Health Informatics

Practical Guide

Seventh Edition

ROBERT E. HOYT MD FACP

WILLIAM R. HERSH MD FACP FACMI

Editors

Health Informatics
Practical Guide
Seventh Edition

Copyright © June 2018 by Informatics Education

Seventh Edition

All rights reserved. No part of this book may be reproduced or transmitted in any form, by any means, electronic or mechanical, including photocopying, recording, or by any information storage and retrieval system, without written permission from the publisher, except for inclusion of brief excerpts in connection with reviews or scholarly analysis.

Disclaimer

Every effort has been made to make this book as accurate as possible, but no warranty is implied. The information provided is on an "as is" basis. The authors and the publisher shall have neither liability nor responsibility to any person or entity with respect to any loss or damages arising from the information contained in this book. The views expressed in this book are those of the authors and do not necessarily reflect the official policy or position of any university or government.

First Edition: June 2007
Second Edition: August 2008
Third Edition: November 2009
Fourth Edition: September 2010
Fifth Edition: January 2012
Sixth Edition: January 2014

Print copy ISBN: 978-1-387-64241-0
eBook ISBN: 978-1-387-82750-3

Editors

ROBERT E. HOYT MD, FACP
Diplomate, Clinical Informatics, American Board of Preventive Medicine
Informatics Education
www.informaticseducation.org
Pensacola, FL

WILLIAM R. HERSH MD, FACP, FACMI
Diplomate, Clinical Informatics, American Board of Preventive Medicine
Professor and Chair, Department of Medical Informatics & Clinical Epidemiology
Oregon Health & Science University
www.billhersh.info
Portland, OR

Contributors

ELMER V. BERNSTAM MD, MSE
Associate Dean for Research
School of Biomedical Informatics
Reynolds and Reynolds Professor of Biomedical Informatics
Professor of Internal Medicine
The University of Texas Health Science Center
Houston, TX

HARRY B. BURKE MD, PHD
Professor of Medicine
Chief of Section of Safety, Quality and Value
Biomedical Informatics
Uniformed Services University of the Health Sciences
Bethesda, MD

TREVOR COHEN MD, MBCHB, PHD
Associate Professor of Biomedical InformaticsMD
School of Biomedical Informatics
The University of Texas Health Science Center
Houston, TX

BRIAN E. DIXON MPA, PHD, FHIMSS
Associate Professor, Department of Epidemiology, IU Richard M. Fairbanks School of Public Health
Research Scientist, Center for Biomedical Informatics
Regenstrief Institute Inc.
Indianapolis, IN

ALISON FIELDS BS MPH
College of Health
University of West Florida
Pensacola, FL

M. CHRIS GIBBONS MD, MPH
Johns Hopkins Medical Institutions
Greystone Group, Inc
Baltimore, MD

JOHN GRIZZARD MD
Associate Professor, Radiology
Director, Non-Invasive Cardiovascular Imaging
Virginia Commonwealth University
Richmond, VA

WILLIAM HERSH MD, FACP, FACMI
Professor and Chair, Department of Medical Informatics & Clinical Epidemiology
School of Medicine
Oregon Health & Science University
Portland, OR

ROBERT HOYT MD, FACP
Diplomate, Clinical Informatics
Informatics Education
Pensacola, FL

TODD JOHNSON PHD
Professor, School of Biomedical Informatics
University of Texas Health Science Center
Houston, TX

HAROLD LEHMANN MD, PHD
Director, Division of Health Sciences Informatics
Professor of Health Sciences Informatics
Johns Hopkins University School of Medicine
Baltimore, MD

STEVE MAGARE MSC
Research Officer, Health Informatics
KEMRI-Wellcome Trust Programme
Nairobi, Kenya

SARITA MANTRAVADI PHD, MS, MPH, CPH, CHES
Pearland, TX

GLEBER NELSON MARQUES PHD, MS
Adjunct Professor of Computational Science
College of Health Sciences and Medicine
Mato Grasso State University
Mato Grasso, Brazil

THOMAS MARTIN PHD
Assistant Professor and Graduate Program Director
College of Public Health
Temple University
Philadelphia, PA

KEN MASTERS PHD
Assistant Professor of Medical Informatics
Medical Education Unit
College of Medicine & Health Sciences
Sultan Qaboos University
Sultanate of Oman

VISHNU MOHAN MD, MBI, MBCS, FACP
Associate Professor of Medical Informatics, General Internal Medicine and Management
Department of Medical Informatics and Clinical Epidemiology
Program Director, Clinical Informatics Sub-Specialty Fellowship
Oregon Health & Science University
School of Medicine
Portland, OR

NAOMI MUINGA MSC
Research Officer, Health Informatics
KEMRI – Wellcome Trust Programme
Nairobi, Kenya

CHRIS PATON BMBS, BMEDSCI, BMA, FACHI
Group Head for Global Health Informatics at the Centre for Tropical Medicine
University of Oxford
Oxford, England

SAURABH RAHURKAR BDS, DRPH
Public Health Informatics Postdoctoral Fellow
Center for Biomedical Informatics
Regenstrief Institute
Indianapolis, IN

JOHN RASMUSSEN MBA
Chief Information Security Officer
MedStar Health
Columbia, MD

INDRA NEIL SARKAR, PHD, MLIS, FACMI
Associate Professor of Medical Science
Center for Biomedical Informatics
Brown University
Providence, RI

YAHYA SHAIHK MD, MPH
Johns Hopkins Medical Institutions
Greystone Group, Inc.
Baltimore, MD

JOHN SHARP MSSA, PMP, FHIMSS
Director, Personal ConnectedHealth Alliance
HIMSS Innovation Center
Cleveland, OH
Adjunct Faculty, Health Informatics
Kent State University
Kent, OH

DALLAS SNIDER PHD
Assistant Professor, Computer Science
Hal Marcus College of Science and Engineering
University of West Florida,
Pensacola, FL

Table of Contents

PREFACE xi

ACKNOWLEDGEMENTS xii

CHAPTER 1: OVERVIEW OF HEALTH INFORMATICS. 1
  Introduction. 1
  Definitions of Informatics and related terms . . . 2
  Background. 4
  Historical highlights 7
  Key users of HIT 8
  Organizations involved with Health Informatics . 8
  Federal Government initiatives. 11
  Public private partnerships 12
  Barriers to HIT adoption 15
  Health Informatics programs and careers . . . 17
  Health Informatics resources 20
  Future trends 21
  References 22

CHAPTER 2: HEALTHCARE DATA, INFORMATION AND KNOWLEDGE . . . 29
  Introduction. 29
  Definitions and concepts 29
  Converting data to information to knowledge . 33
  Clinical data warehouses 35
  What makes informatics difficult? 38
  Complexity of knowledge models 39
  Why Health IT sometimes fail 41
  Future trends 42
  References 44

CHAPTER 3: COMPUTER AND NETWORK ARCHITECTURES 47
  Introduction. 47
  Computers 47
  The Internet and World Wide Web 54
  Web services 56
  Networks 57
  Future trends 62
  References 63

CHAPTER 4: ELECTRONIC HEALTH RECORDS. 67
  Introduction. 67
  Electronic health record justification 68
  National Academy of Medicine's vision for EHRs. 72
  EHR key components. 73
  Computerized physician order entry 77
  Clinical decision support systems 79
  Electronic patient registries. 81
  Practice management integration. 82
  EHR adoption. 83
  EHR challenges. 84
  The HITECH Act and Meaningful Use . . . 88
  The impact of the Meaningful Use program . . 90
  Logical steps to selecting and implementing an EHR. 90
  Recommended reading 91
  Future trends 92
  References 94

CHAPTER 5: STANDARDS AND INTEROPERABILITY 101
  Introduction. 101
  Identifier standards 104
  Transaction standards. 106
  Terminology standards 114
  Recommended reading 126
  References .126

CHAPTER 6: HEALTH INFORMATION EXCHANGE 131
Introduction. 131
History of the US health information network initiatives. 133
Interoperability 135
Impact of the HITECH Act on HIE 136
Health information organizations. 137
Health information organization examples. . . 139
Status of US health information exchange . . . 140
Health information exchange concerns. 141
Newer HIT developments to promote HIE . . . 142
Health information exchange resources 142
Recommended reading 144
Future trends 144
References 145

CHAPTER 7: HEALTHCARE DATA ANALYTICS 149
Introduction. 149
Terminology of analytics 149
Challenges to data analytics 152
Research and application of analytics 152
The role of informaticians in analytics 154
Path forward for analytics 155
Recommended reading 156
References 156

CHAPTER 8 CLINICAL DECISION SUPPORT. 161
Introduction. 161
CDS benefits and goals 163
Organizational proponents of CDS. 164
CDS methodology 166
CDS standards 169
CDS functionality. 170
CDS sharing 173
CDS implementation 174
CDS challenges 174
Lessons learned 177
Recommended resources 178
Future trends 178
References 179

CHAPTER 9: SAFETY, QUALITY AND VALUE 183
Introduction. 183
Safety, quality and value 185
Using the EHR to improve quality, safety and value. 188
The inability to interpret free text has limited safety, quality and value. 190
Safety and quality detection and reporting. . . 192
Clinical decision support systems 197
Future trends 201
References 202

CHAPTER 10: HEALTH INFORMATION PRIVACY AND SECURITY. 213
Introduction. 213
Basic security principles 213
The healthcare regulatory environment . . . 215
HIPAA 215
Other regulations and healthcare privacy and security 219
Business drivers for security and privacy . . . 219
Breaches in the news and consequences 220
Threat actors and types of attacks 221
Tools used to protect healthcare privacy and security 223
Future trends – emerging risks in healthcare 226
Recommended reading 227
References 227

CHAPTER 11: HEALTH INFORMATICS ETHICS 233
Introduction. 233
Informatics ethics. 234
International considerations: ethics, laws and culture 236
Codes of individual countries. 237
Pertinent ethical principles 238
Difficulties applying medical ethics in the digital world 239
Transferring ethical responsibility 241
Electronic communication with patients and caregivers 241
Practical steps. 242

Health Informatics ethics and the medical student 245
Future trends 247
References 248

CHAPTER 12: CONSUMER HEALTH INFORMATICS. 253
Introduction. 253
Definitions 253
Personal health records 254
Patient engagement 256
Patient – Clinician electronic communication. 256
Efficacy of consumer health informatics . . .260
CHI and healthcare reform 261
References 264

CHAPTER 13: MOBILE TECHNOLOGY AND MHEALTH 271
Introduction. 271
History .271
Current mobile technology 272
mHealth in clinical settings. 274
mHealth in the home setting 274
mHealth for wellness and sports 275
mHealth in low and middle-income countries 276
Mobile technology for research and development 276
Regulatory requirements 276
mHealth challenges 277
Future trends in mHealth 277
mHealth resources 277
Recommended reading 278
References 278

CHAPTER 14: EVIDENCE BASED MEDICINE AND CLINICAL PRACTICE GUIDELINES. 283
Introduction. 283
Importance of EBM. 284
The evidence pyramid 285
Risk measures and terminologies. 288
Limitations of the medical literature and EBM. 289
Evidence based health informatics 291

EBM resources 293
Clinical practice guidelines. 294
Developing clinical practice guidelines 294
Initiating clinical practice guidelines. 297
Clinical practice guideline example 297
Electronic clinical practice guidelines 298
Clinical practice guideline resources. 299
Recommended reading 300
Future trends 300
References 301

CHAPTER 15: INFORMATION RETRIEVAL FROM MEDICAL KNOWLEDGE RESOURCES 307
Introduction. 307
Content .309
Indexing. 312
Retrieval 316
Retrieval systems 317
Evaluation. 318
Future directions 322
Recommended reading 323
References 323

CHAPTER 16: MEDICAL IMAGING INFORMATICS. 327
Introduction. 327
Typical PACS workflow. 330
PACS for a hospital desktop computer 331
PACS extensions: web-based image distribution. 331
Medical imaging and mobile technology. . . .333
Digital imaging advantages and disadvantages 334
Imaging informatics education 335
Recommended reading 335
Future trends 335
References 337

CHAPTER 17: TELEMEDICINE 339
Introduction. 339
Teleconsultations 341
Telemonitoring 344
Telemedicine initiatives and resources 347
International telemedicine 349

Recommended reading 349
Barriers to telemedicine 350
Telemedicine organizations. 352
Future trends 352
References .353

CHAPTER 18: BIOINFORMATICS 357
Introduction. 357
Genomic primer. 358
Importance of translational bioinformatics. . .359
Bioinformatics projects and centers 360
Personal genomics 365
Genomic information integrated
    with EHRs. 367
Recommended reading 368
Future trends 369
References .370

CHAPTER 19: PUBLIC HEALTH INFORMATICS. 373
Introduction. 373
The role of informatics in public health 375
Information systems to support public
    health functions 376
Case management systems 381
The Public Health Information Network 382
Meaningful use and public health 383
Geographic information systems 383
Common types and sources of public
    health data 385
Global public health informatics 385
Challenges in global public health
    informatics. 388
Public health informatics workforce 390
The role of clinics, hospitals and
    health systems 390
Recommended reading 390
Future trends 391
References .392

CHAPTER 20: ERESEARCH 397
Introduction. 397
Preparatory to research 397
Study initiation 398
Study management and data management . . .400

Data management systems for FDA
    regulated studies 402
Interfaces and query tools 402
Data analysis 403
Recommended reading 403
Future trends 404
References .404

CHAPTER 21: INTERNATIONAL HEALTH INFORMATICS. 409
Introduction. 409
Health informatics in Europe. 410
Health informatics in Australasia. 414
Health informatics in Africa 416
Health informatics in Asia 420
Health informatics in South America. 424
Health informatics in North America 426
Resources. .427
Challenges and barriers. 428
Future trends 429
References .429

CHAPTER 22: INTRODUCTION TO DATA SCIENCE 439
Introduction. 439
Data basics .441
Statistics basics. 441
Database systems. 444
Data analytical processes 447
Major types of analytics 449
Putting it all together. 455
Natural language processing and
    text mining. 455
Visualization and communication 457
Big data. 458
Analytical software for healthcare workers . .460
Data science education 461
Data science careers 462
Data science resources 462
Data science challenges. 462
Future trends 463
Appendix 22.1 464
References .467

INDEX 471

Preface to the Seventh Edition

The seventh edition comprises many of the chapters included in the sixth edition and supplement with new content in each chapter and multiple new authors. We are honored to have Dr. William Hersh from Oregon Health & Science University as co-editor of this edition and he will assume the role of primary editor of all future editions. He is also a contributor of multiple chapters in the seventh edition.

We will continue the same overall chapter framework. Chapters will begin with learning objectives, followed by an introduction and history of the subject and conclude with challenges/barriers, resources, recommended reading, future trends, key points, conclusions and references. In the seventh edition, we will focus on post-HITECH Act changes in Health Informatics and other interesting developments in the field that have occurred since we published the sixth edition four years ago.

Several new authors have been added to the existing authors. The chapter on electronic health records was co-authored by Dr. Vishnu Mohan from the Oregon Health & Science University. Brian Dixon from Regenstrief Institute authored the chapter on public health informatics. Tom Martin from Temple University co-authored the chapter on telemedicine. John Rasmussen is the new author of the Privacy and Security chapter and Harry Burke is the new author of the chapter on patient safety, quality and value.

The editors and authors have provided the most up-to-date and pragmatic information about health informatics based on the continuous review of medical and lay (grey) literature. The approach taken by the editors and authors is consistent with applied informatics and not theoretical informatics. The textbook is intended to be an expansive review of the field of health informatics that will appeal to both undergraduate and graduate students in multiple fields.

Information for instructors: On our website www.informaticseducation.org we provide information about how to purchase the textbook in its many formats. Under the "Instructor's" tab we explain that instructors can receive a PDF textbook download after registering. If they supply the university course number for the course requiring the textbook, we will also provide access to the PowerPoints and Instructor Manual. The Instructor Manual includes the following sections: background, learning objectives, chapter outline, teaching recommendations, student exercises and sample questions. Sign up for our newsletter under the "About Us tab to learn about any new textbook developments.

Information for students: All chapters include an extensive bibliography section and many chapters have a recommended reading section. In addition, we include web resources in each chapter for supplemental education.

We appreciate feedback regarding how to make this book as user friendly, accurate, up-to-date and as educational as possible.

Robert E. Hoyt MD FACP

William R. Hersh MD FACP FACMI

Acknowledgements

We would like to thank our associate editor Ann Yoshihashi MD for her many contributions to help launch the seventh edition. Without her hard work and dedication, the seventh edition would be just a notion on the drawing board.

We would also like to thank Ghislain Viau of Creative Publishing Book Design for his textbook preparation, design and formatting expertise. info@creativepublishingdesign.com

1

Overview of Health Informatics

ROBERT E. HOYT • ELMER V. BERNSTAM • WILLIAM R. HERSH

"During the past few decades the volume of medical knowledge has increased so rapidly that we are witnessing an unprecedented growth in the number of medical specialties and subspecialties. All these difficulties arise from the present, nearly unmanageable volume of medical knowledge and the limitations under which humans can process information."

—Marsden S. Blois, *Information and Medicine: The Nature of Medical Descriptions*, 1984

LEARNING OBJECTIVES

After reading this chapter the reader should be able to:

- State the definition and origin of health informatics
- Identify the drivers behind health informatics
- Describe the key people and organizations involved in health informatics
- Discuss the impact of the HITECH Act and Affordable Care Act on health informatics
- List the barriers to health information technology (HIT) adoption
- Describe educational and career opportunities in health informatics

INTRODUCTION

Health (or as originally called, medical) informatics emerged as a discipline in the 1960s but has only recently gained recognition as an important component of healthcare. Its emergence is partly due to the multiple challenges facing the healthcare system today. As the quote above indicates, the growth in the volume of medical knowledge and patient information that occurred due to better understanding of human health has resulted in more treatments and interventions that produce more information. Likewise, the increase in specialization has also created the need to share and coordinate patient information. Furthermore, clinicians need to be able to access medical information expeditiously, regardless of location or time of day. Technology has the potential to help with each of those areas.

With the advent of the Internet, high speed computers, voice recognition, wireless and mobile technologies, healthcare professionals today have many more tools at their disposal. However, in general, technology has been advancing faster than healthcare professionals can assimilate it into their practice of medicine. One could also argue that there is a critical limitation of current information technology that manages data and not information. Thus, there is a mismatch between what we need (i.e., tools to help us manage meaningful data = information) and what we have (i.e., effective tools for managing data, but ineffective tools to manage information). Additionally, given the volume of data and rapidly changing technologies, there is a great need for ongoing informatics education of all healthcare workers.

This chapter will present an overview of health informatics with emphasis on the factors that helped create and sustain this new field and the key players involved.

Data, Information, Knowledge and Wisdom Hierarchy

Informatics is the science of information and the blending of people, biomedicine and technology.

1

Individuals who practice informatics are known as informaticians or informaticists. There is an information hierarchy that is important in the information sciences, as depicted in the pyramid in Figure 1.1. Notice that there is much more data than information, knowledge or wisdom. Not all data are meaningful, thus there is more data than information, knowledge or wisdom produced. The following are definitions to better understand the hierarchy:

- Data are symbols representing observations about the world. Data are the plural of datum (singular). Thus, a datum is the lowest level of representation, such as a number in a database (e.g., 5), or packets sent across a network (e.g., 10010100). There is no meaning associated with data; the 5 could represent five fingers, five minutes or have no real meaning at all. Modern computers store, manage, process and transmit data accurately and rapidly.
- Information is meaningful data or facts from which conclusions can be drawn by humans or computers. For example, *five fingers* has meaning in that it is the number of fingers on a normal human hand.
- Knowledge is information that is justifiably considered to be true. For example, an elevated fasting blood sugar level suggests an increased likelihood of diabetes mellitus.
- Wisdom is the critical use of knowledge to make intelligent decisions and to work through situations of signal versus noise. For example, a rising blood sugar can indicate diabetes and other secondary causes of hyperglycemia.

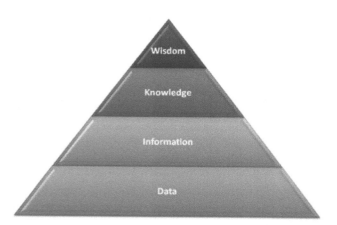

Figure 1.1: Information hierarchy

Ideally, health information technology (HIT) provides the tools to generate information from data that humans (clinicians and researchers) can turn into knowledge and wisdom.[1-2] Thus, enabling and improving human decision making with usable information is a central concern of informaticians. This concept is discussed in much more detail in the chapter on healthcare data, information and knowledge.

The information sciences tend to promote data in formats that can be rapidly transmitted, shared and analyzed. Paper records and reports do not allow this, without a great deal of manual labor. The advent of electronic health records (EHRs) and multiple other healthcare information systems provided the ability and the need to collate and analyze large amounts of data to improve health and financial decisions. Figure 1.2 displays some of the common sources of health data.

Figure 1.2: Health Data Sources
(EHR=electronic health records, PHR=personal health record, HIE=health information exchange)

With ever increasing amounts of health-related data, we have seen the growth of new hardware, software and specialists to handle growing amounts of data. Enterprise systems have been developed that: integrate disparate information (clinical, financial and administrative); archive data; provide the ability to 'mine' data using business intelligence and analytic tools. This is discussed in more detail in the chapter on data mining and analytics and the chapter on data science. Figure 1.3 demonstrates a typical enterprise data system.

DEFINITIONS OF INFORMATICS AND RELATED TERMS

Health informatics is the discipline concerned with management of healthcare data and information through the application of computers and other information technologies. It is more about applying information in the healthcare field than it is about technology per se. That

Chapter 1: Overview of Health Informatics

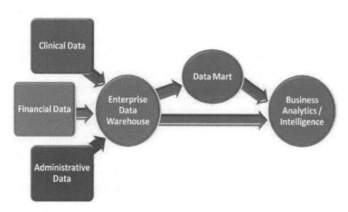

Figure 1.3: Enterprise data warehouse and data mining

is one of the many reasons it is different than pure information technology (IT) in a healthcare organization. Technology merely facilitates the collection, storage, transmission and analysis of data. Health informatics also includes data standards (such as HL7) and controlled medical vocabularies (such as SNOMED) that we will cover in the chapter on data standards. It also addresses issues of usability, clinical workflow, and other aspects of optimizing the use of data and information in healthcare.

Not only are there multiple definitions of health informatics, but there are other expressions of the term that use different adjectives before the noun informatics. For example, the premiere professional association for the field, the American Medical Informatics Association (AMIA), prefers the broader term **biomedical informatics** because it encompasses **bioinformatics** as well as medical, dental, nursing, public health, pharmacy, imaging and research informatics. Hersh uses the term **biomedical and health informatics** to describe the overarching field.[3] (see Figure 1.4) Both AMIA and Hersh then subdivide the field with other terms based on the spectrum from cellular and molecular processes (**bioinformatics**) to the person (**clinical informatics**) to the population (**public health informatics**). **Clinical informatics** also focuses on informatics in the healthcare system and may be further subdivided into the healthcare discipline (nursing informatics, dental informatics, pathology informatics, etc.) or a focus on the consumer or patient (**consumer health informatics**). Hersh also adds broad categories that overarch the spectrum cell-person-population that are focused on images (**imaging informatics**) and research (**research informatics**).

One can also step back and define the word **informatics** itself. The origin of the term is attributed to Dreyfus in 1962.[4] The word originally saw use in Europe, in particular France and Russia (informatique), and often was used synonymously with computer science. Hersh states that informatics is the discipline focused on the acquisition, storage, and use of information in a specific setting or domain.[3] As such, one of the things that distinguishes informatics from information science and computer science is its rooting in a domain. The former School of Informatics at the State University of New York Buffalo defined informatics as the Venn diagram showing the intersection of people, information, and technology. Friedman has stated his *"fundamental theorem"* of biomedical informatics, which states that informatics is more about using technology to help people do cognitive tasks better than about building systems to mimic or replace human expertise.[5]

A historian of the field, Collen, states that the phrase medical informatics was first used in 1974.[6] Although there are other disciplines that use the term informatics (legal informatics, chemoinfomatics, social informatics), its use is probably most prevalent in biomedicine and health.

There are also a number of frequently-cited definitions of biomedical (and health) informatics:

- *"science of information, where information is defined as data with meaning. Biomedical informatics is the science of information applied to or studied in the context of biomedicine. Some, but not all of this information is also knowledge."*[7]
- *"scientific field that deals with biomedical information, data, and knowledge - their storage, retrieval, and optimal use for problem solving and decision making."*[8]

The original term **medical informatics** has been replaced by other terms, including **health informatics**, usually to include the work of non-physician scientists (e.g., biologists) and practitioners (e.g., nurses) and even patients or consumers. The American Medical Informatics Association (AMIA) prefers the broader term **biomedical informatics** defined as *"the interdisciplinary field that studies and pursues the effective*

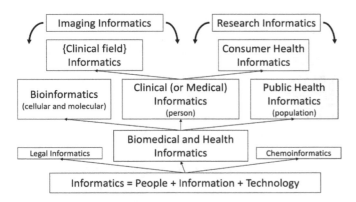

Figure 1.4 Biomedical Informatics

uses of biomedical data, information, and knowledge for scientific inquiry, problem solving and decision making, motivated by efforts to improve human health."[9]

Biomedical informatics is a multidisciplinary field, as it encompasses bioinformatics, genomics, computer science, cellular biology and the social and behavioral sciences. As we move closer to integrating human genetics into the day-to-day practice of medicine, this more global definition is increasingly relevant.

Some informatics-related terms are important to call out or elaborate:

Clinical Informatics is the sub-field of biomedical informatics that focuses on the level of a single individual. Physicians and non-physicians work in this field. For physicians, clinical informatics is the name of the new sub-specialty that allows board certification through the American Board of Preventive Medicine (and the American Board of Pathology for pathologists).[10]

Bioinformatics is the sub-field of biomedical informatics concerned with biological data, particularly that derived from genomics and the other "omics," such as proteomics, metabolomics, and transcriptomics, although such data are increasingly linked to clinical, public health or other data.

There are also several other terms worth noting related to informatics:

Health information technology (HIT or health IT) is defined as the application of computers and technology in healthcare settings.

Health information management (HIM) traditionally focused on the paper medical record and coding. With the advent of the electronic health record (EHR), HIM specialists now must deal with a new set of issues, such as privacy and multiple new concepts such as voice recognition.

For a discussion of the definition, concepts and implications (e.g. distinguishing from other related fields), see the articles by Hersh[3] and Bernstam, Smith and Johnson.[7]

BACKGROUND

Given the fact that most businesses incorporate IT into their enterprise fabric, one could argue that it was just a matter of time before the tectonic forces of medicine and IT collided. As more medical information was published and more healthcare data became available as a result of computerization, the need to automate, collect, archive and analyze data escalated. Also, as new technologies such as EHRs appeared, ancillary technologies such as disease registries, voice recognition and picture archiving and communication systems arose to augment functionality. In turn, these new technologies prompted the need for expertise in health information technology that spawned new specialties and careers.

Health informatics emphasizes *information brokerage*; the sharing of a variety of information back and forth between people and healthcare entities. Examples of medical information that needs to be shared include: lab results, x-ray results, vaccination status, medication allergy status, consultant's notes and hospital discharge summaries. Medical informaticians harness the power of information technology to expedite the transfer and analysis of data, leading to improved efficiencies. The field also interfaces with other fields such as the health sciences, computer sciences, biomedical engineering, biology, library sciences and public health, to mention a few. Informatics training, therefore, must be expansive and in addition to the topics covered in the chapters of this book must include IT knowledge about networks and systems, database management, usability, process re-engineering, workflow analysis and redesign, quality improvement, project management, leadership, teamwork, implementation and training.

HIT facilitates the processing, transmission and analysis of information and interacts with many important functions in healthcare organizations and serves as a common thread (Figure 1.5). This is one of the reasons the Joint Commission created the management of information standard for hospital certification.[11]

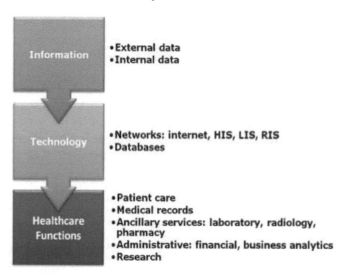

Figure 1.5: Information, information technology and healthcare functions

Many aspects of health informatics noted in Figure 1.5 are interconnected. To accomplish data collection and analysis there is the *hospital information systems*

(HIS) that collects financial, administrative and clinical information and subsystems such as the *laboratory information system* (LIS) and *radiology information systems* (RIS), with the latter often called the *picture archive and communication system* (PACS). As an example, a healthcare organization might be concerned that too many of its diabetic patients are not well controlled and believes it would benefit by offering a disease management web portal. With a portal, patients can upload blood sugar and blood pressure results to a central server so diabetic educators and/or clinicians can analyze the results and make recommendations. They also have the option to upload physiologic parameters via their mobile devices. The following technologies and issues are involved with just this one initiative and covered in other chapters:

- The web-based portal involves consumer (patient) informatics and telemedicine.
- Use of a smart phone is an important type of mobile technology.
- Management of diabetes requires online medical resources, evidence-based medicine, clinical practice guidelines, disease management and an EHR with a disease registry.
- If the use of the diabetic web portal improves diabetic control, clinicians may be eligible for improved reimbursement, known as value-based reimbursement.

There are multiple motivations driving the adoption of health information technology, but the major ones are the need to:

- Increase the efficiency of healthcare (improve physician, nurse and overall healthcare productivity)
- Improve the quality (patient outcomes) of healthcare, resulting in improved patient safety
- Reduce healthcare costs
- Improve healthcare access with technologies such as telemedicine
- Improve communication, coordination and continuity of care
- Improve medical education for clinicians and patients
- Standardization of medical care

These technologies and systems are critical for healthcare to achieve what the Institute for Healthcare Improvement describes as the "Triple Aim" of healthcare, which is to improve the patient's experience, improve the health of populations and reduce the cost of healthcare.[12] Each of these can benefit from intelligent use of data and information.

Over the past 40 years, there has been increasing recognition that some wide variations in practice cannot be justified on objective clinical grounds. For example, patients in some areas of the United States are undergoing more invasive procedures than similar patients in other areas. Thus, there has been a movement to standardize the care of common and expensive conditions, such as coronary artery disease, heart failure and diabetes. *Clinical practice guidelines* are one way to provide advice at the point of care and we will discuss this in more detail in the chapter on evidence-based medicine.

This textbook discusses the driving forces motivating informatics and their inter-relationships. In addition to the motivation to deliver more efficient, safer and less costly healthcare, there is the natural diffusion of technology which also exerts an influence. In other words, as technologies such as wireless and voice recognition become more common, easier to use and less expensive, they will have an inevitable impact or pressure on the practice of medicine. Technological innovations appear at a startling pace as stated by Moore's Law:

"the number of transistors on a chip will double approximately every two years."[13]

Moore's Law describes the exponential growth of transistors in computers. Technology will continue to evolve at a rapid rate, but it is important to realize that it often advances in an asynchronous manner. For example, laptop computers have advanced greatly with excellent processor speed and memory, but their utility is limited by a battery life of roughly 6-8 hours. This is a significant limitation given the fact that most nurses now work eight to twelve-hour shifts, so short battery life is one factor that currently limits the utility of laptop computers in healthcare. This may be overcome with tablet computers or a new battery design.

Healthcare is also subject to shifts in technology. A good example would be mobile technology that was quickly adopted by a large percentage of the world's population and is strongly competing with landlines and desktop PCs. Digital imaging and voice recognition could also be considered evolving technological innovations. We can expect more innovations in the future, and we can only hope they are associated with a lower, not higher price tag than existing technologies.

The EHR, covered in another chapter, could be considered the centerpiece of health informatics with its potential to improve patient safety, medical quality, productivity and data retrieval. EHRs are a focal point of most patient encounters currently. Multiple resources that are currently standalone programs are being incorporated or integrated into the EHR, e.g. electronic prescribing, physician and patient education, genetic profiles, patient portals, disease registries and artificial intelligence, to mention a few. It is anticipated that EHR use will eventually be shown to improve patient outcomes, such as morbidity and mortality because of clinical decision support tools that decrease medication errors and standardize care with embedded clinical guidelines. However, at present, because EHRs do not adequately support clinicians' information needs and workflow, they do little to improve patient care and, in some cases, have been shown to reduce the quality of care.[14] Informaticians will play a major role in helping to reverse this trend. Among other things, it will not be enough to simply store electronic data; it must be shared among disparate partners. We will address HIE (information sharing) in a later chapter.

The Importance of Data

One of the outcomes of EHRs, RISs, mobile technologies, and other systems is the voluminous amount of healthcare data. As pointed out by Steve Ballmer, past CEO of Microsoft, there will be an *"explosion of data"* because of automating and digitizing multiple medical processes.[15] Newer technologies such as electronic prescribing and HIE will produce data that heretofore has not been available. This explains, in part, why technology giants such as Microsoft, Intel and IBM have entered the healthcare arena. As we begin mining medical data from entire regions or organizations we will be able to make much better evidence-based decisions. Increasing data has also created the new buzz word "Big Data" that has several definitions:

- Data so large it can't be analyzed or stored on one computational unit[16]
- Five Vs: the definition started with three Vs but has increased to five:
  - Volume: massive amounts of data are being generated each minute
  - Velocity: data is being generated so rapidly that it needs to be analyzed without placing it in a database
  - Variety: roughly 80% of data in existence is unstructured so it won't fit into a database or spreadsheet. There is tremendous variety, in terms of the data that could potentially be analyzed. However, to do this requires new training and tools.
  - Veracity: current data can be "messy" with missing data and other challenges. Because of the very significant volume of data, missing data may be less important than in the past
  - Value: data scientists now have the capability to turn large volumes of unstructured data into something meaningful. Without value, data scientists will drown in data and not information or knowledge.[17]

The textbook will point out in other chapters that large organizations, such as Kaiser Permanente have the necessary IT tools and expertise, financial resources, leadership and large patient population to be able to make evidence-based decisions in almost all facets of medicine. Pooling data is essential because most practices in the United States are small and do not provide enough information on their own to show the kind of statistical significance we need to alter the practice of medicine.[18]

The US federal government understands the importance of data and information to make evidence-based medical decisions. In 2009, a Presidential Open Government Directive was issued for the heads of the government agencies to promote the publication of government information online, improve the quality of data and to promote transparency.[19] Consistent with that policy Project Open Data and Data.gov were created to share data of interest to multiple communities.[20-21] HealthData.gov is part of this initiative and serves to make datasets from the federal agencies available to a multitude of interested parties, such as healthcare organizations, developers, researchers, etc. Datasets are available through categories: health, state, national, Medicare, hospital, quality, community and inpatient. Because of this initiative, a variety of applications, mashups and visualizations have been developed. As of mid-2017 there were 254 health-related data sets on the site.[22]

The following are examples of other applications or programs producing health-related data:

- Community Health Status Indicators: summarizes the health of the 3,143 counties in the US. Counties are rated as better, moderate or worse, compared to peer counties[23]
- Child Growth Charts: CDC web site that provides percentile charts for children[24]
- Behavioral Risk Factor Surveillance System (CDC): telephone surveys that collect information about risk behaviors, chronic health conditions and preventive services[25]
- Births (CDC): is part of the national vital statistics system collects data from birth certificates[26]
- Mortality and deaths (CDC): lists the number of deaths for leading causes of death[27]
- National Survey of Older Americans: is part of HHS and surveys Americans 60 and older to determine use of elderly services[28]

- State Cancer Profiles: includes interactive maps and graphs of cancer trends at the county, state and national levels.[29]

The federal government continues to add new sources of health-related data available to the public, healthcare professionals and researchers. Health Datapalooza is an annual event launched because of the Health Data Initiative (HDI), sponsored by HHS and the National Academy of Medicine. This public-private partnership brings together disparate users of healthcare data, to improve healthcare quality and safety.[30] Additional data resources are discussed in several other chapters.

The most recent and significant event to affect the health information sciences in the United States was the multiple programs associated with the HITECH Act of 2009, discussed later in this chapter. The programs included substantial financial support for EHRs, HIE and a skilled HIT workforce.

The introduction of information technology into the practice of medicine has been tumultuous for many reasons. New technologies are expensive, they may negatively affect workflow (e.g., interacting with the computer rather than with patients and data entry after work) and require advanced training.[31] Unfortunately, this type of training rarely occurs during medical or nursing school or after graduation. More healthcare professionals who are *bilingual* in technology and medicine will be needed to realize the potential of new technologies. Vendors, insurance companies and governmental organizations will also be looking for the same expertise.

HISTORICAL HIGHLIGHTS

Information technology has been pervasive in the field of Medicine for only about three decades, but its roots began in the 1950s.[32] Since the earlier days the field has experienced astronomical advances in technology, to include, personal computers, high resolution imaging, the Internet, mobile technology and wireless, to mention only a few. In the beginning, there was no strategy or vision as to how to advance healthcare using information technology. Now, we have the involvement of multiple federal and private agencies that are plotting future healthcare reform, supported by health information technology. The following are some of the more noteworthy developments related to health information technology:

- Computers. The first electronic general-purpose computer (ENIAC) was released in 1946 and required 1,000 sq. ft. of floor space. Primitive computers such as the Commodore and Atari appeared in the early 1980s along with IBM's first personal computer, with a total of 16K of memory.[33] Ironically, not everyone saw the future popularity of personal computers. Ken Olson, the president and chairman of Digital Equipment Corporation said in 1977 "*There is no reason anyone would want a computer in their home.*"[34] There has been dramatic growth in global PC sales until 2011 and after that a slow decline, presumably due to widespread use of mobile technology.[35]

Computers were first theorized to be useful for medical diagnosis and treatment by Ledley and Lusted in the 1959 when they published *Reasoning Foundations of Medical Diagnosis*.[36] They reasoned that computers could archive and process information more rapidly than humans. The programming language known as Massachusetts General Hospital Multi-Programming System (MUMPS) was developed in Octo Barnett's lab at Massachusetts General Hospital in the 1970s. MUMPS exists today in the popular EHR known as VistA, used by the Veterans Affairs medical system and Epic Systems Corporation.[37]

- German scientist Gustav Wagner developed the first professional organization for informatics (German Society for Medical Documentation, Computer Science and Statistics) in 1949.[38]
- It is thought that the origin of the term medical informatics dates to the 1960s in France ("Informatique Medicale").[39]
- MEDLINE is the National Library of Medicine's bibliographic database that contains more than 24 million references in the biomedical sciences. In the mid-1960s MEDLINE and MEDLARS were created to organize the world's medical literature. For older clinicians who can recall trying to research a topic using the multi-volume print text *Index Medicus*, this represented a quantum leap forward.[40]
- Artificial Intelligence is a term used when a machine demonstrates learning and/or problem solving. Artificial intelligence (AI) projects in medicine, such as MYCIN (Stanford University) and INTERNIST-1 (University of Pittsburgh), appeared in the 1970s and 1980s.[41] Since the 1960s, AI has had alternating periods where research flourished and where it floundered, known as "*AI winters.*"[14] Natural language processing (NLP) is a subarea of AI that has the potential to intelligently interpret free text.
- Internet. The development of the Internet began in 1969 with the creation of the government project ARPANET.[42] The World Wide Web (WWW or web)

was conceived by Tim Berners-Lee in 1990 and the first web browser Mosaic appeared in 1993.[43-44] The Internet is the backbone for digital medical libraries, HIE and web-based medical applications. Although the terms *Web* and *Internet* are often used interchangeably, the Internet is the *network-of-networks* consisting of hardware and software that connects computers to each other. The Web is a set of protocols (particularly related to HyperText Transfer Protocol or HTTP) that are supported by the Internet. Thus, there are many Internet applications (e.g. email) that are not part of the Web. This is discussed further in the chapter on computer and network architectures. By March 2017 there were more than 3.7 billion Internet users in the world.[45]
- Electronic Health Record. The EHR has been advocated since the 1970s and was formally recommended by the Institute of Medicine (now known as the National Academy of Medicine) in 1991.[46] EHRs will be discussed in more detail in a later chapter.
- Mobile technology. The Palm Pilot personal digital assistant (PDA) appeared in 1996 as the first truly popular handheld computing device.[47] PDAs loaded with medical software became standard equipment for residents in training in the 1990s. They have been supplanted by smartphones, such as the iPhone. Smartphones and tablets will be discussed in more detail in the chapter on mobile technology. The popularity of mobile technology is evidenced by the fact that beginning in 2011 smartphone sales exceeded the sale of personal computers.[48] Gartner, the world's largest information technology research analyst reported that 1.5 billion smartphones were sold in 2016.[49]
- Human Genome Project. In 2003, the Human Genome Project (HGP) was completed after thirteen years of international collaborative research. Mapping all human genes was one of the greatest accomplishments in scientific history. Finalizing a draft of the genome was the first step. What remains is making intelligent use of the data. In other words, we need to understand the difference between data (the code), information (what the code means) and knowledge (what we do with the information).[50] Data from large databases will likely change the way we practice medicine in the future. The HGP will be discussed in the chapter on bioinformatics.

KEY USERS OF HEALTH INFORMATION TECHNOLOGY

HIT is important to all stakeholders in healthcare. HIT generates important data that is transmitted, visualized, analyzed and archived by those in the field of health informatics. There are many important users of HIT:
- Patient – the individual who receives healthcare, often called a consumer or citizen when they are well
- Provider – those who "provide" healthcare, e.g., physicians, nurses, allied health providers
- Purchaser – those who buy healthcare, usually employers or the government
- Payor – those who "pay" the healthcare system, i.e., the insurance companies and government
- Public health – protectors of the public's health

These stakeholders have diverse applications of HIT that they use during the healthcare process, maintaining health, or conducting research.[51]

ORGANIZATIONS INVOLVED WITH HEALTH INFORMATICS

There are many types of organizations involved in health informatics. One way to classify them is professional/trade associations, government, and industry. The latter can be subdivided into healthcare and HIT industries. Both academia and industry also have associations that represent their interests and are typically non-profit. Another category of organization is public-private partnerships.

Professional and Trade Associations

There are many professional organizations in the field of biomedical and health informatics. Probably the premier organization is the **American Medical Informatics Association (AMIA)**. The mission of AMIA is to advance the informatics professions relating to health and disease. To this end, it advances the use of health information and communications technology in clinical care and clinical research, personal health management, public health population, and translational science with the ultimate objective of improving health.[9]

There are several other professional organizations devoted to aspects of biomedical and health informatics. One is the **Healthcare Information and Management Systems Society (HIMSS)**, which is commonly thought to represent industry in the health IT field with a focus on vendors and consultants.[52]

The **American Health Information Management Association (AHIMA)**, represents the health information management (HIM) profession.[53] The **Association of Medical Directors of Information Systems (AMDIS)**, represents physician leaders in HIT[54] and the **Alliance for Nursing Informatics** is a coalition that focuses on nursing informatics.[55]

The **Public Health Informatics Institute** focuses on informatics workforce supply to public health.[56] The **International Society for Computational Biology (ISCB)** focuses on computational biology.[57] Likewise, the **Society for Imaging Informatics and Medicine (SIIM)** focuses on imaging informatics.[58] The **Association for Computing Machinery (ACM)** is the professional association for computer science.[59] **The Medical Library Association (MLA)** focuses on health science librarianship.[60]

There are many specialty societies for healthcare professionals and most have some interest in informatics. For example, the **American Medical Association (AMA)** has several programs that mainly focus on running and maintaining physician practices, including the use of health IT.[61] Likewise, the **American Nurses Association (ANA)** has an interest in informatics issues related to nursing.[62] The **Association of American Medical Colleges (AAMC)**, the professional organization for medical schools, is very active in informatics, as is the **American College of Physicians (ACP)**, the professional society for internal medicine physicians, and the **American Academy of Family Physicians (AAFP)**.[63-65]

Governmental Agencies

Although informatics is carried out at all levels of government, in the US it is predominantly an activity of the federal government. In the US, the **Department of Health & Human Services (HHS)** is the cabinet-level agency that is an umbrella for most of the important government agencies that involve HIT. The **Office of the National Coordinator for Health Information Technology (ONC)** reports directly to the Secretary of HHS and is not an agency per se. Other operating divisions under HHS include:
- National Institutes of Health (NIH)
- Agency for Healthcare Research & Quality (AHRQ)
- Centers for Medicare & Medicaid Services (CMS)
- Centers for Disease Control & Prevention (CDC)
- Health Resources & Services Administration (HRSA)
- Indian Health Service (IHS)
- Food and Drug Administration (FDA)
- Administration on Aging (AOA)[66]

Office of the National Coordinator for Health Information Technology (ONC). The ONC oversees the application of HIT, mostly focused on the adoption of EHRs. It led the implementation of the Health Information Technology for Economic and Clinical Health (HITECH) Act, which provided $30 billion funding through the American Recovery and Reinvestment Act (ARRA) in incentives for EHR adoption. Its main focus now is on standards, interoperability, HIE, and safety of EHRs and their clinical data.[67] The following are the broad goals of the 2015-2020 Federal Health IT Strategic Plan developed by ONC.[68] The specific objectives and strategies are outlined in detail in the plan displayed in table 1.1 on next page.

ONC initially established the Health IT Policy Committee (HITPC) and the Health IT Standards Committee (HITSC), but in the 2015 timeframe both committees were replaced by the Health Information Technology Advisory Committee (HITAC). HITAC recommends to the National Coordinator policies, standards, implementation specifications and certification criteria related to HIT.

There are many other US government agencies involved in aspects of health informatics.

The **National Institutes of Health (NIH)** is the premiere federal agency for biomedical research. One of its institutes is the **National Library of Medicine (NLM)**. NLM serves as the nation's, and really the world's, medical library, but it is also the lead federal funder of research and training in biomedical informatics. The importance of the NLM was reaffirmed recently with the appointment of a new leader, and the incorporation of various data science initiatives within NHS into NLM.[69]

Agency for Healthcare Research and Quality (AHRQ). The AHRQ is *"the lead Federal agency charged with improving the quality, safety, efficiency, and effectiveness of health care for all Americans. As one of 12 agencies within the Department of Health and Human Services, AHRQ supports health services research that will improve the quality of health care and promote evidence-based decision making."* This agency sets aside significant grant money to support healthcare information technology (HIT) research each year. AHRQ also maintains the National Resource Center for HIT, an extensive patient safety and quality section and an extensive HIT Knowledge Library with over 6,000 resources.[70]

Centers for Medicare and Medicaid Services (CMS). CMS is responsible for providing care to 55.3

Table 1.1: Goals of the 2015-2020 Federal Health IT Strategic Plan

Goals	Objectives
Goal 1: Advance person-centered and self-managed health	Objective 1A: Empower individual, family and caregiver health management and engagement
	Objective 1B: Foster individual, provider and community partnerships
Goal 2: Transform healthcare delivery and community health	Objective 2A: Improve health care quality, access and experience through safe, timely, effective, efficient, equitable and person-centered care
	Objective 2B: Support the delivery of high-value care
	Objective 2C: Protect and promote public health and health, resilient communities
Goal 3: Foster research, scientific knowledge and innovation	Objective 3A: Increase access to and usability of high quality electronic health information and services
	Objective 3B: Accelerate the development and commercialization of innovative technologies and solutions
	Objective 3C: Invest in, disseminate and translate research on how health IT can improve health and care delivery
Goal 4: Enhance nation's health IT infrastructure	Objective 4A: Finalize and implement the Nationalwide Interoperability Roadmap
	Objective 4B: Protect the privacy and security of electronic health information
	Objective 4C: Identify, priortize and advance technical standards to support secure and interoperable health information and health IT
	Objective 4D: Increase user and market confidence in the safety and safe use of health IT

million Medicare (2015 data) and 69 million Medicaid patients (2013 data). To improve quality and decrease costs, CMS has information technology pilot projects in multiple areas, to include pay-for-performance demonstration projects that link payments to improved patient outcomes. They reimburse Medicare and Medicaid clinicians for Meaningful Use of certified EHRs. Several informatics-related projects will be discussed in later chapters.[71]

Centers for Disease Control and Prevention (CDC). Although not a primary information technology agency, the CDC has used HIT to promote population health-related issues. Among their programs of interest:

- Public Health Information Network (PHIN), covered in the chapter on public health informatics
- Human Genome Epidemiology Network (HuGENET™) correlates genetic information with public health
- Family History Public Health Initiative is a web site that records family history information and encourages saving it in a digital format, so it can be shared. This is discussed more in the chapter on bioinformatics
- Public Health Image Library contains photos, images and videos on medical topics
- National Health and Nutrition Evaluation Survey (NHANE) program surveys about 10,000 US citizens every two years and shares the results with the public and researchers
- Geographic information systems (GIS) are also covered in chapter on public health informatics
- Podcasts, RSS feeds and apps on medical topics[72]

Health Resources and Services Administration (HRSA) is part of HHS with the primary mission of assisting medical care for the underserved and uninsured in the United States, particularly in rural areas. They support federally qualified health centers (FQHCs) and rural health centers (RHCs). HRSA supports grants for community health centers to include the installation and upgrades of health information technology. They have been a long-term grant supporter of telemedicine. On their web site, they post a variety of health-related data in the HRSA data warehouse. Searchable topics are presented with the ability to present as a table, chart, map or report.[73]

National Committee on Vital and Health Statistics (NCVHS) is a public advisory body to the Secretary of Health and Human Services. It is composed of 18 members from the private sector who are subject matter experts in the fields of health statistics, electronic HIE, privacy/security, data standards and epidemiology. They have been very involved in advising the Secretary in matters related to the HealtheWay (Nationwide Health Information Network), HIPAA, interoperability and other important topics.[74]

National Institute of Standards and Technology (NIST) is a physical science laboratory that is part of the U.S. Department of Commerce and serves to promote and verify measurements and standards. This federal agency makes EHR testing recommendations. The following is a list of some of the pertinent publications related to EHRs:

- (NISTIR 7741) NIST Guide to the Processes Approach for Improving the Usability of Electronic Health Records
- (NISTIR 7742) Customized Common Industry Format Template for Electronic Health Record Usability Testing
- (NISTIR 7743) Usability in Health IT: Technical Strategy, Research, and Implementation
- (NISTIR 7769) Human Factors Guidance to Prevent Healthcare Disparities with the Adoption of EHRs[75]

FEDERAL GOVERNMENT INITIATIVES

The federal government has maintained that HIT is essential to improving the quality of medical care and containing costs; two important aspects of healthcare reform. It is a major payer of healthcare with the following programs: Medicare/Medicaid, Veterans Health Administration, Military Health System, Indian Health Service and the Federal Employees Health Benefits Program. It is therefore no surprise that they are heavily involved in HIT and stand to benefit greatly from interoperability. Agencies such as Medicare/Medicaid and AHRQ conduct HIT pilot projects that potentially could improve the quality of medical care and/or decrease medical costs. The federal government has recognized the importance of technology in multiple areas and as a result has a federal chief technology officer and chief technology officer for HHS.

While the US government has been involved in many aspects of health informatics since the inception of the field, its role increased dramatically with the passage of the American Recovery and Reinvestment Act (ARRA) in 2009.

The most significant governmental initiative that affected the field of Informatics was the HITECH Act. This legislation impacted HIT adoption, particularly EHRs, as well as training and research. HITECH had five broad goals: (a) improve medical quality, patient safety, healthcare efficiency and reduce health disparities; (b) engage patients and families; (c) improve care coordination; (d) ensure adequate privacy and security of personal health information; (e) improve population and public health. Title IV and XIII of ARRA, known as the Health Information Technology for Economic and Clinical Health (HITECH) Act was devoted to funding of HIT programs. The HealthIT.gov website outlines the details of many of the HITECH programs.[67] Readers are encouraged to visit this web site often as HIT policy is subject to frequent change. In addition to the major programs, the following are also important initiatives that were part of HITECH:

- Privacy and HIPAA changes; to be discussed in chapter on privacy and security
- The National Telecommunications and Information Administration's Broadband Technology Opportunities Program. This funded the National Broadband Plan discussed in the chapter on telemedicine
- Indian Health Services HIT programs
- Social Security Administration HIT programs
- Veterans Affairs (VA) HIT programs[76]

The Patient Protection and Affordable Care Act (PPACA) was enacted into law in March 2010 and is commonly known as the Affordable Care Act (ACA, or "Obamacare"). Its primary goals were to increase insurance coverage by expanding private and Medicaid coverage, to reduce healthcare costs, and to improve patient outcomes. The main focus of the legislation so far has been to increase health insurance coverage through the following mechanisms:

- Regulations that prevent insurers from discriminating against people with pre-existing conditions and prohibit lifetime caps on healthcare costs
- The requirement that all individuals have adequate insurance (and thus pay into the system while healthy)
- Subsidies to make that insurance affordable – for the lowest-income families, insurance is provided directly by Medicaid, while for those with higher incomes insurance is subsidized at rates that diminish with increasing income

Other areas within the ACA include:

- Patient Centered Outcomes Research Institute (PCORI) that funds patient-centered and comparative effectiveness research

- The CMS Innovation Center that evaluates healthcare models such as the Accountable Care Organization (ACOs)
- The National Prevention and Health Promotion Strategy
- Independence at Home Demonstration Projects
- Readmission Reduction Program to penalize healthcare systems with excessive readmissions
- Value based reimbursement to hospitals and physicians based on quality measures
- Scholarships and loan repayments for primary care physicians
- Grants for Health Centers to support HIT[77]

Medicare Access and CHIP Reauthorization Act (MACRA) of 2015

The main focus of this law was to replace the sustainable growth rate (SGR) formula, which was slated (but never implemented) to cut Medicare payment for physicians. MACRA also created a new framework for physician reimbursement, aiming to reward them for value and not volume of care provided. A number of other chapters in this book cover aspects of this subject. This new legislation is subject to policy change so should be interpreted with that context.[78]

State Governments and HIT

There are a variety of state-based HIT initiatives, evaluating the adoption of technologies such as EHRs, HIE and e-prescribing. State Medicaid offices are anxious to conduct pilot projects aimed at reducing costs and/or improving quality of care.[79]

International Governments and HIT

This chapter focuses primarily on US health informatics, but the reality is that this is an important and emerging field worldwide. Other countries have less expensive and less fragmented healthcare systems, but they also must deal with aging populations and rising chronic diseases. Meanwhile, technology continues to evolve unabated and in the case of mobile technology is quite affordable. They are therefore looking for healthcare solutions using cost-effective health information technology. Issues such as IT interoperability among European nations and certification are challenges all countries face. In the case of Europe and the European Union they refer to Health IT as eHealth and IT as information and communication technology (ICT).

The Digital Agenda for Europe (DAE) was created to enhance the economic condition in Europe and modernize all industries, to include healthcare. They have also established ICT-related cooperative efforts outside the EU. In 2013, they established ties with the US Department of Health and Human Services to further eHealth cooperation. The established Roadmap focuses on two high priority areas: standards development for interoperability and workforce development to increase skilled health IT workers in Europe. The timeline for this cooperative initiative was 18 months.[80] Multiple other international eHealth initiatives, collaborations and innovations are discussed in other chapters.

International health informatics is a mature sophisticated movement that is supported by multiple countries and international organizations such as the World Health Organization (WHO). The WHO fully supports eHealth with multiple programs and projects. One of their newest collaborations is the WHO Collaborating Centre in Consumer Health Informatics, established to help patients manage their own health. The most prominent international informatics organization is the **International Medical Informatics Association (IMIA)** that supports the International Journal of Medical Informatics. Several international conferences are held to collaborate and support health informatics research efforts. Other international medical informatics associations are discussed in the chapter on International Health Informatics.

PUBLIC-PRIVATE PARTNERSHIPS

National Academy of Medicine (NAM). This prominent organization was formerly known as the Institute of Medicine (IOM). It has published several highly influential reports on HIT over the last couple of decades. These reports have been widely cited, and have been very influential, influencing legislation, such as HIPAA and HITECH. They are available for download via PDF, and hard copies can be purchased on the National Academies Press website.

The first round of IOM reports came out in the 1990s, and into 2000. They focused mainly on identifying problems. The original report, *The Computer-Based Patient Record*, was published in 1991, and then revised in 1997. This was the first volume to bring together all the research identifying problems with paper records, the fact that they're illegible, inefficient, and error prone, and really made the point that the computer-based record was vital to modern health care.[81]

Another influential report was *For the Record*, coming out in 1997, noting that, while there were benefits of EHRs, they would be compromised by inadequate privacy, security, and related problems. This report informed the details of the HIPAA legislation, which now are an integral part of modern US health care.[82]

The *Networking Health* report that came out in 2000 looked at the then much less mature Internet and its role. Of course, this was before the era of smartphones and other ways that we interact online, including social media. But it noted the potential for networked health and addressed one issue at the time, which some people believed what was needed was a separate health Internet, and this report took a contrary view. Another important conclusion of this report was that the availability of the network was more important than its raw bandwidth, that is, the availability so that the Internet could be accessed any time with a reasonable amount of performance was more important than the pure speed of moving content, particularly images, around.[83]

The IOM report that garnered the most press was, *To Err Is Human*. This report brought together research that had previously been done, noting that, first, medical errors are a lot more common than many believed, but even more important that the errors were a systems problem, that you couldn't pin error problems on any one person, or any one segment of the health care industry. Many errors occurred when an error was made somewhere, and it propagated through the system, and the system was not able to identify the error before it happened.[84]

These reports would not be that helpful if they just talked about the problems. As such, the next round of reports began to lay out solutions and started with a vision for a better health care system. The *Crossing the Quality Chasm* report was an early attempt to address issues of quality and the chasm between the way the health care system was and could be. This report argued that health care had to embody safety and quality throughout, that the system needed to be patient-centered and evidence-based, and this report developed a set of aims and rules for what they called high quality 21st century health care.[85]

The *Crossing the Quality Chasm* report developed a set of aims for 21st century health care. It stated that health care should be safe, so avoiding injuries from care intended to help people. It should be affective, so that services provided based on tests and treatments that had justification in the scientific literature and avoiding care that would be unlikely to benefit individuals. Health care in the 21st century should be patient-centered, so respectful of patient's preferences, needs, and values. It should be timely, so that individuals who need care can get it when they need it, and not have delays. Care should also be efficient, so avoiding waste of equipment, supplies, and energy. And it should be equitable, so everyone in the system was able to get care, and it was not denied to them based on any kind of personal characteristics, whether ethnicity, or socio-economic status, or other factors.

The *Crossing the Quality Chasm* report also laid out some rules for 21st century health care. It pointed out that patient needs and values should be the driver of variation in care, and not geography, or access to specialists, or other factors that really don't relate to the individual. Care should be based on continuous healing relationships, so available 24/7, and by all modalities, including online. The patient should be the source of control of their care. There should be shared knowledge among all in health care, from the provider to the patient, to the institution, and so forth. There should be free flow of information and transparency, obviously, though, protecting individual privacy. There should be a focus on anticipating needs, rather than reacting to them, so trying to determine ahead of time how to best allocate resources, rather than react when resources are needed. And decision making should be evidence-based, based on science.

In 2003, *Fostering Rapid Advances in Healthcare...* was published under the guise of the need to foster rapid advances in health care. It called for demonstration projects that implemented the vision of the Quality Chasm report. It looked at some of the information technology issues. It noted that there was lack of interoperability of data, so that data was trapped in different silos. This was in an era where there were very few EHRs. However, we know, even in modern times, we still have this problem with interoperability of data and it being trapped in silos. To unlock information, we need standards and interoperability, so information can move between systems. This report also noted some of the research looking at the misalignment financial incentives when it came to IT, and that those who benefited from the system financially were not necessarily the same as those who were paying for the system.[86]

Further delving into the harm problem of health care, the IOM came out with a report in 2004 on patient safety. There were many recommendations that came out of this report, a number of which focused on information technology. This report restated the case for the National Health Information Infrastructure, as it was called then, the idea that we have a national health information system, that we now call the Nationwide Health Information Network. It called for federal government leadership and public private partnerships. It reiterated the need for standards, so information could more easily move between systems with the patient. It also called for error reporting systems when they did occur that would be protected. Similar to the airline industry, individuals could disclose errors and not face punitive damages if they made attempts to rectify them. This report also made the case for a number of safety initiatives that are carried out in other industries, such as the airline industry, the nuclear power industry, things like adverse event analysis and near-miss analysis.[87]

A further evolution in the view of the IOM reports started to look at using data to learn more about what health care does and try to improve it, to facilitate research. This led to the notion of the *learning health system*, the idea that, like other industries, we would capture data, and analyze it and attempt to improve what we do based on what we learned. The report that defined the learning health system was called, *Knowing What Works in Health Care*, recognizing that doing that requires capacity in information systems and people who know how to implement and use those systems. This report noted again, when EHRs were much less widely used, that the growing amount of data could aid the learning health care system.[88]

Another workshop that came out of this report focused on the infrastructure needed for the learning health system and there was recognized the need for human capacity, including workforce development, including individuals trained in informatics.

By 2009, it was recognized that, even though there was much potential that had been recognized for HIT over a decade through these reports, that progress was not an as good as it could be. The report published in 2009, *Computational Technology for Effective Health Care*, noted that, even though there were a small number of exemplary institutions that were using IT well, we were failing at disseminating those benefits to other institutions. This report called for rethinking some of the approaches, such as focusing on clinical gains in an incremental fashion, not trying to revolutionize the system overnight, and aiming to make gains incrementally. It called for improving the coordination of health care, and improving the way it was financed, moving away from fee-for-service towards things like bundled payments. The report also called for avoiding the monolithic business-oriented systems that were evolving in health care institutions, and instead focusing on federations of systems that would provide value for patients and clinicians.[89]

Further drilling down into the learning health care system led to the realization that there would need to be a digital infrastructure, that the system would need to build a structure of information—capture, and use, and sharing, and analysis—that would protect the rights of individuals. It would still enable the health care system function to perform efficiently. This report identified several themes for the digital infrastructure of the learning health care system and basically took an approach that there needs to be continuous learning. It needs to be integrated into the existing system. Considerations of scale were important, but also that the system must be decentralized and responsive to local needs. There would need to be low barriers to participate and minimizing complexity, all centered around a fabric of trust, so that patients, clinicians, and others would take part.

As more and more IT implementation occurred, especially after the HITECH Act, it was also increasingly recognized that, although IT systems could reduce error and harm, they could also increase it. In 2012, the IOM published its *Health IT and Patient Safety* report, noting that systems that could improve care might also introduce error and cause harm, if not designed and applied properly. This report called for federal oversight.[90]

In late 2012, another IOM report came out, which brought a good deal of the vision and plans for implementation together. This report was called, *Best Care at Lower Cost—The Path to Continuously Learning Health Care in America*. The report started by reviewing the problems in health care, the required action that must be taken to decrease waste, estimated in this report to be $750 billion US dollars out of a $2.5 trillion system, and leading to 75,000 premature deaths. It draws on the analysis of Berwick and others on the sources of waste in the health care system, whether it's services being provided that are not necessary, or any type of service, whether necessary or not, being inefficiently delivered. It notes that, for many things in the health care system, prices are too high relative to the cost. The US health care system also suffers from excess administrative costs, estimated to be as high as 30%. There are missed opportunities for prevention, not only prevention of disease, but prevention of complications once disease has developed and the problems of fraud in the health care system.[91]

In its vision, this report describes a number of components for the learning health care system, things like records being immediately updated and available for use by patients; care being delivered that has been proven reliable at the core, and then tailored to patient needs and preferences at the margins; the patient and family having their needs and preferences being a central part of the decision process; health care functioning as a team, and all of the team members being fully informed of each other's activities at real time—so coordination of care; prices and total costs being fully transparent to all in the care process; incentives for payment being structured to reward outcomes in value, not volume of provided services; promptly identifying errors and correcting them; and routinely capturing patient outcomes, and using those to implement a system of continuous improvement.

The report notes that health care progresses from basic science understanding, which then becomes evidence that it actually works in humans, and then leads to care that's delivered to patients. The current system has many missed opportunities, and wastes, and harm, so in translating from science to evidence, or evidence to care, or care to the patient experience, we have several problems. And instead, we should aim that there be a loop of science to evidence to care, and back informing science. And patients and clinicians and communities are part of that. And the missed opportunities, waste, and harm are minimal. All the above-mentioned IOM/NAM reports can be located and purchased on the National Academies Press web site.[92]

BARRIERS TO HEALTH INFORMATION TECHNOLOGY ADOPTION

Up until the passage of the HITECH Act, the United States was behind many industrialized nations, in terms of per capita payments towards HIT. As of May 2017, CMS (Medicare) paid $9.5 billion for eligible professionals and $24.6 billion for eligible hospitals for adoption and meaningful use of EHRs. CMS (Medicaid) paid out $5.5 billion to eligible professionals and $11.9 billion to eligible hospitals.[93]

Despite these large payments and good adoption statistics for inpatient and ambulatory EHRs, many problems still exist. Health information technology adoption has multiple barriers listed below and discussed in other chapters:

Inadequate time. This complaint is a common thread that runs throughout most discussions of technology barriers. Busy clinicians complain that they don't have enough time to read or learn about new technologies or research vendors. They are also not reimbursed to become technology experts. They usually must turn to physician champions, local IT support, Regional Extension Centers or others for technology advice. A 2016 time-motion study noted that *"for every hour physicians provide direct clinical face time to patients, nearly 2 additional hours is spent on EHR and desk work within the clinic day."*[31] We cannot expect busy clinicians to adopt future technologies until this issue is solved or improved.

Inadequate information. As pointed out earlier in the chapter, clinicians need information, not data. Current HIT systems are data rich, but information poor. This is discussed in detail in the healthcare data, information and knowledge chapter.

Inadequate expertise and workforce. For the United States to experience widespread HIT adoption and implementation, it will require education of all healthcare workers. Hersh emphasizes the need for a work force capable of leading implementation of the EHR and other technologies.[94] Educational offerings will need to be expanded at universities, community colleges and medical, nursing and pharmacy schools. There is a substantial difference between healthcare organizations, in terms of HIT sophistication. The first Work Force for Health Information Transformation Strategy Summit, hosted by the American Medical Informatics Association (AMIA) and the American Health Information Management Association (AHIMA) made several strategic recommendations regarding how to improve the work force.[53] AMIA has been the leader in Health Informatics education, with its 10 x 10 Program.[95] Their goal is to train 10,000 skilled workers over 10 years. The Community College Consortium graduated a significant number of students, but it is too early to know how successful job placement has been. HIT vendors are looking for applicants with both IT and clinical experience, in addition to good people skills and project management experience.[96] In addition to skilled informaticians; we will need to educate residents in training and faculty at medical schools, given the rapidly changing nature of HIT. The APA Summit on Medical Student Education Task Force on Informatics and Technology recommended that instead of CME, we need *"longitudinal, skills-based tutoring by informaticians."*[97] Family Medicine residency programs are generally ahead of other specialty training programs regarding IT training, promoting a longitudinal approach to IT competencies.[98]

Inadequate cost and return on investment data. The literature on the economic aspects of HIT adoption and implementation is mixed and based on different assumptions and methods. An often-cited barrier is a mismatch between costs and benefits of HIT. The clinicians/providers bear the costs (and/or do the extra work), whereas hospitals/insurers/government reap the benefits. For example, clinicians must enter more data into computers so that insurers or government (or researchers) have a clearer understanding of what happened. An article by Bassi and Lau posits a cost-benefit evaluation should have six components: having a perspective, options for comparison, time frame, costs, outcomes and comparison of costs and outcomes for each option. Examples of high quality economic reviews are available in their paper.[99]

High cost to adopt. Technologies such as picture archiving and communications systems (PACS) and EHRs are very expensive. ARRA helped underwrite the initial purchase of some technologies, but long-term support will be a different challenge. There is still limited evidence that most technologies will save money. This is discussed in more detail in the chapter on EHRs.

Change in workflow. Significant changes in workflow will be required to integrate technology into the inpatient and outpatient setting. As an example, clinicians may be accustomed to ordering lab or x-rays by giving a handwritten request to a nurse who places the order. Now they must learn to use computerized physician order entry (CPOE). As with most new technologies, older users have more difficulty changing their habits, even if it will eventually save time or money. Poor usability is also an important impediment to good workflow and we will address this in the chapter on EHRs. As already noted, there is evidence that young physicians are spending more time on the computer and less with the patient, which is disconcerting. According to Dr. Carolyn Clancy, a prior director of AHRQ, *"The main challenges are not technical; it's more about integrating HIT with workflow, making it work for patients and clinicians who don't necessarily think like the computer guys do."*[100]

Privacy concerns. The Health Insurance Portability and Accountability Act (HIPAA) of 1996 was created initially for the portability, privacy and security of personal health information (PHI) that was largely paper-based. HIPAA regulations were updated in 2009, and again in 2013, to better cover the electronic transmission of PHI or (ePHI). This Act has caused healthcare organizations to re-think healthcare information privacy and security. A concern is that HIPAA is open to interpretation. Thus, different organizations have different interpretations of HIPAA and different privacy rules. This will be covered in more detail in the chapter on privacy and security. In the past decade, there have been many serious healthcare security breaches and stolen identities in addition to new threats such as ransom ware, thus adding to the angst.

Legal issues. The Stark and Anti-kickback laws prevent hospital systems from providing or sharing technology such as computers and software with referring physicians. Exceptions were made to these laws in 2006. This is particularly important for hospitals to share EHRs and e-prescribing programs with clinician's offices. Many new legal issues are likely to appear.

Behavioral change. Perhaps the most challenging barrier is behavior. In the Prince by Machiavelli, it was stated "there is nothing more difficult to be taken in hand, more perilous to conduct, or more uncertain in its success, than to take the lead in the introduction of a new order of things."[101] In 1962, Everett Rogers wrote Diffusion of Innovations in which he delineated different categories of acceptance of innovation:

- The innovators (2.5%) are so motivated; they may need to be slowed down
- Early adopters (13.5%) accept the new change and teach others
- Early majority adopters (34%) require some motivation and information from others to adopt
- Laggards (16%) require removal of all barriers and often require a direct order[102]

It is important to realize, therefore, that at least 50% of medical personnel will be slow to accept any information technology innovations and they will be perceived as dragging their feet or being *Luddites*.[103] With declining reimbursement and emphasis on increased productivity, clinicians have a natural and sometimes healthy dose of skepticism. They dread widespread implementation of anything new unless they feel certain it will make their lives or the lives of their patients better. In this situation, selecting clinical champions and conducting intensive training are critical to implementation success.

HIT hype versus fact. The Gartner IT Research Group describes five phases of the hype-cycle that detail the progression of technology from the technology trigger to the peak of inflated expectations to the trough of disillusionment to the slope of enlightenment to the plateau of productivity.[104] Figure 1.6 shows the hype cycle for a variety of emerging technologies for 2017.

Importantly, current studies that evaluate HIT often yield mixed results for multiple reasons contributing to skepticism discussed in these articles.[105-106] Both the RAND Corporation and the Center for Information Technology Leadership reported in 2005 that HIT would save the US about $80-180 billion annually for widespread EHR and HIE adoption.[107] The Congressional Budget Office (CBO), on the other hand, refuted that overly optimistic viewpoint in 2008. They published a monograph entitled *Evidence on the Costs and Benefits of Health Information Technology* that reviewed the evidence on the adoption and benefits of HIT, the costs of implementing, possible factors to explain the low adoption rate and the role of the federal government in implementing HIT. The bottom line for the CBO was *"By itself, the adoption of more health IT is generally not sufficient to produce significant cost savings."*[108] The controversy continues. Several articles call into

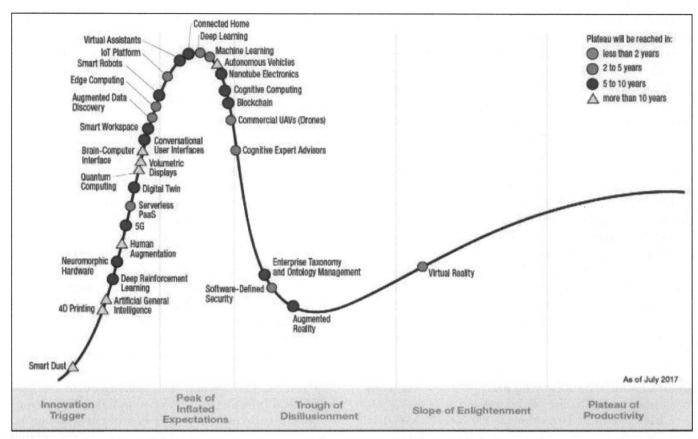

Figure 1.6: Gartner Hype Cycle of Emerging Technology 2017 (Courtesy Gartner.com)

question the presumption that HIT adoption will generate significant cost saving, while another review was more positive.[109-115] Karsh et al. discussed twelve HIT fallacies that added a sober note to the discourse.[116] In the chapter on evidence-based medicine there is a section on evidence based health informatics that sheds light on some of the research deficiencies that challenge the field. Finally, Carol Diamond of the Markle Foundation pointed out that HIT success can't be measured by the number of hospitals that have adopted EHRs or other HIT, but instead whether patient outcomes improve.[117]

Despite a myriad of challenges, some practices have flourished in the HITECH era. The following info box highlights the success of one small primary care practice that maximized HIT to improve the quality of care.

HEALTH INFORMATICS PROGRAMS AND CAREERS

Health Informatics Academic Programs

Historically, health informatics degrees have been offered at the graduate level, usually through a Master of Science (MS) or Doctor of Philosophy (PhD) program. In recent years, health informatics degrees have started to be offered at the undergraduate level, at the bachelor's or even associate degree level. Many options have emerged over the years, from traditional campus-based experiences to online offerings. Another trend has been the development of so-called Graduate Certificate programs, which are usually a subset of a full master's degree. The AMIA Web site maintains a catalog of such educational programs.[95]

Another option for training is a fellowship. Traditionally, fellowships in informatics were funded by the NLM or Veteran's Affairs (VA).[119-120] With the emergence of the clinical informatics subspecialty, new clinical fellowships have emerged, accredited by Accreditation Council on Graduate Medical Education (ACGME). There are nearly 30 such fellowships now, and an up-to-date list is on the AMIA Web site.[95]

Health Informatics Careers

The timing is excellent for a career in health informatics. With the emphasis on increasing adoption of EHRs and other technologies, there has been tremendous interest in health informatics. Healthcare organizations and HIT vendors need workers who are knowledgeable in both technology and healthcare. They are looking for experienced individuals who can hit the ground running, to direct implementation of multiple types of HIT such as EHRs and new standards such as ICD-10. Informaticians will be needed to design, implement and govern many new technologies arriving on the medical scene, as well as train users. Informatics training programs will need to continue the process of designing curricula based on actual needs from the industry. Government reimbursement for EHRs and other technologies has further increased the need for skilled HIT workers. The Health Informatics Forum, HIMSS, American Nurse Informatics, AHIMA and the AMIA web sites list multiple interesting health IT jobs. Other job categories include: nurse and physician informaticists, information directors, chief information officers (CIOs) and chief medical information officers (CMIOs). Recruiting organizations also maintain multiple listings for health IT jobs.

There are a wide variety of jobs available in the informatics realm. The following are just a few common positions in a healthcare organization:

Chief Medical Informatics Officer (CMIO) is usually a physician but could be a nurse who generally reports to the Chief Information Officer (CIO), Chief Executive Officer (CEO) or Chief Medical Officer (CMO). This individual usually works with the CIO to develop a strategic IT plan and to help with the implementation of technologies by clinical staff. CMIOs are less IT oriented and more oriented towards overcoming the barriers to adoption and they provide feedback and education to their staff. They evaluate new technologies that may transform healthcare and along with the CIO they help develop policies that affect privacy and security. They commonly have a Master's degree in one of the information sciences.

In 2002 HIMSS developed a Certified Professional in Health Information Management Systems (CPHIMS) certification and exam. This is primarily aimed at IT professionals who work in healthcare. In 2017, certified professionals were composed of 39% CIOs/VPs, 12% IT or business consultants, 14% directors/managers and the rest staff, nurses and educators. They must have a bachelor's degree and 5 years of information management experience (2 years in healthcare) or a graduate degree and 3 years of information management experience (2 years in healthcare).[121]

Physician CMIOs may also become board certified in Clinical Informatics, discussed in this same section.

Nurse Informaticist (NI) is a nurse who can be the CMIO or an individual who works in the nursing department, IT department or is dual hatted. There are

---

Case Study of Small Primary Care Practice Maximizing HIT

Foresight Family Physicians is located in Grand Junction, Colorado and is staffed by two physicians, a nurse practitioner, a physician assistant, care managers, a behavioral health provider, medical assistants and support staff. They have served about 5300 patients in this location over the past 25 years.

They were early adopters of an EHR and they joined the local health information organization (HIO). Furthermore, they joined two federal initiatives – the Beacon Consortium and the Comprehensive Primary Care Initiative. This provided HIT guidance and enhanced non-visit based payments and new financial opportunities. The office established a QI team that reviewed a variety of electronic reports daily (list of pending patient reminders, Admission Discharge Transfer [ADT] reports, etc.) The HIE generated the ADT reports and every patient who visited an emergency room or was discharged from hospital received a phone call from the practice, resulting in an improvement in ER visits and admissions. Because of depression screening results they hired a behavioral health provider.

Clearly, this could not have occurred without practice leadership and vision, combined with external financial help and expertise.

Discussion: AHRQ posits that four factors must be in place for a practice like Foresight Family Physicians to maximize HIT
- The right practice culture
- High functioning HIT tools
- Clinical team and staff with appropriate knowledge and skills
- Practice processes and workflow that incorporate effective use of HIT

Even with the above, practices need help in the form of financial incentives and transformational assistance.[118]

three million nurses in the United States, compared to about 800,000 physicians so they are a large pool of knowledge workers. Most nurses are trained to think in terms of systems and process improvement. They are therefore extremely valuable for: project management, IT systems managers, data analysts, technology adoption, implementation and training. Nurse Informaticians have had a certification exam since 1995 and published their Scope and Standards in 2008. To take the certification exam, candidates must have a RN degree, a bachelor's degree or higher, at least 2 years of clinical practice, 30 hours of continuing education in informatics in the past three years and other qualifications explained on their web site.[122-123]

Clinician Informatician (CI) is a clinician who may have formal training with a variety of degrees or simply may have extensive on the job experience and an aptitude for technology. As a result, they are usually early adopters and clinician champions who help the clinical staff in a healthcare organization understand and accept transformational technologies.

AMIA helped establish the subspecialty of *clinical informatics*. In September 2011, it was announced that *clinical informatics* was an approved subspecialty, sponsored by the American Board of Preventive Medicine and the American Board of Pathology. The certification is available to physicians who have a primary specialty designated through the American Board of Medical Specialties (ABMS). There will be an additional period of 5 years (2017-2022) in which physicians can be "grandfathered" in without formal informatics education. In the 2009 March/April issue of JAMIA, the core content for this new specialty was spelled out.[124-125] By the end of 2016 about 1400 physicians have become board certified. The following are admission requirements for certification:
- ABMS member board certification in a current specialty
- Attendance at an accredited in the US or Canada or one deemed satisfactory to the Board
- Current license holder
- Completion of one of the following pathways:
  ○ Three years of practice (in the past 5 years) in the clinical informatics field; at least 25% of a FTE
  ○ If a candidate has completed less than 24 months in a non-accredited program, candidates must submit evidence of the training program.

In 2013 AMIA began the process to create a new certification, the Advanced Health Informatics Certification. The goal would be to provide certification for those individuals who are not eligible for the subspecialty of clinical informatics. Most workers in the health informatics field and members of AMIA are not eligible for certification in clinical informatics so this advanced certification should have broad appeal. The certification should have similar requirements to the subspecialty certification and should be at the graduate level.[126]

In 2016 AHIMA announced a new Certified Professional in Health Informatics (CPHI) certification. Certification examination is highly dependent on knowledge of data analysis, data reporting, data management, database management, health informatics training and project management. To be eligible candidates must have Baccalaureate degree and 2 years of HI experience or a Master's degree with 1 year of HI experience or Masters or higher in HI from regionally accredited academic institution.[127]

Although physicians can become chief medical information officers in very large organizations, the reality is that nurses have the greatest potential to be involved with IT implementation and training at the average hospital or large clinic. Larger, more urban clinics may have the luxury of in-house IT staff, unlike smaller and more rural practices.

Table 1.2 lists the salaries of individuals in the information sciences, based on a 2017 review by PayScale, posted on Monster.com.[128] Many of these figures are averages or medians, actual salary will vary depending on location, education, job demand, job scope and size of the organization.[129] The job site Indeed.com provides a search by city, state or zip code with filters for salary estimate, job title, company, location, job type and employer/recruiter.[130] Another site worth mentioning is HealthITJobs.com with a search engine that searches by 33 job descriptions and locations.[131]

Table 1.2: Health Informatics Positions and Salaries

Job Title	Salary Range
Clinical Informatics Specialist (non-physician)	$68,700 average ($45,500-$93,300 range)
Health Informatics Specialist	$61,000 average ($35,450-$91,600 range)
Clinical Informatics Manager	$92,800 average ($59,000-$127,800 range)
Nurse Informatics Specialist	$66,500 average ($51,300-$89,000 range)
Clinical Analyst	$63,800 average ($40,900-$88,100 range)

While there are many IT certifications available, there is no state or federal licensing or credentialing for health informatics. However, nursing already has an informatics specialty certification.

HEALTH INFORMATICS RESOURCES

Because of the rapidly changing nature of technology it is difficult to find resources that are current. It is also difficult to find resources that are not overly technical that would be appropriate for the health informatics neophyte. There are numerous excellent journals, e-journals and e-newsletters that contain articles that discuss important aspects of health information technology. Because health informatics is gaining popularity in the field of medicine many excellent articles can also be found in major medical journals that do not normally focus on technology. As an example, *Health Affairs*, a bimonthly health policy journal features web exclusives, blogs and e-newsletters of interest to informaticians.[132]

Books

- *Handbook of Biomedical Informatics*. Wikipedia Books.[133]
- *Guide to Health Informatics*. Enrico Coiera, third edition 2015[134]
- *Biomedical informatics: Computer Applications in Health Care and Biomedicine*. EH Shortliffe and J Cimino. Fourth edition. 2014[135]
- *Health Informatics: An Interprofessional Approach*. Nelson R, Staggers N. Second edition. 2017[136]

Journals

- *Journal of the American Medical Informatics Association* (JAMIA) is the journal of AMIA and considered by many to be the foremost scientific journal of the informatics field. Formerly published bimonthly, it has now transitioned to an online-only format. JAMIA features peer-reviewed articles that run the gamut from theoretical models to practical solutions. AMIA has launched a second journal, JAMIA Open, which is an open-access journal that provides a global forum for novel research and insights in the major areas of informatics and related areas such as data science, qualitative research, and implementation science[137]
- *International Journal of Medical Informatics* (IJMI) is a monthly journal that covers information systems, decision support, computerized educational programs and articles aimed at healthcare organizations. In addition to standard articles, IJMI publishes short technical articles and reviews.[138]
- *Journal of Biomedical Informatics* was formerly known as *Computers and Biomedical Research*. This bimonthly journal has a focus on biomedical informatics methods.[139]
- *Journal of AHIMA* is published 11 months of the year for its members to stay current in HIM-related issues.[140]
- *Computers, Informatics, Nursing (CIN)* is a bimonthly print journal targeting nursing professionals. It has 12 online issues and 4 print-only issues per year.[141]
- *BMC Medical Informatics and Decision Making* is an open-access online-only journal publishing peer-reviewed research articles. This journal is part of BioMed Central, an online publisher of over 200 online free full text journals.[142]
- *Journal of Medical Internet Research* (JMIR) is the flagship journal of a family of open-access online journals that publishes articles related to medical informatics. JMIR publishes multiple other open access journals, such as *JMIR Medical Informatics* and *JMIR mHealth and uHealth*.[143]
- *Electronic Journal of Health Informatics* (eJHI) is an Australian-based international open access electronic journal that offers open access (no fee) to both authors and readers.[144]
- *Applied Clinical Informatics* is the fee-based e-journal for the International Medical Informatics Association (IMIA) and AMIA.[145]
- *Perspectives in Health Information Management* is the open-access research peer-reviewed journal of AHIMA and is published four times a year.[146]
- *Online Journal of Public Health Informatics* is an open source general interest peer reviewed e-journal published three times annually.[147]

Informatics-Related E-newsletters

- *HealthCareITNews* is available as a daily online, RSS feed or print journal. It is published in partnership with HIMSS and reviews broad topics in HIT. They also publish the online e-journals *NHINWatch*, *MobileHealthWatch* and *Health IT Blog*.[148]
- *Health IT SmartBrief* is a free newsletter e-mailed three times weekly. Articles are posted from a variety of sources.[149]
- *Health Data Management* offers a free daily e-newsletter, in addition to their comprehensive web site. The web site offers 20 categories of IT information, webinars, whitepapers, podcasts and RSS feeds.[150]

Online Resource Sites

- *InformaticsEducation.org* resource center was created to augment this textbook. The site augments this book

with valuable web links organized in a similar manner as the book chapters. It also includes links to excellent informatics newsletters and journals.[151]
- *Agency for Healthcare Research and Quality Knowledge Library* is another excellent resource with over 6,000 articles and other resources that discuss health information technology related issues. (Library was archived in 2013).[152]
- *HIMSS Health IT Body of Knowledge* is a new site to introduce readers to more than 25 topic categories. Articles, tools and guidelines are offered by HIMSS and other resources.[153]
- *HealthIT.gov* is the official web site for the Office of the National Coordinator for Health Information Technology. The site provides valuable information about HIT initiatives and progress throughout the United States. The web site is divided into information for providers and professionals, patients and families and policy researchers and implementers. For providers, there is information about EHRs, Meaningful Use, privacy and security and related resources. Web site includes a dashboard that reports HIT trends, statistics and datasets.[68]
- *AHIMA HIM Body of Knowledge*™ is a searchable database of HIM-oriented material from AHIMA and governmental sources. Multiple articles are hosted. For example, a search for "data analysis" returned 79 articles.[154]
- *OpenClinical* is a not-for-profit organization that supports advanced knowledge management in the following areas: background, research clinical, commercial and public. The site includes resources that are pertinent to many chapters in this textbook.[155]
- *Health Informatics Forum* is an international forum and blog. In addition, the site offers the massive open online course (MOOC) on health informatics, free of charge. This is the same course administered by many community colleges under the HITECH Act funding.[156]
- *Health Services Research Information Central* is part of the National Library of Medicine and has a section devoted to Health Informatics. Sub-sections include: News, Data, Tools and Statistics, Guidelines, Journals, Grants, Funding and Fellowships, Meetings and Conferences and Key domestic and international organizations.[157]

Informatics Blogs

- *HealthIT Buzz Blog* provides HIT updates from the HHS Office of the National Coordinator for Health Information Technology (ONC).[158]
- *Life as a Healthcare CIO* chronicles the work of Dr. John Halamka of Beth Israel Deaconess Medical Center and who is also CIO of Harvard Medical School.[159]
- *Informatics Professor Blog* and provides the insights of Dr. William Hersh, Professor and Chair of the Department of Medical Informatics & Clinical Epidemiology, Oregon Health & Science University. 160 Additional health informatics resources are posted on his website.[161]
- The *Health Care Blog* is hosted by Matthew Holt and considered to be *"a free-wheeling discussion of the latest healthcare developments"* to include health information technology.[162]
- *E-CareManagement* focuses on chronic disease management, technology, strategy, issues and trends. Content is posted by Vince Kuraitis, a HIT consultant for Better Health Technologies.[163]
- *Biological Informatics* was created by Marcus Zillman to compile multiple biomedical informatics sites (100+) into one, as well as a blog. The emphasis is on bioinformatics.[164]

FUTURE TRENDS

Given the relative newness of health informatics it is not easy to predict the future, but some trends seem worth stressing. Many of these points are discussed in more detail in other chapters.

Regardless of the speed of HIT adoption in medicine, the technology itself will continue to evolve rapidly. Many disruptive technologies such as artificial intelligence will present outstanding opportunities. This will require well trained individuals who understand the technology and have the clinical experience to know how it can be applied successfully in the field of medicine.

Clearly, the federal government is moving from volume based to value-based reimbursement. New healthcare delivery models, such as accountable care organizations will be an experiment well worth watching. If they demonstrate cost savings that are strongly supported by HIT, we can expect increased adoption. More research is needed to determine what additions are evidence based, worthwhile and will impact clinical outcomes.

We anticipate more patient centric medical care and associated technologies; for example, more medical apps for smartphones and personalized genetic profiles. Mobile technologies will continue to be an important medical platform for patients and clinicians.

Expect more artificial intelligence in medicine (AIM) to retrospectively and prospectively interpret medical data. As AI improves we can expect real time predictive analytics, alerts and clinical decision support.

Clearly, the new Precision Medicine Initiative (now known as All of Us) that hopes to individualize medical care will require integration and analysis of a variety of phenotypical and genotypical information. This was funded by the 21st Century Cures Act and discussed further in the chapter on Bioinformatics.[165] Health informatics will play a pivotal role in this initiative.

It is unclear if interoperability issues will be solved in the near future. There is some optimism that open APIs and the new FHIR data standard will be a step in the right direction.

Adler-Milstein et al. wrote a very insightful article in 2017 outlining some of the challenges and solutions for patients, clinicians and researchers to "*cross the health IT chasm*" that are very important.[166]

KEY POINTS

- Health informatics focuses on the science of information, as applied to healthcare and biomedicine
- Health information technology (HIT) holds promise for improving healthcare quality, reducing costs and expediting the exchange of information
- The HITECH Act programs have been a major driver of HIT in the United States
- Barriers to widespread adoption of HIT include: time, cost, privacy, change in workflow, legal, behavioral barriers and lack of high quality studies proving benefit
- Many new degree and certificate programs are available in health informatics
- A variety of health informatics resources are available for a wide audience
- Interoperability and health information exchange are a major priority of the federal government but challenged by multiple issues

CONCLUSION

Health informatics is an exciting and evolving field. New specialties and careers are now possible. Despite its importance and popularity, significant obstacles remain. Health information technology has the potential to improve medical quality, patient safety, educational resources and patient - physician communication, while decreasing cost. Although technology holds great promise, it is not the solution for every problem facing medicine today. We must continue to focus on improved patient care as the single most important goal of this new field.

The effects of the multiple programs supported by the HITECH and Affordable Care Acts will likely be both transformational and challenging for the average practitioner.

Research in health informatics is being published at an increasing rate so hopefully new approaches and tools will be evaluated more often and more objectively. Better studies are needed to demonstrate the effects of health information technology on actual patient outcomes and return on investment, rather than observational studies and studies based solely on surveys and expert opinion.

REFERENCES

1. Ackoff RL. From data to wisdom. J Appl Syst Anal. 1989;16:3-9
2. The DIKW Model of Innovation. www.spreadingscience.com (Accessed June 16, 2017)
3. Hersh WR. A stimulus to define informatics and health information technology BMC Medical Informatics and Decision Making. 2009;9. www.biomedcentral.com/1472-6947/9/24 (Accessed November 4, 2009)
4. Dreyfus, Phillipe. *L'informatique.* Gestion, Paris, June 1962, pp. 240–41
5. Friedman CP. A Fundamental Theorem of Biomedical Informatics. JAMIA 2009;16(2):169-170
6. Collen MF. Origins of Medical Informatics. West J Med 1986;145:778-785
7. Bernstam EV, Smith JW, Johnson TR. What is biomedical informatics? Biomed Inform 2010;43(1):104-10

8. Biomedical Informatics: Computer Applications in Health Care and Biomedicine. Shortliffe E, Cimino J. Third Edition. Springer Science.
9. The American Medical Informatics Association (AMIA). www.amia.org (Accessed May 27, 2017)
10. Detmer DE, Shortliffe EH. Clinical Informatics: prospects for a new medical subspecialty. JAMA 2014;311(20):2067-8
11. The Joint Commission www.jointcommission.org (Accessed May 12, 2017)
12. Institute for Healthcare Improvement: The IHI Triple Aim http://www.ihi.org/Engage/Initiatives/TripleAim/Pages/default.aspx (Accessed May 26, 2017)
13. Intel http://www.intel.com/content/ www/us/en/silicon-innovations/moores-law-technology.html (Accessed September 10, 2013)
14. Computational Technology for Effective Health Care. Immediate Steps and Strategic Directions. 2009. National Academies Press. Stead W W and Li HS, editors http://books.nap.edu/openbook.php?record_id=12572&page=R1 (Accessed September 15, 2013)
15. Balmer S. Keynote Address 2007 HIMSS Conference. February 26, 2007
16. Data Mining: Concepts and Techniques. Han J. Elsevier. 2000. Amsterdam, Netherlands
17. Marr B. Why only one of the five Vs of big data really matters. March 19, 2015 http://www.ibmbigdatahub.com/blog/why-only-one-5-vs-big-data-really-matters (Accessed April 3, 2016)
18. Nyweide DJ, Weeks WB, Gottlieb DJ et al. Relationship of Primary Care Physicians' Patient Caseload With Measurement of Quality and Cost Performance. JAMA 2009;302(22):2444-2450
19. Peter Orszag Office of Management and Budget December 8, 2009. http://www.whitehouse.gov/sites/default/files/omb/assets/memoranda_2010/m10-06.pdf (Accessed June 13, 2013)
20. Project Open Data. https://project-open-data.cio.gov/ (Accessed May 21, 2017)
21. Data.gov https://www.data.gov (Accessed May 20, 2017)
22. Health Data.Gov https://www.healthdata.gov (Accessed May 20, 2017)
23. Community Health Indicators https://wwwn.cdc.gov/communityhealth (Accessed May 21, 2017)
24. Child Growth Charts https://www.cdc.gov/growthcharts/ (Accessed May 21, 2017)
25. Behavioral Risk Factor Surveillance System https://www.cdc.gov/brfss/index.html (Accessed May 21, 2017)
26. Births https://www.cdc.gov/nchs/nvss/births.htm (Accessed May 21, 2017)
27. Mortality and deaths https://www.cdc.gov/nchs/fastats/deaths.htm (Accessed May 21, 2017)
28. National Survey of Older Americans https://aoasurvey.org/default.asp (Accessed May 21, 2017)
29. State Cancer Profiles https://seer.cancer.gov/statistics/scp.html (Accessed May 21, 2017)
30. Health Data Palooza. http://www.academyhealth.org/healthdatapalooza (Accessed May 20, 2017)
31. Sinsky C, Colligan L, Li L et al. Allocation of Physician Time in Ambulatory Practice: A Time and Motion Study in 4 Specialties. Ann Int Med 2016;165(11):753-760
32. Sabbatini RME. Handbook of Biomedical informatics. Wikipedia Books. Pedia-Press. 2009. Germany. http://en.wikipedia.org/wiki/Wikipedia:Books/BiomedicaInformatics
33. Computer History Museum Timeline. http://www.computerhistory.org/timeline/?year=1980 (Accessed September 5, 2013)
34. Kansas University Institute of Technology and Telecommunications Center http://www.ittc.ku.edu/~evans/stuff/famous.html (Accessed September 15, 2013)
35. Global PC Sales Fall to Eight-Year Low. I.T. Now. http://itnow.net/blogamazing-computer-sales-statistics-infographic-html/ (Accessed May 26, 2017)
36. Parmar. The History of Health Informatics. Device Talk. July 25, 2014 http://www.mddionline.com/blog/devicetalk/history-health-informatics-072514 (Accessed May 29, 2017)
37. Hersh WR. Informatics: Development and Evaluation of Information Technology in Medicine JAMA 1992;267:167-70
38. Hard Hats. http://www.hardhats.org (Accessed May 29, 2017)
39. Collen MF, Ball Marion J. The History of Medical Informatics in the United States, second edition. Springer. London. 2015.
40. Medline. https://www.nlm.nih.gov/bsd/pmresources.html (Accessed May 29, 2017)
41. Health Informatics. http://en.wikipedia.org/wiki/Medical_Informatics (Accessed May 22, 2017)
42. Howe, W. A Brief History of the Internet http://www.walthowe.com/navnet/history.html (Accessed May 29, 2017)
43. Zakon, R. Hobbe's Internet Timeline v8.1 http://www.zakon.org/robert/internet/timeline (Accessed May 29, 2017)
44. W3C http://www.w3.org/WWW/ (Accessed May 29, 2017)
45. Internet World Stats. http://www.internetworldstats.com/stats.htm (Accessed May 25, 2017)
46. Berner ES, Detmer DS, Simborg D. Will the Wave Ever Break: A Brief View of the Adoption of Electronic Health Records in the United States. JAMIA 2005; 12(1):3-7 http://www.ncbi.nlm.nih.

46. gov/pmc/articles/PMC543824/ (Full text) (Accessed May 29, 2017)
47. Jenkins D. Personal Digital Assistants: A world of information in the palm of your hand. Clin Nurs Spec 2002;16(1):38-39
48. Smart Planet.McKendrick J. Milestone: more smartphones than PCs sold in 2011. ZDNET February 4, 2012 http://www.smartplanet.com/blog/business-brains/milestone-more-smartphones-than-pcs-sold-in-2011/21828. (Accessed May 29, 2017)
49. Gartner https://www.gartner.com/newsroom/id/3609817 (Accessed May 26, 2017)
50. Human Genome Project. https://www.genome.gov/10001772/all-about-the--human-genome-project-hgp/ (Accessed May 29, 2017).
51. Crossing the Quality Chasm: A New Health System for the 21st Century. 2001. National Academies Press. https://www.nap.edu/catalog/10027/crossing-the-quality-chasm-a-new-health-system-for-the?gclid=CjwKEAjwja_JBRD8idHpxaz0t3wSJAB4rXW5jFZT-k4hxp7psFNMIBvFuK_-muGYTlq-iaEH8mVvFRoCDAfw_wcB (Accessed May 29, 2017)
52. Health Information Systems Society (HIMSS) www.himss.org (Accessed September 20, 2017)
53. American Health Information Management Association (AHIMA). www.ahima.org (Accessed September 20, 2017)
54. Association of Medical Directors of Information Systems. www.amdis.org (Accessed September 20, 2017)
55. Alliance for Nursing Informatics. http://www.allianceni.org/ (Accessed September 20, 2017)
56. Public Health Informatics Institute. https://www.phii.org (Accessed September 20, 2017)
57. International Society for Computational Biology. https://www.iscb.org (Accessed September 20, 2017)
58. Society for Imaging Informatics in Medicine. https://siim.org (Accessed September 20, 2017)
59. Association for Computing Machinery. www.acm.org (Accessed September 20, 2017)
60. Medical Library Association. www.mlanet.org (Accessed September 20, 2017)
61. American Medical Association. https://www.ama-assn.org (Accessed September 20, 2017)
62. American Nursing Association http://www.nursingworld.org/ (Accessed October 10, 2017)
63. Association of American Medical Colleges. https://www.aamc.org (Accessed September 20, 2017)
64. American College of Physicians. www.acponline.org (Accessed September 20, 2017)
65. American Academy of Family Physicians. www.aafp.org (Accessed September 20, 2017)
66. Department of Health and Human Services. www.hhs.gov (Accessed May 29, 2017)
67. Federal HIT Strategic Plan. 2015-2020 https://www.healthit.gov/policy-researchers-implementers/health-it-strategic-planning (Accessed May 29, 2017)
68. Office of the National Coordinator for Health Information Technology. http://healthit.gov (Accessed June 15, 2017)
69. Brennan, PF (2016). The National Library of Medicine: accelerating discovery, delivering information, improving health. Annals of Internal Medicine. 165: 808-809.
70. Agency for Healthcare Research and Quality http://www.ahrq.gov/ (Accessed June 15, 2017)
71. Centers for Medicare & Medicaid www.cms.hhs.gov (Accessed June 15, 2017)
72. Centers for Disease Control and Prevention www.cdc.gov (Accessed June 15, 2017)
73. Health Resources and Service Administration www.hrsa.gov (Accessed June 15, 2017)
74. National Committee on Vital and Health Statistics http://ncvhs.hhs.gov (Accessed May 29, 2017)
75. National Institute of Standards and Technology (NIST). https://www.nist.gov (Accessed May 29, 2017)
76. American Recovery and Reinvestment Act of 2009 Public Law 111 – 5. http://www.gpo.gov/fdsys/pkg/PLAW-111publ5/pdf/PLAW-111publ5.pdf (Accessed May 29, 2017)
77. Affordable Care Act. https://www.healthcare.gov (Accessed May 29, 2017)
78. AMA. Making sense of MACRA. https://wire.ama-assn.org/practice-management/making-sense-macra-glossary-new-medicare-terms (Accessed October 10, 2017)
79. State Alliance for ehealth http://www.nga.org/cms/home/nga-center-for-best-practices/center-publications/page-health-publications/col2-content/main-content-list/state-alliance-white-papers.html (Accessed June 15, 2017)
80. European Union. https://europa.eu/european-union/index_en (Accessed June 15, 2017)
81. Dick, RS, Steen, EB, et al., Eds. (1997). The Computer-Based Patient Record: An Essential Technology for Health Care, Revised Edition. Washington, DC, National Academies Press.
82. Anonymous (1997). For the Record: Protecting Electronic Health Information. Washington, DC, National Academies Press.
83. Anonymous (2000). Networking Health: Prescriptions for the Internet. Washington, DC, National Academies Press.
84. Kohn, LT, Corrigan, JM, et al., Eds. (2000). To Err Is Human: Building a Safer Health System. Washington, DC, National Academies Press.

85. Anonymous (2001). Crossing the Quality Chasm: A New Health System for the 21st Century. Washington, DC, National Academies Press.
86. Corrigan, JM, Greiner, M, et al., Eds. (2003). Fostering Rapid Advances in Health Care - Learning from Systems Demonstrations. Washington, DC, National Academies Press.
87. Aspden, P, Corrigan, JM, et al., Eds. (2004). Patient Safety - A New Standard for Care. Washington, DC, National Academies Press.
88. Eden, J, Wheatley, B, et al., Eds. (2008). Knowing What Works in Health Care: A Roadmap for the Nation. Washington, DC, National Academies Press.
89. Stead, WW and Lin, HS, Eds. (2009). Computational Technology for Effective Health Care: Immediate Steps and Strategic Directions. Washington, DC, National Academies Press.
90. Anonymous (2012). Health IT and Patient Safety: Building Safer Systems for Better Care. Washington, DC, National Academies Press.
91. Smith, M, Saunders, R, et al. (2012). Best Care at Lower Cost: The Path to Continuously Learning Health Care in America. Washington, DC, National Academies Press.
92. National Academies Press. https://www.nap.edu (Accessed June 16, 2017)
93. Centers for Medicare and Medicaid. https://www.cms.gov/Regulations-and-Guidance/Legislation/EHRIncentivePrograms/DataAndReports.html (Accessed July 20, 2017)
94. Hersh W. Health Care Information Technology JAMA 2004; 292 (18):2273-441
95. AMIA 10 x 10 Program. http://www.amia.org (Accessed June 15, 2017)
96. Ackerman K. Jury Still Out on Health IT Workforce Training Programs. iHealthBeat. September 6, 2011. www.ihealthbeat.org (Accessed September 6, 2011)
97. Hilty DM, Benjamin S, Briscoe G et al. APA Summit on Medical Student Education Task Force on Informatics and Technology: Steps to Enhance the Use of Technology in Education Through Faculty Development, Funding and Change Management. Acad Psych 2006;30:444-450
98. Recommended Curriculum Guidelines for Family Medicine Residents http://www.aafp.org/dam/AAFP/documents/medical_education_residency/program_directors/Reprint288_Informatics.pdf (Accessed September 15, 2013)
99. Bassi J, Lau F. Measuring value for money: a scoping review on economic evaluation of health information systems. J Am Med Inform Assoc. 2013;20:792-801
100. Interview with Dr. Carolyn Clancy. Medscape June 2005. www.meds-cape.com (Accessed November 4, 2005)
101. Machiavelli N, The Prince Chapter VI www.constitution.org/mac/prince06.htm (Accessed September 15, 2013)
102. Rogers EM, Shoemaker FF. Communication of Innovation 1971 New York, The Free Press
103. Luddite http://www.thefreedictionary.com/Luddite (Accessed September 15, 2013)
104. Gartner hype cycle http://www.gartner.com/newsroom/id/3412017 (Accessed June 15, 2017)
105. Shcherbatykh I, Holbrook A, Thabane L et al. Methodologic Issues In Health Informatics Trials: The Complexities of Complex Interventions. JAMIA 2008; 15:575-580
106. Goldzweig CL, Towfligh A, Maglione M et al. Costs and Benefits of Health Information Technology: New Trends From the Literature. Health Affairs 2009 28 (2) w282-w293 www.content.healthaffairs.org/cgi/content/abstract/28/2/w282-w293 (Accessed February 4, 2009)
107. Girosi, Federico, Robin Meili, and Richard Scoville. 2005. Extrapolating Evidence of Health Information Technology Savings and Costs. Santa Monica, Calif. RAND Corporation http://rand.org/pubs/research_briefs/RB9136/index1.html (Accessed May 20, 2008)
108. Congressional Budget Office Paper: Evidence on the Costs and Benefits of Health Information Technology www.cbo.gov (Accessed September 15, 2013)
109. Kellerman Al, Jones SS. What will it take to achieve the as-yet-unfulfilled promises of health information technology? Health Affairs 2013; 32(1):63-68
110. Himmelstein DU, Wright A, Woolhandler S. Hospital Computing and the Costs and Quality of Care: A National Study. Am J Med 2009; 123(1):40-46
111. Black AD, Car J, Pagliari C et al. The Impact of eHealth on the Quality and Safety of Health Care: A Systematic Overview. PLoS Medicine. Jan 2011. www.plosmedicine.org (Accessed January 21, 2011)
112. Butin MB, Burke MF, Hoaglin MC, Blumenthal D. The Benefits of Health Information Technology: A Review of the Recent Literature Shows Predominately Positive Results. Health Affairs. 2011;30 (3):464-471
113. Laszewski R. Health IT Adoption and the Other myths of Health Care Reform. January 12 2009 www.ihealthbeat.org (Accessed January 12, 2009)
114. Shekelle P, Morton SC, Keeler EB. Costs and Benefits of Health Information Technology. Evidence Reports/Technology Assessments. No. 132. Agency for Healthcare Research and Quality. April 2006. http://www.ahrq.gov/research/findings/evidence-based-reports/hitsys-evidence-report.pdf (Accessed June 13, 2013)

115. Adler-Milstein J, Green CE, Bates DW. A survey analysis suggests that electronic health records will yield revenue gains for some practices and losses for many. Health Affairs. 2013;32(3):562-570
116. Karsh B, Weinger MB, Abbott PA, Wears FL. Health Information technology: fallacies and sober realities. J Am Med Inform Assoc. 2010;17:617-623
117. Diamond CC, Shirky C. Health Information Technology: A Few Years of Magical Thinking? Health Affairs www.healthaffairs.org 2008;27 (5): w383-w390 (Accessed September 3, 2008)
118. Using Health Information Technology to Support Quality Improvement in Primary Care Using Health Information Technology to Support Quality Improvement in Primary Care. AHRQ White Paper. Pub No. 15-0031-EF March 2015
119. National Library of Medicine https://www.nlm.nih.gov/ep/GrantTrainInstitute.html (Accessed July 23, 2017)
120. US Department of Veteran' Affairs https://www.va.gov/oaa/specialfellows/programs/sf_medicalinformatics.asp (Accessed July 23,2017)
121. CPHIMS certification. http://www.himss.org/health-it-certification/cphims (Accessed September 27, 2017)
122. How to become an informatics nurse. Nurse Journal. http://nursejournal.org/nursing-informatics/how-to-become-an-informatics-nurse/ (Accessed May 27, 2017)
123. American Nurses Credentialing Center. http://www.nursecredentialing.org/informatics-eligibility.aspx (Accessed May 27, 2017)
124. Gardner RM, Overhage JM, Steen EB et al. Core Content for the Subspecialty of Clinical Informatics JAMIA 2009;16 (2):153-157
125. Detmer DE, Munger BS, Lehmann CU. Clinical Informatics Board Certification: History, Current Status, and Predicted Impact on the Clinical Informatics Workforce. Appl Clin Inf 2009;1:11-18
126. Kuperman G. Update on Advanced Interprofessional Informatics Certification Activities. August 20, 2103. www.amia.org (Accessed August 20, 2013)
127. CPHI Certification. http://www.ahima.org/certification/CPHI (Accessed May 27, 2017)
128. Monster.com https://www.monster.com/career-advice/article/health-informatics-jobs (Accessed June 30, 2017)
129. University of South Florida. Morsani School of Medicine. http://www.usfhealthonline.com/ (Accessed June 15, 2017)
130. Indeed Job Site. www.indeed.com (Accessed June 15, 2017)
131. HealthITJobs. www.healthitjobs.com (Accessed October 10, 2017)
132. Health Affairs. http://content.healthaffairs.org (Accessed June 15, 2017)
133. Handbook of Biomedical Informatics. Wikipedia Books. http://pediapress.com (Accessed June 15, 2017)
134. Guide to Health Informatics. Enrico Coiera. Third edition. 2015. CRC Press
135. Biomedical Informatics: Computer Applications in Health Care and Biomedicine. EH Shortliffe and J Cimino. Fourth edition. 2014 Springer. New York, NY
136. Health Informatics: An Interprofessional Approach. Nelson R, Staggers N. 2017. Elsevier https://www.elsevier.com/books/health-informatics/nelson/978-0-323-40231-6
137. Journal of the American Medical Informatics Association http://jamia.bmj.com/ (Accessed June 15, 2017)
138. International Journal of Medical Informatics https://www.journals.elsevier.com/international-journal-of-medical-informatics/ (Accessed June 15, 2017)
139. Journal of Biomedical Informatics https://www.journals.elsevier.com/journal-of-biomedical-informatics/ (Accessed June 15, 2017)
140. Journal of AHIMA http://journal.ahima.org (Accessed June 15, 2017)
141. CIN: Computers, Informatics, Nursing www.cinjournal.com (Accessed June 15, 2017)
142. BMC Medical Informatics and Decision Making. https://bmcmedinformdecismak.biomedcentral.com/ (Accessed June 15, 2017)
143. The Journal of Medical Internet Research. http://www.jmir.org/ (Accessed June 15, 2017)
144. Electronic Journal of Health Informatics http://ejhi.net (Accessed June 15, 2017)
145. Applied Clinical Informatics. http://www.schattauer.de/index.php?id=558&L=1 (Accessed June 15, 2017)
146. Perspectives in Health Information Management. http://perspectives.ahima.org (Accessed June 4, 2017)
147. Online Journal of Public Health Informatics. www.ojphi.org (Accessed June 15, 2017)
148. Health Care IT News http://www.healthcareitnews.com/ (Accessed June 13, 2017)
149. Health IT SmartBrief http://www2.smartbrief.com/getLast.action?mode=sample&b=HIT (Accessed June 15, 2017)
150. Health Data Management www.healthdatamanagement.com (Accessed June 15, 2017)
151. Informatics Education. www.informaticseducation.org/resource-center (Accessed June 15, 2017)
152. Agency for Healthcare Research and Quality. Knowledge Library. https://healthit.ahrq.gov/knowledge-library-archive (Accessed June 15, 2017)

153. HIMSS Health IT Body of Knowledge http://www.himss.org/asp/topics_HITBOK.asp (Accessed June 15, 2017)
154. AHIMA HIM Body of Knowledge. http://bok.ahima.org/ (Accessed June 15, 2017)
155. Open Clinical. www.openclinical.org (Accessed June 15, 2017)
156. Health Informatics Forum www.healthinformaticsforum.com (Accessed June 16, 2017)
157. Health Services Research Information Central. https://www.nlm.nih.gov/hsrinfo/informatics.html (Accessed June 16, 2017)
158. Health IT Buzz https://www.healthit.gov/buzz-blog/ (Accessed June 15, 2017)
159. Life as a Healthcare CIO http://geekdoctor.blogspot.com/ (Accessed June 15, 2017)
160. Informatics Professor Blog http://informaticsprofessor.blogspot.com/ (Accessed June 15, 2017)
161. William Hersh Website www.billhersh.info (Accessed June 15, 2017)
162. The Health Care Blog www.thehealth-careblog.com (Accessed June 15, 2017)
163. E-care management http://e-caremanagement.com (Accessed June 15, 2017)
164. Biological Informatics http://www.zillman.us/subject-tracers/biological-informatics/ (Accessed June 15, 2017)
165. Precision Medicine Initiative. https://allofus.nih.gov/about/program-faq (Accessed October 11, 2017)
166. Adler-Milstein J, Embi PJ, Middleton B, Sarkar IN, Smith J. Crossing the health IT chasm: considerations and policy recommendations to overcome current challenges and enable value based care. JAMIA 2017;24(5):1036-1043

2

Healthcare Data, Information, and Knowledge

ELMER V. BERNSTAM • TODD R. JOHNSON • TREVOR COHEN

"...current efforts aimed at the nationwide deployment of health care IT will not be sufficient to achieve the vision of 21st century health care, and may even set back the cause if these efforts continue wholly without change from their present course."[1]

LEARNING OBJECTIVES

After reading this chapter the reader should be able to:

- Define Data, Information, and Knowledge
- Describe how vocabularies convert data to information
- Describe methods that convert information to knowledge
- Distinguish informatics from other computational disciplines, particularly computer science
- Describe the differences between data-centric and information-centric technology

INTRODUCTION

This chapter, will present a framework for understanding informatics. The definitions of data, information, and knowledge were presented in chapter 1 and this chapter will build upon these definitions to answer fundamental questions regarding health informatics. What makes informatics different from other computational disciplines? Why is informatics difficult? Why do some health IT projects fail?

In chapter 1, the fundamental mismatch between available technology (i.e., traditional computers, paper forms) and problems faced by informaticians was mentioned. In this chapter, these ideas are expanded to understand why many health IT (HIT) projects fail. To help organizations appropriately apply HIT, informaticians must understand the limitations of HIT as well as the potential of HIT to improve health.

To illustrate several points, this chapter will begin with a real-world example of challenges at the information level. (See case study on next page.)

DEFINITIONS AND CONCEPTS

Data, Information and Knowledge

In chapter 1, data, information and knowledge (see Figure 1.1) were defined.[4-5] Recall that **data** are observations reflecting differences in the world (e.g., "C34.9"). Note that "data" is the plural of "datum." Thus, "data are" is grammatically correct; "data is" is not correct. **Information** is meaningful data or facts from which conclusions can be drawn (e.g., ICD-10-CM code C34.9 = "Malignant neoplasm of unspecified part of bronchus or lung"). **Knowledge** is information that is justifiably believed to be true (e.g., "Smokers are more likely to develop lung cancer compared to non-smokers"). This relationship is shown in Figure 2.1 and readers will be referred to this diagram later in the chapter.

Data

To understand the relationship between data, information and knowledge in health informatics, readers must understand the relationship between what happens in a computer and the real world. Computers do not represent meaning. They input, store, process and output zero

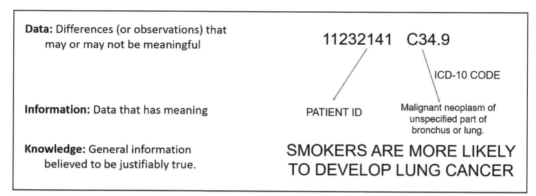

Figure 2.1: Data, information and knowledge

Case Study: The Story of E-patient Dave

In January 2007, Dave deBronkart was diagnosed with a kidney cancer that had spread to both lungs, bone and muscles. His prognosis was grim. He was treated at Beth Israel Deaconess Medical Center in Boston with surgery and enrolled in a clinical trial of High Dosage Interleukin-2 (HDIL-2) therapy. That combination did the trick and by July 2007, it was clear that Dave had beaten the cancer. He is now a blogger and an advocate and activist for patient empowerment.

In March 2009, Dave decided to copy his medical record from the Beth Israel Deaconess EHR to Google Health, a personally-controlled health record or PHR. He was motivated by a desire to contribute to a collection of clinical data that could be used for research. Beth Israel Deaconess had worked with Google to create an interface (or conduit) between their medical record and Google Health. Thus, copying the data was automated. Dave clicked all the options to copy his complete record and pushed the big red button. The data flowed smoothly between computers and the copy process completed in only few moments.

What happened next vividly illustrated the difference between data and information. Multiple urgent warnings immediately appeared, including a warning concerning the prescription of one of his medications in the presence of low potassium levels (hypokalemia) (Figure 2.2). Dave was taking hydrochlorothiazide, a common blood pressure medication, but had not had a low potassium level since he had been hospitalized nearly two years earlier.

Worse, the new record contained a long list of deadly diseases (Figure 2.3). Everything that Dave had ever had was transmitted, but with no dates attached. When the dates were attached, they were wrong. Worse, Dave had never had some of the conditions listed in the new record. He was understandably distressed to learn that he had an aortic aneurysm, a potentially deadly expansion of the aorta, the largest artery in the human body.

Why did this happen? In part, it was because the system transmitted billing codes, rather than doctors' diagnoses. Thus, if a doctor ordered a computed tomography (CT) scan, perhaps to track the size of a tumor, but did not provide a reason for the test, a clerk may have added a billing code to ensure proper billing (e.g., rule out aortic aneurysm). This billing code became permanently associated with the record. To put it another way, the data were transmitted from Beth Israel's computer system to Google's computer system quickly and accurately. However, the meaning of what was transmitted was mangled. In this case, the context (e.g., aortic aneurysm was a billing concept, not a diagnosis) was altered or lost. According to the definitions presented in chapter 1 (and reiterated later in this chapter), meaning is the defining characteristic of information as opposed to data.

After Dave described what happened in his online blog[2] (http://epatientdave.com/), the story was picked up by a number of newspapers including the front page of the Boston Globe.[3] It also brought international attention to the problem of preserving the meaning of data. It became very clear that transmitting data from system to system is not enough to ensure a usable result. To be useful, systems must not mangle the meaning as they input, store, manipulate and transmit information. Unfortunately, as this story illustrates, even when standard codes are stored electronically, their meaning may not be clearly represented.

(off) and one (on). Each zero or one is known as a **bit**. A series of eight bits is called a **byte**. Note that these bits and bytes have no intrinsic meaning. They can represent anything or nothing at all (e.g., random sequences of zeroes and ones).

Bits within computers are aggregated into a variety of **data types**. Some of the most common data types are listed below.
- *Integers* such as 32767, 15 and -20
- *Floating point numbers* (or floats) such as 3.14159, -12.014, and 14.01; the floating point refers to the decimal point
- *Characters* "a," and "z"
- *(Character) Strings* such as "hello" or "ball"

Note that these data types do not define meaning. A computer does not "know" whether 3.14159 is a random number or the ratio of the circumference to the diameter of a circle (known as Pi or π).

Data can be aggregated into a variety of file formats. These file formats specify the way that data are organized within the file. For example, the file header may contain the colors used in an image file (known as the palette) and the compression method used to minimize storage requirements. Common or standardized file formats allow sharing of files between computers and between applications. For example, as long as your digital camera stores photos as JPG files, you can use any program that can read JPG files to view your photos.
- Image files such as JPG, GIF and PNG.
- Text files
- Sound files such as WAV and MP3
- Video files such as MPG

Again, it is important to recognize that neither data types nor file formats define the meaning of the data, except for the purpose of storing or display on a computer. For example, photographs of balloons and microscopes can be stored in JPG files. Nothing about the file format helps us recognize the subject of the photograph.

Informatics vs. Information Technology and Computer Science

Data are largely the domain of information technology (IT) professionals and computer scientists. As computers become increasingly important in biomedicine, biomedical researchers are starting to collaborate with computer scientists. IT professionals and computer scientists concentrate on technology, including computing systems composed of hardware and software as well as the algorithms implemented in such systems. For example, computer scientists develop algorithms to search or sort data more efficiently. Note that *what* is being sorted or searched is largely irrelevant. In other words, the meaning of the data is of secondary importance. It does not matter whether the strings that are being sorted represent proper names, email addresses, weights, names of cars or heights of buildings.

Though they may be motivated by specific applications, computer scientists typically develop general-purpose approaches to classes of problems that involve computation. For example, a computer scientist may design a memory architecture that efficiently stores and retrieves large data sets. The computer science contribution is the

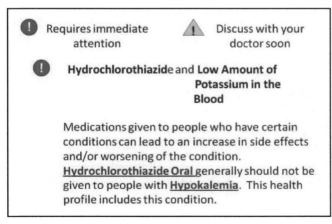

Figure 2.2: Urgent warning in e-patient Dave's record

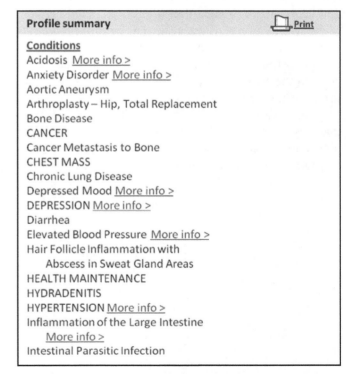

Figure 2.3: e-patient Dave's conditions as reflected in the newly-created personal health record (PHR)

development of the better memory architecture for large data sets; while the memory architecture is not a direct improvement of an EHR per se, it is nonetheless critical to its advancement.

Information and knowledge, on the other hand, are addressed by informatics. To an informatician computers are tools for manipulating information. Indeed, there are many other useful information tools, such as pens, paper and reminder cards. There are significant advantages to manipulating digitized data, including the ability to display the same data in a variety of ways and to communicate with remote collaborators. From an informatics perspective however, one should choose the optimal tool for the information task – often, but not always, the best tool for the task is computer-based.[4,6]

There are areas of overlap between computer science and informatics. For example, information retrieval is widely viewed as a sub-field of computer science and information retrieval researchers often reside in computer science departments. However, we would argue that information retrieval draws on both disciplines. Information retrieval is *"finding material (usually documents) of an unstructured nature (usually text) that satisfies an information need by retrieving documents from large collections (usually stored on computers)."*[7]

Note that information retrieval is concerned with retrieval of information, not data. For example, finding documents that describe the relationship between aspirin and heart attack (myocardial infarction) is an example of an information retrieval task. The central problem is identifying documents that contain certain meaning. In contrast, efficient retrieval of documents (or records) that contain the string "aspirin" can be posed as a database problem (an area of computer science). Importantly, informatics and computer science differ in the problems that they address (see Figure 2.4). It should not be implied that computer science is easier or less intellectually challenging compared to informatics (or vice versa).

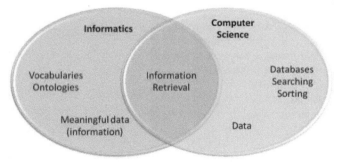

Figure 2.4: Relationship between informatics and computer science in area of information retrieval.

Increasingly, the term "data scientist" is used to refer to professionals engaged in the retrospective analysis of incidentally collected data (such as the online activity of users of a website). With biomedical data, effective analysis often requires the attribution of meaning to such data. So, we would argue, the biomedical data scientist must take both information and raw data into account (i.e., engage in informatics and not solely data analysis).

Artificial Intelligence (AI)

AI is generally considered to be a sub-field of computer science. This is arguably appropriate (particularly for the current crop of AI systems) since the focus is usually on the development of generalizable methods through which a computer can exhibit behavior that appears intelligent. AI is concerned with the development of systems that can do something that previously required human intelligence, such as driving a vehicle in city traffic, winning a game of chess, or solving a logic puzzle. Originally, AI developed in parallel with studies of human expert cognition, and a prevailing notion was that the simulation of intelligent behavior required systems based on knowledge of how expert humans solve problems in a given domain. However, the focus subsequently shifted away from the design of systems that use human-like processes, to the development of systems that can attain human-like performance regardless of how this performance is obtained (i.e., without simulating human cognition or expertise). Recent advances in statistical AI or *"machine learning"* have enabled computers to solve problems that have previously resisted automation. Specifically, *"deep learning"* refers to the use of multi-layer neural networks to learn patterns such as the features of objects in an image. A particularly prominent success in the biomedical domain has been in the field of dermatologic (skin) lesion categorization. Researchers at Stanford University were able to use a very large set of labelled images (129,450) showing various types of skin lesions to train a computer to distinguish specific kinds of malignant (cancerous) lesions from similar benign (non-cancerous) lesions. Importantly, the system was provided only pixel-level data and labels, no attempt was made to provide the system with any knowledge about how to recognize any dermatologic disease. System performance was compared against 21 board-certified dermatologists. The system performed comparably to the dermatologists.[8] This was an impressive and potentially clinically-useful application. This project also illustrates two limitations of deep learning. First, that it requires large sets of labelled data to "train" the system. Second, that the system cannot explain "why" it does something to a human. Other applications of deep learning that may impact biomedicine include natural language processing and speech recognition.

CONVERTING DATA TO INFORMATION TO KNOWLEDGE

We live in the real world that contains physical objects (e.g., *aspirin tablet*), people (e.g., *John Smith*), things that can be done (e.g., John Smith *took* an aspirin tablet) and other concepts. To do useful computation in this context, one must segregate some part of the physical world and create a **conceptual model**. The conceptual model contains only the parts of the physical world that are relevant to the computation. Importantly, everything that is not in the conceptual model is excluded from the computation and assumed to be irrelevant.

The conceptual model is used to design and implement a **computational model**. In Figure 2.5, the real world contains a person, John Smith. There are many other things in the real world including other people, physical objects, etc. There are many things that we can say about this person, they have a name, height, weight, parents, thoughts, feelings, etc. The conceptual model defines what is relevant; everything that is not in the conceptual model is therefore assumed to be not relevant. In our example (Figure 2.5), name and age are chosen. Thus, the height, weight and all other things about John Smith are assumed to be irrelevant. For example, given our conceptual and computational models, one would not be able to answer questions about height. Next a **representation** must be defined. (Figure 2.5). A simple example is that of whole numbers. A representation has three components. The **represented world** is the *information* that one wants to represent (e.g., whole numbers: 0, 1, 2, 3, ...). The **representing world** contains the data that represent the information (e.g., symbols "0", "1", "2", "3", ...). There must be a **mapping** between the represented world and the representing world. In our example, the mapping is the correspondence between whole numbers and symbols that are used to represent them. Note that the data are, in and of themselves, meaningless.

To do anything useful, one must also have rules regarding the mapping (i.e., relationship between the symbols and the real world), and what can be done with the symbols. In our example, these rules are the rules governing the manipulation of whole numbers systems (e.g., addition, multiplication, division, etc.).

The data part of a representational system may also be called its "form", in which case meaning is called its' "content." The word "form" is significant because of its relationship to **formal methods**, which are methods that manipulate data using systematic rules that depend only on form, not content (meaning). These formal methods, including computer programs, depend only on systematic manipulation of data without regard for meaning. Thus, only a human can ensure that the input and output of a formal method (e.g., computer program) correctly capture and preserve meaning. In the skin cancer example, deep learning network described above, the humans who designed or used the network know that the input represents digitized images of skin lesions and that the output represents whether the lesion is cancerous or not. However, the trained network knows nothing about lesions or cancer, it is simply a complex non-linear mathematical function mapping input (digitized images) to output (cancerous vs. non-cancerous). Recent research on deep neural nets has shown that they can be reliably fooled into confidently misclassifying images by adding noise that humans cannot perceive. For instance, researchers can add noise to an image of a panda to create a second image that to humans is indistinguishable from the first, but that causes a deep net to confidently classify the first image as a panda and the second as a gibbon.[10] Of course, there's by now a far larger literature on situations in which human diagnosticians reliably make mistakes.[11] That human and machine diagnosticians reach their conclusions through different processes suggests that they will make different sorts of errors, and that the safest system may be one that takes both of their perspectives into account.

In spite of the fact that formal methods manipulate only form (or data), not meaning, they can be very useful. If the formal method does not violate the rules of the physical world, one can apply the method to solve problems in the real world. For example, a whole number representation can be used to determine how many 8-person boats are needed to transport 256 people across the Nile river (i.e., 256 people divided by 8 people/boat = 32 boats).

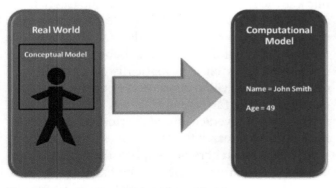

Figure 2.5: Computational framework

However, one must be careful because the formal method (division) can easily violate the rules of the real world. For example, suppose that 250 people are in Cairo and six people are in Khartoum (1,000 miles away) and they must cross at the same time. In this case, 32 boats is the wrong answer since 32 boats are needed in Cairo and another boat is needed at Khartoum. In this example, the real world includes location (Cairo vs. Khartoum), but the conceptual model includes only the number of people; location and distance are ignored. Thus, the computational model (based on the conceptual model) gives an inappropriate answer. It can't be said that the answer is "wrong." Clearly 256/8 = 32; the computer did not malfunction. However, in the case where location is important, the numerical answer is not useful.

The distinction between the real (represented) world, the conceptual model (representing world) and the computational model (that which the computer manipulates) is fundamental to informatics.

When the real world, the conceptual model and the computational model match, it is possible to get useful answers from the computer. When they do not match, such as the case when a critical constraint was left out of the conceptual model, the answers obtained from the computer are not useful. This is what happened in the case of e-patient Dave. Formal methods (computer programs) were developed that linked fields in the Beth Israel Deaconess EHR to fields in Google Health. Data from one were dutifully transferred to the other. However, the meaning (i.e., that the data being transmitted were billing codes, not actual diagnoses) was lost. Further, there was a flaw in the conceptual model, the computational model or both models that prevented dates from being maintained correctly; perhaps because the dates reflected billing dates, rather than the date when a diagnosis was made.

Data to Information

The next step is to convert data into information. Consider the example in Figure 2.1. "C34.9" is, in and of itself, meaningless (i.e., it is a data item or datum). However, ICD-10-CM gives us a way to interpret C34.9 as *"Lung neoplasm, not otherwise specified."* Thus, the vocabulary ICD-10-CM turns the datum into a unit of information.

The computer still stores only data, not information. Thus, only a human can determine whether the meaning is preserved or not. In the case of e-patient Dave, all the computer systems functioned as they were designed. There were no *"computer errors,"* but upon human review, the meaning was mangled.

However, associating ICD-10-CM C34.9 with a patient record labels the patient record (and thus the patient) as having *"Lung neoplasm, not otherwise specified."* Of course, one could design systems that turn data into information without using vocabularies. For example, patient records could be designed that include a bit for each possible diagnosis. Thus, setting the bit corresponding to lung cancer to 1 would be **semantically equivalent** to associating ICD-10-CM C34.9 with the patient's record. Semantically equivalent is simply another way of stating that the meanings are the same.

Transmission of information between computer systems, often referred to as **interoperability**, requires consistency of interpretation in the context of a particular task or set of tasks.[12] The source system (Beth Israel Deaconess EHR for e-patient Dave) and the receiving system (Google Health for e-patient Dave) must share a common way of transforming data into information. However, this is not sufficient. Note that in the case of e-patient Dave, both systems used ICD codes. However, associated information such as dates and most importantly the context: billing code vs. actual diagnosis, was not shared correctly.

Information to Knowledge

Multiple methods have been developed to extract knowledge from information. Note that it would not make sense to directly convert data (which by definition are not meaningful) to knowledge (justified, true belief). Thus, information is required to produce knowledge. Transformation of information (meaningful data) into knowledge (justified, true belief) is a core goal of science.

In the clinical world, most available knowledge is best described as justified (i.e., evidence exists that it is true), rather than proven fact (i.e., it must be true). This is an important distinction from traditional hard sciences such as physics or mathematics.

In this chapter, there is a focus on informatics techniques that are designed to convert clinical information into knowledge. Thus, clinical data warehouses (CDWs) are described that are often the basis for attempts to turn clinical information into knowledge, as well as methods for transforming information into knowledge.

Clinical research informatics is recognized as a distinct sub-field within informatics (see separate chapter on e-research for further information). Clinical research informaticians leverage informatics to enable and transform clinical research.[13-14] By "enable," what is meant is helping researchers accomplish their goals faster and cheaper than is possible using existing methods. For example, searching electronic clinical data may be faster than manually reviewing paper clinical charts. "Transform" means developing methods that allow researchers to do things that they could not do using existing methods. For example, it is not possible to use aggregated clinical data contained in paper records to help clinicians make decisions in real time. One cannot ask, in real-time or near real-time, *"what happened to patients like me, at your institution, who chose treatment A vs. treatment B?"* Although the information required to answer this question is found in the clinical records, a manual chart review cannot be performed in real time. However, to derive knowledge from information and realize the benefits of computerized information, we must ensure that meaning is preserved.

CLINICAL DATA WAREHOUSES (CDWS)

The enterprise data warehouse was introduced in chapter 1. In this chapter, the focus will be on clinical, rather than administrative data, hence the reference to a **clinical data warehouse** or **CDW**.

Increasingly, clinical data are collected via electronic health records (EHRs). Clinical records within EHRs are composed of both **structured data** and **unstructured or (free text)**. Structured data may include billing codes, lab results (e.g., Sodium = 140 mg/dl), problem lists (e.g., Problem #1 = ICD-10-CM C34.9 = "*Lung Neoplasm, Not Otherwise Specified*"), medication lists, etc. In contrast, free text is similar to this chapter – simply human language such as English, called **natural language**. Although templates are often used, key portions of clinical notes are still often dictated and are represented in records as free text.

From an informatics perspective, structured data are much easier to manage – structured data are computationally tractable. Ideally, but not always, these data are encoded using a standard such as ICD-10-CM (previously ICD-10-CM in the United States, see chapter on data standards). Thus, retrieving patients with a particular problem is, theoretically, simply a matter of identifying all records that are tagged with a particular code. As one will see later in this chapter, in practice this does not always work. Further, nuances (e.g., similarity to a previous case) or vague concepts (e.g., light-colored lesion, tall man) may be difficult to convey with a *"one size fits all"* vocabulary.

Similarly, computerized physician order entry (see chapter on electronic health records) can be difficult to implement. If designers allow only structured data, they must anticipate what will be ordered and make choices that constrain the possible inputs. For example, they may choose to use a particular vocabulary for medication orders, allow specific dosing frequencies, etc. Inevitably, however, physicians will want to write unusual orders that will be difficult to accommodate.

Free text, on the other hand, has the advantage of being able to express anything that can be expressed using natural language. On the other hand, it is difficult for computers to process. Indeed, the field of **natural language processing** (NLP) is an active area of research in both computer science and informatics. Within clinical records, the free text notes are critically important. Indeed, as in the case of e-patient Dave, structured data (such as billing codes) may not accurately reflect clinical reality. This is not necessarily anyone's fault. Billing codes were assigned for billing, not for clinical care. Thus, it should not be surprising that using billing codes for a different purpose does not yield the desired result. Over 20 years ago, van der Lei warned:

> *...under the assumption that laws of medical informatics exist, I would like to nominate the first law: Data shall be used only for the purpose for which they were collected. This law has a collateral: If no purpose was defined prior to the collection of the data, then the data should not be used.*[15]

To make sense of clinical records, both structured data and free text must be leveraged. This remains an active area of informatics research.

A clinical data warehouse is a database system that collects, integrates and stores clinical data from a variety of sources including electronic health records, radiology and other information systems. EHRs are designed to support real-time updating and retrieval of individual data (e.g., Joan Smith's age). The general process is shown in Figure 2.6. Data from multiple sources including one or more EHRs are copied into a staging database, cleaned and loaded into a common database where they are associated with **meta-data**. Meta-data are data that describe other data. For example, the notation that a data item is an ICD-10-CM term represents meta-data.

Once loaded into a CDW, a variety of analytics can be applied, and the results presented to the user via a user interface. Examples of simple analytics include summary statistics such as counts, means, medians and standard deviations. More sophisticated analytics include associations (e.g., does A co-occur with B) and similarity determinations (e.g., is A similar to B).

In contrast to EHRs, CDWs are designed to support queries about groups (e.g., average age of patients with breast cancer). Although in principle an EHR may contain the same data as a CDW, databases that support EHRs are designed for efficient real-time updating and retrieval

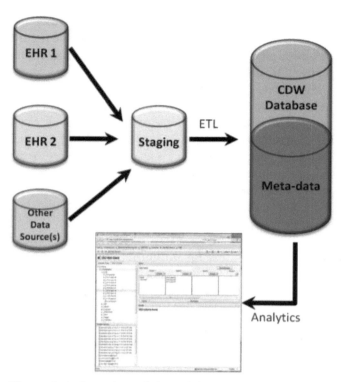

Figure 2.6: Overview of clinical data warehousing (ETL = Extract, transform and load)

quality improvement strategies) in specific patient populations (e.g., retrieve all women who are 40 years old or older who have not had a mammogram in the past year). Similarly, clinical and translational researchers use CDWs to identify trends (e.g., did screening mammograms detect breast cancer at an early stage?).[16] Comparative effectiveness research (CER) or, more broadly, practice-based research, are increasingly important fields that attempt to link research with clinical practice using CDWs. They complement traditional clinical trials that ask very focused questions. For example, a clinical trial might be designed to compare treatment A vs. treatment B in a particular population of patients. In contrast, CER practitioners ask what happened in practice. For example, treatment A has been found to be more effective than treatment B in a clinical trial. What actually happened in practice?

Hospital infection control specialists use CDWs to track pathogens within hospitals. Public health agencies traditionally rely on reporting to conduct surveillance for natural or man-made illnesses (see chapter on public health informatics). However, reporting introduces a delay. Accessing aggregated data at the institutional level can be done much faster using a CDW.

of individual data. Thus, a query across patients rather than regarding an individual may take much more time. Further, since EHRs support patient care, queries about groups may be restricted to ensure adequate performance for clinicians. Another important distinction is that CDWs are usually not updated in real-time. Although update schedules differ, daily or weekly updates of the institutional CDW are typical.

CDWs are rapidly becoming critical resources. They enable organizations to monitor quality by allowing users to query for specific quality measures (see chapter on

One of the most popular clinical data warehousing platforms is the product of the Informatics for Integrating Biology and the Bedside (i2b2) project based at Harvard Medical School.[17] The open source and very modular i2b2 platform was designed to enable the reuse of clinical data for research but can also be very useful for non-research tasks such as quality monitoring. As of August 2017, i2b2 has been implemented at over 100 institutions including academic institutions and commercial entities in the US and abroad.[18]

I2b2 relies on a star schema composed of facts and dimensions (Figure 2.7.). *Facts* are pieces of information that are queried by users (e.g., diagnoses, demographics,

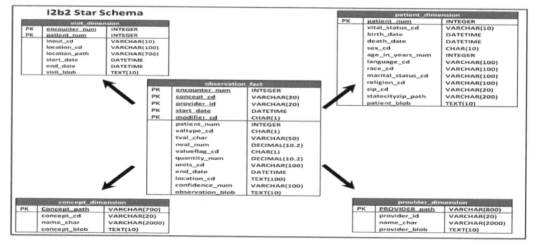

Figure 2.7: i2b2 data model 12

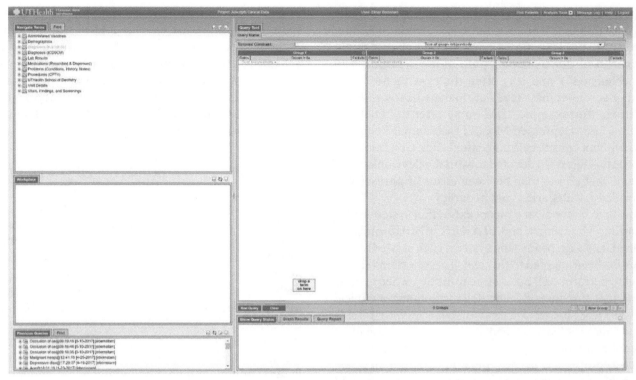

Figure 2.8: i2b2 screenshot showing the result (patient count) of a query for female patients ages 45-64 with ischemic heart disease

laboratory results, etc.) and *dimensions* describe the facts. Note that the data model is organized around facts, rather than individual patients, as would be the case for an EHR. Another benefit of organizing the CDW around observations is that data from multiple sources (e.g., different hospitals) can be aggregated into a common data model – new observations are simply added to the table of facts. Meta-data, such as the vocabulary that was used for encoding the fact, is an important component. Thus, the i2b2 data model by itself is not sufficient to ensure interoperability.

I2b2 also provides a very usable interface to an institutional CDW that can be used by non-informaticians (see Figure 2.8). Users click and drag concepts from the ontology window (upper left) into the query panes (upper right) and obtain results, such as the number of patients fulfilling certain criteria, in lower right. In addition to the basic i2b2 package, specialized modules have been developed for NLP and other tasks.

In short, clinical data are collected via EHRs and archived in CDWs. As EHRs are becoming increasingly common, CDWs are becoming increasingly important. However, to realize the potential of CDWs to improve health, we must do more than archive data. One must turn these data into information and knowledge. Users must be able to "*make sense*" of clinical data; to make clinical data meaningful (data → information) and then learn from aggregated clinical data (information → knowledge). In practice, many of the benefits of EHRs (see chapter 3) actually require a CDW. The transformation of data into information and knowledge is a core concern of informaticians.

Use of Aggregated Clinical Data

To make use of aggregated clinical information, we must be able to recognize records that belong to patients with specific conditions. For example, it is necessary to identify records belonging to patients who have been diagnosed with breast cancer. A simple answer is to rely on billing codes, one of the most common forms of structured data in clinical records. However, as we saw in the case of e-patient Dave, one cannot simply rely on billing codes. Sometimes other structured data are available, problem lists are particularly useful. Unfortunately, problem lists are often out of date or incomplete.[19] Thus, a great deal of interest has focused on extracting information from free text clinical notes.

Concept extraction refers to the problem of identifying concepts within unstructured data, such as discharge summaries or pathology reports. Usually, these concepts are mapped to a controlled vocabulary, such as ICD-10-CM, SNOMED-CT and others. While this may on the surface appear to be a trivial problem, there are many ways in which a single concept might be expressed (for example "high blood pressure" and "hypertension"), and it is often the case that a single word or acronym may have multiple medically relevant meanings (for example "DM" may refer to "Diabetes Mellitus" or "Depressed Mood") that cannot be teased apart without considering contextual cues. Consequently, much effort has been devoted toward the development of systems that aim to map between terms or phrases and controlled vocabularies with accuracy.

Multiple biomedical concept extraction systems exist including MetaMap[20] and cTAKES.[21] Broad-purpose medical language processing systems such as MedLEE,[22] have also been adapted to this end. These systems can be tuned to perform well but require re-tuning when applied to different corpora (e.g., changing institutions) or clinical problems (e.g., breast cancer vs. diabetes mellitus). Table 2.1 summarizes the published performance of these three concept extraction systems; note that the results are not directly comparable to each other due to different tasks, experimental design (e.g., pre-processing), and gold standards (a common limitation).[23-24]

Table 2.1: Published performance of three notable biomedical systems

Concept Extractor	Gold Standard	Precision	Recall	F-score (F1)
cTAKES[21]	Mayo clinic	0.80	0.65	0.72
MetaMap (MMTx)[25]	Proprietary	0.74	0.76	0.748
MEDLEE[26]	Proprietary	0.86	0.77	0.81

Classification refers to the problem of categorizing data into two or more categories. For example, one might want to classify medical records as belonging to patients who have vs. have not been diagnosed with breast cancer. A variety of classification algorithms have been developed, most of which rely on statistical methods. These classification algorithms generally depend on the selection of a set of features, such as the presence or absence of particular terms, concepts or phrases. Once these features have been selected, either manually or through automated methods, medical records can be categorized based on these features. A commonly utilized approach is supervised machine learning, in which an algorithm is used to learn a representation of the features that characterize annotated positive (patients with breast cancer) and negative (patients without breast cancer) cases. New cases can then be categorized automatically based on the extent to which their features are characteristic of previously encountered positive or negative examples. Deep learning using multi-layer neural networks described above, is an example of supervised learning approaches.

WHAT MAKES INFORMATICS DIFFICULT?

Why are some domains highly computerized, while health care and biomedicine resist computerization? Consider the banking system.[4] It is clearly very complex and involves a vast quantities data and meaning. Why do all banks use computers? In contrast to health care, there are no arguments regarding the suitability of computers to track accounts. We argue that in the case of banking, there is a very narrow *"semantic gap"* between data and information. In other words, the correspondence between the data (numbers) and information (account balances) is very direct. As one manipulates the computational model, the meaning of these manipulations follows easily.

Consider the differences between banking data and health care data, such as an account at a bank versus a patient (Table 2.2). One difference is that concepts relevant to health are relatively poorly defined compared to banking concepts. The symbols require significant background knowledge to interpret properly. For example, there are multiple ways that a patient can be "sick" including derangements in vital signs (e.g., extremely high or low blood pressure), prognosis associated with a diagnosis (e.g., any patient with an acute aortic dissection is sick), or other factors. Two clinicians when asked to describe a "sick" individual may legitimately focus on different facts. In contrast, a bank account balance (e.g., $1058.93) is relatively objective and is captured by the symbols. Thus, data-manipulating machines (IT) are much better suited to manipulating bank accounts than clinical descriptors.

In general, if the problem relates strictly to form (data) or is easily reduced to a form-based problem, then computers can easily be applied to solve the problem. Retrieving all abstracts in PubMed containing the string *"breast cancer"* is a question related to data and is easily reducible to a form-based data query. On the other hand, retrieving all documents that report a positive correlation between beta blockers (a class of medications) and weight gain is an information retrieval question that depends on the meaning of the query and the meaning of the text in

Table 2.2: Comparison of health and banking data

	Banking data	Health data
Concepts and descriptions	Precise *Example:* Account 123 balance = $15.98	General, subjective *Example:* sick patient
Actions	Usually (not always) reversible *Example:* Move money A ⊠ B	Often not easily reversible *Example:* Give a medication Perform procedure
Context	Precise, constant *Example:* US $	Vague, variable *Example:* Normal lab values differ by lab
User autonomy	Well-defined and constrained *Example:* What I can do with my checking account = what you can do	Variable and dependent on circumstance *Example:* Clinical privileges depend on training, change over time, depend on circumstances
Users	Clerical staff	Varied, including highly trained professionals
Time sensitivity	Few true emergencies (seconds)	Many time sensitive tasks, highly variable time sensitivity depending on context
Workflow	Well-defined	Highly variable, implicit

the documents. The latter question is not easily reducible to form and is therefore much harder to automate.

Concepts definable with necessary and sufficient conditions are usually relatively easy to reduce to form, and thereby permit some limited automated processing of meaning. However, concepts without necessary and sufficient conditions (e.g., recognizing a sick patient, or defining pain) cannot be easily reduced to data and are much more difficult to capture computationally. Informatics is interesting (and difficult), in part, because many biomedical concepts defy definition via necessary and sufficient conditions.

Blois argued that, to compute upon a system, one must first determine the system's boundaries.[27] In other words, one must define all the relevant components and assume that everything else is irrelevant. However, this is very difficult to do for biological (or human) systems. If the goal is to model the circulatory system, can the renal system be excluded? The endocrine system that includes the adrenal glands (releases epinephrine that constricts blood vessels and raises blood pressure)? The nervous system? And so on. With a bank account, it is easy to draw boundaries around the real-world concepts that affect an accurate account balance. On the other hand, in biomedicine these boundaries are often impossible to precisely define, so our conceptual and computational models are rarely complete and often lead to inaccurate results, such as was seen with e-Patient Dave.

COMPLEXITY OF KNOWLEDGE MODELS

Modeling health care is difficult, but this has not stopped informaticians from trying. Notable modeling attempts include the HL7 Reference Information Model or RIM (see chapter on data standards). Work on the RIM started in 1997 and Release 1 was approved by the American National Standards Institute (ANSI) in 2003. The RIM is one of the major differences between the commonly adopted HL7 version 2.x that has been widely used for decades and version 3, which has not been as widely adopted.[28] One of the problems is that the RIM is very complex (see Figure 2.9) and does not necessarily match all health care environments.

Biomedical informatics is also difficult because biomedical information can be imperfect in several different ways:

- Incomplete information: Information for which some data are missing, but potentially obtainable.
  - Example: What is the past medical history of an unconscious patient who arrives at ED?
- Uncertain information: Information for which it is not possible to objectively determine whether it is true or false. This can also be called epistemic uncertainty, because it arises from a lack of knowledge of some underlying fact. This type of imperfection is addressed by probability and statistics.
  - Example: how many female humans are in the US? Although there is a precise answer to this question at any given moment, we can only estimate the answer using statistics.
- Imprecise information: Information that is not as specific as it should be.
  - Example: Patient has pneumonia. This may be precise enough for some purposes but is not sufficiently precise to determine treatment. For example, antibiotics can treat bacterial pneumonia, but are of little use to a patient with viral pneumonia.
- Vague information: Information that includes elements (e.g., predicates or quantifiers) that permit boundary cases (tall woman, may have happened, large bruise, big wound, elderly man, sharp radiating pain, etc.). Unlike uncertain information, with vague information there is no underlying matter of fact. Even if the age of every female human in the US was known, one could not precisely answer the question of how many mature women were in the US at that time, because "mature" is a term that has boundary cases; there are women who are clearly mature, those who clearly are not, and a number in between for whom one cannot be sure that term applies.
- Inconsistent information: Information that contains two or more assertions that cannot simultaneously hold.
  - Example: Birthdate: 8/29/66 AND 9/17/66

As illustrated in the above examples, all these imperfections may be found in healthcare information. Humans can deal with these imperfections. For example, it can be decided that for clinical purposes, a difference in

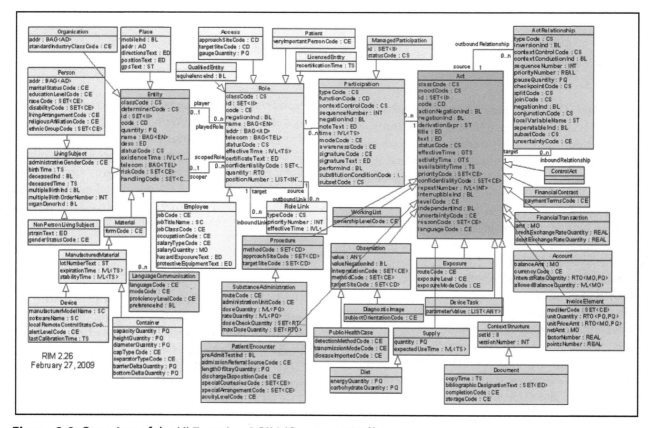

Figure 2.9: Overview of the HL7 version 3 RIM (Courtesy HL7 [29])

patient age of a little over two weeks (a vague statement), is insignificant for clinical purposes. Computers, on the other hand, must be explicitly programmed to make such "judgments." However, the number of possible variances and exceptions is effectively infinite. Thus, they cannot all be anticipated and addressed in advance. This is one reason why clinical decision support often gives advice that is, to a clinician, obviously inappropriate to the current patient situation.

In addition, definitions in health care and biomedicine often change over time. Consider the definition of a gene.[30]

Designing systems that adapt to changes in definition that, in turn, can affect other definitions is difficult. Our computers and programming languages process discrete symbols according to precise formal rules or mathematical expressions. They do not make sense of a highly ambiguous, noisy world or do meaning-based processing. With this background, one can now consider health IT and its various successes and failures in the real world.

WHY HEALTH IT SOMETIMES FAILS

"To improve the quality of our health care while lowering its cost, we will make the immediate investments necessary to ensure that within five years all of America's medical records are computerized. This will cut waste, eliminate red tape, and reduce the need to repeat expensive medical tests... it will save lives by reducing the deadly but preventable medical errors that pervade our health care system." —Barack Obama (Speech on the Economy, George Mason University, January 8, 2009)

Widespread dissatisfaction with health care in America and rapid advancement in information technology has focused attention on Health IT (HIT) as a possible solution. The need for HIT is one of the few topics upon which Democrats and Republicans agree. Both former President Bush and President Obama set 2014 as the goal date for computerizing medical records. To many, HIT seems like an obvious solution to our health care woes. The government's HIT website says that HIT adoption will: improve health care quality, prevent medical errors, reduce health care costs, increase administrative efficiencies, decrease paperwork and expand access to affordable care.[9] However, there is increasing evidence that HIT adoption does not guarantee these benefits. Unmitigated enthusiasm is dangerous for HIT adoption. Similar enthusiasm repeatedly threatens the field of artificial intelligence, resulting in cycles of excitement and disappointment (in artificial intelligence, these cycles are sometimes called "AI winters").

Effects of HIT

HIT is an *"easy sell"* to an American public increasingly dissatisfied with our health care system. Indeed, there is evidence that HIT can improve health care quality, prevent medical errors, and increase efficiency.[31-32] Thus, there is reason for optimism. With the American Recovery and Reinvestment Act (ARRA) of 2009, the US government made a multi-billion-dollar investment in HIT.[33] Similar investments have been made by the governments of Australia,[34] Belgium,[35] Canada,[36] Denmark,[37] and the United Kingdom.[38]

However, many and perhaps even most HIT projects fail.[39] There is also evidence that HIT can worsen health care quality to the point of increasing mortality, increasing errors and decreasing efficiency.[40-42] In November 2011, the Institute of Medicine issued a report entitled *"Health IT and Patient Safety: Building Safer Systems for Better Care"* that concluded: "...*some products have begun being associated with increased safety risks for patients."*[43] There is even a term, "e-iatrogenesis," that refers to the unintended deleterious consequences of HIT.[44] Notably, systems that increase mortality at one institution, do not seem to have the same effect at another institution;[40,45] even though the clinical setting (pediatric intensive care) was similar. Thus, one cannot simply conclude that the system itself is wholly responsible. It is not just the system being implemented, but how it is implemented and in what context that affects the clinical outcomes.

We've Been Here Before: AI Winters

During the 1950s, we were faced with a different problem: the Cold War. Similarly, the government saw IT as a promising (at least partial) solution. If researchers could develop automated translation, we could monitor Russian communications and scientific reports in "real time." There was a great deal of optimism and "...*many predictions of fully automatic systems operating within a few years."*[46]

Although there were promising applications of poor-quality automated translation, the optimistic predictions of the 1950s were not realized. The fundamental problem of context and meaning remains unsolved. This made disambiguation difficult resulting in amusing failures. Humorous examples include: *"the spirit is willing but the flesh is weak"* translated English → Russian → English resulted in the phrase *"the vodka is good but the meat is rotten."*

In 1966, the influential Automatic Language Processing Advisory Committee (ALPAC) concluded that *"there is no immediate or predictable prospect of useful machine translation."*[47] As a result, research funding was stopped and there was little automated translation research in the United States from 1967 until a revival in 1976-1989.[46]

Similarly, there is currently tremendous interest in HIT. Although there is good evidence that HIT can be useful, some will certainly be disappointed. A recent report by the National Research Council (the same body that published the ALPAC report) concluded that "… *current efforts aimed at the nationwide deployment of health care IT will not be sufficient to achieve the vision of 21st century health care, and may even set back the cause if these efforts continue wholly without change from their present course.*"[48] Thus, there is reason for concern that HIT (and perhaps even informatics, in general) may be headed for a bust. Such an "HIT winter" would be unfortunate, since there are real benefits of pursuing research and implementation of HIT.

The Problem: Health Information Technology is Really Health Data Technology

The fundamental problem is that existing technology stores, manipulates and transmits data (symbols), not information (data + meaning). Thus, the utility of HIT is limited by the extent to which data approximates meaning, or more precisely to the ability of HIT to act "as if" it understands the meaning of the data. Unfortunately, in health care, data do not fully represent the meaning. In other words, there is a large gap between data and information. Since the difference between data and information is meaning (semantics), this gap is referred to as the *"semantic gap."*

Social and Administrative Barriers to HIT Adoption. Manipulating data and not information has many consequences for HIT. Note that there is no shortage of computers in hospitals. Hospitals managed their financial data electronically long before they computerized clinical activities. Just like any other organization, many hospitals have functioning e-mail systems and maintain a Web presence. Many clinicians used personal digital assistants,[49] some even communicate with patients using e-mail.

The social and administrative barriers to HIT adoption have been discussed by multiple authors in countless papers. Such barriers include a mismatch between costs and benefits, cultural resistance to change, lack of an appropriately trained workforce to implement HIT and multiple others.[50] To some, clinicians' resistance to computerization appears irrational. However, caution seems increasingly reasonable given the mixed evidence regarding the benefits of poorly-implemented HIT.

FUTURE TRENDS

Significant research problems must be addressed before HIT becomes more attractive to clinicians. Many of these are outlined in a National Research Council report.[48] First, there is a mismatch between what HIT can represent (data) and concepts relevant to health care (data + meaning). This is a very difficult and fundamental challenge that includes multiple long-standing challenges in artificial intelligence (e.g., how computers can be "taught" context or common sense) that have proven very difficult to solve. It seems that until one has true information processing, rather than data processing, technology, the benefits of HIT will be limited.

Second, HIT must augment human cognition and abilities. Friedman expressed this elegantly as the *"fundamental theorem of informatics"*: human + computer > human (humans working with computers should perform better than a human alone).[51] The theorem argues that there must be a clear and demonstrable benefit from HIT. Despite the problems with current HIT, there are clearly situations where HIT can be beneficial. In some ways, human cognition and computer technology are very complementary. For example, monitoring (e.g., waveforms) is much easier for computers than for humans. In contrast, reasoning by analogy across domains is natural for humans but difficult for computers.

How Progress Will Be Made

Researchers are exploring multiple promising paradigm-shifting ideas. Examples of approaches that address some of the fundamental problems described in this chapter can be provided.

One approach is to recognize the complementary strengths of humans and computers. Humans are good at constructing and processing meaning. In contrast, computers are much better at processing data. Users can leverage this understanding to design systems that harness the data-processing power of computers to present (display) data in ways that make it easier for humans to grasp and manipulate meaning. For example, a *word cloud visualization* shows the term frequency in text.[52] The size of the font is proportional to the frequency of the term.

Returning to HIT, one can apply these same principles. For example, Figure 2.10 shows an example of an

EHR that integrates clinical decision support. This is not novel, but this example illustrates what could be done by combining multiple types of information on the same screen with an understanding of the user's task.

Defining scenarios when HIT is beneficial with all relevant parameters and demonstrating that using HIT is *reliably* beneficial in these scenarios remains a research challenge. In its present form, HIT will not transform healthcare in the same way that IT has transformed other industries. This is due in part to the large semantic gap between health data and health information (concepts). In addition, many problems with healthcare require non-technological solutions, such as changes in healthcare policy and financing.

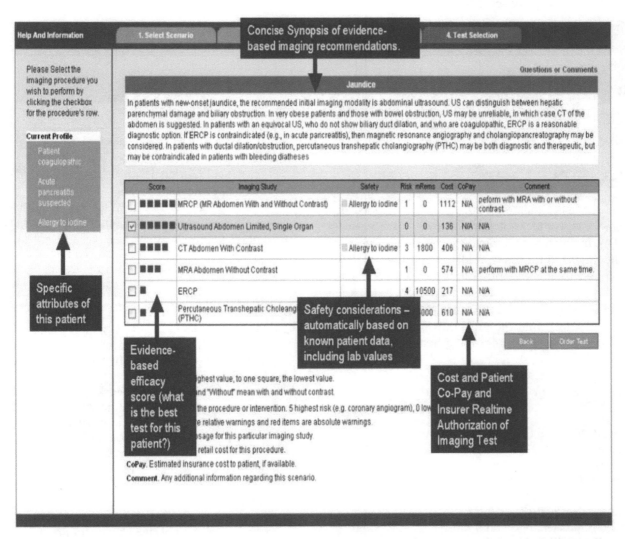

Figure 2.10: EHR screen (from John Halamka) showing integration of decision support into the EHR [53]

KEY POINTS

- Data are observations reflecting differences in the world (e.g., "C34.9") while information is meaningful data or facts from which conclusions can be drawn and knowledge is information that is justifiably believed to be true
- Data are largely the domain of information technology (IT) professionals and computer scientists; information and knowledge are the domains of informatics and informaticians
- Vocabularies help convert data into information
- The transformation of data into information and knowledge is a core concern of informaticians
- When the real world, the conceptual model and the computational model match, we get useful answers from the computer
- Concepts relevant to health are relatively poorly defined compared to e.g. banking concepts
- There is a large "semantic gap" between health data and health information

CONCLUSION

Problems in healthcare are information and knowledge intensive. Current technology is centered on processing data. This mismatch, or semantic gap, between the problems healthcare IT tries to address and the available technology explains the difficulties that informaticians face every day. It also explains the differences between Informatics and Computer Science. Informatics must advance our information and knowledge-processing capabilities to continue improving healthcare through technology.

REFERENCES

1. Stead WW, Lin HS, editors. Computational Technology for Effective Health Care: Immediate Steps and Strategic Directions. National Academies Press; Washington, D.C.: 2009. P. 2
2. E-patient Dave http://epatientdave.com/ (Accessed September 5, 2017)
3. Electronic Health Records Raise Doubt, Boston Globe, April 13, 2009. (Accessed September 5, 2017)
4. Bernstam, E.V., J.W. Smith, and T.R. Johnson. What is biomedical informatics? Journal of biomedical informatics, 2010. 43(1): p. 104-105
5. Floridi, L. Semantic conceptions of information. 2005 October 5, 2005; http://plato.stanford.edu/entries/information-semantic/ (Accessed September 5, 2017)
6. Bernstam, E.V., et al., Synergies and distinctions between computational disciplines in biomedical research: perspective from the Clinical and Translational Science Award programs. Academic Medicine : Journal of the Association of American Medical Colleges, 2009. 84(7): 964-70.
7. Manning CD, Raghavan P, Schutze H. Introduction to information retrieval. Cambridge University Press Cambridge 2008.
8. Andre Esteva A, Kuprel B, Novoa RA, Ko K, Swetter SM, Blau HM, Thrun S. Dermatologist-level classification of skin cancer with deep neural networks. Nature. 2017; 542: 115–118. doi:10.1038/nature21056
9. Miotto R, Wang F, Wang S, Jiang X, Dudley JT. Deep learning for healthcare: review, opportunities and challenges. Briefings in Bioinformatics. 2017 May 6:bbx044.
10. Szegedy C, Zaremba W, Sutskever I, et al. Intriguing properties of neural networks. 2013. Cornell University Library. https://arxiv.org/abs/1312.6199 (Accessed August 30, 2017)
11. Chapman, G.B., & Elstein, A.S. (2000). Cognitive processes and biases in medical decision-making. In G. B. Chapman & F. S. Sonnenberg, (Eds.) Decision-making in health care: Theory, psychology, and applications (pp. 183-210). Cambridge: Cambridge University Press.
12. HealthIT.gov https://www.healthit.gov/buzz-blog/meaningful-use/interoperability-health-information-exchange-setting-record-straight/ (Accessed August 30, 2017)
13. Embi, P.J. and P.R. Payne, Clinical research informatics: challenges, opportunities and definition for an emerging domain. Journal of the American Medical Informatics Association: JAMIA, 2009. 16(3): p. 316-27.
14. Payne, P.R., P.J. Embi, and M.G. Kahn, Clinical research informatics: The maturing of a translational biomedical informatics sub-discipline. Journal of biomedical informatics, 2011.

15. van der Lei, J., Use and abuse of computer-stored medical records. Methods of information in medicine, 1991. 30(2): p. 79-80.
16. Zerhouni, E.A., Translational research: moving discovery to practice. Clin Pharmacol Ther, 2007. 81(1): p. 126-8.
17. Murphy, S.N., et al., Serving the enterprise and beyond with informatics for integrating biology and the bedside (i2b2). Journal of the American Medical Informatics Association: JAMIA, 2010. 17(2): p. 124-30.
18. Informatics for Integrating Biology & the Bedside (I2b2) www.i2b2.org (Accessed September 5, 2017)
19. Szeto, H.C., et al., Accuracy of computerized outpatient diagnoses in a Veterans Affairs general medicine clinic. The American journal of managed care, 2002. 8(1): p. 37-43.
20. Aronson, A.R. and F.M. Lang, An overview of MetaMap: historical perspective and recent advances. Journal of the American Medical Informatics Association: JAMIA, 2010. 17(3): p. 229-36.
21. Savova, G.K., et al., Mayo clinical Text Analysis and Knowledge Extraction System (cTAKES): architecture, component evaluation and applications. Journal of the American Medical Informatics Association: JAMIA, 2010. 17(5): p. 507-13
22. Chen, E.S., et al., Automated acquisition of disease drug knowledge from biomedical and clinical documents: an initial study. Journal of the American Medical Informatics Association: JAMIA, 2008. 15(1): p. 87-98.
23. Chapman, W.W., et al., Overcoming barriers to NLP for clinical text: the role of shared tasks and the need for additional creative solutions. Journal of the American Medical Informatics Association: JAMIA, 2011. 18(5): p. 540-3.
24. Stanfill, M.H., et al., A systematic literature review of automated clinical coding and classification systems. Journal of the American Medical Informatics Association: JAMIA, 2010. 17(6): p. 646-51.
25. Meystre S, Haug PJ. Evaluation of medical problem extraction from electronic clinical documents using MetaMap Transfer (MMTx). Studies in health technology and informatics. 2005 Jan;116:823-8.
26. Friedman, C., et al., Automated encoding of clinical documents based on natural language processing. Journal of the American Medical Informatics Association: JAMIA, 2004. 11(5): p. 392-402.
27. Blois, M.S., Information and medicine: the nature of medical descriptions 1984, Berkeley: University of California Press.
28. Smith, B. and W. Ceusters, HL7 RIM: an incoherent standard. Studies in health technology and informatics, 2006. 124: p. 133-8.
29. HL7 Version 3 RIM http://www.hl7.org/documentcenter/public_temp_B69AB426-1C23-BA17-0CA55CBFEF56C9A3/calendarofevents/himss/2011/HL7%20Reference%20Information%20Model.pdf (Accessed September 5, 2017)
30. Hopkin, K., The Evolving Definition of a Gene. BioScience, 2009. 59(11): p. 928-31.
31. Chaudhry, B., et al., Systematic review: impact of health information technology on quality, efficiency, and costs of medical care. Ann Intern Med, 2006. 144(10): p. 742-52.
32. Bates, D.W., et al., Reducing the frequency of errors in medicine using information technology. J Am Med Inform Assoc, 2001. 8(4): p. 299-308.
33. American Recovery and Reinvestment Act (ARRA) of 2009. ARRA Economic Stimulus Package. Hitech Answers. https://www.hitechanswers.net/about/about-arra/ (Accessed September 5, 2017)
34. HealthConnect Evaluation Department of Health and Ageing (DoHA) 24 August 2009 http://www.health.gov.au/internet/main/publishing.nsf/Content/FAFD8FE999704592CA257BF00020A8CF/$File/HealthConnect.pdf (Accessed September 5, 2017)
35. France, F.R., eHealth in Belgium, a new "secure" federal network: role of patients, health professions and social security services. International Journal of Medical Informatics, 2011. 80(2): p. e12-6.
36. EHRS Blueprint: An interoperable EHR framework. Version 2. March 2006. [cited 2011 December 11]; Available from: https://www.infoway-inforoute.ca/en/component/edocman/resources/technical-documents/391-ehrs-blueprint-v2-full (Accessed September 5, 2017))
37. Protti, D. and I. Johansen, Widespread adoption of information technology in primary care physician offices in Denmark: a case study. Issue brief, 2010. 80: p. 1-14.
38. House of Commons Public Accounts Committee. The National Programme for IT in the NHS: Progress since 2006. Second Report of Session 2008-09. [cited 2011 December 11]; Available from: http://www.publications.parliament.uk/pa/cm200809/cmselect/cmpubacc/153/153.pdf (Accessed September 5, 2017)
39. Littlejohns, P., J.C. Wyatt, and L. Garvican. Evaluating computerized health information systems: hard lessons still to be learnt. BMJ, 2003. 326(7394): p. 860-3.
40. Han, Y.Y., et al., Unexpected increased mortality after implementation of a commercially sold computerized physician order entry system. Pediatrics, 2005. 116(6): p. 1506-12.

41. Levenson, N.G. and C.S. Turner, An Investigation of the Therac-25 Accidents. IEEE Computer, 1993(July): p. 18-41.
42. Koppel, R., et al., Role of computerized physician order entry systems in facilitating medication errors. JAMA, 2005. 293(10): p. 1197-203.
43. Services, C.o.P.S.a.H.I.T.B.o.H.C., Health IT and patient safety: building safer systems for better care, 2011, Institute of Medicine of the National Academies: Washington DC.
44. Weiner, J.P., et al., "e-Iatrogenesis": the most critical unintended consequence of CPOE and other HIT. J Am Med Inform Assoc, 2007. 14(3): p. 387-8; discussion 389.
45. Del Beccaro, M.A., et al., Computerized provider order entry implementation: no association with increased mortality rates in an intensive care unit. Pediatrics, 2006. 118(1): p. 290-5.
46. Hutchins, J., Machine translation: history, in Encyclopedia of language & linguistics, second edition, K. Brown, Editor 2006, Elsevier: Oxford. p. 375-83.
47. ALPAC, Language and machines: computers in translation and linguistics. Report by the Automatic Language Processing Advisory Committee, Division of Behavioral Sciences, National Academy of Sciences, National Research Council., 1966, National Academy of Sciences, National Research Council.: Washington, DC.
48. Computational technology for effective health care: immediate steps and strategic directions, W.W. Stead and H.S. Lin, Editors. 2009, Committee on Engaging the Computer Science Research Community in Health Care Informatics, Computer Science and Telecommunications Board, Division on Engineering and Physical Sciences, National Research Council of the National Academies: Washington, DC.
49. McLeod, T.G., J.O. Ebbert, and J.F. Lymp, Survey assessment of personal digital assistant use among trainees and attending physicians. J Am Med Inform Assoc, 2003. 10(6): p. 605-7.
50. Hersh, W., Health care information technology: progress and barriers. JAMA, 2004. 292(18): p. 2273-4.
51. Friedman, C.P., A "fundamental theorem" of biomedical informatics. J Am Med Inform Assoc, 2009. 16(2): p. 169-70.
52. Visualizations: JAMIA Content. May13, 2009. http://wordcloud.cs.arizona.edu/ (Accessed September 5, 2017)
53. My Life as a CMIO. http://geekdoctor.blogspot.com/2007/11/data-information-knowledge-and-wisdom.html (Accessed September 5, 2017)

3

Computer and Network Architectures

WILLIAM R. HERSH • ROBERT E. HOYT

LEARNING OBJECTIVES

After reading this chapter the reader should be able to:

- Enumerate the most important hardware and software components of a personal computer
- Understand how the Internet and World Wide Web work
- Discuss why web services are used by health information organizations (HIOs)
- List the components of service oriented architecture (SOA)
- Understand the importance of RESTful web services
- Understand the importance of networks in the field of medicine
- Compare wired and wireless local area networks (LANs)
- Describe the newest wireless broadband networks and their significance

INTRODUCTION

The average reader of this book, be they a budding student or seasoned clinician, should understand basic architectures and technologies that are commonly part of health information technology. This chapter will focus on four areas: computers, the Internet, web services and networks.

Although it is commonly said that health informatics is more about information than technology, it is important to understand the basic concepts of computers and especially networks to best determine how the technology can make best use of the data, information, and knowledge.

COMPUTERS

Types of Computers

We may first step back and ask, what exactly is a computer? In the early days of computers, it was clearer. In the early days of computers, they were typically large devices that took up a whole room. With the emergence of personal and desktop computers, they usually took up a desk. These were gradually replaced with so-called laptop computers, which are untethered computers from the office or home setting. And now in modern times, we have computing devices that we carry around, including in the palm of our hand. These mobile devices take the form of tablets and smartphones. Figure 3.1 displays some of the more common computer types.

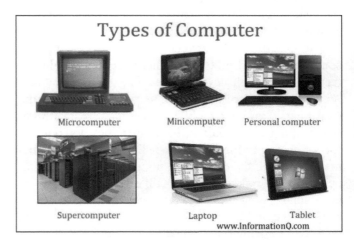

Figure 3.1: Common computer types (Courtesy InformationQ.com)

Let us start from the high end. What are the most powerful computers? In the past, we talked about mainframe computers, which still exist. These are the largest and most expensive computers that often in organizations serve mission-critical needs and many users at the same time. Also, at the very high end are supercomputers, which are powerful computers that are more focused on massive processing and computation of data.

Another category of computers is servers. These are computers that serve many users and handle many transactions across a network. Many of today's modern PCs are powerful enough to serve as servers, but mainframes can serve as servers as well. There is a growing tendency to build server farms consisting of large numbers of relatively commodity style PCs that are easily swapped in and out and maintain a large overall capacity.

Personal computers are probably what most of us think of as computers. These are the desktop and laptop machines, which come in two major categories, the so-called Wintel of mostly Intel-based processors running the Windows operating system, and Apple Macintosh computers running Apple's MacOS. In both categories of Wintel and Apple, laptops now vastly outsell desktops. Whether desktop or laptop, PCs continue to increase in power, especially when they are on networks and have, for the most part, perhaps with the exception of Apple and a few niche Wintel machines, have become essentially a commodity.

Another type of computer that has become very prevalent is the mobile computer. Laptops were the initial incarnation of mobile computers, but now we have computers such as tablets and smartphones that get increasingly smaller. Smartphones are an evolution from a different type of computer that we used to call the Personal Digital Assistant (PDA) and represent the merger of the cell phone with Internet-enabled devices.

The use of mobile devices for health applications is sometimes called mHealth.[1] There are many valuable features that mobile devices have for health interventions, both for healthcare providers, as well as patients and consumers. However, mobile devices also create threats. In particular, organizations have had to develop policies for so-called Bring Your Own Device (BYOD) that make sure that the devices do not undermine computer networks or make data accessible that should not be accessible.[2] A big challenge is to keep mobile computers easy to use and creating innovative applications with them but securing their data and networks.

Tablet computers have undergone an evolution. The tablet computer offers almost the screen size and power of a laptop computer but are even more portable. While the early tablet computers were more rugged, a new genre of lighter and more convenient devices was launched by Apple and followed with devices that use the Google Android operating system. There are also hybrid devices, such as the Microsoft Surface, which can act like a tablet with a touch screen but runs the latest version of Microsoft Windows.

Another type of mobile computer is the smartphone. About one-third of the entire world, and three-quarters of all Americans, own a smartphone.[3] These are hand-held computers that have high portability and are now connected to cellular wireless networks. The two major platforms of smartphones are the iPhone and Android.

The value of mobile devices in healthcare may be most related to their so-called swappable apps, i.e., apps that run on top of a shared data store and can access data collected from sensors, such as GPS, and other sources.[4-5] The most profound impact of smartphones may be in developing countries, where wireless networks are being deployed, in essence bypassing the wired computer networks.[6]

If there has been one certainty about computers, it is that they continue to get increasingly powerful. This has actually been shown quantitatively through Moore's law, which was coined by Intel's Gordon Moore in 1965, who was noticing that everything about computers, whether speed, size, or cost per unit of power, was doubling every 18 months.[7-8] (see Figure 3.2) Examples of Moore's Law can be found by looking at some older devices. A famous picture on the Internet is an IBM computer that was the first computer to ship with a built-in hard disk that was the size of several refrigerators and had five megabytes of storage (https://www.snopes.com/photos/technology/storage.asp). Another comparison comes from a Wikipedia page (https://en.wikipedia.org/wiki/File:Osborne_Executive_with_iPhone_in_2009.jpg), where one of the first commercially successful portable computers, the Osborne Executive from the early 1980s is pictured next to an early iPhone. The Osborne computer is 100 times slower, 100 times heavier, 10 times more expensive, and 485 times bigger than the tiny iPhone.

Many people have predicted the end of Moore's law, and we may finally be reaching the point where computers cannot be made smaller and faster without running into problems of transistors being the size of atoms and being able to pull the heat from them and keep them from overheating. According to a recent article in The Economist, it may well be that the longstanding Moore's law is finally abating, although some disagree.[8-9] Although Moore's law continues to deliver increasingly powerful computers, they do not seem to be getting that much faster. There are aspects of computers that have become much faster over the years, but as Moore's Law provides more computing power, we find ways to take advantage of it so that the overall computing experience is not nearly that much faster. For example, consider graphical user interfaces, which enhance usability but also require more computer power to display the graphics on the screen.

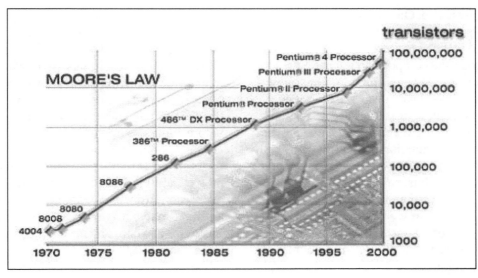

Figure 3.2: Moore's Law (Courtesy Singularity Symposium)

Many lament that we also have "*code bloat*," and that programmers have so much power at their disposal they have less incentive to be efficient in the programs that they write. We also have "*feature creep*," where software producers want to add more features and want to compete with their rivals, even if some of those features are minimally used. There are also other factors, namely congested computer networks, such that faster hardware may just sit idly waiting for information to be delivered over the network. As such, there are a number of mitigating factors for Moore's Law.

Data Storage in Computers

Virtually all modern computers store their data in digital formats. This means that the most fundamental unit of information is discrete. In fact, this fundamental unit is called the binary digit or bit. And an individual bit can assume the value either 0 or 1. Sometimes this is referred to off and on or true or false, but most commonly 0 or 1. Bits are usually grouped together in a sequence of 8, called a byte. A byte can take on 2^8 or 256 different values. For large numbers of bytes, we use prefixes before the name byte, such as a kilobyte, which represents 1,000 bytes, or a megabyte, which represents 1000 kilobytes or a million bytes.

In contrast, analog systems have values on a continuous scale. For example, a clock with hands for the time is an analog device. Even though the underlying electronics may be digital, the clock face itself is analog, in that there is a continuous movement of the hands of the clock through the minutes and the hours. (see Figure 3.3) Many types of data have transitioned from analog to digital, such as music and other sound files. Music has gone from the analog records that were on vinyl to digital forms that was initially on compact disks (CDs) and now is digital in the form of MP3 or other digital formats.

Figure 3.3 Examples of analog vs. digital devices

How is data represented in computers? Essentially, the sequences of bits represent essentially everything, whether numbers, text, the instructions of a computer program, images, video, or genomes. All of these are stored in memory, and the structure by which they're stored determines what they represent, and the computer program that accesses them needs to understand that structure. Probably the most straightforward type of data representation is of numbers. Integers, in particular, are easy to represent in computers. They are essentially straight sequences of bits. Remember that a byte or eight bits, could represent 256 numbers. Depending on the structure chosen to represent the numbers, they can vary from either 0 to 255 or minus 128 to plus 127. Larger integers can be represented by using more bits.

Of course, integers are not the only type of number. There are also floating-point numbers, which have a decimal point. There are different ways that floating-point numbers can be represented. One common approach is to have a sequence for the digit on one side or the other of the decimal point. All of these number issues demonstrate that it is important for the computer to interpret the way the numbers are structured consistently, so that there is proper meaning from them when accessed in memory.

What about representing text in a computer? This is basically done by having a code for each character. The most common representation of English or Roman language characters is ASCII. It represents alphanumeric characters in seven bits, which allows 128 unique characters. That is more than enough to cover the lower case and upper-case letters as well as punctuation points and a few other types of characters. Most PC systems initially extend ASCII to an 8th bit, since computer memory is usually organized in bytes. This allows 256 possible characters. While eight bits per characters may be enough to represent English or other Roman character languages, there are many other languages in the world, some of which have many characters, such as Japanese and Chinese. There are also languages such as Korean, Hebrew, Arabic, Greek, Russian, and more, such that 256 characters is not enough.

For this reason, Unicode was developed (www.unicode.org). Unicode allows representation of much larger character sets by allowing up to 32 bits of representation, allowing literally billions of possible characters. The 8-, 16-, or 32-bit representations are designated by the UTF 8, 16, or 32 designations, which specifies which Unicode set is being used. This is probably sufficient for all human languages, and in fact, Unicode also allows emojis, which are small pictures represented by a single Unicode designation.

Another type of data that commonly represented in computers is images or pictures. Images are represented by picture elements, commonly called pixels. Each pixel, i.e., each element of an image, has one bit in the case of black and white, or more than one in the case of color or grayscale bits at each level of depth. For example, in 8-bit color, there are 256 possible colors or gray scales.

Images take much more memory than text. For example, a 24-bit color, 3000 by 2000 pixel image is 3000 by 2000, or 6 million pixels or 6 megapixels. If there is 24-bit color - 24 bits per pixel – there is a total 144 million bits, or 18 million bytes because there are 8 bits in a byte. This is also 18 megabytes.

Images are commonly compressed, e.g., in the Joint Photographic Experts Group (JPEG) format, an image that is initially 18 megabytes may be compressed substantially. The old adage that a picture is worth a thousand words may well be true with computers. For example, with text, each character takes up one byte so that a single page of text with an average of 50 characters per line over 60 lines takes up about 3000 bytes. But a reasonably high-quality image of that page requires about 1200 by 800 pixels. In black and white, that is 120,000 bytes, although it may be compressed by JPEG or some other compression algorithm. If 8-bit color is added, the picture may get up to 1 million bytes or a megabyte before compression. So indeed, in a computer a picture is worth about a thousand words.

As noted earlier, as amounts of memory get larger, prefixes are used in front of the word "bytes" to make expression easier. For example, if a computer system has a billion bytes, it is much easier to say that it has one gigabyte. The conversions are prefixes that are added at increments of 1,000:

- 1000 bytes = 1 kilobyte (or 1K)
- 1000 kilobytes = 1 megabyte (or 1 meg)
- 1000 megabytes = 1 gigabyte (or 1 gig)
- 1000 gigabytes = 1 terabyte
- 1000 terabytes = 1 petabyte
- 1000 petabtyes = 1 exabyte
- 1000 exabytes = 1 zettabyte
- 1000 zettabytes = 1 yottabyte = 1024 bytes

In the early days of computing, a kilobyte was defined as 1024 bytes or a megabyte as 1024 kilobytes. These were so-called binary conversions. In modern times, these are called kibibytes instead of kilobytes and mebibytes instead of megabyte. But to keep things simpler in the modern era, the standard usage is to use these prefixes that represent increments of 1,000.

The data in computers is held not only in the computer's memory, but also in storage media. There have been many types of storage media over the years that have increased in the amount that they can hold. One of the original storage media for computers was the floppy disk. The largest floppy disks held up to 1440 kilobytes. Larger storage to come after that included CD disks, which hold 640-700 megabytes, which were superseded by the now more common DVD disks that by current standard hold 4.7 gigabytes. Some of the newer standards such as Blu-ray hold up to 50 gigabytes. Of course, these have mostly been superseded by flash memory, which is sometimes called USB memory because it tends to be on so-called *"thumb drives"* that are plugged into USB ports on computers. The capacity of flash memory also continues to grow.

Another important form of storage media is hard disks in computers. Hard disks were originally magnetic but increasingly, there are so-called solid state hard disks, which contain flash memory. It is now typical for magnetic hard disks to be in the range of several terabytes and many flash memory hard disks are approaching that size.

We can look at some digital points of reference to give an idea of the comparative size of different digital objects that might be used in healthcare or in the larger world. A chest x-ray, which is typically about 1200 by 800 pixels at about 8-bit depth is about 7,680,000 bits or 960,000 bytes or about 1 megabyte. Of course, compression usually reduces the size significantly of any type of image. An average size textual book, maybe a medical textbook or something like the Bible is anywhere from one to five megabytes for a typical sized book.

Some have attempted to estimate the size of the digital universe. The EMC Corporation estimated in 2014 that about 4.4 zettabytes of data existed in the world, expanding tenfold by the year 2020 to 44 zettabytes.[10] Healthcare data is actually growing faster than general data, with another report by EMC estimating up to 2.4 zettabytes of health care data by the year 2020.[11]

Computer Hardware and Software

The components of modern computers can essentially be broken down into two categories. There is the hardware, consisting of the physical parts of the computer such as the central processing unit (CPU), memory, auxiliary storage, and input and output devices. This is distinct from software, which is the instructions for the computer and how it acts on the memory. This includes the operating systems, applications or programs, programming languages, and development tools.

With computer hardware, there are some terms that are used that have specific meanings but are not always used properly in the context of those meanings. When talking about the CPU, we generally refer to the chips that are at the core of a computer. These chips typically sit on something called a motherboard that has the main CPU chip and the supporting hardware around it. But sometimes the term CPU is used to describe the entire housing in which a desktop computer sits.

In talking about computer memory, we are typically referring to the active memory that is stored on the chips in a computer. This is technically called random access memory or RAM and is the memory on the motherboard that loses its state when we turn the power off. RAM requires continuous power to maintain it.

Computers have various forms of auxiliary storage where the data is stored, where the memory does not go away when the power is turned off. There are two types of such storage, active and archival. Active storage is used for information that is needed continuously. This data is typically stored on a hard disk directly connected to or a part of the computer. In archival storage, data is less needed less urgently, such as backup information or older information that has exceeded the capacity of the active storage. Archival storage is typically done using magnetic tape, but there are many other ways it can be done, including via optical disks or hard disks, whether they are local or remote on the Internet (increasingly called "cloud storage").

Another important component of computer hardware is input devices. The most common input device that almost every computer has is a keyboard. Sometimes in the case of tiny mobile devices the keyboard is on the screen. With the advent of graphical user interfaces, there are other means to enter data, such as through the use of the mouse and trackpad. Another modality for entering data is speech, with computer recognition of speech and its conversion to language. This is an important application in healthcare.

Also, an important part of computer hardware is output devices. These have typically been monitors and printers. Monitors have improved dramatically over the years as they have gone from cathode ray tubes to flat displays. Their resolution and number of colors has also increased. The original computer monitors, had on the order of 72 to 96 dots per inch, i.e., 72 to 96 individual dots within an inch of the display in each dimension. Printers, on the other hand, have had higher resolution, at a minimum 300 dots per inch, with many printers now having more resolution than that. But monitors have not been standing still either. There are newer higher resolution displays, such as the so-called retina display used by Apple, which ranges from 220 dots per inch on computers to 326 dots per inch on the latest smartphones. Another form of output of computers is computer text to speech capabilities.

Of course, all this computer hardware would not be very useful if there were not software to instruct the computer what to do. Software is typically organized into computer programs, which can be larger programs on large and highly powerful computers, or smaller programs like an app on a smartphone. One way to think of programs is from the title of a classic textbook that is now many decades old, but first described computer programs as consisting of algorithms, the instructions on what to do, and data structures, the data in a structured form on which the algorithms could operate.[12]

In modern times, computer programs are built on so-called layers of abstraction.[13] The idea is that the programmer programs the computer at an appropriate level of detail. For example, an operating system has a great deal of detail in its computer code that move bytes of information around the hardware, but when a programmer is using the operating system level of abstraction he or she may only want to open files or pull data from them without having to know the details of were the bytes go.

Likewise, with graphical user interfaces, there is much detailed computer code that instructs any computer to draw something on a screen in a certain location. But for the graphical user interface, the programmer might want to just put the object in a specific location or drag it around the screen without getting into the details of the computer hardware. The same applies for database systems and networks.

Computer programs also interact with each other within and between computers, sometimes over a network or even the Internet. This interaction is typically done via an application programming interface (API), which defines the rules for different computer programs to interact with each other. APIs are now very important in healthcare, with the growing distribution of data and programs for accessing it.[14]

One important type of computer software is the operating system. The operating system provides value both to the user as well as the programmer. Operating systems typically come with programs that help users manage files, to set up and maintain the computer, and different utilities for various common tasks. For programmers, operating systems provide a standard interface to various services, so they do not need to program the details of how files are accessed on the hardware disc, the pixels of the display, and the bits streaming cross network connections. Programmers can operate at a higher level of abstraction when they work at the operating system level.

There are many well-known operating systems that run on different types of computer hardware. The most widely used PC operating system is Microsoft Windows. There is also the Macintosh Operating System or macOS. Another important operating system is Unix. This system was originally oriented to higher end workstations and servers but is now also popular on PCs in the form of Linux, which is an open source version of the Unix operating system and is viewed by many as an alternative to the more proprietary Windows and Mac OS operating systems. Of course, smartphones and tablets have their operating systems as well. The two market leaders are currently iOS, which is the operating system of the iPhone and the iPad, and Android, the open source operating system from Google that also runs on smartphones and tablets.

Another important concept for operating systems is virtualization, where computers can run so-called virtual machines that have the hardware commands of different machines but are actually represented in software. The virtualization process translates the commands to one machine into the actual commands of the underlying machine. The value of virtualization is that it will isolate the machines from the actual underlying hardware, though sometimes this can be challenging when software has been developed to be well tuned on a specific type of hardware. One examples of virtualization is the ability to run Microsoft Windows on a Macintosh. There are also virtual machines for Linux that enable it to run within a Windows environment.

Another important type of computer software is the programming language. Computer programs give instructions to the hardware telling it how to move and manipulate data. Computer programs are ultimately composed of codes of bits that state what those instructions are to do. The very lowest level instructions are called machine language. When mnemonics are given to those machine language instructions, this is called assembly language. However, assembly language requires far too much detail to attention and makes it difficult for a programmer to deal with higher-level constructs. Thus, there are higher-level computer languages that hide the complexity of assembly language. People sometimes call computer programs in any language, "code."

There are many computer programming languages, which often serve different types of programming better. Among the most commonly used languages include:
- C, C++, C# – used in many modern applications
- MUMPS – used in many early medical applications, now renamed to M and with many modern enhancements
- BASIC – common in early days of PCs
- Python – originated as "scripting" languages for Unix and Web, but achieving larger-scale use, e.g., in machine learning applications
- Java – attempt to create standard language for Web applications
- JavaScript – scripting language for Web browsers
- R – emerging important language for statistics and data analytics

Computer programs are typically written in the programming languages and then run through a compiler, which not only allows them to run faster because the code is translated to machine language, but also has the effect of protecting the intellectual property of the programmer. However, once computer code is compiled, it cannot be modified unless one goes back to the original code. There has been a movement that expresses concern over this and believes that software code should be open and freely available. This type of openly available software is typically called open-source software. The source code is freely available, and there is usually a community that supports, extends, and manages the code, usually under the auspices of some sort of standardizing entity.

Open-source software has been most successful in horizontal markets, such as operating systems, programming languages, Web browsers, and databases.

There is a fair amount of open source software in health care. It is important to remember, however, that even if the source code is free, it still takes resources, time, and/or money, to manage open source software, especially in areas like healthcare. It also requires resources to install the software, train users, and provide support for them. While there are many open-source software programs in healthcare, there has not been widespread use of this model. It is much more prevalent in the biomedical research community, which has a much longer history of sharing software and other tools.

For those who are interested in the availability of open source software, there are a number of sites for general open source software, namely sourceforge.net where many open source projects maintain their code that can be downloaded, and suggestions for enhancements made. Within healthcare, there are some sites that provide links to open source software:
- Source Forge https://sourceforge.net/directory/business-enterprise/enterprise/medhealth
- Wikipedia https://en.wikipedia.org/wiki/List_of_open-source_health_software
- Online Registry of Biomedical Informatics Tools (ORBIT) – https://orbit.nlm.nih.gov/

Database systems

A good deal of what is done with computers in healthcare and other fields is managing data. The systems that manage data are typically called databases. They store large amounts of information, usually in a structured way that makes the data easily accessible by query languages. Most biomedical applications require strong database capabilities.

The most fundamental unit of data in a database is a field that represents a single value of data, such as a heart rate, or a diagnosis, whether it is a diagnosis string of text or a code such as an ICD-10 code. The fields come together in records, which are collections of fields that make up an entry into a table and are a collection of records of the same types. There may be a number of different fields that make up a patient demographics record, such as name date of birth, and ethnicity. Another type of record may be a chemistry panel when the individual tests are done as a group, with the multiple records making up a table. A database usually consists of multiple tables, with a relational structure used to link across the different tables.

A database management systems (DBMS) is a computer program that manages the database and access to it, often using a programming language. The program language is at a fairly high level, such that the programmer can focus on the data and how it is structured, and not the underlying representation in the fields, the records, and the tables.

There are a number of different models for DBMSs. The simplest one is a flat file, where each and every record in the database has the same data. The most common approach to database management systems is the relational database. These are typically queried by structured query language (SQL). Another database model is entity-attribute-value (EAV), where there is a more flexible structure. Instead of having a complete set of tables, each with its set of records, the EAV approach only stores the specific entities, attributes, and values. These types of databases are sometimes called NoSQL, which is based on the MUMPS programming language paradigm that is claimed to be more efficient with data than the relational systems. In all database models, some of the fields are indexed that enables them to be looked up rapidly. If there are large amounts of data, an index will allow access to specific elements of data more quickly than sequentially scanning through the data on the hard disk.

The relational DBMS links records to tables. For example, there are patients who may have many test results or many visits. We only want to have a single record for the patient, but there are multiple records for each set of test results or each visit. This allows efficiency, so that one-time information such as demographics is stored only once, and complex queries across test results or encounters can be done more easily and efficiently. Most query capabilities are based on SQL, which is why relational databases are sometimes called SQL databases.

Figure 3.4 shows an example of a relational database layout for a population of patients. Each individual patient has demographics and other one-time information. But there are many different results and many different encounters. Likewise, the population has more patients, each of which has both the one-time information and then multiple the results and the encounters.

The relational structure enables one to ask common queries of a clinical database relatively efficiently, e.g.:
- Find all patients with Disease X and on Drug Y
- Find all patients who are over Age X and have not had Test Y
- Find all Test X results for patients with Disease Y

Relational databases make these kinds of queries relatively simple, at least simpler than we might undertake with other types of database models.

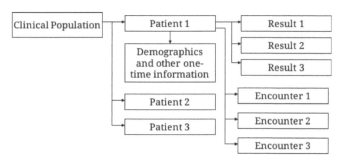

Figure 3.4: Relational Database schema

A final concept about data is that of metadata, which is data about data.[15] In other words, what does a specific data element mean? It may, in the database, carry a simple variable name such as glucose or ethnicity. We may need to know more detail about what the data element actually means, and that is metadata. There are three general types of metadata:

Descriptive – describes data element for discovery or identification

Structural – describes organizational structure of data

Administrative – information on how to manage, such as rights management (who can access and when, how, etc.) and preservation (archiving and storage)

While data is certainly important to computer applications, there must also exist algorithms that use and act on that data. A recent book looked at nine algorithms that *"changed the future."*[16] These are algorithms that have had the most impact on computing and the way that computers interact with our lives. The table of contents of this book lists the algorithms:

- Search Engine Indexing: Finding Needles in the World's Biggest Haystack
- PageRank: The Technology That Launched Google
- Public Key Cryptography: Sending Secrets on a Postcard
- Error-Correcting Codes: Mistakes That Fix Themselves
- Pattern Recognition: Learning from Experience
- Data Compression: Something for Nothing
- Databases: The Quest for Consistency
- Digital Signatures: Who Really Wrote This Software?
- What Is Computable?

THE INTERNET AND WORLD WIDE WEB

Computers must network to exchange data. Computer networks scale from those in a home or office (Local Area Networks or LANs) to massive interconnected networks (an Internet). The Internet is the largest and arguably most important of these large scale international networks. The Internet is a global network-of-networks using the Telecommunications Protocol/Internet Protocol stack (TCP/IP) as the communications standard. The TCP/IP stack allows for layering of different standards and technologies based on the participants in an exchange and the payload being exchanged. The Internet began in the late 1960s as a government project which created a network known as the Advanced Research Projects Agency Network (ARPANET) capable of securely tying together universities and research organizations. The World Wide Web (WWW) operates on top of the Internet and was created by Tim Berners-Lee in 1989. The WWW introduced the web browser, a software program that allows for connection to web servers over the Internet using Hypertext Transfer Protocol (HTTP). The browser can request, retrieve, translate, and render the content from a remote server on the computer screen for users to view. Web pages are written using Hypertext Markup Language (HTML), an implementation of a markup language, or method for defining formatting of text in a document, which has become synonymous with the web. Here is a simple example of html:

```
<html>
  <body>
    <h1>My First Heading</h1>
    <p>My first paragraph</p>
  </body>
</html>
```

Achieving interoperability on the Internet depends on global use of standards. Standards exist for the exchange of data, such as HTTP; the format of data, such as HTML, and the transport of data, such as TCP/IP.

In a TCP/IP network, each device (host) must have an Internet Protocol (IP) address. IP addresses can be distributed amongst different tiers of lower layer networks, or "sub-networks." For IP addressing to function properly in the presence of a sub-network, the machine must both have an IP address and a routing prefix or "subnet mask" (example: IP address of 192.168.10.1 and subnet mask of 255.254.254.0) for it to be considered properly addressable by other network nodes. Two versions of IP addressing exist today, IP version 4 (IPv4) and IP version 6 (IPv6). IPv4 has been around for more than 40 years and is reaching the depletion of its address space. IPv6 is being used to phase out IPv4 before the complete depletion of assignable addresses brings the growth of the Internet to a complete halt. (To determine one's own IP address using a Windows computer, type "ipconfig" in the command line).

Computers are great at thinking in numbers, as that is all they are doing at the lowest level, however communicating a website or computer address in IPv4 or IPv6 to another human is not an easy or issue free process.

To circumvent this, a standard was created known as the Domain Name System (DNS). DNS solves the human address issue by allowing for easier to recognize and remember common language-based addresses to be assigned and mapped to regions of the IP address space. This process which is managed by DNS servers, allows for one to tell someone to visit a website (www.uwf.edu) instead of using its IPv4 or IPv6 address (143.88.3.180). Figure 3.5 demonstrates how this works. Devices can connect to the Internet using a dial-up modem, broadband modem or gateway, Wi-Fi, satellite, and 4/5G cellular data connections.

It is useful to think of the Internet as comprised of two main components, protocols and hardware. The common types of hardware needed are cabling, client computers, servers, hubs, switches, firewalls, gateways, and routers. The client computer is an end point using a network service provided by a server. Each machine addressable on a network is known as a node. Computers connect to the Internet through an Internet Service Provider (ISP) such as Bell South or AT&T. For example, if one uses a web browser (e.g. Chrome, Safari, Opera, Internet Explorer, Firefox) to connect to a web site there are many systems involved in servicing that request. An electronic request for an IP address is sent via the network link provided by the one's ISP to a DNS server. The DNS server then matches the requested domain name and responds with an IP address. The browser is now capable of sending an HTTP GET request (again routed through one's ISP provided link) to the IP address returned from the DNS request. The result of this set of transactions is an HTTP response with an HTML payload from the server. The browser can now render and display the document defined by the HTML response on the user's screen.

For this to occur the message must be sent using small packets of information. Packets can arrive via different routes, useful when there is web congestion, and are reassembled back at one's computer. All traffic sent using TCP/IP (such as phone calls over the Internet (VoIP) and email) are sent using packets. A router is a node which directs the packets on the Internet. The role of ordering these packets and making sure that they make it to their intended recipient in the proper structure is one of the jobs of TCP/IP.

The Uniform Resource Locator (URL) is a specified address to a specific resource. A URL (sometimes also referred to as the Uniform Resource Identifier or URI) can, for example, specify a document provided on the WWW by a web server (e.g. http://www.google.com). The first part of the URL is the protocol identifier, indicating which protocol will be necessary to retrieve the resource. The remainder, known as the resource name, specifies the address of the system to retrieve from as well as the full path to the content to retrieve. The protocol identifier and the resource name are separated by a colon and two forward slashes. As an example, http://uwf.edu/about describes HTTP as the protocol, "uwf.edu" as the server to which the HTTP request will be made, and "/about" as the path to the resource being requested.[17-18] HTTPS is a request over a *"secure socket layer"* which encrypts and decrypts web page requests, in addition to web pages returned by the web server. The most common domains end in .com, .edu, .org, .net, .mil, .gov and .int. Newer top level domains (TLDs) include .biz, .info, .jobs, .mobi, .name and .tel. In addition, extensions exist for countries, such as .br for Brazil.

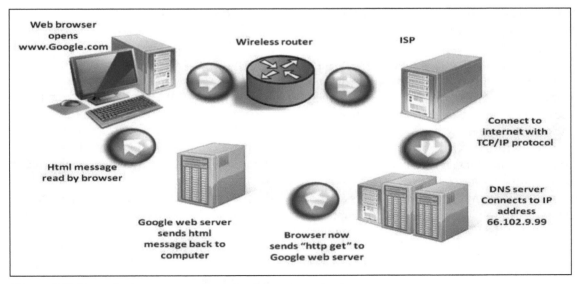

Figure 3.5: How the Internet works to locate web content

WEB SERVICES

Prior to the advent of the Internet, disparate businesses and health care entities were not able to easily exchange data; instead data resided on a local PC or server and controlled communication links (such as via modem) were required to transport that data to another system. *Web services* are task specific applications which are deployed in a platform independent manner via a series of transactions to and from other web-aware applications/services over a network (such as the Internet). Web services can reduce the cost of converting data with external partners, by allowing for a modular component of a larger system to be invoked with little up-front effort.

Web services can be broken down into two categories. Representational State Transfer (or RESTful) services are lightweight services which use existing Internet infrastructure and World Wide Web concepts as their backbone. Simple Object Access Protocol (SOAP) web services utilize a potentially complex series of eXtensible Markup Language (XML)-based ontologies to describe and invoke services over a network. There are obvious pros and cons to each concept, but most often the tradeoff between ease of implementation versus technical depth of field is the main point of comparison struck between the two.

RESTful Services

Representational State Transfer (REST), as a concept, is an aggregate description of the functional model of how HTTP allows for the deployment of the WWW over the Internet. It can be utilized to provide non-WWW content delivery over any application protocol, not solely trapped in the realm of the HyperText Transfer Protocol (HTTP). It is important to realize that REST is an architecture, not a standard. As such, there are endless possibilities as to how REST can be applied to act as a service bus. Even though REST itself is not a standard, many standards are utilized when it is used for service interaction. Communication with a RESTful service is a relatively quick process and can utilize any existing content standard for packaging its messaging. Most commonly, a RESTful service will use XML or JavaScript Object Notation (JSON) for this content delivery. RESTful web services require three basic aspects:
- URI (Uniform Resource Identifier). URI is a set of characters defining a specific object, resource, or location. One of the more common uses for a URI is in providing a Uniform Resource Locator (URL) for an object on the WWW. In a RESTful service, a URI can describe the service being invoked or a component within said service.
- Operation Type (GET, DELETE, POST, PUT). These HTTP methods can be extended past their WWW function to provide four different points of access to a RESTful service. If a URI identifies an object, the HTTP operation type defines an accessor method to that object (e.g. GET a list, POST an update, PUT a new record, DELETE a purged record).
- MIME Type (Multipurpose Internet Mail Extensions). MIME is a means of communicating the content type used within a message transferred over the Internet. Typically, in a RESTful service, this would be XML or JSON, but it could be any other type.

Web Services using SOAP

SOAP is a protocol standard for interacting with web services. These services require a set of standards for content and a service oriented architecture (SOA) stack, a collection of services. The most common standards used in web services transactions are HTTP, as the Internet protocol, with XML as the delivery language. SOAP web services require three basic platform elements:
- SOAP (Simple Object Access Protocol): a communication protocol between applications. It is an XML-based platform-neutral format for the invocation and response of web services functions over a network. It re-uses the HTTP for transporting data as messages.
- WSDL (Web Services Description Language): an XML document used to describe and locate web services. A WSDL can inform a calling application as to the functionality available from a given service, as well as the structure and types of function arguments and responses.
- UDDI (Universal Description, Discovery and Integration): a directory for storing information about web services, described by WSDL. UDDI utilizes the SOAP protocol for providing access to WSDL documents necessary for interacting with services indexed by its directory.

So how does this work? SOAP acts as the means of communicating, UDDI provides the service registry (like the yellow pages) and WSDL describes the services and the requirements for their interaction. One can begin the process acting as a service requester seeking a web service to provide a specific function. The application would search a service directory for a function that meets one's needs using a structured language. There is a service requester seeking a web service. One searches using a search engine that uses a structured language.

Once the service provider is located, a SOAP message can be sent back and forth between the service requester and service provider. A service provider can also be a service consumer, so it is helpful to view web services like the *bus* in a PC, where one plugs in a variety of circuit boards.

Health information organizations (HIOs) often require a Master Patient Index (MPI) service to locate and confirm patients and a Record Locator Service (RLS) to identify documentation on those patients. For connecting multiple HIOs one may also require gateways (a network point that acts as an entrance to another network) and adapters (software that connects to applications).[19-21] A valuable article, *"Improving Performance of Healthcare Systems with Service Oriented Architecture,"* describes how SOA is the logical backbone for HIOs and electronic health records.[22] Another resource for understanding SOA and healthcare was published in by the California HealthCare Foundation, *Lessons from Amazon.com for Health Care and Social Service Agencies*.[23]

The Open Systems Interconnection (OSI) Model

The Open Systems Interconnection (OSI) created a conceptual model in 1984 to help with understanding network architectures. This model divides computer-to-computer communication into seven layers known as the OSI Stack. (See figure 3.6). The Stack's seven layers are divided into upper and lower layers as follows:

- Upper layers
  - 7. Application. This is the layer where applications access network services. Examples, software for database access, email and file transfer and the Internet protocols FTP, HTTP and SMTP.
  - 6. Presentation. This layer translates (formats) the data for the application layer for the network. Examples, data encryption and compression.
  - 5. Session. This layer establishes, maintains and terminates "sessions" between computers.
- Lower layers
  - 4. Transport. This layers deals with error recognition and recovery. It handles message size issues and can reduce large messages into smaller data packets. The receiving transport layer can send receipt acknowledgments. The Internet protocol related to this is TCP.
  - 3. Network. This layer is involved with message control, switching and routing. Translates logical addresses into physical addresses.
  - 2. Data link. This layer packages data from the physical layer into frames (special packets) and is responsible for error free transfer from one computer (node) to another.
  - 1. Physical. This layer deals with the unstructured raw data stream from the other layers. Specifically, it encodes data and decides whether the bits will be sent via a digital or analog mode and decides if the bits will be transmitted as electrical or optical signals. This layer is involved with communication with devices. Examples: USB, Bluetooth and RS-232.

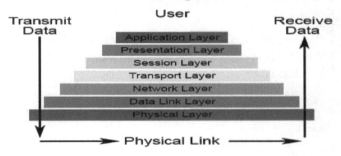

Figure 3.6: OSI Model (Courtesy University of Washington)

NETWORKS

A network is a group of computers that are linked together to share information. Although most medical data reside in silos, there is a distinct need to share data between offices, hospitals, insurers, health information organizations, etc. A network can share patient information as well as provide Internet access for multiple users. Networks can be small, connecting just several computers in a clinician's office or very large, connecting computers in an entire organization in multiple locations.

There are several ways to access the Internet: dial-up modem, wireless fidelity (WiFi), a Digital Subscription Line (DSL), 4G/5G telecommunication, cable modem or T1 lines. The most common type of DSL is Asymmetric DSL (ADSL) which means that the upload speed is slower than the download speeds, because residential users utilize the download function more than the upload function thus allows a segmenting of available bandwidth to give the illusion of greater availability. Symmetric DSL is also available and features similar upload and download speeds. Cable modem networks can either be fully coaxial up to a fiber channel node further upstream or can begin with fiber optic transmission to the building, with coaxial cable run internally. Table 3.1 displays

Table 3.1: Data transfer rates

Transmission method	Theoretical max speed	Typical speed range
Dial-up modem	56 Kbps	56 Kbps
DSL	6 Mbps	1.5-8Mbps downlink/128 Kbps uplink
Cable modem	30 Mbps	3-15 Mbps downlink/1-3 Mbps uplink
Wired Ethernet (Cat 5)	10 Gbps	100 Mbps
Fiber optic cable	100 Gbps	2.5-40 Gbps
T-1 line	1.5 Mbps	1-1.5 Mbps
Wireless 802.11g Wireless 802.11n Wireless 802.11ac	54 Mbps 300 Mbps 7 Gbps	1-20 Mbps 40-115 Mbps 400 Mbps – 1 Gbps
WiMax	70 Mbps	54-70 Mbps
LTE	60 Mbps	8-12 Mbps
Bluetooth	24 Mbps	1-24 Mbps
3G	2.4 Mbps	144-384 kbps
4G	100 Mbps	10-70 Mbps
Satellite	10 Mbps	10 Mbps

data transfer speeds based on the different technologies. Multiple factors influence these speeds, so that theoretical maximum as well as more typical speed ranges are listed.

Information Transmission via the Internet

Given the omnipresent nature of the Internet and faster broadband speeds, the Internet is the network of choice for transmission of voice, data and images. It is important to understand the basics of transmission using packets of information. The Internet Protocol (IP) is a standard that segments data, voice and video into packets with unique destination addresses. Routers read the address of the packet and forward it towards its destination. Transmission performance is affected by the following:

- **Bandwidth** is the size of the pipe to transmit packets (a formatted data unit carried by a packet mode computer network). Networks should have bandwidth excess to operate optimally
- **Packet loss** is an issue because packets may rarely fail to reach their destination. The IP Transmission Control Protocol (TCP) makes sure a packet reaches its destination or re-sends it. The **User Datagram Protocol** (UDP) does not guarantee delivery and is used with, for example, live streaming video. In this case, the user would not want the transmission held up for one packet
- **End-to-end delay** is the latency or delay in receiving a packet. With fiber optics, the latency is minimal because the transmission occurs at the speed of light
- **Jitter** is the random variation in packet delay and reflects Internet spikes in activity

Packets travel through the very public Internet. Encryption techniques defined by the Federal Information Processing Standard (FIPS) encodes the content of each packet so that it can't be read while being transmitted on the Internet. Encryption, however, adds some delay and increase in bandwidth requirements.[24]

Network Types

Networks are named based on connection method, as well as configuration or size. As an example, a network can be connected by fiber optic cable, Ethernet or wireless. Networks can also be described by different configurations or topologies. They can be connected to a common backbone or bus, in a star configuration using a central hub or a ring configuration. In this chapter networks will be described by size or scale.

Personal Area Networks (PANs). A PAN is a close proximity network designed to link phones computers, etc. The most common technology to create a wireless personal area network or WPAN is Bluetooth. Bluetooth technology has been around since 1995 and is designed to wirelessly connect an assortment of devices at a maximum distance of about 300 feet with the most recent Bluetooth devices. Version 5 was released in late 2016 and claims to have quadrupled the range and doubled the speed. It does have the advantages of not requiring much power and connecting automatically. It operates in the 2.4 MHz frequency range. Clearly, the most common application of Bluetooth today is as a wireless headset to connect to a mobile phone, however human interface devices (such as keyboards, mice and fitness apps) are tipping the scales on Bluetooth usage. Many new computers are Bluetooth enabled and if not, a Bluetooth USB adapter known as a dongle can be used or a Bluetooth wireless card. This technology can connect multiple devices simultaneously and does not require "*line of sight*" to connect. In an office, Bluetooth can be used to wirelessly connect computers to keyboards, mice, printers and smartphones. This will avoid the tangle of multiple wires. Bluetooth can connect in one direction (half duplex) or in two directions (full duplex). Security must be enabled since the transmission range is short, and hackers have taken advantage of this common frequency. Bluetooth supports a variety of smart devices, as part of the Internet of Things (IoT), discussed in another section. In addition, faster Bluetooth 5.0 devices are available with speeds in the 2 Mbps range that piggyback on the 802.11 standard.[24] Devices with this standard may transmit for months or years on a coin-type battery.[25]

WPANs can also use other standards: Infrared to connect devices using the IrDA standard, ZigBee networks, Wireless USB and a body area network (BAN). A wireless body area network (WBAN) is also known as a body sensor network which is gaining importance in healthcare with new body sensors being developed continuously.[26] Another wireless sensor network protocol known as ANT™ is available for ultra-low power applications. The proprietary network operates on the 2.4 GHz ISM band. This protocol has wide applicability with wellness, fitness and home monitoring wireless sensors. A variety of chip sets, developer's tools and ANT USB dongles are discussed on the web site.[27]

Local Area Networks (LANs). LAN generally refers to linked computers in an office, hospital, home or close proximity situation. A typical network consists of nodes (computers, printers, etc.), a connecting technology (wired or wireless) and specialized equipment such as hubs, routers and switches.

1. **Wired networks.** To connect several computers in a home or office scenario, a hub or a network switch is needed. Routers direct messages between networks and the Internet; whereas, switches connect computers to one another and prevent delay. Unlike hubs that share bandwidth, switches operate at full bandwidth. Switches are like traffic cops that direct simultaneous messages in the right direction. They are generally not necessary unless multiple computers are running on the same network. To handle larger enterprise demands Gigabit Ethernet LANs are available that are based on copper or fiber optics. Cat5e or Cat6 cables are necessary. Greater bandwidth is necessary for many hospital systems that now have multiple IT systems, such as an electronic medical record and picture archiving and communication systems (PACS). A typical wired LAN is demonstrated in Figure 3.7.

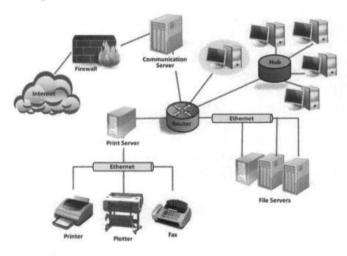

Figure 3.7: Typical wired local area network (Courtesy Department of Transportation)

To connect to the Internet through an Internet Service Provider (ISP) one has several options:

a. Phone lines can connect a computer to the Internet by using a dial-up modem. The downside is that the connection is relatively slow. Digital subscription lines (DSL) also use standard phone lines that have additional capacity (bandwidth) and are a much faster network connection than dial up. DSL also has the advantage over modems of being able to access the Internet and use the telephone at the same time. Home or office networks can use phone lines to connect computers, etc. Newer technologies include frequency-division multiplexing (FDM) to separate data from voice signals. This type of network is inexpensive and easy to install. Speeds of 128 Mbps can be expected even when the phone is in use. Up to 50 computers can be connected in this manner and hubs and routers are not necessary. Each computer must have a home phone line network alliance (PNA) card and noise filters are occasionally necessary. The downside is largely the fact that not all home rooms or exam rooms have phone jacks.

b. Power lines are another option using standard power outlets to create a network. Power line Ethernet products tend to be inexpensive with reasonable transfer speeds. All that is needed is a power outlet in each room.[28]

c. Ethernet is a network protocol and most networks are connected by fiber or twisted-pair/copper wire connections. Ethernet networks are faster, less expensive and more secure than wireless networks. The most common Ethernet cable is category 5e (Cat 5e) unshielded twisted pair (UTP). Cat 6, 6a, 7 and 8 are newer cables that are faster but more expensive.[29]

2. **Wireless (WiFi) networks (WLANs).** Wireless networks are based on the Institute of Electrical and Electronics Engineers (IEEE) 802.11 standard and operate in the 900 MHz, 2.4 GHz and 5 GHz frequencies. These frequencies are "unlicensed" by the FCC and are therefore available to the public. Figure 3.8 shows the radio frequency portion of the electromagnetic spectrum where wireless networks function.

Wireless networks have become much cheaper and easier to install, so many offices and hospitals have opted to go wireless. This allows laptop/tablet PCs and smartphones in exam and patient rooms to be connected to the local network or Internet without the limitations of hardwiring, but it does require a wireless router and access points. If an office already has a wired Ethernet network, then a wireless access point needs to be added to the network router. A wireless router or access point being used as the hub of communications between systems makes the wireless network be in a state known as infrastructure mode. An ad hoc or peer-to-peer mode means that a computer connects wirelessly directly to another computer and through a routing device. In general, wireless is slower than cable and can be more expensive but does not require hubs or switches. The standards for wireless continue to evolve. Most people have used early 802.11 networks that operated on the 2.4 GHz frequency at peak speeds of 54 Mbps with a range of about 100 meters. Keep in mind that this frequency is vulnerable to interference from microwaves, some cordless phones and Bluetooth. 802.11ac is the newest standard that can operate at theoretical speeds up to 7 Gbps with a frequency of 5 GHz and multiple bandwidths of 20, 40, 80 and 120 MHz. This is accomplished with multiple input/multiple output (MIMO) or multiple antennas that send and receive data much faster and at greater distances. Actual data transfer speeds may be slower than the theoretical max speeds for several reasons. Most modern laptop computers have wireless technology factory installed so a wireless card is no longer necessary.

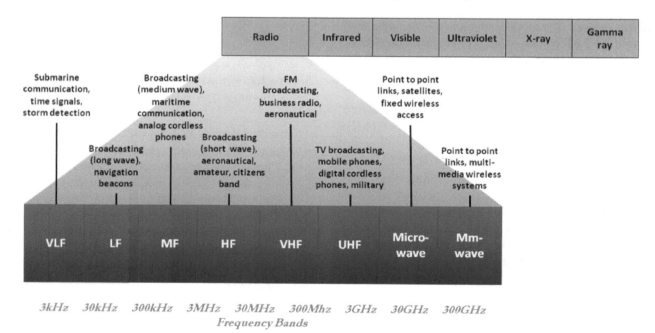

Figure 3.8: Radio frequency spectrums (Courtesy Commission for Communications Regulation).

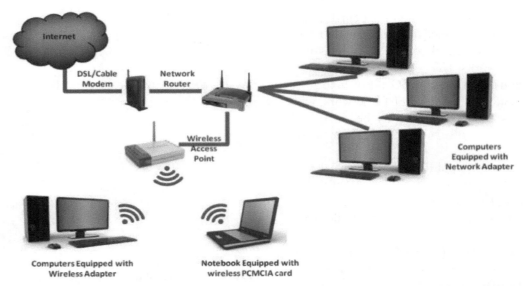

Figure 3.9: Wireless Local Area Network (WLAN) (Courtesy Home-Network-Help.com)

In Figure 3.9 a simple WLAN is demonstrated, with access to the Internet over a cable modem and the possibility of both Ethernet and wireless connectivity to different client computers demonstrated.

A wireless router will connect the computers, server and printers and has a range of about 90 to 120 feet. For a larger office or hospital multiple access points will be necessary. The network router is usually connected to the Internet by an Ethernet cable to DSL or a cable modem. Security must be established using an encryption scheme such as WiFi Protected Access II (WAP2) encryption. Other best-practices for securing a wireless network are the use of a firewall and a unique media access control (MAC) address filtering. Each device on a network has a unique address (MAC) and routers can have security lists which only allow known devices or MACSs into the network.

An emerging trend for hospitals is to use Voice over IP on a wireless network, referred to as VoWLAN. Hospitals can use existing wireless networks to contact nurses, physicians and employees with any wireless enabled device. Devices such as the Nortel VoWLAN phone or Vocera are frequently used. The chief advantage of this approach is saving local and long-distance phone call charges. Using this technology, a patient could directly contact a nurse making rounds so a nurse is not forced to be located near a central nurse-call system. While in the hospital this system could replace landlines, pagers, cell phones and 2-way radios. The downside is that a strong signal is necessary for this system and is more important than that needed with just data.

Another wireless option is wireless mesh networks that rely on a single transmitter to connect to the Internet. Additional transmitters transmit signals to each other over a wide area and only require a power source. Municipalities, airports, etc. are using this type of technology to cover larger defined areas.[30-32]

Wide Area Networks (WANs). Cross city, state or national borders. The Internet could be considered a WAN and is often used to connect LANs together.

Global Area Networks (GANs). GANs are networks that connect other networks and have an unlimited geographic area. The problem with broadband technology is that it is expensive and the problem with WiFi is that it may result in spotty coverage. These shortcomings created an initiative known as Worldwide Interoperability for Microwave Access (WiMax), using the IEEE 802.16 standard. This 4G (fourth generation) network is about 10 times faster than 3G and has greater capacity which is equally important. The network is also known as a global area network (GAN) with operating speeds in the 54-70 Mbps range. The goal is to be faster than standard WiFi and reach greater distances, such that it might replace broadband services and permit widespread wireless access to the Internet by PCs or phones. A user would be able to access the Internet while traveling or from a fixed location. Ironically, the introduction of one 4G network (WiMax) was so slow that major carriers elected LTE, discussed in the next paragraph.

The second 4G wireless network rolled out in US cities was Long Term Evolution or LTE, offered by Verizon, AT&T, Sprint and T-Mobile. As of mid-2017 LTE is available to most US customers.[33] LTE operates in the 700 MHz range and has theoretical maximum download rates of 100 Mbps and upload rates of 50 Mpbs.[34]

4G wireless transports voice, video and data digitally via Internet Protocol (IP) rather than through switches which will reduce delay and latency. 3G phones will not work on 4G networks. LTE Advanced is the latest evolution of 4G and a step closer to 5G. As of June 2017, the United States had the sixth fastest LTE network, with Singapore, South Korea and Hungary as the top 3.[35] 5G cellular coverage will be rolled out in 2018 in some parts of the US.

LTE is now used for public safety networks or (PS-LTE). Given its relatively low cost, high reliability and excellent speed, it can be used by multiple state and federal agencies. It will operate in the upper 700 MHz band.[36]

Virtual Private Networks (VPNs). If a clinician desires access from home to the electronic health records, one option is a virtual private network (VPN). In this case, the home computer is the client and is attached to the network at work by communicating with a VPN server associated with that network. Communicating with nodes over a VPN is akin to working from that network's physical location. The Internet can serve as the means of connection with VPN working over both wired and wireless LANs. Authentication and overall security are key elements of setting up remote access to someone else's computer network. (Figure 3.10) "Tunneling protocols" encrypt data by the sender and decrypt it at the receiver's end via a secure tunnel. In addition, the sender's and receiver's network addresses can be encrypted. The end user can use the VPN option in computers using the Windows operating system. Type VPN into the search window and a set up wizard will create the network.[37]

FUTURE TRENDS

There is a tremendous amount of government and civilian data on the Internet, but it often is stored in formats such as portable document format (PDF) that are largely non-computable. The Semantic Web will find and interpret the data or create a common framework for data sharing. Data will need to be tagged with metadata tags (data that describes data) and known as "linked data." The World Wide Web Consortium (W3C) has promoted the notion of Resource Description Framework (RDF) as the means to describe documents and images. Another specification will be Web Ontology Language. Better definitions will produce better search results. It will also allow for applications run on the Internet to receive and understand data from another application. Sir Timothy Berners-Lee, considered to be the father of the WWW, now promotes the concept of linked data as part of the

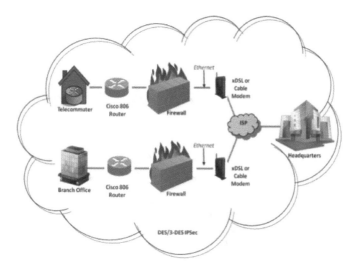

Figure 3.10: Virtual private network diagram (Courtesy Cisco)

RDF. He points out that currently one must have application programming interfaces (APIs) and programs like Excel and PDF to represent data. If the data was linked and encoded by RDF standards, the extra steps would not be necessary. Slowly, organizations such as BestBuy, eBay, BBC and Data.gov have begun participation in web 3.0.[38-39]

Internet2 is a not-for-profit networking consortium of universities, government agencies, researchers and business groups developing applications and a network for the future. The current network is known as *Abilene* and it operates at 100 gigabits per second (Gbps). It has been deployed to almost 100,000 institutions in more than 100 countries.[40]

The Iowa Health System has created a high-speed private network, known as HealthNet connect for medical sharing described in the info box.[41] While much of medical care does not require high speed networks, there are new reasons they may expand healthcare options. The following are some of the reasons high speed networks would be necessary:
- Interactive videos delivering telehealth
- 3D imaging
- Tactile tools (e.g. remote robotic surgery)
- Surgical education
- Implant modeling

The Institute of Electrical and Electronics Engineers (IEEE) released new standards (802.3ba) for 40 and 100 Gigabit Ethernet networks in June 2010. The most recent standard is 802.3bm (2015).[42] It is anticipated that this network will be used by researchers and others like the Department of Energy (ESnet) who need advanced speed. There is talk about bundling 100 Gigabit pipes

> **Ultra-Fast Health Information Exchange**
>
> **HealthNet connect** HealthNet connect is a eHealth consortium that connects clinicians across four states using a high-speed private fiber optic network. The network provides health information exchange to include large images, education, network services, cloud computing, clinical research and telemedicine. They also offer BroadNet connect or ultra-fast fiber optic cable for non-healthcare businesses. [23]

to create a Terabit Ethernet. WiFi will get potentially faster (4.6 Gbps) with 802.11ad routers operating in the 60GHz spectrum. It's predecessor 802.11ac used 2.4GHz or 5GHz frequencies. It is thought that transmissions will be roughly seven times faster than current technologies, but with the drawback of shorter transmission distances.[43]

The Internet of Things (IoT): with so many "smart" devices available to the average consumer/patient that can use both WiFi and Broadband Internet new opportunities arise. Many of these same devices have sensors, such as activity sensors or accelerometers. Other medical sensors might include EEG, EMG, ECG, BP and temperature. In addition, there are "smart homes" and other appliances that have a variety of sensors that could all be connected or linked using the Internet. Therefore, the IoT is a huge network of connected devices raising new possibilities and challenges.[44-45] In the healthcare domain IoT becomes Health IoT and usually consists of the sensors, a handheld device and the network server. This infrastructure could be shared with an available application programming interface (API).[46] It is not difficult to envision a connected home, hospital or an entire city. The amount of data produced would be difficult to comprehend, with the possibility of new software and hardware spin offs. Moreover, the potential for embedded "artificial intelligence" and algorithms would be considerable. The greatest concern is with security of such a wide-open network. Not only would the smart devices require protection, so would the confidentiality, integrity and privacy of data collection. We are still a distance off fully utilizing the IoT in our personal lives and in healthcare.

> **KEY POINTS**
>
> - Informaticians need to understand basic computer technology and network architectures
> - The Internet is the premier network for all industries, to include healthcare
> - Wireless networks have become more attractive due to faster speeds and lower prices
> - Wireless broadband will make Internet access faster and more widely available

CONCLUSION

Computer systems can use TCP/IP to allow for the transmission of data over multiple different protocols to provide content sharing across a network such as the Internet. Disparate services can be integrated by using web services as part of SOA. This platform provides the greatest degree of flexibility for many businesses, to include health information organizations (HIOs).

Hospitals' and clinicians' offices rely on a variety of networks to connect hardware, share data/images and access the Internet. Despite initial cost, most elements of the various networks discussed continue to improve in terms of speed and cost. Many clinicians' offices will require a network expert to ensure proper installation and maintenance. Wireless technology (WiFi) has become commonplace in most medical offices and hospitals with increasing speeds. When wireless broadband (LTE) becomes cost effective it may become the network mode of choice for many. Network security will continue to be an important issue regardless of mode.

REFERENCES

1. Krohn R and Metcalf D, *mHealth: From Smartphones to Smart Systems*. 2012, Chicago, IL: Healthcare Information Management Systems Society.
2. Anonymous, Bring Your Own Devices Best Practices Guide. 2012, Good Technology: Redwood City, CA, http://i.dell.com/sites/doccontent/business/smb/sb360/en/Documents/

good-byod-best-practices-guide.pdf (Accessed September 4, 2017)
3. Poushter J, Smartphone Ownership and Internet Usage Continues to Climb in Emerging Economies. 2016, Pwer Research Center: Washington, DC, http://www.pewglobal.org/2016/02/22/smartphone-ownership-and-internet-usage-continues-to-climb-in-emerging-economies/ (Accessed September 4, 2017)
4. Mandl KD and Kohane IS, Escaping the EHR trap--the future of health IT. New England Journal of Medicine, 2012. 366: 2240-2242.
5. Mandl KD, Mandel JC, and Kohane IS, Driving innovation in health systems through an apps-based information economy. Cell Systems, 2015. 1: 8-13.
6. Hecht J, The bandwidth bottleneck. Nature, 2016. 536: 139-142.
7. Moore GE, Cramming more components onto integrated circuits. Electronics, 1965. 38. ftp://download.intel.com/research/silicon/moorespaper.pdf (Accessed September 8, 2017)
8. Denning PJ and Lewis TG, Exponential laws of computing growth. Communications of the ACM, 2017. 60(1): 54-65.
9. Anonymous, After Moore's Law: Double, double, toil, and trouble, The Economist. March 12, 2016. http://www.economist.com/technology-quarterly/2016-03-12/after-moores-law (Accessed September 8, 2017)
10. Anonymous, The Digital Universe of Opportunities. 2014, EMC Corp.: Hopkinton, MA, http://www.emc.com/collateral/analyst-reports/idc-digital-universe-2014.pdf (Accessed September 9, 2017)
11. Anonymous, The Digital Universe Driving Data Growth in Healthcare. 2014, EMC Corp.: Hopkinton, MA, http://www.emc.com/analyst-report/digital-universe-healthcare-vertical-report-ar.pdf (Accessed September 10, 2017)
12. Wirth N, Algorithms + Data Structures = Programs. 1976, Englewood Cliffs, NJ: Prentice-Hall.
13. Abelson H, Sussman J, and Sussman J, Structure and Interpretation of Computer Programs, 2nd Edition. 1996, Cambridge, MA: MIT Press.
14. Allen A, Open APIs: A nerdy phrase with big meaning for health care, Politico. July 24, 2017. https://www.politico.com/story/2017/07/24/open-apis-a-nerdy-phrase-with-big-meaning-for-health-care-240897 (Accessed September 11, 2017)
15. Riley J, Understanding Metadata: What Is Metadata, and What Is It For? 2017, National Information Standards Organization: Baltimore, MD, http://www.niso.org/publications/understanding-metadata-what-metadata-and-what-it-primer (Accessed September 11, 2017)
16. MacCormick J, Nine Algorithms That Changed the Future: The Ingenious Ideas That Drive Today's Computers. 2012, Princeton, NJ: Princeton University Press.
17. Web basics. Lehigh University. www.lehigh.edu/~jsb4/webbasicsorig.PDF (Accessed August 16, 2017)
18. Mozilla. https://developer.mozilla.org/en-US/docs/Learn/Common_questions/How_does_the_Internet_work Accessed September 26, 2017)
19. Sankaran V. The role of SOA in improving health quality. www.omg.org/news/meetings/workshops/HC-2008/15-02_Sankaran.pdf (Accessed March 26, 2010)
20. Barry, D. K., Web Services. https://www.service-architecture.com/articles/web-services/index.html (Accessed August 16, 2017)
21. Ananthamurthy L. Introduction to web services. DocSlide. https://download.docslide.net/documents/introduction-into-web-services-ws.html (Accessed August 16, 2017)
22. Juneja G, Dournaee B, Natoli J, Birkel S. Improving performance of healthcare systems with service oriented architecture. InfoQ. March 7, 2008. https://www.infoq.com/articles/soa-healthcare (Accessed September 9, 2013)
23. Lessons from Amazon.com for Health Care and Social Service Agencies. March 2009. California Healthcare Foundation www.chcf.org (Accessed August 16, 2017)
24. Gemmill J. Network basics for telemedicine J Telem and Telecare 2005;11:71-76
25. Bluetooth. https://www.bluetooth.com/news/pressreleases/2016/12/07/bluetooth-5-now-available (Accessed August 16, 2017)
26. Ullah S, Higgins H, Braem B et al. A comprehensive survey of wireless body area networks: on PHY, MAC and Network layers solutions. J Med Syst. 2012;36(3):1065-1094
27. ANT Wireless Sensor Network Protocol. www.thisisant.com (Accessed August 15, 2017)
28. Lacoma T. Is wifi too unreliable? Powerline networking may be what you need. May 8, 2016 https://www.digitaltrends.com/computing/everything-you-need-to-know-about-powerline-networking/ (Accessed August 21, 2017)
29. Hastings N. Still need the reliability of wired Internet? Here's how to choose an Ethernet cable. May 1, 2017 https://www.digitaltrends.com/computing/differences-between-ethernet-cables/ (Accessed August 21, 2017)
30. Smith C, Gerelis C. Wireless Network Performance Handbook. McGraw-Hill, Columbus, Ohio, 2003
31. Smith JE. A primer on wireless networking essentials. EDI. www.ediltd.com (Accessed August 15, 2017)

32. Lewis M. A Primer on Wireless Networks. Fam Pract Management. Feb 2004. www.aafp.org (Accessed August 16, 2017)
33. LTE PC World August 2013 www.pcworld.com (Accessed August 16, 2017)
34. LTE. Gizmodo Feb 18, 2009 and Mar 11 2009 www.gizmodo.com (Accessed August 15, 2017)
35. Triggs J. State of the world's 4G LTE networks – June 6, 2017 http://www.androidauthority.com/state-of-4g-lte-2007-777792/ (Accessed August 16, 2017)
36. How-to-guide for LTE in Public Safety https://www.cmu.edu/silicon-valley/dmi/files/howto_guide.pdf (Accessed August 15, 2017)
37. How to set up VPN in Windows. PCWorld. May 2013. http://www.pcworld.com/ article/210562/how_set_up_vpn_in_windows_7.html (Accessed May 27, 2013) (Archived)
38. Nations D. What is Web 3.0 and Is IT Here Yet? March 22, 2017. https://www.lifewire.com/what-is-web-3-0-3486623 (Accessed August 16, 2017)
39. Semantic Web. http://www.w3.org/standards/semanticweb/ (Accessed August 15, 2017)
40. Internet2 www.Internet2.edu (Accessed August 16, 2017)
41. HealthNet connect http://www.hncbnc.com/healthnetconnect-lc.aspx (Accessed August 15, 2017)
42. IEEE www.ieee.org (Accessed August 15, 2017)
43. Chester E. 4.66 Gbps WiFi: How 60 GHz wireless works—and should you use it? Arstechnica. December 15, 2016. https://arstechnica.com/gadgets/2016/12/802-11ad-wifi-guide-review/ (Accessed August 15, 2017)
44. Yeh KH. A Secure IoT-Based Healthcare System with Body Sensor Networks. IEEE Access. 2016;4:10288–99.
45. Morgan J. A Simple Explanation of the Internet of Things. Forbes. May 13, 2014. www.forbes.com (Accessed August 18, 2017)
46. Hu L, Qiu M, Song J, Hossain MS, Ghoneim A. Software defined healthcare networks. IEEE Wirel Commun. 2015;22(6):67–75.

4

Electronic Health Records

ROBERT E. HOYT • VISHNU MOHAN

LEARNING OBJECTIVES

After reading this chapter the reader should be able to:

- Explain the definition and history of electronic health records (EHRs)
- Describe the limitations of paper-based health records
- Identify the benefits of electronic health records
- List the key components of an electronic health record
- Describe the ARRA-HITECH programs to support electronic health records
- Describe the benefits and challenges of computerized order entry and clinical decision support systems
- State the obstacles to purchasing, adopting and implementing an electronic health record
- Enumerate the unintended adverse consequences related to EHRs

INTRODUCTION

There is no topic in health informatics as important, yet controversial, as the electronic health record (EHR). Attempts at developing and promoting EHRs go back over 40 years. However, only in recent years have EHRs become firmly rooted in the US Healthcare system. Despite their widespread recent adoption, they are very much a work in progress.

The Problem Oriented Medical Information System (PROMIS) was developed in 1976 by The Medical Center Hospital of Vermont in collaboration with Dr. Lawrence Weed, the originator of the problem oriented record and subjective, objective, assessment and plan (SOAP) formatted notes. Ironically, the inflexibility of the concept led to its demise.[1] In a similar time frame, the American Rheumatism Association Medical Information System (ARAMIS) appeared. All findings were displayed as a flow sheet. The goal was to use the data to improve the care of rheumatologic conditions.[2] Other EHR systems began to appear throughout the US: the Regenstrief Medical Record System (RMRS) developed at Wishard Memorial Hospital, Indianapolis; the Summary Time Oriented Record (STOR) developed by the University of California, San Francisco; Health Evaluation Through Logical Processing (HELP) developed at the Latter Day Saints Hospital, Salt Lake City and The Medical Record developed at Duke University,[3] the Computer Stored Ambulatory Record (COSTAR) developed by Octo Barnett at Harvard and the De-Centralized Hospital Computer Program (DHCP) developed by the Veterans Administration.[4]

In 1970 Schwartz optimistically predicted, *"clinical computing would be common in the not too distant future."*[5] In 1991, the Institute of Medicine (IOM) (now known as the National Academy of Medicine) recommended EHRs as a solution for many of the problems facing modern medicine.[6] However, following the IOM recommendation, little progress was made for multiple reasons. As Dr. Donald Simborg stated, the slow early acceptance of EHRs was like the *"wave that never breaks."*[7]

The Health Information Technology for Economic and Clinical Health (HITECH) Act that was part of the American Recovery and Reimbursement Act (ARRA) of 2009 was a game changer for EHRs, with incentive programs established by the Centers for Medicare & Medicaid Services (CMS) for the "meaningful use" of certified EHRs utilizing defined criteria to specify eligibility and objectives, as well as other programs that supported EHR education and health information exchange. The EHR incentive program will be discussed in more detail later in this chapter.

In this chapter, we will primarily discuss outpatient (ambulatory) electronic health records, including logical steps to selecting and implementing an EHR.

Electronic Health Record Definitions

There is no universally accepted definition of an EHR. As more functionality is added the definition will need to be broadened. Importantly, EHRs are also known as electronic medical records (EMRs), computerized medical records (CMRs), electronic clinical information systems (ECIS) and computerized patient records (CPRs). Throughout this book, we will use electronic health record (EHR) as the accepted and inclusive term.

Figure 4.1 demonstrates the relationship between EHRs, EMRs and personal health records (PHRs).[8] As indicated in the diagram, PHRs can be part of the EMR/EHR system which may cause confusion.

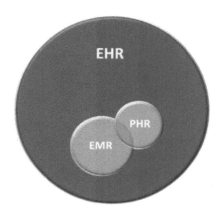

Figure 4.1: Relationship between EHR, PHR and EMR

The National Alliance for Health Information Technology proposed the following definitions to standardize terms:[9]

Electronic Medical Record: *"An electronic record of health-related information on an individual that can be created, gathered, managed and consulted by authorized clinicians and staff within one healthcare organization."*

Electronic Health Record: *"An electronic record of health-related information on an individual that conforms to nationally recognized interoperability standards and that can be created, managed and consulted by authorized clinicians and staff across more than one healthcare organization."*

Personal Health Record: *"An electronic record of health-related information on an individual that conforms to nationally recognized interoperability standards and that can be drawn from multiple sources while being managed, shared and controlled by the individual."*

ELECTRONIC HEALTH RECORD JUSTIFICATION

Some of the most significant reasons why healthcare systems might benefit from the widespread transition from paper to electronic health records include:

Paper Records Are Significantly Limited

Much of the criticism of handwritten prescriptions can also be applied to handwritten office notes. Figure 4.2 illustrates these problems. Even though the clinician in this example used a paper template, the handwriting is illegible and severely limits the ability of other clinicians (and perhaps the clinician generating the note themselves) to extract and use information from the document. Further, the document cannot be electronically shared or stored. The data elements captured in the note cannot be analyzed using computational tools. Other shortcomings of paper records: they are expensive to copy, transport and store; easy to destroy; difficult to analyze and determine who has seen it; and exert a negative impact on the environment. By contrast, electronic patient encounters are legible, information can be easily viewed, transmitted, and (if stored in structured format) analyzed, and storage of electronic records requires a fraction of the space that paper records demand. Almost every industry, from retail to transportation, or banking, is now computerized and digitized for rapid data retrieval and trend analysis.

Figure 4.2: Outpatient paper-based patient encounter form

With the relatively recent healthcare models of patient centered medical home model and accountable care organizations there are new reasons to embrace methods that facilitate data aggregation and reporting, prime amongst them the ability to optimize reimbursement. It is much easier to retrieve and track patient data using EHRs and patient registries than to use labor intensive paper chart reviews. EHRs allow for faster retrieval of lab or x-ray results, and it is likely that EHRs will have an electronic problem summary list that outlines a patient's major illnesses, surgeries, allergies and medications. It is important to note that paper charts are missing during the clinical encounter as much as 25% of the time, according to one study.[10] Even if the chart is available; specific pertinent data elements are missing in 13.6% of patient encounters, according to another study.[11]

Table 4.1 shows the types of missing information and its frequency. According to the President's Information Technology Advisory Committee (PITAC), 20% of laboratory tests are re-ordered because previous studies are not accessible.[12] This statistic has great patient safety, productivity and financial implications.

Table 4.1: Types and frequencies of missing information

Information Missing During Patient Visits	% Visits
Lab results	45%
Letters/dictations	39%
Radiology results	28%
History and physical exams	27%
Pathology results	15%

EHRs allow easy navigation through the entire medical history of a patient. Instead of *pulling paper chart volume 1 of 3* to search for a lab result, it is simply a matter of a few mouse clicks. Another important advantage is the fact that the record is available 24 hours a day, seven days a week and does not require an employee to pull the chart, nor extra space to store it. Adoption of electronic health records has saved money by decreasing full time equivalents (FTEs) associated with ensuring the routine access of patient records, and converting records rooms into more productive space, such as exam rooms. Importantly, electronic health records are accessible to multiple healthcare workers at the same time, at multiple locations. While a billing clerk is looking at the electronic chart, the primary care physician and a specialist can analyze clinical information simultaneously. Moreover, patient information is readily available to physicians on call, so they can review records on patients who are not in their panel. This information may be available off-site; thus, with an EHR a physician can access a patient record from home, instead of having to drive to the office, open the medical records room, and physically search for the patient record.

Furthermore, EHRs improve the level of coding. For example, templates may help remind clinicians to add specific details of the history or physical exam they have performed to justify an appropriate level of coding for the work that they have performed. A study of the impact of an EHR on the completeness of clinical histories in a labor and delivery unit demonstrated improved documentation, compared to prior paper-based histories.[13]

Unlike paper records, EHRs can provide clinical decision support, such as alerts and reminders, which help improve medical decision making. This will be covered later in the chapter.

Another potential advantage of EHRs over paper records is in facilitating clinical research. Not only can the EHR identify eligible patients, it can potentially integrate with research platforms. For example, EHR4CR is a European project involving 35 academic and private partners to create a platform to conduct clinical trials based on EHRs.[14-16]

Need for Improved Efficiency and Productivity

Clinicians want to have patient information available for whenever and wherever they need it. Compared to a paper chart, an EHR allows lab results to be retrieved much more rapidly, thus saving time and money. If lab or x-ray results are frequently missing at the time of the clinical encounter, they are often repeated which adds to this country's staggering healthcare bill. EHRs allow for reduction in duplication of tests; an early study using computerized order entry showed that simply displaying past test results reduced duplication and the cost of future testing by 13%.[17]

EHRs also help to avoid the decrease in efficiency and productivity that occurs due to duplicate prescriptions. It is estimated that 31% of the United States $2.3 trillion-dollar healthcare bill is utilized on administrative tasks.[18] EHRs help to reduce redundant administrative paperwork; for example, they can interface with a billing program that submits claims electronically. Communicating lab results to a patient in the days of paper records often involved a cumbersome communication procedure, but with an EHR, lab results can be forwarded via secure messaging or made available to the patient for viewing via a portal.

Electronic health records can help with efficiency of documentation by utilizing templates and pre-defined macros that generate text. Templates can import relevant data, such as pertinent lab tests, directly into the note. Point-and-click models of navigation, and the use of drop-down menus can reduce documentation time. Of course, one unintended consequence of automating a significant component of patient notes is the introduction of boilerplate language into the clinical record, which adds unneeded text into clinical notes, hindering comprehension.

Embedded clinical decision support is another feature of a comprehensive EHR. Clinical practice guidelines, disease or condition registries, linked educational content and patient handouts can be part of the EHR. This may permit finding the answer to a medical question while the patient is still in the exam room or assist in medical decision making at the point of care.

Clinician workflows can be streamlined by aggregating multiple functions into a single area of the EHR; for example, a physician may be able to sign multiple patient encounters in a single screen of the EHR rather than having to go into each patient chart one by one to sign their note.

EHR dashboards allow clinicians to quickly get a sense of where they stand – not only with parameters of efficiency and productivity (for example a dashboard that displays the number of patients seen each day) but also with patient outcome-related parameters (for example a dashboard that displays to the clinician the degree of control that the diabetic patients in their panel have achieved in comparison to their clinician peers).

However, it should be noted that although EHRs appear to improve overall office productivity, they commonly increase the work of clinicians, particularly with regards to data entry. This will be discussed further in the Loss of Productivity section.

Quality of Care and Patient Safety

As has been previously suggested, an EHR can improve patient safety through multiple mechanisms: (1) Improved legibility of clinical notes, (2) Improved access anytime and anywhere, (3) Reduced duplication, (4) Reminders and clinical alerts (for example a reminder that announces if relevant tests or preventive services are overdue), (5) Clinical decision support that reminds clinicians about drug-drug interactions, known medication allergies, cost and correct dosage of drugs, etc., (6) Electronic problem summary lists (PSLs) provide diagnoses, allergies and surgeries at a glance. Despite the before mentioned benefits, some studies, such as the one by Garrido, have examined quality process measures before and after EHR implementation and failed to show improvement.[19]

To date there has only been one study published that suggested use of an EHR decreased mortality. This EHR had a disease management module designed specifically for renal dialysis patients that could provide more specific medical guidelines and better data mining to potentially improve medical care. The study suggested that mortality was lower compared to a pre-implementation period and compared to a national renal dialysis registry.[20]

It is likely that we are only starting to see the impact of EHRs on quality. Based on internal data Kaiser Permanente determined that the drug Vioxx had an increased risk of cardiovascular events before that information was widely disseminated.[21] Similarly, within 90 minutes of learning of the withdrawal of Vioxx from the market, the Cleveland Clinic queried its EHR to see which patients were on the drug. Within seven hours they deactivated prescriptions and notified clinicians via e-mail.[22] Compare this to the process if paper records were in place – how tedious would it be to go through each patient's paper chart looking to see if Vioxx was included in their medication list, or if patients were noted to be taking the medication in clinical notes. Clearly, electronic clinical quality measure (eCQM) reports are far easier to generate with an EHR compared to a paper chart that requires a chart review. Quality reports can also be generated from a data warehouse or health information organization (HIO) that receives data from an EHR and other sources.[23] Quality reports are the backbone for healthcare reform which will be discussed further in other chapters.

Patient Expectations

The general public has a favorable view of the EHR - according to a Harris Interactive Poll for the Wall Street Journal Online, 55% of adults thought an EHR would decrease medical errors; 60% thought an EHR would reduce healthcare costs and 54% thought that the use of an EHR would influence their decision about selecting a personal physician.[24] The Center for Health Information Technology can make a reasonable case that EHR adoption results in better customer satisfaction through fewer lost charts, faster refills and improved delivery of patient educational material.[25] Patient portals that are part of EHRs are likely to be a source of patient satisfaction as they allow patients access to their records with multiple other functionalities such as online appointing, secure messaging, medication renewals, etc.

Governmental Expectations

EHRs are considered by the federal government to be transformational and integral to healthcare reform. As a result, EHR reimbursement was a major focal point of the HITECH Act. It was the goal of the US Government to have an interoperable electronic health record by 2014. In addition to federal government support, states and payers have utilized initiatives to encourage EHR adoption. CMS is acutely aware of the potential benefits of EHRs to help coordinate and improve disease management in older patients.

Financial Savings

The Center for Information Technology Leadership (CITL) early on suggested that ambulatory EHRs would save $44 billion yearly and eliminate more than $10 in rejected claims per patient per outpatient visit. It should be noted that this optimistic financial projection assumed widespread EHR adoption, health information exchange, interoperability and minor changes in workflow.[26] Several of these processes have not come to fruition.

However, some of the conclusions of this organizations continue to retain their validity. A reasonably articulate case can be made for EHR-related cost savings from eliminated chart rooms and record clerks; as well as a reduction in the need for transcriptionists with the advent of point-and-click templating and voice recognition software, and electronic prescribing has indeed led to fewer callbacks from pharmacists requiring help to decipher physician handwriting. The labor costs of chart pulls are reduced with EHRs, thus saving full time equivalents (FTEs).

Some of the financial savings associated with EHR use are also generated from optimal encounter coding and the increased ability to capture otherwise lost charges. More efficient patient encounters translate to tighter schedules where more patients can be seen each day. Improved savings to payers from medication management are possible with reminders to use generic medications in contrast to more expensive options. EHRs also allow the effective administration of preferred medication lists.

It is not known if EHR adoption will decrease malpractice, hence saving physician and hospital costs. A survey by the Medical Records Institute of 115 practices involving 27 specialties showed that 20% of malpractice carriers offered a discount for having an EHR in place. However, medicolegal risks may be increased during implementation of EHRs – there is an increased risk of errors during the *"implementation chasm"* as clinicians transition from one system to another.[27] These risks include documentation and training gaps, and issues due to software "bugs" and failures. Further, as systems mature the use of email messaging, copy-and-paste models of documentation, and information overload could increase risk.[28] Of course, the presence of EHRs may also be helpful to clinicians – for example, in one study of physicians who had a malpractice case in which documentation was based on an EHR, 55% said the EHR was helpful.[29]

Technological Advances

The time for EHRs is now. The Internet and World Wide Web make the application service provider (ASP) concept for an electronic health record possible. An ASP option means that the EHR software and patient data reside on a remote web server that can be remotely accessed. Computer speed, memory and bandwidth have advanced such that digital imaging is also a reality, so digital image data can be part of an EHR system. Wireless and mobile technologies permit untethered access to the hospital information system and the electronic health record. The unfolding story of the EHR is closely tied to advances in technology that make EHR-related innovations possible.

Need for Integrated and Aggregated Data

Paper health records are standalone, lacking the ability to integrate with other paper forms or information. The ability to integrate health records with a variety of other services and information and to share the information is critical to the future of healthcare reform. Digital healthcare information can be integrated with multiple internal and external applications:

- Integrate with health information organizations (HIOs)
- Integrate with analytical software for data mining to examine optimal treatments, etc. For example, Beth Israel Deaconess Care Organization plans to launch a cloud-based analytics platform that will potentially offer real time population health analytics on EHR based data.[30]
- Integrate genomic data with the electronic health record. Many organizations have begun this journey. There is more information in the chapter on bioinformatics[31]
- Integrate with local, state and federal government information systems for quality reporting and public health issues
- Integrate with algorithms and artificial intelligence. Researchers from the Mayo Clinic could extract Charlson Comorbidity determinations from EHRs, using natural language processing, instead of conducting manual chart reviews.[32]
- Integrate with personal devices, such as activity monitors, glucose monitors, etc.

EHR as a Transformational Tool

It is widely agreed that US Healthcare needs reform in multiple areas. Widespread EHR adoption is a critical part of implementing, maintaining and optimizing a modern healthcare infrastructure. Large organizations such as the Veterans Health Administration and Kaiser Permanente use robust EHRs (VistA and Epic) that generate a significant amount of data for analysis and change the practice of medicine. The integration of data analytics with care has resulted in the improvement in standardization of care, care coordination and population health for these and other similar organizations. In addition, they have begun the process of collecting genomic information for future linking to their electronic records.[33-34]

Need for Coordinated Care

According to a Gallup poll it is very common for older patients to have more than one physician: no physician (3%), one physician (16%), two physicians (26%), three physicians (23%), four physicians (15%), five physicians (6%) and six or more physicians (11%).[35]

Having more than one physician mandates good communication between the primary care physician, the specialist and the patient. This becomes even more of an issue when different healthcare systems are involved. O'Malley et al. surveyed 12 medical practices and found that in-office coordination was improved by EHRs, but the technology was not mature enough to improve coordination of care with external physicians.[36] Electronic health records are being integrated with health information organizations (HIOs) so that inpatient and outpatient patient-related information can be accessed and shared, thus improving communication between disparate healthcare entities. Home monitoring (telehomecare) can transmit patient data from home to an office's EHR also assisting in the coordination of care. We will point out in a later section that coordination of care across multiple medical transitions is part of Meaningful Use.

Figure 4.3 shows the early perceptions of physicians regarding EHR benefits in a 2011 National Center for Health Statistics (NCHS) survey.[37]

NATIONAL ACADEMY OF MEDICINE'S VISION FOR EHRS

The history and significance of the National Academy of Medicine (NAM) (previously known as the Institute of Medicine) is detailed in chapter 1. They have published multiple books and monographs on the direction US

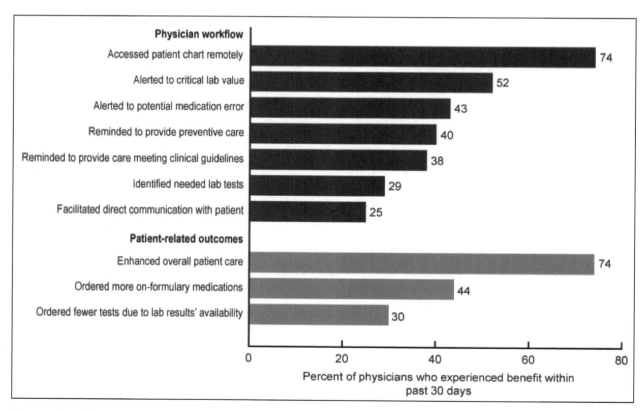

Figure 4.3: Physician's perceptions of EHR benefits

Medicine should take, including *The Computer-Based Patient Record: An Essential Technology for Health Care*. This visionary work was originally published in 1991 and was revised in 1997 and 2000.[6] In this book and their most recent work *Key Capabilities of an Electronic Health Record System: Letter Report* (2003) they outline eight core functions all EHRs should have:

- Health information and data: For the medical profession to make evidence-based decisions, you need a lot of accurate data and this is accomplished much better with EHRs than paper charts; *if you can't measure it, you can't manage it.*
- Result management: Physicians should not have to search for lab, x-ray and consult results. Quick access saves time and money and prevents redundancy and improves care coordination.
- Order management: CPOE should reduce order errors from illegibility for medications, lab tests and ancillary services and standardize care.
- Decision support: Should improve overall medical care quality by providing alerts and reminders.
- Electronic communication and connectivity: Communication among disparate partners is essential and should include all tools such as secure messaging, text messaging, web portals, health information exchange, etc.
- Patient support: Recognizes the growing role of the Internet for patient education as well as home telemonitoring.
- Administrative processes and reporting: Electronic scheduling, electronic claims submission, eligibility verification, automated drug recall messages, automated identification of patients for research and artificial intelligence can speed administrative processes.
- Reporting and population health: We need to move from paper-based reporting of immunization status and biosurveillance data to an electronic format to improve speed and accuracy.[38]

ELECTRONIC HEALTH RECORD KEY COMPONENTS

Many current EHRs have more functionality than the eight core functions recommended by NAM/IOM and this will increase over time. The following components are desirable in any EHR system. One of the advantages of certification for Meaningful Use is that it helped standardize what features were important. The following are features found in most commercial EHRs available currently:

- Clinical decision support systems (CDSS) to include alerts, reminders and clinical practice guidelines. CDSS is associated with computerized physician order entry (CPOE). This will be discussed in more detail in this chapter and the patient safety chapter.
- Secure messaging (e-mail or text messaging) for communication between patients and office staff and among office staff. EHRs will likely include messaging that is part of the Direct Project, explained in the chapter on health information exchange. Telephone triage capability is important.
- Practice management software, scheduling software and patient portals that are embedded or connect with an interface. This feature will handle billing and benefits determination and discussed further in another section.
- Managed care module for physician and site profiling. This includes the ability to track Health plan Employer Data and Information Set (HEDIS) or similar measurements and basic cost analyses
- Referral management feature
- Retrieval of lab and x-ray reports electronically
- Retrieval of prior encounters and medication history
- Computerized Physician Order Entry (CPOE). Primarily used for inpatient order entry but ambulatory CPOE also important. This will be discussed in more detail later in this chapter.
- Electronic patient encounter. One of the most attractive features is the ability to create and store a patient encounter electronically. In seconds, you can view the last encounter and determine what treatment was rendered.
- Multiple ways to input information into the encounter should be available: free text (typing), dictation, voice recognition and templates.
- The ability to input or access information via a smartphone or tablet PC
- Remote access from the office, hospital or home
- Electronic prescribing discussed in a section to follow
- Integration with a picture archiving and communication system (PACS), discussed in a separate chapter
- Knowledge resources for physician and patient, embedded or linked
- Public health reporting and tracking
- Ability to generate electronic clinical quality measures (eCQMs) for reimbursement, discussed in the chapter on quality improvement strategies
- Problem summary list that is customizable and includes the major aspects of care: diagnoses, allergies, surgeries and medications. Also, the ability

to label the problems as acute or chronic, active or inactive. Information should be coded with ICD-10 or SNOMED CT so it is structured data.
- Ability to scan in text or use optical character recognition (OCR)
- Ability to perform evaluation and management (E & M) determination for billing
- Ability to create graphs or flow sheets of lab results or vital signs
- Ability to create electronic patient lists and disease registries.
- Preventive medicine tracking that links to clinical practice guidelines
- Security and privacy compliance with HIPAA standards
- Robust backup systems
- Ability to generate a Continuity of Care Document (CCD) or Continuity of Care Record (CCR), discussed in the data standards chapter
- Support for client server and/or application service provider (ASP) option [39]

The following screen shots reflect the eight basic functions of an EHR as outlined by NAM. Screen shots were derived from the open source EHR LibreHealth.[40]

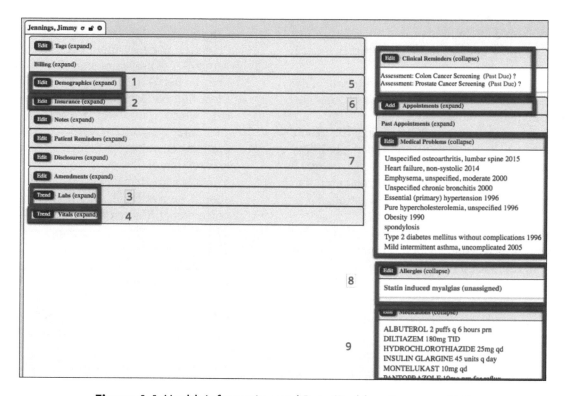

Figure 4.4: Health Information and Data (Problem Summary List)
1. Demographics
2. Insurance information
3. Lab results and graphing capability
4. Vital signs and graphing capability
5. Clinical reminders and alerts
6. Future and past appointments
7. Medical problems
8. Allergies
9. Medications

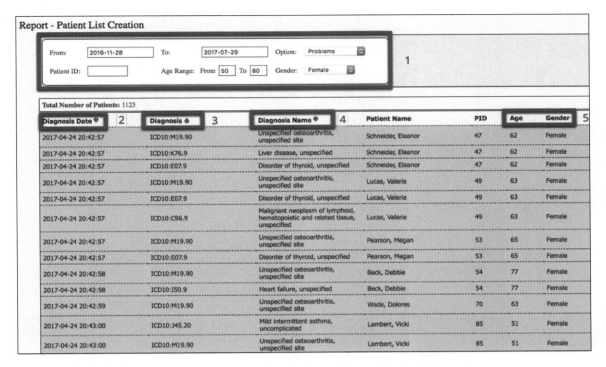

Figure 4.5: Health Information and Data (Patient reports)
1. Search criteria
2. Diagnosis date
3. Diagnosis by ICD-10 code
4. Diagnosis Name
5. Age and Gender

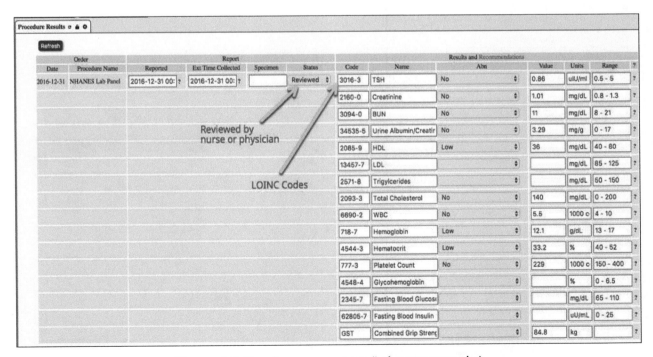

Figure 4.6: Results management (Laboratory results)

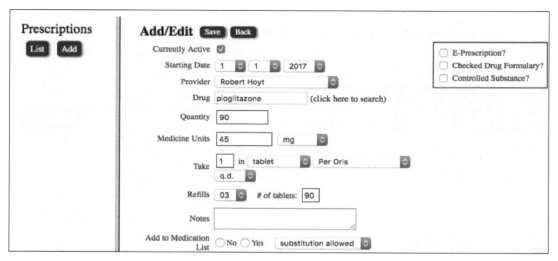

Figure 4.7: Order Management (electronic prescription)

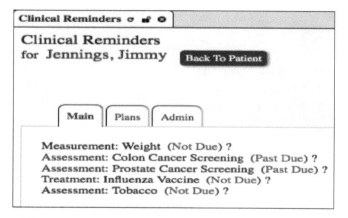

Figure 4. 8: Decision support (alerts and reminders)

Figure 4.10: Patient Support (Patient Education)

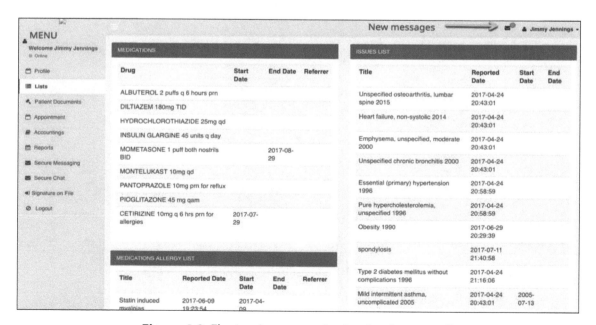

Figure 4.9: Electronic communication (patient portal)

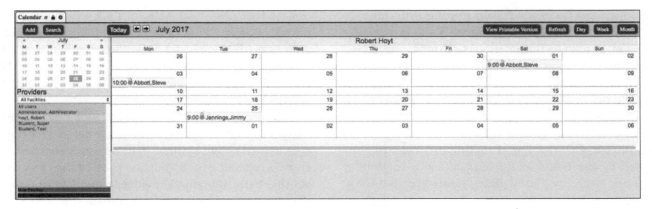

Figure 4.11: Administration (Calendar)

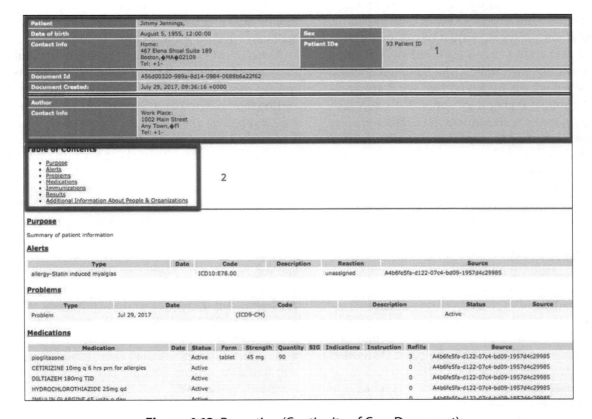

Figure 4.12: Reporting (Continuity of Care Document)
1. Patient demographics and identification
2. Hyperlinked table of contents

COMPUTERIZED PHYSICIAN ORDER ENTRY (CPOE)

CPOE is an EHR feature that processes orders for medications, lab tests, imaging, consults and other diagnostic tests. Many organizations such as the National Academy of Medicine and Leapfrog see CPOE as a powerful instrument of change. However, there is limited evidence that CPOE reduces medication errors, cost and variation of care. This is discussed in the following sections.

Reduce Medication Errors

CPOE has the potential to reduce medication errors through a variety of mechanisms.[41] Because the process is electronic, rules can be embedded that check for allergies, contraindications and other alerts. Koppel et al. note that when compared to paper records, CPOE overcomes the issue of illegibility, is associated with fewer errors associated with ordering drugs with similar names, is more easily integrated with decision support systems than paper, can be easily linked to drug-drug interaction warning, is more likely to identify the prescribing physician, is able to link to adverse drug event (ADE) reporting systems, can avoid medication errors like trailing zeroes, creates data that is available for analysis, can point out treatment and drugs of choice, can reduce under and over-prescribing, and allows prescriptions to reach the pharmacy quicker.[42]

Inpatient CPOE: This functionality was recommended by the NAM as far back as 1991. In 1998, a landmark study by David Bates published in JAMA showed that CPOE can decrease serious inpatient medication errors by a relative risk reduction of 55%. However, this frequently cited article did not show reduction of potential adverse drug events (ADEs).[43] Additionally, many of the studies showing reductions in medication errors using technology were reported by a limited number of academic institutions who had implemented their own home grown EHR and had the infrastructure to allow robust technology support. In contrast, other hospital systems with commercial EHRs were less likely to experience the same optimistic results. A 2008 systematic review of CPOE with CDSS by Wolfstadt et al. only found 10 studies of high quality and those dealt primarily with inpatients. Only half of the studies could show a statistically significant decrease in medication errors, none were randomized and seven were homegrown systems, so results are difficult to generalize.[44]

With the inception of CPOE we are seeing evidence of new errors that result from technology. The term *"e-iatrogenesis"* has been used to describe this phenomenon, the term can be defined as *"patient harm caused at least in part by the application of health information technology."*[45] Others refer to this phenomenon as unintended adverse consequences (UACs). Campbell et al. delineated these unintended consequences as falling within nine types (in order of decreasing frequency): 1) more/new work for clinicians; 2) unfavorable workflow issues; 3) never ending system demands; 4) problems related to paper persistence; 5) untoward changes in communication patterns and practices; 6) negative emotions; 7) generation of new kinds of errors; 8) unexpected changes in the power structure; and 9) overdependence on the technology.[46]

A 2005 article reported that the mortality rate increased 2.8%-6.5% after implementing a well-known EHR.[47] In a 2006 article, also from a children's hospital implementing the same EHR, they found no increase in mortality; perhaps due to better planning and implementation. One of the authors stated that the CPOE system eliminated handwriting errors, improved medication turnaround time and helped standardize care.[48] Nebeker reported on substantial ADEs at a VA hospital following the adoption of CPOE that lacked full decision support, such as medication alerts.[49] On the other hand,, another inpatient study showed a reduction in preventable ADEs (46 vs. 26) and potential ADEs (94 vs. 35) compared to pre-EHR statistics.[50]

A more recent systemic review and meta-analysis suggested that transition from paper-based ordering to commercial CPOE systems in ICUs was associated with an 85% reduction in medication prescribing error rates, but that there was mixed evidence that CPOE reduced ICU mortality. The study concluded *"there is also a critical need to understand the nature of errors arising post-CPOE and how the addition of advanced CDSSs can be used to provide even greater benefit to delivering safe and effective patient care."*[51]

Outpatient CPOE: There is more of a chance for a medication error written for outpatients, because there are far more prescriptions written in the ambulatory setting than in acute care facilities. According to an optimistic report by the Center for Information Technology Leadership, adoption of an ambulatory CPOE system (ACPOE) would eliminate about 2.1 million ADEs per year in the USA. This could potentially prevent 1.3 million ADE-related visits, 190,000 hospitalizations and more than 136,000 life-threatening ADEs.[26] However, a systematic review by Eslami was not as optimistic as he concluded that only one of four studies demonstrated reduced ADEs and only three of five studies showed decreased medical costs. Most showed improved guideline compliance, but it took longer to electronically prescribe and there was a high frequency of ignored alerts (alert fatigue).[52] Kuo et al. reported medication errors from primary care settings. The study concluded that 70% of medication errors were related to prescribing and that 57% of errors might have been prevented by electronic prescribing.[53]

Reduce Costs

Several studies have shown reduced length of stay and overall costs in addition to decreased medication costs with the use of CPOE.[54] Tierney was able to show in an early study an average savings of $887 per admission when orders were written using guidelines and reminders, compared to paper-based ordering that was not associated with clinical decision support.[55]

One study by Nuckols et al. suggested that in comparison to paper records, CPOE saved, on average between $11.6 million and $170 million per hospital in mean lifetime savings in 2012 dollars, depending on their size. The study also suggested that quality-adjusted life-years (QALYs) also increased proportionately, and anticipated increases in CPOE implementation from 2009 through 2015 could save $133 billion and 201,000 QALYs nationwide.[56]

Reduce Variation of Care

There is significant potential for CPOE to facilitate the ability of clinicians to reduce variations in care, and studies have reflected this. For example, one study showed excellent compliance by physicians when the drug ordered by them was changed by a decision support algorithm.[57]

On the surface CPOE implementation seems easy: just replace paper orders with an electronic format. The reality is that CPOE represents a significant change in work flow and is not just the adoption of new technology. Any change in workflow requires extensive training to allow end-users to adapt to the new paradigm. With this in mind, it should be noted that many of the studies that extoll the ability of CPOE to reduce variation in the delivery of clinical care were conducted at medical centers with well-established health informatics programs; where the acceptance level of new technology was unusually high. Several of these institutions such as Brigham and Women's Hospital developed their own EHR and CPOE software.

Compare this experience with that of a rural or critical access community hospital implementing CPOE for the first time with potentially inadequate IT, financial and leadership support. It is likely that smaller and more rural hospitals and ambulatory care offices will have a steep learning curve.

Adoption of CPOE was initially slow, partly because of associated costs and partly because workflow changes were extensive – the process of computerized input is significantly more cumbersome and often slower than scribbling orders on a paper cahrt.[58] Physician resistance to change was on the forefront of implementers minds at the onset of deployment of CPOE; early unsuccessful implementations such as the one in Cedars-Sinai Medical Center in California gave clinical informaticians pause for thought.

Over time, CPOE has become palatable to most physicians. One reason is that the functionality is now well-established in clinical workflows; personnel that used to exclusively exist to enter physician orders, such as inpatient unit secretaries, now no longer perform the task of entering orders. Another reason is that house staff, and not attendings, write most orders in teaching hospitals and academic medical centers, and as a consequence of being the "*EHR generation*" of physicians they do not know any other way to enter orders other than CPOE.

Even though CPOE is in widespread use, it does require forethought, leadership, planning, training and the use of physician champions to implement successfully.

CLINICAL DECISION SUPPORT SYSTEMS (CDSS)

Clinical decision support may be defined as "*any electronic or non-electronic system designed to aid directly in clinical decision making, in which characteristics of individual patients are used to generate patient-specific assessments or recommendations that are then presented to clinicians for consideration.*"[59]

However, clinical decision support is not just an alert or notification. Table 4.2 outlines some of the clinical decision support available today. Calculators, knowledge bases and differential diagnoses programs, which primarily originated as standalone programs, are increasingly being integrated into modern EHR systems.

Table 4.2: Clinical decision support

Type of CDSS	Examples
Knowledge	UpToDate, DynamedPlus
Calculators	eCalcs
Trending/Patient tracking	Flow sheets, graphs
Medications	CPOE and drug alerts
Order sets/protocols	CPGs and order sets
Reminders	Mammogram due
Differential diagnosis	DxPlain
Radiology CDSS	ACR Select
Laboratory CDSS	What lab tests to order
Public health alerts	Infection disease alerts, SMART apps

Sheridan and Thompson have discussed various levels of CDSS: (level 1) all decisions by humans, (level 2) computer offers many alternatives, (level 3) computer restricts alternatives, (level 4) computer offers only one alternative, (level 5) computer executes the alternative if the human approves, (level 6) human has a time line before computer executes, (level 7) computer executes automatically, then notifies human, (level 8) computer informs human only if requested, (level 9) computer informs human but is up to computer and (level 10) computer makes all decisions.[60] Most EHR systems may offer alternatives and provide reminders but make no decisions on their own. With artificial intelligence and natural language processing becoming more sophisticated, this could change in the future.

A majority of studies in the informatics literature regarding CDSSs have been sourced from four institutions – the Veterans Administration (VA), Brigham and Women's Hospital and associated Partners Healthcare, the Regenstrief Institute, and Intermountain Healthcare. These institutions did not use commercial EHRs for most of their history and the CDSS literature reflects this. (Ironically, many of the institutions with early home grown EHRs have transitioned to large commercial EHRs). But, despite this significant limitation, there is some evidence that both commercial as well as home-grown CDSSs improve process measures across a variety of settings.[61] But evidence that suggests improvements in clinical outcomes, efficiency measures and even the case for financial ROI remains less definitive.

There have been some attempts to study why certain types of CDSS succeed while others fail. Roshanov et al. identified several factors that could partially explain this phenomenon. Their paper concluded that success of CDSSs was likely to be greater when they provided advice concurrently to clinicians and patients, or when practitioners were required to provide reasons when over-riding CDSS advice than for systems that did not.[62] On the other hand, CDSSs that presented decision support within electronic charting or order entry systems was associated with a higher likelihood of failure.

No treatise on CDSSs is complete without mentioning the well-cited "Ten Commandment" paper by Bates et al. that delineated the ten principles of effective clinical decision support. This paper specifically references the speed of the CDSS, the need for CDSS to anticipate clinician needs and deliver in real time, and the ability of the CDSS to fit into the user's workflow. The "commandments" also include some pithy observations: that usability matters, that physicians strongly resist suggestions not to carry out an action in the absence of an alternative, instead preferring to change direction rather than stop, and that simple interventions are often more effective than complex guidelines. Bates et al. also note that CDSSs need to ask the end-user for additional information when it is necessary, that CDSS implementers should monitor the impact of their interventions, and that knowledge-based systems should be maintained and curated appropriately.[63]

Knowledge support. Numerous digital knowledge-based resources are being integrated with EHRs. For example, UpToDate[64] has been embedded into several EHRs, and diagnostic (ICD-10) codes can be hyperlinked to further information or you can use *infobuttons*. Infobuttons are a HL7 standard and commonly used to link to important information.[65] Other products such as Dynamed Plus, are available as *infobuttons*, web services API, search box widget, or embedded links.[66]

Calculators. Calculators are now embedded into most commercial EHRs, particularly in the medication and lab ordering sections. Important clinical calculations, such as calculations that assist appropriate antibiotic dosing based on kidney function (creatinine clearance) as well as scoring patient risks such as the 10-year cardiovascular risk, can now be achieved at the point of care.

Flow sheets, graphs and other visual representation of patient data. The ability to view graphic representations of patient data, such as results and vital signs, allows clinicians to visualize data and track trends in a manner that assists in cognitive clinical reasoning.

Medication ordering support. CDSSs help clinicians detect known allergies, identify potential drug-drug interactions, as well as avoid prescribing excessive or ineffective dosages of medications. These interventions have obvious potential in decreasing medication errors and improve patient safety.

Reminders. Computerized reminders that are part of the EHR assist in tracking the yearly preventive health screening measures, such as mammograms. Shea et al. performed a meta-analysis and concluded that there was clear benefit for vaccinations, breast cancer and colorectal screening.[67] A well-designed system should allow for some customization of the reminders as national recommendations change over time. However, it should be noted that reminders are not always heeded by busy clinicians who may choose to ignore them.

Order sets and practice guidelines. Order sets are groups of pre-established related orders that are related to a symptom or diagnosis. For instance, with just a few mouse clicks a provider may place an order set for pneumonia that might include the antibiotic of choice, supplemental oxygen, an order for a chest x-ray, etc.; thus the clinician's workflow is rendered more efficient. Order sets can also reflect best practices (clinical practice guidelines), thus offering optimal patient care.

Differential Diagnosis generators. Dxplain is a differential diagnosis program developed at Massachusetts General Hospital. When the patient's symptoms are entered, it generates a differential diagnosis (a list of stratified diagnostic possibilities). The program has been in development since 1984 and is currently web-based.

A licensing fee is required to use this program. As of 2018 it cannot be integrated into an EHR.[68] In spite of the potential benefit, an extensive 2005 review of CDSSs revealed that only 40% of the 10 diagnostic systems studied showed benefit, in terms of improved clinician performance.[69] Liebovitz offers suggestions as to how future EHRs could improve diagnoses.[70] Artificial intelligence continues to improve so it is likely that EHRs will have the ability to assist with differential diagnosis in the future.

Radiology CDSS. Physicians, particularly those in training, may order imaging studies that are either incorrect or unnecessary. For that reason, several institutions have implemented clinical decision support to improve imaging study ordering. Appropriateness criteria have been established by the American College of Radiologists. Massachusetts General Hospital has had radiology order entry since 2001; a study showed a decline in low utility imaging study orders from 6% down to 2% and this decrease was attributed to the use of decision support.[71]

Beginning on January 1, 2018, the Protecting Access to Medicare Act (PAMA) will require referring clinicians to access and understand appropriate use criteria (AUC) prior to ordering certain diagnostic imaging services: CT, MRI, nuclear medicine exams and PET scans for Medicare patients. These criteria can be standalone or integrated into the EHR. The American College of Radiology has developed the CDSS, known as ACR Select.[72]

Laboratory CDSS. It should be no surprise that clinicians occasionally order inappropriate lab tests, for a variety of reasons. A Dutch study of primary care demonstrated that 20% fewer lab tests were ordered when clinicians were alerted to lab clinical guidelines.[73] Another study showed a decrease in duplicate test ordering when laboratory CDSSs were available.[74]

Public Health Alerts. The New York Department of Health and Mental Hygiene used Epic EHR's *"Best Practice Advisory"* to alert New York physicians about several infectious disease issues. The EHR-based alert also hyperlinked to disease specific order sets for educational tips, lab and medication orders.[75] A newer approach to public health clinical decision support has been to use web services and the new data standard FHIR, discussed in the next section and the chapter on data standards.[76]

Smart apps. Like smartphone apps, new applications were developed that link within the EHR to add more functionality. Fast healthcare interoperability resources (FHIR) is a relatively new HL7 standard that utilizes open standards such as RESTFul APIs. That means, they can communicate with any EHR that has open APIs. Apps can be created in a short period of time for clinical care, public health and research. The SMART App Gallery includes dozens of apps, as of 2018. More apps are described on the HL7 web site.[77-78]

More information about clinical decision support can be found in the chapter on clinical decision support systems and these references.[63, 79-82]

ELECTRONIC PATIENT REGISTRIES

Unlike EHRs that focus on individual patients, registries focus on populations. Patient registries are defined as *"an organized system that uses observational study methods to collect uniform data (clinical and other) to evaluate specified outcomes for a population defined by a particular disease, condition, or exposure, and that serves one or more predetermined scientific, clinical, or policy purposes."*[83]

Modern registries tend to fall into the following categories:
- Chronic disease management: for example, a diabetic educator might have all patients with type 2 diabetes in a registry for management purposes.
- Research registries: if a healthcare system has all e.g. total hip replacement patients in a single registry they can evaluate and compare different outcomes with different prostheses, etc.
- Safety registries: reporting to e.g. the FDA
- Public health registries: reporting immunizations, cancer and biosurveillance
- Quality registries: an option to report performance data to e.g. CMS[84]

Therefore, registries perform multiple functions, to include:
- Natural history of disease: following patients over time is important for both clinicians and researchers to better understand disease progression
- Effectiveness: treatments are better evaluated with larger patient populations, something that would be difficult with a single EHR or clinic
- Safety: studying larger patient populations in a registry is likely to be more valid than a small population
- Quality: to meet value-based reimbursement programs, clinicians will have the option to upload patient results to a registry with batch reports to e.g. CMS

Registry functions are very consistent with the IOM's vision of a *"learning healthcare system"* where treatment is based on constant analysis of patient data to generate the most current and best evidence.

Historically, early patient registries were paper-based, followed by electronic spreadsheets, followed by electronic standalone registries and finally electronic registries integrated with electronic health records. While many electronic health records permit creation of patient lists, most to do not have comprehensive registries with embedded clinical practice guidelines. If not integrated with an EHR, a HIO or a web-based registry, the inputting of patient information would have to be manual, which is not efficient. To create interoperability between EHR and registry there must be syntactic and semantic interoperability discussed in the chapter on health information exchange and this reference.[84]

An interesting recent initiative is the Guideline Advantage, a collaborative program between the American Heart Association, American Cancer Society and the American Diabetes Association. With EHR connectivity, practices can submit performance measures covering heart disease, diabetes and cancer to this qualified clinical data registry (QCDR) that will forward the data to CMS, as part of quality reporting. Participants can view data at the individual, physician and practice level, as well as benchmark their results against national averages.[85]

There is some evidence that the use of registries results in improved care. For example, Han et al. was able to show that patients with type 2 diabetes were more likely to have appropriate laboratory tests done and retinal exams compared to those not in a registry. In addition, patients in a registry were less likely to be admitted or be seen in an emergency room.[86]

PRACTICE MANAGEMENT INTEGRATION

Most medical offices have had computerized practice management (PM) systems for many years, regardless of whether they utilized paper or electronic health records, or a hybrid of these two. While there are many reasons why PM systems have become so prevalent, one primary driving force has been the ability of practice management systems to generate more rapid claims submission and adjudication. Without an electronic system, time and money would be lost on faxes, phone calls and snail mail. The American Medical Association estimated that inefficient claims submission systems lead to about $210 billion annually in unnecessary costs.[87] A PM system is designed to capture all the data from a patient encounter necessary to obtain reimbursement for the services provided. This data is then used to:

- Generate claims to seek reimbursement from healthcare payers
- Apply payments and denials
- Generate patient statements for any balance that is the patient's responsibility
- Generate business correspondence
- Build databases for practice and referring physicians, payers, patient demographics and patient encounter transactions (i.e., date, diagnosis codes, procedure codes, amount charged, amount paid, date paid, billing messages, place and type of service codes, etc.)

Additionally, a PM system provides routine and ad hoc reports so that an administrator can analyze the trends for a given practice and implement performance improvement strategies based on the findings. For example, a medical office administrator can use the PM system to compare different payers with regards to the amount reimbursed for each given service or the turn-around time between claims submission and payment. The results lead to deciding which managed care plans the practice will participate in versus those plans that the practice may want to consider not accepting in the future. Another example is to analyze all payers for a given service performed in the practice to determine if that service is a good use of the practice's clinical time. This analysis provides one aspect of whether the practice should consider continuing to offer a certain service such as case management of a patient who is receiving home health services through an agency. Of course, the administrator must weigh services that aren't profitable against any negative impact on overall patient satisfaction, but the PM system provides a means of analyzing payment performance.

Most PM systems also offer patient scheduling software that further increases the efficiency of the business aspects of a medical practice. Finally, some PM systems offer an encoder to assist the coder in selecting and sequencing the correct diagnosis (International Classification of Diseases, Current revision, clinically modified for use in the United States, or ICD-10-CM) and procedure (Current Procedural Terminology, fourth edition or CPT-4® and Healthcare Common Procedure Coding System or HCPCS) codes. Even when a physician determines the appropriate codes using a *superbill*, (a list of the common codes used in that practice along with the amount charged for each procedure), there are times when a diagnosis or procedure is not listed on the superbill and an encoder makes it efficient to do a search based on the main terms and select the best code. Furthermore, some encoders are packaged with tools such as a subscription to a newsletter published by the American Medical Association (AMA) known as "CPT® Assistant" that help the practice comply with correct coding initiatives which in turn optimize the reimbursement to which the practice is legally and ethically entitled and avoids fraud or abuse fines for improper coding.

Clinical and Administrative Workflow in a Medical Office

Several steps are common to almost any medical practice with regards to treating patients and getting reimbursed properly for the services provided. The steps are subdivided based on whether the patient has been to this practice previously for any type of service. The first step is to get the patient registered. This can be accomplished via a practice website or by the patient calling the office to schedule an appointment. Figure 4.13 demonstrates typical outpatient office workflow.

ELECTRONIC HEALTH RECORD ADOPTION

Outpatient (Ambulatory) EHR Adoption

In 2006, the adoption rate of ambulatory EHRs was reported to be in the 10% to 20% range.[93] The most recent National Ambulatory Medical Care Survey (2015) reported that 86% of office-based respondents had an EHR; 54% with a basic system and 78% with a certified system. The percentage varied by state from a low of 66% to a high of 90%.[88]

Adoption of an EHR does not necessarily indicate that the end-user is using the advanced capabilities of an EHR, as indicated in Table 4.3 from HIMSS Analytics. The results indicate that very few hospital systems have achieved an advanced level of EHR sophistication.[89]

Table 4.3: HIMSS Analytics EMR Adoption Model 4th quarter 2017

Stage	Capabilities	Ambulatory EHR	Inpatient EHR
7	Complete EMR, data analytics to improve care	10.8%	6.4%
6	Documentation templates, full CDSS, closed loop medication administration	21.8%	33.8%
5	Full R-PACS	8.6%	32.9%
4	CPOE, clinical decision support (clinical protocols)	0.8%	10.2%
3	Clinical documentation, CDSS (error checking)	9.4%	12.0%
2	CDR, controlled medical vocabulary, CDS, HIE capable	16.5%	1.8%
1	Ancillaries installed: lab, rad, pharmacy	30.5%	1.5%
0	Ancillaries not installed	1.9%	1.4%

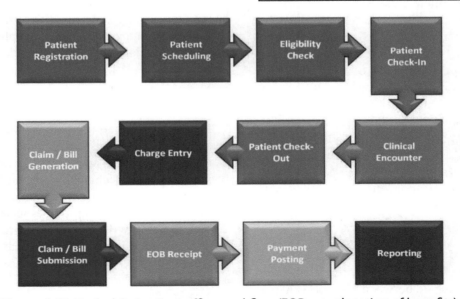

Figure 4.13: Typical Outpatient office workflow (EOB = explanation of benefits)

Inpatient EHR Adoption

The Office of the National Coordinator reported that adoption of a certified inpatient EHR by non-federal acute care hospitals had risen from 72% in 2009 to 96% in 2015 (the most current data).[90]

As anticipated, EHR adoption by rural or small non-teaching hospitals continues to be lower than by larger, urban hospitals and academic medical centers.[91]

International EHR Adoption

Until about a decade ago, the US lagged behind many other developed countries in its adoption of EHRs.[92] A 2009 study showed that the US continued to lag in EHR adoption among primary care physicians in developed countries.[93] A 2015 Organization for Economic Cooperation and Development report on 38 countries stated that *"all but 2 pilot countries reported use by at least half of primary care physicians and many had rates above 75%."*[94]

However, other countries have had their share of implementation failures. As described in the chapter on International Health Informatics, in 2011 the United Kingdom had to dismantle their $17 billion health IT project, the NHS National Programme for IT (NPfIT).[95]

ELECTRONIC HEALTH RECORD CHALLENGES

Many of the same barriers to HIT adoption discussed in Chapter 1 also pertain to EHR adoption and use.

Financial Barriers

Although there are models that suggest significant savings after the implementation of ambulatory EHRs, the reality is that ambulatory EHRs are expensive to implement, particularly for smaller or solo practices. Multiple surveys report that lack of funding is the number one barrier to EHR adoption.[96] In a 2005 study published in *Health Affairs*, initial EHR costs averaged $44,000 (range $14-$63,000) per FTE (full time equivalent) and ongoing annual costs of $8,500 per FTE. These costs included the purchase of new hardware, etc. Financial benefits averaged about $33,000 per FTE provider per year. Importantly, more than half of the benefit derived was from improved coding.[97] This is not a surprise given the fact that studies have shown that physicians often *under-code* for fear of punishment or lack of understanding what it takes to code to a certain level.[98] A 2008 survey reported about one-third of physicians paid between $500-$3,000 per clinician, one-third paid between $3,001-$6,000 and about one-third paid more than $6,000 per clinician for their EHR.[99]

A 2011 study reported on the financial and nonfinancial costs of implementing a commercial EHR in a healthcare network in Texas. They calculated that implementation for a five-physician primary care practice would be about $162,000 with $85,500 in maintenance expenses in the first year. They also estimated that the average end-user would require 134 hours to train and prepare for implementation.[100] Another study reported on 5-year return on investment from 49 practices that were part of the Massachusetts eHealth Collaborative, before and after EHR implementation. The study was prior to CMS reimbursement under the HITECH Act but was similar in that the eHealth Collaborative paid for most costs related to purchase and implementation. They found only 27 percent of practices would achieve a positive five-year return and that a majority would experience a loss. The average projected loss over five years was $43,473 per physician. There were striking differences between the winners and losers of EHR adoption.[101] Eastaugh sampled 62 hospitals and was impressed that most overestimated the ability of EHRs to improve efficiency. The EHR/HIT expenses were 4.3-8.1% of total revenue and 22-39% of available capital. He supports hospitals calculating the total cost of ownership (TCO) to achieve a more realistic appraisal of actual EHR cost.[102]

It is important to consider that integration with other disparate systems such as practice management systems can be very expensive and hard to factor into a cost-benefit analysis. The web-based application service provider (ASP) option is less expensive in the short term and perhaps in the long term, when you factor in the expenses to maintain and upgrade an office client-server network. According to many studies EHR adoption was far higher in large physician practices that could afford the initial high cost.[103]

Physician Resistance

Next to EHR reimbursement lack of support by medical staff was consistently the second most commonly perceived obstacle to adoption.[104] Physicians have to be shown that a new technology is good for their patients, and not just a tool to save time or make money. Often these benefits are hard to prove, especially because EHR implementation will not fix old work flow issues and indeed require clinicians to significantly alter the way they deliver patient care. Change also requires buy-in from all stakeholders, not just from management or a few early adopters.

Loss of Productivity

It is likely physicians will have to work at reduced capacity for several months with gradual improvement depending on training, aptitude, etc. This is a period when physician champions can help maintain morale and momentum with a positive attitude. According to one systematic review CPOE used on central station desktops for CPOE was not time efficient; the weighted average relative time difference across these studies reported an increase in documentation time of 238.4%.[105]

A study of Internal Medicine physicians published in 2014, reported that attending physicians lost about 48 minutes of free time per clinic day, compared to 18 minutes lost per day for trainees.[106] A time motion study of 57 physicians in 4 specialties concluded *"for every hour physicians provide direct clinical face time to patients, nearly 2 additional hours is spent on EHR and desk work within the clinic day."*[107]

Loss of productivity is, in part, due to the change in workflow discussed in the next section.

Work Flow Changes

EHR end users, and indeed all clinicians delivering patient care, will have to change workflows when an EHR is implemented – changes in documentation, the way that patient information is routed between clinicians and ancillary staff, changes in communication patterns both between clinicians and with the patient, new procedures and policy for electronic data privacy and security, even changes in the workflow of how clinical information is handed off to on-call colleagues, consultants and inpatient providers. If these changes to clinical work flows are not anticipated, post-implementation dissatisfaction may increase alarmingly. Work flow analysis will also determine optimal alterations to work flows to ensure uninterrupted and efficient patient care after implementation.

Reduced Physician-Patient Interaction

The addition of the computer in the exam room has heralded a paradigm shift in clinician-patient communication. No longer are clinicians able to maintain eye contact with the patient as often as they did during the days of paper – now clinicians need to divide their time between the patient and the EHR screen, often to the detriment of the patient. Careful attention to workstation placement and ergonomics may mitigate some of these issues, for example the use of a tablet PC may help diminish the time the clinician spends not looking at the patient. The overall effect of exam room technology also depends on the skill of the physician integrating the technology appropriately when they are with the patient at the point of care.[108-110]

Because CPOE and encounter documentation takes longer to complete (on average) compared to paper processes, there is a valid concern that attending physicians or housestaff will be forced to spend more time documenting on the computer and less time with the patient. A study reported in 2013 showed that interns spent only 12% of their time in direct patient-related care, but 40% on the computer.[111] A second report in 2013 reported that emergency room physicians spent 28% of their time in direct patient care but 43% of time with data entry. On average, the total number of mouse clicks for a 10 hour shift approached 4,000.[112] These findings further strain the already negative perception of many patients that they don't have enough face time with their physician.

Usability Issues

Usability has been defined as the *"effectiveness, efficiency and satisfaction with which specific users can achieve a specific set of tasks in a particular environment."*[113]

Ratwani et al. studied the user-centered design of 11 EHR vendors and concluded that there was great variability in usability.[114] A 2017 systematic review by Ellsworth et al identified 120 articles that discussed EHR usability and determined that there was a paucity of quality published studies, as well as a lack of a standard formal means to evaluate usability.[115]

The American Medical Informatics Association (AMIA) board has made a set of recommendations to enhance patient safety and quality of care by improving the usability of EHRs:

- Recommendations for the academic informatics community includes emphasizing usability and human factors research by prioritizing standard use cases, developing a core set of measures specific to adverse events related to health IT use, research and promote best practices for safe EHR implementation
- Policy recommendations include standardizing EHR systems and ensuring interoperability, establishing an adverse event reporting system for health IT, and developing an educational campaign for safe and effective EHR use
- Specific recommendations for EHR vendors include developing a common user interface style guide for certain EHR features, and performing formal usability assessments on their products

- Recommendations for end users suggests adopting best practices for EHR system implementation and ongoing maintenance, and monitoring how IT systems are used with the goal of reporting IT-related adverse events.[116]

Another issue is that commercial EHRs are typically never delivered to healthcare organization as out-of-the-box turnkey installations. Organizations adapt the product to their needs, and the production environment may be somewhat different than the vendor's test environment. Wright et al. recommends testing EHRs in the production environment to mimic how they are used by clinicians. This will allow unintended behavior to be more readily apparent before patient harm can occur.[117]

Integration Issues

Integrating clinical information system elements can be challenging. Best-of-breed solutions require close attention to integration of individual components, and even commercial EHR products require integration with existing software, or with external connections that are required to continue business as usual. Interoperability with other EHRs, registries, health information networks, and data warehouse may also need to be considered. Needless to say, integration can be expensive.

Ultimately, genomic information will also be extensively integrated with most EHRs. Already, several large healthcare systems have begun the journey by linking genotypical information with phenotypical information in the EHR. As pointed out by Hazin et al. and discussed further in the chapter on bioinformatics, there are a host of ethical, legal and social implications and challenges associated with this integration.[118] We are a long way away from having the physician and patient utilize genomic information in any comprehensive manner, and lack clinical decision support to alert patients and providers of significant genomic risk. Additionally, both clinicians and patients may not appreciate the complexity associated with interpreting genomic data. Evidence of this was found in a 2010 survey by CAHG that reported that 90% of physicians polled thought that genomics-based medicine would influence healthcare by "some" to "great extent" but only 8% claimed to be very familiar with genomic medicine and only 16% stated they had training in this area in medical school.[119]

Quality Reporting Issues

EHRs have the potential to generate a variety of data necessary for compliance with meaningful use objectives, to include quality reports. Quality reports have been tied to physician reimbursement in several situations, however, obstacles associated with linking physician compensation to the quality of care they provide remain. In early 2013, two reports from Weill-Cornell Medical College in New York City highlighted issues with quality measure reporting. In one study the accuracy of automated EHR data reporting was low, compared to manual chart review. In another study that examined quality reporting in the Primary Care Information Project in New York it was noted that within the first two years of using an EHR there was no improvement in overall quality, even with high levels of technical assistance.[120-121]

Lack of Interoperability

Data standards are necessary for interoperability, and reimbursement for Meaningful Use mandates that EHRs demonstrate the ability to exchange information. Although we have numerous standards already accepted (discussed in Chapter 5) they will likely need to be updated and new standards added based on use cases. Perhaps the most interesting standard with interoperability ramifications is the HL7 standard known as Fast Healthcare Interoperability Resources (FHIR). A variety of resources have been created to handle common healthcare use cases. FHIR resources are structured data in the form of XML or JSON objects. Each resource has a unique URL. Lab results and other data could be called up using RESTful APIs.[77] Some of the major EHR vendors are already actively involved with this standard. FHIR is discussed further in Chapter 5.

Furthermore, computers are based on data and not information, as discussed in the chapter on healthcare data, information and knowledge.

Privacy Concerns

The HITECH Act of 2009 introduced a new certification process for EHRs sponsored by ONC, in addition to CCHIT certification. This new certification ensures that EHRs will be able to support Meaningful Use and that they also will be HIPAA compliant. ONC certification includes requirements on database encryption, encryption of transmitted data, authentication, data integrity, audit logs, automatic log off, emergency access, access control and accounting of HIPAA releases of information. The HITECH Act also strengthened the prior HIPAA requirements as they relate to EHRs, particularly in the areas of enforcement of HIPAA and notification of breaches. Both civil and criminal penalties for Business Associates (as well as covered entities) were introduced. Civil penalties in their harshest form can range up to 1.5 million dollars. If a data breach of PHI (protected health information) occurs, all affected individuals must be notified. If more than 500 individuals are affected, HHS must be notified as well. Sale of PHI is prohibited.[131] Users of EHRs must:

- Use HIPAA compliant technology
- Provide physical and software security of data systems
- Provide physical and software security of their network(s) including mobile and remote computing
- Provide access control with defined user roles, passwords and user authentication and auditing
- Monitor and manage user behavior
- Have written security policies and procedures
- Have an effective disaster recovery plan[122]

EHRs pose new potential privacy and security threats for patient data, but with proper technology as well as proper health entity and user behavior, these risks can be mitigated. On the bright side, EHRs offer new safeguards unavailable in the paper record world, like audit trails, user authentication, and back-up copies of records. Further details are available in the chapter on privacy and security.

Legal Aspects

A 2010 *Health Affairs* article estimated that malpractice costs in the US are around $55 billion dollars annually (in 2008 dollars) or 2.4% of what we spend on health care.[123] Will EHRs increase or decrease that number? Unfortunately, there is no definitive answer. Most studies suggesting lower malpractice claims after EHR implementation are not designed to prove cause and effect and may not be generalizable to other practices or regions.[124] Arguments can be made for either outcome. On one hand, by increasing the quality of care, theoretically EHRs should reduce malpractice risk. Yet this conclusion assumes that quality and malpractice are related in a linear fashion, which may well not be the case. On the other hand, EHRs that are poorly designed, or that contain bugs, could promote new errors. This risk also points to a need for monitoring and reporting EHR-generated errors with the intention of taking corrective action to avoid "e-iatrogenesis." The Office of the National Coordinator (ONC) for Health IT understands that a system of monitoring and corrective action for EHR-related errors needs to be implemented and outlined its plans in a December 2010 statement.[125]

Two important areas of potential risks and benefits include clinical documentation and clinical decision support. One might expect that the more comprehensive documentation produced by EHRs will improve a physician's defense against malpractice. It certainly may. However the automated way that EHRs carry information forward from one note to the next can also promote errors, for example if a piece of data is recorded incorrectly from the start, yet never corrected.[126] E-discovery laws now allow electronically stored data related to patient records to be considered discoverable for the purpose of malpractice, so the metadata and audit trails that supplement EHR documentation can be used both to defend and to impeach a physician in a malpractice case.[127] Decision support alerts and guidelines embedded into EHRs could potentially provide a defense against malpractice claims if their advice is followed. But what if alerts or guidelines are overridden? There may be very appropriate reasons to do so, but will physicians be expected to document the reason for each and every alert they override? Will they run the risk of being penalized if they don't?

Improved access to information provided by health information exchanges (HIEs) should improve the coordination of care, the quality of medical information that is available, and thus the quality of medical decision making. But, will clinicians overlook key nuggets of clinical information simply because they are overwhelmed by the volume of information they receive? Will ready access to outside information on a patient make a physician more liable if he or she doesn't always actively search for every piece of potentially relevant information? In addition, user errors can arise as users climb a steep learning curve to become proficient with EHRs. Care needs to be taken particularly during the implementation of an EHR to guard against user error.

Finally, as EHRs become the standard of care, will practicing without an EHR become a medicolegal liability? At this point in time it is still undetermined whether EHRs will significantly impact the incidence and expense of malpractice in a positive or a negative way.[128]

Inadequate Proof of Benefit

Successful implementation of HIT at a medical center with a long-standing history of systemic IT support does not necessarily translate to another healthcare organization with less IT support and infrastructure. A systematic review by Chaudry is often cited as proof of the benefits of HIT, but in his conclusion, he states *"four benchmark institutions have demonstrated the efficacy of health information technologies in improving quality and efficiency. Whether and how other institutions can achieve similar benefits and at what costs, are unclear."*[129]

There have been several articles that failed to demonstrate a significant impact of EHRs on medical quality in the US and in Europe.[130-134] A more positive study was published in 2011 of more than 25,000 diabetics in 46 practices that showed achievement of diabetic care was significantly better for practices with EHRs, compared to paper-based practices. They measured intermediate outcomes and not actual patient outcomes, so we don't know the impact on morbidity or mortality.[135] Three additional observational articles measured intermediate outcomes, such as hemoglobin A1c levels, but only one study showed significant benefit.[136-138] A study comparing New York primary care physicians with and without EHRs showed a statistically higher score on nine quality measurement in those clinicians who used an EHR.[139] Another article compared quality measures in those physicians who attested for MU, compared to those who did not and found that MU was associated with marginally better results in 2 measures, worse results for 2 measures and no difference in 3 measures.[140]

A systematic review published in 2012 that looked at the economics of HIT and medication management could find little evidence that CPOE or CDSS were cost effective. Importantly, they noted that the quality of the literature was heterogonous and of poor quality.[141] Another systematic review evaluated the impact of point-of-care computer reminders, as part of CPOE/CDSS on physician behavior and found a very small positive effect. Specifically, the review found that the reminders improved adherence to care by a median of only 4.2%.[142] There has also been a hope and perception that having prior test results readily available in the EHR would reduce testing duplication. In a large retrospective study of before and after EHR implementation, having access to electronic results of lab and imaging results resulted in increased, rather than decreased ordering.[143]

Patient Safety, Reliability, EHRs and Unintended Consequences

Patient Safety. Unfortunately, with implementation of most technologies new problems and issues arise that were not considered initially. EHRs are no exception to this observation and a variety of unintended consequences have been reported. Weiner coined the term *e-iatrogenesis* to mean "*patient harm caused at least in part by the application of health information technology.*"[154] Several studies have shown increased errors after implementing CPOE.[45, 49, 126, 144-146] Campbell et al. outlined nine examples of unintended consequences related to CPOE implementation:

1. "*More work for clinicians*
2. *Unfavorable workflow changes*
3. *Never ending demands for system changes*
4. *Conflicts between electronic and paper-based systems*
5. *Unfavorable changes in communication patterns and practices*
6. *Negative user emotions*
7. *Generation of new kinds of errors*
8. *Unexpected and unintended changes in institutional power structure*
9. *Overdependence on technology*"[147]

Alert fatigue is another common unintended consequence related to CPOE, discussed in more detail in the chapter on patient safety.

The US federal government is keenly aware of the unintended consequences associated with HIT and EHRs after reports by the Joint Commission and the Institute of Medicine.[148-149] Furthermore, the Pennsylvania Patient Safety Authority published a report on errors related to use of default values in 2013. They reported that wrong-time, wrong-dose, inappropriate auto-stops and wrong-route errors were often related to default values that should have been changed.[150]

In response to concerns AHRQ released the monograph *Guide to Reducing Unintended Consequences of Electronic Health Records* in 2011. This Guide discusses unanticipated and undesirable consequences of EHR implementation.[151] In mid-2013, ONC released the report HIT Patient Safety and Surveillance Plan. The plan will make EHR error reporting easier, to include allowing the EHR to generate the report to patient safety organizations (PSOs).[152]

ONC developed a series of EHR risk assessment tools known as Safety Assurance Factors for EHR Resilience (SAFER) guides. The series includes 9 guides covering EHR risk areas, such as system interfaces, system configuration and patient identification. The guides are available on the ONC web site.[153]

Reliability. In spite of successful EHR implementations, we have also seen dramatic failures in 2013, with EHR shutdowns from 1 to 10 days.[154-155] Healthcare organizations must develop backup plans to include temporarily relying on paper-based processes until the EHR is re-established.

With better training or re-design some of the technology-related errors are likely to be overcome. More research is needed to obtain a balanced opinion of the impact of EHRs on quality of care, patient safety and productivity. Furthermore, we will need to study the impact on all healthcare workers and not just physicians.

THE HITECH ACT AND MEANINGFUL USE

Arguably, the most significant EHR-related initiative occurred in 2009 as part of the American Recovery and Reinvestment Act (ARRA). Two major parts of ARRA, Title IV and Title XIII are known as the Health Information Technology for Economic and Clinical Health or HITECH Act.

For clinicians to participate in this program they had to: (1) be eligible, (2) register for reimbursement, (3) use a certified EHR, (4) demonstrate Meaningful Use (MU), and (5) receive reimbursement.

Eligible Professionals (EPs)

Medicare: Medicare defined EPs as doctors of medicine or osteopathy, doctors of dental surgery or dental medicine, doctors of podiatric medicine, doctors of optometry and chiropractors. Hospital-based physicians such as pathologists and emergency room physicians are not eligible for reimbursement. Hospital-based is defined as providing 90% or more of care in a hospital setting. The exception is if more than 50% of a physician's total patient encounters in a six-month period occur in a federally qualified health center or rural health clinic. Physicians may select reimbursement by Medicare or Medicaid, but not both. They cannot receive Medicare EHR reimbursement and federal reimbursement for e-prescribing. They can receive Medicare reimbursement as well as participate in the Physicians Quality Reporting System (PQRs). If they participate in the Medicaid EHR incentive program they can participate in all three programs.

Medicaid: Medicaid EPs are defined as physicians, nurse practitioners, certified nurse midwives, dentists and physician assistants (physician assistants must provide services in a federally qualified health center or rural health clinic that is led by a physician assistant). Medicaid physicians must have at least 30% Medicaid volume (20% for pediatricians). If a clinician practices in a federally qualified health center (FQHC) or rural health clinic (RHC), 30% of patients must be *needy individuals*. The Medicaid program is administered by the states and physicians can receive a one-time incentive payment for 85% of the allowable purchase and implementation cost of a certified EHR in the first year, even before Meaningful use is demonstrated. Medicaid is also different from Medicare in the following: payment over six years does not have to be consecutive and there are no penalties for non-participation.[156]

Registration: Registration began in January 2011. Medicare physicians had to have a National Provider Identifier (NPI) and be enrolled in the CMS Provider Enrollment, Chain and Ownership System (PECOS) and National Plan and Provider Enumeration System (NPPES) to participate.[156]

Certified EHRs: An EHR had to be certified by a recognized certifying organization for a physician or hospital to receive reimbursement. As of mid-2017, there were four organizations that provide certification.[157] Standards and certification criteria are listed, as are the currently certified EHRs. Users can view ambulatory and inpatient EHR categories and search by product name. The search should review who certified the EHR, whether it was for a complete or modular EHR and the EHR certification ID number they would need for reimbursement. The newest 2014 EHR certification is for stage 2 meaningful use.[158]

Meaningful Use (MU): The goals of MU are the same as the national goals for HIT: (a) improve quality, safety, efficiency and reduce health disparities; (b) engage patients and families; (c) improve care coordination; (d) ensure adequate privacy and security of personal health information; (e) improve population and public health. Three processes stressed by ARRA to accomplish this are: e-prescribing, health information exchange and the production of quality reports. Meaningful Use consists of three stages: stage 1 would begin the basic process of data capturing and sharing; stage 2 would require advanced data processes and sharing and stage 3 would aim at improving patient outcomes.

- Stage 1: Meaningful Use mandated a *core set* and a *menu set* of objectives. Participants needed to meet 15 core objectives and five out of 10 menu objectives. They also needed to choose at least one population and public health measure. For each objective, there were reporting measures that must be met to prove Meaningful use. Once a clinician completed two years of MU under stage 1, they moved on to stage 2.[156]
- Stage 2: to align stages 1 and 2, CMS released a modification of MU criteria in October 2015 that gave guidance for the future. 2016 was the last year EPs could enroll in the Medicare MU program. In 2017 EPs attested to new criteria under the Quality Payment Program.[156]
- Stage 3: Medicaid physicians and eligible hospitals will follow new Stage 3 guidelines. EPs would have 10 objectives and eligible hospitals 9 objectives. EHRs must be certified by either 2014 or 2015 guidelines. The reporting period for 2017 was a minimum of any continuous 90-days during the year. The changes to specific Medicaid objectives for 2017 are posted on the CMS web site.[159]

Stage 3 was intended to be implemented in 2018, but instead Medicare providers will transition to Quality Payment Program (QPP) that is part of the Medicare Access and CHIP Reauthorization Act (MACRA). This program offers 2 tracks for medical practices. The first track is the Merit-based Incentive Payment System (MIPS). Clinicians need to report up to 6 quality, 4 improvement and 9 advancing care information measures, for a minimum of 90 days. The quality category replaces PQRS, the improvement activities are new and the advancing care information replaces the Medicare MU program. The second track is the Advanced Alternate Payment Model, which provides more reimbursement (5%) but involves some financial risk. The MU program will change names to Advancing Care Information (ACI) and a new Quality Payment Program (QPP) would be based on value, not volume. For the ACI program there are two routes to submit performance data. One involves 15 measures and the other 11 measures. Details are available on the QPP web site.[160] MACRA will be discussed in multiple chapters but is likely to change over time.

ONC's Health IT Dashboard posts visualizations about most key aspects of meaningful use and health IT. CMS posts information about meaningful use payments. As of December 2017, CMS paid EPs $24.8 billion through the Medicare program and $12.5 billion through the Medicaid program.[156]

THE IMPACT OF THE MEANINGFUL USE PROGRAM

There is little evidence to suggest that meaningful use programs have improved patient outcomes, as opposed to patient processes. In other words, clinicians may be reporting hemoglobin A1c results more often on their patients with diabetes but does that translate into fewer myocardial infarctions or amputatons? While Stage 3 meaningful use and MIPS attempts to be more outcome based, the reality is that the retrospective reporting of processes continues.

In an interesting 2017 article by Goroll entitled *"Emerging from from EHR Purgatory – Moving from Process to Outcomes"* the author posits that EHRs have beome a billing platform where adequate documentation and coding skill is mandatory, yet there is low physician and patient satisfaction with the process.[161] Perhaps with Alternate Payment Models (APMs), the second payment model under MACRA/MPP, they will result in a prospective payment model with more attention to outcomes and less on processes. CMS Innovation Center launched such a program called Comprehensive Primary Care Plus (CPC+) in 2017 with 2866 primary care practices involved. Participants will focus on improving access, continuity and population health.[162] Payment details include:

- **Care Management Fee:** participants will be paid a per-beneficiary-per month (PBPM). The payment is risk-adjusted and paid on a quarterly basis.
- **Performance-Based Incentive Payment:** CPC+ will prospectively also pay based on patient experience measures, clinical quality measures, and utilization measures related to cost
- **Payment under the Medicare Physician Fee Schedule:** Track 1 will continue to bill and receive payment from Medicare fee-for-service (FFS). Track 2 will also continue to bill, but the FFS payment will be reduced to account for CMS shifting part of the FFS payments into CPC+ payments

LOGICAL STEPS TO SELECTING AND IMPLEMENTING AN EHR

EHR implementations are complex affairs. They are not simply IT projects. They are practice transformation projects that should be considered socio-technical-economic initiatives. If EHRs are approached as simply software to be installed and clinicians are similarly approached as users to be trained in using the software, an EHR implementation will undoubtedly falter or even fail. Thus, health care organizations involved in implementing an EHR are wise to plan their process carefully, with attention to the following questions:
- Why are we doing this?
- Who should be involved?
- How will this impact end-users and how do we prepare them?
- What will be the major barriers?
- What should we start doing now to overcome identified barriers?
- Are we ready for change?
- How will the change be managed?

Implementation of an EHR can be divided into three separate, yet intertwined phases: Pre-implementation, implementation and post-implementation.[163] While each phase is distinct, the success of subsequent phases depends upon the thorough planning and execution of the prior stages.

Pre-implementation begins with deciding whether to purchase an EHR (it is rare for a health care organization to create one themselves these days) and ends with signing a contract with a vendor for a specific EHR. This requires a thorough understanding not only of the organization's needs and current state but also of the selected software's abilities and limitations. The main activity in pre-implementation is choosing the EHR that will be used, but several steps that might be done during implementation, such as workflow mapping, may be done and some say should be done, during pre-implementation. Workflow mapping involves a detailed step-by-step description, typically utilizing a flowchart of how a process is accomplished. For example, how are notes created or how are patient messages handled or how are prescription refills managed?[164]

Implementation of the EHR starts with the signing of the contract and ends with the go-live date. Experts in IT implementations often categorize facets of implementation into People, Process, or Technology issues.[165] Alternatively, they can be termed: *Team, Tactics and Technology.*

Issues related to people are particularly important in an EHR implementation. Unless the people issues are managed well from the start, later adoption of the varied functionality inherent in an EHR will likely suffer. Key people issues include leadership, change management, goal establishment and expectation setting. An implementation will have three key types of leaders: a project manager, a senior administrative sponsor, and a clinical champion. The clinical champion will invariably be a physician, but hospital settings will typically have a nurse champion as well. The need for a project manager, someone knowledgeable and experienced in managing a complex IT project with overlapping timelines and multiple stakeholders, is obvious. Senior leadership sponsorship and support is also essential, because an EHR implementation will affect nearly all aspects of a hospital or clinic's operations and thus consistent support from the organization's leader or leaders will be required as inevitable bumps in the road are encountered.

Some healthcare organizations have learned the hard way that implementing an EHR without one or more physician champions can be disastrous. When it comes to clinical matters, physicians rely on other physicians. Because an EHR affects clinical practice in so many ways, respected, supportive, influential clinicians are needed to encourage other physicians to accept and utilize the system effectively.[166]

In inpatient settings, a nurse or clinical champion is essential to ensure that decisions made incorporate all disciplines within the facility. When implementing an EHR it is important to view operations from all perspectives (e.g. physicians, nurses, medical assistants, pharmacists, other support personnel and administrators). Without a nurse champion, decisions made might be solely physician-focused. Additionally, nurses commonly drive the change process in hospitals. Commitment to success, engagement of everyone, and a shared interest in improvement is paramount, so attitude is everything.[167]

Because of the degree of change involved in implementing an EHR for the first time, change management skills are needed. This topic is beyond the scope of this book, but many good resources can be found on it. One good introductory and classic resource is Kotter's book *Leading Change.*[168] An important part of change management is setting goals and establishing expectations.

Many specific process (or tactical) decisions are determined during implementation. How will we use the EHR to redesign our workflows? What is our data entry strategy? Which data will we enter discretely, which will we scan and which (if any) will we leave out of the EHR? Who will do this data entry and when? What order sets will we create? What other information systems will the EHR connect to and what kind of interfaces will it require? Will we follow a *big bang* (all personnel/sites and EHR functions at once) or a phased implementation approach (certain user groups and/or certain sites/departments and or certain EHR functions in sequential order)? How will we conduct user training? What will we do about note templates? How much customization will we allow? How will we utilize super-users be utilized? EHR software does vary in its complexity.

Small practices may adopt EHRs as a subscription service (SaaS) where they only need to maintain an Internet connection and user terminals and everything else is done for them remotely. Large practices may be completely self-contained with their own institutional servers, intranet, backup, terminals and IT staff. Large practice and hospital IT departments will often maintain multiple software environments for the EHR, including distinct and separate production (live), test, and training environments.

Implementation of the EHR is followed by the *post-implementation* phase which remains in effect for the duration of EHR use. This phase involves maintaining, optimizing, reassessing and improving the EHR's content and capabilities, facility workflows/processes, and staff training with a focus on continuous improvement and patient safety. In a sense, EHR implementation is never done. As clinical sites learn more about the software from using it, they often learn how to use the software in previously unanticipated ways. And certainly, as the EHR software is periodically upgraded, new functionality is added that increases efficiencies or opens up new possibilities. Post-implementation can also be referred to as maintenance, sustainment or optimization.

RECOMMENDED READING

- *Efficiency and safety of speech recognition for documentation in the electronic health record* (2017). This article compared standard keyboard documentation with speech recognition (SR) (Dragon) by 35 emergency room physicians using the EHR (Cerner). Participants had prior EHR AND SR training. They were able to demonstrate that task completion was 18% slower with SR and resulted in more errors.[169]

- *Electronic health record adoption in US hospitals: the emergence of a digital "advanced use" divide* (2017). Using 2008-2015 survey data, researchers looked at the prevalance of "basic vs comprehensive" EHRs. New survey questions were sent to establish "performance measurement" and "patient engagement." Their data showed that only 37.5% of hospitals adopted at least 8 of 10 performance measurement functions and 41.7% adopted 8 of 10 paatient engagement functions. They concluded that overall EHR adoption was excellent, but a minority of hospitals had robust systems with performance and patient engagement functionality.[170]

- *HITECH Retrospective: Glass Half-Full or Half-Empty?* (2017). In this blog, Dr. Hersh reviewed several articles in the New England Journal of Medicine that discuss the impact of the HITECH Act. One author points out the increase in HIT adoption as a result of this Act but another bemoans the shorcomings. Dr. Hersh puts this in perspective for the reader.[171]

FUTURE TRENDS

As the practice of healthcare advances EHRs can keep pace by utilizing more sophisticated, integrated and real-time analytics, increasing standardization, enhancing interoperability, and linking tightly with more sophisticated patient portals than those currently available. A desired outcome is that data and information will no longer remain locked in the plethora of EHR silos used by physicians and hospitals, but will electronically flow from one to the other with ease.[172] We can also expect there to be more integration between hospital EHRs and the myriad of medical devices with "smart" attributes, such as IV pumps, blood pressure monitors, glucometers, and other products that may be part of the Internet of Things. Remote patient monitoring, generating data when patients are in their homes or when they are on the move, may also be integrated with the EHR.

Substitutable Medical Applications, Reusable Technologies (SMART) on FHIR is a relatively new initiative that creates apps for EHRs, analogous to smartphone apps. With SMART apps, an API can read data from an EHR and write it to the EHR. As of mid-2017 there were 41 apps posted on a web site., and this number will only grow over time. Figure 4.14 displays a heart failure predictive analytics calculator built using this new paradigm.[173]

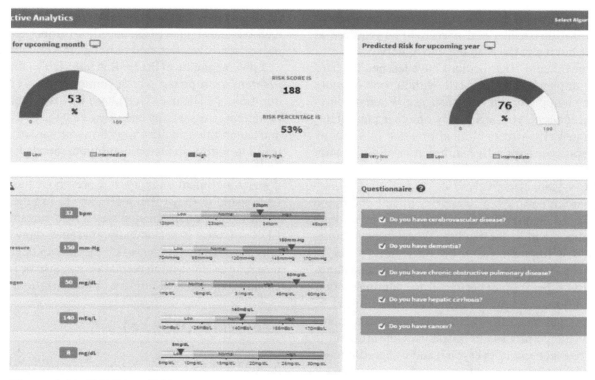

Figure 4.14: SMART on FHIR App

The AMIA EHR-2020 Task Force reported on the status and future direction of EHRs in 2015 with the following recommendations:
- Decrease data entry burden for the clinician, by using patient and other care team input
- Separate data entry from reporting; use natural language processing and new interfaces to produce reports
- EHRs should enable a *"learning healthcare system"* and promote research.
- Regulations should simplify certification, improve interoperability, reduce data re-entry and emphasize patient outcomes.
- Modify reimbursement strategies to support novel EHR innovation
- Enhance EHR certification transparency
- Everyone should be transparent about unintended consequences and best practices to mitigate risk
- EHR vendors should use open APIs, to be open to developers, researchers and patients
- Expand EHR use beyond acute hospital and office care.
- Improve usability[174]

The federal government will continue to look for overcoding and other potential abuses.[175] It is likely there will be new coding guidelines as a result of multiple questions about legitimate EHR billing practices. IT vendors are also being scrutinized, evidenced by the revocation of two EHR certifications in 2013.[176]

Experts suggest several trends, including an increased reliance on cloud computing,[177] and large shared databases used for comparative effectiveness research,[178-179] increased use of natural language processing,[180] more pervasive use of telehealth (virtual visits and consultations),[181] improved clinical decision support, more use of patient registries built into EHR workflow,[182] and greater use and integration of wireless remote outpatient monitoring of patients.[183-184] At least 3 EHR vendors are working on virtual assistants, similar to Amazon's Alexa to save key strokes and time spent looking for results, etc.[185] Rajkomar et al. reported in 2018 their effort to combine FHIR representation, EHR data and deep learning to accurately predict in-hospital mortality and 30 day remission.[186] Clearly, artificial intelligence will be used to improve the EHR's ability to be part of a *"learning healthcare system."*[187]

The Meaningful Use program is transitioning to the QPP, but its future remains uncertain due to budget constraints and a new HHS administration. But no matter how healthcare reimbursement evolves, clearly the direction healthcare is taking is to reimburse for quality and not quantity.

Even today, the future of EHRs is not entirely clear, but one thing is certain – there will be no return to paper records, because much like the advent of the automobile over a century ago rendered the horse-drawn buggy obsolete, the EHR, now used by an entire generation of clinicians who know of no other alternative, will endure.

KEY POINTS

- Electronic health records are central to a modern healthcare system
- Paper-based systems are fraught with multiple shortcomings
- Reimbursement for electronic health records by the federal government dramatically increased EHR adoption
- Despite the potential benefits of electronic health records, obstacles and controversies persist
- Clinical decision support systems are still immature and will likely improve in the future with artificial intelligence
- Advance planning and training is mandatory for successful EHR implementation

CONCLUSION

Without doubt, Medicare and Medicaid reimbursement for EHRs and e-prescribing (the Meaningful Use Program) has been the most significant impetus to promote EHR adoption. However, we lack detailed data regarding EHR failure rates, and are still learning lessons from MU stages 1 and 2.

Enterprise-scale clinical information systems have been transformational for large organizations like the VA, Kaiser-Permanente and the Cleveland Clinic, but the reality is that medicine in this country is mostly practiced by small medical groups, with limited finances and IT support. As a new trend, we are seeing outpatient clinicians opt to re-engineer their business model centered on an EHR. Their goal is to reduce overhead by having fewer support staff and to concentrate on seeing fewer patients per day but with more time spent per patient. When this is combined with secure messaging, e-visits and e-prescribing the goal of the *e-office* is achievable.[188]

REFERENCES

1. Weed LL. Medical Records that guide and teach. NEJM 1968;278:593-600
2. Fries JF, McShane DJ. ARAMIS: a proto-typical national chronic disease data bank. West J Med 1986;145:798-804
3. Atkinson, JC, Zeller, GG, Shah C. Electronic Patient Records for Dental School Clinics: More than Paperless Systems. J of Dental Ed. 2002;66 (5): 634-642
4. National Institutes of Health Electronic Health Records Overview. April 2006. MITRE Corp. http://www.himss.org/files/ HIMSSorg/content/files/Code%20180%20MITRE%20Key%20Components%20of%20an%20EHR.pdfhttp://www.ncrr.nih.gov/publications/informatics/EHR.pdf (Accessed September 23, 2013)
5. Schwartz WB. Medicine and the computer. The promise and problems of change NEJM 1970;283:1257-64
6. The Computer-Based Patient Record: An Essential Technology for Health Care, Revised Edition (1997) Institute of Medicine. The National Academies Press http://www.nap.edu/openbook.php?record_id=5306 (Accessed August 3, 2017)
7. Berner ES, Detmer DE, Simborg D. Will the wave finally break? A brief view of the adoption of electronic medical records in the United States JAIMA 2004;12
8. Stead WW, Kelly BJ, Kolodner RM. Achievable steps toward building a National Health Information Infrastructure in the United States JAMIA 2005;12:113-120
9. Defining Key Health Information Technology Terms. April 28, 2008 www.nahit.org Organization no longer active. (Accessed May 20, 2008)(No longer active)
10. Tang PC et al. Measuring the Effects of Reminders for Outpatient Influenza Immunizations at the point of clinical opportunity. JAMIA 1999;6:115-121
11. Smith PC et al. Missing Clinical Information During Primary Care Visits JAMA 2005;293:565-571
12. The President's Information Technology Advisory Committee (PITAC) http://www.nitrd.gov/pubs/pitac/ (Accessed September 16, 2013)(No longer active)
13. Eden KB, Messina R, Li H et al. Examining the value of electronic health records on labor and delivery. Am J Obstet Gynecol 2008;199; 307.e1-307.e9
14. Electronic Health Records for Clinical Research. http://www.ehr4cr.eu/ (Accessed May 27, 2017)
15. Coorevits P, Sundgren M, Klein GO, Bahr A, Claerhout B, Daniel C, et al. Electronic health records: New opportunities for clinical research. Vol. 274, Journal of Internal Medicine. 2013. p. 547–60.
16. Cowie MR, Blomster JI, Curtis LH, Duclaux S, Ford I, Fritz F, et al. Electronic health records to facilitate clinical research. Clinical Research in Cardiology. 2016;1–9.
17. Tierney WM. Computerized Display of Past Test Results. Annals of Int Medicine 1987;107:569-574
18. Lohr S. Building a Medical Data Network. The New York Times. November 22, 2004. (Accessed December 20, 2005)
19. Garrido T, Jamieson L, Zhou Y, Wiesenthal A, Liang L. Effect of electronic health records in ambulatory care: retrospective, serial, cross sectional study. BMJ 2005;330:1313-1316
20. Pollak, VE, Lorch JA. Effect of Electronic Patient Record Use on Mortality in End Stage Renal Disease, a Model Chronic Disease: A Retrospective Analysis of 9 years of Prospectively Collected Data. Biomed Central http://www.biomedcentral.com/1472-6947/7/38. (Accessed August 3, 2017)
21. US Food and Drug Administration http://www.fda.gov/ola/2004/vioxx1118.html (Accessed Aug 15, 2006)
22. Badgett R, Mulrow C. Using Information Technology to transfer knowledge: A medical institution steps up to the plate [editorial] Ann of Int Med 2005;142;220-221
23. Housman D. Quality Reporting Through a Data Warehouse Pat Safety & Qual Healthcare Jan/Feb 2009;26-31
24. Wall Street Journal Online/Harris Interactive Health-Care Poll www.wsj.com/health (Accessed October 24, 2006)
25. Potential benefits of an EHR. AAFP's Center for Health Information Technology. www.centerforhit.org/x1117.xml (Accessed August 3, 2017)
26. Center for Information Technology Leadership. CPOE in Ambulatory Care. www.citl.org/research/ACPOE.htm (Center no longer active). (Accessed November 9, 2005)
27. Lorenzi NM, Novak LL, Weiss JB, Gadd CS, Unertl KM. Crossing the implementation chasm: a proposal for bold action. J Am Med Inform Assoc 2008;15:290-296
28. Mangalmurti SS, Murtagh L, Mello MM. Medical malpractice liability in the age of electronic health records. N Engl J Med 2010; 363:2060-2067
29. Patient Safety & Quality Healthcare. New Survey Addresses Relationship of EMRs to Malpractice Risk. August 22, 2007. www.psqh.com (Accessed August 23, 2007)
30. BIDCO to launch population health management platform | Health Data Management. 2017 May 25 Available from: https://www.

healthdatamanagement.com/news/bidco-to-launch-population-health-management-platform (Accessed June 1, 2017)
31. Biome Biobank. http://icahn.mssm.edu/research/ipm/programs/biome-biobank (Accessed August 3, 2017)
32. Singh B, Singh A Ahmed A et al. Derivation and Validation of Automated Electronic Search Strategies to Extract Charlson Comorbidities from Electronic Medical Records. Mayo Clinic Proceedings 2012; 87(9):817-824
33. KP Inside. 101 letters to the people of Kaiser Permanente. 2012. George C. Halvorson, Charleston, SC
34. Connected for Health. Using Electronic Health Records to Transform Care Delivery. 2010. Jossy-Bass A Wiley Imprint. Louise L. Liang, Editor. Hoboken NJ
35. Jacobe D. Worried about...the Financial Impact of Serious Illness. Gallup Serious Chronic Illness Survey 2002. www.gallup.com (Accessed January 29, 2006)
36. O'Malley AS, Grossman JM, Cohen GR et al. Are Electronic Medical Records Helpful for Care Coordination? Experiences of Physician Practices. J Gen Int Med 2009 DOI: 10.1007/s11606-009-1195-2 22 December 2009
37. CDC. Physician Adoption of Electronic Health Records: United States, 2011 https://www.cdc.gov/nchs/products/databriefs/db98.htm (Accessed May 25, 2017)
38. Key Capabilities of an Electronic Health Record System: Letter Report. Committee on Data Standards for Patient Safety. https://www.nap.edu/catalog/10781/key-capabilities-of-an-electronic-health-record-system-letter-report (Accessed August 3, 2017)
39. Carter J Selecting an Electronic Medical Records System, second edition, 2008 Practice Management Center. American College of Physicians http://www.acponline.org/pmc"www.acponline.org/pmc (Accessed January 10, 2009)
40. LibreHealth EHR Project. http://librehealth.io/projects/lh-ehr/ (Accessed July 8, 2017)
41. Bates DM, Teich JM, Lee J et al. The impact of Computerized Physician Order Entry on Medication Error Prevention JAMIA 1999;6:313-321
42. Koppel R et al. Role of Computerized Physician Order Entry Systems in Facilitating Medication Errors. JAMA 2005;293:1197-1203
43. Bates DW et al. Effect of computerized physician order entry and a team intervention on prevention of serious medication errors JAMA 1998;280:1311-1316
44. Wolfstadt JI, Gurwitz JH, Field TS et al. The Effect of Computerized Physician Order Entry with Clinical Decision Support on the Rates of Adverse Drug Events: A Systematic Review. J Gen Int Med 2008;23(4):451-458
45. Weiner JP, Kfure T, Chan K et al. "e-Iatrogenesis": The Most Critical Unintended Consequence of CPOE and Other HIT. JAMIA 2007; 14(3):387–388
46. Campbell EM, Sittig DF, Ash JS, Guappone KP, Dykstra RH. Types of unintended consequences related to computerized provider order entry. Journal of the American Medical Informatics Association. 2006 Sep 1;13(5):547-56.
47. Han YY et al. Unexpected increased mortality after implementation of a commercially sold computerized physician order entry system Pediatrics 2005;116:1506-1512
48. Del Beccaro MA et al. Computerized Provider Order Entry Implementation: No Association with Increased Mortality Rate in An Intensive Care Unit Pediatrics 2006;118:290-295
49. Nebeker J et al. High Rates of Adverse Drug Events in a highly computerized hospital Arch Int Med 2005;165:1111-16
50. Holdsworth MT et al. Impact of Computerized Prescriber Order Entry on the Incidence of Adverse Drug Events in Pediatric Inpatients. Pediatrics. 2007;120:1058-1066
51. Prgomet M, Li L, Niazkhani Z et al. Impact of commercial computerized provider order entry (CPOE) and clinical decision support systems (CDSSs) on medication errors, length of stay, and mortality in intensive care units: a systematic review and meta-analysis. JAMIA. 2017;24(2): 413–422
52. Eslami S, Abu-Hanna, A, de Keizer, NF. Evaluation of Outpatient Computerized Physician Medication Order Entry Systems: A Systematic Review. JAMIA 2007;14:400-406
53. Kuo GM, Phillips RL, Graham D. et al. Medication errors reported by US family physicians and their office staff. Quality and Safety in Health Care 2008;17(4):286-290
54. Mekhjian HS. Immediate Benefits Realized Following Implementation of Physician Order Entry at an Academic Medical Center JAMIA 2002;9:529-539
55. Tierney, WM et al. Physician Inpatient Order Writing on Microcomputer Workstations: Effects on Resource Utilization. JAMA 1993;269(3):379-383
56. Nuckols TK, Asch SM, Patel V et al. Implementing computerized provider order entry in acute care hospitals in the United States could generate substantial savings to society. The Joint Commission Journal on Quality and Patient Safety. 2015 Aug 31;41(8):341-AP1.
57. Teich JM et al. Toward Cost-Effective, Quality Care: The Brigham Integrated Computing System.

Pp19-55. Elaine Steen [ed] The Second Annual Nicholas E. Davis Award: Proceedings of the CPR Recognition symposium. McGraw-Hill 1996
58. Ashish KJ et al. How common are electronic health records in the US? A summary of the evidence. Health Affairs 2006;25:496-507
59. Kawamoto, K, Houlihan, CA, Balas, EA, Loback DF. Improving clinical practice using clinical decision support systems: a systematic review of trials to identify features critical to success. BMJ. March 2005. https://www.ncbi.nlm.nih.gov/pmc/articles/PMC555881/ (Accessed June 13, 2007)
60. Sheridan TB, Thompson JM. People versus computers in medicine. In: Bogner MS, (ed). Human Error in Medicine. Hillsdale, NJ: Lawrence Erlbaum Associates, 1994, pp141-59
61. Bright, T. J., Wong, A., Dhurjati, R. et al. Effect of clinical decision-support systems: a systematic review. Annals of internal medicine. 2012;157(1), 29-43.
62. Roshanov, P. S., Fernandes, N., Wilczynski, J. M et al. Features of effective computerised clinical decision support systems: meta-regression of 162 randomised trials. BMJ: British Medical Journal, 2013. 346. http://www.bmj.com/content/346/bmj.f657 (Accessed August 28, 2017)
63. Bates, D. W., Kuperman, G. J., Wang, S. et al. Ten commandments for effective clinical decision support: making the practice of evidence-based medicine a reality. JAMIA. 2003; 10(6):523-530.
64. UpToDate http://www.uptodate.com/home/ehr-interface (Accessed June 20, 2017)
65. HL7 Infobutton. http://www.hl7.org/implement/standards/product_brief.cfm?product_id=208 (Accessed July 9, 2017)
66. DynaMed Plus. http://www.dynamed.com/home/access-options/ehr-access/integration-methods (Accessed July 2, 2017)
67. Shea S, DuMouchel W, Bahamonde I. A meta-analysis of 15 randomized controlled trials to evaluate computer based clinical reminder systems for preventive care in the ambulatory setting. JAMIA. 1996;3:399-409
68. DxPlain. http://www.mghlcs.org/projects/dxplain. (Accessed August 3, 2017)
69. Osheroff JA, Pifer EA, Teich JM, Sittig DF, Jenders. Improving Outcomes with Clinical Decision Support: An Implementer's Guide. HIMSS Publication 2005. http://marketplace.himss.org/OnlineStore/ProductDetail.aspx?ProductId=3318 (Accessed September 16, 2013)
70. Liebovitz D. Next steps for electronic health records to improve the diagnostic process. Diagnosis [Internet]. De Gruyter; 2015 Jan 1 ;2(2):111–6. https://www.degruyter.com/view/j/dx.2015.2.issue-2/dx-2014-0070/dx-2014-0070.xml (Accessed June 20, 2017)
71. Rosenthal DI, Wilburg JB, Schultz T et al. Radiology Order Entry With Decision Support: Initial Clinical Experience. J Am Coll Radiol 2006;3:799-806
72. American College of Radiology https://www.acr.org/Advocacy/Economics-Health-Policy/Clinical-Decision-Support (Accessed June 1, 2017)
73. Van wijk MAM, Van der lei, J, MOssveld, M, et al. Assessment of Decision Support for Blood Test Ordering in Primary Care. Ann Int Med 2001;134:274-281
74. Procop GW, Yerian LM, Wyllie R, Harrison AM, Kottke-Marchant K. Duplicate Laboratory Test Reduction Using a Clinical Decision Support Tool. Am J Clin Pathol .2014; 141(5):718–23. http://www.ncbi.nlm.nih.gov/pubmed/24713745 (Accessed July 5, 2017)
75. Lurio J, Morrison FP, Pichardo M et al. Using electronic health record alerts to provide public health situational awareness to clinicians. JAMIA 2010;17:217-219
76. A Web Services Approach to Public Health Clinical Decision Support | David Raths | Healthcare Blogs. October 22, 2016. https://www.healthcare-informatics.com/blogs/david-raths/population-health/web-services-approach-public-health-clinical-decision-support (Accessed June 30, 2017)
77. 2017 HL7® FHIR® Applications Roundtable http://www.hl7.org/events/fhir/roundtable/2017/03/final.presentations.cfm (Accessed July 15, 2017)
78. SMART App Gallery Available from: https://apps.smarthealthit.org/ (Accessed August 15, 2017)
79. Briggs B. Decision Support Matures. Health Data Management August 15 2005. www.healthdatamanagement.com (Accessed August 20, 2005)
80. M.J. Ball. Clinical Decision Support Systems: Theory and Practice. Springer. 1998
81. Clinical Decision Support Systems in Informatics Review. www.informatics-review.com/decision-support. (Accessed January 23, 2006)
82. Improving Outcome with Clinical Decision Support: An Implementer's Guide. Osheroff, JA, Pifer EA, Teich JM, Sittig DF, Jenders RA. HIMSS 2005 Chicago, Il
83. Gliklich RE, Dreyer NA, Leavy MB. Interfacing Registries with Electronic Health Records. Agency for Healthcare Research and Quality (US); 2014 https://www.ncbi.nlm.nih.gov/books/NBK208625/ (Accessed June 20, 2017)
84. Registries for Evaluating Patient Outcomes: A User's Guide. AHRQ. 2010. Rockville, MD. AHRQ Pub. No.10-EHC049 http://www.effectivehealthcare.

ahrq.gov/ehc/products/74/531/Registries%20 2nd%20ed%20final%20to%20Eisenberg%20 9-15-10.pdf (Accessed August 25, 2017)
85. Guideline Advantage. http://www.guidelineadvantage.org/TGA/ (Accessed August 1, 2017)
86. Han W, Sharman R, Heider A et al. Impact of electronic diabetes registry "Meaningful Use" on quality of care and hospital utilization. J Am Med Informatics Assoc 2016 Mar;23(2):242–7. https://academic.oup.com/jamia/article-lookup/doi/10.1093/jamia/ocv040 (Accessed May 23, 2017)
87. Pulley J. The Claims Scrubbers. November 13, 2008. Healthcare IT News. http://www.healthcareitnews.com (Accessed August 3, 2017)
88. CDC. Ambulatory Health Care Data. https://www.cdc.gov/nchs/ahcd/web_tables.htm#2015 (Accessed August 1, 2017)
89. US Ambulatory EMR Adoption Model. http://www.himssanalytics.org/emram (January 17, 2018)
90. ONC Data Brief No. 23. April 2015 Adoption of Electronic Health Record Systems among US Non-Federal Acute Care Hospitals 2008-2014. https://www.healthit.gov (Accessed August 3, 2017)
91. DesRoches CM, Worzala C, Joshi MS, et al. Small, Nonteaching and Rural Hospitals Continue to be Slow in Adopting Electronic Health Record Systems. Health Affairs. 2012;31 (5):1092-1099
92. Anderson et al. Health Care Spending and use of Information Technology in OECD Countries. Health Affairs. 2006; 25 (3): 810-831.
93. Schoen C, Osborn R, Squires D et al. A Survey of Primary Care Doctors in Ten Countries Shows Progress in Use of Health Information Technology, Less in Other Areas. Health Affairs 2012;31(12):1-12
94. Zelmer J, Ronchi E, Hypponen et al. International health IT benchmarking: learning from cross-country comparisons. JAMIA. 2017;24(2):371-379
95. The Electronic Health Record in the UK. April 3, 2017. Centre for Public Impact. https://www.centreforpublicimpact.org/case-study/electronic-health-records-system-uk/ (Accessed October 2, 2017)
96. Wang SJ et al. A Cost-Benefit Analysis of Electronic Medical Records in Primary Care Amer J of Med 2003;114:397-403
97. Miller RH et al. The Value of Electronic Health Records In solo or small group practices Health Affairs 2005;24:1127-1137
98. King MS, Sharp L, Lipsky M. Accuracy of CPT evaluation and management coding by Family Physicians. J Am Board Fam Pract 2001;14(3):184-192
99. Moore P. Tech Survey: Navigating the Tech Maze. September 2008. Physicians Practice www.physicianspractice.com (Accessed October 4, 2008)
100. Fleming NS, Culler SD, McCorkle R et al. The Financial and Nonfinancial Costs of Implementing Electronic Health Records in Primary Care Practices. Health Affairs 2011. 30(3): 481-489
101. Adler-Milstein J, Green CE, Bates DW. A Survey Analysis Suggests That Electronic Health Records Will Yield Revenue Gains for Some Practices and Losses for Many. Health Affairs 2013;32(3):562-570
102. Eastaugh S. Electronic health records lifecycle cost. J Health Care Finance. 2013 http://hsrc.himmelfarb.gwu.edu/sphhs_hsml_facpubs/11 (Accessed May 25, 2017)
103. 2003 Commonwealth Fund National Survey of Physicians and Quality of Care. http://www.cmwf.org/surveys/surveys_show.htm?doc_id=278869 (Accessed August 3, 2017)
104. Brailer DJ, Terasawa EL. Use and Adoption of Computer Based Patient Records California Healthcare Foundation 2003 www.chcf.org (Accessed February 10, 2005)
105. Poissant L, Pereira J, Tamblyn R et al. The impact of electronic health records on time efficiency of physicians and nurses: a systematic review. JAMIA 2005;12:505-516
106. McDonald CJ, Callaghan FM, Weissman A et al. Use of Internist's Free Time by Ambulatory Care Electronic Medical Record Systems. JAMA Intern Med 2014;174(11):1860. http://archinte.jamanetwork.com/article.aspx?doi=10.1001/jamainternmed.2014.4506 (Accessed May 25, 2017)
107. Sinsky C, Colligan L, Li L et al. Allocation of Physician Time in Ambulatory Practice: A Time and Motion Study in 4 Specialties. Annals Int Med. 2016;165(11):753-760
108. Hsu J, Huang J, Fung V et al. Health Information Technology and Physician-Patient Interactions: Impact of Computers on Communication during Outpatient Primary Care Visits. J Am Med Inform Assoc. 2005;12:474-480.
109. White A, Danis M. Enhancing Patient-Centered Communication and Collaboration by Using the Electronic Health Record in the Examination Room. JAMA 2013;309(22):2327-2328.
110. American Medical Association. Report of the Board of Trustees. May 2013. http://www.ama-assn.org/assets/meeting/ 2013a/a13-bot-21.pdf (Accessed August 3, 2017)
111. Block L, Habicht R, Wu AW et al. In the Wake of the 2003 and 2011 Duty Hours Regulations, How Do Internal Medicine Interns Spend Their Time? J Gen Int Med 2013. DOI: 10.1007/s11606-013-2376-6
112. Hill RG, Sears LM, Melanson SW. 4000 Clicks: a productivity analysis of electronic medical records in a community hospital ED. Amer J Emerg Med 2013. Online September 21, 2013.

113. Boone, E. EMR Usability: Bridging the Gap Between the Nurse and Computer. Nursing Management 2010;41(3): 14-16
114. Ratwani RM, Fairbanks RJ, Hettinger AZ, Benda NC, BN D, SJ D. Electronic health record usability: analysis of the user-centered design processes of eleven electronic health record vendors. J Am Med Informatics Assoc 2015;22(6):1179–82. https://academic.oup.com/jamia/article-lookup/doi/10.1093/jamia/ocv050 (Accessed May 25, 2017)
115. Ellsworth MA, Dziadzko M, O'Horo JC et al. An appraisal of published usability evaluations of electronic health records via systematic review. J Am Med Informatics Assoc 2017;24(1):218–26. http://www.ncbi.nlm.nih.gov/pubmed/27107451 (Accessed May 25, 2017)
116. Middleton, B., Bloomrosen, M., Dente et al. (2013). Enhancing patient safety and quality of care by improving the usability of electronic health record systems: recommendations from AMIA. JAMIA. 2013;e2-e8 https://www.ncbi.nlm.nih.gov/pmc/articles/PMC3715367/ (Accessed August 30, 2017)
117. Wright A, Aaron S, Sittig DF. Testing electronic health records in the "production" environment: an essential step in the journey to a safe and effective health care system. JAMIA. 2016 http://jamia.oxfordjournals.org/content/early/2016/04/22/jamia.ocw039.abstract (Accessed May 25, 2017)
118. Hazin R, Brothers KB, Malin BA et al. (2013) Ethical, legal, and social implications of incorporating genomic information into electronic health records. Genet Med 15: 810-816.
119. Coamey J. MDx and MDs: Is a dose of knowledge the prescription for adoption? www.healthtech.com/uploadedFiles/Conferences/.../CAHG_LandmarkPhysicianStudy.pd... (Accessed August 1, 2017)
120. Kern LM, Malhotra S, Barron Y et al. Accuracy of Electronically Reported "Meaningful Use" Clinical Quality Measures. Ann Int Med 2013;158: 77-83
121. Ryan AM et al. Small physician practices in New York needed sustained help to realize gains in quality from use of electronic health records. Health Affairs 2013;32(1): 53-62
122. HIPAA Survival Guide. www.hipaasurvivalguide.com (Accessed August 3, 2017)
123. Mello, MM et al. National Costs of The Medical Liability System. Health Affairs. 2010; 29, (9):1569-1577
124. Quinn MA, Kats AM, Kleinman K. The relationship between electronic health records and malpractice claims. Arch Int Med 2012;172(15):1187-8
125. David Blumenthal, M.D., M.P.P. Study of Patient Safety and Health Information Technology Institute of Medicine, National Academy of Sciences. December 14, 2010 http://www.iom.edu/ (Accessed October 6, 2011)
126. Robb D, Owens L. Breaking Free of Copy/Paste. OIG work plan cracks down on risky documentation habit. J Ahima March 13, 2013:46-47 www.journal.ahima.org (Accessed August 3, 2017)
127. Mangalmurti, SS. Murtagh L, Mello MM. Medical Malpractice Liability in the Age of Electronic Health Records. New England Journal of Medicine. 2010; 363: 2060-2067
128. Virapongse, A et al. Electronic Health Records and Malpractice Claims in Office Practice. Archives of Internal Medicine. 2008; 168(21):2362-2367.
129. Chaudry B, Wang J, Wu S et al. Systematic Review: Impact of Health Information Technology on Quality, Efficiency and Costs of Medical Care. Ann of Int Med 2006;144:E12-22.
130. Yu FB, Menachemi N, Berner ES et al. Full Implementation of Computerized Physician Order Entry and Medication-Related Quality Outcomes: A Study of 3364 Hospitals. Am J of Qual Meas 2009; E1-9doi:10.1177/106286060933626
131. Study: EHR Adoption Results in Marginal Performance Gains. November 16, 2009. www.ihealthbeat.org No longer active. (Accessed November 23, 2009)
132. EHR IMPACT. The socio-economic impact of interoperable electronic health record (EHR) and ePrescribing systems in Europe and beyond. Final study report. October 2009. www.ehr-impact.eu (Accessed August 3, 2017)
133. Jones SS, Adams JL, Schneider EC et al. Electronic Health Record Adoption and Quality Improvement in US Hospitals. December 2010. www.ajmc.com (Accessed August 3, 2017)
134. Romano MJ, Stafford RS. Electronic Health Records and Clinical Decision Support Systems. Impact on National Ambulatory Care Quality. Arch Int Med 2011;171(10):897-903
135. Cebul RD, Love TE, Kain AK, Hebert CJ. Electronic Health Records and Quality of Diabetic Care. NEJM 2011;365(9):825-833
136. Crosson JC, Obman-Strickland PA, Cohen DJ et al. Typical Electronic Health Record Use in Primary Care Practices and the Quality of Diabetes Care. Ann Fam Med. May/June 2012;10(3). www.annfammed.org (Accessed June 28, 2012)
137. Reed, M, Huang J, Graetz I et al. Outpatient Electronic Health Records and the Clinical Care and Outcomes of Patients with Diabetes Mellitus. Ann Intern Med 2012;157(7):482-489.
138. McCullough JS, Christianson J, Leerapan B. Do Electronic Medical Records Improve Diabetes Quality in Physician Practices? Am J Man Care 2013; 19(2). www.ajmc.com (Accessed February 27, 2013)

139. Kern LM, Barrón Y, Dhopeshwarkar R V., Edwards A, Kaushal R. Electronic health records and ambulatory quality of care. J Gen Intern Med. 2013;28(4):496–503
140. Samal L, Wright A, Healey MJ, Linder JA, Bates DW. Meaningful Use and Quality of Care. JAMA Intern Med 2014;174(6):997. http://archinte.jamanetwork.com/article.aspx?doi=10.1001/jamainternmed.2014.662 (Accessed May 27,2017)
141. O'Reilly D, Tarride JE, Goeree R et al. The economics of health information technology in medication management: a systematic review of economic evaluations. J Am Med Inform Assoc 2012;19:423-438
142. Shojana KG, Jennings A, Mayhew A et al. Effect of point-of-care computer reminders on physician behavior: a systematic review. CMAJ 2010;182(5):E216-E225
143. McCormick D, Bor DH, Woolhandler S et al. Giving Office-Based Physicians Electronic Access To Patients' Prior Imaging and Lab Results Did Not Deter Ordering of Tests. Health Affairs. 2012;31(3):488-496
144. Bates DW. Computerized physician order entry and medication errors: Finding a balance. J of Bioinform 2005;38:259-261
145. Ash JS, Berg M, Coiera E. Some Unintended Consequences of Information Technology in Health Care: The Nature of Patient Care Information System-Related Errors. JAMIA. 2004;11:104-112
146. Berger RG, Kichak, JP. Computerized Physician Order Entry: Helpful or Harmful JAMIA 2004;11:100-103
147. Campbell EM, Sittig DF, Ash JS et al. Types of Unintended Consequences Related to Computerized Order Entry. J Am Med Inform Assoc. 2006;13(5):547-556
148. Sentinel event alert 42. Safely Implementing health information and converging technologies. https://www.jointcommission.org/sentinel_event_alert_issue_42_safely_implementing_health_information_and_converging_technologies/ (Accessed August 3, 2017)
149. Health IT and patient safety building safer systems for better care. November 2008. http://www.iom.edu (Accessed September 17, 2013)
150. Pennsylvania Patient Safety Advisory. September 2013. Vol 10 (3): 92-95 http://patientsafetyauthority.org/Pages/Default.aspx (Accessed September 17, 2013).
151. Guide to Reducing Unintended Consequences of Electronic Health Records. https://www.healthit.gov/unintended-consequences/ (August 3, 2017)
152. HIT Patient Safety and Surveillance Plan. http://www.healthit.gov/sites/default/files/aspa_0392_20130618_onc_fs_hit_safety_plan_v03final.pdf (Accessed August 3, 2017)
153. SAFER Guides https://www.healthit.gov/safer/ (Accessed May 25, 2017)
154. McCann E. Setback for Sutter after $1B EHR crashes. August 26, 2013. www.healthcareitnews.com (Accessed September 5, 2013)
155. Minghella L. Be Prepared: Lessons from an Extended Outage of a Hospital's EHR System. August 30, 2013. www.healthcare-informatics.com (Accessed September 16, 2013)
156. Medicare and Medicaid EHR Incentive Program. https://www.cms.gov/Regulations-and-Guidance/Legislation/EHRIncentivePrograms/index.html?redirect=/EHRIncentiveprograms (Accessed August 15, 2017)
157. ONC Health IT Certification Program. https://www.healthit.gov/policy-researchers-implementers/about-onc-health-it-certification-program (Accessed September 17, 2017)
158. Certified Health IT Product List. CMS. https://chpl.healthit.gov/#/search (Accessed August 3, 2017)
159. CMS. Stage 3 Program requirements for providers attesting to their state's Medicaid EHR Incentive Program. https://www.cms.gov/Regulations-and-Guidance/Legislation/EHRIncentivePrograms/Stage3Medicaid_Require.html (Accessed August 25, 2017)
160. Quality Payment Program. https://qpp.cms.gov/mips/advancing-care-information (Accessed August 25, 2017)
161. Goroll AH. Emerging from EHR Purgatory? Moving from Process to Outcomes. N Engl J Med [2017 May 25;376(21):2004–6. http://www.nejm.org/doi/10.1056/NEJMp1700601 (Accessed May 30, 2017)
162. CMS Innovation Center. https://innovation.cms.gov/initiatives/comprehensive-primary-care-plus (Accessed August 25, 2017)
163. Lorenzi NM, Kouroubali A, Detmer DE et al. How to successfully select and implement electronic health records (EHR) in small ambulatory practice settings. BMC Medical Informatics and Decision Making 2009, 9:15
164. Skolnik, ed. Electronic Medical Records: A practical guide for primary care. New York, NY: Humana Press. 2011
165. Keshavjee, K et al. Best practices in EMR implementation. AMIA: ISHIMR 2006. 1-15.
166. Adler, KG. How to successfully navigate your EHR implementation. Family Practice Management 2007; 14(2): 33-39
167. Adler KG. Successful EHR Implementations: Attitude is Everything. Nov/Dec 2010. www.aafp.org/fpm (Accessed February 20, 2011)

168. Kotter JP. Leading Change. Boston, MA: Harvard Business School Press; 1996.
169. Hodgson T, Magrabi F, Coiera E. Efficiency and safety of speech recognition for documentation in the electronic health record. JAMIA 2017;24(6):1127-1133
170. Adler-Milstein J, Holmgren AJ, Kralovec P et al. Electronic health record adoption in US hospitals: the emergence of a digital "advanced use" divide. JAMIA 2017;24(6):1142-1148
171. Hersh W. HITECH Retrospective: Glass Half-Full or Half Empty? Informatics Professor. October 6, 2017. http://informaticsprofessor.blogspot.com/2017/10/hitech-retrospective-glass-half-full-or.html (Accessed November 6, 2017)
172. O'Malley. Tapping the unmet potential of health information technology. New England Journal of Medicine. 2011; 364(12):1090-1091
173. SMART Apps. https://apps.smarthealthit.org/ (Accessed July 3, 2017)
174. Payne TH, Corley S, Cullen TA, Gandhi TK, Harrington L, Kuperman GJ, et al. Report of the AMIA EHR-2020 Task Force on the status and future direction of EHRs. J Am Med Informatics Assoc 2015 ;22(5):1102–10. https://academic.oup.com/jamia/article-lookup/doi/10.1093/jamia/ocv066 (Accessed August 24, 2017)
175. Schulte F. How doctors and hospitals have collected billions in questionable Medicare fees. Center for Public Integrity. September 15, 2012. www.publicintegrity.org (Accessed August 3, 2017)
176. Health and Human Services, News Release. Certification for electronic health record product revoked. April 25, 2013. www.hhs.gov. (Accessed April 25, 2013)
177. ONC's new query health initiative—what's in it for e-patients http://e-patients.net (Accessed September 20, 2013)
178. Narathe, AS and Conway, PH. Optimizing health information technology's role in enabling comparative effectiveness research. The American Journal of Managed Care. 2010;16(12 Spec No.):SP44-SP47.
179. Jha AK. The promise of electronic records – Around the corner or down the road? JAMA 2011, 306(8):880-881.
180. Kaiser Permanente. http://xnet.kp.org/ future No longer active. (Accessed September 20, 2013)
181. Bates, DW and Bitton, A. The future of health information technology in the patient-centered medical home. Health Affairs. 2010; 29(4):614-621.
182. Topol, E. The future of healthcare: Information technology. Modern Healthcare. www.modernhealthcare.com (Accessed October 1, 2013)
183. Kilian J and Pantuso B. The Future of Healthcare is Social. www.fastcompany.com (Accessed September 20, 2013)
184. Diamond J, Fera B. Implementing an EHR. 2007 HIMSS Conference February 25-March 1. New Orleans
185. Monica K. athenahealth, eClinicalWorks, Epic to Launch Virtual Assistants. January 11, 2018. www.ehrintelligence.com (Accessed February 2, 2018)
186. Rajkomar A, Oren E, Chen K et al. Scalable and accurate deep learning for electronic health records. 2018. Cornell University Library. https://arxiv.org/abs/1801.07860 (Accessed February 2, 2018)
187. Oram A. A learning EHR for a Learning Healthcare System. January 24, 2018. www.emrandhipaa.com (Accessed February 2, 2018)
188. MGMA. Medical Practice Today, 2009. www.mgma.com (Accessed September 20, 2013)

5

Standards and Interoperability

WILLIAM R HERSH

LEARNING OBJECTIVES

After reading this chapter, the reader should be able to:
- Explain the importance of standards and interoperability for health and biomedical data
- Discuss the major issues related to identifier standards, including the debate on patient identifiers
- Describe the various message exchange standards, their explicit roles, and the type of data they exchange
- Discuss the different terminology systems used in biomedicine and their origins, content, and limitations

INTRODUCTION

Standards are critically important in health informatics. They promote consistent naming of individuals, events, diagnoses, treatments, and everything else that takes place in healthcare. They allow better use of data for patient care,[1] as well as re-use of that data, such as for quality assurance, research, and public health.[2] Standards enhance the ability to transfer data among applications, thus leading to better system integration. They also facilitate interoperability among information systems and users. The best resources for learning more about standards come from the book by Benson and Grieve[3] and the Web site of the HL7 International organization (http://www.hl7.org/), whose activities will be described extensively throughout this chapter.

Before we discuss the actual standards of healthcare, let us define what a standard is. According to the International Standards Organization, a standard comes from "*a standard document established by consensus and approved by a recognized body that provides for common and repeated use, rules, guidelines or characteristics for activities or their results, aimed at the optimum degree of order in a given context.*"[4] (Underlining emphasizes critical components.)

Interoperability

Standards facilitate an important process known as interoperability. The original definition of interoperability was published in 1990 by the Institute for Electronic and Electrical Engineers (IEEE), and continues to be widely cited: "*Interoperability being the ability of two or more systems or components to exchange information and to use that information that has been exchanged.*"[5] The IEEE definition of interoperability in health care was extended by the now-defunct organization National Association for Health Information Technology (NAHIT) and was endorsed by 40 other health care organizations.[6] Interoperability in this instance was defined as, "*the ability of different information technology systems and software applications to communicate, to exchange data accurately, effectively, and consistently, and to use the information that has been exchanged.*" IEEE has updated its definition of interoperability and its current definition is now, "*the ability of a system or product to work with other systems or products without special effort on the part of the customer. Interoperability is made possible by the implementation of standards.*" (http://www.ieee.org/education_careers/education/standards/standards_glossary.html)

Some describe levels of interoperability, which has been done for healthcare by Walker et al.[7]
- Level 1 – no interoperability, e.g., mail, fax, phone, etc.
- Level 2 – machine-transportable (structural); information cannot be manipulated, e.g., scanned document, image, PDF
- Level 3 – machine-organizable (syntactic); sender and receiver must understand vocabulary, e.g., email, files in proprietary format

- Level 4 – machine-interpretable (semantic); structured messages with standardized and coded data, e.g., coded results from structured notes, lab, problem list, etc.

Value and Limitations of Standards

The value of standards has been known throughout human history and is not just unique to the computer era. Roman chariots placed the wheels a specific distance apart so that pathways through ancient Roman cities could be used. The emergence of standards for railroad cars enabled different railroads to be built out around various countries, all adhering to a standard of the wheels being a certain distance apart. When telephones started to become international, the emergence of standards allowed calls to be made from one country to another, from one type of phone to another. An early standard with computers was ASCII text. Even though there are some variations between end-of-line characters for different computer operating systems (e.g., Windows, Mac, and Unix), ASCII text is a standard and text written on one computer from one vendor can be used on another computer. Even more recently, standards such as Wi-Fi enable computers, smartphones, tablets, and other devices to connect wirelessly to the Internet. Many global financial transactions are based on standards, with the ubiquitous automated teller machine (ATM) allowing us to access money from our bank accounts almost anywhere in the world.

While there are many benefits of standards, there are also some limitations. Standards can lead to dominance by one segment of the industry and may stifle innovation. They may limit computer applications to a more restricted feature set. For example, the standards for operating systems and productivity applications that emerged from Microsoft in the latter part of the 20th century – Microsoft Windows and Office – have benefits of widespread use but also limitations as well.

Sometimes the world does not fit into standards. A well-known example is the language Esperanto, which was an attempt to create a standard language by which humans could communicate, especially in business, and science, and other types of transactions.[8] Ultimately, however, English prevailed in that function, and Esperanto never caught on. There is also a famous quote, *"The nice thing about standards is that there are so many of them to choose from."*[9] Of course, that defeats the purpose of standards, which is to have a single standard that those working in an area can develop their work around.

Standards Development

Hammond describes four common approaches to developing standards:[10]
- Ad hoc – groups agree to informal specifications
- De facto – single vendor controls industry
- Government mandate – government agency creates standard and mandates its use
- Consensus – interested parties work in open process

The development process for standards is very important. Hammond lists the stages of development of a standard:[10]
- Identification
- Conceptualization
- Discussion
- Specification
- Early implementation
- Conformance
- Certification

If standards are going to be developed in an open and transparent process, there need to be standards bodies to convene those who are developing and, ultimately, using the standards. Typically, these are private nonprofit organizations. In the US, there is an organization called the American National Standards Institute (ANSI, www.ansi.org), which accredits standards development organizations (SDOs) including those that work in health care. There are SDOs that work in health care focused on different standards applied in different areas:
- The Accredited Standards Committee (ASC X12, http://www.x12.org/) focuses on business transactions
- Health Level 7 (HL7 http://www.hl7.org/) focuses on messaging standards
- The American Society for Testing and Materials (ASTM, https://www.astm.org/) is also an SDO and has a committee that develops health IT related standards, Committee E31 on health care informatics

Of course, standards need to interoperate not only in the United States, but also globally. The most important international standards body is the International Organizations for Standardization (ISO, https://www.iso.org/). ISO has many technical committees, one of which is Technical Committee 215 that focuses on health informatics standards. There is also a European standards organization similar to ANSI in the United States, The European Committee for Standardization (CEN, https://www.cen.eu/), which has a Technical Committee 251 in CEN that is focused on health informatics standards. Another standards organization of note is the International Telecommunication Union (ITU http://

www.itu.int/), an agency of the United Nations that is focused on telecommunication standards in general.

In the United States, health information standards have been promoted both by the government as well as private nonprofit organizations. There have been several approaches by the US government over the years that, for the most part, have attempted to identify standards that are ready for use and then to promote their use. Until recently, all health-related standards work was led by the Office of the National Coordinator for Health IT (ONC, https://www.healthit.gov/) Standards Committee. The work of this committee has now been subsumed by the ONC Health Information Technology Advisory Committee (HITAC, https://www.healthit.gov/hitac/).

The health information standards process has been aided by the National Institute for Standards and Technology (NIST, https://www.nist.gov/), which is the US federal agency that leads standards development, not only in health care, but in all aspects. NIST has focused its efforts in health care on supporting ONC, and the National Library of Medicine (NLM, https://www.nlm.nih.gov/) with its focus on terminology standards and making sure that there is appropriate terminology being used within messaging standards, such as HL7.

The ONC Health IT Standards Committee published an interoperability vision that gave an overall vision and framework for how interoperability work should proceed[11] and an interoperability roadmap detailing how to get there.[12] Figure 5.1 gives an overview of the interoperability roadmap, and the specific activities that are laid out starting with core technical standards and functions.

Another outcome of this work is the Standards Advisory, an annual release of a list of standards that are deemed ready for widespread adoption.[13] Some perspectives on adopting standards come from a couple of academic papers. One of which looks at what it takes for standards to become both mature and adoptable,[14] while the other reviews use cases to establish that an electronic health record (EHR) system is open and interoperable.[15]

Another important standards and interoperability activity is Integrating the Health Enterprise (IHE, http://ihe.net/). This is a non-federal effort, so a private nonprofit organization, that identifies and demonstrates solutions to real world interoperability problems. IHE organizes interoperability showcases where they demonstrate various solutions, both in person at various meetings, as well as virtually over the Internet.

One topic that comes up commonly in discussion of health IT standards is, why can't health IT be more like banking? After all, with banking we can literally take our ATM card from our local bank and put it in any ATM around the world and, due to a standard-- ISO 8583-- we can get out local currency and have it charged against

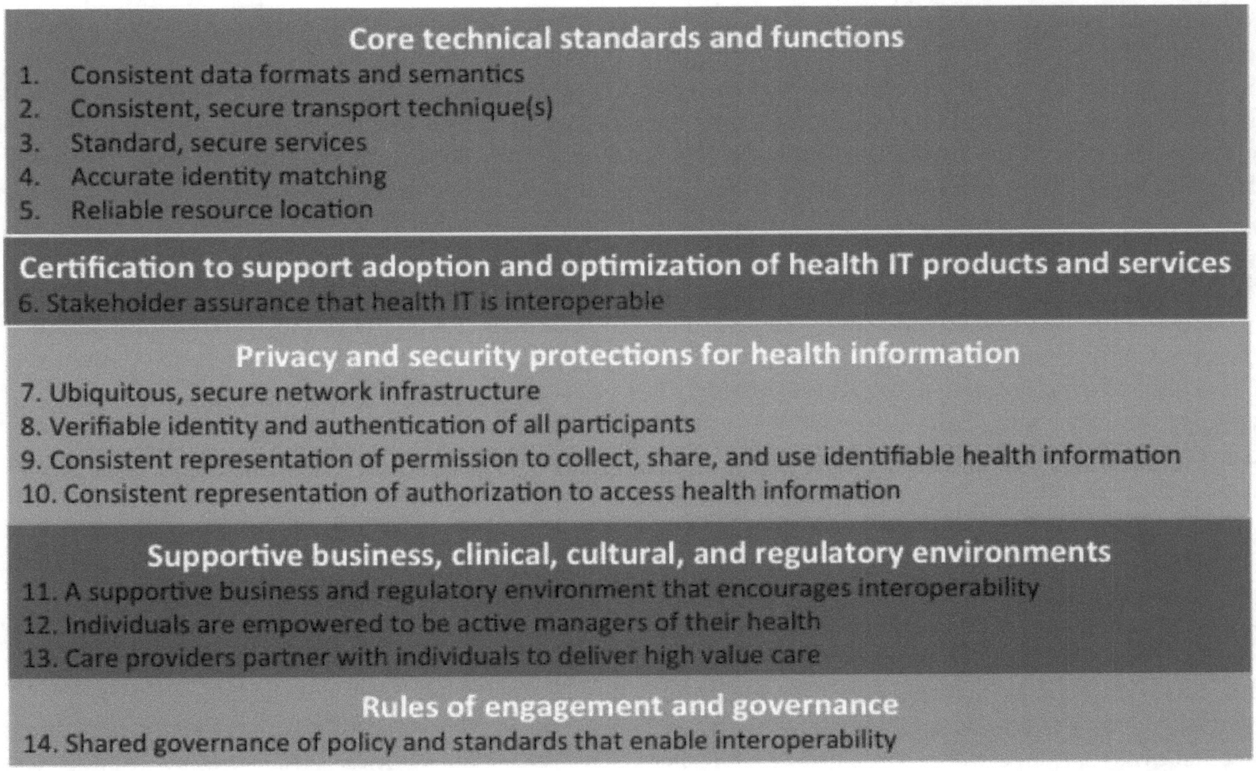

Core technical standards and functions
1. Consistent data formats and semantics
2. Consistent, secure transport technique(s)
3. Standard, secure services
4. Accurate identity matching
5. Reliable resource location

Certification to support adoption and optimization of health IT products and services
6. Stakeholder assurance that health IT is interoperable

Privacy and security protections for health information
7. Ubiquitous, secure network infrastructure
8. Verifiable identity and authentication of all participants
9. Consistent representation of permission to collect, share, and use identifiable health information
10. Consistent representation of authorization to access health information

Supportive business, clinical, cultural, and regulatory environments
11. A supportive business and regulatory environment that encourages interoperability
12. Individuals are empowered to be active managers of their health
13. Care providers partner with individuals to deliver high value care

Rules of engagement and governance
14. Shared governance of policy and standards that enable interoperability

Figure 5.1: Interoperability roadmap

our bank account (ISO 8583-1:2003 – https://www.iso.org/obp/ui/#iso:std:iso:8583:-1:ed-1:v1:en).

If banking can do standards worldwide so easily, why not healthcare? Bernstam and Johnson have discussed why the analogy often does not hold.[16] Banking data is relatively simple, mostly numbers-- it's very precise. The actions, the context, the users, and the workflow of banking are all very simple relative to the complexity of health care. Even though we need to have strong security around banking data, the complexity and variability of health data are much greater. And, again, while we can learn some from banking about standards, it doesn't completely describe why we are not yet there with health IT standards.

In the rest of this chapter, we will cover four types of standards: identifier standards, transaction standards, message exchange standards, and terminology standards.

IDENTIFIER STANDARDS

Identifier standards aim to create identifiers for all entities that participate in healthcare: patients, providers, employers, health plans.

Patient Identifiers

Of these identifiers, probably the most controversial are patient identifiers. Clearly, there is value for having a single, standardized patient identifier. Among these benefits are the easy linkage of records, such that when different records exist for a patient, they can easily be linked together. This facilitates health information exchange, when patients move from one healthcare organization to another. It also potentially reduces errors and costs from when there are duplicate records and they need to be merged together. These same benefits, however, can be risks. The easy linkage of records potentially compromises the privacy and confidentiality for a patient.

Standard identifiers aim to reduce the problems of both *duplicate* and *overlaid* records, as shown in Figure 5.2. A duplicate record occurs when more than one record exists for a patient, whereas an overlaid record takes place when more than one patient is mapped to the same record.

A study over a decade ago identified that errors in patient identifiers compromised the quality of care and can be costly, noting an expense of about $4,500 to correct duplicate patient records in an operating room, taking 325 minutes of work of those in the health care system, and with the cost increasing over the length of time that the error was not identified.[17] More recently it has been shown that duplicate records are likely to be associated with missed abnormal test results.[18]

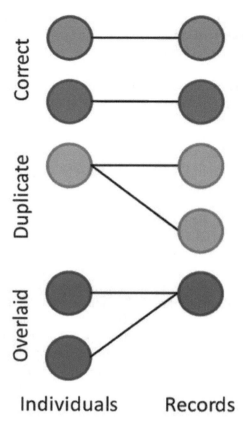

Figure 5.2: Duplicate and overlaid records

In large medical centers, there are many patients who have the same name. One analysis of five large, academic centers found that the occurrence of patients who have the same first and last name was anywhere from 16% to 40%.[19] That was somewhat reduced when the date of birth was added, although there was still a substantial amount of overlap of names and birth dates. This analysis also found that these institutions had highly variable policies for how they prevented duplicate records, how they detected them, and how they removed them once they were found. These institutions also had different approaches to mitigating these errors when they did happen.

What are some of the key attributes we would want in patient identifiers? These were laid out in a report by Connecting for Health published in 2005, and include:
- Unique – only one person has an identifier
- Non-disclosing – discloses no personal information
- Permanent – will never be re-used
- Ubiquitous – everyone has one
- Canonical – each person has only one
- Invariable – will not change over time[20]

While the issue of national health identifiers is a controversial political issue in the United States, it is a non-issue in most other industrialized, developed countries. For example, in New Zealand, there is a National Health Index. Not only is that index used for all health purposes, but there is a Web site that describes why that index exists, why it is important, and what the government does to protect privacy (http://www.health.govt.nz/our-work/health-identity/national-health-index/).

Iceland also has a national identifier in its health sector database.[21] Iceland is also the home of a national genetic database.[22] In Singapore, all citizens have a national registration identity card (NRIC). In addition, all long-term visitors get a foreign identification number (FIN). These national numbers are used for all identification, not just healthcare. Most other Western European countries also use national patient identifiers without much controversy.

Should there be government-issued patient identifiers in the United States? This was mandated by the Health Insurance Portability and Accountability Act (HIPAA) legislation in the mid-1990s. HIPAA mandated that there be patient identifiers for all citizens in the United States. But there was tremendous political pushback, and that requirement of HIPAA was postponed and eventually abandoned.

We already do have a national identifier in the United States, which is the social security number (SSN). For many years, the Veterans' Administration used the SSN as its patient record number. It turns out, however, that the SSA is a poor identifier. So even if the US desired a national health identifier, it should probably not be the SSN.

There are a number of issues related to SSNs.[23] There are many duplicates in the population, estimated to be up to 3% to 5%. There is also no check digit in the SSN that enables a checksum process to validate when the number is electronically transmitted. The SSN is also used for many other purposes, and for that reason many advocate that it not be used for healthcare. It has been shown that the SSN can be used to de-identify individuals in public health data sources.[24]

Some have advocated for voluntary identifiers in the US, with those agreeing voluntarily signing a consent and being assigned a voluntary, national health identifier.[25-26] There has been a standard developed for this (ASTM E2553, http://www.astm.org/Standards/E2553.htm), but there has not been implementation of such a program.

Others have argued that a national health identifier is unnecessary in the US and, because it's politically unfeasible, is *"not worth the fight."*[27] There may be other ways to achieve the goals of a national identifier, such as record linkage, described below. But others have argued back, saying that a unique patient identifier would reduce errors and improve system interoperability in the US.[28-30] The cost of such a system would not be cheap, but could be offset by other improvements in healthcare.

Are there alternatives to a national patient identifier? Probably the most promising approach is the use of probabilistic matching algorithms, where various attributes of the patient, e.g., name, address, date of birth, phone number, and others, are matched in a probabilistic manner. These attributes are not always recorded identically and may change over time, so robust algorithms are required that can provide matching with a high level of confidence. There is a long history of research in this area dating back over a decade and showing that many methods that have been developed show a relatively high level of accuracy in matching patients.[31-36] Some of these methods are used in health information exchange systems, where it is a requirement to match patients across different health systems.[37-38] There are still problems such as non-standardized[39] or dirty and missing data[40] that make these algorithms challenging.

The ONC commissioned a report on the current state of patient records matching.[41] It noted that successful record matching techniques were an imperative from the standpoint of patient safety, care coordination, data quality, and other issues. The report reviewed the current state of the art and noted that, for the most part, it works well, but would benefit from some changes to standards in health care systems, namely, the standardizing of patient identifying attributes in patient records, such as:

- First/given, middle/second given, and last/family names
- Suffix – e.g., Jr./Sr., II/III/etc., MD/RN/PhD, Esq., etc.
- Date of birth – YYYYMMDD, with HHMMSS if available
- Current and historical addresses – in some international format
- Phone number – all known
- Gender – from HL7 value set; M, F, UN

The report noted there would also need to be a process for handling changes in these attributes across the healthcare system, for example, a name change or address or a phone number changed. The report also concluded that the matching algorithms still needed to be improved and advocated further research to evaluate different algorithms, even looking at additional attributes that might be valuable in the matching process. Finally, the report called for widespread dissemination of best practices so that they could be implemented by vendors, health care organizations, and others.

Other Identifiers

Other types of identifiers in healthcare are much less controversial. Few would argue that we should not have identifiers for healthcare providers. The original provider identifier was the Universal Physician Identifier Number (UPIN), which was maintained by the US government for physicians who treated Medicare patients. But since not all physicians treat Medicare patients, this was superseded by the National Provider Identifier (NPI), which is assigned to all physicians in the US. There is a national provider system that issues the NPI, which is a 10-digit number whose last digit serves as a check digit. This allows a checksum process to verify that the identifier is transmitted correctly. The payor for Medicare in the US, the Centers for Medicare and Medicaid Services (CMS), will not process claims without use of the NPI.

Also, unlikely to be controversial, perhaps with the exception of the administrative overhead they may cause, are employer and health plan identifiers. Employers must have a standard Employer Identifier Number (EIN). In addition, the Affordable Care Act requires health plans to have either a Health Plan Identifier (HPI) or an Other Entity Identifier (OEID) that is an identifier for use in transactions.

TRANSACTION STANDARDS

Transaction standards are important for the business of healthcare. There is a set of transaction standards for healthcare called ASC X12N, which were developed to encourage electronic commerce for health claims, simplifying what was previously a situation of over 400 different formats between insurance companies and others for healthcare transactions. The HIPAA legislation mandated the use of the ASC X12N standards for healthcare business electronic data exchange under the guise of *administrative simplification.* The original version of ASC X12 was called 4010. This was superseded by a new version that was released in 2012 called 5010. The use of the 5010 transaction standards is a requirement for payment for any government healthcare related transactions and is also used by many private insurance companies as well.

The major transactions in 5010 and their identifier numbers include:
- Health claims and equivalent encounter information (837)
- Enrollment and disenrollment in a health plan (834)
- Eligibility for a health plan (request 270/response 271)
- Health care payment and remittance advice (835)
- Health plan premium payments (820)
- Health claim status (request 276/response 277)
- Referral certification and authorization (278)
- Coordination of benefits (837)

One of these transaction standards made the front-page news around October 2013 when the healthcare.gov website was launched. There were initially many problems with the website, a significant one of which was the improper implementation of the 834 standard for enrollment and disenrollment in health plans. In fact, a reporter from The Washington Post said that Obamacare's most important number was 834, referring to the problems that many insurers had with inadequate implementation of the standard.[42] (A more technical description of the problem was also described.)[43] There were also many other informatics lessons to be learned from the rollout of healthcare.gov, not just the political issues, but the federal IT procurement issues and the management of large-scale, complex projects.[44]

MESSAGING STANDARDS

There are many message exchange standards that focus on different types of messages and different types of data. On one level, healthcare data standards are almost synonymous with Health Level Seven or "HL7." However, the name HL7 refers to several entities. There is HL7, the organization that develops and supports standards, which is properly called HL7 International. There are also the standards of HL7 themselves, mainly the two different versions of the main HL7 messaging standards. These two standards are substantially different and incompatible with each other. The name HL7 comes from the OSI 7-layer model of network communications.

HL7 Version 2

Version 2 of HL7 is widely used throughout health care. It is a so-called syntactic standard, whereas Version 3 aims for true semantic interoperability. HL7 Version 2 has several versions that have added subsequent refinements, but they are all part of HL7 Version 2. HL7 Version 2 is supported by most vendors of health care information systems for interchange of data. HL7 Version 2 messages use the ASCII format delimit the different fields with the vertical bar character (|).

HL7 Version 2 is mostly a syntax. This means that the sender and the receiver must understand the meaning of the messages. Some of the later versions of HL7 Version 2 add more semantics (or meaning) so that there is some consistency of what the messages mean. Within HL7 Version 2, each message has segments, and each of the segments has a three-character identifier and then values that follow it. Some of these segments and their identifiers include:
- MSH – message header

- EVN – event type
- PID – patient identifier
- OBR – results header
- OBX – result details

Here is an example of HL7 Version 2 message:

MSH|^~\&|||^123457^Labs|||200808141530||
ORU^R01|123456789|P|2.4
PID|||123456^^^SMH^PI||MOUSE^MICKEY||
19620114|M|||14Disney Rd^Disneyland^^^MM1 9DL
PV1|||5N|||||G123456^DR SMITH
OBR|||54321|666777^CULTURE^LN|||20080802|||||||
SW^^^FOOT^RT|C987654
OBX||CE|0^ORG|01|STAU|||||F
OBX||CE|500152^AMP|01||||R|||F
OBX||CE|500155^SXT|01||||S|||F
OBX||CE|500162^CIP|01||||S|||F

We see the message has different segments and each of the segments has a header. The first is the message header, which tells us that this is a report from an entity called Lab 123457. It lists the date and the reference number of the lab test. It also provides patient identifying information. The patient is named Mickey Mouse, with date of birth, gender, address, and city. There is another segment on the provider, which in this case is Doctor Smith with identifier G123456 who's located on Ward 5N of a hospital. There is the observation, which is a swab from the right foot that is being assessed for bacterial culture. The result of the test is an organism, which, in this case, is *Staphylococcus aureus*. As in all microbiological specimens, it is tested for susceptibility to different antibiotics, including ampicillin, trimethoprim-sulfa, and ciprofloxacin. We see that the organism is resistant to ampicillin, but sensitive to trimethoprim-sulfa and ciprofloxacin.

As noted above, there have been different releases of HL7 Version 2, starting with the very first and basic version implemented in 1990. These include:

- V2.1 (1990): First implemented version - very basic, but still used
- V2.2 (1994): Basic enhancements
- V2.3 (1997): Scheduling and Finance messages added
- V2.3.1 (1999): Pathology, Allergies, Referral as Scheduling
- V2.4 (2000): Clinical Focus - Referrals and Discharge Summaries
- V2.5 (2003): Data field lengths standardized
- V2.5.1 (2007): Four data items added due to US regulatory requirements
- V2.6 (2007): Data type changes
  - Coded Element (CE) → Coded With (No) Exceptions (CNE/CWE)
  - Time Stamp (TS) → Date/Time (DTM)
- V2.7 (2011): Collaborative Care Message
- V2.7.1 (2012): Lab Orders/Results features added for US "Meaningful Use"
- V2.8 (2014): Authorization and Ordering information added

HL7 Version 2 has been a highly successful standard and it continues to be widely used. Its major limitation is that it is a syntactic standard. There are only a small number of semantic definitions, so the sender and the receiver need to know the language of the message.

HL7 Version 3

HL7 Version 3 is an attempt to introduce semantics into messaging. Its goal is semantic interoperability so that each HL7 Version 3 message has a specific meaning no matter which system is using it. If one adheres to the standard any system can understand the meaning, at least in principle, of an HL7 Version 3 message. For this to happen, HL7 Version 3 is based on Reference Information Model (RIM). This is an object model of the entities that pass messages. HL7 Version 3 is also implemented in a more modern format, namely Extensible Markup Language (XML). However, HL7 Version 3 is complex, some would say complicated. Others have gone as far as to call it incoherent. HL7 Version 3 works by building messages around the RIM. The RIM is object oriented and there are five abstract classes, with the elements of the message defined in the context of these abstract classes:

- Entity – things in world, e.g., people, organizations, other living subjects, drugs, devices
- Role – capability or capacity, e.g., patient, practitioner
- Participation – role in context of an act, e.g., performer, target
- Act – clinical or administrative definitions, e.g., observation, diagnosis, procedure
- Act relationship – links between acts, e.g., diagnosis act

All clinical, administrative, financial, etc. activities of healthcare can be expressed in "constraints" to the RIM. We can see this complexity by looking at an example of a pulse measured by a physician during an office visit in Figure 5.3. In this scenario, there is a patient and a physician making an observation about the patient's pulse. There is an observation that comes from that, the measurement of the pulse, which is modeled in a format called Entity-Attribute-Value, or EAV. This will ideally be based on a standard terminology, although HL7 Version 3 itself does not specify that. But a pulse rate of 50 beats per minute can be assigned to this example of an HL7 Version 3 message.

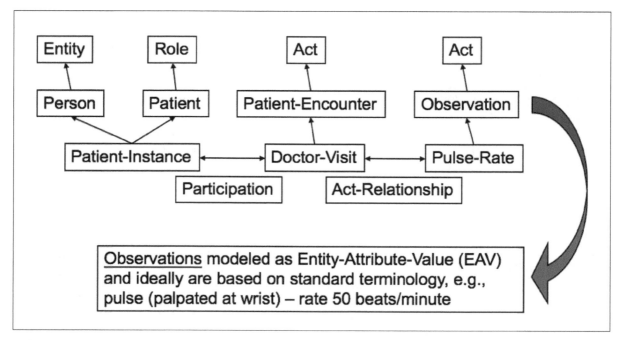

Figure 5.3: Clinical observation with HL7 v.3

Fast Healthcare Interoperability Resources (FHIR)

The complexity of HL7 Version 3 has stymied major use of it. However, with the widespread adoption of electronic health records and other clinical data systems, a new robust interoperability standard is needed. This has led to the Fast Healthcare Interoperability Resources (FHIR) standard. FHIR steps backwards somewhat in terms of the complexity of HL7 Version 3 but moves forward in aiming to allow interoperability to proceed. When FHIR emerged as the leading candidate for interoperability, HL7 International took over its development.

FHIR is best understood by navigating through its Web site. A good starting point is the page that organizes its documentation into modules (http://www.hl7.org/fhir/modules.html). A key component of FHIR is its Resources, which comprise the content of its messages. Figures 5.4 and 5.5 show FHIR resources for patient and medication. There are 6 types of resources: (http://www.hl7.org/fhir/resourceguide.html):

- Clinical: The content of a clinical record
- Identification: Supporting entities involved in the care process
- Workflow: Manage the healthcare process
- Financial: Resources that support the billing and payment parts of FHIR
- Conformance: Resources use to manage specification, development and testing of FHIR solutions
- Infrastructure: General functionality, and resources for internal FHIR requirements

Other HL7 Standards

The HL7 International organization has many other activities. These include the Clinical Context Object Workgroup, which aims to develop standards such as single sign-on and passing of context of the clinical or patient across applications being used, and the Clinical Decision Support Workgroup, which aims to develop standards around clinical decision support applications.

Clinical Document Architecture (CDA)

Another important activity of HL7 is the Clinical Document Architecture (CDA). CDA is important because most health care information is in the form of documents, and these are used to allow humans to read them. But as documents become electronic, it may be desirable for them to also have computable structures. CDA defines a standard structure and the metadata for that structure. A key aspect of CDA is templates. These are re-usable parts of documents that occur across different documents. The unstructured part of documents then can be wrapped in the CDA framework.

The current version of CDA, Version 2, has three levels of interoperability, which are depicted in Figure 5.6:

- Level 1 – general document specification
- Level 2 – adds document types with allowable structures
- Level 3 – adds mark-up expressible in structured form, such as RIM

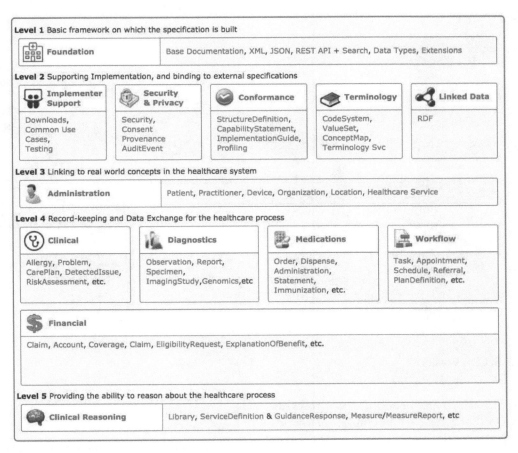

Figure 5.4: FHIR Resources

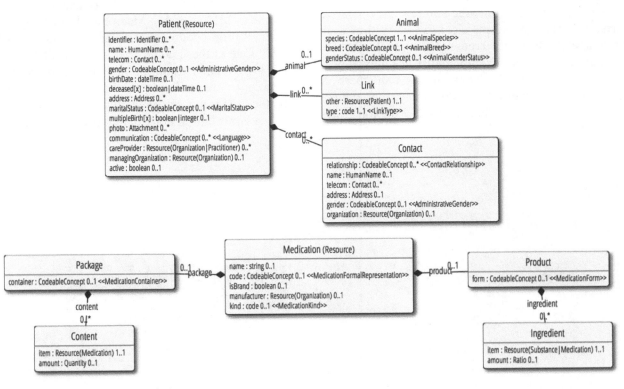

Figure 5.5: FHIR Resource linking

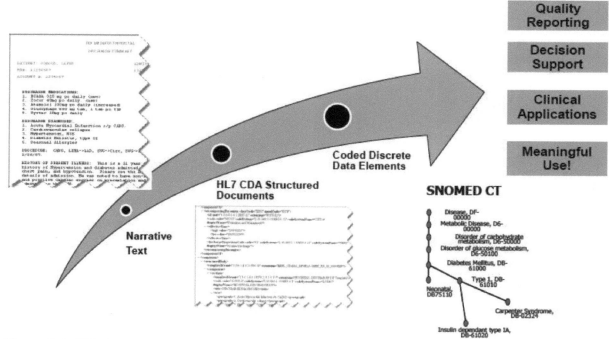

Figure 5.6: CDA Version 2

In recent years, there has been an effort towards Consolidated CDA (C-CDA). C-CDA consists of a series of reusable templates for documents and sections. There are document templates that represent the specific types of documents that are commonly used in medical records, such as clinical notes, discharge summaries, operative reports, and history and physical exam. Within each document template are different sections, each of which contain actual data. For example, allergies may appear in many different types of documents, so that wherever allergies appear in any type of document, they adhere to the allergy section template. Figure 5.7 demonstrates how these section templates may be used in different kinds of documents.

In essence, C-CDA allows building of documents from standardized components, which, in turn, contain standardized information. ONC released an Implementation Guide in 2012 as part of Meaningful Use Stage 2, which defined nine document templates, as shown in Figure 5.8.[45] One of these is the Continuity of Care Document (CCD), which is a summary of the patient as they move from one care setting to another. There are other types of documents that are typically used in patient care, such as the consultation note, diagnostic imaging, discharge

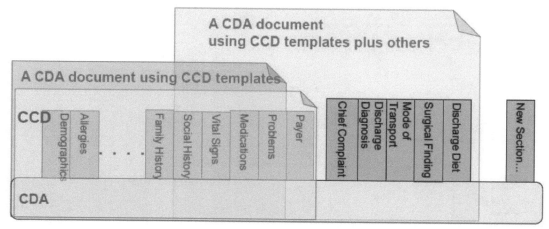

Figure 5.7: C-CDA template sections

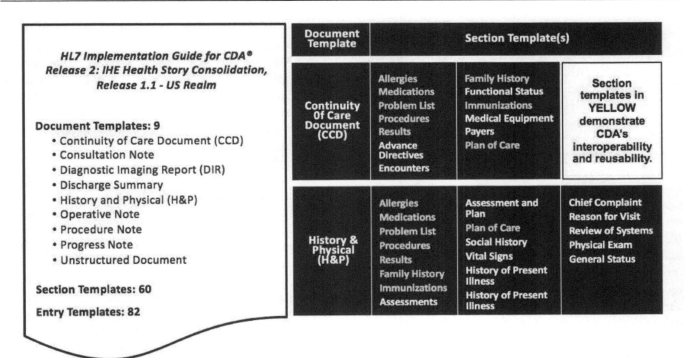

Figure 5.8: Document template types

summary, history and physical, operative note, procedure note, progress note, and another general unstructured document. Each of these documents has reusable section templates.

Imaging Standards

Of course, documents are not the only type of data in healthcare for which interoperability is desired. Another important type of data in health care is image data, for which we may want to move from the devices that capture them into records so that they can be viewed, and then we may want to archive them in various ways. The Digital Imaging and Communications (DICOM) standard is intended for the transport of images. It was developed by the American College of Radiology and the National Electrical Manufacturers Association, and there is a Web site devoted to its details (http://dicom.nema.org/).

DICOM defines how images and the metadata associated with those images is moved between various electronic devices, including information systems. The systems that store images and make them available for a wide variety of healthcare uses are called picture archiving and communication systems (PACSs).

One of the challenges for DICOM is that the ease of moving images around in the modern Internet, being able to display them in Web browsers, has led to a good deal of image transfer that does not use DICOM or take advantage of all the standardization inherent within it. This leads to clinical problems in that the information associated with an image may not be complete.

There are two overall parts to a DICOM message, the header and the actual image data (Figure 5.9). The header data contains information about the patient, the type of image, and how it was captured. It also contains information about the structure and compression of an image, e.g., if it is a JPEG image, how much compression has been used on that image.

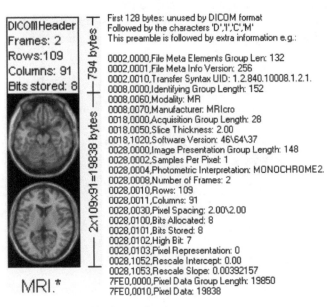

Figure 5.9: DICOM standard

Device Standards

Like images, there has been a proliferation in medicine of devices that capture, send, or receive data. This requires standards for communicating with these devices.[46] According to the US Food, Drug, and Cosmetic Act, Section 201(h), a medical device is defined as *"an instrument, apparatus, implement, machine, contrivance, implant, in vitro reagent, or other similar or related article, including a component part, or accessory that is intended for use in the diagnosis of disease or other conditions, or in the cure, mitigation, treatment, or prevention of disease, in man or other animals, or intended to affect the structure or any function of the body of man or other animals, and which does not achieve any of its primary intended purposes through chemical action within or on the body of man or other animals and which is not dependent upon being metabolized for the achievement of any of its primary intended purposes."*

The original standard for controlling and linking information from medical devices was called the Medical Information Bus (MIB).[47-48] The MIB was developed at Intermountain Healthcare in Utah. Most implementations of the MIB just transfer data, although the standard can issue commands to devices. For example, one could change the settings of an intravenous (IV) fluid pump. The MIB standard achieved standardization through IEEE as the 1073 standard and then later was designated an ISO 11073 standard but was never widely adopted beyond the small number of settings where it was developed.

More recently, there have been other medical device standards that have been implemented and are achieving use. One is the open source Integrated Clinical Environment (OpenICE, https://www.openice.info/), which is a prototype clinical ecosystem that connects medical devices and clinical applications, mostly in the intensive care unit, but in other places around hospitals. OpenICE provides a framework for integrating devices and apps, and it includes another standard called the Medical Device Plug-and-Play (MDPnP, http://www.mdpnp.org) interoperability program that enables devices that adhere to the standard to be plugged in and be ready for usage, much in the way we do with universal serial bus (USB) devices on computers.

Another effort for medical device standards has been the Personal Connected Health Alliance (http://www.pchalliance.org/ , formerly Continua Health Alliance), which is a consortium of companies and organizations that are devoted to interoperability, mostly of personal telehealth devices. Figure 5.10 from the organization's playbook[49] demonstrates the Continua personal health ecosystem. There are a variety of sensors, such as scales for weight, blood glucose monitoring machines, spirometers that measure airflow, and different sensors that may be in homes. The data from these sensors is transferred though wired connections, such as USP or Internet, or may go wirelessly through WiFi.

Ultimately, data is captured by computers that aggregate it. That data is then transferred across a network to applications such as an EHR or some other collector of all the information. The Continua consortium describes this all as the personal health ecosystem and is developing standards for the connection of all these devices and the data transfer between them.

Prescribing Standards

Another important type of message to exchange is the prescription. There is a family of standards around electronic prescribing developed by the National Council for Prescription Drug Programs (NCPDP, www.ncpdp.

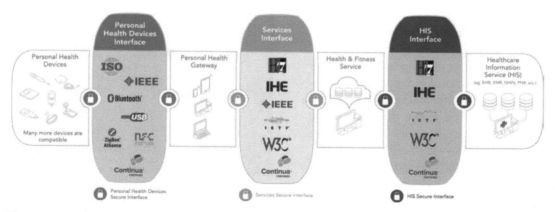

Figure 5.10: Continua personal health ecosystem

com), whose SCRIPT is the communications standard between the prescriber and the pharmacy. SCRIPT is required for use in the Meaningful Use criteria and have led to the widespread adoption of electronic prescribing.

Laboratory Standards

Another relatively common type of message exchange is that of reporting laboratory data. This has led to the development of the EHR Laboratory Interoperability and Connectivity Standard (ELINCS)[50] ELINCS has been operationalized using HL7 Version 2.5.1.

(http://www.hl7.org/implement/standards/product_brief.cfm?product_id=31). The goal of ELINCS is to cover the entire process of outpatient laboratory usage, from ordering by the clinician to collection of the specimen, to sending it to the lab, to the lab having the specimen processed and measured and then sending the results back to the clinician.

Patient Summaries

Another type of information that we may wish to exchange is a patient summary. This was recognized over a decade ago and led to the development of initially the Continuity of Care Record (CCR). The goal for the CCR was to be, "*a set of basic patient information consisting of the most relevant and timely facts about a patient's condition.*"[51] The goal for its use was to be available when the patient was referred or transferred or discharged, either among health care providers or facilities, and it would convey basic information for providing continuity of care.

The original CCR standard, however, was not compatible with any existing standards, and this led HL7 and several vendors to create the Continuity of Care Document (CCD), which would be based on HL7 Version 3, and the Clinical Document Architecture (CDA). There were battles and lawsuits, but eventually the CCD prevailed because it was compatible with other standards.

The CCD has resulted in more standard use of the document.[52] However, many implementations of the standard make errors.[53] There is also some allowable variation within the standard, such that its semantic interoperability has not been fully achieved. Nonetheless, the CCD is an important document that is a patient summary that is easily moved between most EHR and other patient information systems.

Stage 2 of the Meaningful Use criteria set forth the required data set for patient summaries. The so-called required data set for the CCD includes:

- Header
- Purpose
- Problems
- Procedures
- Family history
- Social history
- Payers
- Advance directives
- Alerts
- Medications
- Immunizations
- Medical equipment
- Vital signs
- Functional status
- Results
- Encounters
- Plan of care

Another use of the patient summary is to allow the patient to download his or her summary. This began with the Veteran's Administration (VA) Blue Button Initiative that allowed VA patients to go to the VA portal and download an electronic summary of their medical data. This idea was then taken up by ONC as well as several vendors in their personal health record (PHR) systems. Now this whole activity is called the Blue Button Initiative. The ONC has further facilitated development of the Blue Button Toolkit that enables vendors and others to download patient summaries from within their systems.

Platforms

The final topic for messaging standards is platforms. There has been much criticism that today's EHRs are large, monolithic systems and not platforms on top of which other applications and innovations can be built. Mandl and Kohane called these monolithic systems "traps" and argued the need to move beyond them.[54] Mandl et al. have developed the Substitutable Medical Apps Reusable Technologies (SMART) platform,[55] based on the idea that there should be an underlying platform upon which "apps" can be built that access a common store of data and functions.[56] This builds on an analogy of smartphones that have many underlying functions, such as global positioning system (GPS) that different apps can access. SMART apps may be mobile apps or Web-based. The platform also uses a security standard called OAuth2.

SMART has also adapted FHIR as its API for accessing data, hence the phrase SMART on FHIR.[57] It has been implemented for EHRs and extended to areas like genomics[58] and precision medicine[59] applications. Major vendors, such as Epic, have started to support it.[60] The effort has further partnered with the Clinical Information Modeling Initiative, or CIMI, which is devoted to building standards-based clinical information models and by the Healthcare Services Platform Consortium (HSPC, http://hspconsortium.org/).

Many EHR vendors have also started to take up the transforming their monolithic systems into these kinds of platforms. Surescripts is a company that started as an e-prescribing platform and are extending it to clinical messaging (http://surescripts.com/products-and-services/clinical-direct-messaging) and a national record locator service for all patients who have records in the Surescripts system (http://surescripts.com/products-and-services/national-record-locator-service). Clinical messages then can be sent across this platform. Allscripts is an EHR vendor that has developed an open architecture platform that enables developers to build on top of the core Allscripts system (http://www.allscripts.com/company/partners/allscripts-developer-program-registration). EPIC has developed an open API, adapting FHIR and providing a way to access data and services. The have developed an "app store" to distributed SMART on FHIR apps for their platform (http://open.epic.com).

TERMINOLOGY STANDARDS

The final category of standards we will discuss are terminology standards. The benefits of computerization of clinical data depend upon its "normalization" to a consistent and reliable form so we can carry out tasks such as aggregation of patient data, clinical decision support, and clinical research. However, clinical language is also inherently vague, sometimes by design, and that cam be at odds with the precision of computers. A comprehensive reference on all of the different terminology standards is the book by Giannangelo.[61]

With terminology, there are terms that may mean the same thing, like *cancer* and *carcinoma*. But inside a computer, the ASCII codes for the letters of those words are no more similar than the codes for the words *apple* and *zebra*. Medicine is sometimes criticized for having such vague language. Some have argued that we need "*fewer words and more meaning in medicine*," such as in air traffic control, where the communications that are allowed between pilots and air traffic controllers are much more limited.[62] Another example is the military, where communications on the battlefield use language that is constrained.

Just as we saw at the beginning of this chapter that there is a standard to define what a standard is there is likewise a terminology of terminologies. The term *terminology* itself generally refers to a collection of terms. But the notion of a term is not so simple. Most terminology is based on concepts, which are a thing or an idea that is expressed in one or more terms. Concepts have *synonyms* where different terms describe the same concept. There are also *polysems*, which are terms that mean more than one concept.

We sometimes talk about *dictionaries*, which have concepts plus their meaning. Dictionaries often list some of the different terms that describe a concept, i.e., the synonyms of that concept. A *thesaurus* is a resource that groups synonyms by the concept to which they refer. We also talk about a *vocabulary*, which is a collection of concepts and terms in a domain, such as healthcare, or information technology, or some subset of those. And then there is an *ontology*, which consists of structured concepts and the relationships between them that gives a more formal representation of knowledge.

There are a number of use cases for standardized terminology, including:[63]
- Information capture – documenting findings, conditions, and outcomes
- Communication – transferring information
- Knowledge organization – classification of diseases, treatments, etc.
- Information retrieval – accessing knowledge-based information
- Decision support – implementing decision support rules

Dealing with language and terminology is usually a lot harder for computers than humans. As humans, we understand synonymy and polysemy of terms. For example, there are many ways can we say *common cold*. There are many different synonyms that, as humans, especially those who have some training in medicine, can understand. When we use terms like *cold, upper respiratory infection, URI, laryngitis, bronchitis, rhinitis, viral syndrome*, and more. They are not all quite exactly synonyms, but we understand the similarity between the terms. Computers, on the other hand, just view these as bits in memory and do not really understand that they mean similar things unless we program the computer to do so.

Likewise, for polysemy, we can take a word like *lead*. It can be used in many ways in medicine. It can be used as a verb, as in, *hypertension leads to heart disease*. We can also talk about an *EKG lead, lead poisoning*, and others.

Cimino has elucidated some of the "*desiderata*" for constructing medical or clinical vocabularies.[64] Most vocabularies have a hierarchical structure and then some sort of coding scheme where a code is assigned to every concept. We ultimately want to represent the terms and the concepts as codes and we want to use these codes in information systems.

There are various approaches that can be used for codes. They can be numerical, where they are assigned sequentially or randomly. They might be a mnemonic, such as an abbreviation. There could be a hierarchical code that indicates the level in the hierarchy that the concept exists. There may be juxtaposition of codes where there are composite codes that indicate a concept is consisting of more primitive concepts. And then there is a combination codes where those composites use ordering. In general, as argued by Cimino,[64] we should avoid semantic codes that put meaning into the codes themselves. Concept codes are best represented by an identifier that does not say anything about the meaning.

One of the major uses of terminologies is coding. We assign the codes in a terminology to represent diseases, treatments, findings in a patient, and so forth. Rosenbloom describes three types of terminologies:[65]
- Interface – support data entry[66]
- Processing – optimize natural language processing
- Reference – enable storage, analysis, retrieval

Coding is one of the major activities of the Health Information Management (HIM) profession.[67] With the growth of uses of terminology and technologies to capture it, the HIM field has been changing substantially.[68] Those who do coding often make use of what's called *computer-assisted coding*, which is computer programs that are designed specifically to assist human coders in the coding process.[69]

There are many terminology standards in biomedicine. Some of them have evolved to carry out specific purposes. There are terminology standards for diagnoses, drugs, laboratory findings, procedures, and other aspects of healthcare. There are several terminologies for nursing. There are terminologies for literature indexing and medical devices. There are also a couple comprehensive terminologies that attempt to cover all these areas and link the terms from these different terminology standards together into a comprehensive whole.

International Classification of Diseases (ICD)-9

One of the earliest terminology systems, and still highly important, is the *International Classification of Diseases* (ICD).[70] ICD was first developed in 1893, when it was called the International List of Causes of Death. And the initial primary purpose for ICD was to compile mortality statistics. It was developed in London and eventually passed to the World Health Organization. Along the way, ICD changed in name to International Classification of Diseases, as it has evolved to code diseases more than just the cause of death.

The major use of ICD is modern times is in coding diagnoses for health insurance claims, which is why we sometimes hear data collections with ICD codes called *claims data*. In addition to diagnosis codes, they may include procedure codes and other type of data that is used for health insurance claims.

Until recently, the version of ICD used in the US was ICD-9. ICD-9 was approved by the World Health Organization in 1975. Even though ICD-10 was released in 1990, ICD-9 continued to be used in the US until 2015. ICD-9, the original version from the World Health Organization, was organized hierarchically, with one digit in the code for each level of the hierarchy and having codes up to four digits.

ICD is often extended by different countries and was extended in the US to ICD-9-CM, with CM standing for clinical modifications. This process added more detail and a fifth digit, so there could be up to five-digit codes. ICD-9-CM also had an additional set of codes, V codes, for encounters related to prevention and screening. If a patient is screened for something, they would not get the diagnosis code, but rather the V code that they were being screened for that diagnosis. ICD-9-CM also has G codes that document the provision of specific services, such as those that embodied in quality measures.

The US finally discontinued ICD-9 in October 2015 with the transition to ICD-10-CM, although there is still plenty of data that is encoded in ICD-9-CM. Below is an example of ICD-9-CM for some types of heart disease. One type is acute myocardial infarction, which is listed under diseases of the circulatory system under the subcategory of ischemic heart disease. Acute myocardial infarction has an ICD-9-CM code of 410. The fourth digit is used to indicate the location in the heart of the acute myocardial infarction, e.g., anterior, inferoposterior, or subendocardial. If the clinician has not specified where the myocardial infarction occurs, the last category 410.9, myocardial acute unspecified, is used.

> Diseases of the circulatory system (390-459)
> Ischemic heart disease (410-414)
> (410) Acute myocardial infarction
> (410.0) MI, acute, anterolateral
> (410.1) MI, acute, anterior, NOS
> (410.2) MI, acute, inferolateral
> (410.3) MI, acute, inferoposterior
> (410.4) MI, acute, other inferior wall, NOS
> (410.5) MI, acute, other lateral wall
> (410.6) MI, acute, true posterior
> (410.7) MI, acute, subendocardial
> (410.9) MI, acute, unspecified
> ...

(414) Other forms of chronic ischemic heart disease

...
- o + (414.01) Coronary atherosclerosis, native coronary artery
- o + (414.02) Coronary atherosclerosis, autologous vein bypass graft
- o + (414.04) Coronary atherosclerosis, artery bypass graft

There are other types of ischemic heart disease. Under 414, we see different types of the ischemic heart disease, one of which is coronary atherosclerosis, which itself may occur in the native coronary artery of an individual, or it might occur in a bypass graft, which has come either from a vein or an artery. There are specific codes that go out to the fifth digit.

There are a number of limitations of ICD-9 that limit its usefulness beyond just providing billing codes so that reimbursement can take place.[71] One of the limitations of ICD-9 is the use of Not Otherwise Specified (NOS) codes. These indicate another category that often can be ambiguous, especially when diseases change over time. For example, decades ago clinicians spoke of *non-A, non-B hepatitis*, when medicine did not know about any of the other types of hepatitis. Now there are hepatitis C, hepatitis D, and others. For a patient who might have been coded from the past as non-A, non-B hepatitis, it would be difficult to perform queries in databases that use the newer more specific names.

Another limitation of ICD-9 is Not Elsewhere Classified (NEC). This basically indicates that there is no separate specific code other than what is given. For example, even though there are several codes for so-called major depression, there is only one code for non-major depression, which is 311, Depressive Disorder Not Elsewhere Classified.

There are a number of other limitations of ICD-9.[71] For example, using digits in the codes can be problematic when there might be more than 10 items at a given level of the hierarchy. Another problem with ICD-9 is that the granularity, the level of detail, is often inadequate. For example, there is only one code for most cancers in each location, e.g., 162.4, malignant neoplasm of middle lobe, bronchus, or lung. Of course, there are many different types of neoplasms that can occur in that area and ICD-9 does not provide the ability to specify them. In addition, for the most part, ICD-9 is not extensible, so modifiers cannot be added for more detailed location, severity, and it is unable to indicate any kind of causal relationships.

International Classification of Diseases (ICD)-10

ICD-10 was adopted by the World Health Organization in 1990. There were significant changes made in the structure from ICD-9, allowing more granularity of codes. After numerous delays over several years, it was finally implemented in the US in October, 2015 as ICD-10-CM.[72] Also in the US, inpatient procedure codes were added and called ICD-10-PCS, although CPT-4 is still used for outpatient procedures.

Figure 5.11 illustrates some of the differences between ICD-9-CM and ICD-10-CM. The total number of codes in ICD-10-CM is tripled from ICD-9-CM. In addition, the codes themselves can extend out to seven characters compared to five for ICD-9-CM. In ICD-9-CM, that first character is a number, or it may be the specific V or G or E codes. In ICD-10-CM, the first character is a letter, so any letter. The second character is a number, and then the remaining characters can be letters or numbers.

ICD-9-CM	ICD-10-CM
• 13,000+ codes	• 69,000+ codes
• 3-5 characters	• 3-7 characters
• First character numeric or V/G/E	• First character alpha • Character 2 numeric • Character 3-7 alphanumeric
• Characters 1-3 – category • Characters 4-5 – etiology, anatomic site, or other clinical detail	• Characters 1-3 – category • Characters 4-6 – etiology, anatomical site, or other clinical detail • Character 7 – extension, e.g., initial, subsequent, sequelae

Figure 5.11: ICD-9-CM comparison with ICD-10-CM

For both systems, the first three characters in the code give the category. In ICD-9-CM, the next two characters in the code provide more detail about the etiology, anatomical site, or other clinical detail. That's extended to three characters in ICD-10-CM. There is also a seventh character in ICD-10-CM that's called an extension and it allows the code to be extended in various ways. The most common way that the code is extended is to talk about the visit to the health care system, whether it is the initial encounter, a subsequent encounter, or sequelae that occur from one of those. But there are other extensions as well.

The major difference between ICD-9-CM and ICD-10-CM is the increased granularity on a massive scale, where as many ICD-9-CM codes represent perhaps a single disease or a single condition, ICD-10 adds many modifiers and, as such, has much higher granularity. An example of this is seen with the difference between the single ICD-9 code, 995.29 Unspecified adverse effect of other drug, medicinal and biological substance and the following sample of codes related to adverse drug events:

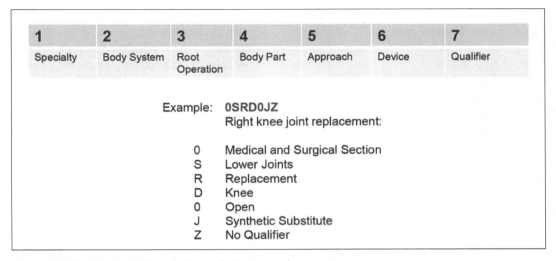

Figure 5.12: ICD-10-PCS code for right knee replacement

T360X5A Adverse effect of penicillins, initial encounter
T361X5A Adverse effect of cephalosporins and other beta-lactam antibiotics, initial encounter
T362X5A Adverse effect of chloramphenicol group, initial encounter
T363X5A Adverse effect of macrolides, initial encounter
T364X5A Adverse effect of tetracyclines, initial encounter
T365X5A Adverse effect of aminoglycosides, initial encounter
T366X5A Adverse effect of rifampicins, initial encounter
T367X5A Adverse effect of antifungal antibiotics, systemically used, initial encounter
T368X5A Adverse effect of other systemic antibiotics, initial encounter
Plus 170 additional codes

By the same token, ICD-10-PCS increases massively the number of procedure codes. ICD-9 had relatively moderate numbers of procedure codes, but now ICD-10-PCS increases that substantially and has a seven-character structure that represents the aspects of procedures listed here, the specialty, of the body system, the root operation, body part, approach, device, and qualifier. Figure 5.12 shows an example here for the ICD-10-PCS code of right knee joint replacement.

Granularity is an issue for ICD-10-PCS as well. As seen in Figure 5.13, a single ICD-9 code for pericardiectomy, the removal of the pericardium, the membrane that surrounds the heart, has many codes in ICD-10-PCS, with different operative approaches and different operative techniques are reflected in different codes. A number of commentators have had fun poking at the excess granularity of ICD-10-CM, whether it reaches a point of absurdity, particularly those who are critical of the bureaucracy in health care.[73] An example is how one

37.31 Pericardiectomy	025N0ZZ Destruction of Pericardium, Open Approach
	025N3ZZ Destruction of Pericardium, Percutaneous Approach
	025N4ZZ Destruction of Pericardium, Percutaneous Endoscopic Approach
	02BN0ZZ Excision of Pericardium, Open Approach
	02BN3ZZ Excision of Pericardium, Percutaneous Approach
	02BN4ZZ Excision of Pericardium, Percutaneous Endoscopic Approach
	02TN0ZZ Resection of Pericardium, Open Approach
	02TN3ZZ Resection of Pericardium, Percutaneous Approach
	02TN4ZZ Resection of Pericardium, Percutaneous Endoscopic Approach

Figure 5.13: Pericardiectomy ICD-9 and ICD-10-PCS codes

might be struck by a falling object on board a watercraft. ICD-10-CM goes to the level of detail of what type of watercraft, e.g.,

V93.40 – Merchant ship
V93.41 – Passenger ship
V93.42 – Fishing boat
V93.43 – Powered watercraft
V93.44 – Sailboat
V93.48 – Unpowered watercraft
V93.49 – Unspecified

About half of ICD-10-CM codes are related to the musculoskeletal system, particularly injuries. This is not surprising, given the combinatorial explosion that occurs from all the different possible injuries in all the different anatomical sites involving all the different anatomical parts of the human body. A quarter of all codes are related to fractures, due to that combinatorial explosion. About a third of codes distinguish laterality, left versus right. Therefore, most impacted by ICD-10 are the medical specialties, or areas, of orthopedics, obstetrics and gynecology, and behavioral health. Primary care has a medium level of impact, with the other medical specialties having a low level of impact.

For many years, implementation of ICD-10 was successively delayed in the US. But finally, in October 2015, ICD-10-CM and ICD-10-PCS were implemented in the US. (The author of the book, *ICD-10 Illustrated*, actually noted there was an ICD-9 code to describe this, 738.42, *delayed milestones*.[74] The implementation of ICD-10-CM was led by the Centers for Medicare and Medicaid Services. And the administrator of CMS noted that the implementation was by and large successful.

Like ICD-9, CMS provides a look-up Web site where one can go type words into a search box and retrieve the appropriate ICD-10 codes (https://www.cms.gov/medicare-coverage-database/staticpages/icd-10-code-lookup.aspx).

There have been several informatics concerns about ICD-10-CM. One of these is the excess granularity, described above. Many advocated that ICD-10 never be adopted, that it just be skipped, and the US move from ICD-9-CM directly to ICD-11.[75] Part of the reason for that is that ICD-11 will be built on a compositional terminology, SNOMED, described below.[76] However, ICD-11 is not yet completed, and it would probably be another two, maybe three, years before the development of ICD-11-CM.

Another informatics concern is that the mappings between ICD-9-CM and ICD-10-CM are somewhat convoluted; they're not just straight-forward mappings.[77] These convoluted mappings range across different medical specialties from relatively few for hematology to many for obstetrics and injuries. A survey of physicians after the implementation of ICD-10-CM found that the majority of them saw little value for ICD-10-CM, or really any type of ICD coding, because it just states a diagnosis and does not present much more detail on the patient.[78] Coders, however, saw more value for ICD-10-CM and hoped to convince physicians and others of its value.

Diagnosis Related Groups (DRGs)

Another terminology standard for diseases is the Diagnosis Related Groups (DRGs). DRGs were originally developed to aggregate ICD-9 codes into groups that could be used for health services research to look at hospital costs. The DRG system consists of several hundred codes that lump hospital illnesses together that are roughly comparable in the resources they should be using. However, not true to its original intention, DRGs were adopted by the predecessor of CMS, the Health Care Financing Authority (HCFA), to be used for the Prospective Payment System for hospitalization under Medicare starting in the 1980s. Since then, all hospitalizations have been classified by their DRG, and that influences the reimbursement that hospitals receive for the hospitalization.

Here, are some examples of DRGs for respiratory diseases:

Respiratory disease w/ major chest operating room procedure, no major complication or comorbidity 75
Respiratory disease w/ major chest operating room procedure, minor complication or comorbidity 76
Respiratory disease w/ other respiratory system operating procedure, no complication or comorbidity 77
Respiratory infection w/ minor complication, age greater than 17 79
Respiratory infection w/ no minor complication, age greater than 17 80
Simple Pneumonia w/ minor complication, age greater than 17 89
Simple Pneumonia w/ no minor complication, age greater than 17 90
Respiratory disease w/ ventilator support 475
Respiratory disease w/ major chest operating room procedure and major complication or comorbidity 538
Respiratory disease, other respiratory system operating procedure and major complication 539

These DRG codes tend to categorize multiple different types of diseases that are in the same general body area and require the same amount of resources. The definitions of these DRGs are laid out in quite explicit detail, and they define how much reimbursement the hospital gets for the patient who is hospitalized with this condition.

Drug Terminology

There are several different code sets around drug terminology that are interrelated, but it is sometimes confusing as to what roles they play. These different code sets are mostly led by the US government. In fact, there is a collaboration among different federal agencies called FedMed, where there has been an agreed set of standards, comprehensive, and freely accessible federal medication terminologies. Included in FedMed are the National Drug Codes (NDC), the Unique Ingredient Identifier (UNII), the VA National Drug File Reference Terminology (NDF-RT), the NCI Thesaurus Structured Product Labeling (NCIt SPL), and RxNorm and RxTerms from the National Library of Medicine.

The NDC is basically a packaging standard. There is an 11-digit code maintained by the US Food and Drug Administration for each and every pharmaceutical preparation (https://www.fda.gov/Drugs/InformationOnDrugs/ucm142438.htm). The first five digits representing the manufacturer; the next four digits representing the product name, strength, and dose forms; and the final two digits being the code for packaging, such as the number of tablets in the bottle. One of the major challenges of NDC is that those middle four digits about product name, strength, and dose vary from different manufacturers. Thus, the same drug from a different manufacturer will have a different middle four digits. This problem is overcome because the NDC codes map into the other terminology systems, but many data aggregations contain NDC codes that can make it challenging to perform queries and analysis of that data.

There are other drug terminology standards that are part of FedMed. The UNII specifies the ingredients in drugs and other compounds, both the active and inert substances. The NDF-RT maintains much detail about drugs' mechanism of actions, physiologic effect, and structural class. The NCIt SPL maintains pharmaceutical dosage form, route of administration, and potency.

RxNorm brings all of these drug terminologies together and is meant to be the semantic structure for all formulations and their components of drugs (https://www.nlm.nih.gov/research/umls/rxnorm/).[77] Within RxNorm is RxTerms, which provides an interface terminology to RxNorm so that the name of drugs can be linked to the more specific details about them.[80] In FedMed, RxNorm and RxTerms are the standards into which other drug terminologies must map. There are tools related to RxNorm. One is RxNav, which provides an API for term look up (https://rxnav.nlm.nih.gov/).[81] There is also RxImageAccess, which is a set of images of pills that one can use to take a picture of an unknown pill and identify which drug it is (http://rximage.nlm.nih.gov/docs/doku.php). Finally, there is RxMix, which allows applications to be built from different APIs around RxNorm, RxTerms, NDFRT, and RxImageAccess (https://mor.nlm.nih.gov/RxMix/).[82]

Figure 5.14 shows the structure and relationships in RxNorm. This structure represents all the information

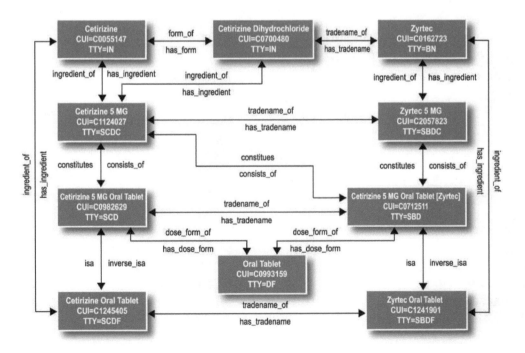

Figure 5.14: RxNorm standard

we would want to know about a drug, its ingredients, its brand name, the dosage that it comes in, the type of packaging, and so forth. And RxNorm gives us a structure to do this for all drugs, so that we can, in a computable way, represent the drug or multiple drugs that a patient is taking.

Finally, we can look at how the different FedMed terminologies come together in Figure 5.15. We see in the center that the active ingredient, the clinical drug, come from RxNorm, the VA NDF-RT adds various other information about the drug. The NLM FDA DailyMed system adds additional information about the dosage form and the drug product. Finally, the NDC code brings all of that together, providing information about the packaged products.

Logical Observation Identifiers Names and Codes (LOINC)

The Logical Observation Identifiers Names and Codes (LOINC) standard started for laboratory tests and names but has been extended into other types of measures as well as languages beyond English.[83-84] LOINC consists of observations and each observation and has several attributes, the main one being the component, or analyte, which is the substance or entity that is being measured or observed. These components may have properties such as mass concentration, numeric fraction. They have a time that they were observed. There is a specimen that usually comes from a system in the body, such as blood or cerebral spinal fluid.

There is a scale by which the observation is measured. It may be qualitative, quantitative, ordinal, or nominal. There may be a method associated with the observation, the procedure that is used to make that observation. Also, in the distribution of LOINC is the Regenstrief LOINC Mapping Assistant (RELMA, https://loinc.org/relma/), which is a Windows program that allows searching of the LOINC database and helps one map their local codes to LOINC codes.

Below are several examples of LOINC codes:
Blood glucose GLUCOSE:MCNC:PT:BLD:QN:
Serum glucose GLUCOSE:MCNC:PT:SER:QN:
Urine glucose concentration
    GLUCOSE:MCNC:PT:UR:QN:
Urine glucose by dip stick
    GLUCOSE:MCNC:PT:UR:SQ:TEST STRIP
Ionized whole blood calcium CALCIUM.
    FREE:SCNC:PT:BLD:QN:

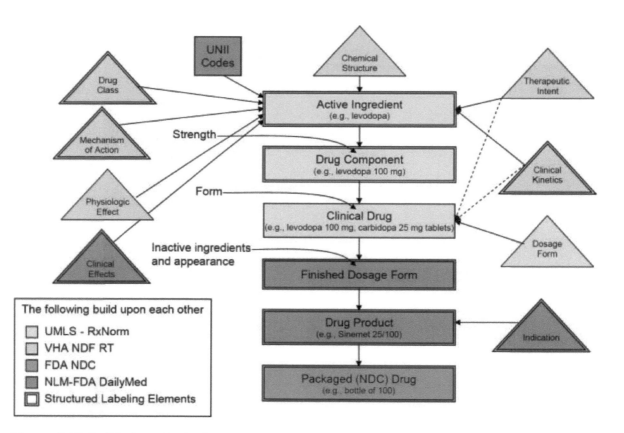

Figure 5.15: FedMed terminologies

24 hour calcium excretion CALCIUM.
   TOTAL:MRAT:24H:UR:QN:
Automated hematocrit
   HEMATOCRIT:NFR:PT:BLD:QN:AUTOMATED COUNT
Manual spun hematocrit
   HEMATOCRIT:NFR:PT:BLD:QN:SPUN
Erythrocyte MCV ERYTHROCYTE
   MEAN CORPUSCULAR
   VOLUME:ENTVOL:PT:RBC:QN:AUTOMATED COUNT
ESR by Westergren method
   ERYTHROCYTE SEDIMENTATION
   RATE:VEL:PT:BLD:QN:WESTERGREN

Current Procedural Terminology (CPT-4)

Another terminology that is important to physicians is Current Procedural Terminology (CPT-4). This is a classification of the procedures that are performed by physicians and usually, in addition to an ICD-9 code, there must be a CPT-4 code reported so that the physician can be reimbursed by the government or a private insurance company. There are also certain types of CPT codes called evaluation and management (E&M) codes, which document the intensity of clinical encounters such as office visits. The CPT-4 system is copyrighted and maintained by the American Medical Association. (https://www.ama-assn.org/practice-management/cpt-current-procedural-terminology)

CPT-4 is part of a larger procedure coding system, the HCFA Common Procedure Coding System (HCPCS). HCPCS provide different levels of procedure coding. The first level contains CPT-4 codes. Then the second level includes other codes for items and supplies and non-physician services. There used to be a third level, which would allow organizations to create their own local codes, and these were abolished under the HIPAA standards rules in 2003.

Systematized Nomenclature of Medicine (SNOMED)

The Systematized Nomenclature of Medicine (SNOMED, http://www.snomed.org/) is a controlled terminology that attempts to cover all of medicine and health care. It is more formally known as SNOMED Clinical Terms (SNOMED CT). SNOMED was originally developed by the College of American Pathologists (CAP), who originally created the Systematized Nomenclature of Pathology (SNOP). In the 1980s, SNOP was extended to cover all of medicine and was renamed to SNOMED. The SNOMED CT moniker comes from the fact that the original SNOMED merged with another terminology system being developed in England called the Clinical Terms Project, and these two merged together into a single system in 2000 to become SNOMED CT.[85]

In 2007, the ownership of SNOMED was transferred to an international standards body, the International Health Terminology Standards Development Organization (IHTSDO). This standards body now maintains SNOMED CT and continues to develop and expand it and translate it into other languages. SNOMED CT is currently available in the US English, UK English, Spanish, Danish, and Swedish, and it is being translated to additional languages.

One of the original limitations of SNOMED CT was a license that restricted its usage. In 2003, the former owner of SNOMED, the CAP, negotiated with the NLM to create what was then a five-year license for all of the United States. The license now has been transferred to IHTSDO and continues to be available. The license allows SNOMED to be used in the US by all public and private entities for any health care, public health, research, educational, or statistical use. There are many other countries that have licensed SNOMED CT for countrywide usage. This allows SNOMED CT to be used to encode patient level data sets and redistribute them to others if large portions of the vocabulary are not extracted and transferred as well.

SNOMED CT has a great deal of documentation on its website, including a starter guide that provides an overview of the system.[86] SNOMED CT is the richest vocabulary for describing clinical observations and findings. Every analysis of different terminology systems has found its coverage to be most extensive.[87-88] Additional resources for SNOMED include:

- Document library – http://snomed.org/doc
- Look-up and browser – http://browser.ihtsdotools.org
- E-learning about SNOMED CT – http://elearning.ihtsdotools.org

One of the key features of SNOMED is the use of what is called a *multiaxial* or *compositional* approach. This means that compound terms can be combined from smaller terms, for example, inflammation of the lung, without requiring a term for inflammation in every body location. In addition, there can be modifiers added to terms, such as *severe* or *worsening*. SNOMED CT contains more than 300,000 concepts, more than one million descriptions or terms that express those concepts, and more than one million relationships between those concepts.

Figure 5.16 from the starter guide gives the big picture of SNOMED, which basically consists of concepts that are organized into hierarchies, like most terminology systems. SNOMED also has attributes that allow

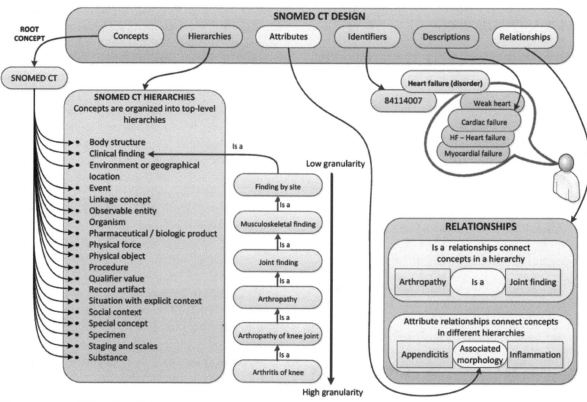

Figure 5.16: SNOMED CT design

relationships to be expressed between the different concepts, which give this multiaxial or compositional nature to terms that can be specified. SNOMED has specific identifiers for each concept. There are also descriptions or synonyms for those concepts, and the ability to create these relationships based on attributes.

Figure 5.17 shows an example of a SNOMED CT description for the concept *myocardial infarction*. There is a preferred term in English, which is myocardial infarction. There is also a fully specified name that gives the official SNOMED CT name. There are other synonyms for this concept as well as an identifier that might be used

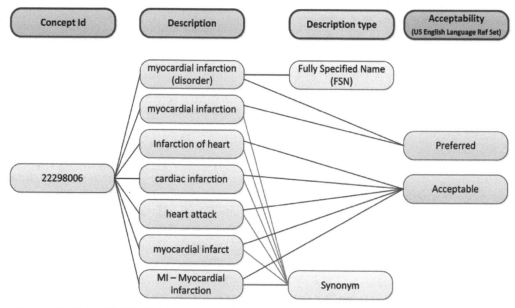

Figure 5.17: SNOMED CT description for myocardial infarction concept

in a computer program that would specify the concept and map to any of its expressions of the concept.

SNOMED CT also has *attributes*, which allow the compositional approach to take place and for more complex terms to be built from more fundamental ones. The categories of attributes in SNOMED include:
- Clinical finding concepts
- Procedure concepts
- Evaluation procedure concepts
- Specimen concepts
- Body structure concepts
- Pharmaceutical/biologic product concepts
- Situation with explicit context concepts
- Event concepts
- Physical object concepts

SNOMED also recognizes that, when using terminology, especially interface terminology, we do not necessarily want to have to construct complex terms from simpler ones, so there are many *pre-coordinated* concepts, where there is a clinically meaningful term that is a combination of more basic terms. An example is shown in Figure 5.18, where *fracture of the tibia* can be modeled into levels of detail as an injury of the tibia, fracture of the lower limb. For the purposes of clinical use, the pre-coordinated expression is used and then mapped to this more underlying set of defining relationships.

Nursing Terminologies

There are also nursing vocabularies, which are designed for nursing practice to capture nursing observations, diagnoses, interventions, and patient outcomes.

There are several different vocabularies, and they suffer from the same problems seen in vocabularies in general, in that the vocabularies are based on irreconcilable information models so are not easily combined with each other. In addition, the terms in the vocabulary are not always expressed in the ways that clinicians express them, so there can be challenges in mapping observations made by clinicians into the terms of the vocabulary. The vocabularies can also be very tedious to use in patient documentation. And these lead to a question of whether the data that has been captured then can be transferred across settings. Table 5.1 shows the different nursing vocabularies that are in use and what categories of concepts they cover. These vocabularies are improved by the ANA and included in terminology collections such as UMLS Metathesaurus (described next).

Unified Medical Language System (UMLS)

As far back as the 1980s, it was recognized that there was a plethora of different vocabulary systems, and to achieve the full value of computerized data, there was a need to reconcile these them. This led to development of the *Unified Medical Language System* (UMLS. https://www.nlm.nih.gov/research/umls/) Project, which was launched by the NLM in the late 1980s and was an attempt to reconcile these vocabularies.[89] There are three components of the UMLS:
- Metathesaurus – the thesaurus based on all the component vocabularies of the UMLS, described in more detail below

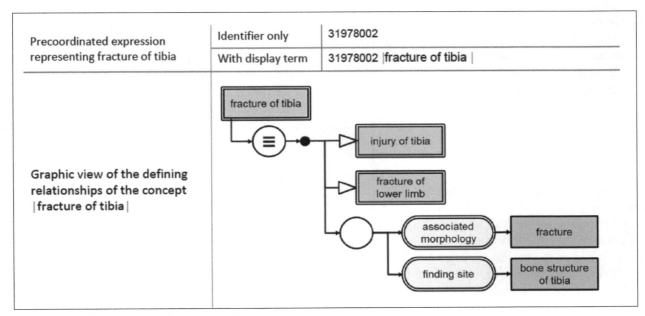

Figure 5.18: SNOMED CT Fracture of tibia

Table 5.1: Nursing Terminologies

Nomenclatures	Diagnoses	Interventions	Outcomes
National Nursing Diagnosis Association (NANDA)	X		
Nursing Interventions Classification (NIC)		X	
Nursing Outcomes Classification (NOC)			X
Omaha System	X	X	X
Clinical Care Classification	X	X	X
International Classification for Nursing Practice	X	X	X

- Semantic Network – maps generic relationships between the semantic types of the concepts that are in the Metathesaurus, such diseases and treatments
- Specialist Lexicon – collection of words and terms mainly designed to assist in natural language processing applications

According to its documentation on the UMLS Web site, the Metathesaurus is a *"database of information on concepts that appear in one or more of the number of different controlled vocabularies and classifications used in biomedicine."* It is designated a "Metathesaurus" in that it identifies equivalent terms across terminologies or vocabularies and links them.[79]

In the Metathesaurus, all terms that are conceptually the same are linked together as a concept. Each concept may have one or more terms, each of which represents an expression of the concept from a source terminology that is not just a simple lexical variant (i.e., differs only in word ending or order). Each term may consist of one or more strings, which represent all the lexical variants that are represented for that term in the source terminologies. One of each term's strings is designated as the preferred form, and the preferred string of the preferred term is known as the canonical form of the concept.

Each Metathesaurus concept has a single concept unique identifier (CUI). Each term has one term unique identifier (LUI), all of which are linked to the one (or more) CUIs with which they are associated. Likewise, each string has one string unique identifier (SUI), which likewise are linked to the LUIs in which they occur. In addition, each string has an atomic unique identifier (AUI) that represents information from each instance of the string in each vocabulary. Figure 5.19 depicts the English-language concepts, terms, and strings for the Metathesaurus concept *atrial fibrillation*. Each string may occur in more than one vocabulary, in which case each would be an atom.) The canonical form of the concept and one of its terms is atrial fibrillation. Within both terms are several strings, which vary in word order and case.

There are number of limitations to the Metathesaurus, which is one of the reasons why its use has been modest. The Metathesaurus only maps one-to-one relationships. There may be a term in one vocabulary that might map to multiple terms in another vocabulary, but the Metathesaurus does not map those many to one or one to many relationships. It only maps one into one relationships. Another limitation of the Metathesaurus is that the only terms in it are those that come from the source vocabularies. There may be other ways to express a term, but if that expression of the term is not in one of the source vocabularies, then it will not be in the Metathesaurus. Another limitation of the Metathesaurus is that there is no unifying hierarchy. It contains descriptions of all the hierarchies of the source vocabularies, but there is no unified hierarchy that exists for the entire Metathesaurus. And finally, it is not extensible like SNOMED, in that terms are individual and atomic and they can't be combined or modified. For all these reasons, the Metathesaurus has relatively modest use, and its value has been more as a repository for vocabularies where system developers who use different vocabularies can go and find information about terms in those vocabularies and perhaps expand their systems by taking advantage of the linkages to other vocabularies.

Other Terminologies and Activities

There are several other healthcare vocabularies
- Common Dental Terminology (CDT) – equivalent of CPT for dental procedures
- Medical Subject Headings (MeSH) – used to index biomedical literature for retrieval, discussed in the chapter on information retrieval

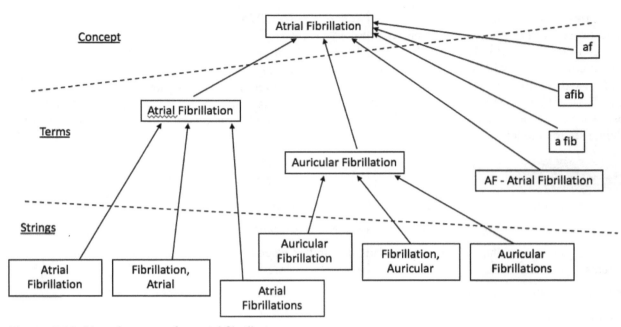

Figure 5.19: Metathesaurus for atrial fibrillation

- Universal Medical Device Nomenclature (UMD) – describes medical devices
- Diagnostic and Statistical Manual of Mental Disorders (DSM) – catalogs psychiatric and psychological conditions
- International Classification of Functioning, Disability, and Health (ICF)
- International Classification of Primary Care (ICPC)

There are several other terminology activities worthy of mention. One consists of efforts to create common data elements (CDEs, https://www.nlm.nih.gov/cde/) for research studies. These have been developed by the US National Institutes of Health and aim to standardize reporting in different types of research studies. Examples include:

- Patient Reported Outcome Measurement System (PROMIS, www.nihpromis.org)
- National Institute of Neurological Disorders and Stroke Common Data Elements Project (www.commondataelements.ninds.nih.gov/)
- Global Rare Diseases Registry (GRDR, www.grdr.info)
- Consensus Measures for Phenotypes and Exposures (PhenX, www.phenx.org)

Another effort aims to model important clinical elements that might be captured in an electronic health record and used for varying purposes from clinical decision support to biomedical research, et cetera. The rationale for the Clinical Element Model (CEM) is to go beyond the so-called *"stack of coded items"* that might be used to represent a clinical finding.[90] An example of a model is shown in Figure 5.20, which represents the measurement of systolic blood pressure and certain qualifications. For example, it was measured in the right arm, and the patient was sitting. An effort called the Clinical Information Modeling Initiative (CIMI, https://www.opencimi.org/) aims to create CEMs for many types of clinical data. CEMs are used extensively at Intermountain Healthcare in Utah.[91]

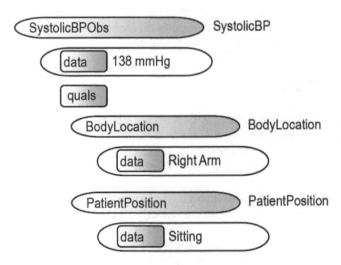

Figure 5.20: Clinical Element Model

There are also a number of commercial efforts in aspects of terminology. One comes from the company Intelligent Medical Objects (IMO, https://www.e-imo.com/), which provides a variety of terminology services such as mapping of free text to control terms of keeping terminologies systems like ICD-10 and SNOMED up to date and providing means to access that terminology. Another commercial effort is Medcin (Medicomp, http://www.medicomp.com/), which is a terminology system that is focused on documentation at the point of care using an EHR. Finally, 3M has developed HDD Access ((https://www.hddaccess.com/), which has been moved to an open source model.

RECOMMENDED READING

Benson, T and Grieve, G (2016). *Principles of Health Interoperability - SNOMED CT, HL7 and FHIR*, Third Edition. London, England, Springer.

Giannangelo, K, Ed. (2015). *Healthcare Code Sets, Clinical Terminologies, and Classification Systems*, Third Edition. Chicago, IL, AHIMA Press.

KEY POINTS

- Standards are important for interoperability of healthcare data and information.
- Identifier standards are important to identify people and organizations in healthcare
- Individual patient identifiers are controversial, demonstrating value and risk to linking individuals in information systems
- Messaging standards are important to move data and information between information systems
- Terminology standards are important for normalizing the representation of all concepts used in health information systems

CONCLUSIONS

This chapter looked at different standards for identifiers, transactions, message exchange, and terminology. The ultimate goal of all these standards is to move towards semantic interoperability, which has been described as a computer utterance in one information system having the same effect in any other[92]. Whether that is a diagnosis or a clinical finding that leads to data aggregation or clinical decision support, the computer utterance should have the same effect in any computer system.

REFERENCES

1. Samal L, Dykes P, Greenberg J, et al. Care coordination gaps due to lack of interoperability in the United States: a qualitative study and literature review. *BMC Health Services Research*. 2016;16:143.
2. Safran C, Bloomrosen M, Hammond W, et al. Toward a national framework for the secondary use of health data: an American Medical Informatics Association white paper. *Journal of the American Medical Informatics Association*. 2007;14:1-9.
3. Benson T, Grieve G. *Principles of Health Interoperability - SNOMED CT, HL7 and FHIR*, Third Edition. London, England: Springer; 2016.
4. Anonymous. *ISO/IEC Guide 2: standardization and related activities – general vocabulary: 2004*. Geneva, Switzerland: International Organization for Standardization; 2004.
5. Anonymous. *IEEE Standard Computer Dictionary: A Compilation of IEEE Standard Computer Glossaries*. Piscataway, NJ: IEEE Press; 1990.
6. Anonymous. *What is Interoperability?* Chicago, IL: National Alliance for Health Information Technology;2005.
7. Walker J, Pan E, Johnston D, Adler-Milstein J, Bates D, Middleton B. The value of health care information exchange and interoperability. *Health Affairs*. 2005;24:w5-10-w15-18.
8. Patterson R, Huff S. The decline and fall of Esperanto: lessons for standards committees. *Journal of the American Medical Informatics Association*. 1999;6:444-446.
9. Tanenbaum A, Wetherall D. *Computer Networks (5th Edition)*. Boston, MA: Pearson Education; 2010.
10. Hammond W, Jaffe C, Cimino J, Huff S. Standards in Biomedical Informatics. In: Shortliffe E, Cimino J, eds. *Biomedical Informatics: Computer Applications in Health Care and Biomedicine*, Fourth Edition. London, England: Springer; 2014:211-254.
11. Anonymous. *Connecting Health and Care for the Nation: A 10-Year Vision to Achieve an Interoperable Health IT Infrastructure*. Washington, DC: Department of Health and Human Services; June 5, 2014 2014.
12. Anonymous. *Connecting Health and Care for the Nation: A Shared Nationwide Interoperability Roadmap Version 1.0*. Washington, DC: Department of Health and Human Services; January 30, 2015.

13. Anonymous. 2017 Interoperability Standards Advisory. Washington, DC: Department of Health and Human Services;2017.
14. Baker D, Perlin J, Halamka J. Evaluating and classifying the readiness of technology specifications for national standardization. *Journal of the American Medical Informatics Association.* 2015;22:738-743.
15. Sittig D, Wright A. What makes an EHR "open" or interoperable? *Journal of the American Medical Informatics Association.* 2015;11:1099–1101.
16. Bernstam E, Johnson T. Why health information technology doesn't work. *Frontiers of Engineering.* 2009;39(4):30-35.
17. Fernandes L, Lenson C, Hewitt J, Weber J, Yamamoto J. *Medical record numbers errors.* Chicago, IL: Initiate; April, 2001.
18. Joffe E, Bearden C, Byrne M, Bernstam E. Duplicate patient records--implication for missed laboratory results. Paper presented at: AMIA Annual Symposium Proceedings 20092009; San Francisco, CA.
19. McCoy A, Wright A, Kahn M, Shapiro J, Bernstam E, Sittig D. Matching identifiers in electronic health records: implications for duplicate records and patient safety. *Quality and Safety in Health Care.* 2013;22:219-224.
20. Anonymous (2005). Linking Health Care Information: Proposed Methods for Improving Care and Protecting Privacy. Washington, DC, Markle Foundation. https://www.markle.org/sites/default/files/linking_report_2_2005.pdf
21. Arnason J. Personal identifiability in the Icelandic health sector database. *The Journal of Information, Law and Technology.* 2002;2.
22. Gulcher J, Stefansson K. The Icelandic healthcare database and informed consent. *New England Journal of Medicine.* 2000;342:1827-1830.
23. Winkler W. Should Social Security numbers be replaced by modern, more secure identifiers? *Proceedings of the National Academy of Sciences.* 2009;106:10877-10878.
24. Acquisti A, Gross R. Predicting Social Security numbers from public data. *Proceedings of the National Academy of Sciences.* 2009;106:10975-10980.
25. Hieb B. The case for a voluntary national healthcare identifier. *Journal of ASTM International.* 2006;3(2).
26. Hieb B. *Designing a Voluntary Universal Healthcare Identification System.* Stamford, CT: Gartner;2008.
27. Ferris N. Why a national health care ID isn't worth the fight. *Government Health IT*2005.
28. Hillestad R, Bigelow J, Chaudhry B, et al. *Identity Crisis: An Examination of the Costs and Benefits of a Unique Patient Identifier for the U.S. Health Care System.* Santa Monica, CA: RAND Corporation;2008. MG-753-HLTH.
29. Detmer D. Activating a full architectural model: improving health through robust population health records. *Journal of the American Medical Informatics Association.* 2010;17:367-369.
30. Aranow M. It's Time for a National Patient Identifier. In: Halamka J, ed. *Life as a Healthcare CIO*2013.
31. Grannis S, Overhage J, Hui S, McDonald C. Analysis of a probabilistic record linkage technique without human review. Paper presented at: Proceedings of the 2003 AMIA Annual Symposium2003; Washington, DC.
32. Grannis S, Overhage J, McDonald C. Real world performance of approximate string comparators for use in patient matching. Paper presented at: MEDINFO 2004 - Proceedings of the Eleventh World Congress on Medical Informatics2004; San Francisco, CA.
33. Tromp M, Ravelli A, Méray N, Reitsma J, Bonsel G. An efficient validation method of probabilistic record linkage including readmissions and twins. *Methods of Information in Medicine.* 2008;47:356-363.
34. Tromp M, Ravelli A, Bonsel G, Hasman A, Reitsma J. Results from simulated data sets: probabilistic record linkage outperforms deterministic record linkage. *Journal of Clinical Epidemiology.* 2011;64:565-572.
35. Li X, Shen C. Linkage of patients records from disparate sources. *Statistical Methods in Medical Research.* 2013;22:31-38.
36. Sayers A, Ben-Shlomo Y, Blom A, Steele F. Probabilistic record linkage. *International Journal of Epidemiology.* 2016:Epub ahead pf print.
37. Kho A, Cashy J, Jackson K, et al. Design and implementation of a privacy preserving electronic health record linkage tool in Chicago. *Journal of the American Medical Informatics Association.* 2015;22:1072-1080.
38. McFarlane T, Dixon B, Grannis S. Client Registries: Identifying and Linking Patients. In: Dixon B, ed. *Health Information Exchange - Navigating and Managing a Network of Health Information Systems.* Amsterdam, Netherlands: Elsevier; 2016:164-182.
39. Randall S, Ferrante A, Boyd J, Semmens J. The effect of data cleaning on record linkage quality. *BMC Medical Informatics & Decision Making.* 2013;13:64.
40. Ong T, Mannino M, Schilling L, Kahn M. Improving record linkage performance in the presence of missing linkage data. *Journal of Biomedical Informatics.* 2014:Epub ahead of print.

41. Morris G, Farnum G, Afzal S, Robinson C, Greene J, Coughlin C. *Patient Identification and Matching - Final Report.* Baltimore, MD: Audacious Inquiry;2014.
42. Kliff S. Obamacare's most important number: 834. *Washington Post.* October 23, 2010, 2013.
43. Laszewski R. The 834 Problem. *The Health Care Blog* 2013.
44. Blumenthal D. Reflections on Health Reform: A Tale of Two IT Procurements. *HuffPolitics Blog* 2013.
45. Anonymous. *Companion Guide to HL7 Consolidated CDA for Meaningful Use Stage 2.* Washington, DC: Department of Health and Human Services;2012.
46. Day B. Standards for Medical Device Interoperability and Integration. *Patient Safety & Quality Healthcare* 2011.
47. Kennelly R, Gardner R. Perspectives on development of IEEE 1073: the Medical Information Bus (MIB) standard. *International Journal of Clinical Monitoring and Computing.* 1997;14:151-154.
48. Kennelly R. Improving acute care through use of medical device data. *International Journal of Medical Informatics.* 1998;48:145-149.
49. Anonymous. *Continua Adoption Playbook.* Arlington, VA: Personal Connected Health Alliance; May, 2017.
50. Sujansky W, Overhage J, Chang S, Frohlich J, Faus S. The development of a highly constrained health level 7 implementation guide to facilitate electronic laboratory reporting to ambulatory electronic health record systems. *Journal of the American Medical Informatics Association.* 2009;16:285-290.
51. ASTM International www.astm.org/COMMIT/E31_ConceptPaper.doc
52. D'Amore J, Sittig D, Ness R. How the continuity of care document can advance medical research and public health. *American Journal of Public Health.* 2012;102:e1-e4.
53. D'Amore J, Mandel J, Kreda D, et al. Are Meaningful Use Stage 2 certified EHRs ready for interoperability? Findings from the SMART C-CDA Collaborative. *Journal of the American Medical Informatics Association.* 2014;21:1060-1068.
54. Mandl K, Kohane I. Escaping the EHR trap--the future of health IT. *New England Journal of Medicine.* 2012;366:2240-2242.
55. Mandl K, Mandel J, Murphy S, et al. The SMART Platform: early experience enabling substitutable applications for electronic health records. *Journal of the American Medical Informatics Association.* 2012;19:597-603.
56. Mandl K, Mandel J, Kohane I. Driving innovation in health systems through an apps-based information economy. *Cell Systems.* 2015;1:8-13.
57. Mandel J, Kreda D, Mandl K, Kohane I, Ramoni R. SMART on FHIR: a standards-based, interoperable apps platform for electronic health records. *Journal of the American Medical Informatics Association.* 2016:Epub ahead of print.
58. Alterovitz G, Warner J, Zhang P, et al. SMART on FHIR Genomics: facilitating standardized clinico-genomic apps. *Journal of the American Medical Informatics Association.* 2015;22:1173-1178.
59. Warner J, Rioth M, Mandl K, et al. SMART precision cancer medicine: a FHIR-based app to provide genomic information at the point of care. *Journal of the American Medical Informatics Association.* 2016:Epub ahead of print.
60. Bloomfield R, Polo-Wood F, Mandel J, Mandl K. Opening the Duke electronic health record to apps: implementing SMART on FHIR. *International Journal of Medical Informatics.* 2017;99:1-10.
61. Giannangelo K, ed *Healthcare Code Sets, Clinical Terminologies, and Classification Systems, Third Edition.* Chicago, IL: AHIMA Press; 2015.
62. Voytovich A. Reduction of medical verbiage: fewer words, more meaning. *Annals of Internal Medicine.* 1999;131:146-147.
63. Chute C. Medical Concept Representation. In: Chen H, Fuller S, Hersh W, Friedman C, eds. *Medical Informatics: Advances in Knowledge Management and Data Mining in Biomedicine.* New York: Springer-Verlag; 2005:162-182.
64. Cimino J. Desiderata for controlled medical vocabularies in the twenty-first century. *Methods of Information in Medicine.* 1998;37:394-403.
65. Rosenbloom S, Miller R, Johnson K, Elkin P, Brown S. Interface terminologies: facilitating direct entry of clinical data into electronic health record systems. *Journal of the American Medical Informatics Association.* 2006;13:277-288.
66. Rosenbloom S, Miller R, Johnson K, Elkin P, Brown S. A model for evaluating interface terminologies. *Journal of the American Medical Informatics Association.* 2008;15:65-76.
67. Scott K. *Medical Coding for Non-Coders, Edition 2.* Chicago, IL: American Health Information Management Association; 2013.
68. Calhoun M, Rudman B, Watzlaf V. Vision 2016 to Reality 2016: building a profession. *Journal of AHIMA.* 2012;83(8):18-23.
69. Tully M, Carmichael A. Computer-assisted coding and clinical documentation: first things first. *Healthcare Financial Management.* Vol 662012:46-49.
70. Chute C. Clinical classification and terminology: some history and current observations. *Journal of*

the American Medical Informatics Association. 2000;7:298-303.
71. Chute C. The Copernican era of health terminology: a recentering of health information systems. Paper presented at: Proceedings of the AMIA 1998 Annual Symposium1998; Orlando, FL.
72. Outland B, Newman M, William M. Health policy basics: implementation of the International Classification of Disease, 10th Revision. *Annals of Internal Medicine.* 2015;163:554-556.
73. Mathews A. Walked Into a Lamppost? Hurt While Crocheting? Help Is on the Way. *Wall Street Journal.* September 13, 2011, 2011.
74. Sklevaski N. *Struck by Orca: ICD-10 Illustrated.* 2014.
75. Chute C, Huff S, Ferguson J, Walker J, Halamka J. There are important reasons for delaying implementation of the new ICD-10 coding system. *Health Affairs.* 2012;31:836-842.
76. Rodrigues J, Robinson D, DellaMea V, et al. Semantic alignment between ICD-11 and SNOMED CT. *Studies in Health Technology and Informatics.* 2015;216:790-794.
77. Boyd A, Li J, Burton M, et al. The discriminatory cost of ICD-10-CM transition between clinical specialties: metrics, case study, and mitigating tools. *Journal of the American Medical Informatics Association.* 2013;20:708-717.
78. Butz J, Brick D, Rinehart-Thompson L, Brodnik M, Agnew A, Patterson E. Differences in coder and physician perspectives on the transition to ICD-10-CM/PCS: a survey study. *Health Policy and Technology.* 2016:Epub ahead of print.
79. Bodenreider O. The Unified Medical Language System (UMLS): integrating biomedical terminology. *Nucleic Acids Research.* 2004;32:D267-D270.
80. Fung K. RxTerms - a New Interface Terminology to RxNorm. *NLM Technical Bulletin.* Bethesda, MD: National Library of Medicine; 2008.
81. Bodenreider O, Nelson S. RxNav: a semantic navigation tool for clinical drugs. Paper presented at: MEDINFO 2004 - Proceedings of the Eleventh World Congress on Medical Informatics2004; San Francisco, CA.
82. Peters L, Mortensen J, Nguyen T, Bodenreider O. Enabling complex queries to drug information sources through functional composition. *Studies in Health Technology and Informatics.* 2013;192:692-696.
83. Vreeman D, Chiaravalloti M, Hook J, McDonald C. Enabling international adoption of LOINC through translation. *Journal of Biomedical Informatics.* 2012;45:667-673.
84. Vreeman D. *LOINC Essentials.* 2017.
85. Spackman K. SNOMED RT and SNOMED CT: promise of an international clinical terminology. *MD Computing.* 2000;17(6):29.
86. Anonymous. *SNOMED CT - Starter Guide.* Copenhagen, Denmark: International Health Terminology Standards Development Organisation; July 31, 2014 2014.
87. Wasserman H, Wang J. An applied evaluation of SNOMED CT as a clinical vocabulary for the computerized diagnosis and problem list. Paper presented at: Proceedings of the AMIA 2003 Annual Symposium2003; Washington, DC.
88. Elkin P, Brown S, Husser C, et al. Evaluation of the content coverage of SNOMED CT: ability of SNOMED clinical terms to represent clinical problem lists. *Mayo Clinic Proceedings.* 2006;81:741-748.
89. Humphreys B, Lindberg D, Schoolman H, Barnett G. The Unified Medical Language System: an informatics research collaboration. *Journal of the American Medical Informatics Association.* 1998;5:1-11.
90. Coyle J, Heras Y, Oniki T, Huff S. *Clinical Element Model.* Salt Lake City, UT: Intermountain Health Care; November 14, 2008 2008.
91. Oniki T, Coyle J, Parker C, Huff S. Lessons learned in detailed clinical modeling at Intermountain Healthcare. *Journal of the American Medical Informatics Association.* 2014;21:1076-1081.
92. Dolin R, Alschuler L. Approaching semantic interoperability in Health Level Seven. *Journal of the American Medical Informatics Association.* 2011;18:99-103.

6

Health Information Exchange

ROBERT E. HOYT • WILLIAM R. HERSH

LEARNING OBJECTIVES

After reading this chapter the reader should be able to:

- Identify the need for and benefits of health information exchange (HIE) and interoperability
- Describe the concept of health information organizations (HIOs) and how they integrate with the national HIE strategy
- Summarize the differences between Direct and eHealth Exchange
- Enumerate the basic and advanced features offered by HIOs
- Detail the obstacles facing health information exchange
- Summarize the newest HIE models

INTRODUCTION

In recent years, there has been substantial growth in the adoption of the electronic health record (EHR) in ambulatory and hospital settings across the United States, fueled largely by incentive funding provided by the Health Information Technology for Economic and Clinical Health (HITECH) Act.[1] One key challenge to effective use of HIT, however, is that most patients in the U.S., especially those with multiple conditions, receive care across a number of settings.[2-3] To enable data to follow patients wherever they receive care, attention has recently focused on health information exchange (HIE), defined as the *"reliable and interoperable electronic sharing of clinical information among physicians, nurses, pharmacists, other health care providers, and patients across the boundaries of health care institutions, health data repositories, States, and other entities who are not within a single organization or among affiliated providers."*[4] The HITECH Act recognized that EHR adoption alone was insufficient to realize the full promise of HIT, allocating $563 million for States or State-designated entities to establish HIE capability among health care providers and hospitals.[5] As a result of HITECH funding, HIE adoption has grown in a parallel though somewhat smaller manner. By 2015, 82% of U.S. hospitals had engaged in some form of HIE.[6]

HIE is also a critical element of Meaningful Use (MU) and integral to the future success of healthcare reform at the local, regional and national level. Exchange of health-related data is important to all healthcare organizations, particularly federal programs such as Medicare or Medicaid, for several reasons. The federal government determined that HIE is essential to improve: the disability process, continuity of medical care issues, bio-surveillance, research and natural disaster responses.[7] As a result, the federal government has been a major promoter of HIE and the development of data standards to achieve interoperability. Electronic transmission of data results in faster and less expensive transactions, when compared to standard paper-based mail and faxes. If the goal of the federal government was only to promote electronic health records, then the result would be electronic, instead of paper silos of information. Instead, they created a comprehensive game plan to share health information among disparate partners.

Chapter 1 discussed programs from the HITECH Act that support HIE and interoperability. HIE is an important part of Meaningful Use, particularly stage 2 and is also integral to accountable care organizations (ACOs), patient centered medical homes (PCMHs) and Medicare Access and Chip Authorization Act (MACRA).

Exchange of patient information is an international issue and not limited to just the United States. A 2012 survey of 10 high income countries asked if physicians could electronically exchange patient summaries and test results outside their own practices. Canada reported a low of 14% and New Zealand reported a high of 55%; the US reported 31%. Furthermore, they found that fewer than 25% of US physicians were notified when one of their patients visited the emergency department and only 16% received information from specialists when changes were made to medications or a care plan.[8]

HIE most commonly involves the exchange of clinical results, images and documents. It is also important to share financial and administrative data among disparate entities as well. Table 6.1 lists some of the common types of health-related data that are important to exchange among the many healthcare partners.

Table 6.1: Common types of health-related data exchanged

Data	Examples
Clinical results	Lab, pathology, medication, allergies, immunizations and microbiology data
Images	Radiology reports; scanned images of paper documentation
Documents	Office notes, discharge notes, emergency room notes
Clinical Summaries	Continuity of Care Documents (CCDs), personal health record extracts
Financial information	Claims data, eligibility checks
Medication data	Electronic prescriptions, formulary status, history
Performance data	Quality measures, such as blood pressure and cholesterol levels
Case management	Management of the underserved, e.g. emergency room utilization
Public health data	Infectious diseases outbreak data, immunization records, etc.
Referral management	Management of referrals to specialists

This chapter will begin with HIE-related definitions and then chronicle of the evolution of local, state and national organizations created for HIE.

Definitions

The following are commonly cited definitions related to health information exchange.

- Health Information Exchange (HIE) is the *"electronic movement of health-related information among organizations according to nationally recognized standards."*[9]
- Health Information Organization (HIO) is *"an organization that oversees and governs the exchange of health-related information among organizations according to nationally recognized standards."*[9]
- Health Information Service Provider (HISP) is an organization that provides services and support for the electronic exchange of health information.[10]
- Health information blocking *"occurs when persons or entities knowingly and unreasonably interfere with the exchange or use of electronic health information."*[11-12]
- Interoperability is defined as *"the ability of two or more systems or components to exchange information and to use the information that has been exchanged."* This implies that the data is computable and that standards exist that permit interoperability.[13]
- Opt-In and Opt-Out refers to patient consent policies; the ability for *content creators* to determine whether the personal health record data they create can be shared as well as with whom. Under an opt-in scenario, no health information can be exchanged unless the patient signs a specific informed consent document permitting the sharing of data. Opt-out assumes that consumers grant permission for the exchange of personal health information as part of the broader informed consent that they sign when they receive care from a clinician and the halting of data sharing must be triggered by an action from the patient.
- Regional Health Information Organization (RHIO) is *"a health information organization that brings together health care stakeholders within a defined geographic area and governs health information exchange among them for the purpose of improving health and care in that community."*[9] Note that the term RHIO is inexact because HIOs do not have to be regional; they can include only one city or an entire state. Furthermore, HIOs can be created for specific populations, such as those on Medicaid or the uninsured. In keeping with these new definitions,

the acronym HIO will be used when addressing health information organizations.

The definition of HIE may vary, depending which country is surveyed. Adler-Milstein et al. published a study on benchmarking from seven developed countries. Four countries defined HIE as a summary of care record, while one defined it as a method to secure messages between physicians and hospitals.[14] It has also been noted that HIE should be used as a "verb" rather than a "noun" to indicate that it is a process and not focused on the organizations that implement HIE.[15]

There are different types of HIE, as described by Williams:[5]

- *Directed* HIE is the direct sending and receiving of information to support planned care. This includes activities such as referral or transfer of patients and is sometimes called *push* HIE, because the information is pushed from one health care entity to another.
- *Query-based* HIE involves a query process to find information that may be unknown to the provider. An example might be when a patient has a visit to an emergency department. This type of HIE is also called *pull* HIE.
- *Consumer-mediated* HIE is where consumers aggregate and control their own information. At the present, this latter type of HIE is uncommon but it may increase in the future.[16]

HISTORY OF THE US HEALTH INFORMATION NETWORK INITIATIVES

In the early 1990s, Community Health Information Networks (CHINs) began appearing across the US. Approximately 70 pilot projects were created but all eventually failed and were terminated.[17] Most were thought to fail due to lack of perceived value and sustainable business plan and immature technology. Despite this early failure, it became apparent that not only would electronic health records (EHRs) need to be adopted, there would be a need for new local and regional health information organizations (HIOs) to exchange data and eventually connect to a national health information exchange.

In April 2004, President Bush signed Executive Order 13335 creating the Office of the National Coordinator for Health Information Technology (ONC) and at the same time calling for interoperable electronic health records within the next decade.[18] ONC determined a decentralized system was needed to exchange patient information along with set of standards, services and policies that direct how the secure exchange of health information would occur.

NHIN Prototype Architecture

In 2005 ONC provided $18.6 million in funding towards the NHIN Prototype Architecture initiative. The purpose of this initiative was to demonstrate that a *network-of-networks* approach without reliance on a centralized network could successfully exchange information between regional HIOs. The goal was to create a Health Internet comprised of services which facilitate the secure exchange of health information. (see Figure 6.1)

NwHIN Exchange

Using the specifications and services developed during the NHIN Trial Implementation period, several federal agencies and private sector organizations began exchanging health information in 2009. These efforts were known as the Nationwide Health Information Network (NwHIN) Exchange. The Social Security Administration (SSA), which requests 15 to 20 million medical records each year as part of disability determinations, was selected as the first federal agency to use the NHIN standards and policies to connect to a non-federal entity. Recognizing that most veterans and active duty service members receive medical care outside their respective systems, the VA and DoD are also involved in the NwHIN Exchange.

eHealth Exchange, HealtheWay and Sequoia Project

In 2012, the NwHIN Exchange was renamed the eHealth Exchange. Additionally, in 2012 a new entity HealtheWay was created to help direct the future of the NwHIN Exchange. HealtheWay is a non-profit public-private organization that promotes open source, open standards-based exchange of health information. Later, in 2012 this initiative became the Sequoia Project.[19] Care Connectivity Consortium supports the Sequoia Project and consists of six large healthcare organizations, such as the Mayo Clinic, that have the vision of shared HIE. The Sequoia Project supports the following initiatives:

- eHealthExchange: is a group (network) of federal agencies and non-federal organizations for improving patient care through interoperable HIE. It connects the private sector, to include state and regional HIOs to federal agencies. In 2016, it connected all 50 states, four federal agencies and almost 50% of US hospitals.
- Carequality: is not a network, it is a *"trust framework"* to connect current and future networks to each other. The goal is to have more than just one network (eHealthExchange).
- RSNA Image Share Validation: tests vendor compliance for exchange of medical images

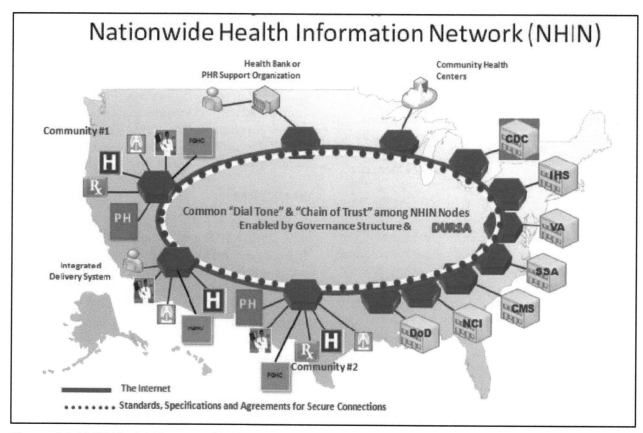

Figure 6.1: NHIN Model (Courtesy ONC)

The Direct Project

The original concept for the NHIN responded to the mobile nature of our society by recognizing the need for healthcare clinicians to have timely access to patient information across multiple organizations and locations. As initially envisioned, this interoperable exchange of patient data between distant and unaffiliated providers would occur through a *network-of-networks* consisting of HIOs and government agencies. By leveraging existing HIOs and the standards with which they were built, it was believed that these tested and reliable core services would speed the development of the NwHIN. The real-world implementation of the NwHIN, however, has been delayed by issues ranging from technical (deciding on how much of the standard to support), to procedural (agreeing upon vocabularies for proper semantic interoperability), to political (reconciling patient privacy and consent laws between locales).

In response to the complexities of building the *network-of-networks*, the NwHIN concept was adjusted by the HIT policy committee's NHIN Working Group to provide more simplistic HIE capabilities via a secure email analogue. This modified version was renamed NwHIN Direct (also referred to by some as NwHIN Lite). The newer model provides a simplified set of standards, policies and services that support the secure exchange of patient data, but in a more lightweight manner. Focusing on the *"email use case"* allowed for a simpler, scalable, more direct exchange to support achieving Stage 2 Meaningful Use criteria.

Launched in 2010, The Direct Project focused on the deployment of functionality using the lowest cost of entry from a technical and operational perspective. The purpose was to supplement traditional fax and mail methods of exchanging health information between known and trusted recipients with a faster, more secure, Internet-based method. In other words, Direct helps **provider A** transmit to **provider B** patient summaries, reconciliation of medications, lab and x-ray results. Use cases include connecting clinician-clinician, clinician-patient, clinician-health organization, and health organization-health organization exchange. An example of Direct is a primary care physician sending a specialist a clinical summary on a patient that is being referred for care.

The system is based on secure messaging that is managed by a health information service provider (HISP). HISPs can be a healthcare entity, an HIO or an IT organization. The role of the HISP, in Direct, is to provide user authentication, message encryption and maintenance of system security for sending and receiving organizations or clinicians. By contracting with an HISP, health entities avoid the need for multiple data use and reciprocal support agreements (DURSAs) or contracts with every provider with whom they exchange data

The Direct Project relies on *push* technology, which refers to sending (pushing) data to a provider. Pushed messages can include attachments, such as referral summary documents. This push process is much simpler than *pull* technology where a health information exchange database is queried (pulled) for matches to the patient and then relevant document results are *pulled*. The HISP can maintain a provider directory, like an email address book or contacts list, containing relevant provider demographics including the direct email address that is used to authenticate both the sender and receiver. This process is less complicated than creating and maintaining master patient indices (MPIs) and record locator services (RLSs) that underlie pull technology.

Open source software has been developed to allow for a Direct Project compliant EHR to receive these secure messages and initiate new messages to other Direct Project participants. Direct Project providers must obtain a Direct Address and a security certificate from a HISP. An example of such a secure Direct Address would be b.wells@direct.aclinic.org. Direct messages can be received and sent by clinicians regardless of whether they have an EHR. However, most EHRs offer Direct support as part of their efforts to achieve stage 2 Meaningful Use certification. These efforts permit messages to appear in the system's email inbox and output such as Continuity of Care documents (CCDs) can be generated and transmitted seamlessly and securely from one EHR to another.

One of the largest HISPs is SureScripts an electronic prescription network provider. In 2017, they reported having connections to 35 state health information networks, and a variety of other large healthcare entities as part of their Direct network, known as SureScripts Health Information Network.[20] In mid-2016 The Direct Project extended secure messaging to patients and other healthcare consumers.[21]

The most recent usage statistics were reported by ONC in 2015. Their data showed that 64% of clinicians had an EHR with messaging capabilities, but only 16% had the ability to view, download and transmit secure messages.[22]

Blue Button Project

Blue button means the presence of a blue button in an electronic application such that a patient can download their healthcare data. (see Figure 6.2) Various organizations such as the Department of Veterans Affairs, Medicare and large payer organizations have taken the lead to make this available. Initially, data was primarily based on administrative claims data and available as an ASCII or PDF formatted file.[23] With increased adoption of electronic health records and Meaningful Use requirements structured clinical documents can be generated and shared. Blue buttons could be part of every patient portal or personal health record that is integrated with a personal health record (PHR) providing patients with easily identified ready access to their record in a portable format.

ONC has promoted the idea that more should be done with this user-friendly initiative and therefore developed the Blue Button Plus project. Blue Button Plus represents the ability to have these records in a human readable and machine-readable format and the ability to send or share them. The end user has the choice whether to print or share them electronically. This also helped eligible professionals meet Meaningful Use stage 2 requirements (view, download and transmit) as Blue Button Plus can leverage consolidated CDAs (see chapter on data standards) and the Direct Project.[24]

Figure 6.2: Blue Button

INTEROPERABILITY

Sucessful HIE is dependent on multiple interactions between disparate partners and organizations. In order to accomplish this, there must be interoperability and it must occur at several levels:

- *Foundational interoperability* refers to the infrastructure or technology required to exchange electronic information between disparate systems. Different systems must adopt standardized communication protocols, such as TCP/IP, HTTP, SOAP and REST
- *Syntactic interoperability* means that messages have a structure and syntax that is understandable by the systems exchanging the messages. This requires standards such as XML and HL7 standards
- *Semantic interoperability* means terminology and coding must be the same for the sending and the receiving organizations.[4]

IMPACT OF THE HITECH ACT ON HIE

The 2009 HITECH Act signaled a major federal commitment to expansion of health information technology. Although the HITECH Act focused on incentivizing the expansion of EHRs, it also encouraged the growth of health information exchange through the authorization and funding of the State HIE Cooperative Agreement Program. This program closed the state and regional HIE gap by awarding $548 million to 56 state agencies.[25] HIE was further supported by incorporating HIE into Meaningful Use stage 2 objectives necessary for EHR reimbursement. The bar was set lower in terms of information sharing in stage 1 because most physicians and hospitals lacked the technology to share.[26] One could argue that the HITECH Act was one of the strongest catalysts for HIE in the US, but also one of its most significant limitations. HIE followed EHR adoption and Meaningful Use objectives. Post-acute and long-term facilities, behavioral health and laboratory providers were not eligible for reimbursement under Medicare or Medicaid and there wasn't a strong business case to develop HIE on their own. Hence, there were gaps in US healthcare where HIE was not encouraged or financially supported. Table 6.2 enumerates the modified stage 2 objectives that have definite HIE implications.

Stage 3 Meaningful Use and HIE

Clinicians who bill Medicaid and hospitals will continue to operate under Modified Stage 2 or 3 Meaningful Use programs. Medicare clinicians who bill Medicare part B, can begin their first performance year in 2017 under MACRA, described in the next section. 2016 was the last year of reimbursement for returning eligible Medicare clinicians.

According to CMS, more than 60% of Stage 3 MU objectives will require HIT interoperability. Importantly, CMS advocates use of application program interfaces (APIs) to improve interoperability. The following Stage 3 objectives (4 out of 8) for eligible Medicaid professionals (EPs) are related to HIT interoperability. Each objective has several measures outlined in the reference.

- Patient electronic access: the EP provides patients with electronic access to their health information

Table 6.2: HIE and Modified Stage 2 Meaningful Use Objectives (EP=eligible physician, EH = eligible hospital)

Stage 2 Objective	Group	HIE Implications
Health Information Exchange: When a patient is referred to another care setting or clinician they must transmit a summary of care record for more than 10% of transitions and referrals	EP, EH	This could be achieved through either the Direct Project or HIO
Patient Access: More than 5 % of unique patients should be able to view, download, or transmit health information to a third party More than 50% of unique patients who are discharged from the hospital or ED should be able to view, download, or transmit health information	EP EH	This would be achieved through either a patient portal/EHR or through a HIO
Public Health Reporting: EPs must comply with two of the following: 1) They must submit immunization data 2) They must submit syndromic surveillance data 3) They must submit specialized registry data to a public health agency	EP, EH	This could be achieved through either an EHR or through a HIO

and patient education. This could involve patient portals and/or an application programming interface (API)
- Coordination of care: use certified EHRs to engage with patients about their care
- Health information exchange: EP must provide a summary of care record for transitions of care and incorporate summary of care records from other clinicians into their EHR. The consolidated clinical document architecture (C-CDA) would be a common format, as well as Direct messages
- Public Health and clinical data reporting: EPs engages with a public health agency or registry to submit electronic public health data, using certified EHRs. Data transmitted would include immunization registry reporting, syndromic surveillance reporting and reportable conditions.[27]

HEALTH INFORMATION ORGANIZATIONS (HIOS)

The late 1990s saw the rise of health information organizations (HIOs) in the United States, largely created with federal startup funds. There was, however, no national game plan as to how to create or maintain them.

According to a 2011 national survey there were 85 operational HIOs (exchanging clinical information) out of 255 reported HIE entities.[28] It is not known, however, how many HIOs started and failed.

Most HIOs begin with a collaborative planning process that involves multiple stake holders in the healthcare community. Participation from a broad spectrum of health care entities is necessary for long term sustainability. Potential participants include: insurers (payers), physicians, hospitals, medical societies, medical schools, health informatics programs, state and local government, employers, consumers, pharmacies and pharmacy networks, ambulatory care providers, business leaders, selected vendors and public health departments.

Social capital or an atmosphere of trust is a prerequisite for HIO success. This is particularly true in highly competitive health care regions, where health systems, physician groups, other providers, and payers distrust the motives of the other parties. HIOs are usually complex organizations in which the governing members must reach consensus on governance structure, privacy and security issues, as well as business, technical and legal aspects of HIE. The building of social capital and trust is necessary for sustainability of the HIO.

Multiple functions need to be addressed by a HIO:

- Financing: what will be the sources of short term startup money and on-going revenue? What is the long-term business plan? What is the pricing structure?
- Regulations: what data, privacy and security standards will be used?
- Information technology: who will create and maintain the actual network? Who will do the training? Will the HIO use a centralized or de-centralized data repository?
- Clinical process improvements: what processes will be selected to improve? Will the analysis use claims data or provider patient data? Who will monitor and report the progress?
- Incentives: what incentives exist for disparate entities to join?
- Public relations (PR): how will information on the benefits of the HIO be spread to healthcare organizations, physicians and the public?
- Consumer participation: how will the HIO reach out to stakeholders and patients for input?

The planning phase generally takes several years and generally relies on federal and/or state grant support. Upon completion of the planning phase, the HIO is ready to focus on building the technical infrastructure. The web-based infrastructure can be built by local IT expertise or an HIE-specific vendor. HIOs start with simple processes such as clinical messaging (test results retrieval) before tackling more complicated functionality.

Several types of data exchange models exist and determine how data is shared and stored. The following are general categories:

- Federated: decentralized approach where data is stored locally on a server at each network node (hospital, pharmacy or lab). Data therefore must be shared among the users of the HIO with an import/export scheme
- Centralized: the HIO operates a central data repository that all entities must access
- Hybrid: a combination of some aspects of federated and centralized model

Further details concerning clinical data exchange models as well as HIOs using these models, are discussed in the article by Just and Durkin.[29]

In addition, HIE tend to fall into three ownership categories. It should be noted that there is little written about enterprise or vendor-related HIE at this point.

- Government HIE: exchange of information primarily with government agencies such as the Social Security Administration
- Community-based HIE: the most common type of HIE where a health information organization (HIO)

is established to exchange information between several healthcare entities
- Private HIE: the exchange is either vendor-based, in which HIE is usually part of a large EHR system or enterprise-based and part of a large healthcare system. In the latter case, the healthcare system might share the same EHR or provide interfaces for disparate EHR systems.

Although HIOs utilize a variety of web-based infrastructures they tend to utilize the following similar shared services:
- **Record locator service (RLS)** directs the inquirer to the physical location of the patient's records based on the patient matching by the MPI. These results can in turn allow for retrieval of the documents to which they relate. One such implementation would be a document registry which serves as an index for content housed in a repository.
- **Master patient index (MPI)** is a database containing the registered patients within the HIO. The MPI assigns a unique patient identifier and uses algorithms to locate the correct patient and any existing records by sorting through a myriad of demographic identifiers. Duplicate records, or poor matching algorithms, can still be a problem for most functioning HIOs. An eMPI is an enterprise MPI or software that gives patients a unique ID so there is no duplication within the system
- **Provider directory** lists the potential data suppliers and users pertinent across the HIO. It is likely to include credentials, address, phone numbers, email addresses and hospital affiliation.
- **Data warehouses** such as document repositories provide the storage of patient data accessible via HIE.

The expectation is that HIOs will save money once they are operational. It is presumed that the network will decrease office labor costs (e.g. costs associated with faxing, etc.), improve medical care and reduce duplication of tests, treatments, and medications. Many people feel that insurers are likely to benefit more from HIE than clinicians. One of the potential benefits of health information exchange is more cost-effective electronic claims submission. As reported by the Utah Health Information Network, a paper claim costs $8, compared with an electronic claim cost of $1 plus the $0.20 charge by the HIO; therefore, a savings of $6.80.[30]

HIOs can be for-profit or not-for-profit, however the clear majority are not-for-profit. Operating capital for HIOs in most cases comes from fees charged to participating hospitals, physician offices, labs and imaging centers. Some HIOs charge clinicians a subscription fee (e.g. a flat fee per physician per month), others charge a transaction fee, while others charge nothing. Several HIOs are very transparent regarding their charges and this reference includes a charge matrix for users.[31] HIOs can address the entire medical arena or simply a sector such as Medicaid patients. HIOs can cover a city, region, an entire state, multiple states or an entire country. Because HIE can be a marketing strategy for newer deliver models, accountable care organizations (ACOs) and integrated delivery networks (IDNs) may adopt HIE faster than traditional HIOs can be created. Importantly, IDNs can rapidly offer HIE to their networks without the long and difficult process of creating governance and trust between disparate and competitive healthcare organizations.

There are at least four HIE business models:
- Not for profit HIOs are usually 501(c) 3 tax-exempt organizations that focus on the patient and community and are funded by federal or state funds and rely on tax advantages. An example would be The Health Collaborative.[32]
- Public utility HIOs are usually created and maintained by state or federal funding. An example is the Delaware Health Information Network.[33]
- Physician and payer collaborative HIOs are created within a defined geographic area and can be either for-profit or not-for-profit. An example is the Inland Northwest Health Services HIE.[34]
- For-profit HIOs focus on the financial benefits of exchanging data. An example is the Strategic Health Intelligence HIE.[35]

HIOs are relatively new so many regions have little experience with the concept and further education is necessary for clinicians and healthcare administrators to convince them to participate in the regional HIO. Studies so far have shown that clinicians and patients are not very knowledgeable about HIOs but support the concept of sharing medical information securely.[36-37]

According to a recent HIO survey, of the 255 HIOs that completed the survey, 24 were termed sustainable: that is, operational, not dependent on federal funding in the past year and at least broke even through operational revenue alone. Approximately half of operational HIOs charged participants a subscription fee, but multiple revenue models exist. Many HIOs were not ready for Meaningful Use but many satisfied at least one MU objective such as the exchange of lab results, care summaries, emergency department (ED) episodes or pharmacy summaries. The survey also found that HIOs were more likely to adhere to an opt-out policy than to a policy where consumers must actively give permission to the exchange of their health records. Depending on the consent model adopted by the HIO, patient choice can be made by provider, by data type (lab, radiology, etc.), encounter type, by sending organization, by data field or by sensitive data (mental health, etc.).[38]

The survey group found multiple challenges facing HIOs: developing a sustainable business plan, defining value for providers and consumers, addressing government mandates (e.g. Meaningful Use), addressing technological issues such as integration, governance issues, addressing privacy and security, engaging potential users and accurately linking patient data. The three most common sources of shared information were hospitals, primary care physicians and community/public health clinics.[38]

Some of the more common HIE functions are listed in Table 6.3.[38]

Table 6.3: Health information exchange functionality (Courtesy eHealthInitiative)

Functionality	Functionality
Results delivery	Quality reporting
Connectivity with EHRs	Results distribution
Clinical documentation	Electronic health record (EHR) hosting
Alerts to clinicians	Assist data loads into EHRs
Electronic prescribing	EHR interfaces
Health summaries	Drug-drug alerts
Electronic referral processing	Drug-allergy alerts
Consultation/referrals	Drug-food allergy alerts
Credentialing	Billing

Statewide Health Information Exchange Cooperative Agreement Program (SHIECAP)

In March 2010, fifty-six states, eligible territories, and qualified State Designated Entities (SDE) were funded to build capacity for exchanging health information within and across state lines. This program was created under the HITECH Act to expand HIE/HIO efforts at the state-level while also supporting nationwide interoperability and Meaningful Use. In some states, existing RHIOs expanded to become statewide entities/SDEs. Approximately, $600 million in federal funding was allocated initially and the last funding year for SHIECAP was 2011.[39]

HEALTH INFORMATION ORGANIZATION EXAMPLES

The following are local, regional or statewide HIOs that are innovative and successful and can serve as examples to follow.

Indiana Health Information Exchange

- One of the oldest, largest and most successful HIOs, linking over 100 hospitals and 38 healthcare systems
- Their strategic plan involves more than just sharing data, they plan to improve patient outcomes (population health) through HIE and analytics
- They have integrated with 35 different EHRs
- OneCare™ provides lab, image and hospital admission results, as well as access to clinical data
- PopCare™ is a suite of population health services, such as care and case management[40]

Utah Health Information Network (UHIN)

- Created in 1993, it has been one of the most financially successful non-profit statewide HIOs in existence.
- 90% of Utah physicians and the state government are connected
- They now connect with HIOs in Arizona and western Colorado
- Clinical Health Information Exchange (CHIE): provides a clinical portal, discharge or ED alerts, and Direct Project functions
- Utransend provide a clearinghouse for administrative (billing and eligibility) services
- CareAchieve is a data warehouse for analytics
- HIE handles dental claims transactions[41]

Nebraska Health Information Initiative (NeHII)

- Statewide roll out began July 2009
- HIE offers a dynamic virtual health record (VHR) for visualization of data
- HIE uses Direct Project services
- HIE provides population health analytics
- Public health gateway to submit reports
- Immunization gateway to submit reports
- Admit, discharge and transfer (ADT) reports/alerts
- Prescription drug monitoring program[42]

Maine Statewide Health Information Exchange (HealthInfoNet)

- One of the largest statewide HIOs
- The network known as HealthInfoNet was launched August 2009 and is now also a Regional Extension Center
- Has ability to create a virtual EHR based on collated data
- Offer analytics reporting by subscription: hospital performance, market share, population risk, 30-day readmission risk and variation management[43]

The Health Collaborative (HealthBridge)

This initiative is based in Cincinnati, OH and was one of the early adopters of HIE. They cover Ohio, and parts of Indiana and Kentucky. Their services include:

- Results back as secure messaging (Direct Project), HL7 integration directly and delivered to fax machine
- Immunization and syndromic surveillance sent to public health agencies
- Data analyzed and sent to health plans
- Analytics for measure reporting, risk of readmission alerts, duplicative radiology alerts and opioid prescription alerts, admission alerts[32]

STATUS OF US HEALTH INFORMATION EXCHANGE

HIO Status. It is difficult to know how many individual and state-wide HIOs are in existence and at what stage of maturity and data exchange capabilities. One helpful resource has been the annual national survey sponsored by eHealth Initiative. They have measured HIO maturity based on a stage 1-7 taxonomy, with state 7 representing *"sustainable and fully operational HIO"*…..they offer *"advanced analytics, quality reporting, clinical decision support, PACS reporting…."*. The following are highlights from the most recent (2013) survey:

- 199 organizations *volunteered* to take the survey. It is unknown how many didn't respond and why.
- Interoperability was a major problem due to the necessity to connect to multiple systems and the fact that creating interfaces with e.g. EHRs was difficult and expensive. They desired standardized integrated products and pricing from vendors
- More than half of respondents support accountable care organizations (ACO) and patient centered medical home (PCMH) models
- Federal funding was still needed for many HIOs, particularly advanced HIOs. Most of these are state-designated entities. Only 52 claimed they received enough revenue from users to cover operating costs.
- Patient engagement was limited: 37 HIOs allowed patients to view their data, 24 supported patient scheduling and 17 permitted patients to submit data.
- HIOs continued to face challenges of sustainability, funding and privacy issues but also faced competition from other HIOs, ACOs and HIE vendors. Sharing often did not occur outside the network.
- Ninety organizations used the push model (Direct Project) for messaging.[38]

HIE Status. A 2013 report on *hospital-based* HIE showed that it grew substantially since 2008. Roughly, 60% of hospitals shared electronic health data with physicians and other hospitals outside their organization.[44] However, another 2013 article reported that only 30% of hospitals and 10% of practices participated with a HIO. Test results were the most frequently shared data (82%), followed by discharge summaries (66%) and outpatient clinical summaries (61%). They also reported that fewer than 25% were financially sustainable and most viewed viability as a major issue. Only 10% of reported HIOs could meet all six stage 1 Meaningful Use criteria for HIE.[45] In the report to Congress by ONC in June 2013 they stated that 39 states had the ability to exchange health data via the push technology, whereas, 25 states had pull technology for HIE.[46]

A national survey of HIE leaders published in 2017 reported that 50% of respondents thought EHR vendors routinely engaged in *"information blocking"* and 25% of respondents reported that healthcare systems routinely do it as well.[47]

Evidence of Benefit Status. In 2014 Rudin et al. conducted a systematic review of HIE and concluded that the evidence for benefit was low quality, but likely showed some benefit in the emergency department. Overall usage of the exchanges occurred in only about 10% of patient encounters.[48] Few of the operational HIOs have been published in the literature so results may not be generalizable. Also, certain regions, such as New York seem to have more mature HIE models, compared to nascent initiatives.

A 2015 systematic review of HIE noted that 57% of studies reported some benefit of HIE, but those of high quality were much less likely to demonstrate benefit. They concluded, that despite the widespread adoption of HIE in the US, there was little evidence that it had a significant impact on cost, utilization or quality of care.[49]

Federal HIE Support Status. In 2015 HHS and ONC released *Connecting Health and Care for the Nation: A Shared Nationwide Interoperability Roadmap*, intending to lay the roadmap for interoperability over the ensuing

decade. The recommendations were based on input from multiple stakeholders from the public and industry. Timeline and goals were as follows:
- 2015-2017 emphasis will be on sending and receiving data to promote quality and health outcomes
- 2018-2020 emphasis will be on interoperable health IT to improve health and reduce cost
- 2021-2024 emphasis will be on nationwide interoperability to support a learning health system

The Roadmap views the biggest driver of HIE being a supportive payment and regulatory environment. The Roadmap provides goals to establish the appropriate policy and technologies. Lastly, the Roadmap seeks to measure outcome for patients: *"access to longitudinal electronic health information, can contribute to that information and can direct it to any electronic location."* The provider outcome is *"workflows and practices include consistent sharing and use of patient information form all available and relevent sources."*[50]

The Office of the National Coordinator published their strategies for accelerating HIE in 2016. Many were based on input received in 2013 from an RFI about how to advance HIE. Some of there initiatives were as follows:
- ONC will work with HHS to ensure that all new policies support HIE at multiple levels
- HHS will support HIT standards for HIE and coordinate with all relevant agencies
- Voluntary certification of HIE products
- HHS to include HIE as a quality measure as part of newer value-based reimbursement
- CMS's Health Care Innovation Awards will make grants available for new service models that include HIE
- Develop open source tool kits for HIE to support secure messaging and ADT alerts[51]

HEALTH INFORMATION EXCHANGE CONCERNS

There are multiple concerns surrounding the creation and sustainment of a health information organization. The following are just few of the reported concerns:
- Each HIO has a different business model. Is there enough data to know which model is preferred?
- It is unclear how HIOs will be funded long term. Will funding come from insurers? Clinicians? Employers? Consumers? Federal or state government?
- Approximately $550 million from the HITECH ACT went towards statewide HIE. Has enough been learned at this point to decrease the failure rate?
- Most medical information is free-text or unstructured. How is the data computable and analyzable?
- Will universal standards be adopted or will different standards for different HIOs prevail?
- Poor cities, states and regions tend to be at a disadvantage. What should be done with geographical gaps in HIOs and what regions should they cover? Should they be based on geography, insurance coverage or prior history?
- Will nationwide exchange of health information be possible with a low number of sustainable HIOs fail and incomplete adoption of EHRs?
- What are the incentives for competing hospitals and competing physicians in the average city or region to collaborate and share information?[52]
- Will HIOs have to comply with FISMA regulations?
- How can the price of interfaces be reduced? It is very expensive to create interfaces that connect each EHR to the HIO. There are interface engines that make this process more affordable. A list of common interface engines is available at this reference.[4]
- Will the newest HIPAA regulations (or state personal health information-related laws) become impediments to HIO implementation and operation?
- Opt-in and opt-out patient consent models vary by locality, region, and state. Will one model become standard?
- How to solve the patient matching and identity problem?
- Is there a strong reason to accredit HIOs?
- How will patient privacy and security rules under Meaningful Use come into play in the HIO domain?
- Very little research has been done to identify which physician specialties are the most frequent requestors of patient data from HIOs. Similarly, little is known about which clinical situations benefit the most from data exchange. This suggests that providers may not value HIE. In the future, will clinicians be comfortable making care decisions based on discrete data elements imported from an external record source?
- Will timely access to patient documentation be realized in the face of technical and procedural hurdles?
- Will physician adoption of the Direct Project standards, to meet Meaningful Use paradoxically decrease adoption of the more formal pull model?
- How can payers be more consistently involved in support of HIOs? Will providers trust an HIO that is sponsored by or involves payers?
- When will there be more quantitative and qualitative studies to document value and return on investment?
- Will Accountable Care Organizations (ACOs) increase or decrease HIO use?

- Is the current HIO model too complex for success, compared to other models of HIE?

A 2016 systematic review by Eden et al. reported on the barriers and facilitators related to HIE and included studies performed outside the United States. They noted that the most common barriers were incomplete information, inefficient workflow and information that did not meet the needs of the users. Facilitators included receiving more complete information, intelligent planning, training, implementation and workflow, and involving users in identifying key HIE functions. They reported that sites that used a proxy person to pull data experienced higher HIE usage. The evidence was inadequate to compare HIE barriers based on architecture (centralized versus federated) 0r push versus pull HIE.[53]

NEWER HIT DEVELOPMENTS TO PROMOTE HIE

Given the substantial failures experienced in HIE at the federal and state levels there have been several interesting alternatives that have been proposed in the past few years.

FHIR. In 2011 Health Level 7 (HL7) developed the data standard known as Fast Healthcare Interoperability Resources (FHIR) largely to encourage healthcare interoperability. HL7 messaging has been used in healthcare for a long time but the most recent Version 3 was found to be entirely too complex and inconsistent. FHIR is based on resources such as "patient" and "documents." Resources are located using the Representational State Transfer (REST) system that uses the HTTP protocol, URLs and uses XML or JSON for resource representations. This approach is agile or light weight and is now used extensively in the World Wide Web for transactions. In this manner, data elements of a document, e.g. PSA result could be requested without the entire lab report or a complete patient summary.[54] The Argonaut Project is the private sector initiative that tries to advance FHIR-based APIs and interoperability standards.[55] More details are available in the chapter on data standards.

Blockchain. A blockchain is a *"distributed transaction ledger."* Each block is a set of transactions and each block has a numeric accounting of its contents (hash) and the hash is dependent of the hash of the block before it. The blockchain is not dependent on a central authority, but instead it is distributed to all nodes in the network. Several techniques in the blockchain validate the blocks that are newly added. Blockchains were first used to handle currency (Bitcoins) but some now believe this approach can be applied to the healthcare domain, as a patient's record is a sequence of events, like financial transactions. Transactions don't include the actual documents, but instead reference FHIR Resources located via Uniform Resource Locators (URLs). These complex steps provide security to the network. Patients could potentially benefit from the blockchain strategy because all that they would have to do is provide access to the specialist to the blockchain. Clinicians and organizations would by definition have access to all records in the blockchain. Technical details are available in this reference.[56] There is significant interest in healthcare today with this new approach, as it might also assist the supply chain, claims adjudication and clinical trials, to mention a few.[57] ONC had enough interest in this approach to launch a Blockchain Challenge in 2016 and as a result received about 70 submissions.[58]

OpenHIE. OpenHIE is a 2013 open-source initiative aimed at the underserved populations of middle to low income countries. It uses a service oriented architecture (SOA) platform to conduct HIE and consists of three layers (component, interoperability and external systems layers). The interoperability layer receives communications from the external services and processes messages among these systems and the component layer. Figure 6.3 displays the architecture. TS = terminology service maintains terminologies standards, such as ICD-10 and SNOMED-CT. CR = client registry is the master patient index. SHR = shared health record or patient-centric records. HMIS = health management information system stores healthcare data for analysis. FR = facility registry that maintains IDs of providers.[59]

EHR Vendor-Based HIE. In regions where there are well defined healthcare networks you may find vendor based HIE. For example, Epic Systems EHR has an interoperability product, known as Care Everywhere. This platform can exchange the consolidated clinical document architecture (C-CDA) between Epic and other non-Epic EHRs, as required by Stage 2 Meaningful Use. As of 2014, Care Everywhere was exchanging about 4.6 million C-CDA documents monthly. They can exchange with other EHRs, HIOs, HISPs and the eHealth Exchange.[60] For additional information regarding vendor-based and enterprise-based HIE, we recommend the article by Everson in the recommended reading section.

HEALTH INFORMATION EXCHANGE RESOURCES

It can be argued that creating the technology architecture is the easy part in the life of a HIO. Far more time must be spent planning the governance and financing.

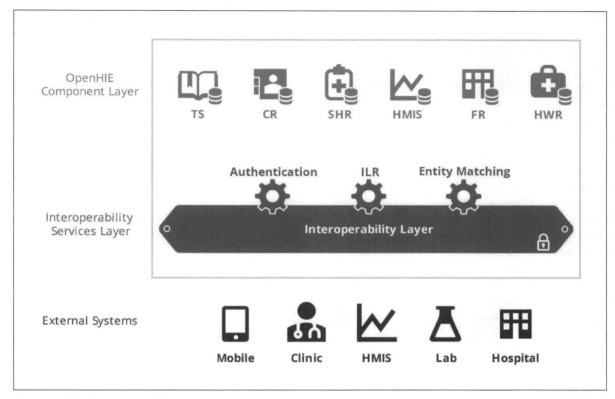

Figure 6.3: OpenHIE architecture

It is therefore critical that localities do their homework to research the lessons learned from others who have successfully built a HIO. The following are valuable resources:

- *Health Information Exchange. Navigating and Managing a Network of Health Information Systems.* This is a contemporary (2016) textbook that covers all aspects of HIE. It also includes five case studies of successful HIOs.4 First Edition. Editor Brian Dixon. 2016. Academic Press. ISBN: 9780128031353. This is a definitive work on HIE and HIOs.
- *ONC Tech Lab* (previously known as the S&I Framework). This ONC site lists HIE pilot projects, standards coordination, testing and utilities and innovation. The overriding goal is to create an environment for consensus building for interoperability and other HIT-related issues.[61]
- *Working Group for the Electronic Exchange of Data (WEDI)* is a non-profit organization consisting of public and private HIE experts who serve as advisors for Health and Human Services. They have multiple working groups and publish information briefs.[62]
- *Integrating the Health Care Enterprise (IHE)* sole purpose is to promote interoperability in healthcare. IHE is a non-profit organization that has members (stakeholders) from a variety of experts in the field. IHE develops IHE profiles that can be used for multiple use cases and have been used by multiple EHR vendors. On their website they list the 11 domains they are currently exploring and developing profiles. For example, the Cardiology domain has 7 committees; one is the cath committee that has the goal of integrating the workflow of this procedure. The profiles are tested and demonstrated at the Connectathon. After the appropriate testing IHE releases the Integration Statements documenting the process supports IHE profiles.[63]
- Commonwell Health Alliance is a non-profit vendor organization that uses existing standards and policies to enable healthcare interoperability. Their service provider RelayHealth uses widely adopted IHE standards and they were among the first to use the FHIR standard for patient identification. Other services include patient enrollment, record location, patient identification and linking to records and data query using FHIR or XCA.[64]
- Chapter 2: Health Information Exchange: Community HIE Efforts in Health Information Technology in the United States, 2015: Transition to a Post-HITECH World. This monograph provides more survey summary data to give readers a good view of current HIE in the US. Robert Wood Johnson Foundation[65]

RECOMMENDED READING

The following articles summarize newer trends and knowledge related to health information exchange:
- *Outcomes from health information exchange: systematic review and future research needs.* Hersh et al provides an overview of the evidence base for value of HIE.[66]
- *Barriers and facilitators to exchanging health information: a systematic review.* This 2016 review reported that the most commonly cited barriers were incomplete information, inefficient workflow and the situation where the exchanged information did not meet the needs of the requesting individual.[53]
- *The implications and impact of 3 approaches to health information exchange: community, enterprise and vendor-mediated health information exchange.* Everson discusses a newer taxonomy of HIE as well as perceived benefits of one model over another. He reinforces the fact that little is written about the newer models and more research is needed.[67]

FUTURE TRENDS

While the success of HIOs continues to be uncertain even with extensive HITECH Act funding, several trends are appearing from the more mature and successful HIOs. First, many facilitated the achievement of Meaningful Use by providing HIE to include quality reporting and other advanced functionalities. Second, clinical messaging is being combined with administrative and financial data to give users more of a dashboard experience, where multiple data sources are aggregated to expose seemingly disparate functions on one web page. It seems likely that eventually integration of EHRs, practice management systems and claims management as core HIO services will occur. This would offer a single platform to conduct all clinical and financial business and the ability to generate a wide range of reports. Third, more efforts to use data secondarily for research and as a means of financially supporting HIOs can be expected. Fourth, data analytics will likely evolve if the need is perceived and the value proven. Fifth, more mergers of HIE vendors and new vendors appearing can be anticipated if accountable care organizations and MACRA continue mandated sharing of health information. Sixth, more interoperability can be expected in the future between electronic health records, home telemedicine monitors and any other devices that generate medical data that should be collated and analyzed into one location for clinician review. Seventh, newer means to identify patients accurately will likely appear and reduce the possibility of duplication of records.

Adler-Milstein and Dixon believe that care coordination, patient-centered care, population health management and a true learning health system will be future drivers of HIE.[68] This assumes that there is either mandates from federal and state governments and a plausible business case for long term sustainability.

KEY POINTS

- Health information exchange is critical for healthcare reform
- Creating the architecture for a Health Information Organization (HIO) is not difficult; developing the long-term business plan is
- Direct is a simple approach to accomplishing HIE
- FHIR and Blockchain are two new potential technologies for HIE

CONCLUSION

Sharing of health-related data is a critical element of healthcare reform and in the United States. Health information exchange among disparate partners is becoming more common in the United States due to evolving HIOs and Meaningful Use objectives. Federal programs support the creation of exchanges as well as the services, standards and policies that make HIE possible. HIOs are proliferating, largely due to government support but they are often impeded by a lack of a sustainable business model, as well as privacy and security issues. The federal government has privatized the Nationwide Health Information Network in an effort to accelerate standards creation and adoption by private sector stakeholders. Similarly, integrated delivery networks are offering health information exchange as a marketing strategy and so they can participate in new healthcare reform delivery models. It is too early to know what a HIO of the future will look like, but it seems clear that more features and better integration can be expected.

REFERENCES

1. Washington, V, DeSalvo, K, et al. (2017). The HITECH era and the path forward. *New England Journal of Medicine*. 377: 904-906.
2. Bourgeois, FC, Olson, KL, et al. (2010). Patients treated at multiple acute health care facilities: quantifying information fragmentation. *Archives of Internal Medicine*. 170: 1989-1995.
3. Finnell, JT, Overhage, JM, et al. (2011). All health care is not local: an evaluation of the distribution of emergency department care delivered in Indiana. *AMIA Annual Symposium Proceedings*, Washington, DC. 409-416.
4. Dixon, B (Ed) (2016). *Health Information Exchange - Navigating and Managing a Network of Health Information Systems*. Amsterdam, Netherlands, Elsevier.
5. Williams, C, Mostashari, F, et al. (2012). From the Office of the National Coordinator: the strategy for advancing the exchange of health information. *Health Affairs*. 31: 527-536.
6. Patel, V, Henry, J, et al. (2016). Interoperability among U.S. Non-federal Acute Care Hospitals in 2015. Washington, DC, Department of Health and Human Services. https://www.healthit.gov/sites/default/files/briefs/onc_data_brief_36_interoperability.pdf
7. Commonwealth Fund. Perspectives on Health Reform. http://www.commonwealthfund.org/publications/perspectives-on-health-reform-briefs/2009/jan/the-federal-role-in-promoting-health-information-technology (Accessed June 27, 2017)
8. Schoen C, Osborn R, Squires D et al. A Survey of Primary Care Doctors in Ten Countries Shows Progress in Use of Health Information Technology, Less in Other Areas. Health Affairs 2012;31(12):1-12
9. Defining Key Health Information Terms. National Alliance for Health Information Technology www.nahit.org No longer active. (Accessed May 21, 2008)
10. Gartner, Summary of the NHIN Prototype Architecture Contracts, A Report for the Office of the National Coordinator for Health IT, May 31, 2007, page 9. http://www.healthit.gov/sites/default/files/summary-report-on-nhin-prototype-architectures-1.pdf (Accessed September 11, 2013)
11. (ONC). Report on Health Information Blocking. 2015;1–39. www.healthit.gov (Accessed June 26, 2017)
12. Adler-Milstein, J and Pfeifer, E (2017). Information blocking: is it occurring and what policy strategies can address it? *Milbank Quarterly*. 95: 117-135.
13. IEEE Standard Computer Dictionary: A Compilation of IEEE Standard Computer Glossaries (New York, NY. 1990)
14. Adler-Milstein J, Ronchi E, Cohen GR, Winn LAP, Jha AK. Benchmarking health IT among OECD countries: better data for better policy. J Am Med Informatics Assoc. 2014;21(1):111–6.
15. Healthcare Information and Management Systems Society Healthcare IT News. 2015. [2015-10-07]. http://www.healthcareitnews.com/news/hie-verb-onc-wants-move-quickly-data-exchange. (Accessed October 20, 2017)
16. Mikk, KA, Sleeper, HA, et al. (2017). The pathway to patient data ownership and better health. *Journal of the American Medical Association*. 318: 1433-1434.
17. Soper P. Realizing the potential of community health information networks for improving quality and efficiency through the continuum of care: a case study of the HRSA community access program and the Nebraska Panhandle partnership for HHS. WHP 023A. December 2001. http://www.stchome.com/media/white_papers/WHP023A.pdf (Accessed September 15, 2011)
18. Executive Order: Incentives for the Use of Health Information Technology and Establishing the Position of the National Health Information Technology Coordinator http://www.whitehouse.gov/news/releases/2004/04/20040427-4.html (Accessed September 11, 2013)
19. The Sequoia Project. www.sequoiaproject.org (Accessed September 15, 2017)
20. Surescripts. www.surescripts.com Accessed (September 11, 2017)
21. Direct Project makes secure messaging available to patients | Health Data Management https://www.healthdatamanagement.com/news/direct-project-makes-secure-messaging-available-to-patients (Accessed June 22, 2017)
22. Office-based Physician Electronic Patient Engagement Capabilities. https://dashboard.healthit.gov/quickstats/pages/physicians-view-download-transmit-secure-messaging-patient-engagement.php (Accessed June 22, 2017)
23. Blue Button Project www.healthit.gov/bluebutton (Accessed June 22, 2017)
24. Blue Button Plus http://bluebuttonplus.org (Accessed June 22, 2017)
25. Kuperman GJ. Health-information exchange: why are we doing it, and what are we doing? J Am Med Inform Assoc 2011;18:678-682.
26. HIE and Meaningful Use state 2 matrix. http://www.himss.org/files/HIMSSorg/content/files/MU2_HIE_Matrix_FINAL.pdf (Accessed September 12, 2013)

27. Stage 3 Program Requirements for Providers Attesting to their States Medicaid EHR Incentive Program. https://www.cms.gov/Regulations-and-Guidance/Legislation/EHRIncentivePrograms/Stage3Medicaid_Require.html (Accessed September 11, 2017)
28. Office of the National Coordinator for Health Information Technology http://www.healthit.gov (Accessed September 12, 2013)
29. Just B, Durkin S. Clinical Data Exchange Models: Matching HIE Goals with IT Foundations. Journal of AHIMA. 2008;79 (2):48-52
30. Sundwall D. RHIO in Utah, UHIN. HIMSS Conference Presentation June 6, 2005
31. Redwood MedNet http://www.redwoodmed net.org/projects/hie/pricing.html (Accessed May 21, 2013)
32. The Health Collaborative. www.healthcollab.org (Accessed November 20, 2017)
33. Delaware Health Information Network. https://dhin.org (Accessed November 20, 2017)
34. Inland Health Services. https://www.inhs.info (Accessed November 20, 2017)
35. Strategic Health Intelligence. http://strategichealthintelligence.com/health-information-exchange/ (Accessed November 20, 2017)
36. Shapiro JS, Kannry J, KushniruK W, Kuperman G. Emergency Physicians Perceptions of Health Information Exchange. JAIMA 2007;14:700-705
37. Wright A, Soran C, Jenter CA et al. Physician attitudes toward HIIE: Results of a statewide survey. JAMIA 2010;17:66-70 doi 10.1197/JAMIA.M5241
38. Ehealth Initiative. HIE Survey. 2013. http://www.ehealthinitiative.org (Accessed September 12, 2013) (Archived)
39. HIE Challenge Grant Program https://www.healthit.gov/providers-professionals/health-information-exchange-challenge-grant-program (Accessed November 20, 2017)
40. Indiana Health Information Exchange. www.ihie.org (Accessed November 21, 2017)
41. Utah Health Information Network http://www.uhin.com/ (Accessed September 12, 2013)
42. Nebraska Health Information Initiative www.nehii.org (Accessed September 12, 2013)
43. Maine Health Information Network. http://www.hinfonet.org/ (Accessed September 12, 2013)
44. Furukawa MF, Patel V, Charles D et al. Hospital Electronic Health Information Exchange Grew Substantially in 2008-2012. Health Affairs 2013;32:1346-1354
45. Adler-Milstein J, Bates DW, Jha AK. Operational Health Information Exchanges Show Substantial Growth, But Long-Term Funding Remains a Concern. Health Affairs;32:1486-1492
46. Update on the Adoption of Health Information Technology and Related Efforts to Facilitate the Electronic Use and Exchange of Health Information. June 2013. A Report to Congress. http://www.healthit.gov/sites/default/files/rtc_adoption_of_healthit_and_relatedefforts.pdf (Accessed July 3, 2013)
47. Alder-Milstein, J, Pfeifer E. Information Blocking: Is It Occurring and What Policy Strategies Can Address It? Milbank Q 2017 Mar;95(1):117-135
48. Rudin RS, Motala A, Goldzweig CL, Shekelle PG. Usage and effect of health information exchange: A systematic review. Ann Intern Med. 2014;161(11):803–11.
49. Rahurkar S, Vest JR, Menachemi N. Despite the spread of health information exchange, there is little evidence of its impact on cost, use, and quality of care. Health Affairs. 2015;34(3):477-483
50. ONC. Connecting Health and Care for the Nation A Shared Nationwide. 2014;1–25. Available from: http://www.healthit.gov/sites/default/files/nationwide-interoperability-roadmap-draft-version-1.0.pdf (Accessed September 7, 2017)
51. ONC. Principles and Strategy for Accelerating Health Information Exchange (HIE). New Engl J Med Heal Serv Res 2016 https://www.cms.gov/ehealth/downloads/Accelerating_HIE_Principles.pdf (Accessed September 7, 2017)
52. Grossman JM, Bodenheimer TS, McKenzie K. Hospital-Physician Portals: The Role of Competition in Driving Clinical Data Exchange. Health Affairs 2006;25(6):1629-1636
53. Eden KB, Totten AM, Kassakian SZ, Gorman PN, McDonagh MS, Devine B, et al. Barriers and facilitators to exchanging health information: a systematic review. Int J Med Inform. NIH Public Access; 2016 Apr;88:44–51. http://www.ncbi.nlm.nih.gov/pubmed/26878761 (Accessed November 20, 2017)
54. Bender D, Sartipi K. HL7 FHIR: An agile and RESTful approach to healthcare information exchange. In: Proceedings of CBMS 2013 - 26th IEEE International Symposium on Computer-Based Medical Systems. 2013. p. 326–31.
55. The Argonaut Project. https://hl7.org/implement/standards/fhir/2015Jan/argonauts.html (Accessed November 20, 2017)
56. Peterson K, Deeduvanu R, Kanjamala P, Boles K. A Blockchain-Based Approach to Health Information Exchange Networks. NIST Work Blockchain Healthc. 2016;(1):1–10.
57. Blockchain's potential use cases for healthcare: hype or reality? | Healthcare IT News http://www.healthcareitnews.com/news/blockchains-potential-use-cases-healthcare-hype-or-reality (Accessed November 20, 2017)

58. Announcing the Blockchain Challenge | Newsroom | HealthIT.gov https://www.healthit.gov/newsroom/blockchain-challenge (Accessed June 23, 2017)
59. OpenHIE. https://wiki.ohie.org/display/documents/OpenHIE+Architecture (Accessed June 23, 2017)
60. Epic Care Everywhere. https://www.epic.com/careeverywhere/ (Accessed June 23, 2017)
61. S & I Framework www.siframework.org (Accessed June 20, 2017)
62. Working Group for the Electronic Exchange of Data (WEDI). http://www.wedi.org/ (Accessed June 23, 2017)
63. Integrating the Healthcare Enterprise https://www.ihe.net/ (Accessed June 24, 2017)
64. Commonwell Health Alliance. www.commonwellalliance.org (Accessed June 24, 2017)
65. Chapter 2: Health Information Exchange: Community HIE Efforts in Health Information Technology in the United States, 2015: Transition to a Post-HITECH World. https://www.rwjf.org/en/library/research/2015/09/health-information-technology-in-the-united-states-2015.html (Accessed June 25, 2017)
66. Hersh, WR, Totten, AM, et al. (2015). Outcomes from health information exchange: systematic review and future research needs. *JMIR Medical Informatics*. 3(4): e39. http://medinform.jmir.org/2015/4/e39/ (Accessed June 23, 2017)
67. Everson J. The implications and impact of 3 approaches to health information exchange: community, enterprise, and vendor-mediated health information exchange. Learn Heal Syst 2017;(December 2016):e10021. http://doi.wiley.com/10.1002/lrh2.10021 (Accessed June 23, 2017)
68. Adler-Milstein J, Dixon, B. Future Directions in HIE. Chapter 16 in Health Information Exchange. Dixon B, Editor. Academic Press. 2016.

7

Healthcare Data Analytics

WILLIAM R. HERSH

LEARNING OBJECTIVES

After reading this chapter the reader should be able to:
- Discuss the difference between descriptive, predictive and prescriptive analytics
- Describe the characteristics of "Big Data"
- Enumerate the necessary skills for a worker in the data analytics field
- List the limitations of healthcare data analytics
- Discuss the critical role electronic health records play in healthcare data analytics

INTRODUCTION

One of the promises of the growing critical mass of clinical data accumulating in electronic health record (EHR) systems is secondary use (or re-use) of the data for other purposes, such as quality improvement and clinical research.[1] The growth of such data has increased dramatically in recent years due to incentives for EHR adoption in the US funded by the Health Information Technology for Economic and Clinical Health (HITECH) Act.[2] In the meantime, there has also seen substantial growth in other kinds of health-related data, most notably through efforts to sequence genomes and other biological structures and functions[3] as well as patient-generated sensor and other data.[4] The analysis of this data is usually called *analytics* (or *data analytics*). This chapter will define the terminology of this field, provide an overview of its promise, describe what work has been accomplished, and list the challenges and opportunities going forward.

TERMINOLOGY OF ANALYTICS

The terminology surrounding the use of large and varied types of data in healthcare is evolving, but the term analytics is achieving wide use both in and out of healthcare. A long-time leader in the field defined analytics as "*the extensive use of data, statistical and quantitative analysis, explanatory and predictive models, and fact-based management to drive decisions and actions.*"[5]

Similarly, IBM defined analytics as "*the systematic use of data and related business insights developed through applied analytical disciplines (e.g. statistical, contextual, quantitative, predictive, cognitive, other [including emerging] models) to drive fact-based decision making for planning, management, measurement and learning.*"[6]

Adams and Klein define three levels and their attributes of the application of analytics[7]:
- Descriptive – standard types of reporting that describe current situations and problems
- Predictive – simulation and modeling techniques that identify trends and portend outcomes of actions taken
- Prescriptive – optimizing clinical, financial, and other outcomes

Much work is focusing now on predictive analytics, especially in clinical settings attempting to optimize health and financial outcomes.

There are several terms related to data analytics. A core methodology in data analytics is *machine learning*, which is the area of computer science that aims to build systems and algorithms that learn from data and will be described in more detail below.[8-9] Related to machine learning is *data mining*, which is defined as the processing and modeling of large amounts of data to discover previously unknown patterns or relationships.[10] A subarea of data mining is *text mining*, which applies data mining techniques to mostly unstructured textual data.[11] Another close but more recent term in the vernacular is *Big Data*, which describes large and ever-increasing volumes of data that adhere to the following attributes:[12]

- Volume – ever-increasing amounts
- Velocity – quickly generated
- Variety – many different types
- Veracity – from trustable sources

Another term is *data science*, defined by Donoho as *"the science of learning from data; it studies the methods involved in the analysis and processing of data and proposes technology to improve methods in an evidence-based manner."*[13] He further states that a data scientist is *"a person who is better at statistics than any software engineer and better at software engineering than any statistician."*

With the increasing digitization of clinical data, hospitals and other healthcare organizations are generating an ever-increasing amount of data. In all healthcare organizations, clinical data takes a variety of forms, from structured (e.g., images, lab results, etc.) to unstructured (e.g., textual notes including clinical narratives, reports, and other types of documents). For example, Kaiser-Permanente estimated in 2013 that its current data store for its 9+ million members exceeded 30 petabytes of data.[14] Other organizations are planning for a data-intensive future. For example, the American Society for Clinical Oncology (ASCO) has been developing its Cancer Learning Intelligence Network for Quality (CancerLinQ).[15] CancerLinQ provides a comprehensive system for clinicians and researchers consisting of EHR data collection, application of clinical decision support, data mining and visualization, and quality feedback.

Another source of large amounts of data is the world's growing base of scientific literature and its underlying data that is increasingly published with journal and other articles (see Chapter on medical information retrieval). One approach to this problem that has generated attention is the IBM Watson project, which started as a generic question-answering system that was made famous by winning at the TV game show Jeopardy![16] IBM has also focused Watson in the healthcare domain[17], although it has been criticized for a paucity of scientific evaluation of its capabilities.[18]

Kumar et al. have noted that the process of data analytics resembles a pipeline and have developed an approach that specifies four major steps in this pipeline, to which data sources and actions on it pertinent to healthcare and biomedicine can be placed.[19] The pipeline begins with input data sources, which in healthcare and biomedicine may include clinical records, financial records, genomics and related data, and other types, even those from outside the healthcare setting (e.g., census data). The next step is feature extraction, where various computational techniques are used to organize and extract elements of the data, such as linking records across sources, using natural language processing (NLP) to extract and normalize concepts, and matching of other patterns. This is followed by statistical processing, where machine learning and related statistical inference techniques are used to make conclusions from the data. The final step is the output of predictions, often with probabilistic measures of confidence in the results. (see Figure 7.1)

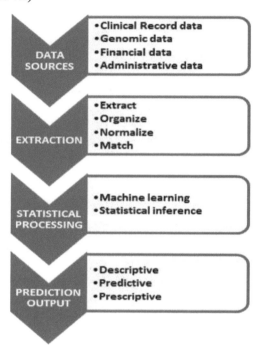

Figure 7.1: From data sources to prediction

The growing quantity of data requires that its users have a good understanding of its *provenance*, which is where the data originated and how trustworthy it is for large-scale processing and analysis.[20] A more peripheral but related term is *business intelligence*, which in healthcare refers to the *"processes and technologies used to obtain timely, valuable insights into business and clinical data."*[7] Another relevant term relevant is the notion promoted by the National Academy of Medicine (previously known as the Institute of Medicine) of the *learning health system*.[21-22] Advocates of this approach note that routinely collected data can be used for continuous learning to allow the healthcare system to better carry out disease surveillance and response, targeting of healthcare services, improving decision-making, managing misinformation, reducing harm, avoiding costly errors, and advancing clinical research.[23]

Another term comes from the call for new and more data-intensive approaches to diagnosis and treatment of disease called *precision medicine*.[24] Advocates for this approach note the inherent complexity of nonlinear systems in biomedicine, with large amounts and varied types of data that need models to enable their predictive value. Technology thought leader O'Reilly also notes that data science is transforming medicine, striving to solve its equivalent of the *Wanamaker dilemma* for advertisers, named after the problem of knowing that half of advertising by merchants does not work, but that the half that does not work is not known.[25]

One of the major motivators for data analytics comes from new models of healthcare delivery, with a focus on value-based care, where reimbursement includes incentives to deliver high-quality care in cost-efficient ways.[26] Healthcare organizations that participate will require a focused IT infrastructure that provides data that can be used to predict and quickly act on excess costs. One of the challenges for healthcare data is that patients often get their care and testing in different settings (e.g., a patient seen in a physician office, sent to a free-standing laboratory or radiology center, and also seen in the offices of specialists or being hospitalized. This has increased the need for development of *health information exchange* (HIE), where data is shared among entities caring for a patient across business boundaries [27]. A well-known informatics blogger has succinctly noted that "ACO = HIE + analytics."[28]

Machine Learning

An essential aspect of data analytics, especially when dealing with large, complex data sets, is machine learning.[8-9] There are many definitions of machine learning, but it is commonly described as the field of computer science that studies how computers learn from data. Machine learning is derived from mathematics and statistics - where relationships are learned from data, and from computer science - where there is emphasis on efficient algorithms, especially those involving large amounts of data.

There are two main categories of machine learning algorithms. In *supervised learning*, algorithms learn to predict a known output that they learn from training data and have their effectiveness evaluated on test data. There is use of training data and test data to avoid over fitting the training data.[29] In *unsupervised learning*, the aim is to find naturally occurring patterns or groupings within the data.

A common application of supervised learning has the aim of predicting a clinical diagnosis from a training data set. For example, an algorithm might process the waves of an electrocardiogram in an effort to interpret them with the goal of learning to make a diagnosis. Another relatively common supervised learning task is the detection of abnormalities on a chest x-ray.

A more specific example of supervised learning is the estimation of risk that comes from large data sets such as the Framingham Heart Study.[30] The risk score from the Framingham study aims to predict the risk of coronary heart disease for an individual. As new types of data become available, the risk score models can be extended to include these new types of data, such as variants in the human genome.

A common use of unsupervised learning is the discovery of new attributes that may be associated with clinical questions such as diagnosis, treatment, or prognosis of disease. An example of unsupervised learning is the genome-wide association study (GWAS), where the goal is to identify genome variants that might be associated with a particular disease.[31] These types of investigations form the basis of precision medicine, where the goal is to identify causes and factors associated with a disease more precisely and develop more targeted treatments.

There are several different tasks for which machine learning is used. The major tasks and their primary uses include:

- Classification – predict class from one or more elements of data
- Regression – predict numerical value from data
- Clustering – group items together
- Density estimation – find statistical values
- Dimensionality reduction – reduce many to few features

There are several approaches used in machine learning for classification. There is a great deal of mathematics that underlie all of them, but they can be understood by looking at what they attempt to do overall. Some of the most common algorithms include:

- k-Nearest Neighbors (kNN) – aim to find category having "closest" number of attributes
- Naïve Bayes – derive conditional probabilities that classify into categories based on Bayes' Theorem
- Support vector machines (SVMs) – for binary classification, draw a "line" that separates one category from others
- Decision trees – develop set of rules that classify into categories

Another approach gaining widespread use in supervised learning are neural networks. When these networks build complex representations from simpler ones, they are referred to as deep learning.[32] A number of applications of deep learning have achieved successful results in making diagnoses and predicting outcomes:

- Diagnosing skin lesions – keratinocyte carcinomas vs. benign seborrheic keratoses and malignant melanomas vs. benign nevi[33]
- Classifying metastatic breast cancer on pathology slide images[34]
- Predicting longevity from CT imaging[35]
- Predicting cardiovascular risk factors from retinal fundus photographs[36]
- Detecting arrhythmias comparable to cardiologists[37]

There are also different approaches to regression. Linear regression is a common approach of predicting a numerical value from a set of data. There are various techniques that improve upon the basic algorithm of linear regression, such as locally weighted, ridge, and lasso techniques. Another type of regression is logistic regression, where the aim is to classify data into binary categories.

The typical steps taken in a machine learning application include:
- Collect data
- Prepare input data – "wrangle" to get in usable format and clean
- Analyze input data
- (If supervised learning) Train algorithm
- Test algorithm
- Use algorithm

CHALLENGES TO DATA ANALYTICS

There are, of course, challenges to data analytics. One concern is that data generated in the routine care of patients may be limited in its use for analytical purposes.[38] For example, such data may be inaccurate or incomplete. It may be transformed in ways that undermine its meaning (e.g., coding for billing priorities). It may exhibit the well-known statistical phenomenon of *censoring*, i.e., the first instance of disease in record may not be when it was first manifested (left censoring) or the data source may not cover a sufficiently long-time interval (right censoring). Data may also incompletely adhere to well-known standards, which makes combining it from different sources more difficult.

There are several other challenges for using clinical data for analytics. Data quality and accuracy is often not a top priority for busy clinicians.[39] Patients receive care at different locations and the data from such care might not be readily available.[40-41] A good deal of clinical data is "locked" in text.[42] Even structured data not usable purely automated for clinical score calculation.[43] Experience has shown that electronic records of patients at academic medical centers are not easy to combine for aggregation.[44] There are also substantial quantities of data, e.g., the average pediatric ICU patient generates 1348 information items per 24 hours.[45]

Others have noted larger challenges around analytics and big data. Boyd and Crawford have expressed some "*provocations*" for the growing use of data-driven research.[46] They note that research questions asked of the data tend to be driven by what can be answered, as opposed to prospective hypotheses. They also note that data are not always as objective as we might like, and that "*bigger*" is not necessarily better. Finally, they raise ethical concerns over how the data of individuals is used, how it is collected, and the possible divide between those who have access to data and those who do not. Similar concerns focused specifically on healthcare data by Neff, who describes a myriad of technical, financial, and ethical issues that must be addressed before making use of big data routinely for clinical practice and other health-related purposes.[47] These challenges also create ethical issues, such as who owns data and who has privileges to use it.[48] Another ethical concern is "*biased algorithms*" that result from explicit or implicit assumptions about how aspects of the world are modeled or operated upon by algorithms.[49-50]

It has also been noted that for a health system to have robust data analytics capability, human and technical infrastructure is key.[51] Elements of the infrastructure include:
- Stakeholder engagement
- Human subjects research protection
- Protection of patient privacy
- Data assurance and quality
- Interoperability of health information systems
- Transparency
- Sustainability

Krumholz notes that new models of thinking and training are needed as well, not only for data analysts and informaticians but clinicians and administrators as well.[52] Other have noted that new tools are required, e.g., a "green button" to help clinicians aggregate data into a local EHR.[53]

RESEARCH AND APPLICATION OF ANALYTICS

The research base around applying analytics to improve healthcare delivery is still in its early stages. There is an emerging base of research that demonstrates how data from operational clinical systems can be used to identify critical situations or patients whose costs are outliers. There is less research, however, demonstrating how this data can be used to actually improve clinical outcomes or reduce costs.

Bates has hypothesized the most important use cases for application of data analytics:[54]
- High-cost patients – looking for ways to intervene early
- Readmissions – preventing
- Triage – appropriate level of care
- Decompensation – when patient's condition worsens
- Adverse events – awareness
- Treatment optimization – especially for diseases affecting multiple organ systems

Studies using EHR data for clinical prediction have been proliferating. One early area of focus was the use of data analytics to identify patients at risk for hospital readmission within 30 days of discharge. The importance of this factor came from the US Centers for Medicare and Medicaid Services (CMS) Readmissions Reduction Program that penalized hospitals for excessive numbers of readmissions.[55] This led to a number of research groups using EHR data attempting to predict patients at risk for readmission.[56-61]

Many other studies have attempted to identify cases or cohorts of patients:
- Identifying patients who might be eligible for participation in clinical studies[62-63]
- Identification of children with asthma[64]
- Improving on ICD-9 to identify patients with hepatocellular carcinoma[65]
- Automated phenotyping using EHR data and machine learning[66]

Other studies have applied analytics to predict healthcare utilization or outcomes:
- Predicting healthcare utilization 7-68 or primary care panel size[69]
- Risk-adjusting hospital mortality rates[70]
- Detecting postoperative complications[71-72]
- Measuring processes of care[73]
- Determining five-year life expectancy[74]
- Predicting risk of suicide and ability to do so better than clinicians[75]
- For cancer patients, detecting potential delays in cancer diagnosis[76] or predicting trajectory of disease using NLP[77]
- Severity of illness in ICU based on usual physiological models[78]
- Cardiovascular event risk prediction[79]

Several analyses have demonstrated ability to predict mortality with EHR data, such as:
- Risk-standardized mortality after acute myocardial infarction[80]
- Mortality in the intensive care unit, augmented with patient similarity data[78], during hospitalization[81], or one year after hospitalization[82]

Other researchers have also been able to use EHR data to replicate the results of randomized controlled trials (RCTs). One large-scale effort has come from the Health Maintenance Organization Research Network's Virtual Data Warehouse (VDW) Project.[83] Using the VDW, for example, researchers were able to demonstrate a link between childhood obesity and hyperglycemia in pregnancy.[84] Another demonstration of this ability has come from United Kingdom General Practice Research Database (UKGPRD), a repository of longitudinal records of general practitioners. Using this data, Tannen et al. were able to demonstrate the ability to replicate the findings of the Women's Health Initiative [85-86] and RCTs of other cardiovascular diseases.[87-88] Likewise, Danaei et al. were able to combine subject-matter expertise, complete data, and statistical methods emulating clinical trials to replicate RCTs demonstr-ating the value of statin drugs in primary prevention of coronary heart disease.[89]

These large repositories have been used for other research purposes. For example, the UKGPRD has been used for determining risk factors for pancreatic cancer[90] and gastroesophageal cancer.[91] Another large data repository in the US allowed replication of prospective cohort studies for risks of venous thromboembolic events in a manner much more efficient that historical retrospective analyses.[92] In addition, the Observational Medical Outcomes Partnership (OMOP) was to apply risk-identification methods to records from ten different large healthcare institutions in the US, although with a moderately high sensitivity vs. specificity tradeoff.[93] Finally, a case report demonstrated a situation where a clinical research database was queried to help make a decision whether to anticoagulate a child with systematic lupus erythematosus (SLE), a question for which no scientific literature existed to answer.[94]

Additional approaches use more novel methods. Denny and colleagues have developed methods for carrying out genome-wide association studies (GWAS) that associate specific findings from the EHR (the "phenotype") with the growing amount of genomic and related data (the "genotype") in the Electronic Medical Records and Genomics (eMERGE) Network.[95] eMERGE has demonstrated the ability to validate existing research results and generate new findings,[96] being able to identify genomic variants, among others, associated with atrioventricular conduction abnormalities, [97] red blood cell traits,[98] while blood cell count abnormalities,[99] and thyroid disorders.[100] More recent work has "inverted" the paradigm to carry out phenome-wide association studies (PheWAS) that associated multiple phenotypes with varying genotypes.[101-102]

Clearly a large and growing body of research demonstrates that EHR and other clinical data can be used to predict outcomes, including adverse ones, as well as diagnoses and eligibility for research studies. The next step in research is to find evidence that such methods lead to improved patient or healthcare system outcomes. There are unfortunately a smaller number of studies, and their results are mixed. One study showed that a readmission reduction tool applied to an existing case management approach helped reduce readmissions,[103] while another found that use of a Bayesian network model embedded in EHR to predict hospital-acquired pressure ulcers led to a tenfold reduction in such ulcers as well as a reduction by one-third in intensive care unit length of stay for such patients.[104] Another study found that a readmission risk tool intervention reduced risk of readmission for patients with congestive heart failure but not those with acute myocardial infarction or pneumonia.[105] An additional study found that an automated prediction model integrated into an existing EHR was successful in identifying patients on admission who were at risk for readmission within 30 days of discharge, but its use had no effect on 30-day all-cause and 7-day unplanned readmission rates in the 12-month period after it was implemented.[106]

Some newer studies include:
- Use of EHR-based acuity score allowed intervention that reduced in-hospital mortality from 1.9% to 1.3%[107]
- A randomized controlled trial of a tool to reduce delay in cancer diagnosis led to earlier diagnosis for colorectal and prostate cancer[108]
- Use of a predictive report based on a NLP tool reduced time in discharge planning meetings and 30-day all-cause mortality although not cost or readmissions[109]

One leader in application of data analytics has been Geisinger Health System, where the development and use of a universal data architecture has led to successes,[110] such as early detection and treatment of sepsis, monitoring and control of surgery costs and outcomes, and closing the loop on appropriate treatment and lack of follow-up.[111]

Another application has focused on children with cerebral palsy, where the implementation of a learning health system led to 43% reduced hospital days, 30% reduction in emergency department visits, and 210% reduction in healthcare costs. [112]

THE ROLE OF INFORMATICIANS IN ANALYTICS

Although much has been written extolling the virtues of analytics and big data, little of it focuses on the human experts who will carry out the work, to say nothing of those who will support their efforts in building systems to capture data, put it into usable form, and apply the results of analysis. Many of those who collect, analyze, use, and evaluate data will come from the workforce of biomedical and health informatics. To this end, we must ask questions about the job activity as well as the education of those who work in this emerging area that analytics thought leader Davenport calls the "*sexiest job of the 21st century.*"[113]

In the worlds of healthcare and biomedicine, the field poised to lead in data science is informatics. After all, informatics has led the charge in implementing systems that capture, analyze, and apply data across the biomedical spectrum from genomics to health care to public health.[114] From basic biomedical scientists to clinicians and public health workers, those who are researchers and practitioners are drowning in data, needing tools and techniques to allow its use in meaningful and actionable ways.

Data science is more than statistics or computer science applied in a specific subject domain. Dhar notes that a key aspect of data science, in particular what distinguishes it from statistics, is an understanding of data, its varying types, and how to manipulate and leverage it.[115] He points out that skills in machine learning are key, based upon a foundation of statistics (especially Bayesian), computer science (representation and manipulation of data), and knowledge of correlation and causation (modeling). Dhar also notes a challenge to organizational culture that might occur as organizations moved from "intuition-based" to "fact-based" decision-making.

It is also clear that there are two types of individuals working with analytics and big data. A report by the McKinsey consulting firm states a need in the US for 140,000-190,000 individuals who have "*deep analytical talent.*"[116] Furthermore, the report notes there will be need for an additional 1.5 million "*data-savvy managers needed to take full advantage of big data.*"[116] Analyses from the UK find similar results. An analysis by SAS estimated that by 2018, there will be over 6400 organizations that will hire 100 or more analytics staff.[117] Another report found that data scientists currently comprise less than 1% of all big data positions, with more common job roles consisting of developers (42% of advertised positions), architects (10%), analysts (8%) and administrators (6%).[118] It was also found that the technical skills most commonly required for big data positions were NoSQL, Oracle, Java and SQL. While these estimates are not limited to healthcare, they also do not include other countries that will have comparable needs to the US and the UK for such talent.

A report from IBM Global Services noted healthcare organizations are lagging behind in hiring individuals who are proficient in both "*numerate*" and business-oriented skills.[119] An additional report from IBM Global Services lists "expertise" among the critical attributes in organizations that are needed to complement technology. This expertise includes the supplementation of business knowledge with analytics knowledge, establishing formal career paths for analytics professionals, and tapping partners to supplement skills gaps that may exist.[120] Another US-based report by PriceWatersCoopers on health IT talent shortages noted that healthcare organizations wanting to keep ahead needed to acquire talent in Systems and data integration, data statistics and analytics, technology and architecture support, and clinical informatics.[121]

The US National Institutes of Health (NIH) also recognizes that big data skills will be important for conducting biomedical research. In 2013, NIH convened a workshop on enhancing training in big data among researchers.[122] Like the healthcare domain, participants called for skills in quantitative sciences, domain expertise, and ability to work in diverse teams. The workshop also noted a need for those working in big data to understand concepts of managing and sharing data. Trainees should also have access to real-world data problems and real-sized data sets to solve them. Longer-term training would be required for those becoming experts and leaders in data science.

What do biomedical and health informaticians working in analytics and big data need to know? An emerging consensus can be drawn from the reports above indicates that a combination of skills will be required:[123-124]

1. Programming - especially with data-oriented tools, such as SQL and statistical programming languages
2. Statistics - working knowledge to apply tools and techniques
3. Domain knowledge - depending on one's area of work, bioscience or health care
4. Communication - being able to understand needs of people and organizations and articulate results back to them

This to be relevant, informatics educational programs will need to introduce concepts of analytics, big data, and the underlying skills to use and apply them into their curricula. There will be a need for appropriate coursework for those who will become the "*deep analytical talent*" as well as higher breadth, perhaps with lesser depth, for the order of magnitude more individuals who will apply the results of big data analytics in healthcare and biomedical research.

PATH FORWARD FOR ANALYTICS

Clearly there is great promise ahead for healthcare driven by data analytics. The growing quantity of clinical and research data, along with methods to analyze and put it to use, can lead to improve personal health, healthcare delivery, and biomedical research. However, there is also a continued need to improve the completeness and quality of data as well as conduct research to demonstrate that it can be applied to solve real-world problems and how that is best done.

One recent analysis laid out recommendations for operational use of clinical data.[125] Although focused on comparative effectiveness research, the recommendations can be applied for almost any data analytics task. The authors called for:

- Adherence to best practices for use of data standards and interoperability
- Processes to evaluate availability, completeness, quality, and transformability of data
- Toolkits and pipelines to manage data and its attributes
- Challenges and metrics for assessing "research grade" of operational data
- Standardized reporting methods for operational data and its attributes
- Adaptation of "best evidence" approaches to use of operational data
- Appropriate use of informatics expertise to assist with optimal use of operational data and to develop published guidelines for doing so
- Research agenda to determine biases inherent in operational data and to assess informatics approaches to improve data

The "*best evidence*" approach is modeled on the framework of evidence-based medicine (EBM), applying the four basic steps of EBM to clinical data instead of scientific studies:

- Ask an answerable question – can question be answered by the data we have?
- Find the best evidence – in this case, the best evidence is the EHR data needed to answer the question
- Critically appraise the evidence – does the data answer the question? Are there confounders?
- Apply it to the patient situation – can the data be applied to this setting?

RECOMMENDED READING

A number of books have been published in recent years on the application of data analytics to healthcare:
1. Gensinger, RA, Ed. (2014). *Analytics in Healthcare: An Introduction*. Chicago, IL, Healthcare Information Management Systems Society.
2. Marconi, K and Lehmann, H, Eds. (2014). *Big Data and Health Analytics*. Boca Raton, FL, CRC Press.
3. Reddy, CK and Agarwal, CC, Eds. (2015). *Healthcare Data Analytics*. Boca Raton, FL, Chapman & Hall.
4. Natarajan, P, Frenzel, JC, et al. (2017). *Demystifying Big Data and Machine Learning for Healthcare*. Boca Raton, FL, CRC Press.

KEY POINTS

- Healthcare data has proliferated greatly, in large part due to the accelerated adoption of EHRs
- Data analytic platforms examine data from multiple sources, such as clinical records, genomic data, financial systems, and administrative systems
- Analytics is necessary to transform data to information and knowledge
- Accountable care organizations and other new models of healthcare delivery will rely heavily on analytics to analyze financial and clinical data
- There is a great demand for skilled data analysts in healthcare; expertise in informatics will be important for such individuals

CONCLUSION

There is a great deal of promise for data analytics to help health and health care in several ways. But we need other aspects of informatics for it to succeed. This robust EHRs and other clinical data sources, adherence to data standards and interoperability, HIE, and usability of clinical systems to make it easy for clinicians to enter good data. We also need improved completeness and quality of data. We need research demonstrating how data analytic techniques are best applied to improve health and outcomes. And finally, we need the human expertise to apply those tools and disseminate both their output, but also the understanding of what the data means.

REFERENCES

1. Safran C, Bloomrosen M, Hammond W, et al. Toward a national framework for the secondary use of health data: an American Medical Informatics Association white paper. *Journal of the American Medical Informatics Association*. 2007;14:1-9.
2. Washington V, DeSalvo K, Mostashari F, Blumenthal D. The HITECH era and the path forward. *New England Journal of Medicine*. 2017;377:904-906.
3. Tenenbaum J. Translational bioinformatics: past, present, and future. *Genomics Proteomics Bioinformatics*. 2016;14:31-41.
4. Li X, Dunn J, Salins D, et al. Digital health: tracking physiomes and activity using wearable biosensors reveals useful health-related information. *PLoS Biology*. 2017;15(1):e2001402.
5. Davenport T, Harris J. Competing on Analytics: The New Science of Winning. Cambridge, MA: Harvard Business School Press; 2007.
6. Anonymous. The value of analytics in healthcare - From insights to outcomes. Somers, NY: IBM Global Services;2012.
7. Adams J, Klein J. Business Intelligence and Analytics in Health Care - A Primer. Washington, DC: The Advisory Board Company; August 22, 2011 2011.
8. Harrington P. Machine Learning in Action. Greenwich, CT: Manning Publications; 2012.
9. Deo R. Machine learning in medicine. *Circulation*. 2015;132:1920-1930.
10. Bellazzi R, Zupan B. Predictive data mining in clinical medicine: current issues and guidelines. *International Journal of Medical Informatics*. 2008;77:81-97.
11. Cohen A, Hersh W. A survey of current work in biomedical text mining. *Briefings in Bioinformatics*. 2005;6:57-71.
12. Zikopoulos P, Eaton C, deRoos D, Deutsch T, Lapis G. Understanding Big Data: Analytics for Enterprise Class Hadoop and Streaming Data. New York, NY: McGraw-Hill; 2011.
13. Donoho D. 50 years of Data Science. Princeton NJ: Tukey Centennial Workshop; September 18, 2015 2015.

14. Gardner E. The HIT Approach to Big Data. *Health Data Management.* Vol 212013:4.
15. Sledge G, Miller R, Hauser R. CancerLinQ and the future of cancer care. *2013 ASCO Educational Book.* 2013:430-434.
16. Ferrucci D, Brown E, Chu-Carroll J, et al. Building Watson: an overview of the DeepQA Project. *AI Magazine.* 2010;31(3):59-79.
17. Ferrucci D, Levas A, Bagchi S, Gondek D, Mueller E. Watson: beyond Jeopardy! *Artificial Intelligence.* 2012;199-200:93-105.
18. Ross C, Swetlit I. IBM pitched its Watson supercomputer as a revolution in cancer care. It's nowhere close. *STAT* 2017.
19. Kumar A, Niu F, Ré C. Hazy: making it easier to build and maintain big-data analytics. *Communications of the ACM.* 2013;56(3):40-49.
20. Buneman P, Davidson S. *Data provenance – the foundation of data quality.* Pittsburgh, PA: Carnegie Mellon University Software Engineering Institute; September 1, 2010 2010.
21. Friedman C, Wong A, Blumenthal D. Achieving a nationwide learning health system. *Science Translational Medicine.* 2010;2(57):57cm29.
22. Smith M, Saunders R, Stuckhardt L, McGinnis J. *Best Care at Lower Cost: The Path to Continuously Learning Health Care in America.* Washington, DC: National Academies Press; 2012.
23. Okun S, McGraw D, Stang P, et al. *Making the Case for Continuous Learning from Routinely Collected Data.* Washington, DC: Institute of Medicine; April 15, 2013 2013.
24. Collins F, Varmus H. A new initiative on precision medicine. *New England Journal of Medicine.* 2015;372:793-795.
25. O'Reilly T, Loukides M, Steele J, Hill C. How Data Science Is Transforming Health Care - Solving the Wanamaker Dilemma. Sebastopol, CA: O'Reilly Media; 2012.
26. Burwell S. Setting value-based payment goals - HHS efforts to improve U.S. health care. *New England Journal of Medicine.* 2015;372:897-899.
27. Kuperman G. Health-information exchange: why are we doing it, and what are we doing? *Journal of the American Medical Informatics Association.* 2011;18:678-682.
28. Halamka J. The "Post EHR" Era. *Life as a Healthcare CIO* 2013.
29. Brownlee J. Overfitting and Underfitting With Machine Learning Algorithms. *Machine Learning Mastery* 2016.
30. Kengne A, Masconi K, Mbanya V, Lekoubou A, Echouffo-Tcheugui J, Matsha T. Risk predictive modelling for diabetes and cardiovascular disease. *Critical Reviews in Clinical Laboratory Sciences.* 2014;51:1-12.
31. Atanasovska B, Kumar K, Fu J, Wijmenga C, Hofker M. GWAS as a driver of gene discovery in cardiometabolic diseases. *Trends in Endocrinology and Metabolism.* 2015;26:722-732.
32. Goodfellow I, Bengio Y, Courville A. *Deep Learning.* Cambridge, MA: MIT Press; 2016.
33. Esteva A, Kuprel B, Novoa R, et al. Dermatologist-level classification of skin cancer with deep neural networks. *Nature.* 2017;542:115-118.
34. Liu Y, Gadepalli K, Norouzi M, et al. Detecting cancer metastases on gigapixel pathology images. *arXivorg.* 2017:arXiv:1703.02442.
35. Oakden-Rayner L, Carneiro G, Bessen T, Nascimento J, Bradley A, Palmer L. Precision radiology: predicting longevity using feature engineering and deep learning methods in a radiomics framework. *Scientific Reports.* 2017;7:1648.
36. Poplin R, Varadarajan A, Blumer K, et al. *Predicting Cardiovascular Risk Factors from Retinal Fundus Photographs using Deep Learning.* Arxiv.org; September 21, 2017.
37. Rajpurkar P, Hannun A, Haghpanahi M, Bourn C, Ng A. *Cardiologist-Level Arrhythmia Detection with Convolutional Neural Networks.* Arxiv.org; July 6, 2017.
38. Hersh W, Weiner M, Embi P, et al. Caveats for the use of operational electronic health record data in comparative effectiveness research. *Medical Care.* 2013;51(Suppl 3):S30-S37.
39. deLusignan S, vanWeel C. The use of routinely collected computer data for research in primary care: opportunities and challenges. *Family Practice.* 2005;23:253-263.
40. Bourgeois F, Olson K, Mandl K. Patients treated at multiple acute health care facilities: quantifying information fragmentation. *Archives of Internal Medicine.* 2010;170:1989-1995.
41. Finnell J, Overhage J, Grannis S. All health care is not local: an evaluation of the distribution of emergency department care delivered in Indiana. Paper presented at: AMIA Annual Symposium Proceedings2011; Washington, DC.
42. Hripcsak G, Albers D. Next-generation phenotyping of electronic health records. *Journal of the American Medical Informatics Association.* 2012;20:117-121.
43. Aakre C, Dziadzko M, Keegan M, Herasevich V. Automating clinical score calculation within the electronic health record - a feasibility assessment. *Applied Clinical Informatics.* 2017;8:369-380.
44. Broberg C, Sklenar J, Burchill L, Daniels C, Marelli A, Gurvitz M. Feasibility of using electronic medical record data for tracking quality indicators in adults with congenital heart disease. *Congenital Heart Disease.* 2015;10:E268-E277.

45. Manor-Shulman O, Beyene J, Frndova H, Parshuram C. Quantifying the volume of documented clinical information in critical illness. *Journal of Critical Care.* 2008;23:245-250.
46. Boyd D, Crawford K. *Six Provocations for Big Data.* Cambridge, MA: Microsoft Research; September 21, 2011.
47. Neff G. Why big data won't cure us. *Big Data.* 2013;1:117-123.
48. Trotter F. Who Owns Patient Data? *The Health Care Blog*2012.
49. O'Neil C. Weapons of Math Destruction: How Big Data Increases Inequality and Threatens Democracy. Crown Publishing Group; 2016.
50. Knight W. Biased Algorithms Are Everywhere, and No One Seems to Care. *MIT Technology Review*2017.
51. Amarasingham R, Patzer R, Huesch M, Nguyen N, Xie B. Implementing electronic health care predictive analytics: considerations and challenges. *Health Affairs.* 2014;33:1148-1154.
52. Krumholz H. Big data and new knowledge in medicine: the thinking, training, and tools needed for a learning health system. *Health Affairs.* 2014;33:1163-1170.
53. Longhurst C, Harrington R, Shah N. A 'green button' for using aggregate patient data at the point of care. *Health Affairs.* 2014;33:1229-1235.
54. Bates D, Saria S, Ohno-Machado L, Shah A, Escobar G. Big data in health care: using analytics to identify and manage high-risk and high-cost patients. *Health Affairs.* 2014;33:1123-1131.
55. Anonymous. Readmissions Reduction Program. Washington, DC: Center for Medicare and Medicaid Services; October 2, 2013 2013.
56. Amarasingham R, Moore B, Tabak Y, et al. An automated model to identify heart failure patients at risk for 30-day readmission or death using electronic medical record data. *Medical Care.* 2010;48:981-988.
57. Donzé J, Aujesky D, Williams D, Schnipper J. Potentially avoidable 30-day hospital readmissions in medical patients: derivation and validation of a prediction model. *JAMA Internal Medicine.* 2013;173:632-638.
58. Gildersleeve R, Cooper P. Development of an automated, real time surveillance tool for predicting readmissions at a community hospital. *Applied Clinical Informatics.* 2013;4:153-169.
59. Hebert C, Shivade C, Foraker R, et al. Diagnosis-specific readmission risk prediction using electronic health data: a retrospective cohort study. *BMC Medical Informatics & Decision Making.* 2014;14:65.
60. Shadmi E, Flaks-Manov N, Hoshen M, Goldman O, Bitterman H, Balicer R. Predicting 30-day readmissions with preadmission electronic health record data. *Medical Care.* 2015;53:283-289.
61. Tabak Y, Sun X, Nunez C, Gupta V, Johannes R. Predicting readmission at early hospitalization using electronic clinical data - an early readmission risk score. *Medical Care.* 2015;55:267-275.
62. Voorhees E, Hersh W. Overview of the TREC 2012 Medical Records Track. Paper presented at: The Twenty-First Text REtrieval Conference Proceedings (TREC 2012)2012; Gaithersburg, MD.
63. Wu S, Liu S, Wang Y, et al. Intra-institutional EHR collections for patient-level information retrieval. *Journal of the American Society for Information Science & Technology.* 2017:in press.
64. Afzal Z, Engelkes M, Verhamme K, et al. Automatic generation of case-detection algorithms to identify children with asthma from large electronic health record databases. *Pharmacoepidemiology and Drug Safety.* 2013;22:826-833.
65. Sada Y, Hou J, Richardson P, El-Serag H, Davila J. Validation of case finding algorithms for hepatocellular cancer from administrative data and electronic health records using natural language processing. *Medical Care.* 2016;54:e9-e14.
66. Halpern Y, Horng S, Choi Y, Sontag D. Electronic medical record phenotyping using the anchor and learn framework. *Journal of the American Medical Informatics Association.* 2016:Epub ahead of print.
67. Haas L, Takahashi P, Shah N, et al. Risk-stratification methods for identifying patients for care coordination. *American Journal of Managed Care.* 2013;19:725-732.
68. Charlson M, Wells M, Ullman R, King F, Shmukler C. The Charlson comorbidity index can be used prospectively to identify patients who will incur high future costs. *PLoS ONE.* 2014;9(12):e112479.
69. Rajkomar A, Yim J, Grumbach K, Parekh A. Weighting primary care patient panel size: a novel electronic health record-derived measure using machine learning. *JMIR Medical Informatics.* 2016;4(4):e29.
70. Escobar G, Gardner M, Greene J, Draper D, Kipnis P. Risk-adjusting hospital mortality using a comprehensive electronic record in an integrated health care delivery system. *Medical Care.* 2013;51:446-453.
71. FitzHenry F, Murff H, Matheny M, et al. Exploring the frontier of electronic health record surveillance: the case of postoperative complications. *Medical Care.* 2013;51:509-516.
72. Tien M, Kashyap R, Wilson G, et al. Retrospective derivation and validation of an automated electronic search algorithm to identify post operative cardiovascular and thromboembolic complications. *Applied Clinical Informatics.* 2015;6:565-576.
73. Tai-Seale M, Wilson C, Panattoni L, et al. Leveraging electronic health records to develop

measurements for processes of care. *Health Services Research.* 2013;49:628-644.
74. Mathias J, Agrawal A, Feinglass J, Cooper A, Baker D, Choudhary A. Development of a 5 year life expectancy index in older adults using predictive mining of electronic health record data. *Journal of the American Medical Informatics Association.* 2013;20(e1):e118-e124.
75. Tran T, Luo W, Phung D, et al. Risk stratification using data from electronic medical records better predicts suicide risks than clinician assessments. *BMC Psychiatry.* 2014;14:76.
76. Murphy D, Laxmisan A, Reis B, et al. Electronic health record-based triggers to detect potential delays in cancer diagnosis. *BMJ Quality & Safety.* 2014;23:8-16.
77. Jensen K, Soguero-Ruiz C, Mikalsen K, et al. Analysis of free text in electronic health records for identification of cancer patient trajectories. *Scientific Reports.* 2017;7:46226
78. Lee J, Maslove D, Dubin J. Personalized mortality prediction driven by electronic medical data and a patient similarity metric. *PLoS ONE.* 2015;10:e0127428.
79. Weng S, Reps J, Kai J, Garibaldi J, Qureshi N. Can machine-learning improve cardiovascular risk prediction using routine clinical data? *PLoS ONE.* 2017;12(4):e0174944.
80. McNamara R, Wang Y, Partovian C, et al. Development of a hospital outcome measure intended for use with electronic health records - 30-day risk-standardized mortality after acute myocardial infarction. *Medical Care.* 2015;53:818-826.
81. Khurana H, Groves R, Simons M, et al. Real-time automated sampling of electronic medical records predicts hospital mortality. *American Journal of Medicine.* 2016;129:688-698.
82. vanWalraven C, Forster A. The HOMR-Now! model accurately predicts 1-year death risk for hospitalized patients on admission. *American Journal of Medicine.* 2017:Epub ahead of print.
83. Hornbrook M, Hart G, Ellis J, et al. Building a virtual cancer research organization. *Journal of the National Cancer Institute Monographs.* 2005;35:12-25.
84. Hillier T, Pedula K, Schmidt M, Mullen J, Charles M, Pettitt D. Childhood obesity and metabolic imprinting: the ongoing effects of maternal hyperglycemia. *Diabetes Care.* 2007;30:2287-2292.
85. Tannen R, Weiner M, Xie D, Barnhart K. A simulation using data from a primary care practice database closely replicated the Women's Health Initiative trial. *Journal of Clinical Epidemiology.* 2007;60:686-695.
86. Weiner M, Barnhart K, Xie D, Tannen R. Hormone therapy and coronary heart disease in young women. *Menopause.* 2008;15:86-93.
87. Tannen R, Weiner M, Xie D. Replicated studies of two randomized trials of angiotensin-converting enzyme inhibitors: further empiric validation of the 'prior event rate ratio' to adjust for unmeasured confounding by indication. *Pharmacoepidemiology and Drug Safety.* 2008;17:671-685.
88. Tannen R, Weiner M, Xie D. Use of primary care electronic medical record database in drug efficacy research on cardiovascular outcomes: comparison of database and randomised controlled trial findings. *British Medical Journal.* 2009;338:b81.
89. Danaei G, Rodríguez L, Cantero O, Logan R, Hernán M. Observational data for comparative effectiveness research: An emulation of randomised trials of statins and primary prevention of coronary heart disease. *Statistical Methods in Medical Research.* 2011;22:70-96.
90. Stapley S, Peters T, Neal R, Rose P, Walter F, Hamilton W. The risk of pancreatic cancer in symptomatic patients in primary care: a large case-control study using electronic records. *British Journal of Cancer.* 2012;106:1940-1944.
91. Stapley S, Peters T, Neal R, Rose P, Walter F, Hamilton W. The risk of oesophago-gastric cancer in symptomatic patients in primary care: a large case-control study using electronic records. *British Journal of Cancer.* 2013;108:25-31.
92. Kaelber D, Foster W, Gilder J, Love T, Jain A. Patient characteristics associated with venous thromboembolic events: a cohort study using pooled electronic health record data. *Journal of the American Medical Informatics Association.* 2012;19:965-972.
93. Ryan P, Madigan D, Stang S, Overhage J, Racoosin J, Hartzema A. Empirical assessment of methods for risk identification in healthcare data: results from the experiments of the Observational Medical Outcomes Partnership. *Statistics in Medicine.* 2012;31:4401-4415.
94. Frankovich J, Longhurst C, Sutherland S. Evidence-based medicine in the EMR era. *New England Journal of Medicine.* 2011;365:1758-1759.
95. Denny J. Mining Electronic Health Records in the Genomics Era. *PLOS Computational Biology.* 2012;8(12):e1002823.
96. McCarty C, Chisholm R, Chute C, et al. The eMERGE Network: a consortium of biorepositories linked to electronic medical records data for conducting genomic studies. *BMC Genomics.* 2010;4(1):13.
97. Denny J, Ritchie M, Crawford D, et al. Identification of genomic predictors of atrioventricular conduction:

using electronic medical records as a tool for genome science. *Circulation*. 2010;122:2016-2021.
98. Kullo L, Ding K, Jouni H, Smith C, Chute C. A genome-wide association study of red blood cell traits using the electronic medical record. *PLoS ONE*. 2010;5(9):e13011.
99. Crosslin D, McDavid A, Weston N, et al. Genetic variants associated with the white blood cell count in 13,923 subjects in the eMERGE Network. *Human Genetics*. 2012;131:639-652.
100. Denny J, Crawford D, Ritchie M, et al. Variants near FOXE1 are associated with hypothyroidism and other thyroid conditions: using electronic medical records for genome- and phenome-wide studies. *American Journal of Human Genetics*. 2011;89:529-542.
101. Denny J, Bastarache L, Ritchie M, et al. Systematic comparison of phenome-wide association study of electronic medical record data and genome-wide association study data. *Nature Biotechnology*. 2013;31:1102-1111.
102. Newton K, Peissig P, Kho A, et al. Validation of electronic medical record-based phenotyping algorithms: results and lessons learned from the eMERGE network. *Journal of the American Medical Informatics Association*. 2013;20(e1):e147-154.
103. Gilbert P, Rutland M, Brockopp D. Redesigning the work of case management: testing a predictive model for readmission. *American Journal of Managed Care*. 2013;19(11 Spec No. 10):eS19-eSP25.
104. Cho I, Park I, Kim E, Lee E, Bates D. Using EHR data to predict hospital-acquired pressure ulcers: A prospective study of a Bayesian Network model. *International Journal of Medical Informatics*. 2013;82:1059-1067.
105. Amarasingham R, Patel P, Toto K, et al. Allocating scarce resources in real-time to reduce heart failure readmissions: a prospective, controlled study. *BMJ Quality & Safety*. 2013;22:998-1005.
106. Baillie C, VanZandbergen C, Tait G, et al. The readmission risk flag: Using the electronic health record to automatically identify patients at risk for 30-day readmission. *Journal of Hospital Medicine*. 2013;8:689-695.
107. Rothman M, Rimar J, Coonan S, Allegretto S, Balcezak T. Mortality reduction associated with proactive use of EMR-based acuity score by an RN team at an urban hospital. *BMJ Quality & Safety*. 2015;24:734-735.
108. Murphy D, Wu L, Thomas E, Forjuoh S, Meyer A, Singh H. Electronic trigger-based intervention to reduce delays in diagnostic evaluation for cancer: a cluster randomized controlled trial. *Journal of Clinical Oncology*. 2015;33:3560-3567.
109. Evans R, Benuzillo J, Horne J, et al. Automated identification and predictive tools to help identify high-risk heart failure patients: pilot evaluation. *Journal of the American Medical Informatics Association*. 2016;23:872-878.
110. Erskine A, Karunakaran P, Slotkin J, Feinberg D. How Geisinger Health System Uses Big Data to Save Lives. *Harvard Business Review*. 2016.
111. Graf T, Erskine A, Steele G. Leveraging data to systematically improve care: coronary artery disease management at Geisinger. *Journal of Ambulatory Care Management*. 2014;37:199-205.
112. Lowes L, Noritz G, Newmeyer A, Embi P, Yin H, Smoyer W. 'Learn From Every Patient': implementation and early results of a learning health system. *Developmental Medicine & Child Neurology*. 2017;59:183-191.
113. Davenport T, Patil D. Data Scientist: The Sexiest Job of the 21st Century. *Harvard Business Review*. Cambridge, MA: Harvard Business Publishing; 2012.
114. Hersh W. A stimulus to define informatics and health information technology. *BMC Medical Informatics & Decision Making*. 2009;9:24.
115. Dhar V. Data science and prediction. *Communications of the ACM*. 2013;56(12):64-73.
116. Manyika J, Chui M, Brown B, et al. Big data: The next frontier for innovation, competition, and productivity. McKinsey Global Institute;2011.
117. Anonymous. An assessment of demand for labour and skills, 2012-2017. London, England: SAS;2013.
118. Anonymous. Adoption and employment trends, 2012-1017. London, England: SAS;2013.
119. Fraser H, Jayadewa C, Mooiweer P, Gordon D, Piccone J. *Analytics across the ecosystem - A prescription for optimizing healthcare outcomes*. Somers, NY: IBM Global Services;2013.
120. Balboni F, Finch G, Rodenbeck-Reese C, Shockley R. Analytics: A blueprint for value. Somers, NY: IBM Global Services;2013.
121. Anonymous. Solving the talent equation for health IT. PriceWaterhouseCoopers; March, 2013 2013.
122. Anonymous. Report of Workshop on Enhancing Training for Biomedical Big Data. National Institutes of Health; July 29-30, 2013 2013.
123. Otero P, Hersh W, Ganesh A. Big Data: Are Biomedical and Health Informatics Training Programs Ready? In: Lehmann C, Séroussi B, Jaulent M, eds. *Yearbook of Medical Informatics 2014*. Stuttgart, Germany: Schattauer; 2014:177-181.
124. Garfield J. Top Skills and Education Requirements for Data Scientists. *Paysa Blog* 2017.
125. Hersh W, Cimino J, Payne P, et al. Recommendations for the use of operational electronic health record data in comparative effectiveness research. *eGEMs (Generating Evidence & Methods to improve patient outcomes)*. 2013;1:14.

8

Clinical Decision Support

ROBERT E. HOYT • HAROLD P. LEHMANN

"Clinical decision support systems link health observations with health knowledge to influence health choices by clinicians for improved health care."

—Robert Hayward, Centre for Health Evidence 2004

LEARNING OBJECTIVES

After reading this chapter the reader should be able to:

- Define electronic clinical decision support (CDS)
- Enumerate the goals and potential benefits of CDS
- Discuss the government and private organizations supporting CDS
- Discuss CDS taxonomy, functionality and interoperability
- List the challenges associated with CDS
- Enumerate CDS implementation steps and lessons learned

INTRODUCTION

Definition

The above definition by Dr. Hayward is widely quoted and general in nature. A more recent definition by the Office of the National Coordinator for HIT states *"Clinical decision support (CDS) provides clinicians, staff, patients or other individuals with knowledge and person-specific information, intelligently filtered or presented at appropriate times, to enhance health and health care."*[1] To be complete, one could argue that any resource that aides the clinician, other healthcare team members or patients with medical decision making should be considered CDS. For the purposes of this chapter, electronic CDS only will be discussed. CDS, most commonly that is part of electronic health records (EHRs) and computerized physician order entry (CPOE), is discussed in the chapter on EHRs. It should be emphasized that CDS tools potentially assist more than just physicians; nurses, pharmacists, radiologists, patients, etc. have the same need for knowledge management. Also, we define clinical decision support systems (CDSS) as the information technology systems that support electronic CDS.

Early in its evolution CDS was discussed in terms of alerts and reminders to clinicians, but as pointed out in the chapter on EHRs, it should be broadened to include many other tools available to clinicians and others at the point of care. This would include diagnostic help, cost data, up-to-date information about emerging clinical problems such as Zika virus, data from disease registries and so forth. The vision is for all CDS data to be electronic, structured and computable, which is frequently not the case.

Despite extensive online medical resources and robust search engines available to all members of the healthcare team, questions concerning correct diagnoses and optimal treatments still arise frequently.[2] For that reason, many experts have strongly promoted CDS as a driving force towards improved patient care and safety. Therefore, CDS and quality are closely related.

The Five Rights of CDS

CDS subject matter experts have stressed that for CDS to be effective it must include five rights:

1. The right information (what): should be based on the highest level of evidence possible and adequately

referenced. There should be good internal and external validity; that is, the recommendations are based on high quality studies and they can be applied to a similar population of patients. These issues are discussed in the chapter on evidence-based medicine and clinical practice guidelines.
2. To the right person (who): that is the person who is making the clinical decision, be that the physician, the patient or some other member of the healthcare team.
3. In the right format (how): should the information appear as part of an alert, reminder, infobutton or order set? That depends on the issue and clinical setting.
4. Through the right channel (where): should the information be available as an EHR alert, a text message, email alert, etc.?
5. At the right time (when) in workflow: new information, particularly in the format of an alert should appear early in the order entry process so clinicians are aware of an issue before they complete, e.g. an electronic prescription. This adds to the usability of the technology.[3]

Historical Perspective

Interest in clinical decision support is not new. In 1959, Ledley and Lusted published an article on medical reasoning in which they predicted computers would be used for medical decision making.[4] Early experience with CDSSs began in the 1970s with several well-known initiatives. These employed a variety of systems and knowledge bases that will be discussed in more detail in a later section. Except for the HELP system, these programs were standalone programs, not integrated with any other systems, such as the EHR, and most are no longer available.

De Dombal's system for acute abdominal pain: was developed at Leeds University in the UK in the 1970s. It was based on Bayesian probability to assist in the differential diagnosis of abdominal pain.[5]

Internist-1: was a diagnostic CDS program developed at the University of Pittsburg in the 1970s that relied on production rules (IF-THEN statements). It used patient observations to generate possible diagnoses. Despite being cumbersome, the knowledge base was used in the subsequent system known as QMR.[6]

Mycin: was a rule-based system developed by Dr. Edward (Ted) Shortliffe and others at Stanford University in the 1970s to diagnose and treat infections.[7]

Later, in the 1980s several initiatives achieved some commercial success.

DxPlain: was developed in 1984 by Massachusetts General Hospital as both a medical reference system and diagnostic CDSS. Based on clinical findings (signs, symptoms, laboratory data) the program generated a ranked list of diagnoses related to the clinical manifestations. DxPlain justifies the diagnoses, suggests further steps and describes atypical manifestations. As a reference, it describes over 2400 diseases, the signs and symptoms, causes and prognosis, as well as references. Institutions can lease the program annually.[8]

QMR: Quick Medical Reference was a diagnostic CDDS consisting of an extensive knowledge base of diagnoses, symptoms and lab findings. It evolved from Internist-1 in the 1980s and discontinued about 2001.[9]

HELP: Health Evaluation Through Logical Processing (HELP) was developed in the 1980s by the University of Utah as a hospital information system and later as a diagnostic CDSS. It provides alerts and reminders, data interpretation, diagnostic help, management suggestions and clinical practice guidelines. They developed the Antibiotic Assistant that was later commercialized into TheraDoc.™ It is still in use today in Intermountain Healthcare hospitals as part of their EHR system. It has been modernized and is now known as HELP2. They have developed thousands of medical logic modules (MLMs) used for CDS and discussed later in the chapter.[10]

Iliad: is a diagnostic CDSS and reference system for professionals developed by the University of Utah in the 1980s and discontinued about 2000. The last version (CD-ROM) covered about 930 diseases and 1500 syndromes with ICD-9 codes for each diagnosis.[11-12]

An extensive list of early initiatives is archived on the web site OpenClinical.[13] The following are a sample of more recent CDS tools:

Isabel: created in 2002, is a differential diagnosis web-based tool available for worldwide use. Signs and symptoms are inputted as free text or imported from the EHR. and a diagnostic checklist is generated as a standalone tool or integrated with the EHR (SNOMED-CT coded). The inference engine relies on natural language processing with a database of about 100,000 documents and 40 proprietary algorithms. It is offered as a paid subscription and a mobile app is available.[14]

SimulConsult: is a diagnostic program based on Bayesian networks with its strength in pediatrics, genetics, neurology and rheumatology. For the knowledge base, it includes about 5,500 diagnoses, and 3000+ genetic variants. A search is conducted using the patient's age and gender and then symptoms, signs, lab tests, images, etc. are added to narrow the differential diagnosis. The program also includes recommended additional testing and articles to read. It is particularly useful for a child with genetic variants and unusual physical findings where a differential diagnosis is important.[15]

SnapDx: a free mobile app that is a diagnostic CDS for clinicians. It is based on positive and negative likelihood ratios (LRs) derived from the medical literature. As of early-2018 the program covered about 50 common medical conditions (Apple iOS only).[16] (See figure 8.1)

CDS BENEFITS AND GOALS

Electronic health records alone are electronic filing cabinets of healthcare data. Without additional tools, they could also be referred to as electronic data management systems (EDMSs). It is when they include helpful information and guidance that they have the potential to positively impact patient care.

Some of the potential benefits and goals of CDS based on several expert sources are listed in Table 8.1.[3, 17-18]

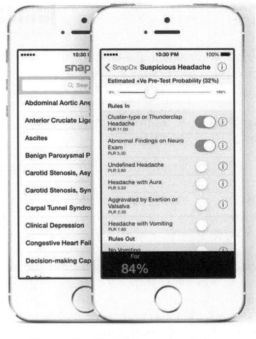

Figure 8.1: SnapDx on smartphone

Table 8.1: CDS Benefits and Goals

Benefits and Goals	Details
Improvement in patient safety	· Medication alerts · Improved ordering
Improvement in patient care	· Improved patient outcomes · Better chronic disease management · Alerts for critical lab values, drug interactions and allergies · Improved quality adjusted life years (QALY) · Improved diagnostic accuracy
Reduction in healthcare costs	· Fewer duplicate lab tests and images · Fewer unnecessary tests ordered · Avoidance of Medicare penalties for readmission for certain conditions · Fewer medical errors · Increased use of generic drugs · Reduced malpractice · Better utilization of blood products
Dissemination of expert knowledge	· Sharing of best evidence · Education of all staff, students and patients
Management of complex clinical issues	· Use of clinical practice guidelines, smart forms and order sets · Interdisciplinary sharing of information · Case management
Monitoring clinical details	· Reminders for preventive services · Tracking of diseases and referrals
Improvement of population health	· Identification of high-cost/needs patients · Mass customized messaging
Management of administrative complexity	· Supports coding, authorization, referrals and care management
Support clinical research	· Ability to identify prospective research subjects

ORGANIZATIONAL PROPONENTS OF CLINICAL DECISION SUPPORT

Multiple governmental and non-governmental organizations have been proponents of CDS. The following are known leaders in the field.

National Academy of Medicine (previously known as the Institute of Medicine (IOM)

The IOM has been a strong proponent of the use of information technologies to improve access to clinical information and support clinical decision making. In Crossing the Quality Chasm, they stated *"Development of decision support tools to assist clinicians and patients in applying the evidence, evidence - based care processes, supported by automated clinical information and decision support systems, offer the greatest promise of achieving the best outcomes from care for chronic conditions."* In many of their subsequent publications they also mentioned CDS as an evolutionary tool.[19]

American Medical Informatics Association (AMIA)

In 2006, A Roadmap for National Action on CDS was published under the auspices of AMIA. The Roadmap outlined three pillars of CDS:

1. *"Best Knowledge Available When Needed: the best available clinical knowledge is well organized, accessible to all, and written, stored and transmitted in a format that makes it easy to build and deploy CDS interventions that deliver the knowledge into the decision-making process*
2. *High Adoption and Effective Use: CDS tools are widely implemented, extensively used, and produce significant clinical value while making financial and operational sense to their end-users and purchasers*
3. *Continuous Improvement of Knowledge and CDS Methods: both CDS interventions and clinical knowledge undergo continuous improvement based on feedback, experience, and data that are easy to aggregate, assess, and apply. "*

AMIA has also established a CDS Working Group to carry out the vision of the Roadmap and to offer a forum for the advancement of CDS.[20-21]

Office of the National Coordinator (ONC)

ONC funded research by the RAND Corporation and Partners Health Care/ Harvard Medical School in a project known as "Advancing Clinical Decision Support." Four tasks have been outlined:

- *"Task 1: Distill best practices for CDS design and CDS implementation, preparing resources on best practices for broad dissemination through a variety of online channels.*
- *Task 2: Distill best practices and standards for sharing CDS knowledge and produce an open online platform for sharing CDS knowledge artifacts (such as alerts, order sets, etc.) among EHR vendors and/or provider organizations.*
- *Task 3: Develop a "clinically important" drug-drug interaction (DDI) list, as well as a legal brief about the liability implications of using the clinically important DDI list.*
- *Task 4: Develop a process that engages specialty bodies in weighing performance gaps vs. CDS opportunities to select targets for meaningful use of CDS by specialists"*

ONC also has a special section on CDS sharing which is important if CDS is expected to become widespread and vetted by multiple organizations. Sharing is intended to be part of Task 4 above and should include these activities:

- *"Identified key requirements and features for a Knowledge Sharing Service (KSS)*
- *Proposed important attributes of a governance model and editorial policy*
- *Proposed an architecture based on leveraging existing standards where available*
- *Proposed standard XML schemas for commonly deployed intervention types, which could be imported into a vendor system*
- *Identified gaps in standards that present barriers to progress and make recommendations to appropriate standards development organizations*
- *Deployed a first-generation KSS*
- *Populated the KSS with illustrative XML CDS interventions targeting Meaningful Use and implemented custom style sheets so the XML files could be viewed in human-readable form"*

ONC established the Health eDecisions (HeD) project which ran from 2012-2014 to develop and harmonize CDS standards. They developed an XML schema for representing knowledge and a Virtual Medical Record (vMR) for representing patient data. In 2014 ONC launched the Clinical Quality Framework Initiative to further harmonize those standards that would integrate guidelines for electronic clinical quality measures (eCQMs) and CDS. ONC has also been instrumental in the development of the Quality Data Model that includes clinical concepts necessary for eCQMs, as part of the meaningful use program.[22-23]

Agency for Healthcare Research and Quality (AHRQ)

AHRQ funded several CDS initiatives that began in 2008:

- The Clinical Decision Support Consortium (CDSC) is an initiative to develop and test web-based delivery of CDS, primarily for EHRs. The goal of the CDS Consortium is "*to assess, define, demonstrate, and evaluate best practices for knowledge management and clinical decision support in healthcare information technology at scale – across multiple ambulatory care settings and EHR technology platforms.*"[24]

Access to newly developed CDS is through the public CDSC Knowledge Management Portal that supports search and retrieval of CDS interventions. CDS can be unstructured or highly structured with encoded logic. Brigham and Women's Hospital was the major partner with participating organizations located throughout the US. As of early 2015 the portal contained over 100 CDS tools consisting of: alerts, reminders, templates, order sets, reference information, Infobuttons, and value sets. The tools can be viewed in XML format or in human-readable style sheets. In mid-2014 the Consortium was placed on hiatus, awaiting additional funding.

- Guidelines into Decision Support (GLIDES). Yale School of Medicine was charged with translating CPGs into structured data for the outpatient treatment of common diseases. (Example: childhood obesity and asthma). Guidelines would be disseminated at multiple sites using different EHRs. This generated the Guideline Elements Model (GEM), a knowledge model for guidelines that incorporates a set of more than 100 tags to categorize guideline content.
- CDS eRecommendations project used 45 recommendations from the US Preventive Services Task Force and 12 recommendations relevant to stage 1 meaningful use and created a framework for use with EHRs. Their final report was published in 2011.
- CDS Key Resources: Two CDS white papers and other monographs are included on the AHRQ web site and are available in the resource section of this chapter.
- US Health Information Knowledgebase (USHIK) is an AHRQ initiative to support knowledge "artifacts" using HL7 Health eDecision (HeD) schema. The XML versions can be downloaded and shared by healthcare organizations or EHR vendors seeking to meet meaningful use CDS requirements.[24-26]

Health Level 7 (HL7)

This international standards development organization (SDO) has a working group dedicated to advancing electronic CDS with the following goals:

- *"Work on CDS standards for knowledge representation, such as, Infobuttons and order sets*
- *Work on patient-centered monitoring such as alerts and reminders*
- *Work on population-centric monitoring and management, such as disease surveillance*
- *Work on representation of CPGs*
- *Develop a data model for clinical decision support*
- *Identify existing HL7 messages and triggers for CDS"*

HL7 has also developed the Fast Healthcare Interoperability Resources (FHIR) standard that can be used for CDS and will be discussed in another section.[27]

National Quality Forum (NQF)

The NQF CDS Expert Panel met in 2010 and developed a CDS taxonomy with four components: triggers, input data, intervention and action steps. Their goal was to create the taxonomy for future quality performance measures and to map to the quality data set (QDS) model, necessary for electronic clinical quality measures (eCQMs).[28]

Leapfrog

This patient safety organization has long promoted computerized physician order entry (CPOE). As part of their approach they developed a CPOE Evaluation Tool that tests a hospital's EHR with multiple mock scenarios, to include 12 CDS categories, such as therapeutic duplications.[29-30]

Healthcare Information and Management Systems Society (HIMSS)

HIMSS published one of the best-known textbooks on CDS implementation.[2] Moreover, they created the Electronic Medical Record Adoption Model (EMRAM) that rates the different levels of US EHR adoption from 1 to 7. Full use of CDS would qualify as a stage 6 level. By mid- 2017, of the participating healthcare organizations 32.7% of hospitals had achieved level 6, but only 20.4% by ambulatory facilities.[31]

Centers for Medicare and Medicaid Services (CMS) Meaningful Use Program

CMS is responsible for reimbursing eligible physicians and hospitals for meaningful use of certified EHRs. CMS views clinical decision support to be integral to quality measures and the improvement of patient care. In stage 1 the following objectives were related to CDS:

- Capturing clinical data in a standard, coded manner.
- Utilizing computerized provider order entry.

- Implementing drug-drug, drug-allergy, and drug-formulary checks.
- Setting patient reminders per patient preference.
- Performing medication, problem, and medication allergy reconciliation at transitions of care.

Stage 2 meaningful use required greater use of CDS tools. In 2014, eligible professionals had to report on 9 out of 64 total CQMs, while eligible hospitals and critical access hospitals had to report on 16 out of 29 total CQMs. Specifically, to comply with MU core measure 6 they were required to do the following:

- *"Implement five CDS interventions related to four or more CQMs, if applicable, at a relevant point in patient care for the entire EHR reporting period"*
- *"Enable the functionality for drug-drug and drug-allergy interaction checks for the entire EHR reporting period"*[32]

Despite support from multiple agencies and organizations, widespread implementation of CDSSs has been slow. This will be discussed in more detail in the section on CDS Challenges.

CDS METHODOLOGY

CDS entails two phases: Knowledge Use and Knowledge Management.

Knowledge Use involves several typical steps demonstrated in figure 8.2 which is the taxonomy created by the National Quality Forum. Triggers are an event such as an order for a medication by a user. Input data refers to information within, for example the EHR, that might include patient allergies. Interventions are the CDS actions such as displayed alerts. The action step might be overriding the alert or canceling an order for a drug to which the patient is allergic.[33] Managing this knowledge-based process is called *knowledge engineering*. Because most decision support is performed through tools and processes set by vendors, most knowledge engineers do not call themselves by this explicit name but perform the tasks implicitly.

Figure 8.2: CDS Use Phases

Most of these decision support systems are table driven or depend on software coding that embeds clinical, regulatory, or administrative knowledge into the programming driving the system. Focusing on the technology underlying CDSSs, Berner articulates 2 high-level classes where the knowledge is separable from the clinical-system programming: knowledge based (KB) and non-knowledge based (NKB).[34] While systems based on these architectures (algorithms) are uncommon in clinical operations, we expect them to become more prevalent as the clinical-data environment stabilizes and developers figure out how best to use these more advanced methods in the operational environment.

Knowledge Management, in turn, involves knowledge acquisition, knowledge representation, and knowledge maintenance.

Knowledge Acquisition includes expert-based knowledge or data-based knowledge. The former may come from clinical practice guidelines external to the organization and from clinical expertise from within the organization. Data-based knowledge may also come from models built on data from outside the institution (e.g., APACHE scoring[35]) or from data mining from within the institution. Specific methods are described shortly.

Knowledge Representation has many forms. The choice of representation should depend on the problem at hand, the level of expertise of the knowledge engineer, the resources available, and the commitment of the institution to knowledge maintenance. The classical knowledge-based CDS consists of a knowledge base (evidence-based information), an inference engine (software to integrate the knowledge with patient-specific data) and a means to communicate the information to the end-user, such as a pop-up alert in the EHR. There are several different types of knowledge representations in these KB CDSSs:

- **Configuration**: Here, knowledge is represented by the choices specified by the institution, which may be context sensitive. Thus, a formulary may "know" the insurance status of a patient. Knowledge acquisition involves having the committee in charge of the clinical system (e.g., pharmacy) involved, and knowledge maintenance, reviewing those choices on some frequency or as problems arise.
- **Table based**: Many systems represent their rules for care in tables. For instance, drug–drug interactions have one basic rule: If the patient is on one drug in a row in the table, and an order is placed for a second drug in the same row, and the level of alerting is above the threshold specified by the institution, then the ordering clinician will be alerted. The rows of drug pairs, along with their level of

danger, represent drug-drug-interaction knowledge. Knowledge may be acquired from a vendor (in the case of drug-drug interactions) or from the CDS Committee. Knowledge maintenance involves vendor updates and periodic reviews, in light of patient-quality indicators or patient-safety events.
- **Rules based**: Also known as production rules and "expert systems." The knowledge base consists of IF-THEN statements, such as: if the patient is allergic to sulfa derivatives and a sulfa drug is prescribed, then an alert will be triggered.[7] In contrast to tables, the rules may be of arbitrary complexity. The attraction of rules is that they should be *modular*, in that any single rule can be debugged in isolation of others. The classic example of rules-based CDS is MYCIN, which, while demonstrating many features necessary for CDS, was not adopted into practice, primarily because the system could not have its needed in-use data fed to it by the host EHR.[36] The rule-based architecture is preferable to code-driven software, because to modify the latter means to modify software that directly runs the clinical system. Knowledge acquisition here is generally internal; efforts to share rules have often floundered on the problem that MYCIN faced: how to get data of specific EHRs of different environments into the same rules.[37] See **CDS Standards**, below.
- **Bayesian networks**: These structures use forms of Bayes' Theorem (conditional probabilities) to calculate the (posterior) probabilities of diseases (or other state of concern), based on the pretest probability, prevalence of each disease, P(Disease), conditioned on patient-specific data (such as symptoms). For example, the probability of a disease given a positive test using Bayes' Formula would be:

$$P(Disease|Test+) = \frac{P(Test+|Disease) \times P(Disease)}{P(Test+)}$$

P(Disease|Test+) (also called positive predictive value) and P(Test+|Disease) (also called sensitivity) are conditional probabilities. To deal with multiple findings (tests, signs, symptoms), two assumptions are often made; each finding is independent of other findings and the findings can be grouped into categories, such as present or absent. The results can be both surprising and enlightening.[38] (See example posted by Cornell University in the infobox). Bayesian systems in use involve networks of diseases and findings; knowledge acquisition means to specify the many conditional probabilities in the network. The odds form of Bayes theorem is Posterior Odds (Disease|Test Positive) = Prior Odds (Disease) x Likelihood Ratio Positive; the log form of this has the log prior odds added to the log likelihood ratio (LLR); the LLR then has the semantics of *"weight of evidence."* A variety of free Bayesian calculators are available on the Internet.

Examples of Bayesian CDS are: Iliad, Gideon, SimulConsult and De Dombal's system.[5,11,15,39] Systems that are pseudo-Bayesian (that is, use weighing of evidence in non-Bayesian ways) include Dxplain.[8]

> **Bayesian Probability Example**
>
> The prevalence of breast cancer is about 1% in women (ages 40-50) and a woman with breast cancer has a 90% chance of a positive test (mammogram). In addition, there is a 10% chance of a false positive test (mammogram is positive, but woman doesn't have cancer).
>
> Based on this information, what is the probability that a lady age 45 has breast cancer based on a positive mammogram?
>
> **Answer**: 9 in 108 or about 8% [38]

Knowledge maintenance means there is a need to keep knowledge up to date, from the level of the program through the committees in charge and to track changes and reasons. This maintenance has proven challenging to most organizations.[40]

Non-Knowledge based CDS

When the knowledge representation is a model derived from data mining, the system is called a non-knowledge based CDS. Data mining methods may involve artificial intelligence (AI), (e.g. neural networks, machine learning) or more traditional statistical methods, like linear or logistic regression. These are *data-based* systems, that require the models be developed and validated prior to being used in clinical operation. An open-source program for all these methods is the WEKA environment.[40] The power of these approaches is that the methods first can analyze large data sets, looking for new trends or patterns at the population level. Then, the resulting model can derive recommendations specific to the patient at hand and lie at the heart of predictive analytics.[41] AI machine learning has also been used for *"pattern recognition"* which has become relatively routine in medical diagnostic devices (interpreting images, electrocardiograms, etc.). This approach does not require patient symptoms or physiological findings for the interpretation. For AI models that do require such data, there is no agreement in the field about the amount of validation that is needed to incorporate a model developed elsewhere into a local system. Data mining and predictive modeling can be categorized as supervised or unsupervised machine learning.

Supervised machine learning assumes that the user knows ahead of time what classes or categories exist. A training sample is used that has target (dependent) variables and input (independent) variables. With multiple training sessions, the goal is to narrow the gap between observed and expected observations. The predictive model can be further classified as a classification model if the target involves discrete (categorical or nominal) data (can only be certain values, such as blood type) and a regression model if the target involves continuous data (data that can be anything numerical in a range, such as a patient's weight).

An important aspect of supervised learning is classification or the mapping of data into predefined classes. Techniques to accomplish this are:

- The computational model known as **neural networks** was first reported in 1943 by McCulloch and Pitts.[40] Neural networks are a popular approach and are capable of both supervised and unsupervised machine learning. The networks are arranged in layers and each layer is an array of processing elements or neurons. The input layer receives multiple inputs and in the hidden layer signals are processed and an output is generated to the output layer. Using the supervised model, the outputs are compared to the target output and training with input-output pairs is repeated until the trained output and desired target output are similar. (See figure 8.3). The results are dependent on a good training set, otherwise the results may not reflect a larger population.[41]

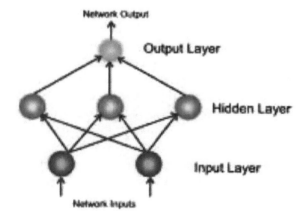

Figure 8.3: Neural networks

- **Logistic regression** is used to analyze data where the output is binary, (for example, the patient has or doesn't have hypertension). The independent variables or input are computed against the output/outcome and odds ratios are produced. The end result is a prediction model and this approach has been used for many years by healthcare researchers.[42]
- **Decision trees**. There are two types of "decision trees" used in knowledge engineering. One is a *classification tree*, where the attributes of an individual case (patient, population, setting) are input, and the algorithm classifies to one class or the other. (see Figure 8.4) Typical algorithms here are C4.5 and recursive tree partitioning, both available in many statistical and machine learning packages.[41,46-47] The "tree" here has to do with use of the attributes, generally in a binary way and recursively (i.e., the same operation is done at each step). The other type of decision tree is used in *decision analysis* to derive an optimal action, or event, a flowchart of action.[48] The "tree" here lays out the space of possible outcomes, which have more to do with unfolding over time than with attributes. They generally consist of decision nodes (squares), chance nodes (circles) and terminal nodes or outcomes (triangles). Probabilities are assigned to the path branches, while costs or other measures of value or preference are assigned to the outcomes. Figure 8.4 displays a decision tree for deciding if patients need a soft, hard or no contact lens. In this example, the decision tree calculates that the first logical branch is tear production. These decision trees are a subset of the wider set of *decision models*. While mostly used to inform policy, a few decision models have been implemented to provide patients real-time decision support as part of *shared decision making*.[49]

Unsupervised machine learning analyzes data without any classes, so the learning system develops new classes or patterns (clustering). This is a valuable approach if the user desires an exploratory data analyses, not based on prior classifications. There are several different types of unsupervised learning.

- **Cluster analysis** is a technique that organizes a large data set into distinct groups. Most commonly they are grouped as a hierarchy. Cluster analysis has been used to group gene sequences, for example identifying clusters of genes associated with breast cancer. Although this technique is fast, it is challenged by clusters that may be difficult to interpret.[50] An example of how this approach is used can be found in an article by Newcomer et al. that evaluated a cohort of Kaiser-Permanente patients that were in the top

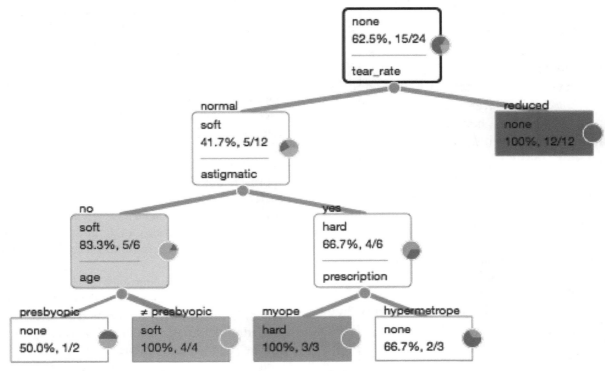

Figure 8.4: Decision Tree for Contact Lens Recommendations

20% for cost of care over two years. This approach identified discrete groups of chronically ill patients that would benefit from care management.[51] This is one of the many approaches healthcare organizations, particularly accountable care organizations, might adopt to lower costs.[52]

- **Association Rules** generate probabilities that look like IF-THEN statements. Association rules have been used for "market basket analysis", for example, if a man buys diapers, there is a 90% chance he will buy beer.

CDS STANDARDS

Much medical knowledge is independent of the specific institution where it is used: calcium and bicarbonate should *never* be in the same IV line, because limestone will precipitate in the line and in the vein; patients surviving myocardial infarction should be treated with beta-blockers, because the best evidence shows a decrease in subsequent mortality.[53] Patients with type 1 diabetes do better if they have close control; however, implementing this directive depends on local resources. Thus, for several decades, developers of CDS have struggled with how to share the medical knowledge that rises above local concerns, while enabling local committees to modify the rules to be in accord with local realities. The following are some of the significant standards developed and certified to address these concerns.

Arden Syntax: this standard was developed in 1999 and is now an ANSI/HL7 standard (v2.7-2008). The core representations are Medical Logic Modules (MLMs), which encode information for a single medical decision (text file), so it can be shared. There are a standard set of categories (*maintenance, library, knowledge*) and slots (e.g., within *knowledge, data* and *logic*). It is an open standard that can be used by individuals, healthcare organizations and vendors for development of clinical rules. The problem is that MLMs can't be shared and this is known as the *"curly braces problem"* because the assignment of local data to the rules variables, in the *data* slot, depends on institution-specific programming code placed between curly braces ({}) in that slot. Note in figure 8.5 hematocrit is contained within a pair of curly braces. Although several EHR vendors use this standard, overall usage has been low.[54-55]

GELLO: is a class-based object-oriented language that can create queries to extract and manage data from EHRs, to create decision criteria. GELLO is an attempt to address the curly braces problem by inputting the patient data required from a "virtual EMR," which a vendor would then be responsible for linking to. It was also developed by HL7 as a standard query and expression language. It is part of HL7 version 3 and provides the framework for manipulation of clinical data for CDS.[57]

data:
>   blood count storage := event
>   {'complete blood count');
>   hematocrit := read last ({'hematocrit'}:
>   previous_hct := read last ({'hematocrit'})
>   where it occurred before the
>   time of hematocrit);;

A patient's hematocrit may be stored as "hematocrit" in the EHR, or as "Hct" or as a database SELECT instruction. The specifics comprise the content of the curly braces.

Figure 8.5: Data slot of an MLM [56]

GEM: is an international ASTM standard developed so clinical practice guidelines, the source of much of the knowledge, could be stored and shared in a XML format. GEM addresses the lack of uniformity on the source of knowledge that clinical guidelines are written as text, which requires much interpretation to turn into machine-directing rules. Yale University developed this standard and the GLIDES initiative discussed in an earlier section. Another feature is the "GEM Cutter", an XML editor that facilitates guideline markup.[58]

Guideline Interchange Format (GLIF): was developed by several biomedical informatics programs to enable sharable and computer-interpretable guidelines at the knowledge level.[59]

Clinical Quality Language (CQL): The quality-improvement community discovered that population-level imperatives look like clinically-directed rules. CQL is a draft HL7 standard being evaluated as a new language to represent eCQMs and CDS. It facilitates human and machine-readable representations in XML.[60]

Infobuttons: Infobuttons refers to the function of linking patient data to general information; they are often implemented as Web-linked icons that permit downloading of guidelines, articles, monograph entries, or other canonical information. Infobuttons are associated with an HL7 Infobutton Management Standard that standardizes context-sensitive links embedded within EHR systems. The OpenInfobutton project is an open-source initiative by the Veterans Health Administration (VHA) and the University of Utah. As an example, the content provider UpToDate connects ICD-9 codes, lab results and medication information of individual patients via EHR infobuttons to their online knowledge base.[61-62]

Fast Healthcare Interoperability Resources (FHIR): is a HL7 standard that holds great promise for healthcare sharing. The JASON Task Force that was part of the HIT Standards and Policy Committee recommended that FHIR be used and public application programming interfaces (APIs) be adopted by EHR vendors to promote interoperability. FHIR is a RESTful API that is a http-based standard that uses either XML or JSON for data representation, OAuth for authorization and ATOM for queries. This same web services approach is used by Facebook and Google. FHIR is data and not document centric so EHR (A) can place a http request for data from EHR (B); a clinician could request just the results for one lab test, instead of a consolidated CDA document. This approach would also facilitate interactions between an EHR and a CDS database/server. In addition, this strategy permits apps to be created that will interface with EHRs. Figure 8.6 demonstrates an actual CDS app for monitoring bilirubin levels in infants used by Intermountain Healthcare.

The Argonaut Project was created to promote and test this strategy with leaders in the healthcare and EHR vendor industries. HL7 has created a variety of clinical and administrative "resources" (structured data) that can be requested. Examples of resources might be a medication lists, allergies, and problem lists and these can be combined in a message. Each resource would use available standards such as LOINC, RxNorm, etc. This approach is faster, simpler and more flexible than other attempts at interoperability.

CDS is one of the FHIR resources, such that a query can be initiated for a decision based on parameters. For example, FHIR could be used to identify a drug-drug interaction.[27, 63-64] As of early-2018 fifty-four SMART apps are listed in the app gallery. [64]

CDS FUNCTIONALITY

There are several other ways to classify CDSS, in addition to knowledge based and non-knowledge-based systems. For example, Osheroff et al. developed a taxonomy of functional intervention types that was related to the type and timing of tasks.[3]

Importantly, CDS can be located within the EHR or link to a remote CDS program. CDS can activate before, during or after a patient encounter. Activation can be automatic or "on demand." Alerts can be non-interruptive or interruptive (e.g. the process stops until the physician responds to the alert).

Table 8.2 lists a taxonomy of CDS that is based on CDS goals, stated earlier. There is overlap among the categories, as sections such as patient safety and patient care are interrelated.[3, 17-18]

Wright and colleagues and the Provider Order Entry Team (POET) at Oregon Health & Science University elicited a taxonomy from 11 CDS experts and validated the taxonomy with 9 vendors and 4 CDSS developers.[65] Based on function, they arrived at 3 high-level categories: Order facilitators, point-of-care alerts/reminders, and relevant information displays.

Ordering Facilitators. While prescribing support is the main focus of order entry CDS, it is important to realize that clinicians (and patients) need guidance regarding appropriate lab and imaging orders, as well as cost data. With healthcare reimbursement heading in the direction of value-based purchasing, there needs to be greater emphasis on ordering the drugs, labs and images of choice that are evidence based (more comments on this in the chapter on EHRs).

Order sets: are special orders in EHRs that are customized to provide guidance for common problems encountered with hospital admissions. As an example, for community acquired pneumonia (CAP) an order set might recommend a certain antibiotic, when the follow-up chest X-ray will be ordered, oxygen use, etc. This is an effort to standardize care based on evidence-based medicine. Order sets are researched, adopted and modified by local clinicians.

Order sets should make the process of writing admission orders faster with choices provided with drop down lists. The order set standard is an HL7 standard that aims to standardize order set libraries.[66] There has not been a great deal of research in this area. An earlier report by the Veterans Administration (VA) and a report in 2010 summarized their experience with order sets in seven disparate healthcare systems. Both studies confirmed that creating and maintaining order sets is labor intensive, but they save time and improve care. They also point out that many of the order sets were underutilized so ongoing scrutiny is needed.[67-68]

There are commercial providers of order sets, such as Zynx that offers more than 800 order sets capable of being integrated with many EHR vendors.[69]

Therapeutic support: programs such as Theradoc® provide a clinical surveillance program that includes an *Infection Control Assistant* (monitors hospital acquired infections) and *Pharmacy Assistant* that provides drug surveillance of antibiotics and anticoagulants (blood thinners). This platform integrates with clinical EHR data.[70] It is not known how many commercial EHR vendors offer clinical calculators for common medical issues and risk prediction. Allscripts EHR has fully integrated a battery of calculators (68) in their workflow.[71] Calculators can be purchased or home-grown. For example, the Cleveland Clinic wanted to improve prescribing the blood thinner (heparin) so they created an online calculator that was eventually integrated into their EHR.[72] The new SMART platform with FHIR will make it possible to develop apps that integrate with commercial EHRs with open APIs in the near future. A variety of CDS therapeutic tools should become available.[64]

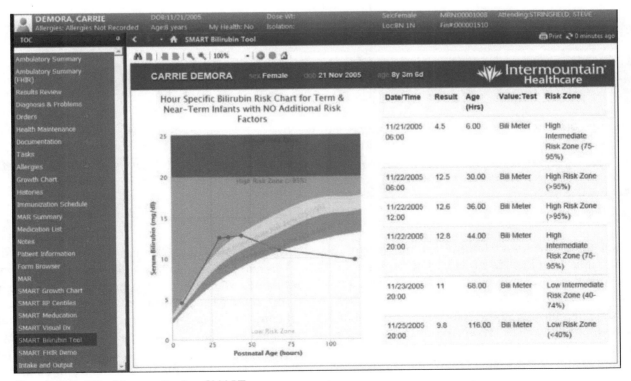

Figure 8.6: Bilirubin monitoring SMART app

Table 8.2: CDS Taxonomy

Function	Examples
Patient Safety	· Medication alerts · Critical lab alerts · Ventilator support alerts · Improved drug ordering for warfarin and glucose · Infusion pump alerts · Risk calculation · Improved legibility · Diagnostic aids
Cost	· Reminders to use generic drugs or formulary recommendations · Fewer duplications · Reminders about costs of drugs, lab tests and imaging studies · Reduce Medicare penalties for readmissions · Reduced medication errors · Reduced malpractice claims · Better utilization of blood products
Patient Care	· Embedded clinical practice guidelines, order sets, and clinical pathways · Better chronic disease management · Identify gaps in recommended care · Immunization aids · Diagnostic aids · Sepsis alerts (**see Case Study infobox**) · Antibiotic duration alerts · Prognostic aids · Patient reminders · Pattern recognition for images, pulmonary function tests and EKGs, blood gases, Pap smear interpretation
Disseminating Expert Knowledge	· Use of infobuttons for clinician and patient education · Provide evidence-based medicine with embedded clinical practice guidelines and order sets
Managing Complex Clinical Issues	· Reminders for preventive care for chronic diseases · Care management · Predictive modeling based on demographics, cost, and clinical parameters
Managing Complex Administrative Issues	· Decision modeling · Research recruitment

Smart forms or smart templates: are electronic encounter notes that include structured questions pertinent to selected diseases, augmented with CDS. For example, a smart form for diabetes would include questions that center on care of the diabetic patient with the usual co-morbidities. The advantage of this approach is that it is relatively fast, and the output is primarily structured data. Also, the smart form can have recommended diagnostic tests and treatments embedded in the form. Partners HealthCare has been a pioneer in smart forms and has designed them for acute respiratory tract infections, diabetes and coronary artery disease. One of the downsides of this approach is that it generates a "robotic" note that does not provide a historical narrative. It is unknown how many templates combine a structured approach with free text (hybrid approach). Templates are generally used for very mild common illnesses such as sore throat, disability determinations, etc. For a

> **Case Study: Prediction of ICU sepsis using minimal EHR data**
>
> Researchers used machine learning software (InSight) to predict sepsis based on a validated ICU data set (MIMIC-III) of 40,000+ patients. The predictors were: vital signs, oxygen saturation, Glasgow Coma Score and age; all easily obtainable from most EHRs.
>
> Using the test dataset with a prevalence of sepsis of 11.3% machine learning classification performance was compared with the existing sepsis scoring systems SIRS, SOFA and MEWS. The results using receiver operating characteristic (ROC) curves and area under the curve (AUC) measures, InSight outperformed the other risk scoring systems. Additionally, they randomly deleted 60% of the data and still outperformed the other scoring systems.
>
> It should be noted that this was a retrospective and not prospective study, so further research will be necessary to see if eventual patient morbidity and mortality is reduced by utilizing this machine learning model in the ICU setting
>
> *Desautels T, Calvert J, Hoffmann J et al. Prediction of Sepsis in the Intensive Care Unit With Minimal Electronic Health Record Data: A Machine Learning Approach. JMIR 2016;4(3):e28*

discussion of the issues related to structured notes versus free text narratives, see the article by Rosenbloom et al.[73]

Alerts and Reminders: alerts serve to warn the clinician about potential problems, during or after the patient visit. Common alerts appear as pop-up boxes and are triggered by electronic prescribing, ordering of labs or images and return of results. The alerts can be interruptive, forcing the clinician to provide a response or non-interruptive, providing information or advice during a patient encounter. Reminders, as the name implies, remind a clinician or patient about preventive medicine measures such as "mammogram is due." These are often triggered before the patient is seen and may or may not appear as pop-up boxes.

Relevant Information Displays:

- **Infobuttons, hyperlinks, mouse-overs:** infobuttons were covered in the section on CDS standards. These tools can be used by any member of the healthcare team or patients. Depending on the EHR vendor, users can connect to a variety of embedded internal and external educational resources. For example, the medical resource UpToDate offers an embedded link for nine major commercial EHRs.[74]
- **Diagnostic support:** non-integrated differential diagnosis support is available through Simulconsult and DxPlain.[8,15] Isabel, discussed earlier, is capable of automatically pulling coded symptom and sign data from the EHR.[14] Ultimately, natural-language-processing programs will be more routine, able to pull in all types of data and fully integrate with the EHR, so that support tools are auto-populated with demographics and clinical data, thus saving the clinician time.
- **Dashboards:** are an infrequently discussed type of CDS, because they are aimed at *population-based* decision making, rather than individual clinical decision making. Defined broadly, dashboards can include any type of patient summary, flow chart of lab results or vital signs or a disease registry. Excellent data representation helps the clinician track and trend common medical problems.

CDS SHARING

Currently, there is no single way that CDS knowledge is represented and shared. Importantly, there doesn't seem to be a universal incentive for disparate healthcare organizations to share CDS tools and other HIT. Various CDS standards exist as discussed in a prior section. Ideally, standards would allow for sharing among healthcare organizations and EHR vendors, but in 2018, we do not have CDS interoperability. Thus far, two approaches have been taken: structure the knowledge base for sharing, using a standard vocabulary or access a central repository of CDS using service-oriented architecture (SOA).[3] Both ONC and AHRQ have sponsored CDS sharing initiatives in the past, as outlined in the prior chapter section *Clinical Decision Support by Organizations*.

Several other CDS initiatives should be mentioned. The Socratic Grid is an open-source SOA platform that supports CDS of clinical data and utilizes more than just production rules and offers a variety of prediction models.[75] OpenCDS is another open-source open-standards SOA platform that provides CDS tools and resources. This public-private collaboration uses open patient data as input and conclusions/recommendations are delivered as output.[76]

ONC convened an expert panel on CDS sharing in 2013 and they listed the key principles for sharing: *"(1) prioritize and support the creation and maintenance of a national CDS knowledge sharing framework; (2) facilitate the development of high-value content and tooling, preferably in an open-source manner; (3) accelerate the development or licensing of required, pragmatic standards; (4) acknowledge and address medicolegal liability concerns; and (5) establish a self-sustaining business model."*[77]

It seems likely FHIR will become a new approved standard for healthcare data sharing, including CDS, in the near future.

CDS IMPLEMENTATION

Multiple excellent resources are available for CDS implementation. Table 28.3 will outline some of the logical steps used for the management of most CDS implementation projects. The implementation specifics provided came from the following references.[4,23,78-80]

CDS CHALLENGES

General

Robert Greenes aptly stated *"the difficulty in deploying and disseminating CDS is in large part due to the lack of recognition of how hard the job is and lack of availability of tools and resources to make this job easier."* Furthermore, not all CDS is the same; diagnostic CDS is more complicated than simple allergy alerts.[81]

CDS is clearly challenged by the knowledge explosion in medicine, big data from digitized medicine and the addition of genomics into clinical practice. Currently, incentives to use CDS is aligned with meaningful use payments but this program does not apply to all physicians and what happens when the program goes away?

Organizational support: Clearly, for CDSS to be successful there has to be support from the administrative, IT and clinical leadership. It should be part of the mission and vision of the healthcare organization. This implies that CDS is a priority, there are incentives and it is achievable. Only large healthcare organizations with a track record of innovation and advanced IT support can create and maintain a sophisticated CDDS, as we currently know them. It is still not clear that CDS has enough value or a strong business case as it currently stands. Table 8.3 outlines CDS implementation steps.

CDS Reports and Reviews: Much has been written about the impact of CDS on medical and administrative processes, as well as patient outcomes. This section will summarize several of the major reviews on this topic. It should be noted that there have been many reviews of the impact of health information technology (HIT) on medical and administrative processes, as well as reviews on computerized physician order entry (CPOE) that includes CDS. Moreover, many reviews were written before the HITECH Act and the EHR Incentive program, such that studies performed before the widespread adoption of EHRs may have different results from the post-adoption period.[82] The following are reviews that specifically evaluate CDSSs:

- Two systematic reviews by Garg et al. and Kawamoto et al. in 2005 discussed early CDS impact. According to Garg et al. 64% of the studies showed improvement in clinician performance and 13% showed improvement in actual patient outcomes.[83] Kawamoto et al. similarly found that 68% of the studies demonstrated improvement in clinical practice. They identified the following success factors: CDS was integrated into clinical workflow; CDS was electronic, not paper based; CDS provided support at the right time and location and CDS provided actionable recommendations.[84]

- Berner wrote a CDS State of the Art paper for AHRQ in 2009. She made the point that many studies focused on inpatients and many were done at academic medical centers with home grown EHRs and strong IT support. This may yield different results from studies based on commercial EHRs in community settings. She also noted that many of the studies were not randomized controlled trials and most focused on processes and not patient outcomes. Additionally, there were methodological limitations noted in most studies. The point was made that qualitative studies are also needed to better understand why some CDSS implementations work and others do not.[34]

- Black et al reported in 2011 on the impact of health information technology (eHealth) on healthcare quality and safety. Like others, this effort focused on systematic reviews from 1997 to 2010. Overall, they felt the studies were of poor quality. CDS was a category of their review but they concluded that there was a large gap between the potential and proven benefits of all HIT, to include CDSSs. Evaluations tended to focus on benefits with little attention to unintended adverse consequences and cost. The authors included a table of studies demonstrating CDS benefit and a table for evidence of risks.[85]

Table 8.3: Logical Steps for CDS Implementation

Logical Steps	Details
Project initiation	• Ensure clinical and non-clinical leadership are onboard and have a shared vision • Ensure CDS is synched with organizational goals, patient safety/quality measures and meaningful use objectives • Determine the business case/value of CDS for the organization • Determine feasibility from a manpower and financial standpoint and acceptance by clinicians • Ensure objectives are clear and attainable • Identify key stakeholders and assess buy-in • Understand that the CDS needs of specialists are different from primary care • Assess readiness, EHR capability and IT support • Assess the clinical information systems (CISs) involved • Assess knowledge management capabilities • Assemble the CDS team: clinical leaders, CMIO, administrative and nursing leaders, managers, EHR vendor and IT experts • Identify clinical champions • Develop CDS charter
Project planning	• Consider a SWOT analysis (strengths, weaknesses, opportunities and threats) • Utilize standard planning tools such as Gantt charts and swim lanes • Develop timeline • Decide whether to build or buy CDS content • CDS committee should select CDS interventions that fit their vision • Be sure to follow the 5 Rights of CDS • Map the different processes involved with CDS and be sure they integrate with the clinician's workflow • Determine whether you will measure structure, processes and/or outcomes • Plan the intervention: triggers, knowledge base, inference engine and communication means • Educate staff and gain their input • Design the CDS program for improvement over baseline performance in an important area for the organization. In other words, be sure you can measure outcomes and compare with baseline data • Investigate the needed CDS standards required • Follow the mandates of change management, e.g. John Kotter's Eight Step Model • Communicate goals of CDS project to all affected
Project execution	• Provide adequate training and make CDS training part of EHR training • Develop use cases • Test and re-test the technology: unit, integration and user acceptance testing • Decide on incremental roll-out or "big bang" • Provide a mechanism for feedback in the CDS process, as well as formal support
Project monitoring and control	• Use data from feedback, override logs, etc. to modify the system as needed • Compare the alert and override rates with national statistics • Measure percent of alerts that accomplished desired goals • Communicate the benefits and challenges to the end-users as they arise • Use tools such as the AHRQ Health IT Evaluation ToolKit • Knowledge management maintenance; are guidelines unambiguous and up to date? Who will maintain the content?

- Lobach et al. published an evidence report on CDS and knowledge management for AHRQ in 2012. Their review covered the period 1976 through 2010, to include 148 randomized controlled trials. They concluded that commercially and locally developed CDSSs (inpatient and outpatient) improved preventive services, ordering clinical studies and prescribing therapies. They identified six new success factors: integration with CPOE, promotion of action, lack of need of additional data entry by clinicians, evidence based CDS, local involvement and CDS for patients and clinicians. Only 20% of trials reported on actual patient outcomes and only 15% reported on cost.[86]
- Jasper et al. reported a systematic review on prior CDS systematic reviews (SRs) in 2011. They used the AMSTAR rating system for systematic reviews and only examined SRs with a score of 9 or above. In spite of using this scoring system, only 17 out of 35 SRs were of high quality. They concluded that 57% of SRs showed practitioner improvement and 30% showed a positive impact on patient outcomes. They made the point that simple CDS, such as drug alerts and preventive care reminders are likely successful because they require minimal patient data.[87]
- Bright et al., in another systematic review of the literature from 1976 to 2011, reported their results in 2012. The authors were the same as in the review by Lobach. Like other reviews they noted the majority reported on the effect of CDS on processes and not outcomes or cost. They commented that few mentioned unintended adverse consequences, workflow or efficiency issues and studies were heterogeneous, making comparisons difficult.[88]
- Jones et al. reported a systematic review of HIT, with a focus on meaningful use covering the time period from 2010 to 2013. Fifty-seven percent of the reports evaluated CDS, with 65 percent being positive, 17 mixed, 11 neutral and 7 negative. The authors concluded that most reports were positive in regard to the effects on quality, safety and efficiency, but there was a paucity of information about contextual and implementation factors that would have been helpful to determine why some implementations were not successful.[89]

To summarize, there is ample evidence to show CDS improves a variety of processes but there is much less evidence to show a positive impact on clinical outcomes, such as mortality, length of stay, health-related quality of life and adverse events. The same holds for measuring economic outcomes.[4]

Unintended adverse consequences (UACs): As with all HIT implementations, unanticipated and undesirable side effects are discovered over time. Ash et al. convened an expert panel to analyze UACs related to CDSSs. They concluded that there were two major UAC patterns noted: problems associated with CDS content and the presentation of information on the computer screen. Content issues were related to shifting of human roles, outdated CDS and misleading CDS. Presentation issues were related to rigidity, alert fatigue and a variety of potential errors such as incorrect auto-completes and timing issues.[90]

Alert fatigue is perhaps the most publicized UAC related to CDSSs. van Der Sijs et al. performed a review of this issue in 2006 with several important observations. Physicians override alerts between 49-96% of the time but most are for good reasons. Only 2-3% of the time does overriding result in a true adverse drug event (ADE). Reviewers tended to agree with the overrides 95.6% of the time. These data suggest that too many alerts are unnecessary and poorly written.[91] High override rates (81-87%) were also reported out of the VA system, however a study in Boston showed an acceptance rate of 67% for interruptive alerts.[92-93] Clearly, how well the alerts are written, how they appear in workflow and how well the innovation is accepted plays a role.

Medico-legal: because CDSSs make recommendations that may alter patient outcomes, they have legal ramifications. CDS needs to be based on the best recognized evidence and must be updated regularly. Ignoring EHR generated alerts or treatment recommendations also has legal implications. This raises the question, do healthcare organizations need to archive alerts and responses, in case there are negative outcomes?[94] For the above reasons the FDA has suggested that they regulate HIT that has patient safety implications. As of early 2018, the FDA has not provided guidance regarding CDS regulations and this may have a dampening effect on CDS innovation and development.[95]

In addition, product liability for CDS malfunctioning may discourage innovation by EHR vendors.

Clinical: CDS must fit into the workflow of all members of the healthcare team. CDS needs to be integrated within the EHR and adhere to the five rights of CDS, covered earlier in the chapter. Currently, most CDS is part of meaningful use objectives, but to be truly successful, other decisions must be supported, shared, updated and measured. For example, medication CDS must eventually include adjustments for the patient's weight, gender, renal and liver function, mental status, age, etc.

Technical: as discussed in other sections, uniform standards and an acceptable interoperability platform are necessary so that CDS can be standardized, updated, shared and stored. Also, for CDS to be well received and promulgated it must be well designed and match usability criteria. Furthermore, the creation of local CDS (not vendor generated) is complicated and labor intensive, requiring a multi-disciplinary team, coupled with maintenance and evaluation. CDS is therefore an example of a "socio-eco-technical" implementation.

CDS will likely be more complicated in the future as we move beyond meaningful use objectives and consider incorporating genetic data into decision support. Highly complex CDS will magnify alert fatigue and the need for near constant updating of rules and algorithms.

Lack of interoperability: the lack of interoperability has been one of the most challenging obstacles to EHRs and HIT in general. Ideally, CDS should be shared so that healthcare organizations do not have to reproduce work that has already been done. As discussed in prior sections, multiple other organizations are working on a solution.

Long term CDS benefits: Durability remains a question mark. Multiple studies have shown that if there is scrutiny of a systematic change, like CDS, there seems to be improvement; but when the oversight diminishes, so do the results.[96]

LESSONS LEARNED

There is a wealth of literature on avoiding the pitfalls of CDS design, implementation and maintenance. We will summarize some of the best-known lessons learned and recommendations from a variety of resources into Table 8.4.[4, 34, 70-71, 97-98]

Many of the studies were conducted pre-HITECH Act so better data may be available as CDS interventions are generated and reported as part of meaningful use. For example, the study by Roshanov et al. suggested that standalone CDS programs were linked to a higher success rate than those what were part of CPOE.[99] Will this hold up with newer data? We also don't know if CDS content that is purchased will be a better or worse choice for organizations.

Table 8.4 Lessons Learned

Lesson Learned	Comments
Project initiation • Healthcare organizations have competing priorities • CDS cannot come from external mandate	Ensure the organization can support a new CDS initiative. Even if CDS is intended to match meaningful use, it must be embraced by all and match organizational goals
Project planning • Customization of content and workflow is important • One size CDS does not fit all • CDS must match the 5 Rights of CDS • Make CDS as non-intrusive and non-interruptive as possible • Ideally, there should be recommendations for clinicians and patients • Interventions should include a reason for overrides • Intervention should make recommendation and not just assessment • "Do CDS with users, not to them" • EHR data must be up to date for triggers to work correctly	Customization is desirable but labor intensive and not available at smaller organizations. Specialists and primary care clinicians have different needs. Clinicians do not want to stop and speed is important. Table 8.1 CDS Taxonomy
Project execution • Feedback buttons in CDS work well • Include CDS training into EHR training • CDS must be tested for UACs and patient safety	User feedback is critical
Project monitoring and control • Knowledge management is time consuming • Be sure intervention content is up to date	There may have to be a separate knowledge management committee

Executing CDS the way it comes out of the box is not enough. Given that alert fatigue is so prevalent (up to 95% ignoring alerts), implementers have the obligation of looking at the entire decision making/therapeutic process and not just the physician's interaction with the system.

RECOMMENDED RESOURCES

Textbooks

- Berner ES. Clinical decision support systems theory and practice. Clinical decision support systems theory and practice. 2007.[100]
- Greenes Robert A. Clinical Decision Support: The Road to Broad Adoption. Second Edition. 2014 [81]
- Osheroff Improving Outcomes with Clinical Decision Support. An Implementer's Guide. Second Edition. 2012 HIMSS. Chicago, IL[4]

Web sites

- AHRQ CDS Initiatives [26]
- ONC CDS Resources [23]
- Harvard Library of Evidence. Website includes 236 clinical decision rules with Oxford level of evidence. The (IF-THEN) rules area available for download as FHIR or XML. [101]

White papers

- Lobach et al. Enabling Health Care Decision making Through Clinical Decision Support and Knowledge Management: Evidence Rep. 2012 [86]
- Berner E. Clinical Decision Support Systems: State of the Art (PDF) 2009 [34]
- Karsh BT. Practice Improvement and Redesign: How Change in Workflow Can Be Supported by CDS (PDF). AHRQ 2009.[102]
- Tcheng JE, Bakken S, Bates DW et al. Optimizing Strategies for Clinical Decision Support. Summary of a Meeting Series. April 2018. National Academy of Medicine, Wash. DC. [103]

Articles

- Felcher AH, Gold R, Mosen DM et al. Decrease in unnecessary vitamin D testing using clinical decision support tools: making it harder to do the wrong thing. Researchers used three CDS tools in the EHR to improve vitamin D testing. Sixth months after implementation, screening rates decreased from 74 to 24 tests per 1000 members, and the proportion of appropriate screening increased from 56.2% to 69.7%.[104]
- Hay SN, Brecher ME. Evidence-based clinical decision support. This article was written from the perspective of improving laboratory testing. The article was divided into pre-analytical, analytical and post-analytical testing with examples of how CDS can be utilized in each phase.[105]

FUTURE TRENDS

In the section on CDS sharing it was reported that subject matter experts felt that the US needed a national CDS knowledge sharing framework. Presumably, this would be the responsibility of ONC or AHRQ. To accomplish this, there would likely need to be more data standards created and a business case for widespread CDS adoption. Clearly, Meaningful Use objectives have provided a business case but the future of healthcare reform is uncertain.[94, 106]

A newer direction appears to be using FHIR for interoperability and the creation of CDS applications that integrate with commercial EHRs. Using this approach there would be new incentives for developers to create a variety of decision support tools for clinicians and patients.

We can anticipate newer methods to deliver CDS in the future. For example, one speech recognition vendor (M*Modal) in 2017 embeds simple CDS into the dictated notes to point out deficiencies. For example, if a clinician dictates heart failure but fails to specify acute or chronic, systolic versus diastolic, this is recognized through "natural language understanding" and a pop-up appears. This might improve coding with ICD-10. Other issues related to note quality could also trigger alerts.[107] To our knowledge this approach has not been objectively evaluated yet.

KEY POINTS

- With evolving technologies and strategies, the definition of CDS should be broad
- There is widespread support for CDS implementation as part of healthcare reform
- In spite of widespread support, CDS faces multiple challenges
- The evidence, thus far, suggests CDS benefits processes more than patient outcomes

CONCLUSION

Busy clinicians often do not incorporate the best evidence into their medical practices and the hope is that robust CDS, that is part of modern EHRs, will offer them the best evidence within their natural workflow.[102] The Institute of Medicine (National Academy of Medicine) has promoted the idea of the "learning health system" that is part of US healthcare of the future. Their third recommendation towards reform promoted *"Decision support tools and knowledge management systems should be routine features of healthcare delivery to ensure that decisions made by clinicians and patients are informed by current best evidence."*[108]

The concept of clinical decision support is not in its infancy, but the use of electronic CDS that is fully integrated with EHRs is new. Therefore, we still need better research as to what works and what does not. To date, attempts to deliver the best evidence via CDS has produced mixed results. It is likely we will see further innovations arising that will make CDS more usable and more acceptable by the average busy clinician.

REFERENCES

1. Clinical Decision Support. Office of the National Coordinator. http://www.healthit.gov/policy-researchers-implementers/clinical-decision-support-cds (Accessed January 15, 2015)
2. Ely JW, Osheroff JA, Maviglia SM et al. Patient-Care Questions That Physicians Are Unable to Answer. JAMIA 2007;14(4):407-414
3. Ledley RS, Lusted LB. Reasoning Foundations of Medical Diagnosis. Science. 1959;130(3366):9-21
4. Improving Outcomes with Clinical Decision Support: An Implementer's Guide Jerome A. Osheroff, Jonathan M. Teich, Donald Levick, Luis Saldana, Ferdinand T. Velasco, Dean F. Sittig, Kendall M. Rogers, Robert A. Jenders: Second Edition. 2012 HIMSS. Chicago, IL
5. De Dombal FT, Leaper DJ, Staniland JR, McCann AP, and Horrocks JC. Computer-aided diagnosis of acute abdominal pain. British medical journal. 1972;2(5804):9.
6. Parker RC, and Miller RA. Creation of realistic appearing simulated patient cases using the INTERNIST-1/QMR knowledge base and interrelationship properties of manifestations. Methods Inf Med. 1989;28(4):346-51. Data Model update Jan 2015 http://www.healthit.gov/sites/default/files/qdm_4_1_2.pdf (Accessed June 16, 2015)
7. Shortliffe EH, Davis R, Axline SG, Buchanan BG, Green CC, and Cohen SN. Computer-based consultations in clinical therapeutics: Explanation and rule acquisition capabilities of the MYCIN system. Computers and Biomedical Research. 1975;8(4):303-320. doi:10.1016/0010-4809(75)90009-9.
8. DxPlain http://www.mghlcs.org/projects/dxplain (Accessed February 12, 2015)
9. Quick Medical Reference (QMR) http://www.openclinical.org/aisp_qmr.html (Accessed February 15, 2015)
10. Health Evaluation Through Logical Processing (HELP) http://intermountainhealthcare.org/isannualreport/2009/Website/help2.html (Accessed February 9, 2015)
11. Iliad 4.5 http://www.ramex.com/title.asp?id=1292 (Accessed February 9, 2015)
12. Illiad http://www.openclinical.org/aisp_iliad.html (Accessed February 9, 2015)
13. OpenClinical http://www.openclinical.org/aisinpractice.html (Accessed February 9, 2015)
14. Isabel http://www.isabelhealthcare.com/home/default (Accessed February 9, 2015)
15. SimulConsult www.simulconsult.com (Accessed February 1, 2015)
16. SnapDx http://www.snapdx.co/ (Accessed February 9, 2015)
17. Perreault LE, Metzler JB. A pragmatic framework for understanding clinical decision support. Journal of Healthcare Information Management. 1999;13(2):5-21.
18. Gilmer TP, O'Connor PJ, Sperl-Hillen JM, Rush WA, Johnson PE, Amundson GH, Asche SE, and Ekstrom HL. Cost-effectiveness of an electronic medical record based clinical decision support system. Health Serv Res. 2012;47(6):2137-58. doi:10.1111/j.1475-6773.2012.01427.x.
19. Corrigan JM. Crossing the quality chasm. Building a Better Delivery System. 2005. National Academies Press. www.nap.edu
20. Osheroff JA, Teich JM, Middleton B, Steen EB, Wright A, and Detmer DE. A roadmap for national action on clinical decision support. Journal of the American Medical Informatics Association: JAMIA. 2009;14(2):141-5. doi:10.1197/jamia.M2334 http://www.pubmedcentral.nih.gov/articlerender.fcgi?artid=2213467&tool=pmcentrez&rendertype=abstract (Accessed February 15, 2015)
21. American Medical Informatics Association (AMIA) www.amia.org (Accessed January 28, 2015)
22. Advancing Clinical Decision Support. RAND corporation. http://www.rand.org/health/projects/clinical-decision-support.html (Accessed January 29, 2015)

23. Office of the National Coordinator. Clinical Decision Support. http://www.healthit.gov/policy-researchers-implementers/clinical-decision-support-cds (Accessed January 29, 2015)
24. CDS Portal http://cdsportal.partners.org (Accessed January 29, 2015)
25. CDS Consortium http://www.cdsconsortium.org/ (Accessed February 1, 2015)
26. Clinical Decision Support Initiatives http://healthit.ahrq.gov/ahrq-funded-projects/clinical-decision-support-cds-initiative (Accessed January 9, 2015)
27. HL7 www.hl7.org (Accessed January 8, 2015)
28. Quality Forum. Driving Quality and Performance Measurement—A Foundation for Clinical Decision Support: A Consensus Report. NQF. Washington DC 2010. http://www.qualityforum.org/Publications/2010/12/Driving_Quality_and_Performance_Measurement_-_A_Foundation_for_Clinical_Decision_Support.aspx (Accessed January 8, 2015)
29. Kilbridge PM, Welebob EM, and Classen DC. Development of the Leapfrog methodology for evaluating hospital implemented inpatient computerized physician order entry systems. Qual Saf Health Care. 2006;15(2):81-4. doi:10.1136/qshc.2005.014969.
30. Leapfrog Group. www.leapfroggroup.org (Accessed January 29, 2015)
31. HIMSS Electronic Medical Record Adoption Model http://www.himssanalytics.org/emram/emram.aspx (Accessed December 20, 2017)
32. Centers for Medicare and Medicaid Services. www.cms.gov (Accessed January 29, 2015)
33. Driving Quality and Performance Measurement—Foundation for Clinical Decision Support. NQF. http://www.qualityforum.org/Publications/2010/12/Driving_Quality_and_Performance_Measurement__A_Foundation_for_Clinical_Decision_Support.aspx. (Accessed February 28, 2015)
34. Berner ES. Clinical decision support systems: state of the art. AHRQ Publication. 2009;(09-0069):4-26.
35. Wong DT, Knaus WA. Predicting outcome in critical care: the current status of the APACHE prognostic scoring system. Can J Anaesth 1991; 38:374-83
36. Buchanan BG, Shortliffe EH. Rule-Based Expert Systems: The MYCIN Experiments of the Stanford Heuristic Programming Project. Reading, MA: Addison-Wesley; 1984
37. Introduction to Knowledge Systems. Mark Stefik. Morgan Kaufmann Publishing. 1995. San Francisco, CA.
38. Baye's Formula. Cornell University. http://www.math.cornell.edu/~mec/2008-2009/TianyiZheng/Bayes.html (Accessed February 10, 2015)
39. Gideon. http://www.gideononline.com/ (Accessed July 25, 2016)
40. Geissbuhler A, Miller RA. Distributing knowledge maintenance for clinical decision support systems: the knowledge library model. Proc AMIA Symp 1999:770-774
41. WEKA Machine Learning. http://www.cs.waikato.ac.nz/ml/weka/ (Accessed July 25, 2016)
42. Predictive Analytics. http://www.predictiveanalyticstoday.com/what-is-predictive-analytics/ (Accessed July 26, 2016)
43. McCulloch WS, and Pitts W. A logical calculus of the ideas immanent in nervous activity. The bulletin of mathematical biophysics. 1943;5(4):115-133.
44. Sordo M. Introduction to Neural Networks i Healthcare. 2002. Open Clinical http://www.openclinical.org/docs/int/neuralnetworks011.pdf (Accessed February 2, 2015)
45. What is logistic regression? Statistics Solutions. http://www.statisticssolutions.com/what-is-logistic-regression/ (Accessed July 26, 2016)
46. C4.5 Programs for machine learning. J. Ross Quinlan. 1993. Morgan Kaufmann Publishers, San Mateo, CA.
47. Wang F. Adaptive semi-supervised recursive tree partitioning: The ART towards large scale patient indexing in personalized healthcare J. Bio Inform 2015; 55:41-54
48. Hunink MGM. Decision Making in Health and Medicine: Integrating Evidence and Values. Cambridge; Cambridge University Press; 2001
49. Gambhir SS, Shepherd JE, Shah BD et al. Analytical Decision Model for the Cost Effective Management of Solitary Pulmonary Nodules. J Clin Onc 1998;16(6):2113-2125
50. Hardin JM, Chhieng DC. Data mining and clinical decision support systems. Chapter 3 in Berner ES. Clinical decision support systems theory and practice. 2007. Springer. Second edition.
51. Newcomer SR, Steiner JF, and Bayliss EA. Identifying subgroups of complex patients with cluster analysis. The American journal of managed care. 2010;17(8): e324-32.
52. Colak C, Karaman E, Turtay MG et al. Application of knowledge discovery processes on the prediction of stroke. Compute Methods Programs Biomed 2015 119(3):181-185
53. AHRQ Beta-Blockers for Acute Myocardial Infarction. http://archive.ahrq.gov/clinic/commitfact.htm (Accessed February 8, 2016)
54. Arden Syntax. HL7. http://www.hl7.org/implement/standards/product_brief.cfm?product_id=2 (Accessed February 1, 2015)
55. Jenders RA. Decision Rules and Expressions. Chapter 15 in Greenes RA. Clinical decision support the road to broad adoption. 2014.

56. Hripcsak G. Writing Arden Syntax Medical Logic Modules. Comput Biol Med. 1994 Sep;24(5):331-63.
57. GELLO. HL7. http://www.hl7.org/implement/standards/product_brief.cfm?product_id=5 (Accessed February 2, 2015)
58. GEM. ASTM. http://www.astm.org/Standards/E2210.htm (Accessed February 3, 2015)
59. GLIF. Biomedical Informatics Research and Development Center. University of Rochester. http://www.birdlab.org/research-glif.cfm (Accessed February 2, 2015)
60. Strasberg H. Clinical Quality Language. www.solutions.wolterskluwer.com January 12, 2015 (Accessed March 1, 2015)
61. OpenButton Project http://www.openinfobutton.org/ (Accessed February 3, 2015)
62. UpToDate. Infobutton. http://www.uptodate.com/home/hl7 (Accessed February 3, 2015)
63. FHIR HL7. http://hl7-fhir.github.io/index.html (Accessed February 10, 2016)
64. SMART on FHIR http://smarthealthit.org/smart-on-fhir/ (Accessed June 2, 2016)
65. Wright A, Sittig DF, Ash JS et al. Development and evaluation of a comprehensive CDS taxonomy: comparison of front-end tools in commercial and internally developed electronic health record systems. J Am Med Assoc 2011;18(3):232-242
66. HL7 Order Set Standard. http://www.hl7.org/implement/standards/product_brief.cfm?product_id=287 (Accessed March 8, 2015)
67. Payne T, Hoey P, Nichol P et al. Preparation and Use of Pre-constructed Orders, Order Sets and Order Menus in a Computerized Provider Order Entry System. Journal of the American Medical Informatics Association. 2003;10(4):322-329
68. Wright A, Sittig EF, Carpenter JD et al. Order Sets in Computerized Physician Order Entry Systems: An Analysis of Seven Sites. AMIA 2010 Symposium Proceedings. :892-896
69. Zynx Health. www.zynxhealth.com (Accessed February 7, 2016)
70. Theradoc. www.theradoc.com (Accessed March 21, 2015)
71. eCalcs. http://www.galenhealthcare.com/products-services/products/ecalcs/ (Accessed March 6, 2015)
72. Butterfield S. Let the computer do the math. ACP Hospitalist August 2014. www.acphospitalist.org (Accessed March 7, 2015)
73. Rosenbloom ST, Denny JC, Xu H, Lorenzi N, Stead WW, and Johnson KB. Data from clinical notes: a perspective on the tension between structure and flexible documentation. Journal of the American Medical Informatics Association. 2011;18(2):181-186
74. UpToDate www.uptodate.com (Accessed March 4, 2015)
75. Socratic Grid www.socraticgrid.org (Accessed March 24, 2015)
76. Open CDS. www.opencds.org (Accessed March 24, 2015)
77. Kawamoto K, Hongsermeier T, Wright A, Lewis J, Bell DS, and Middleton B. Key principles for a national clinical decision support knowledge sharing framework: synthesis of insights from leading subject matter experts. Journal of the American Medical Informatics Association. 2013;20(1):199-207 http://www.ncbi.nlm.nih.gov/pmc/articles/PMC3555314/
78. Project Management www.project-management.com (Accessed March 24, 2015)
79. Byrne C, Sherry D, Mercincavage L et al. Advancing Clinical Decision Support. Key Lessons in Clinical Decision Support Implementation. Westat Technical report. 2011. http://www.healthit.gov/sites/default/files/acds-lessons-in-cds-implementation-deliverablev2.pdf (Accessed January 20, 2015)
80. Kotter J. Leading change. Boston Mass.: Harvard Business School Press; 1996.
81. Greenes RA. Clinical decision support. The road to broad adoption. 2014. Second edition. Elsevier.
82. Romano MJ, Stafford RS. Electronic Health Records and Clinical Decision Support Systems. Impact on Ambulatory Care Quality. Arch Int Med 2011; 10:897-903
83. Garg AX, Adhikari NK, McDonald H, Rosas-Arellano MP, Devereaux PJ, Beyene J, Sam J, and Haynes RB. Effects of computerized clinical decision support systems on practitioner performance and patient outcomes: a systematic review. JAMA. 2005;293(10):1223-38. doi:10.1001/jama.293.10.1223.
84. Kawamoto K, Houlihan CA, Balas EA, and Lobach DF. Improving clinical practice using clinical decision support systems: a systematic review of trials to identify features critical to success. Bmj. 2005 ;330(7494) :765. doi: http://dx.doi.org .
85. Black AD, Car J, Pagliari C, et al. The Impact of eHealth on the Quality and Safety of Health Care: A Systematic Overview. PLoS Medicine. January 2011; 8(1). www.plosmedicine.org (Accessed June 4, 2015)
86. Lobach D, Sanders GD, Bright TJ, Wong A, Dhurjati R, Bristow E, Bastian L, Coeytaux R, Samsa G, and Hasselblad V. Enabling health care decision making through clinical decision support and knowledge management. 2012
87. Jaspers MWM, Smeulers M, Vermeulen H, and Peute LW. Effects of clinical decision-support systems on practitioner performance and patient outcomes: a synthesis of high-quality systematic review findings. Journal of the American Medical Informatics Association: JAMIA. 2011;18(3):327-334.

88. Bright TJ, Wong A, Dhurjati R, Bristow E, Bastian L, Coeytaux RR, Samsa G, Hasselblad V, Williams JW, Musty MD, Wing L, Kendrick AS, Sanders GD, and Lobach D. Effect of clinical decision-support systems: a systematic review. Ann Intern Med. 2012;157(1):29-43. doi:10.7326/0003-4819-157-1-201207030-00450
89. Jones SS, Rudin R, Perry T et al. Health Information Technology: An Updated Systematics Review with Focus on Meaningful Use. Ann Intern Med. 2014 ;160(1) :48-54
90. Ash J. S., Sittig, D. F., Campbell, E. M., Guappone, K. P., & Dykstra, R. H. (2007). Some unintended consequences of clinical decision support systems. AMIA ... Annual Symposium Proceedings / AMIA Symposium. AMIA Symposium, 26–30.
91. Van der Sijs H, Aarts J, Vulto A, Berg M. Overriding of drug safety alerts in computerized physician order entry. JAMIA 2006; 13:138-147.
92. Lin C-P, Payne TH, Nichol WP, Hoey PJ, Anderson CL, and Gennari JH. Evaluating clinical decision support systems: monitoring CPOE order check override rates in the Department of Veterans Affairs' Computerized Patient Record System. Journal of the American Medical Informatics Association. 2008;15(5):620-626
93. Shah NR, Seger AC, Seger DL, Fiskio JM, Kuperman GJ, Blumenfeld B, Recklet EG, Bates DW, and Gandhi TK. Improving acceptance of computerized prescribing alerts in ambulatory care. Journal of the American Medical Informatics Association. 2006;13(1):5-11.
94. Greenberg M, and Ridgely MS. Clinical decision support and malpractice risk. JAMA: the journal of the American Medical Association. 2011;306(1):90-91
95. Wicklund E. FDA urged to clarify clinical decision support regulations. mHealth Intelligence. February 26, 2016. www.mhealthintelligence.com (Accessed September 16, 2016)
96. Gerber JS, Prasad PA, Fiks AG et al. Durability of benefits of an outpatient antimicrobial stewardship intervention after discontinuation of audit and feedback. Research Letter. December 17 2014 JAMA;2014312(23):2569-2570
97. Bates DW, Kuperman GJ, Wang S et al. Ten Commandments for Effective Clinical Decision Support: Making the Practice of Evidence-based Medicine a Reality. Journal of the Amer Med Info Assoc 2003;10(6):523-530
98. Eichner J and Das M. Challenges and Barriers to Clinical Decision Support Design and Implementation Experienced in the Agency for Healthcare Research and Quality CDS Demonstrations. AHRQ Pub. No.10-0064-EF March 2010
99. Roshanov PS, Fernandes N, Wilczynski JM et al. Features of effective computerized clinical decision support systems: meta-regression of 162 randomized trials. BMJ 2013;346: f657 doi: 10.1136/bmj. f657
100. Clinical Decision Support Systems. Theory and Practice. Second Edition. Eta S. Berner, Editor. 2007. Springer. New York, NY. http://www.unimasr.net/ums/upload/files/2012/Mar/UniMasr.com_ad22bb3650b5a1fa7e31e56a8e03f3a0.pdf (Accessed June 3, 2016)
101. Harvard Library of Evidence. http://libraryofevidence.med.harvard.edu/app/library (Accessed December 12, 2017)
102. Karsh BT. Practice Improvement and Redesign: How Change in Workflow Can Be Supported by CDS (PDF). AHRQ 2009 http://healthit.ahrq.gov/sites/default/files/docs/page/09-0054-EF-Updated_0.pdf (Accessed January 5, 2016)
103. Tcheng JE, Bakken S, Bates DW et al. Optimizing Strategies for Clinical Decision Support. Summary of a Meeting Series. April 2018. National Academy of Medicine, Wash. DC. https://nam.edu/optimizing-strategies-clinical-decision-support/ (Accessed April 10, 2018)
104. Felcher AH, Gold R, Mosen DM et al. Decrease in unnecessary vitamin D testing using clinical decision support tools: making it harder to do the wrong thing. JAMIA. 2017;24(4):776-780
105. Hay SN, Brecher ME. Evidence-based Clinical Decision Support. November 21, 2016. www.Laboratory-manager.advanceweb.com (Accessed November 11, 2017)
106. Stage 3 Proposed Rule https://s3.amazonaws.com/public-inspection.federalregister.gov/2015-06685.pdf (Accessed June 2, 2016)
107. M*Modal https://mmodal.com/ (Accessed November 9, 2017)
108. Best Care at Lower Price: The Path to Continuously Learning Health Care in America. 2012 http://www.iom.edu/Reports/2012/Best-Care-at-Lower-Cost-The-Path-to-Continuously-Learning-Health-Care-in-America.aspx (Accessed March 25, 2015)

9

Safety, Quality and Value

HARRY B. BURKE

LEARNING OBJECTIVES

After reading this chapter the reader should be able to:
- Define safety, quality, near miss, and unsafe action
- List the safety and quality factors that justified the clinical implementation of EHR systems
- Discuss three reasons why the EHR is central to safety, quality, and value
- List three issues that clinicians have with the current EHR systems and discuss how these problems affect safety and quality
- Describe a specific electronic patient safety measurement system and a specific electronic safety reporting system
- Describe two integrated CDSSs and discuss how they improve safety and quality

INTRODUCTION

From a safety and quality perspective, health informatics is the electronic acquisition, storage, and use of medical information to improve medical care. Health informatics has profoundly changed the practice of medicine and significantly improved the safety and quality of clinical care.

Health informatics, formerly medical informatics, has a long and honorable history. In the 1960s hospitals started using mainframe computers for their accounting and billing, and for storing laboratory test results which were printed and put in the patient's paper chart. As early as 1972 the National Library of Medicine, a part of the National Institutes of Health, began funding informatics training programs whose main purpose was to train individuals to apply computer and information science to medicine.[1]

The federal government enacted legislation and established entities designed to develop health informatics in order to improve safety and quality. The Agency for Health Care Policy and Research was established under the Omnibus Budget Reconciliation Act of 1989 (103 Stat. 2159). It was reauthorized with its name changed to the Agency for Healthcare Research and Quality (AHRQ) under the Healthcare Research and Quality Act of 1999. Its mission is to "*produce evidence to make health care safer, higher quality, more accessible, equitable, and affordable.*"[2] During this time health informatics, as a discipline, began to appreciate the centrality of the patient's medical record and to understand that if medicine was to improve its safety and quality it would have to computerize the medical record.

The Evolution of the EHR

At the turn of the 20th century, medical records were handwritten on index cards and stored in envelopes.[3] As the century progressed medicine became more complex and medical records became more complex. In addition, the demand for accurate medical information increased, which required that physicians write more detailed and complete notes.[4-6] These factors increased the size and scope of the medical record, thus it became a large, paper-based loose-leaf collection of clinical notes, laboratory values, radiology reports, and consultations.

As early as the mid-1960s, there were calls for computerizing outpatient clinic records.[7] Yet it was not until 1991, with the publication of the Institute of Medicine's *CPR Report – Computer-based Patient Record,* that there was an in-depth analysis of some of the potential benefits of EHRs, a discussion of issues related to implementation barriers including privacy and cost, and a national call for the adoption of a computer-based patient record.⁸ In 1996, President Clinton signed the Health Insurance Portability and Accountability Act (HIPAA) into law (Public Law 104 - 191). It was designed to make health insurance more affordable and accessible and it included important provisions to address the transmission and privacy issues related to electronic personal health information. It focused national attention on health information technology and the use of EHRs.

The increased national awareness of, and interest in, safety and quality, led the Institute of Medicine (now known as the National Academy of Medicine) to publish a series of landmark reports that shaped the national dialogue on healthcare safety, quality, and emphasized the importance of the electronic medical record:

- Published in 2000, To Err Is Human, Building a Safer Health System, focused on the safety and quality of care. It claimed that almost 100,000 hospital deaths were caused by medical errors. It asserted that that the problem was that good clinicians were working in a dysfunctional system and it set forth "a national agenda...for reducing medical errors and improving patient safety through the design of a safer health system."9
- Published in 2003, Key Capabilities of an EHR System discussed the basic functions of an EHR, database management, and data standards.10-11 It called for clinicians to abandon paper-based charts and move to electronic health systems and computer-aided decision support systems.
- Published in 2004, Patient Safety: Achieving a New Standard for Care, asserted that in order to prevent errors and to learn from the errors that do occur, a new health care delivery system was needed. It advocated for a radical restructuring of the medical system that had evolved over the last 100 years. The new system would be based on a culture of safety and the implementation of electronic information systems. In addition, it proposed the development of health care data standards for the exchange, reporting, and analysis of safety data.12

Yet, in the 13 years after the 1991 publication of *CPR Report – Computer-based Patient Record*, EHRs had not been widely adopted in clinical medicine.[13] There were few computers in outpatient examination rooms and hospital rooms, so there was little ability to use EHRs in the day-to-day practice of medicine.[14]

The widespread implementation and clinical use of electronic medical records in the United States began with President George Bush's 2004 State of the Union Address in which he said, *"By computerizing health records, we can avoid dangerous medical mistakes, reduce costs, and improve care."*[15] This speech was followed by Executive Orders, by several major legislative initiatives, and by implementation rules and regulations:

- In 2004, the Office of National Coordinator for Health Information Technology was established by Executive Order. It was a national office whose mission was to promote and oversee the development of health information technology.
- In 2009, the American Recovery and Reinvestment Act was passed (H.R. 1, Pub. L. 111-5). It included the Health Information Technology for Economic and Clinical Health (HITECH) Act which authorized the use of financial incentives to promote the meaningful use of EHRs to improve safety and quality. It also mandated that the National Coordinator for Health Information Technology oversee the implementation of EHRs.16
- In 2010, the Patient Protection and Affordable Care Act (known colloquially as Obamacare) (H.R. 35-90, Pub. L. 111-148) included mandated financial incentives for hospitals and clinicians for improvements in the quality of care of Medicare patients.
- In 2015, the Medicare Access and CHIP Reauthorization Act (MACRA), (H.R. 2, Pub. L. 114–10), institutionalized the use of health informatics to assess quality, improve clinical care, and lower costs. It allowed the Centers for Medicare and Medicaid Services to financially incentivize clinicians and medical organizations to adopt electronic medical record systems and to demonstrate their "meaningful use."
- In 2017, the Merit-Based Incentive Program System (MIPS) was introduced along with the Alternative Payment Models (APMs), which replaced several prior initiatives including meaningful use, in order to better monitor and improve safety, quality, and value.

These federal initiatives have driven the transition from paper-based, to electronic, health records but several issues related to the EHR remain unresolved including: record portability, the transmission of records between EHR systems, and clinician acceptance. The EHR will be the basis for many of the advances in safety, quality, and value.

In parallel with the federal clinical improvement initiatives, the National Library of Medicine redoubled its support of health informatics education and training. It expanded its mission from creating computer-related medical applications to supporting sixteen graduate level biomedical informatics training programs in the areas of translational bioinformatics, clinical research informatics, healthcare informatics, and public health informatics.[17]

Many private and quasi-private organizations have been created to improve safety and quality. For example, the **National Quality Forum's** mission *"is to lead national collaboration to improve health and healthcare quality through measurement. We strive to achieve this mission by: Convening key public- and private-sector leaders to establish national priorities and goals to achieve healthcare that is safe, effective, patient-centered, timely, efficient, and equitable; Working to ensure that NQF-endorsed standards will be the primary standards used to measure and report on the quality and efficiency of healthcare in the United States; and By Serving as a major driving force for and facilitator of continuous quality improvement of American healthcare quality."*[18] In addition, every major medical center has a robust safety and quality program.

Health Informatics Has Made Great Progress Over The Last Decade

The rise of health informatics in two domains has significantly improved safety and quality:
1. Clinical activities – (a) widespread installation of computers in outpatient examination rooms and hospital rooms, (b) adoption of EHR systems and their use during the clinical encounter, and (c) implementation and use of CDSSs; and
2. Monitoring and improvement activities – (a) auditing of EHRs in order to assess clinician performance and monitor patient safety, and (b) implementation of improvement initiatives.

The rest of this chapter explores these domains from the perspective of safety and quality.

The Enormity And Complexity Of Medicine In The United States

Improving safety and quality is a very difficult because of the pervasiveness of disease, the complexity of medicine, and the importance of health in everyone's life. Disease pervasiveness is reflected by the fact that 83.6% of adults had contact with a health care professional in 2015, there were 884.7 million physician office visits in 2015, 125.7 million hospital outpatient visits in 2011, 141.4 million emergency department visits in 2014, and 35 million hospital admissions in 2015.[19-23] Many of these patient encounters were complex and each provided multiple opportunities for clinicians, healthcare systems, and even patients, to make mistakes. These mistakes can have many causes, including: the rarity and/or complexity of the medical conditions and procedures, clinician time pressures and distractions, miscommunication and misunderstanding, patient personality characteristics, and defects in the healthcare system's delivery of medical care. The sheer number of encounters, and the fact that clinicians and patients are human, means that there will always be medical safety and quality issues. The job of health informatics is to minimize these issues by: (1) providing real-world simulations and other forms of training to improve clinician safety and quality performance, (2) the real-time monitoring and detection and clinician notification of activities that place patients at risk of a safety or quality event and the detection and clinician notification of the occurrence of safety and quality events, and (3) modifying the healthcare system in order to prevent the recurrence of unsafe actions or inactions. Health informatics plays an important role in safety, quality, and value.

QUALITY, SAFETY, AND VALUE

Quality And Safety

Safety and quality are related in the sense that a lapse in safety almost always lowers quality, but safe medicine is not always high-quality medicine.

From the physician's perspective, quality, safety, and value are: do good work, don't mess up, and don't charge a lot.

From the patient's perspective, quality, safety, and value are: (1) medical care that they understand, that takes into account their preferences and expectations, and that they agreed to; (2) that is appropriate for the medical condition and performed properly; and (3) that they can afford.[24]

From the Institute of Medicine's perspective, quality is a set of six aspirational goals: medical care should be safe, effective, timely, efficient, patient-centered, and equitable.[25] In this set of goals, safe and effective focuses on the physician; timely, efficient and patient-centered focuses on the healthcare system; and equitable focuses on societal virtues.

From the Center for Medicare and Medicaid Services' perspective, quality is a set of six aspirational goals: "*(1) make care safer by reducing harm caused in the delivery of care; (2) strengthen person and family engagement as partners in their care; (3) promote effective communication and coordination of care; (4) promote effective prevention and treatment of chronic disease; (5) work with communities to promote best practices of healthy living; and (6) care affordable.*" As Medicaid states, this is *"better health, better care, lower cost through improvement."* [26-27]

From the Agency for Healthcare Research and Quality's perspective, quality is *"the degree to which health care services for individuals and populations increase the likelihood of desired health outcomes and are consistent with current professional knowledge.*"[28]

From the clinical outcome perspective, quality, safety, and value are the ability of medical professionals to provide safe, affordable and appropriate care that achieves the expected clinical outcomes. Outcomes are usually risk adjusted and disease specific, and they take into account the dangers lurking in the side-effects of treatment.

Value

Value is the how important something is to us. It can be expressed in monetary terms; how much a person or organization will pay, or accept, for something they want to receive or give. In this sense, value is usually comparative because resources are almost always finite. At other times, it is expressed in qualitative terms, for example, how much a person "values" health, how much a patient "values" recovering from an illness, or how much an organization "values" providing high quality care.

There are at least four perspectives on medical value: (1) the physician and/or healthcare system that delivers the care (provider of care), (2) the patient and sometimes the patient's family (recipient of care), (3) the employer or government agency that either directly or indirectly pays for the care (payer for care), and (4) the norms of the society within which the care occurs. It should come as no surprise that these four perspectives are not identical. For example, in a University of Utah survey, conducted by Leavitt Partners in 2017, of 5,031 patients, 687 physicians, and 538 employers, they found that 88% of physicians equated value with the quality of care, 60% of employers ranked cost as the key component of value, and 45% of patients said that value was affordable out-of-pocket expenses.[29] Thus, there is a major conflict between the quantitative and qualitative views of value, depending on whether the respondent is the deliverer, recipient, or payer of medical care.

From the patient perspective, patients value health but the monetary value they place on healthcare varies with necessity. For example, patients usually attach little monetary value to prevention. In addition, they usually do not want to pay out of pocket, so requiring payments from patients can be a significant disincentive to their utilizing medical services. But when patients are sick, they are usually willing to pay more for their medical care and, when someone else is paying for that care, patients and their families tend to want everything done for them, no matter what the cost.

Another perspective is to frame value in terms of the net clinical benefit. For example, in terms of the net benefit of the treatment group in a clinical trial, where the next benefit is the sum of the clinical benefit and toxic effects, compared to the controls.[30] In cancer, the idea is that at least 20% of the control patients must still be alive and there must be at least a 50% improvement due to the treatment.[31-32] Of course, this is an oversimplification since it does not take into account the magnitude/duration of the improved outcome. Furthermore, it assumes a measurement outcome precision that does not exist, it may be unrealistically stringent, and it may not correspond to either the patient's or clinician's perception of a clinical benefit, especially since their expectations are rarely accurate.[33-35] Another type of net clinical benefit relates to low-value testing. For example, the view that there is little net clinical benefit in imaging patients who present with acute uncomplicated back pain.[36] Although reducing imaging utilization may appear to be a simple idea, it may have unintended consequences. For example, patients who are refused imaging may go to another clinician, they may experience a catastrophic outcome that could have been avoided by imaging, and they may no longer trust their clinician – which can lead to poor clinician-patient relationships and can affect other patient medical problems.

From a monetary perspective, there is little accurate information regarding the cost of care. Although 76% of physicians consider cost when making treatment decisions, it is not clear what "cost" they are considering.[29] Is it the direct expense involved in delivering medical care to a specific patient? If so, this cost is rarely known by anyone in healthcare because it is rare that there is accurate patient-level cost accounting. Is it the payments that employers and government agencies contractually make? Other than Medicare and Medicaid, these contract payments are considered a secret by the third-party payers, and physicians rarely know these payments. Other than the physician's own prices, it is unusual for a physician to know the list prices or payments for care. Patients almost never know the cost of care until they receive an explanation of benefits, reflecting the charges and payments for care. Finally, there is the list price of care, which is the price that a patient pays if he or she is not covered by a third-party contract or government agency. Historically, the patients who could not afford health insurance were the patients who have paid the most for medical care.

Another perspective is to employ a valuation method that combines the payments for specific services/therapies with predictions of patient future states. For example, one way to assess value is to use quality-adjusted life years (QALY), based EQ-5D utility estimates, either directly or through simulations, to determine value.[37] If the treatment does not produce sufficient quality-adjusted life years, then it is not of value. Although normative systems are interesting they have several problems. First, the individual patient predictions are rarely accurate. Second, populations may provide different utility estimates.[37] Third, individual patients are quite variable in what they value and how much they value it, and their valuations can change over time.[38-39] Fourth, many patients with limited life expectancies do not even want to discuss their life expectancy.[40] Finally, patients rarely reject a treatment simply because it has a low quality-adjusted life year estimate.

There have been attempts to define value in terms of what a life is worth, i.e., how much society should pay to add one year to a person's life. But no one has been able to justify a specific number, for example, no one has been able to successfully argue that we should pay $50,000 to add one year but that we should not pay $51,000 to add one year to a person's life. When a patient is dying, these numbers appear meaningless to both the patient and the family.

Another monetary approach is to create equations, where value is equal to the [service/treatment] times [quality of the service/treatment] divided by [price of providing the service/treatment]. It is not clear that these are the only terms that should be in the equation. Furthermore, it turns out that the numerator and denominator are difficult to measure and that there is no method for weighting the importance of each of the terms in the equation. Finally, attempts have been made in the U.S. to rank order services and treatments from high value to low value, but none have been able to set and maintain a threshold below which they will not pay because; inevitably, more and more exceptions are made until the system ceases to effectively ration care.[41]

Interestingly, one can ask the question, what is the value of health informatics? For example, what is the economic benefit of a clinical decision support system (CDSS) that is used for cardiovascular disease prevention? Jacob et al. recently undertook a systematic review to answer this question. He found that, *"The symposium noted the difficulty in transitioning from judgments of economic value at the level of specific implementations to a judgment about the aggregate of the implementations: costs and benefits have to be summed over implementations with different organizational contexts, technologies, functions, outcomes, scales, and scope. This systematic economic review of one type of health information technology, namely CDSS, encountered similar difficulties among others in synthesizing the economic evidence from various implementation instances."*[42] In other words, because of a paucity of information regarding the drivers of cost and benefit, and a lack of cost metrics, they were unable to determine if CDSSs were cost-beneficial or cost-effective for cardiovascular disease prevention. The problems Jacob et al. describe are not limited to CDSSs, they apply to most medical economic benefit calculations.

From a societal perspective, potential medical costs are almost infinite because there is no limit to the amount of money that can be spent trying to achieve perfect clinical outcomes for all patients and all conditions. Therefore, all societies ration care, sometimes the rationing is explicit but usually it is implicit and hidden from view.[43] For example, the United Kingdom National Health Service practices explicit rationing and, as British government's spending has declined, the rationing has increased.[44-45] Another form of rationing practiced by both the U.K. and Canada is to make people wait until they either give up, die, or go to the private healthcare system. In Canada, after a referral from a general practitioner, there is a two and one-half month wait to see a specialist and another two and one-half month wait to receive treatment.[46] Rationing priorities, what societies are willing spend their healthcare money for, is based on: (1) the medical conditions it is willing to treat, (2) the number of patients with those conditions, and (3) the cost, and sometimes efficacy of the treatment. Society achieves value by maximizing the impact of its resources on what it considers important; by treating the patients that society deems should receive medical care, by delivering that care at the lowest possible cost given the political situation, and by achieving the largest possible medical benefit for the most patients. In other words, societies choose who to include and who to exclude from care.

In the U.S., the Centers for Medicare and Medicaid Services (CMS) is using financial incentives and disincentives to regulate the practice of medicine. The Centers for Medicare and Medicaid Services has established a "value-based" payment system that aims to get value for its money by eliminating *"inappropriate and unnecessary"* care and by applying quality metrics to improve the quality of care.[47] The main thrust of this approach is to reduce costs; it assumes that improving safety and quality will reduce costs. But value-based payment systems, as they are currently implemented and without significant assistance from health informatics, may not result in significant reductions in the cost of care.[48-49] Furthermore, it is not clear that CMS's anticipated cost reductions will significantly offset the cost of improving safety and quality. For example, in a population of high-cost Medicare patients, it was found that only 4.8% of the spending was preventable.[50]

To be clear, safety and quality cost money, they require increased expenditures on clinician safety and quality training, they require that clinicians spend time on quality improvement activities rather than seeing patients, and they require investments in healthcare personnel and information systems. On the other hand, in the future, health informatics and technology may be able to improve the efficiency of healthcare while, at the same time, enhance safety and quality, by monitoring and assessing care and reporting that information to clinicians in order to prevent unsafe actions from occurring and by directing clinicians to perform those actions (cognitive and procedural) that improve safety and quality – which should also improve value.

USING THE EHR TO IMPROVE QUALITY, SAFETY AND VALUE

An EHR system consists of a graphical user interface for entering and viewing information, a sophisticated relational database that can acquire, store and retrieve information, and powerful, extensible auditing and reporting systems. Prior to President Bush's speech there had been a great deal of interest in EHRs and there were many small, primitive by today's standards, EHR systems – none of which were in widespread use. The main barrier to their acceptance was that few physicians used computers in their clinical practice, even fewer had computers in their exam rooms, and almost none used them routinely during their interactions with patients. The reason for this situation was that the paper chart had been optimized by clinicians over the previous hundred years and it was an extremely efficient clinical data acquisition, storage, and retrieval system.[51]

But paper records had at least five disadvantages. First, it could be difficult to read the handwriting in some notes, there could be errors and there could be non-standard abbreviations. Second, occasionally the patient's current chart would be checked out of medical records and it could take several hours to retrieve it. Third, charts were local and could not be accessed remotely. Fourth, paper records took up a great deal of clerical personnel time and a large amount of space. Fifth, and the most important disadvantage, was that it was very expensive to manually review charts.[51] For example, the cost of a manual review varied from $74 to $350 per chart, depending on the amount of information extracted.[52] This meant that, except for peer review, charts were not routinely audited to determine physician and nurse performance, and hospitals were not being evaluated and compared in terms of their performance. The EHR is easily readable, it is always available (except when it isn't), it takes up very little space (except in the exam room where it takes up a great deal of space), and it can be used to aggregate and analyze clinician and healthcare system performance.

Within a year of President Bush's speech, CMS began rolling out programs to financially encourage clinicians and hospitals to purchase computers and EHR programs and, later, to meaningfully use them. These programs were usually based on financial incentives and penalties. The American Recovery and Reinvestment Act of 2009 significantly increased physician adoption of EHRs and the Centers for Disease Control and Prevention reported that, by 2015, 87% of office-based physicians were using an EHR.[53-55] In addition, CMS began introducing quality metrics that clinicians and hospitals had to meet in order to continue to receive financial incentives and not incur financial penalties.

The relationship between EHRs and clinical quality has been investigated, usually in cross-sectional studies of process measures.[56-63] The results have been equivocal. No study has demonstrated a benefit across all its quality measures. Some have shown a partial benefit,[59-61, 64] while others have not demonstrated a significant clinical impact on quality.[56-58, 61-62, 65] Furthermore, one of the few retrospective longitudinal studies that used the quality measure hemoglobin A1c and compared before and after the introduction of an EHR did not find significant quality improvement attributable to the use of an EHR.[66] At the time President Bush spoke there were not, and to this day there have not been, any large scale randomized prospective studies that demonstrate significant improvements in safety and quality directly attributable to just the use of EHRs. Furthermore, it is unlikely that one will be conducted since it would require half of the physicians in the study to return to handwritten notes.

Although EHRs may not have a direct effect on safety and quality, they do allow administrative personnel to, for the first time, monitor clinician and healthcare system performance. Prior to EHRs individual physician charts were audited (usually as part of the peer review process) but there was no aggregation of a physicians' medical records, or healthcare system records, and, therefore, no assessment of their performance. Furthermore, the EHR allows for indirect safety and quality improvements related to: 1) determining the nature, frequency, and severity the safety and quality issues, 2) assessing quality and safety issues and implement solutions, and 3) determining if the implemented solutions had, in fact, improved medical care. Furthermore, adding CDSSs to EHRs has the potential to directly improve safety and quality.

Clinician Problems with EHRs

The conversion from paper charts to an EHR system has created several problems. One issue is that moving from paper to the EHR or moving from one EHR to another, usually results in the loss of most of a patient's past medical information. Commonly, only medications, allergies, and problem lists are transferred to the new system. Access to the old system usually continues for a brief period of time, but the cumbersome use of two parallel systems usually ends rather quickly. In addition, the use of a new EHR requires a great deal of clinician training, but most transitions provide relatively little clinician training, and the training they do receive usually consists of a short didactic related to the major features of the system and workflow "cheat sheets", followed by clinician trial and error during patient encounters. The loss of patient medical histories and the inept use of EHR systems have created significant safety and quality issues.

Clinicians are currently experiencing problems related to their use of the EHR. One problem is clinician time and efficiency. In the past, clinicians quickly reviewed the patient's history, including the clinical information that had accrued since the last clinical note; they interacted with the patient; and they wrote a new narrative note that contained a summary of the patient's progress, an assessment of the clinical encounter, and a description and explanation of the clinical plan for the patient. Currently, clinicians sign into the computer, access the EHR, open the patient's record, and search the relevant patient information using pull-down menus, click boxes, and opening windows. Either during or after the patient encounter, clinicians add information to the patient's record using the same pull-down menus, click boxes, and opening windows. Furthermore, clinicians usually manually type the patient information into the EHR. Finally, physicians may not be allowed to write free text notes, they may be required to use template-based notes.[67] All of these activities require time, present a fragmented view of the patient, are opportunities for mistakes, and reduce clinician productivity.

Currently, physicians spend at least 5.9 hours out of an 11.4-hour work day using their EHR.[68-69] Typing into a computer and accessing information during the clinical encounter reduces the amount of time spent interacting with patients and distracts both the patient and clinician during the clinical encounter.[70] Over one-third of patients believe that the physician's use of a computer in the clinical encounter negatively affects physician-patient communication.[71] This means that clinicians may be less productive, they may see fewer patients, they may generate fewer relative value units (RVUs) and, due to computer-based distractions, they may find it difficult to communicate with their patients.

It is true that the EHR does improve the narrative note quality; physician notes are more complete.[72-73] But, in order to save time, clinicians have resorted to a shortcut that the EHR makes available to them. An unintended consequence of the EHR is that it allows clinicians to cut-and-paste from previous notes into the current note. This adds old, out-of-date information to the encounter note, it reduces the amount of current information in the note, and it makes it difficult to rapidly obtain an accurate understanding of the patient's current status.[74] Cutting and pasting does save time, but at the expense of the clinical quality of the note.

It is well known in human factors research that technology can cause human mistakes and there has been a growing recognition since 2005 that EHRs can cause mistakes.[75-88] Issues with system functionality, for example, poor user interfaces and fragmented displays, have delayed the delivery of care – which is a safety issue. There have been an increasing number of safety professional liability claims related to EHRs, including the use of copy and paste, insufficient area for documentation, poor drop-down menus, and improper templates.[88-89] In a recent health information technology review, 53% of the studies found that health information technology problems were associated with patient harm and death, and near-miss events were reported in 29% of the studies.[90] It is ironic that one of the main reasons for adopting EHR was because they would improve safety by eliminating mistakes, only to find that they have been replaced by health information technology-related mistakes. Furthermore, whereas handwriting mistakes were relatively easy to detect and correct, health information technology-related mistakes have been much more difficult to detect and correct.[90] Some have advocated for proactive detection of health information technology-related problems while others have advocated for the redesign of health information technology systems to reduce human error.[91-93] Both solutions are necessary, but both are expensive to implement.

A recent RAND report stated that, *"the current state of EHR technology significantly worsened professional satisfaction in multiple ways. Poor EHR usability, time-consuming data entry, interference with face-to-face patient care, inefficient and less fulfilling work content, inability to exchange health information between EHR products, and degradation of clinical documentation were prominent sources of professional dissatisfaction."*[67] Furthermore, physicians complain that, because of shortcomings in the design and implementation of health information technology systems, current EHRs do not deliver sufficient clinical value to compensate for their difficulty and expense.[94] It was suggested in a recent review that we should rethink the definition of meaningful use, reduce EHR difficulty, and improve their clinical utility.[95] In other words, EHRs may be necessary but they not sufficient, for increasing the safety and quality of medical care.

Finally, EHR systems, and their related hardware and software are, in their current form, very expensive to buy, maintain, and upgrade. In addition, they are very expensive to use because they slow down clinicians in terms of finding the relevant information, they require checking of boxes, and they necessitate typing into a computer. One of the primary goals of health informatics is to improve the usability and efficiency of EHRs.

Societal Perspective

Both the good and the bad aspects of clinicians being required to use EHRs have been discussed, but there is also a societal perspective. *"With rapid consolidation of American medicine into large-scale corporations, corporate strategies are coming to the forefront in health care delivery, requiring a dramatic increase in the amount and detail of documentation, implemented through use of EHRs (EHRs). EHRs are structured to prioritize the interests of a myriad of political and corporate stakeholders, resulting in a complex, multi-layered, and cumbersome health records system, largely not directly relevant to clinical care. Drawing on observations conducted in outpatient specialty clinics, we consider how EHRs prioritize institutional needs manifested as a long list of requisites that must be documented with each consultation. We argue that the EHR enforces the centrality of market principles in clinical medicine, redefining the clinician's role to be less of a medical expert and more of an administrative bureaucrat, and transforming the patient into a digital entity with standardized conditions, treatments, and goals, without a personal narrative."*[96] Health informatics can be viewed as dehumanizing patients and the clinicians; it can be viewed as part of a larger effort to advance the interests of corporations and governments, at the expense of patients and clinicians. One way to humanize health informatics is to demonstrate its positive benefits to patients and clinicians, for example, by showing: (1) its ability to improve the delivery of medical services, (2) its ability to assist in the selection of the best therapy for an individual patient, and (3) its ability to improve patient outcomes.

THE INABILITY TO INTERPRET FREE TEXT HAS LIMITED QUALITY, SAFETY AND VALUE

In addition to the many problems clinicians encounter when using EHRs, administrators and researchers have their own problems with EHRs. They want to automatically extract meaningful information so that they can assess the safety, quality and value of medical care. "Meaningful" in this context is medical information that tells us what we want to know about the patient. For example, we may want to know if a patient was injured, the cause of the injury, where the injury is located, if the patient is in pain, what treatment was provided, and whether the treatment was effective. Furthermore, we may want to aggregate all the patients with this particular injury in order to perform an observational study of the effectiveness of a treatment. In other words, administrators and researchers want to automatically extract information from EHRs so they no longer have to perform manual chart reviews in order to determine how patients are being treated and their outcomes.

The problem is that the clinical information they are interested in is being typed into the EHR as free text and there is currently no automated way to read these notes with sufficient accuracy so that the extracted data are highly reliable. Because the clinician's narrative text cannot be automatically read with high accuracy, administrators have resorted to requiring that clinicians check boxes and fill in structured fields in the EHR – because this information can be automatically extracted and analyzed. In addition, administrators have used "administrative" data, usually data from the EHR that is used for billing, to assess performance. Unfortunately, administrative data does not contain important clinical information, it contains biases related to payer reimbursement, and there are significant challenges related to its use in assessing safety and quality.[97]

Binary Data: Check Boxes

There are three kinds of information in most EHRs; check boxes, structured and semi-structured text, and free text. Check boxes are labeled binary fields that the clinician checks. They can be one item or a list of items, any one or more of which may be relevant to the clinical problem. The boxes may be related to prevention, signs and symptoms, diagnosis, prognosis and treatment, outcomes. Check boxes are useful to administrative personnel because: (1) once the boxes are checked, the data can be easily and inexpensively acquired, (2) the data is already organized in terms of category/subcategory, e.g., prevention or the diagnosis and treatment of disease, and (3) the data can be aggregated.

Check boxes require a great deal of checking. For example, for each symptom, the clinician has to check some or all of the boxes related to onset, duration, frequency, location, setting, alleviating/aggravating factors, quality, intensity, severity, temporal trends, and unique manifestations and, in terms of pain, paroxysmal pain (shooting, sharp, electric, hot, and radiating), superficial pain (itchy, cold, numb, sensitive, and tingling), and deep pain (aching, heavy, dull, cramping, and throbbing), and so forth. Furthermore, clinicians have to check the relevant boxes that show the reasons for: (1) working up the patient, (2) determining the diagnosis, and (3) selecting a specific treatment. They must also describe the outcome of the treatment. In addition, they must somehow communicate (1) their reasoning, (2) who the patient is and what his or her values are, and (3) why the patient and clinician, using shared decision making, selected a particular treatment plan. Finally, because of the many boxes, there are many possible box-checking errors.

In other words, check boxes can be useful for discrete, simple information. Unfortunately, they are of little use for more complex information because there is currently no way to combine and organize large disparate collections of check boxes in order to create meaningful clinical information. The more complex and detailed the meanings of interest, the more hierarchal and detailed the boxes must be, the more boxes that must be checked, and the more difficult it is to put all the box information together into an integrated, coherent and clinically useful medical description of the patient.

Alphanumeric Data: Structured and Semi-structured Fields

When we advance from binary (existence/nonexistence) data to alphanumeric data, we move from check boxes to structured and semi-structured data. Fully structured fields are fields that take specific values, for example, laboratory values and prescription orders. Structured data can also be exact text, for example, drop down menus that contain all the possible diagnoses in urological pathology. In this situation, every item on the menu has a corpus of text with a specific meaning. When an item is selected the exact same text is always inserted into the EHR.

Semi-structured fields are usually domain-specific and have limited, pre-specified vocabularies, for example, radiology reports. Although the text may vary slightly, the predefined vocabulary establishes the meaning. Because the words and phrases are already known, a key word or key phrase search can be used to find the meanings in the text.

In both structured and semi-structured fields, the type of information is already known by the label of the field, and what we want to know is the token for the patient. For example, for laboratory data, the type is already known, e.g., the field is labeled HbA1c, and we want to know the token, namely, the patient's numeric value in the A1c field. In other words, we search a specific field and the information in the field is the meaning.

Alphanumeric Data: Unstructured Fields

The problem with using just check boxes and structured fields is the paucity of information they to provide. Check boxes and structured fields cannot fully represent the complexity and individuality of patients, their diseases, and their treatments. Unstructured fields allow clinicians to generate free text, so that they can properly describe the patient and the patient's condition, explore possible diagnoses and the reasons for selecting one diagnosis over another, and justify the treatment that was selected for an individual patient. The need to automatically find the meanings expressed in free text has long been recognized as one of the most important goals of health informatics. The automatic search for meaning in free text is the province of natural language processing.

Natural Language Processing

Natural language processing (NLP) is a computer program that takes as its input the clinician's electronic free text and it returns as its output the meaning of the text. In order to determine the meaning of text one needs to label (named entities) the medical concepts relating to patient signs and symptoms, tests and imaging, risk assessment, diagnosis, prognosis, treatment, and outcome and then determine the relationship between the named entities one is interested in. For example, if one wanted to know if the patient had had chest pain [named entity] associated with his myocardial infarction [named entity], one would determine if both terms were present (usually but not always in the same sentence) and if a relationship between the two was described in the free text.

The main problems with finding meaning in text is that: (1) there are many ways to say the same meaning, (2) the meaning can depend on the context, and (3) much of the medical free text is written in a telegraphic style that does not obey the rules of English. Furthermore, meaning can cross sentence boundaries, for example, there can be co-reference.[98]

There are many approaches to natural language processing; the three most commonly used are key word, rule-based, and machine learning. A recent review of natural language processing methods found that 24% used key word, 67% used rule-based, and 9% used statistical methods.[99] Since that review the number of statistical methods has increased dramatically. Common information extraction methods include: MedLEE, HITEx, and cTAKES.[100-102]

Key word, or key phrase, are searches for an exact word or phrase in the text. Since the meaning of the key word is known, if the word is found in the text, then the meaning of the text is also known. It can also search for the root form of the word, for variations of the word including abbreviations and plurals, and for synonyms. The major drawback of this approach is that it is too specific. One must perform a search for every way the word can be written. Furthermore, (1) it does not include modifiers such as "not," (2) it does not include predicates, and (3) it does not take context into account. In other words, just knowing that a word is present is not usually sufficient for determining the meaning of the text.

Rule-based systems are more flexible. They allow for searching text in terms of subject-predicate statements; if X is in the text, then search for Y, if both X and Y are found then you have found the meaning of the text. For example, if the term "pain" is in the text, then search for an anatomic location, such as leg. If you find both, then you infer that the patient has leg pain. Clearly, this is superior to a key word search, but many of the problems inherent in the key word search remain. For example, (1) you must specify all possible variations of the subject and predicate, (2) it does not take into account modifiers, and (3) its view of context is limited to the predicate.

Machine learning (based on statistical methods) defines the search for meaning in free text as a classification task. The ways that a meaning can be written are called patterns and patterns can be learned. For example, the relationship between a patient having an A1c and the value of the A1c can be written many different ways including, *"the patient's A1c is 6.7,"* and *"the patient's A1c is well controlled at 6.7."* Words sets are coded as patterns and trained on text that both contains these patterns (positive) and text that does not contain these patterns (negative). The better machine learning is at distinguishing positive patterns from negative patterns (and non-patterns) the more accurate it is. The idea is that during training machine learning will learn to create classes of patterns and to generalize these classes to patterns it has not seen before, where the generalized pattern is not an exact set of words, rather, it is the features in the text that indicate the existence or nonexistence of the searched for meaning. In other words, the trained model will be able to use a set of features to define the target meaning and it will use this set to detect the existence of the target meaning in the sentence. Many machine learning algorithms have been used, including support vector machines, Bayesian conditional probability models, and artificial neural networks (also called deep learning).

Currently, no key word, rule-based, or machine learning method has performed with a sufficiently high accuracy that it can be used to reliably find any meaning in any medical free text.

SAFETY AND QUALITY DETECTION AND REPORTING

In most situations, an EHR system will not, by itself, improve the safety and quality of medical care but the clinical use of an EHR is a necessary prerequisite for quality and safety improvement. Furthermore, EHR systems assist and are, many times, essential for: (1) detecting and reporting of adverse events, (2) safety and quality improvement initiatives, and (3) prevention programs.

Focus On Actions Rather Than Events

An adverse event is *"An unexpected and undesired incident directly associated with the care or services provided to the patient."*[103] In other words, an adverse event is a safety event that reached the patient. *"A near miss is any event that could have had an adverse patient consequence but did not and was indistinguishable from a full-fledged adverse event in all but outcome."*[104] A near miss event and an adverse event can have the same cause and this can result in their being confounded. Essentially, a near miss event and an adverse event only differ in that one reached the patient and the other did not.[105]

Historically, safety has focused on events, such as near miss events and adverse events, rather than on the unsafe actions that cause these events – even though almost every near miss event and adverse event is caused by at least one unsafe action. If we are to prevent the occurrence of safety events, we must focus on the causes of the events, i.e., the unsafe actions or unsafe inactions that give rise to the event because, once an unsafe event occurs, it is too late to prevent it. All that can be done is to ameliorate any resulting patient harm and try to prevent that specific adverse event from recurring. If we can prevent near misses by preventing unsafe actions, then we can prevent adverse events. An unsafe action is any action (or inaction) that has the potential to cause an adverse event. Once an unsafe action occurs steps must be taken to eliminate its recurrence and the recurrence of any unsafe actions associated with the proximate unsafe action. Furthermore, unsafe conditions are usually caused by unsafe actions. If an unsafe condition is observed, it must be eliminated and the unsafe actions associated with it must be identified and eliminated.

Although we normally think of unsafe actions as being produced by clinicians in hospitals, everyone involved in a patient's care, including the patient, can produce unsafe actions and these actions can occur in clinics, independent living situations (with or without home care), and residential care, including assisted living and nursing facilities. In addition, although we have been examining safety from the clinician and healthcare system perspective, it is important to understand that patients can feel unsafe due to actions that the clinician and healthcare system may not recognize as unsafe.[106]

When we move away from events and to actions, we can recognize the importance of unsafe inactions. Unsafe inactions are missed care.[107] Missed care is *"any aspect of required care that is omitted either in part or in whole or delayed."*[108] They are not always mistakes, many times they are actions that are selectively not performed, usually due to time pressure, because it is believed that the inaction will not create a safety risk. In addition, primary care physicians may omit patient teaching, follow-up, emotional support, and mental health needs because of time constraints and administrative burdens.[109] Inactions can create safety risks and they must be recognized and incorporated in a safety program.[107,109] Currently, it is very difficult to detect unsafe inactions but, in the future, it may be possible for CDSSs to be trained to detect unsafe inactions.

In terms of hospitals, it is well known that unsafe actions occur frequently throughout hospitals.[110] Many are not reported, and from a system perspective, they go unnoticed.[111] The essential questions are: how are unsafe actions to be detected, which unsafe actions should be reported, and how are unsafe actions to be prevented? One approach to detecting unsafe actions is for an automatic clinician decision support system to monitor performance. Such a system requires: (1) an electronic safety detection system, (2) all actions are entered into the system in real time, and (3) the system analyzes the entered information and reports safety issues in real time. With this system in place an unsafe action that was not noticed by the performing clinician will be detected and reported by the system. The system will send a safety message to the clinician and to safety personnel that an unsafe action either has, or is, occurring. This will provide an opportunity for clinicians to truncate, and perhaps even prevent, the unsafe action. A limited version of this system exists in pharmacy CPOEs.

Reporting Unsafe Events and Actions

Currently, it is the responsibility of individuals to detect, report, and correct most unsafe actions. How good are individuals at reporting unsafe actions? Westbrook found that of the 218.9/1,000 clinically important prescribing errors, only 13.0/1,000 errors were reported by the clinical staff.[112] Another study also found that few adverse event medication errors were recorded in the EHR.[113] Furthermore, two-thirds of near miss events were reported by a witness and one-third were self-reported.[114] Given that most near miss events are probably not witnessed, this suggests that many, if not most, near miss events are not reported. Further support for not reporting near misses comes from residents who preferred to discuss an adverse event with their supervisor and at department-led conferences, rather than reporting the event.[115]

The current safety systems are retrospective, they operate on a case-by-case basis, they detect few errors, their incident evaluation and resolution process can miss the correct causes, the process takes a long time and can be expensive, and long-delayed corrective action can be ineffective. Furthermore, other than the aggregation of individual event reports, the current system does not have the ability to perform systematic, patient-level safety assessments across the medical system and it does not have the ability to aggregate that information to detect systemic problems.

Currently, the proximate individual is usually blamed for an unsafe action. It is said that the individual made a mistake, forgot, exhibited poor communication, did not comply with policies and procedures, and much more.[114,116] In reality, a properly trained medical professional has safely performed that action many times in the past. But people are not perfect; there are random mistakes in human performance. In other words, on any given day, every individual has a probability of making a mistake. On this day, the mistake was made by this person – on another day it may be made by another person. The real problem is that there was no recognition of the fallibility of man and of the risk inherent in each medical activity and, as a consequence, there were no processes in place to prevent most mistakes from occurring.

One approach to mistakes is punitive – blame the individual. For example, administrative personnel may believe that they must educate the "offending attending physician and his or her staff."[117] But competent individuals feel that it is unfair to blame them for the mistake and denigrating the physician and his or her staff is counterproductive.[91] Although punitive measures can lead to anger, resentment, and a negative culture – playing the "blame game" is still prevalent in many healthcare systems. Most researchers who have assessed the utility of the punitive approach have rejected it. Instead, they have called for the option of anonymous reporting; for an expert, objective, systematic standardized process to analyze and understand unsafe actions; and for feedback regarding that changes were made in the system so that clinicians can feel that they are a part of the safety improvement process.[37, 91, 115, 118-125]

Near miss events and adverse events are often blamed on individuals because their unsafe actions are usually the proximate cause. Previous unsafe actions related to the near miss or adverse event, and the conditions surrounding the event, may be noted but are typically not considered part of the primary cause of the event. Furthermore, in an investigation of a near miss it is rarely recognized that the individual is part of a healthcare system and that it is the system that allowed the individual to produce an unsafe action. Finally, there is a failure to understand that the real cause of the unsafe action may be: (1) inadequate training and/or supervision by the system, (2) overwork and stressful conditions within the system, or (3) the system's inability to manage the clinical environment. For example, in hospital settings, interruptions are associated with more than 80% of the orders entered into the wrong EHR.[118] Unfortunately, by far the most common response to an unsafe action is to try to change people rather than to improve the system.[114,116, 126-127] There is a critical need for health information systems that can monitor clinician actions and report problems before they occur, so we do not blame competent health care professionals.

Finally, in most systems, the safety personnel are usually reactive, they spend most of their time filling out patient safety reports, investigating events, and providing documentation. It is the frontline supervisors and clinicians who must find the time and resources required to proactively improve safety.[128] It must be recognized that the safety personnel and the frontline clinicians need improved health informatics systems in order to efficiently and effectively perform their jobs.

Activity Related to Near Miss Events

When a near miss has been detected several things can happen. According to Jeffs there are three possible responses to a near miss. There is the "*quick fix*," where the effect of the near miss is dealt with but nothing else is done. There is the "*going into a black hole*," where the near miss is dealt with and reported to the system, but clinicians never learn if it was fixed and, if so, how it was fixed. There is the "*closing off the Swiss-cheese holes*," where the near miss is dealt with and reported to the system, the system takes corrective action to prevent its recurrence, and the relevant information is returned to the clinicians.[110]

It is difficult to select and mount a systematic response to near miss events because, although there are approaches to categorizing the severity of close calls and adverse events for comparative analysis, for example, the Safety Assessment Code Matrix, there is currently no consensus regarding which near miss events should be reported and how they should be dealt with.[129-131] Furthermore, even the reporting or harm is fraught with difficulty. The Agency for Healthcare Research and Quality released version 1.2 of its Harm Scale in April 2012. It has a two-part harm assessment process for harm, namely, the degree and duration of harm. Degree of harm consists of a five-point scale: death, severe harm, moderate harm, mild harm, and no harm. Duration of harm consists of a two-point scale: permanent (at least one year) and temporary harm.[132] A consistent problem in safety is the creation of scales that have low interrater agreement. For example, the AHRQ Harm Scale v1.2 has kappa's of around 0.50 and raters have a great deal of difficulty distinguishing between severe, moderate, and mild harm.[133]

One reason for the quick fix is that it is the expedient solution. Another reason for the prevalence of quick fixes is that the person who produced the unsafe action does not want to be blamed for it, so fixing but not reporting it becomes the preferred solution. In addition, clinicians are over committed, and they have competing priorities; taking the time to report an unsafe action may not be their highest priority.[110]

For reporting and acting on unsafe actions, one can take a Safety Assessment Code Matrix approach, namely, to prioritize unsafe actions in terms of the combination of their probability of causing an adverse event and the degree of severity of a resulting adverse event.[129] But the scoring system should not be based on subjective judgments, rather, it should be an evidence-based quantitative assessment. Using a data-driven expert system for guidance, some low risk unsafe actions can be quickly fixed, while more serious unsafe actions require a report, systemic corrective action, and feedback to the clinicians. Furthermore, reporting should be electronic and standardized so that the reported information can be properly analyzed, effectively acted upon, and electronically transmitted across the healthcare system. Finally, there must be: (1) a non-punitive response to the report, (2) an effective organizational response including change management (learning) within a facility and across the healthcare organization, and (3) feedback to leadership, safety personnel, and clinicians regarding the organizational response.[119,123]

In addition to assessing risk, health informatics can assist in evaluating and eliminating unsafe actions by collecting the necessary data. These data should be analyzed as a ratio, where the numerator is the number of detected unsafe actions (and inactions) and the denominator is the opportunity for an unsafe action (and inaction), over a specified time interval. The opportunity for an unsafe action is, for a properly trained person, the product of the complexity of the action, the complexity of the activity within which the action occurs, the frequency of the action, the frequency of the activity, and person's activation, over the specified time interval. Person activation is his or her arousal.[134] The expected level of activation is equal to 1.0 when the performer has a normal arousal, to <1.0 when the performer has too low an arousal (usually when the task is repetitive and/or boring), and to >1.0 when performer has too high an arousal (usually when under a great deal of stress). In other words, every medical activity will have a denominator, which may be somewhat imperfect, but which allows for the identification of those activities that have the highest chance of unsafe actions and which adjusts for the observed rate of unsafe actions, thus placing the observed unsafe actions in the context of their probability of occurrence. This should be calculated by an expert system and the results should be the targets of a learning healthcare system.

Patient Safety Systems

Currently, medical personnel report safety events by manually filling in either a paper-based or electronic reporting form, for example, in the Patient Safety Reporting System.[135] The safety event information is sent to safety personnel where the event is documented and, if it is a Joint Commission sentinel event, and sometimes even if it isn't, it is investigated and reported.[136] The safety report can be deficient in several ways: (1) it may lack standardization of data, (2) it may not include of all the relevant data, and (3) it may contain analysis biases.[137] That said, patient safety reporting programs have been successful in medicine.[119, 125, 138-139] But attempts to adapt industry safety approaches to medicine have resulted in numerous practical problems.[140] An important problem is that most industry systems use a total reporting approach which, in medicine, means that *"any unintended or unexpected incident that could have or did lead to a harm"* must be reported.[141] This is based on the belief that increased reporting will increase safety. This assumption can lead to a focus on quantity rather than quality. To take notice of every event, to mandate that each one must be properly reported, and to require that corrective action be taken for each reported event, will overwhelm most safety programs.[110] For example, an oncology practice implemented a reporting program and in its first three years it received 688 reports, each of which had to undergo a *"plan, do, study, act"* quality improvement cycle.[37] In a radiation department, over a two-year period 1,897 near misses were reported though their voluntary, electronic incident system. This represented an average of one near miss for every patient treated.[122] In a diverse group of primary care practices, over a nine-month period, 632 near misses were reported but only 32 quality improvement projects could be initiated.[121] It is well known that *"the frequency of near misses in daily practice does make it impractical for clinicians to report every near miss, or for the organization to respond to every near miss."*[110]

Most safety programs are not integrated into the EHR system. Their reports are usually not automatically filled in and the investigation is not automatically conducted, and the investigation results written. Health informatics needs to develop and implement an automated safety reporting system that is a part of the EHR. In the future, a large part of quality and safety will be triggered electronically, much of the data will auto populate the form, much of the investigation will be performed by a data-driven expert system, and the report will be automatically written.

Root Cause Analysis

A root cause analysis is usually a reactive process that attempts to discover the prior causal event or events that gave rise to a specific safety event. In addition, it attempts to determine how to prevent the safety event from recurring. The usual methodology is to assess the safety event and infer the cause of the event. The problem is that this approach commits the *post hoc, ergo propter hoc* (*"After this, therefore because of this"*) logical fallacy. As David Hume pointed out in *Of Miracles*, the effect does not contain within it its cause; for any effect, there are many possible causes and the effect cannot prove its cause. Furthermore, there are almost always a cascade of causes that result in an observed effect.[142] The outcome of most root cause analyses is to point to the proximate clinician, even though there are usually several unsafe actions leading up to the clinician being involved in the safety event. One way to solve this problem is to use health informatics. One can use information in the EHR to: (1) determine all antecedent actions, (2) create a decision model that uses these antecedents, the unsafe action, and other relevant EHR information, and (3) run the model to determine the probably of each of the possible sequences causing the safety event. In other words, set-up the variables, simulate the situation, find the most probable causes, and fix those causes.

Safety and Quality Measurement

There are many quality measurement systems. One of the most utilized is the National Committee for Quality Assurance's Healthcare Effectiveness Data and Information Set (HEDIS) measures.[143] In 2017, it consisted of a set of 91 measures that assessed how well patients were being cared for by clinicians and healthcare systems. The system uses defined and structured field searches, surveys, and self-reporting. In addition, the Centers for Medicare and Medicaid Services currently has 271 quality measures that are used to justify payment.[144] Finally, the Agency for Healthcare Research and Quality's National Guideline Clearinghouse is a database that contains thousands of clinical practice guidelines, most of which measure quality of care.[145] Unfortunately, the explosion of measures has led to inefficiencies and imbalances, including poorly defined measures, duplicate and overlapping measures, and an over representation of measures in some areas.[146]

Measurement is a necessary component of all quality improvement projects. Unfortunately, too many projects are *ad hoc* and local. The editors of JAMA Internal Medicine critically appraised the quality improvement studies that they receive for publication and they found that many of them were of poor quality for the following reasons: (1) they were not generalizable, the problem existed only at one center or the intervention was only performed at a limited number of centers, (2) many studies only focused on changes in health care processes, use, or cost rather than on clinical outcomes, (3) they did not assess, in addition to benefits, potential adverse effects, (4) value, in terms of cost savings, did not reflect the costs associated with the intervention, (5) it was rare for there to be a control group, (6) no attempt was made to use statistical methods that approximated randomization, and (7) even when blinding was possible it was not done.[147]

Most medical organizations are collecting EHR data and manual data that can be used to construct process and outcome measures.[148] Although process measures are not very reliable, they tend to be more reliable than outcome measures because the healthcare system directly controls processes, but outcomes are affected by patient behaviors.[149-153] In fact, it has been suggested that outcomes, at least in cancer, are not a good measure of quality.[154]

These data are periodically aggregated in order to assess and improve the organization's performance. The aggregate results can be presented numerically or graphically, or a combination of the two. The display of these measures is usually called a dashboard. For example, a Veterans Integrated Services Network, in order to improve the quality, safety, and value of its care of veterans, developed 300 dashboards and reports.[155]

Although apparently a very simple task, in reality, the communication of actionable safety and quality information is devilishly difficult. What to display, how to display it, and what it means are very challenging issues. In the past, the development of dashboards and reports has been primarily *ad hoc*. Administrators usually targeted specific measures for performance assessment and the targeted measures drove the creation of dashboards. Typically, there was no explicit plan regarding how to operationalize the organization's safety and quality objectives in terms of an integrated set of aggregated measures and there was little recognition whether the dashboard is for strategic, tactical and operational use.[156] Furthermore, there was no evaluation method to determine if the organization's dashboard goals had been met. For example, Karami et al. identified seven evaluative categories for dashboards, namely, user customization, knowledge discovery, security, information delivery, alerting, visual design, and integration and system connectivity – few of which are systematically evaluated during the development, and use of, a dashboard.[157] In addition, there was little understanding of significant intellectual, financial, and personnel resources necessary to create an effective dashboard.[156] Finally, the evidence for the utility of dashboards is slight. A recent review found that most of the dashboard literature consisted of dashboard descriptions and individual case reports rather than empirical studies.[156] In other words, it is not yet known whether dashboards, as opposed to other methods of understanding safety and quality results, are effective at improving safety and quality. Information technology is at its best when it operates in real time and when its information drives immediate actions that prevent or ameliorate an unsafe action.[158]

The federal government, as one of the largest U.S. medical payers, has long been interested in knowing the safety of the care provided to its beneficiaries by its payee hospitals. In 2001, the Centers for Medicare and Medicaid Services created the Medicare Patient Safety Monitoring System, which, in 2009, was transferred to the Agency for Healthcare Research and Quality. The Medicare Patient Safety Monitoring System performs manual chart reviews to determine the national rates for 21 types of adverse events and it creates a baseline for evaluating national patient safety initiatives.[159] Shortly after its creation, in 2003, the Agency for Healthcare Research and Quality developed its 27 item Patient Safety Indicators that screen for adverse events that are likely to be preventable.[160]

Individual reporting and Patient Safety Indicators underreport safety events. They fail to detect approximately 90% of the hospital events.[161] The Global Trigger Tool, developed by the Institute for Healthcare Improvement in 2003, assesses the safety of care provided by individual hospitals. It can detect up to 90% of adverse events, in comparison to approximately 1% using voluntary reporting systems and 9% using the Patient Safety Indicators.[161]

The Global Trigger Tool process involves randomly selecting ten discharged patient medical records every two weeks at a hospital. Two reviewers independently review the same charts for the presence of one or more of 53 "*triggers*," which are entries in the medical record that require further investigation to determine whether an adverse event occurred and, if so, its severity. Regardless of the size of the chart and the complexity of the patient's medical problems, each chart review is limited to 20 minutes. The two reviewers arrive at a consensus regarding triggers, adverse events, and severity. A physician adjudicator, the final arbitrator, and the reviewers then come to a final determination regarding the number, type, and severity of events. The physician does not review the records; he/she only assesses the reviewers' results.

The Global Trigger Tool has improved the safety event detection process by defining a set of triggers and providing a systematic process for their evaluation. There is a substantial body of evidence that supports the fact that the Global Trigger Tool significantly improves safety.[161-175] But the manual Global Trigger Tool does have important limitations: (1) because it is not risk adjusted, it cannot be used to compare different types of hospitals, (2) it is very labor intensive and expensive, (3) it exhibits low abstractor agreement, (4) it does not examine all inpatients, and (5) a physician must adjudicate the abstractor's findings for each putative adverse event.

Many organizations have partially implemented an electronic version of the Global Trigger Tool, using information from the EHR, including check boxes and structured fields.[176] The problem with this approach is that it generates a huge number of triggers, each of which must be assessed by a reviewer for the existence of adverse events, which is very time consuming and expensive. Furthermore, many of the triggers are not captured by the check boxes and structured fields, so this approach does not eliminate the need for a reviewer checking the medical records.

In addition to the Global Trigger Tool's lack of comprehensiveness, it does not assess all hospitalized patients, it employs an *ad hoc* search process for the triggers, each reviewer examines the chart in his or her own way. This may be one of the reasons for the low agreement between raters in terms of the triggers and for the adverse events.[177] The low agreement means that there is a substantial amount of error in the number of triggers and which triggers are detected. Another issue is that the review of the patient's medical record is limited to 20 minutes. It is well known that the longer the patent is in the hospital the greater the chance of mistake, thus the limited chart review underestimates the number of errors and biases the types of errors detected.[178]

The Agency for Healthcare Research and Quality has begun development of the Quality and Safety Review System to replace the Medicare Patient Safety Monitoring System.[179] The new system is designed to overcome one of the main limitations of previous measurement systems, namely, the *ad hoc* search for safety event information in the chart. The Quality and Safety Review System directs reviewers to look for specific information based on questions automatically generated by its evidence-based expert system. The system uses existing information in the EHR, including age, sex, diagnoses, procedures, and potential adverse events as the basis for asking reviewers to acquire additional information from the chart. Based on what the reviewer reports, the expert system may ask additional questions before determining whether an adverse event had occurred. The expert system uses explicit, standardized definitions of the variables and of adverse events, and it uses human generated, validated rule-based (if-then) algorithms to ask questions and detect adverse events. The Quality and Safety Review System has a broad scope, its goal is to detect most of the adverse events that occur in hospitals, i.e., to measure "*all cause harm*." This standardized approach will allow reported rates to be compared across hospitals because they will be based on the same definitions and a standardized methodology.

Clearly, the major limitation of the Quality and Safety Review System is its reliance on human reviewers. What is needed in order for it to be maximally effective is for it to have the ability to detect meaning in free text. When this occurs, the system will become automatic, inexpensive, highly reliable, comprehensive (it will scan all patients), and accurate. Furthermore, it will operate in real time and have the capability of notifying clinicians regarding potentially unsafe actions, so they can be prevented from becoming adverse events.[179]

CLINICAL DECISION SUPPORT SYSTEMS

Although it was asserted that the elimination of the paper chart would significantly reduce errors and improve quality, that claim was made before the widespread adoption of the EHR. Since its implementation it has become clear that, although there is no longer any illegible handwriting and the chart is readily available, the EHR produces its own errors, the cut-and-paste function has made the available patient record less intelligible, and clinicians are having a hard time using EHR systems. But all is not lost, for the EHR is a precondition for the development and use of most CDSSs. In the last decade, advances in safety and quality have largely been due to CDSSs that have been built into, and rely upon, the electronic record.

CDSSs have been defined as, *"any software designed to directly aid in clinical decision making in which characteristics of individual patients are matched to a computerized knowledge base for the purpose of generating patient-specific assessments or recommendations that are then presented to clinicians for consideration."*[180] Full-fledged CDSSs have a graphical user interface, contain an algorithm, and display the output of the algorithm. The algorithm can be a human-constructed, rule-based system or, more recently, a trained statistical/probabilistic model. It takes as its input individual patient clinical information and provides as its output predictions regarding an individual patient's: risk of disease including prevention, or diagnosis, or prognosis including treatment and outcome.[181] The basic idea is that CDSSs can be used to prevent or ameliorate unsafe actions or inactions and provide information that can improve the quality of care. CDSSs have also been called expert systems and the use of trained statistical/probabilistic programs to make predictions has been called predictive analytics.

In terms of data acquisition, there are two main types of CDSSs.
1. Free-standing systems: they operate independently of the EHR. They require that an individual manually input the data and receive the results. The main type of autonomous systems has been for diagnosis.
2. Integrated systems: they interact directly with the patient's EHR. They access and analyze the clinical data and report their results. Currently, the main types of integrated systems are: (a) computerized provider order entry for medications, laboratory and radiographic tests and, (b) clinician alerts, reminders, and checklists. In addition, automated detection systems can discover potentially unsafe actions in order to prevent their occurrence or ameliorate their effects. An integrated system can operate in one of two modes, batch, where it periodically accesses, analyzes and reports its results and real-time, where it continuously accesses, analyzes, and reports its results.

The early CDSSs were free-standing because there was no EHR. Other than the importation of laboratory data, all data entry was performed manually by clinicians. With the advent of the EHR, CDSSs can automatically acquire data from the health record database but, because free text is usually not reliably read, these data are usually acquired from checkboxes and structured fields.

Currently, CDSSs are limited to demographic and anatomic/cellular data, both of which have limited predictive power. Molecular biomarkers, which are constative of the disease process, have the potential to allow us to better understand, predict, and treat disease. In the future, when molecular biomarkers (which include gene expression) are routinely included in the patient's medical record, CDSSs will become much more powerful – and the era of personalized medicine (also known as precision medicine) will have begun.[181]

Free-Standing Systems

The early CDSSs were free-standing. Clinicians manually entered the patient's data into the system and received the patient's risk of disease, or a rank order list of possible diseases, or a rank order list of treatments for a disease. The earliest CDSSs were diagnostic, and many supported some form of text data. Although an obvious target, the selection of diagnosis was unfortunate because the most difficult predictions in medicine tend to be those related to the diagnosis of disease.

The first diagnostic CDDS to gain widespread attention was the Internal Medicine diagnostic system Internist-1.[182-183] It consisted of a set of branching if-then rules and contained 570 diseases. There were two issues with Internist-1 and with similar free-standing systems. First, it did not solve a medical problem. In other words, Internal Medicine physicians were perfectly capable of making these diagnoses without Internist-1 and the program did not improve on their diagnostic accuracy. Second, it could take hours to manually input the clinical data into the program and no one wanted to perform that task. Internist-1 morphed into Quick Medical Reference, which was more of an information tool that a diagnostic program. It allowed clinicians to review the diagnostic information in the program's knowledge base. It contained 700 diseases and 5,000 signs, symptoms, and laboratory values. It could function as a textbook and it could generate a rank order list of possible diagnoses.[184-185] Unfortunately, it is no longer commercially available. Additional diagnostic programs included: DXplain, which contains 2,000 diseases, 5,000 clinical manifestations, and uses a modified form of Bayesian statistics; Iliad, which contains 930 diseases, 1,500 syndromes, 13,900 disease manifestations, and 90 simulated cases; and Isabel, which contains 11,000 diagnoses and 4,000 drugs and heuristics.[186-188]

In some diagnostic situations, the primary source of data is the image. In 2006, Tleyjeh et al. created a program, VisualDx, into which clinicians entered descriptors and lesion morphologies and it provided a dermatologic differential diagnosis. Tleyjeh et al. suggested that the program could increase *"clinician awareness of, knowledge about, and skills in the recognition of chemical warfare, bioterrorism, and radiation injuries."*[189] More recently, they developed an app that can build a dermatologic differential diagnosis based on images.[190] The clinician takes a picture of the dermatologic lesion and enters relevant factors such as age, travel, medical and social history, and the location, distribution, and appearance of the lesion into the program. The program, based on a simple matching criterion, lists the possible diagnosis in rank order by likelihood. The program claims to contain more than 2,800 conditions and more than 40,000 images. Chou found that VisualDx could improve the dermatologic diagnostic accuracy of medical students and residents by 19%.[191]

Over the last 10 years there has been a great deal of interest in reducing diagnostic errors. In 2015, the National Academies of Sciences, Engineering and Medicine published *Improving Diagnosis in Health Care*. It described many of the current diagnostic problems and it recommended ways to improve diagnostic accuracy. CDSSs were an integral part of their diagnostic improvement strategy. They stated that *"Diagnostic decision support tools can provide support to clinicians and patients throughout each stage of the diagnostic process, such as during information acquisition, information integration and interpretation, the formation of a working diagnosis, and the making of a diagnosis."*[192] In order to achieve the envisaged automated clinical decision support system an effective, operational natural language processing system will have to be in place.

Integrated Systems

With the advent of the EHR, CDSSs could operate autonomously on check boxes, and on laboratory, radiology, and pathology information without human data entry. Furthermore, the clinical decision support system can be running in the background in real time during the patient encounter, assessing and responding to the information the clinician enters into the EHR. The clinical decision support system's real-time monitoring and response system is what makes it a powerful safety and quality tool.

Computerized Provider Order Entry Systems (CPOEs)

Published in 2007, the Institute of Medicine's *Preventing Medication Errors,* presented information regarding the incidence and cost of medication mistakes, and it offered strategies for reducing them.[193] They pointed out that paper-based prescribing was one of the most common sources of medical mistakes and adverse events. These mistakes were due to many factors, including: (1) illegible handwriting and the use of abbreviations in prescription orders, (2) incomplete and incorrect prescriptions (e.g., incorrect dose calculation, drug name confusion, restarting a discontinued medication), (3) adverse drug-drug interactions, and (4) prescribing a medication that the patient was allergic to.[194-195]

CPOEs are electronic systems that are integrated into the EHR and that allow physicians to electronically order medications. They contain expert systems that evaluate the safety of the order using rules and information in the patient's EHR and they transmit alerts to the clinician when a potentially unsafe action (medication order) is occurring. CPOEs have been shown to reduce duplicate medications, drug overdoses, adverse drug-drug interactions, and the prescribing medications that patients are allergic to.[196-201] A recent systematic review and meta-analysis of CPOEs in the intensive care unit found an 85% reduction in medication errors by clinicians and a 12% reduction in mortality associated with CPOEs.[202] But computer-based systems are not perfect, they can make mistakes.[90] CPOEs can create duplicate prescriptions, miss wrong dose and wrong drug, generate mistakes related to drop down menus, and alerts can malfunction.[194, 203-206] It had been thought that adding additional clinical decision support capabilities to a computerized provider order entry system would offer additional safety and quality benefits, but it did not provide any additional benefit.[207-210] Furthermore, it is not always the case that all aspects of a safety solution need be electronic. A common outpatient medication mistake is dispensing a medication to the wrong patient, which occurs in 1.22 per 1,000 dispensed prescriptions.[200] Simple measures, such as checking the prescription with the patient at the point of sale, can reduce these mistakes by 56%.[211]

Another form of CPOE deals with the ordering of laboratory tests. Whereas, ordering medications dealt with safety, laboratory test ordering systems deal with reducing unnecessary testing in order to reduce the volume of tests and the cost of testing. A recent study used a CDDS to detect tests with a high repetition probability, or great complexity, or which were mutually incompatible within the same order.[212] The system would either cancel the test with no recourse or cancel it but allow the test after a written justification. They found that the provider order entry system reduced testing by 16% and costs by 17%. In a similar manner, radiology testing was reduced when CDSSs reviewed information in the patient's chart and denied testing.[213] Unfortunately, clinicians complained that the system was: (1) not easy to use, (2) too slow, (3) presented a high risk of error, and (4) required frequent interactions between the clinical staff. The investigators concluded that user acceptance and satisfaction were critical to system success. If clinicians did not find that the system benefited them, then they would either not use the system or they would use it in a suboptimal manner.

How well the computer-based clinical systems are implemented can have a profound effect on their acceptance and use.[131, 214-216] For example, in an odd twist of fate, the CPOE system used at a major teaching hospital in France crashed and they had to return to a paper-based order system.[217] The residents were given a satisfaction and user survey for both the electronic and paper order systems. They were almost four times more satisfied with the paper than the electronic system and they did not detect an increase in errors. In other words, computer-based systems that are not user-friendly, not efficient, and do not add clinical value can be detrimental to medical practice. User feedback should be solicited, and the acquired information acted upon, in the creation and deployment of computer-based clinical systems.[131, 218] Finally, because CDSSs have become part of the clinician work-flow, it is important to design them so that they seamlessly integrate into the clinician-patient clinical encounter.[219]

Clinician Alerts, Reminders and Checklists

Alerts are very useful in reducing medication mistakes. Most alerts are for drug-drug interactions. Unfortunately, currently there are far too many alerts, which blunts their effectiveness. Frequent alerts regarding co-administration incompatibilities negatively influenced adherence to the alerts – which resulted in many alerts being either ignored or overridden.[220-226] Alert fatigue has significantly reduced clinician enthusiasm for medication alerts.[226] In other words, there can be too much of a good thing when it comes to safety.

Another kind of alert, a patient-specific electronic reminder, occurs less frequently and has been shown to be an effective safety tool. Reminders that were integrated into an EHR increased clinician adherence to recommended care for diabetes and coronary artery disease.[227] In addition, a recent systematic review showed that reminders were effective in increasing clinician's ordering diabetes testing in women with a history of gestational diabetes.[228] But not all reminders are equally effective. Reminders for appropriate laboratory monitoring had no impact on rates of receiving appropriate testing for creatinine, potassium, liver function, renal function, or therapeutic drug level monitoring.[229] It appears that the efficacy of a reminder depends, in part, on whether there is a clinical problem that the reminder solves.

Safety checklists are activity-specific ordered lists of the actions that must be performed to successfully accomplish the task.[230] They have been useful in reducing preventable medical mistakes.[230-234] They are used in situations where an obligatory sequence of actions must be performed and where, if an action is omitted or an incorrect action is added, there is the potential for an unsafe inaction or action to occur – which could result in an adverse event. Checklists are especially useful in situations where several clinicians are performing coordinated actions on a patient in a complex, multi-stimuli environment. Checklists sequentially focus clinicians' attention on specific tasks. Recently, computer-based interactive, dynamic, adaptive safety checklists have been developed, many of which are linked to EHRs.[235-239] Interactive means that when an item is checked as completed, the system is updated, dynamic means that the checklist advances as the items on the checklist are completed, and adaptive means that the checklist can change based on changing conditions in the clinical workflow. These capabilities are based on the checklist's if-then algorithms and data-driven expert systems. The major limitations of checklists are: (1) they can tell if an action was done or not done but cannot tell if what was done was what was supposed to be done and cannot tell whether it was done correctly, (2) checklists are time consuming, and (3) checklists can disrupt an established workflow. Finally, checklist adherence tends to drop off over time. For example, the use of a childbirth checklist declined from 100% initially, to 72.8% at 2 months, to 61.7% at 12 months.[240]

Real-Time Systems

Although medication ordering and reminder systems operate in real time, they are just the initial steps in real time CDSSs. A long-term system-level step that is necessary to improve safety is to develop and implement sophisticated real-time CDSSs.

1. Clinician-patient clinical encounter: This system takes as its input the real time natural language information written into the EHR by the clinician during the clinician-patient interaction and existing check box and structured data already in the medical record. The CDSS continuously monitors this input in real time, in order to detect unsafe actions and conditions and to report unsafe actions and conditions to the clinician while the interaction is in progress, so a safety event can be prevented. This means that clinicians will have to write in their EHR during the clinical encounter.[241]
2. Medical procedure: This system takes as its input audiovisual information produced in real time during the procedure and existing check box and structured data already in the medical record. The CDSS continuously monitors this input in real time, in order to detect unsafe actions and conditions and to report unsafe actions and conditions to the clinician while the procedure is in progress so that a safety event can be prevented.

Decision Aids

CDSSs can be, and should be, used as decision aids. They provide predictions regarding the risks and benefits of a treatment for an individual patient. These estimates can be discussed with the patient as part of shared decision making. For example, they have been used to decide whether the patient should undergo an elective joint replacement.[242] The use of decision aids has been shown to reduce the rate of hip and knee surgery, thus reducing medical utilization and costs.[243-244]

Regulatory Environment and a Cautionary Note

Although the Centers for Medicare and Medicaid Services and the Agency for Health Care Research and Quality has been the driving forces behind the implementation and use of EHRs and related systems, the U.S. Food and Drug Administration is the federal agency responsible for the regulation of medical devices, including software. The Food and Drug Administration has been interested in medical software for many years, including CDSSs. It held hearings and provided guidance in 1998, 1999, and 2002.[245-247] In 2016, Congress passed the 21st Century Cures Act (Public Law No. 114-255, FDCA § 520(o)(1)(E)), which exempts from regulation software designed for: *"(i) displaying, analyzing, or printing medical information about a patient or other medical information (such as peer-reviewed clinical studies and clinical practice guidelines); (ii) supporting or providing recommendations to a health care professional about prevention, diagnosis, or treatment of a disease or condition; and (iii) enabling such health care professional to independently review the basis for such recommendations that such software presents so that it is not the intent that such health care professional rely primarily on any of such recommendations to make a clinical diagnosis or treatment decision regarding an individual patient."* Many of the current health informatics tools existed without FDA approval. This law not only makes them legal, but it also opens the field to additional innovation.

FUTURE TRENDS

Health informatics will likely develop an accurate natural language processing program. This advance will allow for the automatic detection of meaning in free text. Furthermore, it will drive the redesign of the EHR, thus reducing the number of check boxes and structured data and resurrecting the narrative clinical note. It will be easier for clinicians to use the EHR and for administrators to aggregate and analyze medical data.

Safety and quality will consist of systematic, evidence-based electronic detection and reporting programs. Furthermore, CDSSs will operate in real-time and alert clinicians to actions that may place patients at risk of harm.

KEY POINTS

- It is clear that safety, quality and value are becoming defining characteristics of medical practice
- The U.S. government has, and will continue to, drive safety, quality and value
- Safety must shift its focus from safety events to the unsafe actions that cause the events, for it is only then that we have a real chance to prevent safety events
- The EHR is necessary but not sufficient for improving safety and quality
- In order to achieve their full potential, EHR systems must have the capability to read free text
- Health informatics will be building real-time CDSSs into the EHR and these systems will significantly improve safety, quality and value

CONCLUSIONS

Health informatics must be properly socialized within the medical community. Currently, is exists mostly by fiat. The usual situation is that clinicians are told that a health information technology product is being implemented and they better get used to it. But this mode of implementation does not have long term viability. A recent editorial in Lancet Oncology stated, *"What has become evident over the past two decades or longer is that vast amounts of data have now infiltrated every aspect of our daily lives. From data analytics to artificial intelligence, to predictive modelling and machine learning, we are now seeing these systems being incorporated into all aspects of health, including those found in oncology. But as now shown by the JAMIA study, physicians are not always willing to accept the changes that these systems bring. Although big data offers the promise of easing workflows, ensuring treatment adherence according to guidelines, the analysis of large datasets, maintenance of a centralised records system, improving diagnostic accuracy, and monitoring disease or drug safety surveillance—all of which could be hugely beneficial for the future of health care—clearly a delicate balance is needed when integrating those promises into the clinical decision-making process."*[248]

Health informatics must not just be an administrative endeavor designed to reduce costs. It must become an integrated strategy supported by clinicians and focused on improving safety, quality, and value.

REFERENCES

1. Braude RM. A descriptive analysis of national library of medicine-funded medical informatics training programs and the career choices of their graduates. Med Decis Making 1991;11:33-37.
2. Agency for Healthcare Quality and Research. https://www.ahrq.gov/cpi/about/index.html (Accessed December 4, 2017)
3. Kuenssberg EV. Volume and cost of keeping records in a group practice. Br Med J 1956;1(Suppl 2681):341–3.
4. Dearing WP. Quality of medical care. Calif Med 1963;98:331–5.
5. Weed LL. Medical records that guide and teach. N Engl J Med 1968;278:593–600.
6. Soto CM, Kleinman KP, Simon SR. Quality and correlates of medical record documentation in the ambulatory care setting. BMC Health Serv Res 2002;2:22.
7. Levy RP, Cammarn MR, Smith MJ. Computer handling of ambulatory clinic records. JAMA 1964;190:1033–7.
8. Dick RS, Steen EB, Detmer DE. (Eds.) CPR Report – Computer-based Patient Record. https://www.ncbi.nlm.nih.gov/books/NBK233047/ (Accessed December 4, 201
9. Kohn LT, Corrigan JM, Donaldson MS. (Eds.) To Err Is Human: Building a Safer Health System. Washington, DC: National Academy Press, 2000. https://www.nap.edu/resource/9728/To-Err-is-Human-1999--report-brief.pdf (Accessed December 4, 2017)
10. Institute of Medicine. Key Capabilities of an EHR System, 2003. http://www.nationalacademies.org/hmd/Reports/2003/Key-Capabilities-of-an-Electronic-Health-Record-System.aspx . (Accessed December 4, 2017)
11. Bates DW1, Ebell M, Gotlieb E, Zapp J, Mullins HC. A proposal for electronic medical records in U.S. primary care. JAMIA 2003;10(1):1–10.
12. Aspden P, Corrigan JM, Wolcott J, and Shari M. Erickson SM. (Eds.) Patient Safety: Achieving a New Standard for Care. https://www.nap.edu/catalog/10863/patient-safety-achieving-a-new-standard-for-care . (Accessed December 4, 2017).
13. Embi PJ, Yackel TR, Nowen JL, et al. Impacts of computerized physician documentation in a teaching hospital: perceptions of faculty and resident physicians. J Am Med Inform Assoc 2004;11:300–9.
14. Burke HB, Hoang A, Becher D, Fontelo P, Liu F, Stephens M, Pangaro LN, Sessums LL, O'Malley P, Baxi NS, Bunt CW, Capaldi VF, Chen JM, Cooper BA, Djuric DA, Hodge JA, Kane S, Magee C, Makary ZR, Mallory RM, Miller T, Saperstein A, Servey J, Gimbel RW. QNOTE: An instrument for measuring the quality of EHR clinical notes. J Am Med Inform Assoc 2014;21(5):910-916.
15. Bush GW. State of the Union address, January 2004. Available at, https://georgewbush-whitehouse.archives.gov/stateoftheunion/2004/ . (Accessed December 4, 2017).
16. Jha AK. Meaningful use of EHRs. *The Journal of the American Medical Association* 2010;304:1709–1710.
17. National Library of Medicine, https://www.nlm.nih.gov/ep/GrantTrainInstitute.html . (Accessed December 4, 2017).
18. National Quality Forum. http://www.qualityforum.org/about_nqf/mission_and_vision/ . (Accessed December 4, 2017).
19. Summary Health Statistics Tables for U.S. Adults: National Health Interview Survey https://www.cdc.gov/nchs/fastats/physician-visits.htm (Accessed December 4, 2017).

20. National Ambulatory Medical Care Survey https://www.cdc.gov/nchs/fastats/physician-visits.htm (Accessed December 4, 2017).
21. National Hospital Ambulatory Medical Care Survey https://www.cdc.gov/nchs/data/ahcd/nhamcs_outpatient/2011_opd_web_tables.pdf (Accessed December 4, 2017).
22. National Hospital Ambulatory Medical Care Survey: 2014 Emergency Department Summary https://www.cdc.gov/nchs/fastats/emergency-department.htm (Accessed December 4, 2017)
23. American Hospital Association Annual Survey http://www.aha.org/research/rc/stat-studies/fast-facts.shtml (Accessed December 4, 2017).
24. Brandt WS, Isbell JM, Jones DR. Defining quality in the surgical care of lung cancer patients. J Thorac Cardiovasc Surg. 2017 Oct;154(4):1397-1403.
25. Institute of Medicine. Crossing the Quality Chasm: A New Health System for the 21st Century. Washington, DC: The National Academies Press, 2001. https://www.nap.edu/catalog/10027/crossing-the-quality-chasm-a-new-health-system-for-the . (Accessed December 4, 2017).
26. CMS Quality Strategy. https://www.cms.gov/Medicare/Quality-Initiatives-Patient-Assessment-Instruments/QualityInitiativesGenInfo/Downloads/CMS-Quality-Strategy.pdf . (Accessed December 4, 2017).
27. Medicaid Quality of Care. https://www.medicaid.gov/medicaid/quality-of-care/index.html . (Accessed December 4, 2017).
28. Agency for Healthcare Research and Quality. https://www.ahrq.gov/professionals/quality-patient-safety/quality-resources/tools/chtoolbx/understand/index.html . (Accessed December 4, 2017).
29. Ross C. 'Value' is medicine's favorite buzzword. But whose definition are we using? https://www.statnews.com/2017/11/29/value-medicines-buzzword/(Accessed December 4, 2017).
30. Schnipper LE, Schilsky RL. Are value frameworks missing the mark when considering long-term benefits from immune-oncology drugs? JAMA Oncol December 28, 2017
31. Schnipper LE, Davidson NE, Wollins DS, Tyne C, Blayney DW, Blum D, Dicker AP, Ganz PA, Hoverman JR, Langdon R, Lyman GH, Meropol NJ, Mulvey T, Newcomer L, Peppercorn J, Polite B, Raghavan D, Rossi G, Saltz L, Schrag D, Smith TJ, Yu PP, Hudis CA, Schilsky RL, American Society of Clinical Oncology. American Society of Clinical Oncology statement: a conceptual framework to assess the value of cancer treatment options. J Clin Oncol 2015;33(23): 2563-2577.
32. Schnipper LE, Davidson NE, Wollins DS, Blayney DW, Dicker AP, Ganz PA, Hoverman JR, Langdon R, Lyman GH, Meropol NJ, Mulvey T, Newcomer L, Peppercorn J, Polite B, Raghavan D, Rossi G, Saltz L, Schrag D, Smith TJ, Yu PP, Hudis CA, Vose JM, Schilsky RL. Updating the American Society of Clinical Oncology value framework: revisions and reflections in response to comments received. J Clin Oncol 2016; 34(24):2925-2934.
33. Abel G, Saunders CL, Mendonca SC, Gildea C, McPhail S, Lyratzopoulos G. Variation and statistical reliability of publicly reported primary care diagnostic activity indicators for cancer: a cross-sectional ecological study of routine data. BMJ Qual Saf. 2018 Jan;27(1):21-30.
34. Ben-Aharon O, Magnezi R, Leshno M, Goldstein DA. Association of immunotherapy with durable survival as defined by value frameworks for cancer care. JAMA Oncol December 28, 2017.
35. Hoffmann TC, Del Mar C. Clinicians' Expectations of the Benefits and Harms of Treatments, Screening, and Tests: A Systematic Review. JAMA Intern Med 2017;177(3):407-419.
36. Hong AS, Ross-Degnan D, Zhang F, Wharam JF. Clinician-Level Predictors for Ordering Low-Value Imaging. JAMA Intern Med 2017;177(11):1577-1585.
37. Arnold M. Simulation modeling for stratified breast cancer screening – a systematic review of cost and quality of life assumptions. BMC Health Services Research 2017;17:802.
38. Mehrez A, Gafni A. Quality-adjusted life years, utility theory, and healthy-years equivalents. Med Decis Making 1989;9:142-149.
39. WichmannAB, Adang EMM, Stalmeier PFM, KristantiS, Van den Block L, Myrra JFJ Vernooij-Dassen MJFJ, Engles Y. The use of Quality-Adjusted Life Years in cost-effectiveness analyses in palliative care: Mapping the debate through an integrative review. Palliative Medicine 2017; 31(4):306 – 322.
40. Schoenborn NL, Lee K, Pollack CE, Armacost K, Dy SM, Xue QL, Wolff AC, Boyd C. Older Adults' Preferences for When and How to Discuss Life Expectancy in Primary Care. J Am Board Fam Med. 2017 Nov-Dec;30(6):813-815.
41. Perry P, Hotze T. Oregon's experiment with prioritizing public health care services". AMA J Ethics 2011;13 (4): 241–247.
42. Jacob V, Thota AB, Chattopadhyay SK, Njie GJ, Proia KK, Hopkins DP, Ross MN, Pronk NP3, Clymer JM. Cost and economic benefit of CDSSs for cardiovascular disease prevention: a community guide systematic review. J Am Med Inform Assoc. 2017 May 1;24(3):669-676.
43. Scheunemann LP, White DB. The Ethics and Reality of Rationing in Medicine. Chest 2011;140;1625-1632.
44. Westaby S. Open Heart. New York: New York, Basic Books, 2017.

45. Editorial. The NHS: failing to deliver on Beveridge's promise? Lancet Oncol Vol 19 January 2018
46. Fraser Institute, Waiting your turn: Wait times for health care in Canada. 2017. https://www.fraserinstitute.org/studies/waiting-your-turn-wait-times-for-health-care-in-canada-2017 . (Accessed December 13, 2017).
47. Appari A, Johnson ME, Anthony DL. Health IT and inappropriate utilization of outpatient imaging: A cross-sectional study of U.S. hospitals. Int J Med Inform 109 (2018) 87–95.
48. Roberts ET, Zaslavsky AM, McWilliams JM The Value-Based Payment Modifier: Program Outcomes and Implications for Disparities. Ann Intern Med. 2017 Nov 28. doi: 10.7326/M17-1740.
49. Frakt AB, Jha AK. Face the Facts: We Need to Change the Way We Do Pay for Performance. Ann Intern Med. 2017 Nov 28.
50. Figueroa JF, Joynt Maddox KE, Beaulieu N, Wild RC, Jha AK. Concentration of Potentially Preventable Spending Among High-Cost Medicare Subpopulations: An Observational Study. Ann Intern Med. 2017 Nov 21;167(10):706-713.
51. Burke HB, Hoang A, Becher D, Fontelo P, Liu F, Stephens M, Pangaro LN, Sessums LL, O'Malley P, Baxi NS, Bunt CW, Capaldi VF, Chen JM, Cooper BA, Djuric DA, Hodge JA, Kane S, Magee C, Makary ZR, Mallory RM, Miller T, Saperstein A, Servey J, Gimbel RW. QNOTE: An instrument for measuring the quality of EHR clinical notes. J Am Med Inform Assoc 2014;21(5):910-916.
52. Medicare Patient Safety Monitoring System. https://www.ahrq.gov/downloads/pub/advances/vol2/Hunt.pdf . (Accessed December 4, 2017).
53. American Recovery and Reinvestment Act of 2009, Pub L, 111–5, 123 Stat 115.
54. Steinbrook R. Health care and the American recovery and reinvestment act. N Engl J Med 2009;360:1057–1060.
55. Office of the National Coordinator of Health Information Technology. Physician adoption of EHRs by 2015. https://dashboard.healthit.gov/quickstats/quickstats.php . (Accessed December 4, 2017).
56. Linder JA, Ma J, Bates DW, Middleton B and Stafford RS. EHR use and the quality of ambulatory care in the United States. Arch Intern Med 2007;167:1400–1405.
57. Keyhani S, Hebert PL, Ross JS, Federman A, Zhu CW and Siu AL. EHR components and the quality of care. Med Care 2008;46:1267–1272.
58. Zhou L1, Soran CS, Jenter CA, Volk LA, Orav EJ, Bates DW et al. The relationship between EHR use and quality of care over time. JAMIA 2009;16:457–464.
59. Friedberg MW, Coltin KL, Safran DG, et al. Associations between structural capabilities of primary care practices and performance on selected quality measures. Ann Intern Med 2009;151:456–463.
60. Walsh MN, Yancy CW, Albert NM, Curtis AB, Stough WG and Gheorghiade M et al. EHRs and quality of care for heart failure. J Am Heart Assoc 2010;159:635–642.
61. Poon EG, Wright A, Simon SR, Jenter CA, Kaushal R and Volk LA et al. Relationship between use of EHR features and health care quality: results of a statewide survey. Med Care 2010;48:203–209.
62. Romano MJ and Stafford RS. EHRs and CDSSs: impact on national ambulatory care quality. Arch Intern Med 2011;171:897–903.
63. Kern LM, Barrón Y, Dhopeshwarkar RV, Edwards A, Kaushal R and HITEC Investigators. EHRs and ambulatory quality of care. J Gen Intern Med 2013;28:496–503.
64. Sequist TD, Gandhi TK, Karson AS, Fiskio JM, Bugbee D and Sperling M et al. A randomized trial of electronic clinical reminders to improve quality of care for diabetes and coronary artery disease. J Am Heart Assoc 2005;12:431–437.
65. Matheny ME, Sequist TD, Seger AC, Fiskio JM, Sperling M and Bugbee D et al. A randomized trial of electronic clinical reminders to improve medication laboratory monitoring. JAMA 2008;15:424-429.
66. Burke HB, Becher DA, Hoang, A, Gimbel RW. The adoption of an EHR did not improve A1c values in Type 2 diabetes. J Innov Health Inform 2016;23(1):433-438.
67. Friedberg MW, Chen PG, Van Busum KR, Aunon F, Pham C, Caloyeras J, Mattke S, Pitchforth E, Quigley DD, Brook RH, Crosson FJ, Tutty M. Factors Affecting Physician Professional Satisfaction and Their Implications for Patient Care, Health Systems, and Health Policy. Rand Health Q. 2014 Dec1;3(4):1.http://www.rand.org/content/dam/rand/pubs/research_reports/RR400/RR439/RAND_RR439.pdf . (Accessed December 4, 2017).
68. Sinsky C, Colligan L, Li L, Prgomet M, Reynolds S, Goeders L, Westbrook J, Tutty M, Blike G. Allocation of Physician Time in Ambulatory Practice: A Time and Motion Study in 4 Specialties. Ann Intern Med. 2016 Dec 6;165(11):753-760.
69. Arndt BG, Beasley JW, Watkinson MD, Temte JL, Tuan WJ, Sinsky CA, Gilchrist VJ. Tethered to the EHR: Primary Care Physician Workload Assessment Using EHR Event Log Data and Time-Motion Observations. Ann Fam Med. 2017 Sep;15(5):419-426.
70. Montague E, Asan O. Dynamic modeling of patient and physician eye gaze to understand the effects

of EHRs on doctor-patient communication and attention. Int J Med Inform 2014;83:225-234.
71. Shaarani I, Taleb R, Antoun J. Effect of computer use on physician-patient communication using a validated instrument: Patient perspective. Int J Med Inform 2017;108:152-157.
72. Burke HB, Hoang A, Becher D, Fontelo P, Liu F, Stephens M, Pangaro LN, Sessums LL, O'Malley P, Baxi NS, Bunt CW, Capaldi VF, Chen JM, Cooper BA, Djuric DA, Hodge JA, Kane S, Magee C, Makary ZR, Mallory RM, Miller T, Saperstein A, Servey J, Gimbel RW. QNOTE: An instrument for measuring the quality of EHR clinical notes. J Am Med Inform Assoc 2014;21(5):910-916.
73. Burke HB, Sessums LL, Hoang A, et al. EHRs improve note quality. J Am Med Inform Assoc 2015 Jan;22(1):199-205.
74. Cohen R, Aviram I, Elhadad M, Elhadad N. Redundancy-aware topic modeling for patient record notes. PLoS One. 2014 Feb 13;9(2):e87555.
75. Koppel R, Metlay JP, Cohen A, et al. Role of computerized physician order entry systems in facilitating medication errors. JAMA 2005;293:1197–1203.
76. Weiner JP, Kfuri T, Chan K, et al. "e-Iatrogenesis": the most critical unintended consequence of CPOE and other HIT. J Am Med Inform Assoc 2007;14:387–388.
77. Karsh BT, Weinger MB, Abbott PA, et al. Health information technology: fallacies and sober realities. J Am Med Inform Assoc 2010;17:617–23.
78. Magrabi F, Ong MS, Runciman W, et al. An analysis of computer-related patient safety incidents to inform the development of a classification. J Am Med Inform Assoc 2010;17:663–70.
79. Sittig DF, Classen DC. Safe EHR use requires a comprehensive monitoring and evaluation framework. JAMA 2010;303:450–451.
80. Myers RB, Jones SL, Sittig DF. Review of reported clinical information system adverse events in US Food and Drug Administration databases. Appl Clin Inform 2011;2:63–74.
81. Harrington L, Kennerly D, Johnson C. Safety issues related to the electronic medical record (EMR): synthesis of the literature from the last decade, 2000–2009. J Healthc Manag 2011;56:31–43.
82. Sittig DF, Singh H. EHRs and national patient-safety goals. N Engl J Med 2012;367:1854–1860.
83. Meeks DW, Takian A, Sittig DF, et al. Exploring the sociotechnical intersection of patient safety and EHR implementation. J Am Med Inform Assoc 2014;21:e28–34.
84. Bellwood P, Borycki EM, Kushniruk AW. Awareness of technology induced errors and processes for identifying and preventing such errors. Stud Health Technol Inform 2015;208:61–65.
85. Castro G B, Buczkowski L, Hafner J, et al. Investigations of Health IT related Deaths, Serious Injuries, or Unsafe Conditions. 2015. Final Report, The Joint Commission.
86. Magrabi F, Ong MS, Runciman W, et al. Using FDA reports to inform a classification for health information technology safety problems. J Am Med Inform Assoc 2012;19:45–53.
87. Meeks DW, Smith MW, Taylor L, et al. An analysis of EHR–related patient safety concerns. J Am Med Inform Assoc 2014;21:1053–59.
88. Graber ML, Siegal D, Riah H, Johnston D, et al. EHR– related events in medical malpractice claims. J Patient Saf 2015. PMID: 26558652
89. Troxel DB. EHRs closed claims study. The Doctors Company, 2017. http://www.thedoctors.com/ecm/groups/public/documents/print_pdf/con_id_013553.pdf (Accessed December 4, 2017).
90. Kim MO, Coiera E, Magrabi F. Problems with health information technology and their effects on care delivery and patient outcomes: a systematic review. JAMIA 2017;24(2):246-250.
91. Menon S, Singh H, Giardina TD, Rayburn WL, Davis BP, Russo EM, Sittig DF. Safety huddles to proactively identify and address EHR safety. J Am Med Inform Assoc 2017;24:261-267.
92. Pelayo S, Ong M. Human Factors and Ergonomics in the Design of Health Information Technology: Trends and Progress in 2014. Yearb Med Inform. 2015 Aug 13;10:75-78.
93. Kushniruk A, Nohr C, Borycki E. Human Factors for More Usable and Safer Health Information Technology: Where Are We Now and Where do We Go from Here? Yearb Med Inform. 2016;1:120-125.
94. Verdon DR. EHRs: The real story. Why a national outcry from physicians will shake the health information technology sector. Med Econ 2014;91(3):18–27.
95. Kellerman AL and Jones SS. What it will take to achieve the as-yet-unfulfilled promises of health information technology. Health Affairs 2013;32(1):63–68.
96. Hunt LM, Bell HS, Baker AM, Howard HA. EHRs and the disappearing patient. Med Anthropol Q 2017;31:403-421.
97. Lucyk K, Tang K, Quan H. Barriers to data quality resulting from the process of coding health information to administrative data: a qualitative study. BMC Health Services Research 2017; 17:766.
98. Cohen KB, Lanfranchi A, Choi MJ, Bada M, Baumgartner WA Jr, Panteleyeva N, Verspoor K, Palmer M, Hunter LE. Coreference annotation and resolution in the Colorado Richly Annotated Full Text (CRAFT) corpus of biomedical journal articles. BMC Bioinformatics. 2017 Aug 17;18(1):372

99. Ford E, Carroll JA, Smith HE, Scott D, Cassell JA. Extracting information from the text of electronic medical records to improve case detection: a systematic review. J Am Med Inform Assoc 2016;23(5):1007-15
100. Friedman C, Alderson PO, Austin JH, Cimino JJ, Johnson SB. A general natural- language text processor for clinical radiology. J Am Med Inform Assoc 1994;1(2):161–174.
101. Zeng QT, Goryachev S, Weiss S, Sordo M, Murphy SN, Lazarus R. Extracting principal diagnosis, co-morbidity and smoking status for asthma research: evaluation of a natural language processing system. BMC Med Inform Decis Mak 2006;6(1):30.
102. Savova GK, Masanz JJ, Ogren PV, Zheng J, Sohn S, Kipper-Schuler KC, Chute CG. Mayo clinical Text Analysis and Knowledge Extraction System (cTAKES): architecture, component evaluation and applications. J Am Med Inform Assoc 2010;17(5): 507–513
103. Davies J, Hebert P, Hoffman C, editors. The Canadian Patient Safety Dictionary. Calgary: Royal College of Physicians and Surgeons of Canada and Health Canada. 2003. http://www.royalcollege.ca/portal/page/portal/rc/common/documents/publications/patient_safety_dictionary_e.pdf (Accessed December 4, 2017).
104. Grober ED, Bohnen JMA. Defining medical error. Can J Surg. 2005 Feb; 48: 39–44.
105. Barach P, Small SD. Reporting and preventing medical mishaps: lessons from non-medical near miss reporting. *BMJ* 2000;320:759-63.
106. Hays R, Daker-White G, Esmail A, Barlow W, Minor B, Brown B, Blakeman T, Sanders C, Bower P. Threats to patient safety in primary care reported by older people with multimorbidity: baseline findings from a longitudinal qualitative study and implications for intervention. BMC Health Services Research (2017) 17:754
107. Sasso L, Bagnasco A, Aleo G, Catania G, Dasso N, Zanini MP, Watson R. Incorporating nursing complexity in reimbursement coding systems: the potential impact on missed care. BMJ Qual Saf. 2017 Nov;26(11):929-932.
108. Kalisch BJ, Williams RA. Development and psychometric testing of a tool to measure missed nursing care. J Nurs Adm 2009;39:211–9.
109. Poghosyan L, Norful AA, Fleck E, Bruzzese JM, Talsma A, Nannini A. Primary Care Providers' Perspectives on Errors of Omission. J Am Board Fam Med. 2017 Nov-Dec;30(6):733-742.
110. Jeffs L, Berta W, Lingard L, Baker GR. Learning from near misses: from quick fixes to closing off the Swiss-cheese. BMJ Qual Saf 2012;21:287-294.
111. Ginsburg LR, Chuang YT, Richardson J, Norton PG, Berta W, Tregunno D, Ng P. Categorizing errors and adverse events for learning: a provider perspective. Healthc Q 2009;12:154-60.
112. Westbrook JI, Li L, Lehnbom EC, Baysari MT, Braithwaite J, Burke R, Conn C, Day RO. What are incident reports telling us? A comparative study at two Australian hospitals of medication errors identified at audit, detected by staff and reported to an incident system. Int J Qual Health Care 2015;27:1-9.
113. De Hoon SE, Hek K, van Dijk L, Verheij RA. Adverse events recording in EHR sysetms in primary care. BMC Me Inform Decis Making 2017;17:163.
114. Speroni KG, Fisher J, Dennis M, Daniel M. What causes near-misses and how are they mitigated? Plast Surg Nurs. 2014;34:114-9.
115. Hatoun J, Suen W, Liu C, Shea S, Patts G, Weinberg J, Eng J. Elucidating Reasons for Resident Underutilization of Electronic Adverse Event Reporting. Am J Med Qual. 2016;31:308-14.
116. Heideveld-Chevalking AJ, Calsbeek H, Damen J, Gooszen H, Wolff AP. The impact of a standardized incident reporting system in the perioperative setting: a single center experience on 2,563 'near-misses' and adverse events. Patient Saf Surg. 2014;8:46. doi: 10.1186/s13037-014-0046-1.
117. Yoon RS, Alaia MJ, Hutzler LH, Bosco JA 3rd. Using "near misses" analysis to prevent wrong-site surgery. J Healthc Qual. 2015;37:126-32.
118. Arnold A. Building a learning culture and prevention of error - to near miss or not. J Med Radiat Sci. 2017;64:163-164.
119. Petsching W, Haslinger-Baumann E. Critical incident reporting system (CIRS): a fundamental component of risk management in health care systems to enhance patient safety. Safety in Health 2017;3:9.
120. Hatoun J, Suen W, Liu C, Shea S, Patts G, Weinberg J, Eng J. Elucidating Reasons for Resident Underutilization of Electronic Adverse Event Reporting. Am J Med Qual. 2016;31:308-14.
121. Crane S, Sloane PD, Elder N, Cohen L, Laughtenschlaeger N, Walsh K, Zimmerman S. Reporting and Using Near-miss Events to Improve Patient Safety in Diverse Primary Care Practices: A Collaborative Approach to Learning from Our Mistakes. J Am Board Fam Med. 2015;28:452-460.
122. Kusano AS, Nyflot MJ, Zeng J, Sponseller PA, Ermoian R, Jordan L, Carlson J, Novak A, Kane G, Ford EC. Measurable improvement in patient safety culture: A departmental experience with incident learning. Pract Radiat Oncol. 2015;5:e229-237.
123. Burlison JD, Quillivan RR, Kath LM, Zhou Y, Courtney SC, Cheng C, Hoffman JM. A Multilevel

Analysis of U.S. Hospital Patient Safety Culture Relationships With Perceptions of Voluntary Event Reporting. J Patient Saf. 2016 Nov 3. PMID: 27820722
124. Deraniyagala R, Liu C, Mittauer K, Greenwalt J, Morris CG, Yeung AR. Implementing an Electronic Event-Reporting System in a Radiation Oncology Department: The Effect on Safety Culture and Near-Miss Prevention. J Am Coll Radiol. 2015;12:1191-1195.
125. Howell AM, Burns EM, Hull L, Mayer E, Sevdalis N, Darzi A. International recommendations for national patient safety incident reporting systems: an expert Delphi consensus-building process. BMJ Qual Saf. 2017;26:150-163.
126. Adelman JS, Kalkut GE, Schechter CB, et al. Understanding and preventing wrong-patient electronic orders: a randomized controlled trial. J Am Med Inform Assoc 2013;20:305-310.
127. Lavin JM, Boss EF, Brereton J, Roberson DW, Shah RK. Responses to errors and adverse events: The need for a systems approach in otolaryngology. Laryngoscope. 2016;126:1999-2002.
128. Gibson R, Armstrong A, Till A, McKimm J. Learning from error: leading a culture of safety. Br J Hosp Med (Lond). 2017;78:402-406.
129. Safety Assessment Code (SAC) Matrix, https://www.patientsafety.va.gov/professionals/publications/matrix.asp (Accessed October 16, 2017).
130. Arnold A. Building a learning culture and prevention of error - to near miss or not. J Med Radiat Sci. 2017;64:163-164.
131. Kim MO, Coiera E, Magrabi F. Problems with health information technology and their effects on care delivery and patient outcomes: a systematic review. JAMIA 2017;24(2):246-250
132. AHRQ Harm Scale. http://www.ashrm.org/pubs/files/white_papers/SSE-2_getting_to_zero-9-30-14.pdf . (Accessed December 4, 2017).
133. Williams T, Szekendi M, Pavkovic S, Clevenger W, Cerese J. The reliability of AHRQ Common Format Harm Scales in rating patient safety events. J Patient Saf. 2015 Mar;11(1):52-9.
134. Yerkes RM, Dodson JD. The relation of strength of stimulus to rapidity of habit-formation. Journal of Comparative Neurology and Psychology 1908;18:459–482.
135. NASA. Patient Safety Reporting System, https://psrs.arc.nasa.gov . (Accessed December 4, 2017).
136. The Joint Commission. https://www.jointcommission.org/sentinel_event.aspx . (Accessed December 4, 2017).
137. Noble DJ, Pronovost PJ. Underreporting of patient safety incidents reduces health care's ability to quantify and accurately measure harm reduction. J Patient Saf. 2010;6:247-50.
138. Weiss BD, Scott M, Demmel K, Kotagal UR, Perentesis JP, Walsh KE. Significant and sustained reduction in chemotherapy errors through improvement science. J Oncol Pract. 2017;13:e329-e336.
139. Mitchell I, Schuster A, Smith K, Pronovost P. Patient safety incident reporting: a qualitative study of thoughts and perceptions of experts 15 years after 'To Err is Human'. BMJ Qual Saf. 2016;25:92-99.
140. Macrae C. The problem with incident reporting. BMJ Qual Saf. 2016;25:71-75.
141. National Patient Safety Agency. Seven steps to patient safety. London: NHS National Patient Safety Agency, 2004.
142. Of Miracles. Harvard Classics. Bartleby.com textbooks. www.bartleby.com (Accessed January 2, 2018)
143. National Committee for Quality Assurance. http://www.ncqa.org/hedis-quality-measurement . (Accessed December 4, 2017).
144. Quality Payment Program. https://qpp.cms.gov/mips/quality-measures . (Accessed December 4, 2017)
145. National Guideline Clearinghouse. https://www.ahrq.gov/cpi/about/otherwebsites/guideline.gov/index.html . (Accessed December 4, 2017).
146. Schuster MA, Onorato SE, Meltzer DO. Measuring the cost of quality measurement: a missing link in quality strategy. JAMA 2017;318(13):1219-1220
147. Grady D, Redberg RF, O'Malley PG. Quality Improvement for Quality Improvement Studies. JAMA Intern Med 2017 Nov 27.
148. Schall MC Jr1, Cullen L, Pennathur P, Chen H, Burrell K, Matthews G. Usability Evaluation and Implementation of a Health Information Technology Dashboard of Evidence-Based Quality Indicators. Comput Inform Nurs 2017;35(6):281-288.
149. Rubin HR, Pronovost P, Diette GB. The advantages and disadvantages of process-based measures of health care quality. Int J Qual Health Care 2001;13:469–74.
150. Marang-van de Mheen PJ, Shojania KG. Simpson's paradox: how performance measurement can fail even with perfect risk adjustment. BMJ Qual Saf 2014;23:701–5.
151. Bruckel J, Liu X, Hohmann SF, et al. The denominator problem: national hospital quality measures for acute myocardial infarction. BMJ Qual Saf 2017;26:189–99.
152. Abel G, Saunders CL, Mendonca SC, Gildea C, McPhail S, Lyratzopoulos G. Variation and statistical reliability of publicly reported primary care diagnostic activity indicators for cancer: a

cross-sectional ecological study of routine data. BMJ Qual Saf. 2018 Jan;27(1):21-30.
153. Tinmouth J. Unpacking quality indicators: how much do they reflect differences in the quality of care? BMJ Qual Saf. 2018 Jan;27(1):4-6.
154. Shulman LN, Palis BE, McCabe R, Mallin K, Loomis A, Winchester D, McKellar D. Survival as a quality metric of cancer care: use of the National Cancer Data Base to assess hospital performance. J Oncol Pract. 2017 Nov 1
155. Carmichael JM, Meier J, Robinson A, Taylor J, Higgins DT, Patel S. Leveraging electronic medical record data for population health management in the Veterans Health Administration: Successes and lessons learned. Am J Health Syst Pharm 2017;74(18):1447-1459.
156. Buttigieg SC, Pace A, Rathert C. Hospital performance dashboards: a literature review. J Health Organ Manag. 2017 May 15;31(3):385-406
157. Karami M, Langarizadeh M, Fatehi M. Evaluation of Effective Dashboards: Key Concepts and Criteria. Open Med Inform J 2017;11:52-57
158. Franklin A, Gantela S, Shifarraw S, Johnson TR, Robinson DJ, King BR, Mehta AM, Maddow CL, Hoot NR, Nguyen V, Rubio A, Zhang J, Okafor NG. Dashboard visualizations: Supporting real-time throughput decision-making. J Biomed Inform 2017;71:211-221.
159. Hunt DR, Verzier N, Abend SL, Lyder C, Jaser LJ, Safer N, Davern P. Fundamentals of Medicare Patient Safety Surveillance: Intent, Relevance, and Transparency. https://www.ahrq.gov/downloads/pub/advances/vol2/Hunt.pdf . (Accessed December 4, 2017).
160. Patient Safety Indicators, http://qualityindicators.ahrq.gov/downloads/modules/psi/v31/psi_guide_v31.pdf . (Accessed December 4, 2017).
161. Classen DC, Resar R, Griffin F, et al. "Global Trigger Tool" shows that adverse event in hospitals may be ten times greater than previously measured. Health Aff. 2011;30(4): 581-589.
162. Rozich JD, Haraden CR, Resar RK. Adverse drug event trigger tool: a practical methodology for measuring medication related harm. BMJ Quality Safety. 2003;12(3):194-200.
163. Resar RK, Rozich JD, Classen D. Methodology and rationale for the measurement of harm with trigger tools. BMJ Quality Safety. 2003;12(suppl 2):ii39-ii45.
164. Resar FK, Rozich JD, Simmonds T, Haraden CR. A trigger tool to identify adverse events in the intensive care unit. Joint Commission J Quality Patient Safety. 2006;32(10):585-590.
165. Adler L, Denham CR, McKeever M, et al. Global Trigger Tool: implementation basics. J Patient Safety. 2008;4(4):245-249.
166. Griffin FA, Classen DC. Detection of adverse events in surgical patients using the Trigger Tool approach. BMJ Quality Safety. 2008;17:253-258.
167. Classen DC, Lloyd RC, Provost L, Griffin FA, Resar R. Development and evaluation of the Institute for Healthcare Improvement Global Trigger Tool. J Patient Safety. 2008;4(3):169-177.
168. Griffin FA, Resar RK. IHI Global Trigger Tool for Measuring Adverse Events. 2nd ed. Innovation Series white paper. Cambridge, MA: Institute for Healthcare Improvement; 2009. http://www.IHI.org.
169. Naessens JM, Campbell CR, Huddleston JM, et al. A comparison of hospital adverse events identified by three widely used detection methods. Int J Quality Health Care. 2009;21(4):301-307.
170. Naessens JM, O'Byrne TJ, Johnson MG, Vansuch MB, McGlone CM, Huddleston JM. Measuring hospital adverse events: assessing inter-rater reliability and trigger performance of the Global Trigger Tool. Int J Quality Health Care. 2010;22(4):266-274.
171. Lander L, Roberson DW, Plummer KM, Forbes PW, Healy GB, Shah RK. A trigger tool fails to identify serious errors and adverse events in pediatric otolaryngology. Otolaryngology–Head Neck Surg. 2010;143(4):480-486.
172. Landrigan CP, Parry GJ, Bones CB, Hackbarth AD, Goldmann DA, Sharek PJ. Temporal trends in rates of patient harm resulting from medical care. N Engl J Med. 2010;363:2124-2134.
173. Milbrath C, Pries G, Howard P, Huseth G. Applying the IHI Global Trigger Tool to pediatric and special needs populations. http://www.psqh.com/julyaugust-2011/902-applying-the-ihi-global-trigger-tool-to-pediatric-and-special-needs-populations.html . Published July/August 2011. (Accessed September 5, 2013).
174. Good VS, Saldana M, Gilder R, Nicewander D, Kennerly DA. Large-scale deployment of the Global Trigger Tool across a large hospital system: refinements for the characterization of adverse events to support patient safety learning opportunities. BMJ Quality Safety. 2011;20(1):25-30.
175. Sharek PJ, Parry G, Goldmann D, et al. Performance characteristics of a methodology to quantify adverse events over time in hospitalized patients. Health Serv Res. 2011;46(2):654-678.
176. Snow D. Kaiser Permanente Experience with Automating the IHI Global Trigger Tool (Text Version). Presented at AHRQ 2010 Annual Conference, Bethesda, Maryland, 26–29 September 2010. https://archive.ahrq.gov/news/events/conference/2010/snow/index.html . (Accessed December 4, 2017).

177. Schildmeijer, K, Nilsson L, Arestedt K, Perk J. Assessment of adverse events in medical care: lack of consistency between experienced teams using the global trigger tool. BMJ Quality Safety. 2012;21(4):307-314.
178. Rothschild JM, Landrigan CP, Cronin JW, Kaushal R, Lockley SW, Burdick E, Stone PH, Lilly CM, Katz JT, Czeisler CA, Bates DW. The Critical Care Safety Study: The incidence and nature of adverse events and serious medical errors in intensive care. Crit Care Med. 2005 Aug;33(8):1694-700.
179. Quality and Safety Review System. https://www.ahrq.gov/news/blog/ahrqviews/new-system-aims-to-improve-patient-safety-monitoring.html . (Accessed December 4, 2017).
180. Hunt DL, Haynes RB, Hanna SE, Smith K. Effects of computer-based CDSSs on physician performance and patient outcomes: a systematic review. JAMA. 1998;280(15):1339-46.
181. Burke HB. Predicting clinical outcomes using molecular biomarkers. Biomarkers in Cancer 2016:8:89-99.
182. Miller RA, Pople HE Jr, Myers JD. Internist-1, an experimental computer-based diagnostic consultant for general internal medicine. N Engl J Med. 1982;307(8):468-76.
183. Masarie FE Jr, Miller RA, Myers JD. INTERNIST-I properties: representing common sense and good medical practice in a computerized medical knowledge base. Comput Biomed Res. 1985;18(5):458-79.
184. Miller RA, McNeil MA, Challinor SM, Masarie FE Jr, Myers JD. The INTERNIST-1/QUICK MEDICAL REFERENCE project--status report. West J Med. 1986;145(6):816-22.
185. Miller R, Masarie FE, Myers JD. Quick medical reference (QMR) for diagnostic assistance. MD Comput. 1986;3(5):34-48.
186. Barnett GO, Cimino JJ, Hupp JA, Hoffer EP. DXplain. An evolving diagnostic decision-support system. JAMA 1987;258(1):67-74. (http://dxplain.org/dxp2/dxp.sdemo.asp and http://www.mghlcs.org/projects/dxplain (Accessed December 4, 2017)
187. Warner HR Jr1, Bouhaddou O. Innovation review: Iliad--a medical diagnostic support program. Top Health Inf Manage. 1994;14(4):51-8. http://www.openclinical.org/aisp_iliad.html (Accessed December 3, 2017)
188. Ramnarayan P, Roberts GC, Coren M et al. Assessment of the potential impact of a reminder system on the reduction of diagnostic errors: a quasi-experimental study. BMC Med Inform Decis Mak. 2006;6(1):22. http://www.openclinical.org/aisp_isabel.html (Accessed December 4, 2017)
189. Tleyjeh IM, Nada H, Baddour LM. VisualDx: decision-support software for the diagnosis and management of dermatologic disorders. Clin Infect Dis. 2006 Nov 1;43(9):1177-84.
190. Goldsmith LA. (Ed.) VisualDx. Rochester, NY: VisualDx, 2013.
191. Chou WY1,2, Tien PT2,3, Lin FY4, Chiu PC1,2. Application of visually based, computerised diagnostic decision support system in dermatological medical education: a pilot study. Postgrad Med J. 2017 May;93(1099):256-259.
192. National Academies of Sciences, Engineering, and Medicine. Improving Diagnosis in Health Care. Washington, DC: The National Academies Press, 2015. https://www.nap.edu/catalog/21794/improving-diagnosis-in-health-care . (Accessed December 4, 2017).
193. Aspden P, Wolcott J, Bootman JL, Cronenwett LR. (Eds.) Preventing Mediation Errors. https://www.nap.edu/catalog/11623/preventing-medication-errors . (Accessed December 4, 2017).
194. Armada ER, Villamanan E, Lopez-de-Sa E, Rosillo S, Rey-Blas JR, Testillano ML, Alvarez-Sala R, López-Sendón J. Computerized physician order entry in the cardiac intensive care unit: effects on prescription errors and workflow conditions. J Crit Care 2014;29:188–93.
195. Warrick C, Naik H, Avis S, Fletcher P, Franklin BD, Inwald D. A clinical information system reduces medication errors in paediatric intensive care. Intensive Care Med 2011;37(4):691–4.
196. Kaushal R, Shojania KG, Bates DW. Effects of computerized physician order entry and CDSSs on medication safety: a systematic review. Arch Intern Med 2003;163(12):1409–1416.
197. Ammenwerth E, Schnell-Inderst P, Machan C, Siebert U. The effect of electronic prescribing on medication errors and adverse drug events: a systematic review. J Am Med Inform Assoc 2008;15:585–600.
198. van Rosse F, Maat B, Rademaker CM, van Vught AJ, Egberts AC, Bollen CW. The effect of computerized physician order entry on medication prescription errors and clinical outcome in pediatric and intensive care: a systematic review. Pediatrics 2009;123:1184–90.
199. Nuckols TK, Smith-Spangler C, Morton SC, Asch SM, Patel VM, Anderson LJ, Deichsel EL, Shekelle PG. The effectiveness of computerized order entry at reducing preventable adverse drug events and medication errors in hospital settings: a systematic review and meta-analysis. Syst Rev 2014;3:56.
200. Georgiou A, Prgomet M, Paoloni R, Creswick N, Hordern A, Walter S, Westbrook J. The effect of CPOEs on clinical care and work processes in emergency departments: a systematic review of the quantitative literature. Ann Emerg Med 2013;61(6):644–53.

201. Jozefczyk KG, Kennedy WK, Lin MJ, Achatz J, Glass MD, Eidam WS, Melroy MJ. Computerized prescriber order entry and opportunities for medication errors: comparison to tradition paper-based order entry. J Pharm Pract 2013;26(4):434–7.
202. Prgomet M, Li L, Niazkhani Z, Georgiou A, Westbrook JI. Impact of commercial computerized provider order entry (CPOE) and CDSSs (CDSSs) on medication errors, length of stay, and mortality in intensive care units: a systematic review and meta-analysis. J Am Med Inform Assoc. 2017;24:413-422.
203. Shulman R, Singer M, Goldstone J, Bellingan G. Medication errors: a prospective cohort study of hand-written and computerised physician order entry in the intensive care unit. Crit Care 2005;9(5):R516–21.
204. Colpaert K, Claus B, Somers A, Vandewoude K, Robays H, Decruyenaere J. Impact of computerized physician order entry on medication prescription errors in the intensive care unit: a controlled cross-sectional trial. Crit Care 2006;10(1):R21.
205. Korb-Savoldellia V, Boussadic A, Durieuxc P, Sabatiera B. Prevalence of Computerized Physician Order Entry Systems–Related Medication Prescription Errors: A Systematic Review. Int J Med Inform, online 28 December 2017.
206. Wright A, Ai A, Ash J, Wiesen JF, Hickman TT, Aaron S, McEvoy D, Borkowsky S, Dissanayake PI, Embi P, Galanter W, Harper J, Kassakian SZ, Ramoni R, Schreiber R, Sirajuddin A, Bates DW, Sittig DF. Clinical decision support alert malfunctions: analysis and empirically derived taxonomy. J Am Med Inform Assoc. 2017 Oct 16.
207. Fernandez Perez ER, Winters JL, Gajic O. The addition of decision support into computerized physician order entry reduces red blood cell transfusion resource utilization in the intensive care unit. Am J Hematol 2007;82(7):631–3.
208. Rana R, Afessa B, Keegan MT, et al. Evidence-based red cell transfusion in the critically ill: quality improvement using computerized physician order entry. Crit Care Med 2006;34(7):1892–7.
209. Adams ES, Longhurst CA, Pageler N, Widen E, Franzon D, Cornfield DN. Computerized physician order entry with decision support decreases blood transfusions in children. Pediatrics 2011;127(5):e1112–9.
210. Pageler NM, Franzon D, Longhurst CA, Wood M, Shin AY, Adams ES, Widen E, Cornfield DN. Embedding time-limited laboratory orders within computerized provider order entry reduces laboratory utilization. Pediatr Crit Care Med 2013;14(4):413–9.
211. Cohen MR, Smetzer JL, Westphal JE, Comden SC, Horn DM. Risk models to improve safety of dispensing high-alert medications in community pharmacies. J Am Pharm Assoc 2012;52:584–602.
212. Bellodi E, Vagnoni E, Bonvento B, Lamma E. Economic and organizational impact of a clinical decision support system on laboratory test ordering. BMC Med Inform Dec Making 2017;17:179.
213. Goldzweig CL, Orshansky G, Paige NM, Miake-Lye IM, Beroes JM, Ewing BA, Shekelle PG. EHR-based interventions for improving appropriate diagnostic imaging: a systematic review and meta-analysis. Ann Intern Med. 2015 Apr 21;162(8):557-65.
214. Ohmann C, Boy O, Yang Q. A systematic approach to the assessment of user satisfaction with health care systems: Constructs, models and instruments. Stud Health Technol Inform. 1997;43B:781–5.
215. Ash JS, Fournier L, Stavri PZ, Dykstra R. Principles for a successful computerized physician order entry implementation, AMIA Ann Symp Proc 2003;36–40.
216. Kruse CS, Goetz K. , Summary and frequency of barriers to adoption of CPOE in the US. J Med. Syst 2015;39:15.
217. Griffon N, M. Schuers M, Joulakian M, Bubenheimg M, Leroy J-P, Darmoni SJ. Physician satisfaction with transition from CPOE to paper-based prescription. International Journal of Medical Informatics 103 (2017) 42–48.
218. Hoonakker PLT, Carayon P, Walker JM. Measurement of CPOE end-user satisfaction among ICU physicians and nurses. Appl Clin Inf 2010;1: 268–285.
219. Miller A, Moon B, Anders S, Walden R, Steven Brown S, Montella D. Integrating computerized CDSSs into clinical work: A meta-synthesis of qualitative research. Int J Med Inform 2015;84(12):1009-18.
220. Page N, Baysari MT, Westbrook JI A systematic review of the effectiveness of interruptive medication prescribing alerts in hospital CPOE systems to change prescriber behavior and improve patient safety. Inter J of Med Inform 2017;105: 22–30.
221. van der Sijs H, Aarts J, Vulto A, et al: Overriding of drug safety alerts in computerized physician order entry. J Am Med Inform Assoc 2006;13:138–147
222. Carspecken CW, Sharek PJ, Longhurst C, et al: A clinical case of EHR drug alert fatigue: Consequences for patient outcome. Pediatrics 2013; 131:e1970–e1973
223. Genco EK1, Forster JE2, Flaten H3, Goss F3, Heard KJ3, Hoppe J3, Monte AA3.Clinically Inconsequential Alerts: The Characteristics of Opioid Drug Alerts and Their Utility in Preventing Adverse Drug Events in the Emergency Department. Ann Emerg Med 2016;67:240-248.

224. Straichman YZ, Kurnik D, Matok I, Halkin H, Markovits N, Ziv A, Shamiss A, Loebstein R. Prescriber response to computerized drug alerts for electronic prescriptions among hospitalized patients. Int J Med Inform 2017; dpi.org/10.1016/j.ijmedinf.2017.08.008

225. Kane-Gill SL, O'Connor MF, Rothschild JM, Selby NM, McLean B, Bonafide CP, Cvach MM, Hu X, Konkani A, Pelter MM, Winters BD. Technologic Distractions (Part 1): Summary of Approaches to Manage Alert Quantity With Intent to Reduce Alert Fatigue and Suggestions for Alert Fatigue Metrics. Crit Care Med. 2017;45:1481-1488.

226. Lee S, Lee S, Kim D. Physicians' and pharmacists' perceptions on real-time drug ultilization review system: a nationwide survey. Int J Qual Health Care 2017;doi.org/10/1093/intqhc/mzx085

227. Sequist TD, Gandhi TK, Karson AS, Fiskio JM, Bugbee D, Sperling M, Cook EF, Orav EJ, Fairchild DG, Bates DW. A randomized trial of electronic clinical reminders to improve quality of care for diabetes and coronary artery disease. JAMA 2005;12:431–437.

228. Middleton Pl, Crowther CA. Reminder systems for women with previous gestational diabetes mellitus to increase uptake of testing for type 2 diabetes or impaired glucose tolerance. Cochrane Database Syst Rev. 2014 Mar 18;(3):CD009578.

229. Matheny ME, Sequist TD, Seger AC, Fiskio JM, Sperling M, Bugbee D Bates DW, Gandhi TK. A randomized trial of electronic clinical reminders to improve medication laboratory monitoring. J Am Heart Assoc 2008;15:424-429.

230. Haynes AB, Weiser TG, Berry WR, Lipsitz SR, Breizat A-HS, Dellinger EP, Herbosa T, Joseph S, Kibatala PL, Lapitan MCM, Merry AF, Moorthy K, Reznick RK, Taylor B, Gawande Aa. A surgical safety checklist to reduce morbidity and mortality in a global population. N Engl J Med. 2009;360(5):491–9.

231. Thomassen Ø, Storesund A, Søfteland E, Brattebø G. The effects of safety checklists in medicine: a systematic review. Acta Anaesthesiol Scand. 2014 Jan;58(1):5-18

232. de Jager E, McKenna C, Bartlett L, Gunnarsson R, Ho YH. Postoperative Adverse Events Inconsistently Improved by the World Health Organization Surgical Safety Checklist: A Systematic Literature Review of 25 Studies. World J Surg. 2016 Aug;40(8):1842-58.

233. Garland NY, Kheng S, De Leon M, Eap H, Forrester JA, Hay J, Oum P, Sam Ath S, Stock S, Yem S, Lucas G, Weiser TG. Using the WHO Surgical Safety Checklist to Direct Perioperative Quality Improvement at a Surgical Hospital in Cambodia: The Importance of Objective Confirmation of Process Completion. World J Surg. 2017 Dec;41(12):3012-3024.

234. Pysyk CL. A change to the surgical safety checklist to reduce patient identification errors. Can J Anaesth. 2017 Nov 2.

235. Avrunin GS, Clarke La, Osterweil LJ, Goldman JM, Rausch T. Smart checklists for human-intensive medical systems. In: IEEE/IFIP International Conference on Dependable Systems and Networks Workshops (DSN 2012). IEEE. 2012. p. 1–6.

236. Pageler NM, Longhurst Ca, Wood M, Cornfield DN, Suermondt J, Sharek PJ, Franzon D. Use of electronic medical record-enhanced checklist and electronic dashboard to decrease CLABSIs. Pediatrics. 2014;133(3):738–46.

237. Thongprayoon C, Harrison aM, O'Horo JC, Berrios RaS, Pickering BW, Herasevich V. The Effect of an Electronic Checklist on Critical Care Provider Workload, Errors, and Performance. J Intensive Care Med. 2014. doi:10.1177/0885066614558015.

238. Nan S, Van Gorp P, Korsten HH, Vdovjak R, Kaymak U, Lu X, Duan H. Tracebook: a dynamic checklist support system. In: Computer-Based Medical Systems (CBMS), 2014 IEEE 27th International Symposium On. IEEE. 2014. p. 48–51.

239. Nan S, van Gorp P, Lu X, Kaymak U, Korsten H, Vdovjak R, Duan H. A meta-model for computer executable dynamic clinical safety checklists. BMC Med Inform Decision Making (2017) 17:170.

240. Semrau KEA, Hirschhorn LR, Marx Delaney M, Singh VP, Saurastri R, Sharma N, Tuller DE, Firestone R, Lipsitz S, Dhingra-Kumar N, Kodkany BS, Kumar V, Gawande AA; BetterBirth Trial Group. Outcomes of a Coaching-Based WHO Safe Childbirth Checklist Program in India. N Engl J Med. 2017 Dec 14;377(24):2313-2324.

241. Porat T, Delaney B, Kostopoulou O. The impact of a diagnostic decision support system on the consultation: perceptions of GPs and patients. BMC Med Inform Decis Mak. 2017 Jun 2;17(1):79.

242. Ibrahim SA. Decision aids and elective joint replacement - how knowledge affects utilization. N Engl J Med. 2017 Jun 29;376(26):2509-2511.

243. Arterburn D, Wellman R, Westbrook E, et al. Introducing decision aids at Group Health was linked to sharply lower hip and knee surgery rates and costs. Health Aff (Millwood) 2012; 31: 2094-104.

244. Stacey D, Légaré F, Lewis K, et al. Decision aids for people facing health treatment or screening decisions. Cochrane Database Syst Rev 2017; 4: CD001431.

245. FDA 1. Guidance for FDA Reviewers and Industry, Guidance for the Content of Premarket Submissions for Software Contained in Medical Devices, May 29, 1998.

246. FDA 2. Guidance for Industry, FDA Reviewers and Compliance on Off-the-Shelf Software Use in Medical Devices, September 9, 1999.
247. FDA 3. General Principles of Software Validation; Final Guidance for Industry and FDA Staff, January 11, 2002.
248. Editorial. Clinical decision making: more than just an algorithm. Lancet Oncol 2017;18(12):1553.

10

Health Information Privacy and Security

JOHN RASMUSSEN

LEARNING OBJECTIVES

After reading this chapter the reader should be able to:

- Explain the importance of confidentiality, integrity, and availability as it pertains to health information privacy and security
- Describe the regulatory environment and how it drives information privacy and security programs within the health care industry
- Recognize the importance of data security and privacy as related to public perception, particularly regarding data breach and loss
- Identify different types of threat actors and their motivations
- Identify different types of controls used and how they are used to protect information
- Describe emerging risks and how they impact the health care sector

INTRODUCTION

Information security and privacy have changed immeasurably during the last 15 years. Prior to the year 2000 the motivation of the computer hacker was mainly directed toward intruding into systems and acquiring data for bragging rights. When criminals determined that there was financial opportunity to be had by stealing data and controlling systems, the game had changed.

Opportunities abound in the healthcare space for individuals with malicious or criminal intent. Information systems at hospitals, insurance and pharmaceutical companies, medical device manufacturers, and small provider practices are an inviting and rich target for these individuals. The data maintained by the health care industry can be personally identifiable, clinical, financial, and valuable intellectual property.

This chapter will explore the foundational elements of information security and describe the regulatory environment that is driving the healthcare sector to implement controls to protect data. It will demonstrate the types of breaches that have occurred, and risks associated with the breach of information. The chapter will also describe the different types of controls and their application to help prevent security and privacy incidents from occurring. In addition, individuals who work within healthcare play an important role in securing information and this chapter will describe how they can help. Finally, the chapter will discuss emerging risks to information security and privacy within healthcare.

BASIC SECURITY PRINCIPLES

Electronic information is everywhere. As technology spreads and new technologies develop, the healthcare sector faces many new challenges in protecting the security and privacy of that data. The increased adoption of electronic health records, coupled with personal health records and health information exchanges creates some monumental challenges in the coordination of protection for that data. How does a hospital, clinic, or insurance company secure the most sensitive personal data of individuals and still maintain the ability to use that information for the provision of quality care to the patient?

Some 2013 findings indicate that a little over 12% of participants had withheld information from a healthcare provider because of security concerns.[1] This lack of communication could have dire consequences on the provider/patient relationship and essentially the patient's health as a whole. But without better assurances and solutions by vendors, insurers and health care organizations, it may be difficult to win and keep the public trust.

To understand health information privacy and security it is important to understand the discipline of information security through some basic concepts and to understand where privacy lands as part of that discipline. This section covers some basic security principles and describes how those principles apply to privacy.

Data Classification and Privacy

Information stored, transmitted, or created by electronic systems can come in many flavors. One component of that information can be elements of personal data that can be construed as sensitive information. Any company dealing with data will have a data classification policy that determines the sensitivity of data and how that organization may choose to protect it.

A data classification program should be broken down by Security Objective and Potential Impact if that data is compromised.[2] The Federal Information Processing Standards (FIPS), Standards for Security Categorization of Federal Information and Information Systems (FIPS PUB 199) presents a guideline for classifying data through the Security Objectives (confidentiality, integrity, or availability) to determine if the Potential Impact is low, medium, or high. Once Impact is determined, proper controls can be put into place. There are many different factors that contribute to the impact ranking, including specific data elements and regulatory requirements to protect that data.

There are a myriad of regulatory requirements dealing with the protection of confidential information and protecting the privacy of individuals whom the data references. Those regulations will be discussed later in this chapter.

The privacy of health data falls squarely in the "confidentiality" Security Objective. There are many elements of health data that would be considered sensitive or harmful to the subject of the information if it is accessed, used, or disclosed in an unauthorized manner. Examples of sensitive data elements include Social Security Number, credit card number, sensitive diagnosis, family medical history, etc.

Confidentiality, Integrity, and Availability

According to the International Information Systems Security Certification Consortium (ISC2), among others, there are three pillars of information security (confidentiality, availability, and integrity) that are fundamental to protecting information technology solutions such as health information technology (HIT). In FIPS they are also referred to as "Security Objectives."[3]

Security measures are instituted collectively to meet one or more of these primary goals, with the result being one where confidentiality, availability and integrity are all covered. These terms are ubiquitous in the practice of information security across all industries. Whether in banking, manufacturing, or healthcare, these terms are commonly used when developing and maintaining information security programs.

- Confidentiality refers to the prevention of data loss, and is the category most easily identified with HIPAA privacy and security within healthcare environments. Encryption, access control, and secure authentication are typical controls used to protect confidentiality.
- Availability refers to system and network accessibility, and often focuses on power loss or network connectivity outages. Loss of availability may be attributed to natural, accidental, or intentional disasters. Natural disasters can include tornados, earthquakes, hurricanes or fire. Accidental incidents such as the loss of a personal thumb drive, misconfiguration of a system, or a backhoe digging up a fiber optic cable are the most common availability issues. Availability can also be impacted by individuals or groups intentionally attacking systems or data. These attacks can include denial of service attacks (DoS) which target an organization and bombard it with data until legitimate data can no longer be received, or through ransomware which infects a computer with a virus that will encrypt the user's data until a ransom is paid. To address the risk to availability organizations will use redundant systems, backup batteries and generators, separate data centers, and data backups to protect their data.
- Integrity describes the trustworthiness and permanence of data, an assurance that the lab results or personal medical history of a patient is not modifiable by unauthorized entities or corrupted by a poorly designed process. Database best practices, data loss solutions, and data backup and archival tools are implemented to prevent data manipulation, corruption, or loss; thereby maintaining the integrity of patient data. Organizations also use audit logging for access to systems and databases as a means of protecting the integrity of the data. One strategy employed by hackers is to delete audit logs of systems they intrude upon, thus erasing any evidence of changes they made to the system.

Defense in Depth

With the protection of confidentiality, integrity, and availability as key objectives, and with the help of data classification to know what is important to protect, the organization must then develop a set of controls to protect its data. In most cases one control will not be adequate to protect the data. An organization must employ several different kinds of controls to protect its data from compromise. Building several layers of security controls around data is often referred to as "Defense in Depth."[4]

This concept acknowledges that one layer of defense may not be able to contain the risk to data being compromised and by adding additional layers the level of risk can be reduced significantly. A common analogy to describe this concept is that of a castle. The castle is usually built in a strategic location with a wide plane of view, so the inhabitants can see any threat from far off and enact other defenses. Once the attacker gets close enough to the castle they are in danger from arrows reaching far beyond the walls. If the attacker gets close to the castle they will have to cross a moat and then scale the walls or breach the front gate of the castle. If the attacker is fortunate enough to get over the first wall it is likely that they will see additional walls and additional defensive objects, each more difficult to overcome. While not foolproof, this defensive strategy will make the attacker think twice about the costs associated with attacking the castle.

Healthcare organizations will use a combination of technical, administrative, and physical safeguards to protect their systems and information. Each of these may have additional controls in place to further reduce risk.

An example of defense in depth in healthcare:
1. A healthcare organization has a set of policies and procedures around data privacy and security that the employee receives training on and acknowledges annually – Administrative Safeguard
2. New employees are given a criminal background check, drug screen and credit check prior to employment – Administrative Safeguard
3. The new employee is given a badge with their photo and the ability to access restricted areas through a proximity sensor – Physical and Technical Safeguard
4. Each employee has their own user name and password to access the entities electronic resources – Technical Safeguard
5. Access is given to only the systems that the employee needs to access, minimum necessary – Technical and Administrative Safeguard
6. The computer that the employee uses is encrypted, has patches automatically pushed to the device, and runs anti-virus software – Technical Safeguard
7. The email system the employee accesses has anti-virus installed on the server, has patches updated by the technology team, and is encrypted. It also uses an email gateway to filter out spam, phishing (this is a social engineering technique that tricks the individual into taking an action like clicking on a malicious link, downloading an attachment, or providing their credentials), and emails containing known malware – Technical Safeguard
8. The network that the email system, and computers operate on is located behind a firewall, intrusion prevention system, and switches and routers which employ access control lists (ACLs) to limit the types of traffic allowed on the network – Technical Safeguard.

The examples above are very high-level examples of the controls applied to the healthcare environment. These are standard techniques that start with the individual (these controls are also applied to patients and visitors) and work their way all the way up the technical stack to data centers and the Internet.

THE HEALTHCARE REGULATORY ENVIRONMENT

The healthcare industry is a heavily regulated industry. Given the nature of the enterprise it is not surprising that state, federal, and industry actors want a stake in protecting the safety of the patient population. This section will provide an in-depth review of the Health Insurance Portability and Accountability Act (HIPAA) and identify other industry standards and regulatory requirements as they pertain to healthcare privacy and security.

HIPAA

HIPAA was originally created to promote the portability of health insurance but gained additional prominence as the key regulation affecting the industry by including requirements to protect patient privacy and fulfill certain patient rights. In the 20 years since its passage it has evolved and been updated to reflect new threats and to provide clarification and interpretation to language that has been part of the law. Updates came in the form of the Health Information Technology for Economic and Clinical Health (HITECH) Act, part of the American Recovery and Reinvestment Act of 2009.

HITECH imposed new requirements for breach notification and imposed stiffer penalties for non-compliance with HIPAA.[5] In addition, it added new patient rights to HIPAA.

The HIPAA Privacy and Security Rule (45 CFR Parts 160, 162, and 164) is enforced by the Office for Civil Rights (OCR), which operates under the Office of the Secretary of the Department of Health and Human Services.[6] The OCR began enforcing HIPAA in 2003. Their enforcement actions include the investigation and resolution of patient privacy complaints as well as investigation of breaches of protected health information (PHI).

First and foremost, HIPAA only applies to organizations defined as "covered entities."[7] If an organization does not meet the definition of a covered entity then it is not subject to the law. Types of organizations that *do not* have to follow HIPAA include but are not restricted to the following: independent research organizations, life insurers, employers, many school districts, many law enforcement agencies, and worker's compensation carriers.

Covered entities include:
- Health Plans – which includes health insurers, HMOs, company health plans, and government programs such as Medicare and Medicaid.
- Healthcare clearinghouses
- Healthcare Providers – which includes most doctors, clinics, hospitals, psychologists, chiropractors, nursing homes, pharmacies, and dentists "who transmit any health information in electronic form in connection with a transaction covered by this subchapter" (45 CFR § 160.102). This means the provider must transmit data between two parties to "carry out financial or administrative activities related to health care." [8] For example, a dentist submitting a billing claim to an insurance carrier would meet this definition. If a provider does not meet this definition, i.e. only takes cash payment and does not conduct any "transaction" as defined in HIPAA, ("Transaction means the transmission of information between two parties to carry out financial or administrative activities related to health care"), they are not considered a covered entity and therefore not subject to HIPAA.

For those organizations that are required to abide by HIPAA, patient data and personal information must be protected according to the Security Rule. Protections apply to all protected health information (PHI), whether in hard copy records, electronic protected health information (ePHI) stored on computing systems, or even verbal discussions between medical professionals. Covered entities must put safeguards in place to ensure data is not compromised, and that it is only used for the intended purpose. The HIPAA rules are not designed to and should not impede the treatment of patients.[9]

Covered entities must comply with certain consumer rights. Specifically, a patient may:
- Request and receive a copy of their health records
- Request an amendment to their health record
- Receive a notice that discusses how health information may be used and shared, the Notice of Privacy Practices
- Request a restriction on the use and disclosure of their health information
- Receive a copy of their "accounting of disclosures"
- Restrict disclosure of the health information to an insurer if the encounter is paid for out of pocket
- File a complaint with a provider, health insurer, and/or the U.S. Government if patient rights are being denied or health information is not being protected.[10]

Protected Health Information (PHI)

The term Protected Health Information is defined as *"individually identifiable health information"* with some exclusions. Breaking this down, individually identifiable health information is:
- Information created by a covered entity
- And "relates to the past, present, or future physical or mental health or condition of an individual"
- Or identifies the individual or there is a reasonable basis to believe that the individual can be identified from the information.[11]

The first elements and one of the other two elements must be present to be considered PHI.

Permitted Uses and Disclosures of PHI

HIPAA allows covered entities to use or disclose PHI for treatment, payment, or healthcare operations. "Use" refers to the internal use of PHI within an organization and "disclosure" refers to the release of information outside of the organization. These uses and disclosures are limited to the *"minimum necessary"* to prevent misuse or to protect privacy.[12]

For example, a health system may have two software programs for their electronic health record, one which contains treatment notes and the other billing information. When applying the concept of *"minimum necessary"* a doctor would have access limited to the treatment notes and a billing specialist would be limited to access to the billing system used by the hospital.[13]

Covered Entity Permitted Uses and Disclosures of patient data according to the Privacy Rule:

- To the individual
- For treatment, payment or health care operations
- Uses and disclosures with opportunity to agree or object
  o Facility directories
  o For notification and other purposes
- Public interest and benefit activities
  o Required by law
  o Public health activities
  o Victims of abuse, neglect or domestic violence
  o Health oversight activities
  o Judicial and administrative proceedings
  o Law enforcement purposes
  o Decedents
  o Cadaveric organ, eye, or tissue donation
  o Research
  o Serious threat to health or safety
  o Essential government functions
  o Workers' compensation
- Limited data set – this can include health information and dates; no other individual identifiers are included in a limited data set
- De-identified data – this is data where all individually identifiable information has been removed.14

The 18 individual identifiers named in the rule include:
1. Names
2. All geographic subdivisions smaller than a state, including street address, city, county, precinct, ZIP Code, and their equivalent geographical codes, except for the initial three digits of a ZIP Code if, according to the current publicly available data from the Bureau of the Census:
   a. The geographic unit formed by combining all ZIP Codes with the same three initial digits contains more than 20,000 people.
   b. The initial three digits of a ZIP Code for all such geographic units containing 20,000 or fewer people are changed to 000.
3. All elements of dates (except year) for dates directly related to an individual, including birth date, admission date, discharge date, date of death; and all ages over 89 and all elements of dates (including year) indicative of such age, except that such ages and elements may be aggregated into a single category of age 90 or older
4. Telephone numbers
5. Facsimile numbers
6. Electronic mail addresses
7. Social security numbers
8. Medical record numbers
9. Health plan beneficiary numbers
10. Account numbers
11. Certificate/license numbers
12. Vehicle identifiers and serial numbers, including license plate numbers
13. Device identifiers and serial numbers
14. Web universal resource locators (URLs)
15. Internet protocol (IP) address numbers
16. Biometric identifiers, including fingerprints and voiceprints
17. Full-face photographic images and any comparable images
18. Any other unique identifying number, characteristic, or code, unless otherwise permitted by the Privacy Rule for re-identification[15]

The Business Associate

Not all services can be managed by a hospital or clinic. Some services may need to be outsourced to a partner who has the expertise that the healthcare entity does not have in-house. Some of these partners or vendors will need to view, access, process, or create PHI on behalf of the covered entity. These organizations would then become a Business Associate (BA) to the Covered Entity (CE). Business associates can be an electronic health record (EHR) software company, a third party that assists with billing and claims, or a transcription service.[16]

A BA is required to have a business associate agreement (BAA) in place with the CE under HIPAA. The BAA ensures that the BAs meet the requirements of the Security Rule under HIPAA and places the proper protections around PHI. A BA is also responsible for obtaining BAAs with any other organizations that sub-contract to it and those organizations are also required to follow the Security Rule. This cascades the responsibility of protecting PHI to all of the vendors and sub-contractors who work with that information. This requirement was added to HIPAA as part of the HITECH Act as was the ability for the OCR to enforce HIPAA among business associates. The business associate can now be penalized for violating HIPAA through civil action by the OCR and criminally by the Department of Justice.[17]

The Security Rule

HIPAA requires that covered entities apply safeguards to protect patient information. There are three categories of safeguards identified in HIPAA that have different sets of "required" and "addressable" controls. All required controls must be put in place. An addressable control is one that is also required, however, compensating controls can be used if the addressable control is not available to the covered entity. HIPAA takes into account the ability of a covered entity to apply safeguards based upon their size and financial position and allows for a *"flexibility of approach"* in order for the covered entity to meet the security requirements of the Security Rule.[18]

Safeguard areas and examples of controls:
- Administrative Safeguards – policy or procedures used to provide security and governance to privacy and security.
  o Security management processes to reduce risks and vulnerabilities
  o Security personnel responsible for developing and implementing security policies
  o Information access management - minimum access necessary to perform duty
  o Workforce training and management
  o Evaluation of security policies and procedures
- Physical Safeguards – physical security measures taken to protect information. Typically, they are in the form of locks, security cameras, guards, or badge access to restricted areas.
  o Facility access and control limiting physical access to facilities
  o Workstation and device security policies and procedures covering transfer, removal, disposal, and re-use of electronic media
- Technical Safeguards – technical tools implemented to protect data. These can be firewalls, anti-virus, automatic logoff, session timeouts, intrusion detection, or a wide variety of other technical controls.
  o Access control that restricts access to authorized personnel
  o Audit controls for hardware, software, and transactions
  o Integrity controls to ensure data is not altered or destroyed
  o Transmission security to protect against unauthorized access to data transmitted on networks and via email[19]

Breach Requirements under HIPAA

Subpart D, 45 CFR 164.4XX of HIPAA deals entirely with the breach of protected health information and was added as a result of the HITECH Act to strengthen the privacy protections of HIPAA and outline the requirements for breach notification.

A "breach" is a complicated thing to define. Simply, a breach is an unauthorized acquisition, access, or use of PHI with a number of exceptions.[20]

Exceptions:
1. Data is encrypted. This is considered a safe harbor; or
2. *"Any unintentional acquisition, access, or use of protected health information by a workforce member or person acting under the authority of a covered entity or a business associate, if such acquisition, access, or use was made in good faith and within the scope of authority and does not result in further use or disclosure"*; or
3. *"Any inadvertent disclosure by a person who is authorized to access protected health information at a covered entity or business associate to another person authorized to access protected health information at the same covered entity or business associate, or organized health care arrangement in which the covered entity participates, and the information received as a result of such disclosure is not further used or disclosed"*; or
4. *"A disclosure of protected health information where a covered entity or business associate has a good faith belief that an unauthorized person to whom the disclosure was made would not reasonably have been able to retain such information."*[21]

If the incident does not meet one of the exceptions above a breach risk assessment must be done to determine if the incident exceeds a low probability of compromise. This is determined by looking at four factors together:
1. The nature and extent of the PHI involved. A sensitive diagnosis or the release of a Social Security Number would exceed a low probability for this factor.
2. The person who used the information or who it was disclosed to. If an employee stole PHI for personal gain this would exceed low probability for this factor.
3. Whether the PHI was actually acquired or viewed. A laptop containing PHI, if unencrypted, is considered "acquired" and would exceed low probability for this factor.
4. The extent to which the risk to PHI has been mitigated. If email is sent to an unauthorized individual and no attempt is made to retrieve or delete the email, this would exceed low probability for this factor.[22]

If a breach is determined, the covered entity must notify the individual(s) impacted by the breach. They must inform them within 60 days of when the breach is identified. The notification must include:
- A description of what happened
- A description of the type of PHI that was breached
- Steps the individual can take to protect themselves
- What the covered entity is doing to investigate the breach and mitigate harm

- Contact information for the individual to contact the covered entity[23]

If a breach exceeds 500 individuals, the covered entity must notify the media and must report the breach to the Office for Civil Rights (OCR).

Regardless of the number of individuals impacted by a breach, all breaches must be reported to the OCR annually.[24]

OTHER REGULATIONS AND HEALTHCARE PRIVACY AND SECURITY

While HIPAA is the best known regulatory requirement affecting healthcare, there are a number of other regulations or industry standards that deal with privacy and security protections. Table 10.1 outlines security standards and laws.

BUSINESS DRIVERS FOR SECURITY AND PRIVACY

In addition to the regulatory requirements governing privacy and security within healthcare, the business must calculate cybersecurity as part of its overall risk profile. Healthcare entities are very familiar with calculating the risk of medical malpractice as litigation for malpractice can cost millions to an organization and harm their reputation. As more and more technology is introduced into the healthcare space there is more opportunity for risk that must be considered as part of doing business. Cybersecurity concerns are foremost of these concerns as they can harm the patient, visitors, or employees of a healthcare entity.

A heavy fine from a regulatory entity can impact the reputation of an organization. The OCR has a website known as the *"Wall of Shame"* which shows all breaches currently under investigation and a second site that shows all *"Resolution Agreements."*[25-26] These sites are easy for consumers and media to reference when they want to call attention to a certain entity and can have an impact on the reputation of that entity, though it is very difficult to measure the impact on reputation.

Another business driver for healthcare is loss of market share or value. This ties directly to reputation but is not dependent upon a negative reputational impact. One example of this occurred in 2016 when a short selling financial firm named Muddy Waters bought information about a security vulnerability in a pacemaker produced

Table 10.1: Security Standards and Laws

Security Standard/Law	Brief Description
ISO 20000/27000	International IT Governance and IT Security standards
COBIT	IT Governance framework
ITIL	Information Technology Infrastructure Library, IT service management
NIST SP 800-53	National Institute of Standards and Technology, IT security controls
SOX	Sarbanes–Oxley Act; Public company accounting law
PCI-DSS	Payment Card Industry Data Security Standard – applies to all credit card merchants
FISMA	Federal Information Security Management Act
FERPA	Family Educational Rights and Privacy Act – applies to student records. If a covered entity is also an academic medical center, student health records would be protected under this regulation
State Laws	Almost every state in the United States has laws that require notification in case of a breach. If a covered entity has patients in several different states it will need to consider the breach notification laws for each state if there is an incident involving patients who reside in those states
GDPR	General Data Protection Regulation – this is a European Union regulation that requires companies that process personal information for EU citizens, including medical data, to protect that data.

by St. Jude Medical Inc. from a cyber security firm named MedSec Holdings Inc. and then shorted the stock before publicly disclosing the vulnerability. The announcement caused St. Jude shares to fall by 4.96 percent in one day, also producing a 7.4 percent discount to Abbott Laboratories which was working to acquire St. Jude.[27]

To respond to these risks insurance companies have been building new products that address these concerns or working to modify existing products, so they can fill the needs of this new risk area. Companies like Beazley offer breach response insurance packages that include coverage for notification to individuals impacted by the breach, fines and corrective actions, or incident response.[28]

Privacy And Security Roles And Governance In The Healthcare Organization

Information privacy and security are evolving in organizations as their role becomes more and more important. The roles of Information Security Officer and Privacy Officer are identified in HIPAA as a requirement for covered entities.

The top positions for security are typically called Chief Information Security Officer (CISO) and Chief Privacy Officer (CPO). Their reporting relationships can differ depending on an organization's size, industry, compliance mandates and laws, technology initiatives, maturity, private or public status, and even profit model.

In some models the CISO and CPO report to the Chief Compliance Officer or Chief Risk Officer of an organization.

Other models have a Chief Security Officer role that reports directly to the president of an organization and has a CISO and CPO that report directly to them.

The most common model is to have the CPO report to a Chief Compliance Officer and the CISO report directly to the Chief Information Officer (CIO).

Information security policy is usually established under the direction of the CIO, but it is more common to see an information security committee chartered with the responsibility of creating these policies. Likewise, privacy policies are usually established under the Chief Compliance Officer but can be driven by a committee.

Depending on resources, the information technology teams may consist of network, system administration, security and data personnel, or could be the very same technical staff relied upon for all office or clinic IT needs. No matter the titles, this supporting staff is often tasked to defend key systems, networks, and patient data from risk, and assist with any investigations resulting from a data breach.

BREACHES IN THE NEWS – AND CONSEQUENCES

Healthcare data breaches are frequent and with the strengthening of penalties under the HITECH Act they make a big splash when they hit the news. While the examples listed below may list the financial consequences of a violation of regulatory requirements, they are not inclusive of the costs of a corrective action plan (CAP) which usually accompanies a civil penalty or settlement with the OCR. These costs can be many times higher than the actual penalty as an entity must apply expensive technology to their environment, hire additional security and compliance staff, rewrite policy, and retrain staff all within the window defined under the CAP.

- Advocate Medical Group (2013/2016) – In July 2013, Advocate reported that patient health and identity data for 4 million patients was at risk due to theft. The data were contained on 4 unencrypted company computers stolen from their administrative building and contained names, addresses, birth dates and personal health data. While this is historically the second largest breach, what is most notable is that this is their second large breach (over 500 patients).[29] Less than a month after the announcement of this breach there were two class action lawsuits addressing Advocate's "failure to take the necessary precautions required to safeguard patients' protected health information" and claiming that the computers were stolen from an "unmonitored" room with "little to no security."[30] The suits also cite negligence, invasion of privacy, consumer fraud, and intentional infliction of emotional distress.[31] The OCR settled the investigation with Advocate in 2016 when Advocate agreed to pay the settlement amount of $5.55 million and adopt a corrective action plan. This is the largest settlement to date.
- Oregon Health & Science University (OHSU), (2013/2016) – In March of 2013 OHSU experienced a breach when an unencrypted laptop was stolen from a surgeon who was on vacation in Hawaii. The laptop contained information for over 4000 patients, including surgery schedules. Notification was provided to the individuals affected.[32] Later that year a second breach notification was sent out to an additional 3000 patients for a disclosure of PHI on Google Drive by residents who were using these services to keep a spreadsheet of patients.[33] Since there was no business associate agreement in place with Google, the breach notification was required.

These incidents, occurring in the same year, led to the finding by the OCR that OHSU did not perform adequate risk analysis and apply timely controls to mitigate risk. OHSU settled for $2.7 million and a three-year corrective action plan.34
- TRICARE (2011/2014) - The largest breach in history occurred in 2011 and was reported to have affected between 4.9 and 5.1 million military active duty service members, retirees, and their families within the TRICARE health system.35-36 The breach was in the form of unencrypted backup tapes stolen from the vehicle of an employee of Science Applications International Corp (SAIC), a TRICARE contractor.37 The data was expansive and covered those cared for in military facilities between 1992 and September 2011. Information that was contained on the tapes included names, addresses, birth dates, social security numbers, and personal health data. There was no financial data such as credit card or bank account information contained on the tapes; however, with the level of personal information that was obtained financial ramifications have been reported by the affected patients. Four people initially filed a single $4.9 billion federal lawsuit against TRICARE and SAIC in 2011, but by the close of 2012 the suits grew to eight that were consolidated into one to be heard and handled by the U.S. District Court in Washington, D.C. On May 9th, 2014, the U.S. District Court in Washington, D.C. threw out most of the case stating that it was "speculative" that the plaintiffs may suffer harm from the breach.38 This illustrates one a challenge with the multitude of breaches of personally identifiable information, that it is very hard to prove that harm, damage, or loss was a direct result of the breach.
- Affinity Health Plan, Inc. (2010/2013) – This breach occurred in 2009 but went unreported until 2010 and affected more than 300,000 patient records. While not the most significant in terms of novelty or number of records, the distinguishing feature of this breach is how the data was breached. Affinity had returned seven photocopy machines they had leased long term. Unfortunately, the copiers were then sold to media giant CBS News as part of an investigative report on data security risks. The units had not been wiped before return and confidential patient information remained on their storage hard drives. Three hundred pages of documents from one copier contained personally identifiable information and included sensitive medical test results, cancer diagnoses, and prescription drug information.39 In August of 2013, the U.S. Department of Health and Human Services (HHS) announced a settlement agreement that included a fine of over $1.2 million and a Corrective Action Plan (CAP) that required Affinity to use its best efforts to obtain all hard drives from previously leased machines and to take specific measures to safeguard their patient's health information.40 This story highlights the importance of understanding the comprehensive nature of patient health data storage and exploring non-traditional avenues through which breaches may occur.

THREAT ACTORS AND TYPES OF ATTACKS

It is important to understand the different types of threat actors and their motivations to understand why information from the healthcare industry is targeted. The actors identified below have malicious intent. It is also important to note that an additional threat, albeit inadvertent, to the environment is the individual employee. The regular employee can accidentally initiate a security incident through misuse of computer resources, like visiting a malicious webpage and downloading malware, or they can cause downtime through improperly following change control processes. Some information technology employees may try to initiate a change to the configuration of a system in the production environment without first testing the change in a test instance and creating a back out plan if the change fails. This can lead to all manner of availability issues and can extend the downtime if the change was not formally documented and there is a need to systematically troubleshoot the problem.[41]

- Insiders – these are employees or other individuals within the organization who have direct access to electronic resources.
  o Motivation – financial gain, or revenge.
  o Consequences – these individuals may steal information to resell it or to build their own information. If financial gain is their goal they will target business confidential information or intellectual property. The individual may also hold a grudge against the organization for some perceived or real slight against them. In this case, the individual may be out to cause damage to the organization by deleting information, sharing information without authorization, or causing damage to electronic systems.[42]
- Hacktivists – these are individuals or groups who do not have direct access to the organization's electronic resources.

- o Motivation – hacktivists usually have a grudge or political agenda against the organization.
- o Consequences – they are out to disrupt business or to embarrass the organization. They will deface websites or publicly post personal information about employees to spur additional action and draw attention to their cause.[42]
- Organized Crime – the most common and high profile of all threat actors.
  - o Motivation – financial gain.
  - o Consequences – the goal of organized crime is to earn money from their activities. This is a low risk – high reward endeavor as many countries will not extradite individuals who are accused of stealing data or money. One grand example of a cyber heist is the $81 million theft from the Bangladesh Bank on February 4, 2016.[42-43] Hackers used credentials stolen from 4 employees to initiate transfers to banks in the Philippines. The attack was disrupted and $850 million in additional transfers were stopped. This is an extreme example of organized crime, more commonly these criminals are trying to get smaller payouts by targeting individuals with schemes like Nigerian 219 emails where the individual receives an email from an exiled leader and can receive millions of dollars if they can help that leader transfer money out of a bank where it is held.[44]
- Nation States – can be nation states or hackers sponsored by nation states.
  - o Motivation – national interest, espionage, cyber warfare, financial gain.
  - o Consequences – Nation States are the most nefarious of threat actors because they have the full power and financing of the nation behind their activities. They are out to seek advantages in the international arena through any means necessary and may be responsible for some very costly attacks. If espionage or theft of intellectual property is the motivation a nation state will become an *"Advanced Persistent Threat"* (APT), meaning that they will intrude on a system and lay in wait until they discover what they need.[45-46] They will be very stealthy and the victim will have no idea that the attacker is there until it is too late.
    - § Sony hack – in October 2014 Sony Pictures employees were greeted by an image of a red skeleton and the words "#Hacked by #GOP," on their login screens. The attackers stole many internal documents and other intellectual property and posted some of this information on the Internet to embarrass Sony. The attack has been attributed to North Korea and the motivation was outrage over the movie "The Interview" which depicted an assassination plot against Kim Jong Un.[47]
    - § Electric grid in Ukraine – December 2015 a cyberattack shut down the electrical grid to 250,000 Ukrainians. The attack has been attributed to Russia who had been in a state of conflict with Ukraine for years.[48]
    - § Anthem breach – in February 2015 hackers stole the information of 80 million current and former members of Anthem insurance. This personal information may have been targeted for financial gain or for espionage purposes. Personally identifiable information could be used to create new identities or sign up for financial services. Any health information, if sensitive, could potentially be used for blackmail purposes if the foreign government wanted to have some leverage over a person. It is suspected that China is responsible for this attack.[49]

Types of attacks

There are many different ways a threat actor can attack their target. This section will describe some of the methods of attack.

- Social Engineering – the weakest link in the organization is going to be the end user. Individuals susceptible to manipulation by things that seem familiar and friendly to them but are actually malicious in nature. Social engineering attacks attempt to get the end user to do something they would not normally do, thus allowing the threat actor access to their system or information. Social Engineering is the most frequently used form of attack.[50] There are many different techniques for social engineering, here are some examples:
  - o Phishing - typically in the form of email but can also come in a telephone call or text message, that poses as a legitimate person or institution in order to lure the victim into giving away sensitive information like usernames, passwords, or financial information. They can also lure the victim to download and install malware on their computer or phone. Phishing may also:
    - § May appear to come from someone you know or your place of work
    - § Displays a sense of urgency, such as "invoice overdue"
    - § Appears to be too good to be true

§ Could contain hyperlinks or attachments leading to fake login pages or posing as legitimate documents.[51]
o Shoulder surfing - the attacker merely looks over the victim's shoulder and views restricted information or copies down their password.
o Tailgating - the attacker uses someone else's access to a restricted area by following closely behind the individual who has already used their ID badge to unlock access to that area.
o The promise of free hardware - there is no regular term for this, but it is a highly effective form of social engineering where a threat actor will leave USB sticks seeded in a parking lot or laying on a counter for an unsuspecting employee to pick up. If the employee picks up the USB, plugs it in and clicks on a file on the drive they could infect their machine with malware. One study by Google's anti-abuse research team discovered that 98 percent of 297 dropped USB sticks were picked up and those individuals clicked on files on 45 percent of those recovered sticks.[52]

Other direct attacks that do not involve social engineering, but may occur as the result of successful social engineering attacks include:

- Denial of Service (DoS) - a DoS attack is intended to stop traffic from reaching a certain website or destination through flooding that site with bogus requests, so legitimate traffic cannot get through.[53] On October 21, 2016 Dyn, a company that helps route Internet traffic through management of Domain Name Service (DNS), was hit by - two massive DoS attacks. Since Dyn managed DNS for many parts of the internet the DoS attack not only brought Dyn to a standstill, it impacted traffic routed to many of Dyn's customers and slowed or stopped their traffic.[54]
- Brute Force attack - this is an attack where the threat actor will attempt to throw random credentials at an application login page to see if they can gain access. If an organization uses a strong password policy and a technical control that locks the account after a certain number of unsuccessful login attempt this type of attack can be easily thwarted. However, there are certain devices or systems that come with default passwords set up as administrators for the system. Many people do not change these default passwords so a brute force attack against this type of device can be easily successful after throwing a few passwords at the system like "admin," "password," "123password," "administrator," etc.[55]
- Doxing - this is a relatively new attack technique that has a substantial impact on privacy. Doxing gathers information about the victim and publishes that information in order to embarrass or harass the individual. The more sensitive the data, the more powerful the attack. This attack technique would be typically employed by a hacktivist.[56]

TOOLS USED TO PROTECT HEALTHCARE PRIVACY AND SECURITY

Earlier in this chapter Defense in Depth was defined. This section will describe a few of the tools and operational techniques to prevent the unauthorized access, use, or disclosure of information created or maintained by a healthcare entity. This section is by no means a comprehensive list of solutions that can be used to protect privacy and security. This is a high-level description of some of the tools that encompass the technical, administrative, and physical controls referenced in HIPAA.

There are many technical controls available to healthcare entities to employ in their defenses. These can be applied at different layers through the organization's IT infrastructure.

Client protection

- Patching: Applications and operating systems (OSs) may arrive with vulnerabilities present. These vulnerabilities can be identified through regular use or through individuals testing the software or OS. Some of these vulnerabilities can be exploited to introduce security risks. In order to reduce these risks, the vendors will patch periodically. Patches may also contain feature enhancements or bug fixes. Vendor updates may occur monthly or much less frequently. To prevent the exploitation of these vulnerabilities an organization must patch their systems or applications as soon as practical after the patch is released. The longer a system remains unpatched, the more vulnerable it becomes. Most exploits for systems utilize known vulnerabilities.
- Anti-virus: This is software developed for the computer or device that detects and blocks viruses from executing and infecting the device. Many traditional anti-virus solutions are signature based and will only work if the product is already aware of the virus. Anti-virus solutions must receive updates on a periodic basis to be effective. New variants of viruses are released daily, and the anti-virus solution must be updated as the threats change. Newer solutions of anti-virus software are called next-generation anti-virus. This software catalogs system processes and uses algorithms to detect and block virus behavior.[57]

- Encryption: This technology renders information unreadable without the appropriate key to unlock that information. When applied to clients such as personal computers, laptops, or smart phones, encryption can be file-based or whole disk. File-based encryption does not touch the operating system, it encrypts each of the files on the device. Whole disk encryption renders the entire disk unreadable without the key. Both technologies are considered encryption for data at-rest, meaning that when an individual is logged in or the device is powered on the data is readable. Once the device is turned off or rebooted a key is necessary to unscramble the data.[58]

Application and database protection
- Strong authentication: Keeping applications and databases secure relies on limiting access to the system to those who are authorized to use the system and using an authentication method that proves the individual is who they say they are. Authentication methods can include a password, token, or biometric. A combination of two of these authentication methods, dual-factor authentication, can be used to strengthen the security around these systems. The factors fall into three categories – something one knows, something one has, or something that one is.[59]
    - Password: this is the most widely used method of authentication where a user types in a password to gain access to the system. Passwords can be a PIN number like those used for ATM withdrawals or can be a longer combination of letters, numbers, and special characters.
    - Token: a token can come in the form of a smart card, key fob, or an application on a smartphone that generates a one-time PIN number for the user to enter to access the system.
    - Biometric: this is an authentication method that uses a part of the body to provide access. These can include retina scanning, hand geometry, and fingerprint.
- Privileged account management (PAM): used to protect accounts with elevated access to a system. These systems are meant to limit the time an elevated credential can be used to prevent these accounts being used against the organization. Software programs exist that catalog privileged accounts and change their passwords frequently. To use an administrative level password a technician would be required to log into the PAM system and check out the password they need. The password would be generated, given a short time to live, and signed out to the user.[60]
- Backup and continuity of operations: there are times when a system may become unavailable due to hardware or software failure as well as times when data may be corrupted such as during a ransomware attack. Organizations must have backup and recovery strategies in place which are regularly tested to ensure data and systems can be brought back up in a timely manner. Some organizations will employ a strategy of using several different data centers that mirror each other's data in case there is a power outage or other emergency. These data centers will have redundant power, internet connectivity, and backup power in the form of batteries and generators. Data will also be backed up at these sites to tape or disk. When data is backed up to tape it is typically stored off site, so it would not be damaged or lost if there was a fire or flood at the data center.
- OWASP: Open Web Application Security Project (OWASP) is an organization focused on improving the security of software. Application security begins with good coding and security practices built into the project.[61] OWASP identifies ten proactive controls that should be included in every software development project:
    - Verify for Security Early and Often
    - Parameterize Queries
    - Encode Data
    - Validate All Inputs
    - Implement Identity and Authentication Controls
    - Implement Appropriate Access Controls
    - Protect Data
    - Implement Logging and Intrusion Detection
    - Leverage Security Frameworks and Libraries
    - Error and Exception Handling[62]
- Vulnerability analysis and penetration testing: applications and systems should be checked periodically to see if new vulnerabilities have been discovered or introduced into a system. This can be done by hiring third parties to test or through the organization conducting routine tests against their own systems specially designed to identify vulnerabilities. Once identified the organization will need to develop a mitigation strategy to prevent those vulnerabilities from being exploited.[63]

Border defenses
- Firewalls: these are network devices or software products that allow or reject traffic to different network zones.[64] Firewalls can be used to define different levels of trust on different areas of the

network. For example, the main database for an electronic health record system would not be directly connected to the internet. It would be protected by a firewall that would only allow certain traffic or users to access it.
- Intrusion detection: this technology is used to detect anomalies within the system or network that may not be caught by other layers of protection like firewalls or antivirus software. These solutions monitor network traffic, system logs, and system usage to determine abnormal behavior. Intrusion detection can be network based or host based to provide a more granular view of potentially harmful activity.[65]
- Web filtering: most users in an organization will have a need to access the internet as part of their work. Work policies may also allow a user to access the Internet for personal purpose. Whether for business or personal use, the Internet can be a dangerous place and protections need to be put into place to prevent the user from inadvertently infecting their workstations or allowing a threat into the organization. A web filter will catalog and block access to sites known to contain malicious software. These solutions are also referred to as "content control software" as their secondary purpose may be to limit the users from accessing sites that may not be appropriate for work such as those related to pornography, gambling, or hate speech.[66]
- Virtual Local Area Networks (VLANs): are networks that attach objects from one or more LANs in a way that makes them appear as if they are all on the same LAN. They take advantage of logical connections rather than physical connections and can assist in creating segments that help secure different areas of the network.[67] Within the healthcare environment they can be used to limit access to objects that may be vulnerable to internal or external threats. VLANs can be used to segregate biomedical devices from the rest of the network so they are less likely to be compromised.
- Encryption in transit: earlier in this section encryption for data "at rest" was described. Encryption for data in transit takes data from the source system and renders it unreadable using a set of keys and then sends the data on to its destination where the receiving system decrypts the data using a shared key. This allows information to transit the internet or the network without being captured and read by a threat actor monitoring the traffic.[58] The most visible example of encryption in transit is the Hyper Text Transfer Protocol Secure (HTTPS) used for visiting web sites.

Administrative controls can be a highly effective tool for protecting data. From an organizational perspective data governance is a necessity in reducing risk to the organization.
- Policy - Policies that define the appropriate use of applications and services provided by the organization are fundamental in helping to resolve human resource issues. If an employee is terminated for browsing pornographic websites at work a policy had better been in place prior to the termination or the organization may face a wrongful termination lawsuit. Most individuals would intrinsically understand inappropriate workplace behavior but if there are no rules written down and disseminated to staff, the terminated employee would likely win their wrongful termination suit.
- Contracting - Another administrative control that is highly useful is a contract review process that includes a review for provisions on information security and privacy. It is not enough to have a business associate agreement in place, an organization must define the lowest level of security they are willing to accept from a vendor or a partner to secure their data. This is a risk-based decision that the organization may need to make if the vendor is not willing to budge on contracting provisions.
  o Ensure that the vendor addresses security and privacy controls within a contract. Information should be a shared responsibility, but many contracts place all responsibility for product security on the customer.
  o If the vendor uses hosted solutions make sure there is language that covers these downstream services.
  o If the language in the contract is weak, work with the legal team and the information security team in the organization to build template language for negotiating contracts.
  o Be aware of the risks of "*click through agreements.*" These are contract agreements for services that usually come in the form of a webpage when paying for a service online.[68] If "I agree" is clicked then all the terms and conditions have been agreed to by the organization. These are non-negotiable agreements and could place patient privacy or security at risk if they do not include protective language. They will certainly benefit the vendor and place all the risk on the customer. Employees of a healthcare entity should also be aware that they are accepting risk on behalf of the organization if they click these agreements. Organizations will likely have a signature authority policy in place

that identifies who in the organization can sign contracts.

Physical controls are very important in protecting the security and privacy of data and individuals who are part of the healthcare environment including employees, visitors, and patients. In addition to having guards, gates, cameras, and door locks there are other controls that protect the environment that are not so obvious.

- Proximity cards are used widely in healthcare to restrict access to certain areas. These cards are typically an ID badge worn by the employee and given rights to access areas where they can go. In addition to unlocking doors, proximity cards can be used to unlock workstations or cabinets.
- RFID (radio-frequency identification) uses electromagnetic fields to identify and track items.69 RFID tags can be attached to pieces of expensive mobile equipment like infusion pumps, wheelchairs, or beds to help locate these devices and deter theft. Additionally, these tags could be included in an armband for a newborn and hooked to a security system that automatically locks doors if there is an attempted abduction.
- Cameras can be used to monitor movement in the hallways or main areas of a hospital or clinic. They can also be used to monitor more sensitive areas like pharmacies. Additionally, cameras can be integrated into systems that help reduce fall risk. Some technologies draw virtual borders around beds or chairs and if the patient breaks that boundary it sends an alert to the main desk in the unit. This can be very effective in preventing falls from patients who are fall risks or in preventing a patient from eloping.

FUTURE TRENDS - EMERGING RISKS IN HEALTHCARE

Given the nature of the healthcare industry, it will continue to be viewed as a rich target for malicious actors. The provider segment of the industry within the United States operates in a manner which allows generous access to patients and visitors and relies on the free flow of information to safely provide care. This dynamic places data and computers at a higher-level risk for the theft of devices or data. The growth in the use of health technology only exacerbates this concern by increasing the attack surface and increasing the amount of sensitive data that could be infiltrated. However, this is not the limitation of emerging risks within healthcare.

Health technology not only stores and transmits sensitive information, it can be used to provide direct care to a patient. These technologies assist the nurses and doctors by automating certain tasks. Biomedical devices include many different types of technology that directly touch the patient; including infusion pumps that automatically deliver medications, ventilators which assist in breathing, surgical robots, and pacemakers. While these devices are extremely effective in assisting patients, they use computer code to ensure they function properly. Many of these devices may reside on an operating system similar to what is used on a desktop computer. When this is the case they have the same vulnerabilities that a desktop computer does, and they should also be patched and updated to reduce the risk of compromise.

Biomedical technology is typically managed by a clinical engineering department in a hospital. Traditionally the information technology department has not had a large role in managing these devices, so it is up to biomed to ensure that these devices are operating in a safe manner. This requires them to understand when and how the device should be patched or how the device should be secured on the network. More and more the biomed and IT departments are needing to cooperate to ensure these devices are appropriately secured.

There are several challenges to securing these devices. For one, these devices are reviewed and approved by the Food and Drug Administration (FDA) for use on patients. They go through a long application and review process before they are finally permitted for use. Biomedical device manufacturers want to ensure that the devices are operating as designed and so may be very proprietary about how they patch and support these devices. If a vendor has a restrictive support contract around these devices it can be challenging to receive timely updates. Vendors are allowed to patch these devices without going through the FDA approval process, as long as the security updates do not impact the functionality of the device. However, vendors may choose to bundle their patches and may take a considerable amount of time to test the patches to ensure there are no changes to functionality. The FDA has issued post market guidance on the cybersecurity of medical devices to help ease this process but there is a lot left in the hands of the manufacturer regarding security patching.[70]

Older devices can run on operating systems that are no longer supported by the vendor, thus compounding the issue. If a patch for a vulnerability is not available, the organization must evaluate different methods of securing the device.

The threat is very real. In May 2017 a virus named WannaCry impacted biomedical devices in the US and in Europe with ransomware that made the devices unusable for patient care.[71]

This virus also impacted environmental control systems. These too are an emerging avenue of compromise for hospitals and other care facilities. Environmental control systems manage the temperature and airflow within these buildings and are vital in keeping the appropriate temperature for care and maintaining proper airflow to reduce the risk of infection.

Environmental control systems, biomedical devices, and any other device connected to a healthcare network are going to increase the attack surface. The more vulnerable points on a network, the more chances an attacker must pivot and identify a higher profile target. If a biomedical device can be reprogrammed by an attacker, this device could be used to harm a patient. If an environmental control system can be hacked, and that system is shut down or reprogrammed, this too can have an impact on patient safety. This happened in Arlington, Texas in 2009 when an employee broke into 14 computers, installing malware that allowed others to control the computers. One of these computers controlled the heating and air conditioning for the hospital. The hacker was sentenced to nine years in federal prison for his crime.[72]

While patient privacy and the protection of sensitive information is very important, the ultimate risk lies in the potential physical harm that hacked systems could have on patients.

RECOMMENDED READING

Privacy and security are constantly evolving. There are new vulnerabilities, threats, breaches, and regulatory interpretations on a daily basis. To keep current on some of these issues please check out the following resources:

Wired magazine. https://www.wired.com/category/security/. In-depth security articles.

Krebs on Security. https://krebsonsecurity.com/. Security blog.

Ars Technica. https://arstechnica.com/. Technology news site.

The Register. https://www.theregister.co.uk/security/. International news site.

Health and Human Services. https://www.hhs.gov/hipaa/newsroom/index.html. HIPAA releases.

KEY POINTS

- Confidentiality, integrity and availability are key concepts to understand healthcare information privacy and security
- ARRA and the HITECH Act were designed in part to supplement the administrative, physical and technical safeguards implemented by HIPAA
- Healthcare workers must be able to identify different types of threat actors as well as the appropriate controls
- Security measures will continue to improve but so will the efforts of criminals who seek illicit access to protected health data and identify theft

CONCLUSIONS

This chapter has touched the surface of what it means to protect the privacy and security of health information. It has covered some basic concepts of information security, the healthcare privacy and security regulatory environment, business drivers and organizational structure of information security in healthcare, real healthcare breaches and consequences, threat actors, types of attacks, tools used to protect information, and some emerging security risks to the industry. This broad overview of privacy and security is just a primer to a segment of the healthcare industry that continues to evolve and mature.

Currently there is no way to predict what impacts the industry will face regarding security threats. Who would have predicted in January 2017 that National Security Agency cyber warfare tools would be leaked and then used with great impact against healthcare in the WannaCry attack of May 2017?[73]

To be effective in protecting privacy and security in healthcare it is important for all of those who work in the industry to be aware of the potential vectors of attack, type of information exposed, and the tools that can be used to prevent an attack. Ultimately, the information security and privacy practitioners of an organization rely on everyone in the workforce to be a partner in protecting our information assets and systems.

REFERENCES

1. Agaku IT, Adisa AO, Ayo-Yusuf OA, Connolly GN. Concern about security and privacy, and perceived control over collection and use of health information are related to withholding of health

1. information from healthcare providers. J Am Med Inform.2014;21(2):374-378
2. National Institute of Standards and Technology. FEDERAL INFORMATION PROCESSING STANDARDS PUBLICATION: Standards for Security Categorization of Federal Information and Information Systems. http://nvlpubs.nist.gov/nistpubs/FIPS/NIST.FIPS.199.pdf. Page 6. February 2004. (Accessed Sept 22, 2017)
3. National Institute of Standards and Technology. FEDERAL INFORMATION PROCESSING STANDARDS PUBLICATION: Standards for Security Categorization of Federal Information and Information Systems. http://nvlpubs.nist.gov/nistpubs/FIPS/NIST.FIPS.199.pdf. Page 2. February 2004. (Accessed Sept 22, 2017)
4. US-CERT, United States Computer Emergency Readiness Team. Defense in Depth. https://www.us-cert.gov/bsi/articles/knowledge/principles/defense-in-depth. September 13, 2005. (Accessed Sept 22, 2017)
5. U. S. Department of Health & Human Services. HIPAA Enforcement. https://www.hhs.gov/hipaa/for-professionals/special-topics/HITECH-act-enforcement-interim-final-rule/index.html. June 16, 2017. (Accessed Sept 22, 2017)
6. U. S. Department of Health & Human Services. HIPAA Enforcement. https://www.hhs.gov/hipaa/for-professionals/compliance-enforcement/index.html. July 25, 2017. (Accessed Sept 22, 2017)
7. U. S. Department of Health & Human Services. Guidance Materials for Consumers. https://www.healthit.gov/providers-professionals/implementation-resources/ocr-guidance-materials-consumers. N.D. (Accessed Sept 22, 2017)
8. U. S. Department of Health & Human Services. Covered Entities and Business Associates. https://www.hhs.gov/hipaa/for-professionals/covered-entities/index.html. June 16, 2017. (Accessed Sept 22, 2017)
9. U. S. Department of Health & Human Services. Won't the HIPAA Privacy Rule's minimum necessary restrictions impede the delivery of quality health care by preventing or hindering necessary exchanges of patient medical information among health care providers involved in treatment? https://www.hhs.gov/hipaa/for-professionals/faq/208/wont-minium-necessary-restriction-impede-delivery/index.html. March 14, 2006. (Accessed Sept 22, 2017)
10. Privacy Rights Clearinghouse. The HIPAA Privacy Rule: Patients' Rights. https://www.privacyrights.org/consumer-guides/hipaa-privacy-rule-patients-rights. September 23, 2017. (Accessed Sept 22, 2017)
11. U. S. Department of Health & Human Services. HIPAA Administrative Simplification. pg 16. https://www.hhs.gov/sites/default/files/hipaa-simplification-201303.pdf. March 26, 2013. (Accessed Sept 22, 2017)
12. U. S. Department of Health & Human Services. HIPAA Administrative Simplification. pg 84. https://www.hhs.gov/sites/default/files/hipaa-simplification-201303.pdf. March 26, 2013. (Accessed Sept 22, 2017)
13. U. S. Department of Health & Human Services. HIPAA Administrative Simplification. pg 78. https://www.hhs.gov/sites/default/files/hipaa-simplification-201303.pdf. March 26, 2013. (Accessed Sept 22, 2017)
14. U. S. Department of Health & Human Services. HIPAA Administrative Simplification. § 164.502 pg 77-80. https://www.hhs.gov/sites/default/files/hipaa-simplification-201303.pdf. March 26, 2013. (Accessed Sept 22, 2017)
15. U.S. Department of Health and Human Services National Institutes of Health. How can covered entities use and disclose protected health information for research and comply with the Privacy Rule. http://privacyruleandresearch.nih.gov/pr_08.asp. February 7, 2007. (Accessed Sept 22, 2017)
16. U. S. Department of Health & Human Services. Business Associates. https://www.hhs.gov/hipaa/for-professionals/privacy/guidance/business-associates/index.html. July 26, 2013. (Accessed Sept 22, 2017)
17. U. S. Department of Health & Human Services. Business Associate Contracts. https://www.hhs.gov/hipaa/for-professionals/covered-entities/sample-business-associate-agreement-provisions/index.html. June 16, 2017. (Accessed Sept 22, 2017)
18. U. S. Department of Health & Human Services. HIPAA Administrative Simplification. pg 63. https://www.hhs.gov/sites/default/files/hipaa-simplification-201303.pdf. March 26, 2013. (Accessed Sept 22, 2017)
19. U. S. Department of Health & Human Services. HIPAA Administrative Simplification. pg 64-66. https://www.hhs.gov/sites/default/files/hipaa-simplification-201303.pdf. March 26, 2013. (Accessed Sept 22, 2017)
20. U. S. Department of Health & Human Services. HIPAA Administrative Simplification. pg 71. https://www.hhs.gov/sites/default/files/hipaa-simplification-201303.pdf. March 26, 2013. (Accessed Sept 22, 2017)
21. U. S. Department of Health & Human Services. HIPAA Administrative Simplification. pg 71. https://www.hhs.gov/sites/default/files/

21. hipaa-simplification-201303.pdf. March 26, 2013. (Accessed Sept 22, 2017)
22. U. S. Department of Health & Human Services. HIPAA Administrative Simplification. pg 71. https://www.hhs.gov/sites/default/files/hipaa-simplification-201303.pdf. March 26, 2013. (Accessed Sept 22, 2017)
23. U. S. Department of Health & Human Services. HIPAA Administrative Simplification. pg 71-72. https://www.hhs.gov/sites/default/files/hipaa-simplification-201303.pdf. March 26, 2013. (Accessed Sept 22, 2017)
24. U. S. Department of Health & Human Services. HIPAA Administrative Simplification. pg 72-73. https://www.hhs.gov/sites/default/files/hipaa-simplification-201303.pdf. March 26, 2013. (Accessed Sept 22, 2017)
25. U. S. Department of Health & Human Services. Breach Portal: Notice to the Secretary of HHS Breach of Unsecured Protected Health Information. https://ocrportal.hhs.gov/ocr/breach/breach_report.jsf. N.D. (Accessed Sept 22, 2017)
26. U.S. Department of Health and Human Services. Office for Civil Rights. https://www.hhs.gov/hipaa/for-professionals/compliance-enforcement/agreements/index.html. May 23, 2017. (Accessed Sept 22, 2017)
27. Finkle, J. and Burns, D. St. Jude stock shorted on heard device hacking fears; shares drop. http://www.reuters.com/article/us-stjude-cyber/st-jude-stock-shorted-on-heart-device-hacking-fears-shares-drop-idUSKCN1101YV. August 25, 2016. (Accessed Sept 22, 2017)
28. Beazley. Beazley Breach Response (BBR) in Technology, media & business services: Healthcare. https://www.beazley.com/documents/Factsheets/beazley-bbr-coverage-factsheet-us.pdf. N.D. (Accessed Sept 22, 2017)
29. McCann E. Behemoth breach sounds alarm for 4M. http://www.healthcareitnews.com/news/behemoth-hipaa-breach-sounds-alarms. August 26, 2013. (Accessed Sept 22, 2017)
30. McCann E. Advocate health slapped with lawsuit after massive data breach. http://www.healthcareitnews.com/news/AdvocateHealth-slapped-with-lawsuit-after-massive-data-breach. September 6, 2013. (Accessed Sept 22, 2017)
31. Conn J. Advocate health care sued following massive data breach. http://www.modernhealthcare.com/article/20130906/NEWS/309069953. September 6, 2013. (Accessed Sept 22, 2017)
32. Ouellette, P. Oregon Health and Science University reports data breach. https://healthitsecurity.com/news/oregon-health-and-science-university-reports-data-breach. March 26, 2013. (Accessed Sept 22, 2017)
33. Ouellette, P. OHSU alerts patients of Google cloud security concerns. https://healthitsecurity.com/news/ohsu-alerts-patients-of-google-cloud-security-concerns. July 29, 2013. (Accessed Sept 22, 2017)
34. U.S. Department of Health and Human Services. Office for Civil Rights. Widespread HIPAA vulnerabilities result in $2.7 million settlement with Oregon Health & Science University. http://wayback.archive-it.org/3926/20170127185938/https://www.hhs.gov/about/news/2016/07/18/widespread-hipaa-vulnerabilities-result-in-settlement-with-oregon-health-science-university.html. July 18, 2016. (Accessed Sept 22, 2017)
35. U.S. Department of Health & Human Services. Breaches affecting 500 or more individuals. https://www.hhs.gov/hipaa/for-professionals/breach-notification/index.html. Updated July 26, 2013. (Accessed Sept 22, 2017)
36. Privacy Rights Clearinghouse. Data breaches: a year in review. https://www.privacyrights.org/blog/data-breaches-year-review. December 16, 2011. (Accessed Sept 22, 2017)
37. Vockley M. Safe and secure? Healthcare in the Cyberworld. Biomedical Instrumentation & Technology. 2012; 164-173 https://doi.org/10.2345/0899-8205-46.3.164. (Accessed Nov 25, 2013)
38. Kern, C. Judge Tosses Most Claims in DoD, TRICARE Data Breach Case. https://www.healthitoutcomes.com/doc/judge-tosses-most-claims-in-dod-tricare-data-breach-case-0001. May 16, 2014. (Accessed Sept 22, 2017)
39. Conn J. HHS wants photocopy machines examined as part of data security. http://www.modernhealthcare.com/article/20130815/NEWS/308159953. August 15, 2013. (Accessed Sept 22, 2017)
40. Office of Civil Rights. HHS settles with health plan in photocopier breach case. http://wayback.archive-it.org/3926/20170127183806/https://www.hhs.gov/about/news/2013/08/14/hhs-settles-with-health-plan-in-photocopier-breach-case.html. August 14, 2013. (Accessed Sept 22, 2017)
41. Giandomenico, A. Byline: Know Your Enemy: Understanding Threat Actors. https://blog.fortinet.com/2017/07/13/byline-know-your-enemy-understanding-threat-actors. July 13, 2017. (Accessed Sept 22, 2017)
42. Recorded Future. Proactive Defense: Understanding the 4 Main Threat Actor Types. https://www.recordedfuture.com/threat-actor-types/. August 23, 2016. (Accessed Sept 22, 2017)
43. Zetter, K. That Insane, $81M Bangladesh Bank Heist? Here's What We Know. https://www.wired.com/2016/05/

insane-81m-bangladesh-bank-heist-heres-know/. May 17, 2016. (Accessed Sept 22, 2017)
44. Australian Competition & Consumer Commission. Nigerian Scams. https://www.scamwatch.gov.au/types-of-scams/unexpected-money/nigerian-scams. N.D. (Accessed Sept 22, 2017)
45. Walls, M. Nation-State Cyberthreats: Why They Hack. https://www.darkreading.com/informationweek-home/nation-state-cyberthreats-why-they-hack-/a/d-id/1318522?. January 8, 2015. (Accessed Sept 22, 2017)
46. Barreiro, A. Defending against Advanced Persistent Threats. http://www.techrepublic.com/blog/it-security/defending-against-advanced-persistent-threats/. April 16, 2012. (Accessed Sept 22, 2017)
47. Peterson, A. The Sony Pictures Hack, Explained. https://www.washingtonpost.com/news/the-switch/wp/2014/12/18/the-sony-pictures-hack-explained/?utm_term=.a9563cad386d. December 18, 2014. (Accessed Sept 22, 2017)
48. Greenberg, A. How an Entire Nation Became Russia's Test Lab for Cyberwar. https://www.wired.com/story/russian-hackers-attack-ukraine/. June 20, 2017. (Accessed Sept 22, 2017)
49. Harwell, D. and Nakashima, E. China Suspected in Major Hacking of Health Insurer. https://www.washingtonpost.com/business/economy/investigators-suspect-china-may-be-responsible-for-hack-of-anthem/2015/02/05/25fbb36e-ad56-11e4-9c91-e9d2f9fde644_story.html?utm_term=.80c71d4081ab. February 5, 2015. (Accessed Sept 22, 2017)
50. Lord, M. What is Social Engineering? Defining and Avoiding Common Social Engineering Threats. https://digitalguardian.com/blog/what-social-engineering-defining-and-avoiding-common-social-engineering-threats. July 27, 2017. (Accessed Sept 22, 2017)
51. Phishing.org. What is Phishing? http://www.phishing.org/what-is-phishing. N.D. (Accessed Sept 22, 2017)
52. Armasu, L. Spreading Malware Through Dropped USB Sticks Could Be Highly Effective Research Finds. http://www.tomshardware.com/news/dropped-usb-sticks-spreads-malware,32391.html. August 4, 2016. (Accessed Sept 22, 2017)
53. US-CERT, United States Computer Emergency Readiness Team. Security Tip (ST04-015) Understanding Denial-of-Service Attack. https://www.us-cert.gov/ncas/tips/ST04-015. February 6, 2013. (Accessed Sept 22, 2017)
54. Hilton, S. Dyn Analysis Summary of Friday October 21 Attack. https://dyn.com/blog/dyn-analysis-summary-of-friday-october-21-attack/. October 26, 2016. (Accessed Sept 22, 2017)
55. University of Virginia Computer Science. Blocking Brute Force Attacks. http://www.cs.virginia.edu/~csadmin/gen_support/brute_force.php. 2007. (Accessed Sept 22, 2017)
56. C.S-W. The Economist. What doxxing is, and why it matters. https://www.economist.com/blogs/economist-explains/2014/03/economist-explains-9. March 10, 2014. (Accessed Sept 22, 2017)
57. Johnson, B. What is Next-Generation Antivirus (NGAV)? https://www.carbonblack.com/2016/11/10/next-generation-antivirus-ngav/. November 10, 2016. (Accessed Sept 22, 2017)
58. Raglione, A. Best Practices: Securing Data at Rest, in Use, and in Motion. https://www.datamotion.com/2015/12/best-practices-securing-data-at-rest-in-use-and-in-motion/. December 1, 2015. (Accessed Sept 22, 2017)
59. Harris S, Ouellet E. Security+ Certification All-in-One Exam Guide. Berkeley, CA:McGraw-Hill/Osborne; 2003.
60. Ismail, N. Access all areas? Tracking and managing the privileged users. http://www.information-age.com/access-areas-tracking-managing-privileged-users-123465270/. March 27, 2017. (Accessed Sept 22, 2017)
61. OWASP. Welcome to OWASP the free and open software security community. https://www.owasp.org/index.php/Main_Page. March 13, 2017. (Accessed Sept 22, 2017)
62. OWASP. OWASP Top 10 Proactive Controls 2016. https://www.owasp.org/index.php/OWASP_Proactive_Controls. September 1, 2017. (Accessed Sept 22, 2017)
63. Secureworks. Vulnerability Assessments Versus Penetration Tests. https://www.secureworks.com/blog/vulnerability-assessments-versus-penetration-tests. April 8, 2015. (Accessed Sept 22, 2017)
64. Palo Alto Networks. What is a Firewall? Firewalls and Their Evolution. https://www.paloaltonetworks.com/cyberpedia/what-is-a-firewall. N.D. (Accessed Sept 22, 2017)
65. SANS Institute InfoSec Reading Room. Understanding Intrusion Detection Systems. https://www.sans.org/reading-room/whitepapers/detection/understanding-intrusion-detection-systems-337. 2001. (Accessed Sept 22, 2017)
66. Kaspersky Lab. What is a Web Filter? https://usa.kaspersky.com/resource-center/definitions/web-filter. 2017. (Accessed Sept 22, 2017)
67. Cisco. Chapter: Understanding and Configuring VLANs. https://www.cisco.com/c/en/us/td/docs/switches/lan/catalyst4500/12-2/25ew/configuration/guide/conf/vlans.html. N.D. (Accessed Sept 22, 2017)
68. Wilmer Hale. Are "Click Through" Agreements Enforceable? https://www.wilmerhale.

com/pages/publicationsandNewsDetail. aspx?NewsPubId=86850. March 22, 2000. (Accessed Sept 22, 2017)
69. PC Mag. Definition of: RFID. https://www.pcmag.com/encyclopedia/term/50512/rfid. N.D. (Accessed Sept 22, 2017)
70. U.S. Food and Drug Administration. Cybersecurity. https://www.fda.gov/MedicalDevices/DigitalHealth/ucm373213.htm. August 30, 2017. (Accessed Sept 22, 2017)
71. Fox-Brewster, T. Medical Devices Hit By Ransomware For The First Time In US Hospitals. https://www.forbes.com/sites/thomasbrewster/2017/05/17/wannacry-ransomware-hit-real-medical-devices/#19a26c9a425c. May 17, 2017. (Accessed Sept 22, 2017)
72. Moscaritolo, A. Texas hospital hacker sentenced to nine years. https://www.scmagazine.com/texas-hospital-hacker-sentenced-to-nine-years/article/558753/. March 21, 2011. (Accessed Sept 22, 2017)
73. Nakashima E. The NSA has linked the WannaCry computer worm to North Korea. https://www.washingtonpost.com/world/national-security/the-nsa-has-linked-the-wannacry-computer-worm-to-north-korea/2017/06/14/101395a2-508e-11e7-be25-3a519335381c_story.html?utm_term=.1d32be0c6a5d. June 14, 2017. (Accessed Sept 22, 2017)

11

Health Informatics Ethics

KEN MASTERS

"It is immaterial for the experiment whether it is done with or against the will of the person concerned."
—Dr. Karl Brandt, Final Statement, Nuremberg Trials, 19 July 1947[1]

LEARNING OBJECTIVES

After reading this chapter, the reader should be able to:

- Describe the 20th century medical and computing background to health informatics ethics
- Identify the main sections of the IMIA Code of Ethics for Health Information Professionals
- Describe the complexities in the relationship between ethics, law, culture and society
- Describe different views of ethics in different countries
- Summarize the most pertinent principles in health informatics ethics
- Discuss the application of health informatics ethics to research into pertinent areas of health informatics
- Discuss appropriate health informatics behaviour by medical students

INTRODUCTION

As is obvious from the subject in this text book, health informatics combines themes from medical fields and from informatics fields. It is to be expected, then, that health informatics ethics will combine information from medical ethics and from informatics ethics. This section details the recent history of these two fields so that the reader can understand the context within which modern health informatics ethics is to be discussed. This chapter will first examine some of the historical context in which medical ethics should be understood.

The Road from Nuremberg

Nuremberg (alternate spelling *Nuernberg*, German spelling *Nürnberg*) is a town in Germany. Before and during World War II, Germany's National Socialist Party (Nazi Party) had controlled Germany and had occupied much of Europe. During this time, at least 11 million people (mostly Jews, Poles, Romani ["Gypsies"], Eastern Europeans, and others regarded by the Nazis as "sub-humans" or "undesirables")[2] were systematically murdered in what is now referred to as "The Holocaust." Many of the victims were murdered in large camps called concentration camps.

At the end of World War II, a series of legal trials was held in Nuremberg and other cities to examine crimes against humanity that had been committed in Germany and German-occupied countries.[3] To its shame, the German medical profession had cooperated with the Nazi Party on such a scale that medical and other health professionals were tried separately, and the abridged transcripts of these trials (referred to as the "Medical Case") make up more than 1,300 pages of testimony and supporting documentation.[1,4]

Crimes committed by medical professionals had included widespread euthanasia and sterilization of mentally and physically handicapped people (referred to as "useless eaters" (*unnützen Esser*) and lives "unworthy of life" [*lebensunwerten Lebens* or *Lebensunwerts*),[4-7] and also a large number of medical and biological experiments conducted on concentration camp inmates. Victims included men, women and children. (See Figures 11.1 and 11.2.)

Figure 11.1: Children's Memorial, Mauthausen Concentration Camp

Figure 11.2: Dissection Room, Mauthausen Concentration Camp

Most died extremely painfully as a result of the experiments, and many of those who survived were later murdered by the camp authorities. Permission or consent for the medical experiments was almost never obtained from the inmates. Where "permission" was obtained, it was usually only as an alternative to death, or with the promise of release. None of the surviving victims was ever released by their captors, nor were any death sentences commuted.[4] Ironically, many of the experiments would have been illegal if they had been conducted on animals, as the Nazis had introduced strict laws governing the use of animals in medical experiments.[1]

At the Nuremberg medical trial, a code of conduct, which later became known as "The Nuremberg Code," was presented. The Nuremberg Code was in direct response to the medical crimes.[1] (See the NIH site at https://history.nih.gov/research/downloads/nuremberg.pdf for the Code). The Code emphasised the need for experimental subjects' voluntary consent to the experiment, regard for their safety (including mental suffering), balance of risks, and right to withdraw from the experiment if they wish. In addition, the Code noted that the responsibility for performing the experiment lay with the qualified medical experimenter, and this responsibility could not easily be transferred.

As the prosecution at Nuremberg had noted, the medical professionals who had performed these procedures had violated a basic medical principle of "First, do no harm" (*primum non nocere*).[4] Because the Code was in direct response to the Nazi medical experiments, it focuses on the rights of patients and experimental subjects; it is not a broader code dealing with general ethics of medical practitioners in other situations.

World Medical Associations' (WMA) Declaration of Helsinki

After the publication of the Nuremberg Code, several countries reviewed their medical ethics (See below for some discussion of national influences). In 1964, the World Medical Association drew up the first version of the "Declaration of Helsinki" (DoH), which broadened the concerns of the Nuremberg Code, and was to be applicable across the globe. Since then, the DoH has been through several reviews, and the current version was adopted in Fortaleza, Brazil in 2013.[8]

Although it is not the same as the Nuremberg Code, the DoH is similar to it, as it deals with the human subjects' safety, consent, risks, and right of withdrawal. Amongst the significant additions (significant in the light of health informatics) is the right to "*privacy of research subjects and the confidentiality of their personal information*" (Article 10, 23 and 24). These issues will be addressed later.

INFORMATICS ETHICS

Although one may argue that the history of informatics ethics begins with the ancient Greeks, it is only in the latter half of the 20th century that machine-based information and ethics were viewed together for the first time.[9] At roughly the same time that the Nuremberg Code was being developed, Norbert Wiener first published his book *The Human Use of Human Beings* in which he considered the social and ethical implications in the relationship between machines and humans.[10] From the

1970s onwards, work by people like Kostrewski and Oppenheim[11] and Robert Hauptman[12] dealt with ethical questions in information research. In 1986, Mason introduced the PAPA acronym of Privacy, Accuracy, Property, and Accessibility as part of a *"social contract among people in the information age"* to *"enhance the dignity of mankind."*[13] In 1997, Severson introduced 4 principles of Information Ethics: (1) Respect for intellectual property; (2) Respect for privacy; (3) Fair representation; (4) Non-maleficence (or "doing no harm").[14]

As computers have developed further, codes of ethics for professional organisations have also evolved, and the field of informatics ethics (or data ethics[15]) has grown. Two examples of professional ethics codes are the Association for Computing Machinery's *Code of Ethics and Professional Conduct*[16] and the Canadian Information Processing Society's *Code of Ethics and Professional Conduct*.[17] These codes also refer directly to honouring the rights of the individual, respecting privacy and confidentiality, and doing no harm. (see Figure 11.3)

With similarities between the principles of ethics in medicine and ethics in informatics (especially in areas of respect for subjects, privacy, and doing no harm), it is to be expected that these would be issues contained in health informatics codes that were developed in the late 20th and early 21st century. Indeed, these

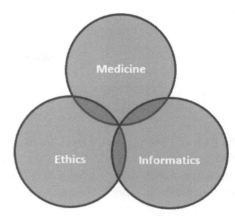

Figure 11.3: Health Informatics Ethics formed from Medicine, Ethics and Informatics

principles are contained in the International Medical Informatics Association's (IMIA) Code of Ethics for Health Information Professionals.[18] Although the code is aimed at health informatics personnel, it should be remembered that any medical person working with electronic data will also be a Health Informatics Professional (HIP) or a Clinical Informatics Professional (CIP).

The IMIA code is extensive, going much further than the Nuremberg Code and the Declaration of Helsinki, and has the following components (Table 11.1)

Table 11.1: IMIA Code

IMIA Code
Part I: Principles
A. Fundamental Ethics Principles: autonomy, equality and justice, beneficence, non-malfeasance, impossibility (recognising that some things are impossible), and integrity.
B. General Principles of Information Ethics: information-privacy and disposition, openness, security, access, legitimate infringement, least intrusive alternative, and accountability.
Part II: Rules of Ethical Conduct
A. Subject-centred duties: these focus on electronic records and are aimed at ensuring that subjects of electronic records are protected from abuse of their information.
B. Duties towards other Health Care Professionals (HCPs): these focus on proper support, keeping HCPs informed of relevant information, maintaining standards of data storage, and intellectual property.
C. Duties towards institutions/ employees: these include integrity, loyalty, ensuring the safety of the institution's data, evaluation of systems' security, alerting and informing the institution of problems in good time and working within their scope of competence.
D. Duties towards society: these include the proper collection, storage and safe-guarding of appropriate data, informing the public, and not participating in work that violates human rights.
E. Self-regarding duties: these include recognising one's own limitations, maintaining competency and avoiding conflict.
F. Duties towards the profession: these include not bringing the profession into disrepute, impartiality, and assisting and maintaining standards of professionalism amongst colleagues.

The IMIA Code is continually under review and serves as a useful guide for all people who work in health informatics fields.

INTERNATIONAL CONSIDERATIONS: ETHICS, LAWS AND CULTURE

The first part of this chapter describes the medical ethics' developments from World War II to the present day, and then the development of health informatics ethics. The impression is one of a great tragedy created by one country's lapse of medical ethics, internationally punished for breaking the widely-accepted ethical practices, resulting in a neat and linear path towards a set of near-perfect and internationally-accepted codes of ethics in the medical and health informatics fields.

While useful, this impression is a deceptive over-simplification, and the student of health informatics needs to be aware of greater complexities, especially with regards to national and international practices, and the relationship between ethics and the law. Part of the reason for the conflict is that ethics in general is strongly influenced by a country's laws and culture, but the relationship between ethics, law, culture and society is unclear, is not fixed internationally, and may be fluid even within a given country over time.

Different Views of Ethics

While there are many theories of ethics, for our purposes, there are three broad views regarding the relationship between ethics, law, culture and society:
- Ethics does not exist outside the law and exists only for the good of a properly ordered and legal society. Therefore, a society's needs and the prevailing laws define ethical behaviour.
- Ethics is usually strongly informed by the law, society, and the prevailing culture, and are extensions of these. There are ethical requirements that are not necessarily required by law, but what is ethical can never conflict with what is legally required.
- Ethics exists entirely outside of the law and is a matter of personal conscience. Because ethics grows from within social practices, there is usually correspondence between ethics and the law; where there is conflict, the ethical viewpoint must always prevail.

Significance of Different Views

In the codes and activities outlined above, one can see the different views being expressed. When these views are translated into practises, the difficulties of implementing ethics become more apparent. Some examples are:
- Among some Western doctors, there was the feeling that the Nuremberg Code was useful for "barbarians," but unnecessary for civilised physicians.[19]
- Many of the principles in the Nuremberg code were not universally followed as standard procedures, even in prosecuting countries. The Nuremberg Code had not existed before World War II; rather, it emerged as a response to the atrocities witnessed. Part of the defence was that, at the time, there had been many international medical experiments performed on condemned prisoners (including conscientious objectors), who received no pardon or reduction in sentence, and it was also questionable whether all international medical experiment subjects (or their parents, in the case of minors) had given their permission.[1,20] Some of the most important medical experiments (such as those by Edward Jenner) had been performed without any indication of consent.
- The medical experiments carried out by the Nazi doctors were almost always in compliance with the law and legal instructions from superiors and were usually meticulously documented in reports.[1,4] A strong argument for supporting the Nazi medical experiments was for the good of society, especially considering the saving of soldiers' lives during war.[1,4] The counter-argument was that, even if a legal order were received, the physician should refuse to obey an order that he believes to be morally unjustified.[1]
- Medical staff from Japan had also conducted many experiments on the Chinese population and had used live prisoners in training procedures for their doctors.[21-22]
- In more than 31 US States, until the late 1970s, Eugenics Boards routinely sterilised people for various reasons, including being poor, or "feeble-minded," or young girls who had been raped. In North Carolina alone, an estimated 7,500 people were sterilised under this program.[23-25]
- Other countries also performed compulsory sterilization; in the former Czechoslovakia, involuntary sterilization (mainly of Romani women) was widely performed from 1973-1991, and other allegations of forced sterilization as late as 2008 have been reported.[26-28]
- Although not widespread, other countries, including the USA, had conducted medical experiments on humans who were not fully informed, and so, could not have given informed consent. Some had been conducted before the war, but many continued well after.

At the very time of the Nuremberg trials, the Tuskegee Syphilis Experiment was being conducted in the USA and ended only in 1972 when it was reported in the press.[29-31] (see Figure 11.4). Other experiments included the U.S. syphilis tests in Guatemala,[32] the Sonoma State Hospital experiments on disabled children,[33] and the radiation experiments on American citizens.[34] In all of these cases, the central tenets of the Nuremberg Code had not been followed. (For further reading on this topic, see Anthony Clare's *Medicine Betrayed*.)[35]

- There are many instances where a second person's rights might override the confidentiality rights of a patient.[36]
- Although the Declaration of Helsinki says that local laws must be respected (Article 10), it points out that "No national or international ethical, legal or regulatory requirement should reduce or eliminate any of the protections for research subjects set forth in this Declaration." This can be a meaningless contradiction
- A code of ethics is only a code of ethics. It carries no legal weight at all. If a person is found to be acting unethically, then their organisations and institutions may take actions such as revoking licenses, and refusing permission to practice, but that is the extent of their powers. A person must be guilty of committing a crime in order to be punished in a court of law. The codes of ethics referred to always place their ethics in the context of law (for example, the confidentiality requirement exists, unless otherwise demanded by law).[16-17]

CODES OF INDIVIDUAL COUNTRIES

The American Medical Informatics Association (AMIA)

With the international differences in medical ethics, it is to be that expected there are also differences in health informatics ethics. Indeed, there are, and several individual countries have their own health informatics codes. A complication is that much of the activity covered by health informatics may also be conducted in other fields, so different codes may exist for workers in those fields. This section highlights a few.

The American Medical Informatics Association (AMIA) has a code of ethics for its members.[37] It is significantly shorter than the IMIA code, but also looks at the ethical relationship between doctor and patient (and the patients' family), colleagues, institutions and employers, and society in general. Regarding patients, there is an emphasis on confidentiality and security of information and using all information for the intended

Figure 11.4: Doctor examining a Tuskegee Syphilis Experiment subject (Source: United States National Archives and Records Administration)

purpose only. In the area of research, the code notes that researchers must ensure "*that the greatest good for society is balanced by ethical obligations to individual patients,*"[37] although there is no specific mention of issues like informed consent and right of withdrawal from the experiment or trial.

The AMIA document is also cognizant of difficulties, however, and makes it clear that the code "*is not intended to be prescriptive or legislative; it is aspirational*"[37]

United Kingdom

In the United Kingdom, the UK Council for Health Informatics Professions (UKCHIP) has the *UKCHIP Code of Conduct,* which "*sets out the standards of behaviour required of health informatics professionals registered with the*" UKCHIP.[38] The code has four short sections, dealing with "*Working to professional standards,*" "*Respecting the rights and interests of others,*" "*Protecting and acting in the interests of patients and the public,*" and "*Promoting the standards and standing of the profession.*" In addition, the UK's General Medical Council (GMC) has guidelines in its Good Medical Practice.[39]

European Parliament Directives

Similarly, there are several European Parliament Directives, such as (95/46/EC) of 1995 [40] and others [41-43] which are binding on member countries of the European Union, deal with the protection of data, and cover a wide range of issues from privacy to security. (Note that much of the material contained in these directives has been updated by the April 2016 General Data Protection

Regulation,[44] which will be enforceable from 25 May 2018). The most pertinent principles have been synthesised by de Lusignan et al.[45] into these:

1. Personal data shall be processed fairly and lawfully.
2. Personal data shall be obtained and processed for one or more specified and lawful purposes and not in any manner incompatible with those purposes.
3. Personal data shall be adequate, relevant and not excessive in comparison to the purpose that it was collected for.
4. Personal data shall be accurate and up to date where necessary.
5. Personal data should not be kept longer than is deemed necessary.
6. Personal data shall be processed in accordance with the rights of individuals as set out in the act.
7. Personal data shall have appropriate security measures in place.
8. Personal data shall not be transferred outside of the European Economic Area (EEA) unless adequate protections exist for the rights and freedoms of data subjects.

While this is a useful guide, the actual legally applicative directive is difficult for the lay person to understand and appears extremely difficult to apply. For example, Article 8 of the 95/46/EC directive states that "*Member States shall prohibit the processing of personal data revealing racial or ethnic origin, political opinions, religious or philosophical beliefs, trade-union membership, and the processing of data concerning health or sex life.*" It is then followed by a series of exceptions where this does *not* apply and is also followed by the statement that "*Subject to the provision of suitable safeguards, Member States may, for reasons of substantial public interest, lay down exemptions in addition to those laid down in paragraph 2 either by national law or by decision of the supervisory authority.*" This means that member states may make laws that override the main paragraph of the article.

There are also several international guides developed by different international medical associations that deal with use of specific health informatics activities, such as electronic health records and email communication between doctor and patient. These guides cover ethical issues such as data privacy and protection. Some examples are American Medical Association's (AMA) *Guidelines for Physician-Patient Electronic Communications*[46] and the *Guide to Australian electronic communication in health care.*[47]

Finally, there are many ethics guides from other disciplines that impact upon ethics in health informatics.[48]

PERTINENT ETHICAL PRINCIPLES

In this rather strange mixture of ethics, laws, and cultural influences, there are some principles that appear to be common. Given the complexities outlined above, it is useful for the student of health informatics to have a summary of the most pertinent ethical points, and that summary is supplied here. Principles such as the right to privacy, informed consent in research, and the non-transferability of ethical responsibilities (accountability) will be discussed. With the understanding that the importance of each of these will be viewed differently in different circumstances, these are useful guides from medical ethics and health informatics ethics. In these descriptions, any reference to patients would refer to research subjects and to their families. In health informatics education, these will extend to students.

- Right to privacy. The patient has a right to privacy, which means that information that the HIP has obtained must not be shared with others unless there is reason to believe that it is in the best interests of the patient.
- Guard against excess: there should be safeguards against excessive personal data collection; only data specifically needed should be collected.
- Security of data. The right to privacy, and maintaining patient safety, also means that there is a responsibility on the researcher to keep the data as secure as possible, in order to prevent unauthorised access to it. As an extension of this and incorporating ethical operations of the institutions in which the HIP works, if the HIP becomes aware of security problems, even those that are not under his/her direct control, the responsible persons must be informed. The emphasis on security, however, must not be so strong that it impedes the patient's right to access that data.
- Integrity of data. This is also related to security, and the HIP has to ensure that data are kept current and accurate. In addition, data cannot be presented in such a way that it presents an untrue picture of reality or is designed to mislead the reader.
- Informed consent: while the patient should be aware of what is to happen, that awareness can be complete only if the patient is informed. Similarly, the researcher may do only what has been consented to, and, if the researcher wishes to do anything else (e.g. use data for any other purpose), then new consent must be obtained. Crucial aspects of informed consent are: "(1) competence of the subject to consent..., (2) disclosure of information, (3) the subject's understanding of the information being disclosed,

(4) volition or choice in the consent, and (5) authorization of the consent."[49]

- Laws: the HIP needs to be aware of the laws that apply. Where there is a conflict between the law and the professional ethics, the HIP will have to make difficult decisions. In addition to the discussion above, this issue is explored further a little later in this chapter.
- Medical ethics. Health informatics ethics is a sub-set of medical ethics. This means that all issues that apply to medical ethics in general, such as the physical and psychological safety of the patient, also apply to health informatics ethics.
- Sharing data. If it is necessary to share the data with anyone else (e.g. for further research, temporary or permanent storage, or data transmission), then the HIP must be sure that all the above principles are also being followed.
- Wider responsibility: HIPs have ethical responsibilities towards their employers and the wider community regarding protecting data and maintaining professional standards.
- Implicit in all these are the principles of beneficence and non-maleficence. This means that the ethics must be beneficial to the patient and must be consciously aimed at preventing harm to the patient.
- Non-transferability: the responsibility and accountability for adhering to these rests with the HIP and cannot simply be transferred.

DIFFICULTIES APPLYING MEDICAL ETHICS IN THE DIGITAL WORLD

The previous section of this chapter traced the recent history of health informatics ethics, and showed the various principles involved. At this stage, it is obvious that the issues are extremely complex. It is now time to turn to some practical examples to see how some of these principles can be applied.

Ethical Issues with Large Databases: Informed Consent and Confidentiality

A difficult issue when conducting research on large databases, including hospital databases of Electronic Health Records or Electronic Medical Records (EHRs or EMRs), is how to obtain informed consent for the use of patient data.

One way of obtaining informed consent for use in research is to obtain *"broad informed consent"*[50] at the time that the information is gathered. This is an idea borrowed from the study of large biological samples and is regarded as the most practical and economically viable approach for researchers. A variation is to grant consent for the database to be used for specific types of research only.

Some databases may be small (such as from a clinic or hospital) while others may be large (a national database), and several countries are grappling with problems of informed consent for researchers while protecting patients from abuse.[51-52] In one instance, Iceland created a national database with *"presumed consent,"* but allowed individuals to opt out of the program, thereby removing their information from the database. This solution is not always simple, and the legal relationship between presumed and informed consent, especially around issues of identifiable genetic material, continues today.[52-53]

Any approach will be influenced by the national culture, so will differ from country to country, and may also differ depending on the nature and purpose of the database.[51] Because obtaining general or presumed consent could be open to abuse, it is extremely important that the researcher ensures that the research does not conflict with other areas of ethics, such as exposing the patients to stress or exposing any identifying information.[50,54]

One should also guard against corporate ownership of such databases, as these organisations may work to different ethical guidelines, and there may be conflicts of interest in the research and research outcomes. Where such ownership cannot be avoided, then researchers should not be unethically influenced in their work.[55]

As addressed in the chapter on Privacy, Health Care Workers in the US need to be aware of the HIPAA Privacy Rule,[56] as this applies to all doctors who transmit any patient data electronically.

Research on electronic postings: privacy and disclosure

The Internet is full of information simply waiting to be analysed. One area that has received attention has been online environments in which users create postings in conversations. These might be in discussion lists (sometimes called "listservs"), forums, bulletin boards, and social media or social networking sites, such as Facebook and Twitter.

In many of these sites, medical information, sometimes very personal, is exchanged. Even in sites where personal information is not exchanged, the context may be a medically-oriented site. The prime ethical questions for the researcher researching these sites revolve around informed consent and the privacy of the information that is shared on these sites. In short, the question is this: are these electronic postings to be treated with the same level of confidence and anonymity that one would apply to patients in a self-help group?

Resolving the issue depends on whether the 'human-subject' model or the 'textual' model is applied. These are explained further below.

- The human subject model. Briefly, the human subject model is an extension of the medical view of patient information, and it views the electronic postings as reflective of real people, and so all the ethical rules regarding informed consent, privacy, and ensuring that there is no psychological or physical harm to participants must apply.[57-58] This means that, before quoting from or referring to a site, the researcher must obtain informed consent.
- The textual object model. An opposing view is that a posting in a bulletin board is merely a piece of text and is subject only to the laws and ethics that govern any piece of text. These might include rules regarding plagiarism and copyright, but do not involve anything to do with a human patient. The text has been placed into a publicly-accessible area (the Internet) and any expectation of privacy and confidentiality is unwarranted. As Walther argues, this is much like a conversation in a park, and that "people do not expect to be recorded or observed although they understand that the potential to do so exists."[59]

If the person has not posted personally identifiable information, then there is even less concern regarding privacy; after all, the only problem that might exist is that the person can be identified. For example, in the USA, a *"Human subject means a living individual about whom an investigator (whether professional or student) conducting research obtains*
  o *data through intervention or interaction with the individual, or*
  o *identifiable private information."*[60]

If people are concerned about identity, however, they can create pseudonyms and usernames that make it difficult to identify them. One may argue that, if they have not taken such precautions, then they are willing to have themselves publicly identified.

Finally, the textual object model is supported by much 20-century literary theory[61-62] which clearly separates any discussion of text from the discussion of the author or even the author's intention.[63-64]

Problems With the Textual Object Model

There are several problems with applying the textual model to medical research, and some of these are:

- The arguments are frequently based on traditions from other fields (e.g. sociology, or literary theory). When working in a medically-related field (specifically, health informatics), so one should give greater credence to ethical rules in that field of study.
- Although a specific posting may not contain information that can identify a person, when many of these postings are combined, it may be easy to form a picture that can be used to identify the person. There is a strong tradition in medicine that, even when objects are researched, they are not specifically identified.
- Based on the many examples above, laws should not be taken as a standard of ethics. At best, they may set a minimum, from which the ethical HIP works.

The Difficulties and Disadvantages of Applying the Human Subject Model to Electronic Postings

Having said this, the researcher must be aware that there are difficulties with applying the human subject model when performing research on electronic postings. This sub-section identifies some of these and suggests solutions.

- Establishing informed consent can be difficult, if not impossible. With a group of several thousand, where people join and leave continually, how does one establish informed consent?
- One approach is to attempt to determine whether these are necessary. This depends largely upon the rules of registration and public access to the list.[57,65-67] If the list is very tightly controlled, where members have to be a member of a particular organisation, and have to supply corroborating evidence of their identity, then the researcher is advised to obtain full informed consent. If, however, the list is large, registration requires only an arbitrary user name and password (and, perhaps an email address), and the site is searched and indexed by general search engines, then informed consent is less important.
- How does one preserve privacy and anonymity? Again, one can be guided by the amount of privacy that is assumed in the group. In addition, however, unless informed consent from individuals has been obtained, the researcher should avoid referring to specific postings or individuals. The researcher should even avoid anonymous quoting, as this can be used through a search engine to identify the original piece of text. Rather, the research should use aggregated data (i.e. totals, means, etc.) to give an overall impression.

- Finally, if the researcher wishes, she/he may wish to disguise the site. This is discussed in more detail below.

Manipulating postings as part of an experiment

All the examples given so far deal with the researcher as a passive observer with no direct intervention. A common process is scientific research, however, is for the researcher to introduce a change in order gauge reaction. In online work with large database, this practice can be fraught with difficulties.

Perhaps the most famous instance was research conducted by Adam Kramer *et al.* who conducted a massive-scale (N=689,003) research project by intentionally manipulating Facebook News Feeds in order to gauge emotional responses to postings in electronic networks.[68] Although the research was scientifically informative, it released a storm of protest because there had been no informed consent, and it was felt that the emotional changes had caused harm.[69]

TRANSFERRING ETHICAL RESPONSIBILITY

A tempting route to reducing the researchers' ethical responsibilities would be to transfer the responsibility for ethical behaviour to others, allowing the researchers to concentrate on the task at hand: the research. This might be done in three ways:
- As long as the researchers are obeying the law, they are safe from prosecution, as the laws of the State are there as a guide.
- If the researchers work at an institution that has an Ethics Committee or an Institutional Review Board (IRB), then they submit a protocol that describes the research beforehand, and then receive ethics approval for the research. The researchers may feel that they are now 'covered' and so can do whatever they like, as long as they stick to the protocol.
- Keeping data secure is a complex technical process, so one should simply have a database manager who takes full responsibility for the data. If the data are then mistakenly made public, it is the database manager who has to deal with the problem.

These, however, are not effective solutions:
- Handing this responsibility over to the law or State is not an acceptable solution as highlighted in the Nuremberg Trials, since laws do not establish ethics.
- Because of the newness of the field of health informatics, IRBs may not have representatives that are fully aware of the ethical issues and technical applications (e.g. that simply searching on a quotation from a forum allows one to find that forum immediately), or the extent to which informed consent is required.
- Ultimately, HIPs are responsible for their data. Technical staff may be responsible for the storage systems, but the overall responsibility for the material cannot be transferred to anyone else. In cases where data breaches have occurred, all parties (including the institution) may face legal prosecution, as was the case in which a clinic's data regarding HIV patients was compromised because of peer-to-peer file sharing software on their computers.[72]

ELECTRONIC COMMUNICATION WITH PATIENTS AND CAREGIVERS

The advent of the "e-patient"[73] had led to a situation where medical practitioners are on unfamiliar territory and need to be taught the skills specifically required for online interaction with patients and ethical issues need to be considered.[74-75]

A common activity of e-patients is to search for further information about doctors, and this is considered perfectly fair. In the case of doctors searching the Internet

---

Researcher's Responsibility for Data Security

In July 2009, The University of North Carolina at Chapel Hill discovered that a computer had been hacked as far back as 2007. The data from the Carolina Mammography Registry containing some 180,000 mammography records (including 114,000 social security numbers) had been potentially exposed.[70]

One of the prime concerns was of responsibility and culpability.

- The chief researcher, Professor Bonnie C. Yankaskas argued that the university IT security staff were responsible for the security of the file server.
- The university argued that the chief researcher was to blame.

Initially, Prof. Yankaskas was demoted and had her salary reduced. After a legal fight costing Prof. Yankaskas some $350,000, her position and salary were restored, but she was forced to retire, effective at the end of 2011.[71]

for further information about their patients, however, the ethical issues are far from clear,[76] and practising health professionals need to tread carefully, especially if their institutions do not have specific guidelines.

For example, all medical students know that it is relatively easy to find most people's email addresses. In the case of a practising doctor, the name, work address and telephone number will already be known to patients. Using that information, finding the email address is a small step. Because of the convenience of email communication (to both the doctor and patients), patients will wish to email the doctor on a range of topics. One of the most important benefits is that instructions given be clearly laid out and can be referred to later by the patient; this greatly reduces the risks to the patient.

There are, however, ethical issues that need to be considered when medical personnel interact online with patients. Two guides that have already been mentioned, the AMA's *Guidelines for Physician-Patient Electronic Communications*[46] and the *Guide to Australian electronic communication in health care*[47] have useful information for the practising doctor. In addition, the AMA Code of Medical Ethics refers specifically to email and social media usage in Chapter 2.[77]

The AMA's guide begins by explaining the value of email communication, and then gives useful advice about setting up the communication channels and some medico-legal issues. This includes things like making the patient aware of who is reading the email, the types of email topics that are acceptable, use of language, and tips for the patients to ensure they can quickly reference the relevant emails. In addition, the guide advises that the physician should not use email communication with new or prospective clients with whom no personal contact has yet been established, should maintain the same ethical standards that apply to other areas of medicine, should ensure that permission has been obtained for email communication, and should ensure that the email has a disclaimer dealing with breaches of security and privacy, identity of corresponding parties and possible delays in responses.

PRACTICAL STEPS

Measures to Ensure Consent Forms and Other Documents Are Understood

This chapter has discussed informed consent at length, and research usually has to be accompanied by a consent form that is signed by the research participant. But what is the certainty that the participant has actually understood the contents of the consent form and other documents (e.g. survey forms)? In face-to-face research, the researcher can pose questions to ensure that the information has been understood; in online research, this is not always possible. (Even in face-to-face research, the use of questions can be embarrassing to the research subject, and time-consuming.) A useful approach is to reduce the complexity level of the language so that the person can understand the form.

For English, there are several tests used to determine the complexity of language in a text, although the most popular are the *Flesch Reading Ease Test*, and the *Flesch-Kincaid Test*.[78] Readers can find out more about these tests, but, essentially, the tests check various characteristics of a document, such as the average number of words in a sentence and the number of syllables in each word. The *Flesch Reading Ease Test* assigns a value of 1 – 100 (where 1 is most difficult, and 100 is easy), and the *Flesch-Kincaid Test* assigns a number that corresponds to the US school grade. This means that a document with a Flesch-Kinkaid Test score of 8 could be understood by an 8th-grader, while a score of 14 would be at university level.

There are several computer applications that can perform the test. If Microsoft Office is used, the test can be implemented in MS-Word, by making the following changes to the settings:

- MS-Word 2007:
  o Click on the **Office Button**
  o Select **Word Options**
  o Select **Proofing | When correcting spelling and grammar in Word**
  o Select **Show readability statistics**
- MS-Word 2010/2013/2016:
  o Click on **File**
  o Select **Options**
  o Select **Proofing | When correcting spelling and grammar in Word**
  o Select **Show readability statistics**

From now on, when a spelling and grammar check is run, the final dialog box will display the readability scores, such as the example in Figure 11.5:

In addition to the percentage of passive sentences, the dialog box will also give the Flesch-Reading Ease and the Flesch Kincaid Grade scores. These statistics can be used to modify documents, and the tests re-run until the results are satisfactory.

Note that it is easy to 'trick' the tests. The aim is not to trick the tests, but rather to use the results of the test as a guide for your own research. For instance, if the document referred to in Figure 9.5 were a document for university students, it would probably be suitable. If it were a consent form to be given to children, it would probably require extensive revision.

Chapter 11: Health Informatics Ethics

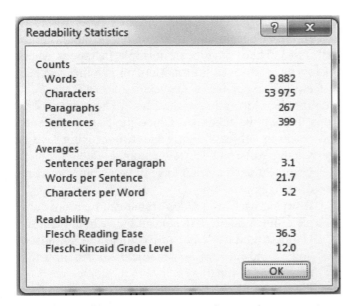

Figure 11.5: Readability statistics from a document in MS-Word 2016

Simple Data Protection

Security breaches at medical facilities occur on an almost daily basis.[79] While network administrators will probably implement several strategies to assist with security, users can also do their part. This is particularly important if one is using a computer in a shared location (such as at home) or using a laptop, which has a high risk of being stolen.

- Several encryption programs encrypt entire disk partitions, folders or individual files. VeraCrypt is a free encryption tool for Windows, Mac and Linux available from https://www.veracrypt.fr/en/Home.html. Combined with standard encryption is the practice of encrypting folders and files that contain nothing of value at all, designed to act as red herrings to lure would-be snoopers away from the real material, and waste their time. One should be aware that users may be legally compelled to reveal their encryption passwords.[80]
- Passwords and document encryption: most operating systems allow users to password-protect their computer. In addition, most word processing, spread sheet and database programs have in-built password and/or encryption protection.
- A computer anti-virus program should be used and kept up to date. There are several good, free anti-virus programs, such as AVG Free (https://www.avg.com/en-ww/homepage) and Avast (https://www.avast.com/free-antivirus-download).
- Anti-spyware and malware software should also be installed on every computer. A good, free anti-spyware program is SUPERAntiSpyware Free Edition (http://www.superantispyware.com/). Malwarebytes Anti Malware (free, available from http://www.malwarebytes.org/) is good anti-malware software.
- Before the computer is discarded or given away, ensure that all data are properly removed. Normal deletions, and removal from the Trash are not good enough, as these files can easily be restored. Users may wish to use several methods in combination, including defragmentation, formatting, and using a free tool like Eraser (http://eraser.heidi.ie/).
- Finally, one might wish to use virtual private networks (VPNs) or any of the mail services that offer protected email, or a plugin, such as Mailvelope (free), which uses encryption for email. These solutions, however, are sometimes costly or technically difficult, and also offer no guarantee, especially in light of recent revelations regarding encryption cracking.[81] If used, they should be seen as an extra layer of security only.

Professionalism in Social Networking sites

The general atmosphere of social networking sites like Facebook and Twitter is one of friendliness, community, joviality. Indeed, the very structure of a social network relies on the free flow of information.[82]

Health professionals, however, need to be aware that they have a professional presence, and that needs to be maintained. They should not be lulled in a false sense of security when social networks promise secure areas and privacy settings, as these secure areas are less secure than one might believe, privacy settings can change at a whim, and the end-user agreement usually allows the social network free access to the material. Research indicates that young medical professionals are not applying privacy restrictions or ensuring that potentially damaging personal information is not posted.[83] Material posted, even when "deleted," is both persistent and searchable. It is important to understand that information is currency and content; if it were entirely private, the social network would cease to exist.

In addition, social networks are not homogeneous, and what is considered acceptable in one network is not necessarily so in another. Even within networks, differences appear. For example, an acceptable posting in one area (subreddit) of Reddit might be unacceptable in another.

When using social media sites, these rules will help to guide users:

- If the decision is made to use social networking sites for personal information, be aware that patients will be able to see this and may react inappropriately. Physicians may choose to use these sites exclusively for professional purposes, as is frequently recommended,75 but even this route has difficulties.[84-85]
- Health professionals may deal with patients at their most vulnerable. Online descriptions and photographs of such situations should not be posted into social networking site, as was done by a Portland nursing assistant.[86-87] It is important to remember that such postings are generally not protected by free-speech laws.[88] (It should be noted, however, that patients frequently audio-record doctors, often surreptitiously.)[89] Your medical facility or system may have policies about photographs and videos. If they do not, two useful guides are that by the UK's General Medical Council[39] and the Doncaster and Bassetlaw Hospitals Guide.[90]
- If any videos are produced for YouTube or similar sites, ensure that all copyright rules are followed. Also insert subtitles (so that hard-of-hearing persons can access them), either by using YouTube's Caption feature, or software like Jubler (free) and Avidemux (free) to add sub-titles.
- Assume that all the rules of patient and research subject confidentiality apply and be aware of other laws such as those concerning libel and copyright. This will apply even if one re-tweets somebody else's tweet. See Scanlon 2012[91] for some of the general legal risks of using Twitter in the UK. Other countries will have similar laws.
- There have been several documented cases of employers requiring staff to reveal their Facebook and other passwords. One needs to be aware of the ethical obligations and the relevant laws in your country regarding the response to such demands.

Removal Of Identifying Material From Electronic Files And Databases

While electronic files are invaluable for medical and health research, great care should be taken to remove identifiable information from these files and databases. This is always important, but even more so when working with conditions (such as HIV/AIDS or psychiatric conditions) that have social stigmas.[54] There are steps to be taken to anonymize data. Unfortunately, the anonymization process is not fool-proof, and the processes of "de-anonymizing" by combining various snippets of information from different databases are also sophisticated.[13,92-95]

Some information (e.g. a patient's face, sound of voice), might be obvious to remove, but others might be less so. There have been multiple instances where doctors have removed information, only to find that lack of technical expertise or experience has left traces of identifiable information in the files.[96] There are several steps that can be taken to reduce the risk of patients' and research subjects' being identified, using free or inexpensive software. These include:
- Using Paint.Net (free) to remove or blur out parts of images.
- Blurring parts of a video frame can be more complex, but Avidemux (free) can blur whole frames.
- Using Audacity (free) to disguise the sounds of voices and remove non-medical information that may identify a patient.
- Using Easy Exif Delete (free) to remove exif metadata (e.g. author, longitude and latitude) that is automatically embedded in .jpg photograph files).
- Ensuring filenames are anonymous. Researchers usually take care to ensure that the patient cannot be identified; for easy classification, however, file names often contain patients' names, and these should also be anonymized.

Limiting Collection Of Visitor Data To Your Website

If an individual or organisation has a website, it is useful to gather information about your visitors. This can be accomplished either by creating your own tracking cookies, or by using some of the many third-party tracking tools. This can be done to collect a wide range of visitor data, usually without consent or even notification, and easily distribute the data to third-party vendors who are not held accountable.

Although this is widely practised, even by well-respected world medical organisations,[97] it violates several laws and ethics' principles. If visitor information is gathered, one needs to obtain the visitors' consent, ensure that only exactly what is needed is gathered, and stipulate clearly:
- What information will be gathered
- How it will be stored and secured
- With whom it will be shared
- For how long it will be kept until being destroyed.

If the website is being run by a third party, one should ensure that these requirements are all met. This can be tested using a free browser plugin like *Ghostery*, to analyse the various data collection systems that are running on the web site

Disguising Of Web Sites Or Bulletin Boards Researched

Unfortunately, no matter how much one tries to protect research subjects, there will be some people who will attempt to discover their true identity.[58] This may because they view it as their right, or do not work to the same ethical model, or who simply see it as a detective game to be played. In medical research, there is a tradition of intentionally changing information in order to prevent people using it to identify the subjects. For example, one may change people's names, cities or even experiences and medical conditions, as long as it does not directly impact on the nature of the research. This is considered "heavy disguise,"[63] and is also employed by some researchers who research web sites.[98–100]

If research is being performed on bulletin boards in a web site, and there is a wish to disguise the research web site so that it has little chance of being discovered, one may wish to create a dummy web site (or "Maryut site"[58]) that is specifically designed to lure investigators away from the real site. The ethics of making Maryut sites, in the interests of safe-guarding the non-disclosure, may be a point of discussion by IRBs and ethics' committees. A danger with this method is that the Maryut site may inadvertently point the detectives to a valid secondary site that has nothing at all to do with the research.

Ensure Irbs, Ethics Committees And Other Administrative Structures Are Aware Of Health Informatics Ethics Issues

This chapter has already referred to the fact that representatives on IRBs may not be familiar with ethical issues in health informatics. It is the responsibility of researchers and other HIPs to inform their IRBs of the health informatics ethics issues. This will allow the IRB to better understand the preventative actions taken by HIPs, and also to understand the motivations behind such actions.

This information-sharing can take the form of workshops and reports, supported by practical implementation when applying for research and grant approval.

HEALTH INFORMATICS ETHICS AND THE MEDICAL STUDENT

The final portion of this chapter deals with medical students. Medical students are already health professionals, simply at the early stages of their careers. Medical students are usually bound by the rules of their national medical professional organisations; in the same way, they should feel bound by the ethical rules of health informatics.

Online Behaviour in Social Networking and Other Interactive Sites

Generally, medical schools' social media policies lag behind technical reality, although some schools have produced some guidelines.[101] As health professionals, students need to remember that everything they post online may stay online for a long time – even if it is deleted, it may be stored in electronic archives. With this in mind, students should be extremely careful about online comments and photographs of themselves, colleagues and patients. (This applies to all students, not only medical students).[102]

A survey amongst US medical schools recently found that as many as 60% reported incidents in which students had posted inappropriate material online. Students were usually given warnings, but, in some cases, were dismissed.[103] Figure 11.6 is a Twitter post by a student who received a poor mark on an app project and reacted by threatening the lecturer with death. (The Arabic reads: *"[Lecturer's Name], We worked on that App for more than 2 hours and the final result is like that!"*). A posting like this might have dire consequences.

Figure 11.6: Student tweet threatening a lecturer for a poor project grade. (Source: Twitter)

There is something of a tradition of medical students' posing with their cadavers, sometimes merely as illustrations for teaching, but very often for other purposes, such as humour.[104] Times have changed, however, and the posting of such photographs on to the Internet is generally not accepted by medical schools.[105] (see Figures 11.7 and 11.8)

Figure 11.7: Dissecting Room, Jefferson Medical College, Philadelphia. (Source: US Library of Congress)

Figure 11.8: Students and teachers of Civil Medical School, Constantinople. (Source: US Library of Congress)

Other Student Activities

Use of mobile devices with cameras. While performing clinical studies, and in the presence of patients, one may wish to look up medical information on the mobile device. Although users are simply using the reference software, be aware that patients may have concerns of privacy if seeing a device with a camera being used. When using such devices, ensure that the camera faces away from any patients; if possible, stand at an angle that would allow them to see the device screen. If there is any doubt, let the patient know that a general disease database is being accessed.

Research projects. Students may be involved in several research projects. Some of these may be small projects performed by one or more students (e.g. surveys of fellow students), or they may be parts of larger projects set by other researchers. In all cases, it is important to remember that one is bound by the same ethical rules that are raised in this chapter. In cases of doubt, speak to the staff, one's supervisor, or members of the Ethics Committee or IRB.

Plagiarism. Similarly, professional conduct extends to plagiarism. Although plagiarism is not specific to health informatics, because of the ease with which information can be copied-and-pasted, there is a temptation to plagiarise others' material. Many universities submit papers to software programs that detect plagiarism. At most institutions, students found guilty of plagiarism may be expelled from their institutions.

Use of Paper Mills. It may be tempting to make use of "Paper Mills." Paper Mills are web sites that allow students to submit assignment details, and somebody else will write the assignment (for a fee). Again, however, if students are caught, they are usually expelled immediately.

Manipulation of electronic files. Electronic files (whether text, audio, video, or still graphics) are easy to manipulate. There are many acceptable reasons for doing so (see above). When performing such manipulation, one must ensure that copyright or other laws are not broken. In addition, one must ensure that the finished product does not present false information.

Recording of lectures and other class activities. It may be tempting to video- or audio-record classes or lectures. This may be useful for students, so that they can watch and/or listen to the lectures afterwards. Before doing this, one must ensure that the lecturer's consent is obtained. In most cases, the permission from fellow students will be needed. As a guide, refer to the discussion on informed consent above.

Using pirated, 'cracked' or other illegal digital files. In most countries, the use of pirated, 'cracked' software or other digital files (e.g. downloaded from 'torrent' sites) is illegal. Being caught performing such operations or with such software will usually lead to prosecution and expulsion. Frequently, such software is also a back-door for viruses and other unwanted software.

Accessing documents illegally. Frequently, while performing research, one will find an abstract to an article, and will wish to read that article. Unfortunately, a great deal of information is available in books or journals that charge subscription fees (i.e. the requestor has to pay to access the journal or individual articles or books). In most cases, a university library will already have paid a fee, and will be able to grant legal access to these resources. In some instances, however, they have not, and so requestors do not have legal access to the resource.

Because students (and qualified doctors) want access to these resources, they are tempted to use sites that share such articles.[106-107] Alternately, they use websites that share usernames and passwords to library databases.[108] The justification for doing so is that, ultimately, patients will benefit from the knowledge that the health professional has gained. Unfortunately, this practice is usually both illegal and unethical.

There are other, both legal and ethical, methods that one may wish to try. These include:
- Searching for the resource in a legitimate site. Frequently, publishers allow authors to place copies of their articles on their own web sites, and in publicly-accessible repositories. These can be searched and the articles freely downloaded.
- Contacting the authors. Authors are usually permitted to send copies of their articles to a limited number of people who request them. Students can contact the author and make such a request. (The author's contact details will usually be visible on the same page that showed the abstract).

FUTURE TRENDS

Because health informatics ethics relies on practices from diverse and continually-evolving fields, it is difficult to make predictions about future trends. That said, however, based on the recent history, there are a few likely scenarios:
- Because medicine and informatics are diverse fields, balancing the ethical practices of health informatics will always be difficult for the HIP. The various codes of ethics will be continually updated to take technological developments into account but will always lag some way behind these developments.
- Digitised medical data will play an increasingly important role as a commodity in patients' lives.
- Because of these changes and the continual emergences of new technologies, HIPs (including students) will be faced with new ethical challenges. They will need to use the basic principles as guides, and their consciences where the principles do not take these developments into account.
- The tension between ethics, culture and law will not become easier in the short term.
- Health informatics ethics is likely to emerge as a field of study by itself.

KEY POINTS

- Health informatics ethics stems from medical ethics and informatics ethics and combines principles from both fields.
- The IMIA Code of Ethics for Health Information Professionals contains guidelines in a range of categories, namely: fundamental ethics principles, general principles, subject-centred duties, and duties towards HIPs, institutions/employees, society, the profession and oneself.
- The relationship between ethics, law, culture and society is extremely complex and fluid, and varies internationally and chronologically. The HIP must become acquainted with the issues that have a direct bearing on his or her practices.
- The most pertinent ethical principles in health informatics ethics relate to: right to privacy, guarding against excess, security and integrity of data, informed consent, data sharing, wider responsibilities, beneficence and non-maleficence and non-transferability of responsibility. All these must be seen within the legal and medical ethics context.
- There are several examples of the application of the principles to research and other situations. These applications can be used as a guide for the HIP, beginning with the HIP as a student.

CONCLUSION

At Nuremberg, a total of 23 defendants (of whom 20 were medical doctors) were tried for medically-related crimes. Seven were acquitted of all charges. Of the 16 found guilty, seven were sentenced to death. These seven, including Dr. Karl Brandt, were executed on 2nd June 1948.

From one of the darkest periods of medical history, codes of ethics evolved. From these codes and codes in informatics ethics, health informatics ethics codes have further evolved. Although these codes have varying degrees of effectiveness, they do provide essential principles for the medical student and health informatics professional who will work with electronic data. It is essential that these principles are understood and applied as conscientiously as possible.

REFERENCES

1. Nuernberg military tribunals. trials of war criminals before the nuernberg military tribunals under control council law no. 10, volume ii: "the medical case" and "the milch case." washington, dc: us government printing office; 1949.
2. Nuernberg Military Tribunals. Trials of war criminals before the Nuernberg Military Tribunals under Control Council Law No. 10, Volume IV: "The Einsatzgruppen Case" and "The Rusha Case." Washington, DC: US Government Printing Office; 1949.
3. The Library of Congress. Nuremberg Trials. 2014 [cited 2017 Jul 11]. Available from: https://www.loc.gov/rr/frd/Military_Law/Nuremberg_trials.html
4. Nuernberg Military Tribunals. Trials of war criminals before the Nuernberg Military Tribunals under Control Council Law No. 10, Volume I: "The Medical Case." Washington, DC: US Government Printing Office; 1949.
5. Hohendorf G, Rotzoll M, Richter P, Eckart W, Mundt C. Die Opfer der nationalsozialistischen „Euthanasie-Aktion T4". Nervenarzt. 2002;73:1065–74.
6. Mostert M. Useless Eaters: Disability as genocidal marker in Nazi Germany. J Spec Educ. 2002;36(3):155–68.
7. Proctor R. Racial Hygiene: medicine under the Nazis. Cambridge: Harvard UP; 1988.
8. World Medical Association. World Medical Association Declaration of Helsinki - Ethical Principles for Medical Research Involving Human Subjects [Internet]. 2013 [cited 2017 Jun 21]. Available from: https://www.wma.net/policies-post/wma-declaration-of-helsinki-ethical-principles-for-medical-research-involving-human-subjects/
9. Capurro R. Towards an ontological foundation of information ethics. Ethics Inf Technol. 2006;8(4):175–86.
10. Wiener N. The Human Use of Human Beings: Cybernetics and Society. Cambridge, Massachusetts: Da Capo Press (Originally published in 1950); 1988.
11. Kostrewski BJ, Oppenheim C. Ethics in information science. J Inf Sci [Internet]. 1980;1(5):277–83. Available from: http://jis.sagepub.com/cgi/doi/10.1177/016555157900100505
12. Hauptman R. Ethical challenges in librarianship. Phoenix, Arizona: Oryx Press; 1988.
13. Mason RO. Four Ethical Issues of the Information Age. MIS Q. 1986;10(1):5–13.
14. Severson R. The Principles of Information Ethics. New York: M.E. Sharp; 1997.
15. Floridi L, Taddeo M. What is data ethics? Phil Trans R Soc A. 2016;374:20160360.
16. Association for Computing Machinery. ACM Code of Ethics and Professional Conduct [Internet]. 1992 [cited 2017 Jul 11]. Available from: http://www.acm.org/about-acm/acm-code-of-ethics-and-professional-conduct
17. Canadian Information Processing Society. Code of Ethics and Professional Conduct [Internet]. 2007 [cited 2017 Jul 11]. Available from: http://www.cips.ca/?q=system/files/CIPS_COE_final_2007.pdf
18. International Medical Informatics Association. The IMIA Code of Ethics for Health Information Professionals [Internet]. 2016 [cited 2017 Jul 11]. Available from: http://imia-medinfo.org/wp/wp-content/uploads/2015/07/IMIA-Code-of-Ethics-2016.pdf
19. Katz J. The consent principle of the Nuremberg Code: Its significance then and now. In: Annas G, Grodin M, editors. The Nazi doctors and the Nuremberg Code: human rights in human experimentation. New York: Oxford UP; 1992.
20. Harkness JM. Nuremberg and the issue of wartime experiments on US prisoners. J Am Med Assoc. 1996;276(20):1672–5.
21. Drea E, Bradsher G, Hanyok R, Lide J, Petersen M, Yang D. Researching Japanese War Crimes Records. Washington, DC: National Archives and Records Administration for the Nazi War Crimes and Japanese Imperial Government Records Interagency Working Group; 2006.
22. Lafleur W, Böhme G, Shimazono S. Dark Medicine: Rationalizing Unethical Medical Research. Bloomington: Indiana UP; 2007.
23. N.C. Justice for Sterilization Victims Foundation. North Carolina Eugenics Informational Brochure [Internet]. 2010 [cited 2017 Jul 11]. Available from: https://www.documentcloud.org/documents/272335-north-carolina-eugenics-informational-brochure.html
24. Schoen J. Choice & Coercion. Birth control, sterilization and abortion in Public Health and Welfare. Chapel Hill & London: The University of North Carolina Press; 2005. 353 p.
25. Severson K. Thousands Sterilized, a State Weighs Restitution. The New York Times. 2011;
26. Motejl O. Final Statement of the Public Defender of Rights in the Matter of Sterilisations Performed in Contravention of the Law and Proposed Remedial Measures [Internet]. 2005. Available from: https://www.ochrance.cz/fileadmin/user_upload/ENGLISH/Sterilisation.pdf
27. European Roma Rights Centre. Coercive and Cruel: Sterilisation and its Consequences for Romani Women in the Czech Republic (1966-2016) [Internet]. 2016. Available from: http://www.compas.org.uk/wp-content/uploads/2017/03/ERRC-Sterilisation-and-its-Consequences-for-

Romani-Women-in-the-Czech-Republic-1966-2016.pdf
28. US Department of State. Country Reports on Human Rights Practices 2009, Vol. 2: Europe and Eurasia, Near East and North Africa. Washington, DC: US Government Printing Office; 2012.
29. Reverby S. Could It Happen Again? Postgrad Med J. 2001;77:553–4.
30. Reverby S. A new lesson from the old "Tuskegee" study. Huffington Post. 2009.
31. Reverby S. Examining Tuskegee: the infamous syphilis study and its legacy. Chapel Hill: University of North Carolina Press; 2009.
32. McNeil D. U.S. Apologizes for Syphilis Tests in Guatemala [Internet]. New York Times. 2010 [cited 2017 Jul 11]. Available from: http://www.nytimes.com/2010/10/02/health/research/02infect.html
33. Leung R. A Dark Chapter in Medical History: Vicki Mabrey On Experiments Done On Institutionalized Children [Internet]. 60 Minutes. 2010 [cited 2017 Jul 11]. Available from: http://www.cbsnews.com/news/a-dark-chapter-in-medical-history-09-02-2005/
34. Advisory Committee on Human Radiation Experiments. Final Report [Internet]. Washington, DC; 1995. Available from: https://fowlchicago.files.wordpress.com/2014/02/advisorycommitte00unit.pdf
35. Clare A. Medicine Betrayed: The Participation of Doctors in Human Rights Abuses. London: Zed Books; 1992.
36. Richards E. Transcript of Tarasoff v. Regents of University of California, 17 Cal. 3d 425, 551 P.2d 334, 131 Cal. Rptr. 14 (Cal. 1976) [Internet]. 1976. Available from: http://biotech.law.lsu.edu/cases/privacy/tarasoff.htm
37. Goodman KW, Adams S, Berner ES, Embi PJ, Hsiung R, Hurdle J, et al. AMIA's Code of Professional and Ethical Conduct. J Am Med Informatics Assoc. 2013;20(1):141–3.
38. UK Council for Health Informatics Professions. UKCHIP Code of Conduct [Internet]. 2017 [cited 2017 Jul 11]. Available from: http://www.ukchip.org/?page_id=1607
39. General Medical Council. Making and using visual and audio recordings of patients [Internet]. The Journal of audiovisual media in medicine. 2011. Available from: http://www.ncbi.nlm.nih.gov/pubmed/12554297
40. European Parliament. Directive 95/46/EC of the European Parliament and of the Council of 24 October 1995. Brussels; 1995.
41. European Parliament. Directive 2006/24/EC of the European Parliament and of the Council of 15 March 2006. Brussels; 2006.
42. European Parliament. Directive 2009/136/EC of the European Parliament and of the Council of 25 November 2009. Brussels; 2009.
43. European Commission Article 29 Data Protection Working Party. Opinion 04/2012 on Cookie Consent Exemption. Brussels; 2012.
44. The European Parliament, The European Council. General Data Protection Regulation. Off J Eur Union. 2016;L119:1–88.
45. de Lusignan S, Chan T, Theadom A, Dhoul N. The roles of policy and professionalism in the protection of processed clinical data: A literature review. Int J Med Inform. 2007;76(SUPPL. 1):261–6.
46. AMA. Guidelines for Physician-Patient Electronic Communications [Internet]. 2008 [cited 2017 Jul 12]. Available from: https://www.google.com/url?sa=t&rct=j&q=&esrc=s&source=web&cd=1&cad=rja&uact=8&ved=0ahUKEwjn5NX7mIPVAhWMChoKHTQcDY4QFggnMAA&url=http%3A%2F%2Fwww.tennlegal.com%2Ffiles%2F430%2FFile%2Fama_Guidelines_for_Physician_electronic_communication.doc&usg=AFQjCNGv-2T
47. Standards Australia. Guide to Australian electronic communication in health care. Sidney, Australia: Standards Australia; 2007.
48. European Commission. Research ethics and social sciences. 2010.
49. Waltz C, Strickland O, Lenz E. Measurement in Nursing and Health Research. 4th ed. New York: Springer; 2010.
50. Petrini C. "Broad" consent, exceptions to consent and the question of using biological samples for research purposes different from the initial collection purpose. Soc Sci Med. 2010;70(2):217–20.
51. Chadwick R, Berg K. Solidarity and equity: new ethical frameworks for genetic databases. Nat Rev Genet. 2001;2(April):318–21.
52. Gulcher J, Stefánsson K. The Icelandic healthcare database and informed consent. New Engl J Med Engl J Med. 2000;342:1827–30.
53. Gertz R. An analysis of the Icelandic Supreme Court judgement on the Health Sector Database Act. SCRIPT-ed. 2004;1(2):241–58.
54. Simon GE, Unutzer J, Young BE, Pincus HA. Large medical databases, population-based research, and patient confidentiality. Am J Psychiatry. 2000;157(11):1731–7.
55. Bernard G, Artigas A, Brigham K, Carlet J, Falke K, Hudson L, et al. The American-European Consensus Conference on ARDS. Am J Respir Care. 1994;149:818–24.
56. United States Department of Health & Human Services. Summary of the HIPAA Privacy Rule [Internet]. 2013 [cited 2017 Jul 12]. Available from:

https://www.hhs.gov/hipaa/for-professionals/privacy/laws-regulations/index.html

57. Eysenbach G, Till JE. Ethical issues in qualitative research on internet communities. Br Med J. 2001;323:1103.
58. Masters K. Non-disclosure in Internet-based research: the risks explored through a case study. Internet J Med Informatics. 2009;5(2):1–12.
59. Walther JB. Research ethics in Internet-enabled research: human subjects issues and methodological myopia. Ethics Inf Technol. 2002;4(4):205–16.
60. Office For Human Research Protection. Protection of Human Subjects. Code Fed Regul. 2009;
61. Wimsatt W, Beardsley M. The Intentional Fallacy. In: The Verbal Icon: Studies in the Meaning of Poetry. Lexington: University of Kentucky Press; 1954. p. 3–18.
62. Barthes R. Image Music Text. New York: Hill and Wang; 1977.
63. Bruckman A. Studying the amateur artist: A perspective on disguising data collected in human subjects research on the Internet. Ethics Inf Technol. 2002;4(3):217–31.
64. Bassett EH, O'Riordan K. Ethics of Internet Research: Contesting the Human Subjects Model. Ethics Inf Technol. 2002;4(1):233–47.
65. Eysenbach G, Wyatt J. Using the Internet for surveys and health research. J Med Internet Res. 2002;4(2):76–94.
66. Frankel MS, Siang S. Ethical and Legal Aspects of Human Subjects Research on the Internet. Washington, DC; 1999.
67. Jankowski N, van Selm M. Research ethics in a virtual world. Guidelines and illustrations. In: Carpentier N, Pruulmann-Vengerfeldt P, Nordenstreng K, Hartmann M, Vihalemm P, Cammaerts B, et al., editors. Media technologies and democracy in an enlarged Europe. Tartu: Tartu University Press; 2007.
68. Kramer ADI, Guillory JE, Hancock JT. Experimental evidence of massive-scale emotional contagion through social networks. Proc Natl Acad Sci. 2014;111(24):8788–90.
69. Arthur C. Facebook emotion study breached ethical guidelines, researchers say. The Guardian. 2014;
70. University of North Carolina. The Carolina Mammography Register Server Compromise: Frequently Asked Questions.
71. Ferreri E. Breach costly for researcher, UNC-CH. News & Observer. 2011;
72. Circuit Court For the 16th Judicial Circuit KC Illinois. Complaint: John Doe 1 et al. v. Open Door Clinic of Greater Elgin [Internet]. 2010. Available from: http://oldarchives.courthousenews.com/2010/03/04/Medical.pdf
73. Ferguson T, e-Patient Scholars Working Group. E-Patients: How They Can Help Us Heal Healthcare. Patient advocacy Heal care Qual [Internet]. 2007; Available from: http://e-patients.net/e-Patient_White_Paper_2015.pdf
74. Masters K. AMEE Guide: Preparing Medical Students for the e-Patient. Med Teach. 2017;39(7):681–5.
75. Herron PD. Opportunities and ethical challenges for the practice of medicine in the digital era. Curr Rev Musculoskelet Med. 2015;8(2):113–7.
76. Genes N, Appel J. The ethics of physicians' web searches for patients' information. J Clin Ethics. 2015;26(1):68–72.
77. AMA. Chapter 2: Opinions on Consent, Communication & Decision Making. In: AMA Code of Medical Ethics [Internet]. AMA; 2016. Available from: https://www.ama-assn.org/sites/default/files/media-browser/code-of-medical-ethics-chapter-2.pdf
78. Kincaid JP, Fishburne RP, Rogers RL, Chissom BS. Derivation of New Readability Formulas (Automated Readability Index, Fog Count and Flesch Reading Ease Formula) for Navy Enlisted Personnel. Research Branch Report 8-75. Memphis, TN: Chief of Naval Technical Training: Naval Air Station Memphis; 1975.
79. DataBreaches.net. DataBreaches.net: [Search Medical] [Internet]. 2017 [cited 2017 Jul 12]. Available from: https://www.databreaches.net/?s=Medical&searchsubmit=
80. Criminal Case No. 10-cr-00509-REB-02. USA vs. Ramona Camelia Fricosu. District of Colorado, USA. 2012.
81. Ball J, Borger J, G G. Revealed: how US and UK spy agencies defeat internet privacy and security. The Guardian. 2013;
82. Papacharissi Z, Gibson PL. Fifteen Minutes of Privacy: Privacy, Sociality, and Publicity on Social Network Sites. In: Reinecke L, Tepte S, editors. Privacy Online: Theoretical Approaches and Research Perspectives on the Role of Privacy in the Social Web. New York: Springer; 2011. p. 75–89.
83. MacDonald J, Sohn S, Ellis P. Privacy, professionalism and Facebook: A dilemma for young doctors. Med Educ. 2010;44(8):805–13.
84. Farnan JM, Sulmasy LS, Worster BK, Chaudhry HJ. Online Medical Professionalism : Patient and Public Relationships : Policy Statement From the American College of Physicians and the. Ann Intern Med. 2013;158(8):620–7.
85. DeCamp M, Koenig TW, Chisolm MS. Social Media and Physicians' Online Identity Crisis. JAMA J Am Med Assoc. 2013;310(6):581–2.
86. Boroff D. Portland nursing assistant sent to jail after posting dying patient's buttocks on

Facebook [Internet]. New York Daily News. 2012 [cited 2017 Jul 12]. Available from: http://www.nydailynews.com/news/national/portland-nursing-assistant-jail-posting-dying-patient-buttocks-facebook-article-1.1035207

87. King R. Female nursing assistant is jailed and banned from using websites for posting dying patient's buttocks on Facebook [Internet]. Mail Online. 2012 [cited 2017 Jul 12]. Available from: http://www.dailymail.co.uk/news/article-2111935/Nursing-assistant-patients-buttocks-Facebook-jailed.html

88. 09-6008 ON. Nina Yoder v. University of Louisville. U.S. Court of Appeals for the Sixth Circuit: Western District of Kentucky at Louisville. 2011.

89. Rodriguez M, Morrow J. Ethical implications of patients and families secretly recording conversations with physicians. JAMA J Am Med Assoc. 2015;313(16):1615–6.

90. Doncaster and Bassetlaw Hospitals (NHS). Photography and Video Policy : to Govern Clinical and Non-clinical Recordings [Internet]. NHS; 2013 [cited 2017 Jul 12]. Available from: https://www.dbth.nhs.uk/wp-content/uploads/2017/07/PAT-PA-14-v.4-Photography-and-Video-Policy-FINAL.pdf

91. Scanlon L. Twitter and the law: 10 legal risks. The Guardian. 2012;

92. McCallister E, Grance T, Scarfone KA. Guide to protecting the confidentiality of Personally Identifiable Information (PII). Gaithersburg, MD: National Institute of Standards and Technology; 2010.

93. Narayanan A, Shmatikov V. De-anonymizing social networks. Proc - IEEE Symp Secur Priv. 2009;173–87.

94. Ohm P. Broken Promises of Privacy: Responding to the Surprising Failure of Anonymization. UCLA Law Rev. 2010;57(6):1701–77.

95. Seeney L. Computational Disclosure Control : A Primer on Data Privacy Protection [PhD Dissertation]. Massachusetts Institute of Technology; 2001.

96. Patrick R. St. Louis-area women sue surgeon after she puts photos of their breasts on the Web. St Louis Post-Dispatch. 2012;

97. Masters K. The gathering of user data by national Medical Association websites. Internet J Med Informatics. 2012;6(2):http://ispub.com/IJMI/6/2/14386

98. Turkle S. Constructions and Reconstructions of Self in Virtual Reality. Mind, Cult Act. 1994;1(3):158–67.

99. Turkle S. Life on the Screen: Identity in the Age of the Internet. New York: Simon & Schuster; 1995.

100. Turkle S. Multiple subjectivity and virtual community at the end of the Freudian century. Sociol Inq. 1997;67(1):72–84.

101. Harrison B, Gill J, Jalali A. Social Media Etiquette for the Modern Medical Student : A Narrative Review. Int J Med Students. 2014;2(2):64–7.

102. Krebs B. Court rules against teacher in MySpace "Drunken Pirate" case. Washington Post [Internet]. 2008; Available from: http://voices.washingtonpost.com/securityfix/2008/12/court_rules_against_teacher_in.html

103. Chretien KC, Greysen SR, Chretien J-P, Kind T. Online Posting of Unprofessional Content by Medical Students. Jama. 2009;302(12):1309–15.

104. Warner JH, Edmonson JM. Dissection: photographs of a rite of passage in American medicine, 1880-1930. New York: Blast Books; 2009.

105. Einiger J. Cadaver photo comes back to haunt resident. ABC Eyewitness News [Internet]. 2010; Available from: http://abc7ny.com/archive/7253275/

106. Masters K. Opening the non-open access medical journals: Internet-based sharing of journal articles on a medical website. Internet J Med Informatics. 2009;5(1):http://ispub.com/IJMI/5/1/6971

107. Masters K. Articles shared on a medical web site – an international survey of non-open access Background. Internet J Med Informatics. 2009;5(2):http://ispub.com/IJMI/5/2/13669

108. Masters K. Opening the closed-access medical journals : Internet- based sharing of institutions ' access codes on a medical website. Internet J Med Informatics. 2009;5(2):http://ispub.com/IJMI/5/2/6358

12

Consumer Health Informatics

WILLIAM R. HERSH • M. CHRIS GIBBONS • YAHYA SHAIHK • ROBERT E. HOYT

LEARNING OBJECTIVES

After reading this chapter, readers should be able to:

- Identify the origins of consumer health informatics (CHI)
- Discuss consumer health informatics tools
- Enumerate the features and format of personal health records
- Identify patient to physician electronic communication tools
- Outline CHI barriers and challenges
- Discuss future trends of CHI

INTRODUCTION

Consumer health informatics (CHI) is the area of health informatics focused on the interaction of consumers, patients, and others with health information systems and applications. This area of informatics emerged with the confluence of widespread availability of the Internet and online information resources with the consumer movement that aimed to empower those who were ill (patients) and not yet ill (consumers) with information to maintain and improve their health, as well as engage in the treatment of their disease. This chapter will define the terminology of CHI and then describe the key applications of CHI, with a focus on the personal health record (PHR) and patient-clinician communication. It will also discuss the efficacy of CHI tools and interventions as well as describe future models of CHI-driven healthcare.

DEFINITIONS

As with many informatics-related terms, there are not only a variety of terms that describe aspects of CHI but diverging definitions of CHI.[1] Some recent overviews of the tools and models have been published.[2-3] According to the Web site of the American Medical Informatics Association (AMIA), CHI is *"the field devoted to informatics from multiple consumer or patient views. These include patient-focused informatics, health literacy and consumer education. The focus is on information structures and processes that empower consumers to manage their own health--for example health information literacy, consumer-friendly language, personal health records, and Internet-based strategies and resources. The shift in this view of informatics analyzes consumers' needs for information; studies and implements methods for making information accessible to consumers; and models and integrates consumers' preferences into health information systems. Consumer informatics stands at the crossroads of other disciplines, such as nursing informatics, public health, health promotion, health education, library science, and communication science."*[4]

Recent years have seen the emergence of many additional terms related to CHI. A term that has been in use for several years is *mHealth*, which generally is an abbreviation for mobile health and typically refers to use of smartphones and other mobile devices.[5-6] Another term gaining traction is *digital health*, which has been defined in ways that are related to the integration of genomics and digital devices or to medical devices from the perspective of the US Food and Drug Administration.[7-8] An additional term that is frequently used is *participatory medicine*, which is defined by the society that advocates for it as *"a movement in which patients and health professionals actively collaborate and encourage one another as full partners in healthcare."*[9]

Finally, a more healthcare-centric term is *patient engagement*, which is defined by the Office of the National Coordinator for Health IT (ONC) as consisting of systems and functions.[10] According to the ONC, patient engagement systems are mostly focused around providing patients an electronic copy of their discharge instructions, providing patients an electronic copy of their health record, and identifying and providing patient-specific educational resources. ONC also notes various functions, such as patients being able to view information from their health or medical record online, the ability to download information from their record, the ability to electronically transmit care and referral summaries to a third party, the ability to pay bills online, to request an amendment, to change or update their record, to schedule an appointment online, to request refills for prescriptions online, and to be able to submit patient-generated data to the health system, things such as blood glucose and weight measurements.

It has been known for over a decade that consumers want access to their health information online. A study by Deloitte in 2008 found that 60% of individuals surveyed wanted physicians to provide online access to their medical records and test results as well as online appointment scheduling.[11] One in four said they would actually pay more for that service. Three in four consumers expressed a desire to have expanded use of in-home monitoring devices, and online tools that would allow them to become more active in their care. Another question asked was, how many understood their health insurance coverage? And only 52% volunteered that they did. At that point in time, they asked, how many maintained a personal health record, either paper based or electronic, with only one in four responded affirmatively that they did.

PERSONAL HEALTH RECORDS

One important consumer health application is the personal health record (PHR). There have been many definitions of the PHR. The Markle Foundation provided an early definition in 2004, defining it as *"an electronic application through which individuals can access, manage, and share their health information, and that of others for whom they are authorized, in a private, secure, and confidential environment."*[12] The American Health Information Management Association (AHIMA) describes the PHR as an electronic lifelong resource of health information needed by individuals to make health decisions guided by these principles, individuals owning and managing their information, which comes from healthcare providers and the individual themselves.[13] It should be maintained in a secure private environment. The individual determines the rights of access to the information. And the PHR does not replace the legal record of the healthcare provider. It is believed that the first mention in the medical literature advocating for patient access to their medical records came in a paper in the *New England Journal of Medicine* in 1973.[14] There have also been some recent systematic reviews of research about the PHR.[15-16]

Tang categorized three types of PHRs.[17] The first and probably most common at this time is the *tethered* PHR, which is an extension of the healthcare provider's EHR. It provides access to some, usually not all, of the information for the individual from the EHR. It will also often allow communication with the provider. Some tethered PHRs may allow the patient to add information into the record, but most do not. Another term used to describe what is essentially a tethered PHR is *patient portal*. A typical patient portal is a subset of the PHR, in that it may not provide patient's access to their healthcare data.[18]

A second type is the *standalone* PHR, which is an isolated application. It may be on a mobile device or a website, but its key defining aspect is that it does not take information from other sources and only contains the information that the patient enters into it. The third type is the *interconnected* or *integrated* PHR. This is a separate application, but it has the ability to interact with one or possibly more provider EHRs. And it will allow test results, scheduling, data collection, etc.[19-20] Figure 12.1 shows a screen shot of a (tethered) PHR.

There are a number of policy issues related to the PHR. The Markle Foundation advocates that the PHR should be controlled by the individual who decides what can be accessed by whom, and for how long.[12] The PHR should contain information from one's lifetime and from all providers. It should be accessible anywhere, at any time, but also private and secure. Information should be transparent. As such, whoever entered or viewed information should be captured and viewable. There should also be easy exchange with other health information systems and professionals. The American College of Physicians (ACP), on the other hand, advocates that physicians should only be responsible for reading and acting on the tethered portion of the PHR and should not be obligated to read or act on the non-tethered part. The ACP also advocates that the physician should be compensated for time spent interacting with the PHR.[22]

A more recent development in the PHR is giving patients access to their clinical notes. This effort is called *OpenNotes* and aims to provide patients with access to the entirety of their medical record, including clinical notes.[23] OpenNotes was initially implemented in three academic centers across the US.[24] Depending on the center, anywhere from half to almost all of patients who signed up for the service actually accessed it. The initial research analysis found that a large majority felt they were in more control of their care. There was also found to be increased adherence to medication by patients using OpenNotes. About one-quarter expressed privacy concerns, and a small number (1-8%, depending on the center) felt confusion, worry, or offense as to what was in the notes. For the most part, physicians taking care of these patients did not perceive any increased time outside of visits, reports of changing content in notes, or requiring more time to write notes. Almost all patients and physicians who started using the system, continued doing so after the initial study ended.

Additional research has been published on the patient perspective from this project. One patient, in particular, a university professor who developed a serious illness, wrote a commentary noting an appreciation of the transparency and access to information.[25] Another study further looked into the patient concerns about privacy. About one-third of patients had privacy concerns both before and after using OpenNotes, but this did not deter usage. Patients were also found to be willing to share their notes with others in their families and other caregivers.

The OpenNotes system has also been implemented and studied in the VA system. It began with the study in one VA center. Patients who used the system found that it improved communication with their providers, it enhanced their knowledge of their health and self-care, it allowed for them to have greater participation in their decision making, and overall benefited their care.[26] When OpenNotes was rolled out across the entire VA, used by tens of thousands of veterans, there was similar positive acceptance.[27] There have been efforts across the United States to encourage wider use of OpenNotes, such as the Northwest OpenNotes Consortium, which is encouraging all healthcare providers to make their notes available in the Pacific Northwest.[28] Many PHR systems now support this kind of functionality when it is enabled.

An increasingly advocated view goes beyond patients viewing their data and notes to actually providing them ownership and control of their data. There is growing consensus that patients should be the owners and stewards of their personal health and healthcare data and also have the right to control access to it for chosen healthcare professionals, institutions, and researchers.[29] Current systems do not facilitate this point of view, as data is for the most part stored in the siloed EHR and other systems of the places where they obtain care. If we accept the view that patients own their data and can control access to it, how do we facilitate the transition from vendor-centric to patient-centric data storage? Does each person have a cloud-based store of their data, to which they grant access to their healthcare providers and others?

This change would have major implications for EHR systems. The current large monolithic systems would need to give way to those that access data in the

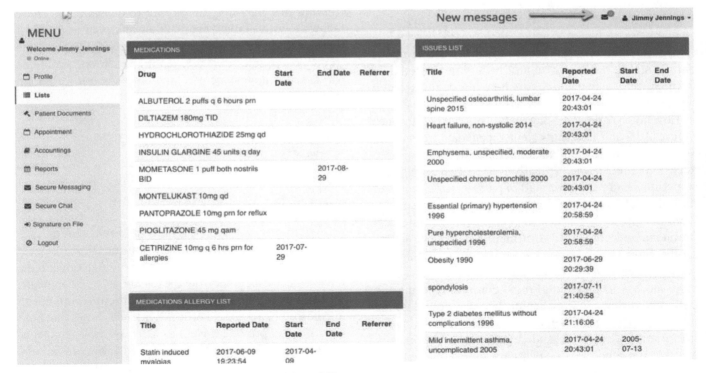

Figure 12.1: Tethered PHR (Courtesy LibreHealth EHR)[21]

cloud-based store. Instead of accessing data from within its own stores, the EHR would instead pull from and push back data to the patient's cloud-based store. There would also need to be integrity of the data, so it could not be altered as well as a model for how this approach would be paid for. Mikk lays out three necessary components for this vision to succeed:

- Standard data elements
- Standard data receipt for each clinical encounter with the push of the encounter into the patient's data store
- A contract that sets the rules for access and control for such a system[30]

PATIENT ENGAGEMENT

As noted above, an overview of patient engagement comes from a statistical brief from ONC that breaks patient engagement into systems and functions.[10] These systems are mostly focused around the meaningful use criteria of providing patients an electronic copy of their discharge instructions, providing patients an electronic copy of their health record within three business days, and identify and provide patient-specific educational resources. They also note various functions. So, for example, patients being able to view information from their health or medical record online, the ability to download information from their record, the ability to electronically transmit care and referral summaries to a third party, the ability to pay bills online, to request an amendment, to change or update their record, to schedule an appointment online, to request refills for prescriptions online, and to be able to submit patient-generated data to the health system, things such as blood glucose and weight measurements.

How much patient engagement is actually being done? Figures 12.2 and 12.3 show some recent data from ONC. We see that more than half of patients are now offered access to their data, typically one to two times, but sometimes more, and over half of patients have accessed their data at least once. Most commonly accessed are laboratory results, but also lists of health and medical problems and current lists of medications. The figure on the right shows various information activities that have been carried out by patients when they have been given access to their data, most commonly using the online record to monitor health, but also downloading information to a computer or mobile device, sending the information elsewhere, and sharing information with at least one other party, such as a family member, or healthcare provider, or someone else who is involved in care. We see that most commonly the sharing is with a family member.

The American Health Information Management Association has published a consumer health information bill of rights that advocates:[31]

- Look at your health information and/or get a paper or electronic copy of it
- Accurate and complete health information
- Ask for changes to your health information
- Know how your health information is used or shared and who has received it
- Ask for limitations on the use and release of your health information
- Expect your health information is private and secure
- Be informed about privacy and security breaches of your health information
- File a complaint or report a violation regarding your health information

Another concern is access to information versus privacy risk. A survey of over 2000 US consumers showed some preferences, with about half believing that the ability to access records electronically outweighed the risk of privacy invasion. A great majority believed that control over health information is important, but about half of them believed that they had little or no control. And concern about privacy invasion for health data was viewed as comparably to concerned about privacy invasion around banking and online shopping.[32]

Another study looked at preferences of the elderly, finding that they may be willing to delegate and share access to information, but that they also wanted granular control with some information being visible and some not.[33] Of course, that creates management problems in terms of how systems make only some information available.

PATIENT – CLINICIAN ELECTRONIC COMMUNICATION

Email Communication

An important part of patient engagement is a patient-clinician electronic communication. A great deal of healthcare has always been provided outside the face-to-face encounter from the telephone, or other means by which patients communicate with clinicians not in the hospital or clinician office. Evidence shows that both patients and providers want to use Internet-based technologies, but the success depends on a variety of factors, such as integration with the electronic health record and healthcare payment models that support their use.[34]

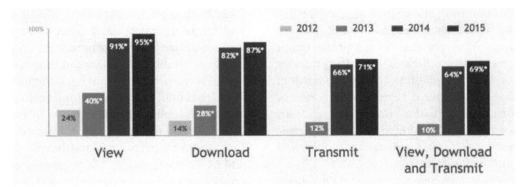

Figure 12.2: Proportion of US hospitals that allow patients to view, download, and/or transmit data[10]

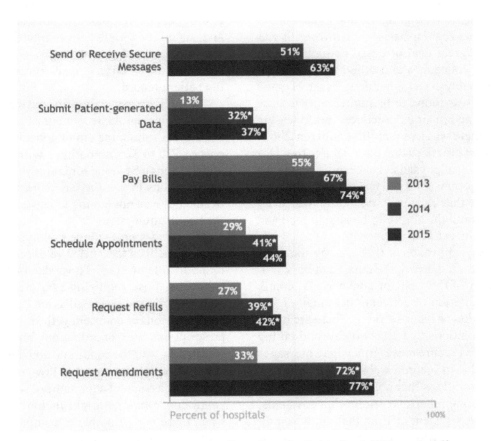

Figure 12.3: Patient engagement functions available from US hospitals[10]

Patients do want to communicate electronically with their clinicians and the healthcare system. A survey from the California Health Care Foundation found that 8% of consumers had emailed their healthcare provider. Of those who were PHR users, 69% were interested in emailing their providers, but had not done so. And even those not using PHRs, slightly over half, still expressed interest in emailing their providers. For those who had used the email, the most common reasons for doing so were to schedule an appointment, re-filling prescriptions, and asking about test results with the significant majority over two-thirds finding it useful and easy.[35] The desire for email with physicians was found in another survey lower than for other non-health Internet tasks, but was higher among women, those who are older, and those who have higher levels of education and income.[36]

More recent studies have shown higher interest in patient-physician email communication. Seth et al. found three-quarters of patients in an urban family medicine clinical in Ontario would be interested in receiving test results and other communications by email, especially among those who were frequent users of email.[37] There are some instances when patients do not prefer email notification of test results, which is when they convey potentially bad news.[38-39] A qualitative study by Giardina et al. verified that patients have negative emotions not only with abnormal test results but sometimes with normal ones as well. They concluded that communications must help patients interpret and manage their online test results and not just report them.[40]

There are, however, some physicians who have been less enthusiastic about email with patients. There were two surveys carried out in Florida before the HITECH era, which found that relatively small numbers used email to communicate with patients, and even smaller numbers did so on a regular basis.[41-42] However, it was found there was a great deal of email communication that went on with colleagues and others in their practices, which is probably even higher now. Usage of email among physicians was found to be higher among those who were younger and in larger practices, and, at least in these early days, there was less than 10% adherence to at least half of the best practices that have been defined for email. Another study of physician concerns included the overload of messages, no ability to triage them (although most systems have that capability now), the insecurity of standard email, and the ability to be able to read and reply to messages in a timely manner.[43]

There are many approaches that can be used for patient-clinician email. Clearly the email can be stand-alone, i.e., not part of the EHR or conventional email. Of course, one limitation of conventional email is that there is no encryption or other security. There are many encrypted email solutions that are available and taking a level higher there is secure messaging where messages are sent and stored by a secured website.

There are also solutions that are integrated with the EHR. Most of the comprehensive vendors have systems, such as the MyChart system of Epic that allows email integrated with the EHR, which, of course, has the advantage of tracking the messages and keeping them secure.[44] There are also secure messaging platforms that can be used, which may also include the ability to triage the message to appropriate individuals in the office.

A number of guidelines for patient-physician electronic communication have been developed over the years. In a critical appraisal of them, Lee et al. noted there was inadequate evidence to back up their recommendations and excess focus on technical and administrative concerns instead of how to use the process optimally.[45]

The American Medical Association established guidelines for physician-patient electronic communications. They were initially established in 2001, revised in 2008, and, for the most part, are still relevant at the present time. The guidelines for communication recommended that practices establish a turnaround time for when messages would be responded to, having patients exercise caution in urgent email messages, retaining all of the communications, establishing the types of transactions allowed by emails and the sensitivity of information within them that would be permitted, and having automated reply, and new messages for completions of requests.[46] There are also guidelines for medical, legal, and administrative issues-- one being the development of informed consent for use, and also the education and acceptable use of email, describing security to patients, encouraging them to use encryption if possible, not sharing email messages and capability with family, marketers, and so forth, and the proper archiving of messages-- ideally, incorporating them into the electronic health record, so they're part of the patient record.

Another longstanding discussion is whether email consultations should be reimbursable. A growing number of insurers reimburse email consultation, typically at a cost of $20 to $25, sometimes with a $5 to $10 co-pay, and there may be value for employers in terms of e-mail consultations in that there may be reduced absenteeism in an employee not having to leave work to consult with the healthcare system.

The American College of Physicians established a set of guidelines for when e-mail consultation should be reimbursable or not.[47] Those deemed not reimbursable would be those requiring two minutes or less, those reporting normal test results that don't require management decisions or counseling, the routine renewal of drug prescriptions or other orders, any kind of brief discussion concerning a stable condition, and questions concerning preparation for some sort of low-risk test that might be ordered as a result of a reimbursed service. Reimbursable email consultations might include the reimbursement when there is a principle of comparability to an actual visit to the physician, services involving a new diagnosis or new treatment, follow-up of maintenance services, reporting of test results that require significant change in treatment or further testing, and also extended counseling for situations requiring urgent contact where delay would lead to the patient being harmed.

There have been a number of studies looking at outcomes with patient-physician email, some of these were in very early days, and an early implementer of patient-physician email was the University of California Davis. They found that physician productivity increased with implementation of email,[48] and that physicians were able to have more visits and achieve more relative value units - a measure of physician productivity per day - when email was implemented.[49] A follow-on study found the telephone volume fell about 18%.[50]

Another early implementation of email came from California Blue Cross Blue Shield, where it was found there was an overall decrease in spending of $1.92 per patient per month for office visits, and $3.69 per patient per month overall.[51] Kaiser Permanente has been a heavy adopter of patient-physician email. An early study there showed that in clinics where it was implemented, there were 9.7% fewer office visits and 13.7% fewer phone contacts.[52] A subsequent study later on found there was increased quality of care, as measured by HEDIS measures, hemoglobin A1c, serum cholesterol, and blood pressure.[53] A more recent study found that 42% of patients who had access to email found it reduced their phone contacts, 36% found it reduced in-person office visits, and 32% felt it led to overall improved health.[54]

Telephonic and Audio-visual Communication

Electronic visits (e-visits or virtual visits) are an example of telehealth or telemedicine where medical care is delivered remotely (telemedicine is covered in much more detail in another chapter). Virtual visits are available as a continuum of care (Figure 12.4).

Figure 12.4: Remote patient communication continuum

A Price Waterhouse study estimated that 20% of outpatient visits could be eliminated by using e-visits.[55] Virtual visits have the advantages of much better security and privacy and the ability to have a third party involved in the billing process. The consensus is that minor complaints can be dealt with more efficiently electronically, thereby allowing sicker patients to be seen in person. Furthermore, patients miss less time from work for minor issues. It has also been pointed out that if the patient provides a history during the e-visit and still has to be seen face-to-face, the physician has the advantage of knowing why the patient is there, therefore saving time. Numerous vendors such as Intuit Health Patient Portal provide the platform for e-visits in addition to their patient portal features. Guidelines need to be established to define what constitutes an e-visit in order for insurance companies to reimburse for the electronic visit.

Questions remain about e-visits, regarding reimbursement (who will pay for what), privacy and what if initiating e-visits causes a drop in office visits leading to reduced office income?

The concept of virtual visits has spawned innovation in the delivery of healthcare as evidenced by the new e-visit vendors described below:

- TelaDoc began as a US-based telephonic consult service intended to supplement the care delivered by the primary care physician. It has since added audio-video visits and the service can be accessed via the web, smartphones or mobile app. The median wait time for service is 10 minutes. They cover common acute medical problems, dermatological and behavioral health issues, sexual health and tobacco cessation. They claim to have twenty million members and offer services 24/7. They have 3100 board-certified physicians in the service and have achieved an outstanding certification in quality from the NCQA. They offer their services also to employers and health plans.[56] A 2014 study of 3000+ TelaDoc patients noted that (compared to patients who had a face to face visit for a similar condition) TelaDoc patients were younger and less likely to have used health care before and less likely to have a follow-up visit to any setting.[57]

- American Well. Although there are similarities with TelaDoc, this vendor is offering a telemedicine platform. With a software development kit (SDK) healthcare systems can modify the American Well app to match their own organization. They also offer health information exchange so that medical information can be shared among disparate partners. Because of these enhancements and the fact that they can connect to 12 specialties, this platform can handle more than urgent care visits. One of their offerings is a healthcare kiosk where patients can contact American Well and this service could be part of an employer, retailer, or healthcare system. Multiway video permits up to 8 physicians or care givers to join the e-visit. They offer 24/7 access for patients from home and aim to coordinate care with the primary care clinician (PCM) and insurance company.[58]

- MDLIVE provides services that are very similar to TelaDoc and treats the same three categories of medical problems. They accept insurance payment and the maximum charge is $59.[59]

- Doctor on Demand is another telemedicine service that offers primarily acute care visits, but unlike TelaDoc and American Well they also offer limited chronic care, preventive care and laboratory tests.

Like the other programs they encourage memberships by retailers, employers and healthcare plans.[60]

Little has been reported about the medical value of e-visits. A 2017 article demonstrated that e-visits improve access to medical care but increase utilization and cost.[61]

An example of an e-visit service tethered to the enterprise EHR, is displayed in the Infobox below.

Virtual Visits at OHSU

Oregon Health & Science University (OHSU) began offering virtual visits for urgent care in 2017. Patients can establish a live video chat using their computer, tablet or smartphone. This service is offered from 7 am to 10 pm seven days a week and that includes children over age 1. A list of the common disorders treated is on the service Web site. (https://www.ohsu.edu/xd/health/services/virtual-care/). Appointment availability is posted, and patients just need to confirm the time. Consent forms are also available to complete before the visit.

The service is staffed by nurse practitioners and physician assistants. This service is only for residents of Oregon due to licensing laws. All or part of the virtual visit is covered by most insurance companies with a maximum charge of $49. All visits take place through the patient portal MyChart that is part of OHSU's Epic EHR system. The patient can see a summary of their visit afterwards and the same summary is forwarded to their primary care provider.

EFFICACY OF CONSUMER HEALTH INFORMATCS

There have been other studies of broader aspects of patient engagement other than e-mail. One systematic review looked at asynchronous patient-provider communication, and studies showed a variety of benefits, including self-efficacy for the patient and improved health outcomes, but the results of these studies were not unequivocal in that some showed benefit and others did not.[62] Another systematic review assessed patient access to medical records and clinical outcomes and found that the association was equivocal between access and improved outcomes.[63]

A survey done of over 2000 US adults with an established relationship with the physician found a number of findings, including the perceived value of their physician using an EHR. A growing number relative to previous surveys had online access to information from their physician, with about half reporting it available. Although there was a desire for additional functionality beyond that which was offered-- there was desire for things like email, appointments scheduling, accessing results, which not all patients had, but wanted. Another interesting finding is that there was less concern about privacy as the functionality of the system increased, as well as the trust in the system increased.[64]

There have been other studies of outcomes of patients having EHR data access. A study from Beth Israel Deaconess Medical Center in Boston found that 22% of all patients that were seen within the health system, enrolled in a portal. Of those who enrolled in the portal 37% sent at least one message to a physician, so 8.4% of all patients. Physicians saw a near tripling of email messages that they received from patients, but the number of messages per 100 patients per month stabilized at an average of about 18.9.[65] In another national survey, it was found that about 30% of patients have been provided access to their EHR data. And for those given access, about 46% actually accessed their records one or more times. And 70% of those who accessed their record found it to be valuable.[66]

A number of health systems now offer virtual or e-visits, and these have been studied. The Health Partners Health System in Minnesota and Wisconsin has an online clinic called virtuwell, and this is made available to patients for simple conditions. The first three years of the system's saw over 40,000 cases, and the resolution of most clinical conditions was comparable to a convenience walk-in clinic that was available to the patient. The analysis found a cost savings of about $88 per episode, which amounts to significant savings when involving tens of thousands of cases. And patient satisfaction was quite high, with almost all patients saying they would recommend the online clinic to others.[67]

Another study was more focused looking at sinusitis and urinary tract infection at the University of Pittsburgh, finding lower costs and higher satisfaction, but also increased use of antibiotics, which raises concerns that there may be an easier ability to prescribe antibiotics, which can be effective, but also lead to antibiotic resistance through e-visits.[68]

There have also been studies assessing the efficacy of PHRs. One study found that a medication review tool within the PHR leads to fewer medication discrepancies.[69] Another study found that patient-entered family history was more valid, than that obtained by the provider in a busy visit.[70] Another study showed that usage of PHRs was associated with improved adherence to well child care and immunization.[71] And finally, in the elderly, PHR usage has been shown to be associated with improved reconciliation of medications and recognition of side effects, but no difference in the use of appropriate medications or adherence measures.[72]

There are, however, some concerns about PHRs. One systematic review found that PHRs require a wide range of health literacy demands on patients and healthcare providers.[73] Patients' abilities to make best use of PHRs is dependent upon level of education and computer literacy, attitudes about sharing health information, and understanding of spoken and written language. The authors conclude that strategies for patients to meet the high health literacy demands of PHRs will be important in their large-scale adoption. One possible way to improve patient understand of notes is to provide easy-to-understand definitions of terms. Chen et al. have prototyped a system that applies natural language processing to notes with the aim of providing definitions of terms that patients can understand.[74]

Turvey et al. assessed the outcomes from giving patient's access to the VA Blue Button to enable them to download their CCDs and share them with providers outside the VA system. Those who were explicitly trained to download their CCD were much more likely to share it with non-VA providers, which resulted in a significant reduction in redundant laboratory tests although it had no impact in medication list concordance.[75]

Other studies have looked at efficacy of patient portals. One study addressed the concern of whether patients viewing test results online may have negative reactions, but this study found that a majority of patients have positive reactions when able to access their laboratory results online.[76] A systematic review looked at the whole spectrum of results from patient portals and found that the results were highly variable and depended on a number of factors. There were personal factors, such as patient age, ethnicity, education level, health literacy, health status as role as a caregiver. There were also healthcare delivery factors, such as provider endorsement and patient portal usability.[77]

Of course, the connection of patients and others to healthcare and health in general goes beyond patient engagement. Many are advocating more patient direct reporting of symptoms, findings, and other data directly into the EHR.[78] Use of these patient-reported outcomes is small but growing, and key questions revolve around whether it can improve care processes, efficiency of care delivered, or ideally patient outcomes.[79]

There are also all of the wearable devices we now wear and collect data from. Can or should we upload this data to the EHR? Physiological sensor data has been shown to be pertinent to care in some conditions, such as inflammatory diseases and diabetes but larger questions remain about how physicians and others in the healthcare system will be able to cope as well as be legally responsible for the increasing amounts of data that can be entered into the EHR by patients and their devices.[80]

CHI AND HEALTHCARE REFORM

The CHI field is evolving at an exceedingly fast pace. Providers and other stakeholders are using digital health tools. Furthermore, the use of technologies such as smart phones, social networks and apps are providing innovative ways to monitor health and well-being, providing greater access to information as well as leading to a convergence of people, information, technology and connectivity to improve health care and health outcomes.[81]

To fully understand the context in which these digital tools are proliferating, several issues must be taken into consideration. For example, it is widely understood that average lengths of stay at US hospitals have dropped significantly in recent years.[82-83] The reasons for these declines are complex but related to economic factors, policy and regulatory forces, disease epidemiology and practice changes. The result is that for the last two decades there has been a decline in care being provided in the hospital and more care provision in the home and community setting for many conditions. Secondly, retail healthcare outlets are rapidly growing in number and popularity.[84] Recent studies show that patients using these facilities have high levels of patient satisfaction, shorter waiting times, lower costs and care quality that is equal to or better than similar care provided in traditional healthcare settings.[85] There is even some evidence that they do a better job of reaching medically underserved populations than historic outlets.[86] Another factor that is important to consider when attempting to understand the current and future impact of digital health tools is that a significant amount of morbidity and mortality is caused by the hospitals themselves. It is widely accepted that many people who go to the hospital, get sick from medical errors and illnesses they did not bring to the hospital.[87] A recent study found that such hospital acquired problems are so common that they are actually the third leading cause of death![88] Obviously, a tremendous amount of good is done within the context of the current healthcare system, yet for most individuals, these findings are still very troubling. Perhaps the most important factors underpinning the growth of digital health are the advances in the computer sciences and broadband networks that are fueling a revolution in medical device innovation and enabling hospital-confined medical devices to become miniaturized, handheld, ingestible, wearable, mobile and operable anywhere there is a broadband connection.[89-94] In fact, some patients who 20 years ago required stays in the intensive care unit followed by lengthy hospitalizations prior to discharge are now are able to go home with small portable devices that do the work the ICU-based machines did just 2 decades ago![95-96] While the development and use of these devices is still growing, it is not difficult to understand the potential impact these devices will likely have on hospitalizations and lengths of stay as noted above. In addition, advances in robotics (a form of digital health) are now enabling surgeons located in one place to operate on patients located across town or across the globe.[97-99] The spectrum of surgeries performed this way will likely increase in the future given the early results and global need for providers and trained medical assistants. Similarly, advances in telemedicine and telehealth are enabling physicians to see, talk to, examine and monitor their patients remotely, lowering the need for inconvenient visits to the doctor's office or unnecessary visits to the ER or hospital.[100-108] Historically, hospitals were in part developed to centralize resources thereby reducing financial costs or improving opportunity costs.[109] In the future however, hospitals may not be able to reduce costs below that of ambulatory care and retail healthcare providers. Also, because of technological advances built on broadband networks described above, there may, no longer be significant opportunity costs associated with centralization of medical infrastructure. The final trend that is important to consider in the context of understanding the digital health potential is the emergence of artificial intelligence, and cognitive computing. Both of which are helping to provide unprecedented levels of data tracking and analytic capacity which in turn is facilitating the generation of insights that are instantly available to medical providers, patients and caregivers alike anytime at the point of need.[110-112]

Given these realities then, why would patients, in the future choose to stay in a hospital? Why would payors insist that covered beneficiaries obtain care in higher risk hospitals when lower risk and lower cost options (like the home) exist with comparable outcomes and higher patient satisfaction levels are available? From this perspective, it is plausible to believe that healthcare systems and in particular hospitals, as they currently exist, will continue to face considerable pressures, that are only increased by the growth of the digital health sector. These pressures pose significant threats to the viability and existence of traditional hospitals. It is becoming increasingly clear that only those hospital systems that proactively embrace the opportunities that the emerging digital health/consumer health informatics sectors provide and innovate on the very notion of what a hospital is, will ultimately be best able to successfully overcome the challenges, provide value to patients and remain financially stable.

Several experts have set forth concepts that attempt to describe potential future organization and delivery of healthcare systems.[113-117] While each attempt at a future vision of healthcare organization and delivery certainly adds to the debate and provides useful thoughts and perspectives, they each suffer from significant limitations and none attempts to provide a comprehensive vision that considers the implications of the consumer health informatics/digital health sector.

The authors briefly describe a vision of future healthcare delivery that, at a conceptual level, attempts to provide such a comprehensive vision. While, we do not believe this, nor any other model, is perfect, it does represent a significant conceptual advance over previous formulations precisely because of its more comprehensive nature. Figure 12.5 summarizes the major components of this vision. Briefly, in the future, it is likely that there will be a significant contraction in the volume of inpatient hospital services. This contraction is likely to be so significant that its impact will be fatal to many currently existing hospitals and result in so significant a restructuring of other hospitals that the majority of hospitals that do survive may bear little resemblance to the hospitals of today, they may no longer be called hospitals. These surviving institutions will focus on patients who are of the highest acuity, critically ill and medically complex, who need procedures and therapies that for other reasons cannot be provided in an alternate, less controlled setting. Given the fact that many conditions that years ago could only be treated in intensive care units that today are managed in part with technology in the ambulatory and home-based setting, we do not see this type of care comprising a large proportion of the total volume of care that will be provided in the future. The care in these facilities is likely to be driven by physician providers, as it often is today. Unlike the healthcare systems of today however, this type of care will likely represent the smallest proportion of the total volume of care provided nationally. Perhaps it will comprise 10-15% of the total volume of care at any given time. Indeed, there will likely be other patients who could benefit from inpatient care services, but otherwise do not need to be in a hospital. Due to advances in telemedicine and telehealth, mobile, wearable, embeddable and cloud computing, as well as advances in artificial intelligence and cognitive computing, these lower acuity patients will increasingly be in the ambulatory care setting in the community or at home. A growing body of literature is demonstrating the value and role of so called "Hospital at Home" models of care delivery.[118-123] It is likely that advances in broadband enabled health technologies will further contribute to the value and potential cost effectiveness of these models of care delivery. Perhaps as much as 15-25% of the total volume of care delivered in the future will be some variation of this hospital at home model of care.

Recent reports detail how several companies from non-healthcare sectors are beginning to make significant advances in the healthcare sector. These include IBM, Microsoft, Google, Apple, Amazon to name a few from the tech sector. However, sectors as varied as the automobile industry and the residential building sector are also exploring opportunities in the health sector.[124-126] While not yet a reality, it suggests that in the not too distant future, whole residential communities including automobiles could be optimized for health and low acuity medical care. One can envision such a model of care become part or supplanting the current notion of Accountable Care Organizations (ACO's) which are geographic areas for which a virtual health system is responsible for the care of consumers living within the area. It may be that these centers could be optimized for post-acute care and chronic disease self-management. To help keep costs down, it is possible that they would be primarily directed by Nurse Practitioners and nurses, who oversee larger numbers of enabling services staff, including patient Navigators, Community Health Workers, Promatoras etc.[127-131] This form model of healthcare delivery may account for as much as 30% of the total volume of care in the future.

SmartCare

Finally, with advances in digital health and consumer health informatics continuing to grow and evolve in society, it is likely that care delivery that is mediated at least in part through these mechanisms will become the largest model of care delivery accounting for as much as 40-50% of the total volume of care that is provided in the future. It is likely that for healthcare systems and providers to effectively stay in contact and manage their patients in a value-based healthcare environment, the amount of care that is delivered by these means will have to substantially increase. Collectively we call this model of care Smartcare, in part to distinguish it from other primarily marketing (mHealth, Connected Care etc.) terms that are in common use. In addition, however, the term is also used to highlight the fact that increasingly, due to advances in computer processing, data storage and miniaturization, computing power will be pushed to the edges of the network and therefore enable consumer devices to become increasingly smart. That is to say that they will operate autonomously to detect, decide and appropriately react to issues based

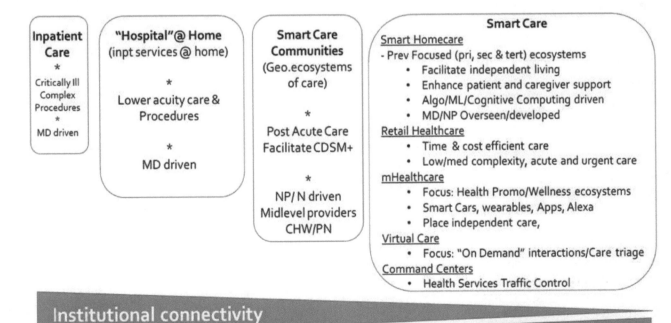

Figure 12.5: The future organization of healthcare delivery

on predetermined algorithms without the direct human input of a healthcare provider. In addition, these solutions will not be individual hardware products, but rather be built into the walls of building, and under the hoods of cars etc. to ultimately be woven seamlessly into the fabric of society in a way that does not require a patient's active or conscious action to capture needed information nor respond in an appropriate manner. Such so called passive interventions have been recognized by public health and medical experts as among the most powerful health intervention possible. Consider the health implications of water fluoridation, iodination of salt and airbags. Their health impact is unquestioned. Consequently, it can be envisioned how homes (as opposed to whole residential communities) and automobiles could themselves become smart and integral components of a broader consumer health ecosystem that is always on, follows the consumer wherever they are and thereby enhances an individuals' ability to live independently, safely, focus on wellness and prevention and help to manage minor health issues and low acuity medical complaint. It is reasonable to surmise that retail healthcare organizations might choose to become major players in exploring and advancing this model of care delivery as could virtual reality and augmented reality designers and innovators. Finally, as this model of care delivery grows there will likely arise the need for health technology "air traffic controllers" and "control centers." These individuals and organizations would have primary responsibility for optimizing data and information flows as well as optimizing the utilization of human resources in the community setting. Large health systems are already thinking about these possibilities and moving in a direction to be prepared for these eventual realities. This is evidenced by the fact that Mercy Hospital system has already developed the first operational "Hospital without beds" that is completely focus on providing and optimizing care at a distance to patients within its network.[132]

The role of the Internet and broadband connectivity cannot be overstated in enabling this vision of the organization and delivery of care to be possible. As the two arrows at the bottom of the graphic illustrate, if we focus on institutional connectivity primarily or preferentially to consumer connectivity, it may result in a situation where healthcare organization have good connectivity, yet some consumer may still have little to no connectivity and therefore in some cases, little to no ability to access available health goods and services. On the other hand, if we additionally prioritize consumer access and work to ensure that all consumers have access to adequate broad band connectivity, it is tantamount to ensuring that all consumers will have access to at least some forms of health care goods and services.

KEY POINTS

- Modern consumers demand a voice at the healthcare market place
- We are moving from standalone personal health records to PHRs tethered to EHRs
- Virtual visits (e-visits) are becoming more mainstream in US healthcare
- Electronic patient-physician communication augments face-to-face communication
- CHI and digital healthcare have the potential to impact healthcare reform

CONCLUSION

It is likely that patients and consumers will increasingly interact with the healthcare system in electronic ways. Questions remain about a number of issues, such as where data will reside, who will control it, and how it will be entered, accessed, and used. Continued research must inform the optimal ways in which data and information systems can be used to improve patient health and treatment of disease as well as the delivery of healthcare.

REFERENCES

1. Flaherty, D, Hoffman-Goetz, L, et al. (2015). What is consumer health informatics? A systematic review of published definitions. *Informatics for Health and Social Care.* 40: 91-112.
2. Abaidoo, B and Larweh, BT (2014). Consumer health informatics: the application of ict in improving patient-provider partnership for a better health care. *Online Journal of Public Health Informatics.* 6: e188. http://ojphi.org/article/view/4903/4570 (Accessed January 1, 2018)

3. Faiola, A and Holden, RJ (2015). Consumer health informatics: empowering healthy-living-seekers through mHealth. *Progress in Cardiovascular Diseases.* 59: 479-486
4. AMIA. https://www.amia.org/applications-informatics/consumer-health-informatics (Accessed January 12, 2018)
5. Krohn, R and Metcalf, D (2012). mHealth: From Smartphones to Smart Systems. Chicago, IL, Healthcare Information Management Systems Society.
6. Marcolino, MS, Oliveira, JAQ, et al. (2018). The impact of mHealth interventions: systematic review of systematic reviews. *JMIR Mhealth Uhealth.* 6(1): e23. http://mhealth.jmir.org/2018/1/e23/ (Accessed February 1, 2018)
7. Sonnier, P (2017). The Fourth Wave: Digital Health. https://fourthwavebook.com/
8. US Food and Drug Administration. https://www.fda.gov/medicaldevices/digitalhealth/ (Accessed January 3, 2018)
9. Society for Participatory Medicine. https://participatorymedicine.org/what-is-participatory-medicine/ (Accessed January 20, 2018)
10. Henry, JW, Pylypchuk, Y, et al. (2016). Electronic Capabilities for Patient Engagement among U.S. Non-Federal Acute Care Hospitals: 2012-2015. Washington, DC, Department of Health and Human Services. https://dashboard.healthit.gov/evaluations/data-briefs/hospitals-patient-engagement-electronic-capabilities-2015.php (Accessed December 20, 2018)
11. Anonymous (2008). 2008 Survey of Health Care Consumers. Washington, DC, Deloitte. http://www.deloitte.com/assets/Dcom-UnitedStates/LocalAssets/Documents/us_chs_ConsumerSurveyExecutiveSummary_200208.pdf (Accessed December 10, 2017)
12. Anonymous (2004). Connecting Americans to Their Healthcare - Final Report of the Working Group on Policies for Electronic Information Sharing Between Doctors and Patients. Washington, DC, Markle Foundation. http://www.markle.org/publications/1250-connecting-americans-their-health-care (Accessed December 8, 2017)
13. Anonymous (2006). The Value of Personal Health Records - A Joint Position Statement for Consumers of Health Care. Bethesda, MD, American Medical Informatics Association. http://www.amia.org/inside/releases/2006/ahima-amiaphrstatement.pdf (Accessed January 28, 2018)
14. Shenkin, BN and Warner, DC (1973). Giving the patient his medical record: a proposal to improve the system. *New England Journal of Medicine.* 289: 688-692
15. Bouayad, L, Ialynytchev, A, et al. (2017). Patient health record systems scope and functionalities: literature review and future directions. *Journal of Medical Internet Research.* 19(11): e388. http://www.jmir.org/2017/11/e388/ (Accessed February 3, 2018)
16. Roehrs, A, AndrédaCosta, C, et al. (2017). Personal health records: a systematic literature review. *Journal of Medical Internet Research.* 19(1): e13. http://www.jmir.org/2017/1/e13/ (Accessed February 3, 2018)
17. Tang, PC, Ash, JS, et al. (2006). Personal health records: definitions, benefits, and strategies for overcoming barriers to adoption. *Journal of the American Medical Informatics Association.* 13: 121-126.
18. ONC. Health IT. Gov. https://www.healthit.gov/providers-professionals/faqs/what-patient-portal (Accessed February 10, 2018)
19. Detmer, D, Bloomrosen, M, et al. (2008). Integrated personal health records: transformative tools for consumer-centric care. *BMC Medical Informatics & Decision Making.* 8: 45. http://www.biomedcentral.com/1472-6947/8/45 (Accessed February 3, 2018)
20. Tang, PC and Lee, TH (2009). Your doctor's office or the Internet? Two paths to personal health records. *New England Journal of Medicine.* 360: 1276-1278.
21. LibreHealth EHR. https://librehealth.io/ (Accessed February 20, 2018)
22. Anonymous (2006). Personal Health Records Policy Statements Adopted by the American College of Physicians. Philadelphia, PA, American College of Physicians. http://www.acponline.org/hpp/phr.pdf (Accessed January 28, 2018)
23. Delbanco, T, Walker, J, et al. (2010). Open Notes: doctors and patients signing on. *Annals of Internal Medicine.* 153: 121-125.
24. Delbanco, T, Walker, J, et al. (2012). Inviting patients to read their doctors' notes: a quasi-experimental study and a look ahead. *Annals of Internal Medicine.* 157: 461-470.
25. Meltsner, M (2012). A patient's view of OpenNotes. *Annals of Internal Medicine.* 157: 523-524.
26. Woods, SS, Schwartz, E, et al. (2013). Patient experiences with full electronic access to health records and clinical notes through the My HealtheVet Personal Health Record Pilot: qualitative study. *Journal of Medical Internet Research.* 15(3): e65. http://www.jmir.org/2013/3/e65/ (Accessed January 20, 2018)
27. Nazi, KM, Turvey, CL, et al. (2014). VA OpenNotes: exploring the experiences of early patient adopters with access to clinical notes. *Journal of the American Medical Informatics Association*: Epub ahead of print.
28. We can do better. https://www.wecandobetter.org/what-we-do/northwest-opennotes-consortium/ (Accessed February 8, 2018)

29. Hersh, WR (2017). From Vendor-Centric to Patient-Centric Data Stores. Informatics Professor. https://informaticsprofessor.blogspot.com/2017/11/from-vendor-centric-to-patient-centric.html (Accessed December 20, 2017)
30. Mikk, KA, Sleeper, HA, et al. (2017). The pathway to patient data ownership and better health. *Journal of the American Medical Association.* 318: 1433-1434.
31. Anonymous (2015). Health Information Bill of Rights. Chicago, IL, American Health Information Management Association. http://bok.ahima.org/PdfView?oid=107674 (Accessed January 28, 2018)
32. Anonymous (2014). Electronic Medical Record (EMR) Benefits Outweigh Privacy Invasion Risk for Chronically Ill. Chicago, IL, Accenture. https://www.accenture.com/t00010101T000000__w__/au-en/_acnmedia/PDF-3/Accenture-Consumers-with-Chronic-Conditions-Electronic-Medical-Records-v2.pdf (Accessed December 12, 2017)
33. Crotty, BH, Walker, J, et al. (2015). Information sharing preferences of older patients and their families. *JAMA Internal Medicine.* 175: 1492-1497.
34. Dixon, RF (2010). Enhancing primary care through online communication. *Health Affairs.* 29: 1364-1369.
35. Undern, T (2010). Consumers and Health Information Technology: A National Survey. Oakland, CA, California Health Care Foundation. http://www.chcf.org/publications/2010/04/consumers-and-health-information-technology-a-national-survey (Accessed February 1, 2018)
36. Wakefield, DS, Kruse, RL, et al. (2012). Consistency of patient preferences about a secure Internet-based patient communications portal: contemplating, enrolling, and using. *American Journal of Medical Quality.* 27: 494-502.
37. Seth, P, Abu-Abed, MI, et al. (2016). Email between patient and provider: assessing the attitudes and perspectives of 624 primary health care patients. *JMIR Medical Informatics.* 4(4): e32. http://medinform.jmir.org/2016/4/e42/ (Accessed January 4, 2018)
38. Choudhry, A, Hong, J, et al. (2015). Patients' preferences for biopsy result notification in an era of electronic messaging methods. *JAMA Dermatology.* 151: 513-521.
39. Friedman, EM (2016). You've got mail. *Journal of the American Medical Association.* 315: 2275-2276.
40. Giardina, TD, Baldwin, J, et al. (2018). Patient perceptions of receiving test results via online portals: a mixed-methods study. *Journal of the American Medical Informatics Association*: Epub ahead of print.
41. Brooks, RG and Menachemi, N (2006). Physician use of e-mail with patients: factors influencing electronic communication and adherence to best practices. *Journal of Medical Internet Research.* 8(1): e2. http://www.jmir.org/2006/1/e2/ (Accessed January 27, 2018)
42. Menachemi, N, Prickett, CT, et al. (2011). The use of physician-patient email: a follow-up examination of adoption and best-practice adherence 2005-2008. *Journal of Medical Internet Research.* 13(1): e23. http://www.jmir.org/2011/1/e23/ (Accessed January 27, 2018)
43. Antoun, J (2016). Electronic mail communication between physicians and patients: a review of challenges and opportunities. *Family Practice.* 33: 121-126.
44. Epic MyChart. http://www.epic.com/software (Accessed February 10, 2018)
45. Lee, JL, Matthias, MS, et al. (2017). A critical appraisal of guidelines for electronic communication between patients and clinicians: the need to modernize current recommendations. *Journal of the American Medical Informatics Association*: Epub ahead of print.
46. AMA H-478.997 Guidelines for patient physician electronic mail http://hosted.ap.org/specials/interactives/_documents/patient_physician_email.pdf (Accessed January 15, 2018)
47. Gorden, MS and DuMoulin, JP (2003). The Changing Face of Ambulatory Medicine: Reimbursing Physicians for Computer-Based Care. Philadelphia, PA, American College of Physicians. https://www.acponline.org/acp_policy/policies/ambulatory_medicine_reimbursements_computer_based_care_2003.pdf (Accessed January 16, 2018)
48. Liederman, EM and Morefield, CS (2003). Web messaging: a new tool for patient-physician communication. *Journal of the American Medical Informatics Association.* 10: 260-270.
49. Liederman, EM, Lee, JC, et al. (2005). The impact of patient-physician Web messaging on provider productivity. *Journal of Healthcare Information Management.* 19: 81-86.
50. Liederman, EM, Lee, JC, et al. (2005). Patient-physician web messaging. The impact on message volume and satisfaction. *Journal of General Internal Medicine.* 20: 52-57
51. Baker, L, Garber, AM, et al. (2003). The RelayHealth webVisit Study: Final Report. Emeryville, CA, Relay Health. http://www.relayhealth.com/rh/general/ourservices/studyResults.aspx (Accessed January 15, 2018)
52. Zhou, YY, Garrido, T, et al. (2007). Patient access to an electronic health record with secure messaging: impact on primary care utilization. *American Journal of Managed Care.* 13: 418-424.

53. Zhou, YY, Kanter, MH, et al. (2010). Improved quality at Kaiser Permanente through e-mail between physicians and patients. *Health Affairs*. 29: 1370-1375.
54. Reed, M, Graetz, I, et al. (2015). Patient-initiated e-mails to providers: associations with out-of-pocket visit costs, and impact on care-seeking and health. *American Journal of Managed Care*. 21: e632-e639.
55. Healthcast 2010: Smaller world, bigger expectations. PWC. November 1999. www.pwc.com
56. TelaDoc. www.teladoc.com (Accessed February 28, 2018)
57. Uscher-Pines L, Mehrotra A. Analysis of TelaDoc use seems to indicate expanded access to care for patients without prior connection to a provider. Health Affairs. 2014; 33(2). https://www.healthaffairs.org/doi/full/10.1377/hlthaff.2013.0989 (Accessed February 28, 2018)
58. American Well. www.americanwell.com (Accessed February 28, 2018)
59. MDLIVE. www.mdlive.com (Accessed February 28, 2018)
60. Doctor on Demand. www.doctorondemand.com (Accessed February 28, 2018)
61. Ashwood JS, Mehrotra A, Cowling D, Uscher-Pines L. Directo to consumer telehealth may increase access to care but does not decrease spending. Health Affairs. 2017;36(3):485-491
62. deJong, CC, Ros, WJ, et al. (2014). The effects on health behavior and health outcomes of Internet-based asynchronous communication between health providers and patients with a chronic condition: a systematic review. *Journal of Medical Internet Research*. 16: e19. http://www.jmir.org/2014/1/e19/ (Accessed December 7, 2017)
63. Davis-Giardina, T, Menon, S, et al. (2014). Patient access to medical records and healthcare outcomes: a systematic review. *Journal of the American Medical Informatics Association*. 21: 737-741.
64. Anonymous (2014). Engaging Patients and Families: How Consumers Value and Use Health IT. Washington, DC, National Partnership for Women & Families. http://nationalpartnership.org/issues/health/HIT/patients-speak.html (Accessed December 20, 2017)
65. Crotty, BH, Tamrat, Y, et al. (2014). Patient-to-physician messaging: volume nearly tripled as more patients joined system, but per capita rate plateaued. *Health Affairs*. 33: 1817-1822.
66. Patel, V, Barker, W, et al. (2014). Individuals' Access and Use of their Online Medical Record Nationwide. Washington, DC, Department of Health and Human Services. http://www.healthit.gov/sites/default/files/consumeraccessdatabrief_9_10_14.pdf (Accessed December 15, 2017)
67. Courneya, PT, Palattao, KJ, et al. (2013). HealthPartners' online clinic for simple conditions delivers savings of $88 per episode and high patient approval. *Health Affairs*. 32: 385-392.
68. Mehrotra, A, Paone, S, et al. (2013). A comparison of care at e-visits and physician office visits for sinusitis and urinary tract infection. *JAMA Internal Medicine*. 173: 72-74.
69. Schnipper, JL, Gandhi, TK, et al. (2012). Effects of an online personal health record on medication accuracy and safety: a cluster-randomized trial. *Journal of the American Medical Informatics Association*. 19: 728-734.
70. Murray, MF, Giovanni, MA, et al. (2013). Comparing electronic health record portals to obtain patient-entered family health history in primary care. *Journal of General Internal Medicine*. 28: 1558-1564.
71. Tom, JO, Chen, C, et al. (2014). Personal health record use and association with immunizations and well-child care visits recommendations. *Journal of Pediatrics*. 164: 112-117.
72. Chrischilles, EA, Hourcade, JP, et al. (2013). Personal health records: a randomized trial of effects on elder medication safety. *Journal of the American Medical Informatics Association*. 21: 679-686.
73. Hemsley, B, Rollo, M, et al. (2017). The health literacy demands of electronic personal health records (e-PHRs): an integrative review to inform future inclusive research. *Patient Education and Counseling*. 101: 2-15.
74. Chen, J, Druhl, E, et al. (2018). A natural language processing system that links medical terms in electronic health record notes to lay definitions: system development using physician reviews. *Journal of Medical Internet Research*. 20(1): e26. http://www.jmir.org/2018/1/e26/ (Accessed February 1, 2018)
75. Turvey, CL, Klein, DM, et al. (2017). Patient education for consumer-mediated HIE - a pilot randomized controlled trial of the Department of Veterans Affairs Blue Button. *Applied Clinical Informatics*. 7: 765–776.
76. Christensen, K and Sue, VM (2013). Viewing laboratory test results online: patients' actions and reactions. *Journal of Participatory Medicine*. 5 http://www.jopm.org/evidence/research/2013/10/03/viewing-laboratory-test-results-online-patients-actions-and-reactions/ (Accessed December 15, 2017)
77. Irizarry, T, DeVitoDabbs, A, et al. (2015). Patient portals and patient engagement: a state of the science review. *Journal of Medical Internet Research*. 17(6): e148. http://www.jmir.org/2015/6/e148/ (Accessed December 15, 2017)

78. Rotenstein, LS, Huckman, RS, et al. (2017). Making patients and doctors happier — the potential of patient-reported outcomes. *New England Journal of Medicine*. 377: 1309-1312.
79. Lavallee, DC, Chenok, KE, et al. (2016). Incorporating patient-reported outcomes into health care to engage patients and enhance care. *Health Affairs*. 35: 575-582.
80. Li, X, Dunn, J, et al. (2017). Digital health: tracking physiomes and activity using wearable biosensors reveals useful health-related information. *PLoS Biology*. 15(1): e2001402. http://journals.plos.org/plosbiology/article?id=10.1371/journal.pbio.2001402 (Accessed December 20, 2017)
81. The Food and Drug Administration. Digital Health. https://www.fda.gov/medicaldevices/digitalhealth/ (Accessed September 6, 2017)
82. Mardis R, Brownson K. Length of stay at an all-time low. Health Care Manag (Frederick) 2003 Apr;22(2):122-7.
83. DeFrances CJ, Hall MJ. 2005 National Hospital Discharge Survey. Adv Data 2007 Jul 12;(385):1-19.
84. Mehrotra A, Lave JR. Visits to retail clinics grew fourfold from 2007 to 2009, although their share of overall outpatient visits remains low. Health Aff (Millwood) 2012 Sep;31(9):2123-9.
85. Mehrotra A, Liu H, Adams JL, Wang MC, Lave JR, Thygeson NM, et al. Comparing costs and quality of care at retail clinics with that of other medical settings for 3 common illnesses. Ann Intern Med 2009 Sep 1;151(5):321-8.
86. Mehrotra A, Wang MC, Lave JR, Adams JL, McGlynn EA. Retail clinics, primary care physicians, and emergency departments: a comparison of patients' visits. Health Aff (Millwood) 2008 Sep;27(5):1272-82.
87. Van Den Bos J, Rustagi K, Gray T, Halford M, Ziemkiewicz E, Shreve J. The $17.1 billion problem: the annual cost of measurable medical errors. Health Aff (Millwood) 2011 Apr;30(4):596-603.
88. Makary MA, Daniel M. Medical error-the third leading cause of death in the US. BMJ 2016 May 3;353:i2139.
89. Gibbons MC, Wilson RF, Samal L, Lehmann CU, Dickersin K, Lehmann HP, et al. Consumer health informatics: results of a systematic evidence review and evidence based recommendations. Transl Behav Med 2011 Mar;1(1):72-82.
90. Gibbons MC. A historical overview of health disparities and the potential of eHealth solutions. J Med Internet Res 2005 Oct 4;7(5):e50.
91. What is telehealth? How is telehealth different from telemedicine? https://www.healthit.gov/telehealth (Accessed September 2, 2017)
92. Finkelstein J, Knight A, Marinopoulos S, Gibbons MC, Berger Z, Aboumatar H, et al. Enabling patient-centered care through health information technology. Evid Rep Technol Assess (Full Rep) 2012 Jun;(206):1-1531.
93. Yilmaz T, Foster R, Hao Y. Detecting vital signs with wearable wireless sensors. Sensors (Basel) 2010;10(12):10837-62.
94. Darwish A, Hassanien AE. Wearable and implantable wireless sensor network solutions for healthcare monitoring. Sensors (Basel) 2011;11(6):5561-95.
95. Kiourti A, Nikita KS. A Review of In-Body Biotelemetry Devices: Implantables, Ingestibles, and Injectables. IEEE Trans Biomed Eng 2017 Jul;64(7):1422-30.
96. Kiourti A, Psathas KA, Nikita KS. Implantable and ingestible medical devices with wireless telemetry functionalities: a review of current status and challenges. Bioelectromagnetics 2014 Jan;35(1):1-15.
97. Hussain A, Malik A, Halim MU, Ali AM. The use of robotics in surgery: a review. Int J Clin Pract 2014 Nov;68(11):1376-82.
98. Avgousti S, Christoforou EG, Panayides AS, Voskarides S, Novales C, Nouaille L, et al. Medical telerobotic systems: current status and future trends. Biomed Eng Online 2016 Aug 12;15(1):96.
99. Ballantyne GH. Robotic surgery, telerobotic surgery, telepresence, and telementoring. Review of early clinical results. Surg Endosc 2002 Oct;16(10):1389-402.
100. Catarinella FS, Bos WH. Digital health assessment in rheumatology: current and future possibilities. Clin Exp Rheumatol 2016 Sep;34(5 Suppl 101):S2-S4.
101. Kvedar J, Coye MJ, Everett W. Connected health: a review of technologies and strategies to improve patient care with telemedicine and telehealth. Health Aff (Millwood) 2014 Feb;33(2):194-9.
102. Ekeland AG, Bowes A, Flottorp S. Effectiveness of telemedicine: a systematic review of reviews. Int J Med Inform 2010 Nov;79(11):736-71.
103. Khan N, Marvel FA, Wang J, Martin SS. Digital Health Technologies to Promote Lifestyle Change and Adherence. Curr Treat Options Cardiovasc Med 2017 Aug;19(8):60.
104. Hollis C, Falconer CJ, Martin JL, Whittington C, Stockton S, Glazebrook C, et al. Annual Research Review: Digital health interventions for children and young people with mental health problems - a systematic and meta-review. J Child Psychol Psychiatry 2017 Apr;58(4):474-503.
105. Bhattarai P, Phillips JL. The role of digital health technologies in management of pain in older people: An integrative review. Arch Gerontol Geriatr 2017 Jan;68:14-24.
106. Kaufman N, Khurana I. Using Digital Health Technology to Prevent and Treat Diabetes. Diabetes Technol Ther 2016 Feb;18 Suppl 1:S56-S68.

107. Widmer RJ, Collins NM, Collins CS, West CP, Lerman LO, Lerman A. Digital health interventions for the prevention of cardiovascular disease: a systematic review and meta-analysis. Mayo Clin Proc 2015 Apr;90(4):469-80.
108. Thomas JG, Bond DS. Review of innovations in digital health technology to promote weight control. Curr Diab Rep 2014;14(5):485.
109. Starr P. The Social Transformation of American Medicine. Basic Books Pubs; 1982.
110. Chen Y, Elenee Argentinis JD, Weber G. IBM Watson: How Cognitive Computing Can Be Applied to Big Data Challenges in Life Sciences Research. Clin Ther 2016 Apr;38(4):688-701.
111. Wu H, Yamaguchi A. Semantic Web technologies for the big data in life sciences. Biosci Trends 2014 Aug;8(4):192-201.
112. Martin-Sanchez F, Verspoor K. Big data in medicine is driving big changes. Yearb Med Inform 2014 Aug 15;9:14-20.
113. Charlesworth K, Jamieson M, Butler CD, Davey R. The future healthcare? Aust Health Rev 2015 Sep;39(4):444-7.
114. Nagle LM, Pitts BM. Citizen perspectives on the future of healthcare. Healthc Q 2012;15(2):40-5.
115. Mezghani E, Da SM, Pruski C, Exposito E, Drira K. A perspective of adaptation in healthcare. Stud Health Technol Inform 2014;205:206-10.
116. Sklar DP. Delivery system reform--visualizing the future. Acad Med 2013 Jul;88(7):905-6.
117. Kimball B, Joynt J, Cherner D, O'Neil E. The quest for new innovative care delivery models. J Nurs Adm 2007 Sep;37(9):392-8.
118. Wilson A, Parker H, Wynn A, Spiers N. Performance of hospital-at-home after a randomised controlled trial. J Health Serv Res Policy 2003 Jul;8(3):160-4.
119. Wilson A, Wynn A, Parker H. Patient and carer satisfaction with 'hospital at home': quantitative and qualitative results from a randomised controlled trial. Br J Gen Pract 2002 Jan;52(474):9-13.
120. Jones J, Wilson A, Parker H, Wynn A, Jagger C, Spiers N, et al. Economic evaluation of hospital at home versus hospital care: cost minimisation analysis of data from randomised controlled trial. BMJ 1999 Dec 11;319(7224):1547-50.
121. Wilson A, Parker H, Wynn A, Jagger C, Spiers N, Jones J, et al. Randomised controlled trial of effectiveness of Leicester hospital at home scheme compared with hospital care. BMJ 1999 Dec 11;319(7224):1542-6.
122. Richards SH, Coast J, Gunnell DJ, Peters TJ, Pounsford J, Darlow MA. Randomised controlled trial comparing effectiveness and acceptability of an early discharge, hospital at home scheme with acute hospital care. BMJ 1998 Jun 13;316(7147):1796-801.
123. Shepperd S, Harwood D, Gray A, Vessey M, Morgan P. Randomised controlled trial comparing hospital at home care with inpatient hospital care. II: cost minimisation analysis. BMJ 1998 Jun 13;316(7147):1791-6.
124. Stoaks U. The Next Big Digital Health Platform For Entrepreneurs To Build On: Your Car. Forbes . 9-1-2015. https://www.forbes.com/sites/unitystoakes/2015/09/01/the-next-big-digital-health-platform-for-entrepreneurs-to-build-on-your-car/#b668eec68956 (Accessed September 1, 2017)
125. Tso R. Smart homes of the future will know us by our heartbeats. Wired . 12-5-2016. https://www.wired.com/insights/2014/10/smart-homes-of-the-future/ (Accessed September 24, 2017)
126. Karten S. Beyond the Smart Home: The Health Hub of the Future. HIT Consultant . 3-9-2016. http://hitconsultant.net/2016/03/09/beyond-the-smart-home-the-health-hub-of-the-future/
127. Gibbons MC, Tyus NC. Systematic review of U.S.-based randomized controlled trials using community health workers. Prog Community Health Partners 2007;1(4):371-81.
128. Whiteman LN, Gibbons MC, Smith WR, Stewart RW. Top 10 Things You Need to Know to Run Community Health Worker Programs: Lessons Learned in the Field. South Med J 2016 Sep;109(9):579-82.
129. Eckenrode J, Campa MI, Morris PA, Henderson CR, Jr., Bolger KE, Kitzman H, et al. The Prevention of Child Maltreatment Through the Nurse Family Partnership Program: Mediating Effects in a Long-Term Follow-Up Study. Child Maltreat 2017 May;22(2):92-9.
130. Olds DL, Robinson J, Pettitt L, Luckey DW, Holmberg J, Ng RK, et al. Effects of home visits by paraprofessionals and by nurses: age 4 follow-up results of a randomized trial. Pediatrics 2004 Dec;114(6):1560-8.
131. Olds DL, Robinson J, O'Brien R, Luckey DW, Pettitt LM, Henderson CR, Jr., et al. Home visiting by paraprofessionals and by nurses: a randomized, controlled trial. Pediatrics 2002 Sep;110(3):486-96.
132. Pepitone J. The $54 million hospital without any beds. http://newsok.com/article/feed/1073187 NewsOK. September 12, 2006. (Accessed September 27 2017)

13

Mobile Technology and mHealth

JOHN SHARP • ROBERT E. HOYT

"Think mHealth as personal health reform"
—Jane Sarasohn-Kahn, Health Populi 2010

LEARNING OBJECTIVES

After reading this chapter the reader should be able to:

- Define mHealth and mobile technologies
- Discuss three mobile technology uses cases in a clinical setting
- Discuss the shortcomings of medical apps for smartphones
- Enumerate the challenges of mHealth in low and middle-income countries
- Identify the software development kits (SDKs) for the iPhone and Android OSs

INTRODUCTION

Mobile health (mHealth) is a term used for the practice of medicine and public health supported by mobile devices. The term mHealth has been defined by the Global Observatory for e-Health of the World Health Organization as *"medical and public health practice supported by mobile devices, such as mobile phones, patient monitoring devices, personal digital assistants and other wireless devices."*[1]

Health care providers use mobile health technology to: access clinical information (e.g., through mobile health apps and mobile-enabled electronic health records), collaborate with care teams (e.g., with secure text messaging), communicate with patients (e.g., through patient portals and text messaging), offer real-time monitoring of patients, conduct research and provide health care remotely, also called telemedicine.

Patients use mobile health technology to: track their own health data through mHealth apps and devices, like the Fitbit®, access their clinical records through mobile-enabled patient portals, and communicate with their providers (e.g., through HIPAA compliant e-mail and secure text messaging).

HISTORY

The exponential growth of mobile technology in the past few years has created opportunities for mHealth to flourish. A 2016 Pew study noted: *"The vast majority of Americans – 95% – now own a cellphone of some kind. The share of Americans that own smartphones is now 77%, up from just 35% in Pew Research Center's first survey of smartphone ownership conducted in 2011."* According to this same survey, adoption of smartphones is about equal among men and women and similar among ethnic groups. Clearly adoption rates drop with age but increase with education and income.[2] (See Figure 13.1)

Early hand-held computers with wireless connections included the Apple Newton and Palm Pilot in the early 1990s. With limited or no wireless networks available, these products depended on synchronizing with desktop computers.

According to Thomas Friedman, *"The year 2007 was a major inflection point: the release of the iPhone, together with advances in silicon chips, software, storage, sensors, and networking, created a new technology platform."*[3] Transmission technologies which enable mobile health applications and devices also include the development

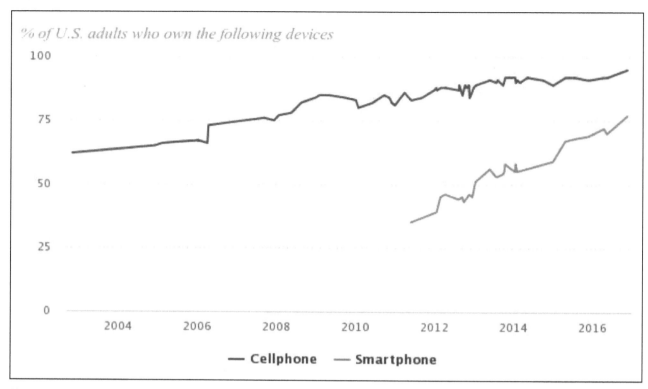

Figure 13.1: Growth of cellphones and smartphones 2004-2016 (Courtesy Pew Research)

of 3G, 4G and soon to be 5G, as well as SMS texting and low energy Bluetooth connections. The wireless protocol 802.11b became widely adopted for wireless transmission methods. The combination of connectivity fed the growth of smartphone apps and connected devices through Bluetooth. Mobile operating systems, such as, the Apple iOS and Google's Android made for common development platforms. This convergence of technology and the opening of the Apple App Store and Google Play, created a marketplace for apps of all kinds. Soon health and fitness apps were deployed at a dizzying pace, but with little quality control.

Another development which spurred the growth of mobile health was the miniaturization of sensors installed in smartphones (e.g., GPS, accelerometers, etc.) and in Bluetooth connected devices, such as, watches, glucose monitors, weight scales and others.

CURRENT MOBILE TECHNOLOGY

Smartphone apps serve as a platform for several mHealth categories. First, text messaging (SMS) can be used for a variety of patient reminders. Second, mHealth apps can inform both patients and providers. Third, smartphone sensors e.g. accelerometers can measure activity with the likelihood that more sensors are on the horizon. Fourth, the smartphone can serve as an intermediary platform that collects data that uploads to the cloud where more elaborate tools can be hosted. Sensors are now available both within smart phones and devices including smart watches, fitness trackers, and devices to monitor blood pressure, continuous glucose monitors, heart monitoring and others which have been enabled by miniaturization. For instance, the accelerometer can count steps and flights of stairs climbed. The touch screen can be used to monitor movement disorders.[4]

Smart phones and tablets work as a platform to not only provide data within apps associated with wireless devices but also transfer that data to the cloud. This process requires well-tested interoperability standards like the Continua Guidelines.[5] (See Figure 13.2)

These guidelines and testing tools outline the transfer of data generated by personal health devices through a personal health device interface, such as, Bluetooth, through a personal health gateway (most commonly today a mobile phone), then through a services interface to a service and then a health information system interface to a health information service (such as, an electronic medical record). Each of these steps is secured and standardized to ensure the data transfer is both safe from hacking and accurate. For instance, something as simple as a date and time stamp for a blood pressure reading from a personal health device must be standardized throughout each step so that it arrives accurately in the electronic health record (EHR) workflow.

Current technologies ranges from simple solutions to complex. For instance, the use of SMS texting can be one of the simplest mobile health interventions. One-way texting, such as, sending wellness reminders via text has been used successfully for people newly diagnosed with diabetes and those who do not own smartphones but have mobile phones with texting available. Text4Baby has been one of the most effective apps to give health reminders to pregnant women.[6] Text2Quit using texting to help smokers quit the habit.[7] SMS texting has also been shown to be effective in sending appointment reminders including for vaccinations.[8] Text messaging can also be personalized to promote healthy activities for those with chronic conditions with positive outcomes.[9] Two-way texting is more challenging but provides the opportunity for the patient to ask questions or report on the current status of their condition.[10]

With the broader use of wireless networks in the home and availability of Bluetooth enabled health devices, the use of remote monitoring has expanded. For instance, the use of wireless scales and wireless blood pressure measurement has enabled the monitoring of conditions like heart failure in the home. To monitor effectively, not only do the devices need to operate reliably in the home and transmit data through a home network but the healthcare team (often a care manager as the first line of defense) needs decision support systems to analyze potential decompensation of those patients.[11]

A requirement of successful use of patient generated health data (PGHD) coming from devices or apps, the data made available to the patient and care team must be analyzed and displayed in a meaningful manner. For instance, a fitness tracker will have an app that displays steps, calories burned and progress toward daily goals. These displays may be graphic visualizations or numeric representations, even comparing statistics from previous days or weeks, and representing a basic version of a personal health dashboard. For care teams, one of the main concerns is being overwhelmed by the amount of data from a population of patients using remote monitoring.[12] Additionally, analysis of home or remote data is not likely to be reimbursed so the interest by the clinician might be low. One of the solutions to this problem is to use analytics to make the data meaningful and actionable. As an example, a care team may receive multiple blood pressure readings or blood glucoses per day. Their concern is when to take action, such as when a reading is out of range, either too high or too low. These provider dashboards can also use dashboard visualizations, such as, warning lights (green, yellow, red) to indicate patients to click through on to get more detailed information. These analytic tools can also help to sort out signal versus noise problems. For instance, with cardiac monitors in the home, a reading could be too low or too high but caused by equipment malfunction or interference in transmission.[13] Some remote monitoring projects are beginning to use machine learning and artificial intelligence to analyze data from remote monitoring devices in real time and offer actionable data based on these analytics.[14] (See Figure 13.3)

It was reported in 2016 that there were 259,000 health apps available globally and the trend is now slowing. Only 24% of healthcare apps recorded more than 50,000 downloads, most likely due to too many choices. In general, health apps do not generate that much income so not all will survive. Consumer apps are the most popular type of healthcare app, and about half deal with chronic

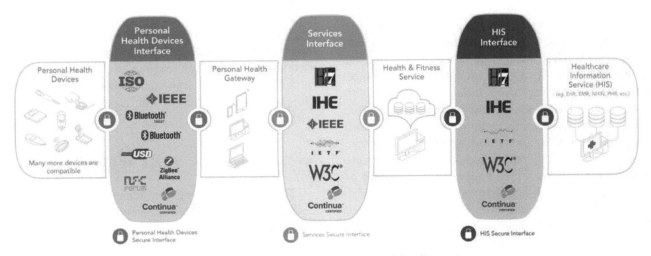

Figure 13.2: Continua Design Guidelines (Courtesy Continua Health Alliance)

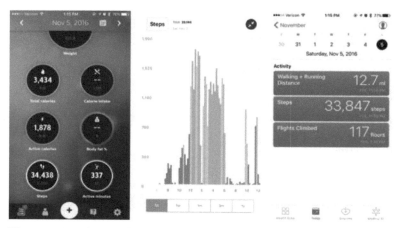

Figure 13.3: Smartphone Apps (Courtesy Philips Healthcare)

disease and about a third deal with wellness and fitness. About 58% use an application programming interface (API) with Apple HealthKit being the most popular API.[15]

According to AppBrain, there are 44,882 Android medical apps 10% were paid apps and only 3% had more than 50,000 downloaded. For "health & fitness" app category in 2018 they recorded 106,4277 apps with 7% as paid apps and only 5% with more than 50,000 downloads.[16] It is known that there are even more apps available on the iTunes Store but a similar evaluation by AppBrain was not available. It is important to note that search terms such as "medical apps" is overly broad and includes apps only distantly related to medical issues.

mHEALTH IN CLINICAL SETTINGS

In both hospitals and outpatient clinics, communication is the primary use of mobile devices. For instance, most EHRs can now be viewed on a mobile device, however, because of the complexity of the data and navigation required for these tools, tablet computers are more likely to be used than smart phones. This mobility has the advantage of allowing the user, whether physician, nurse or other staff, to move from room to room, patient to patient, without dependence on a desktop computer. Nursing have been using *"workstations on wheels"* (with wireless connectivity) for several years now to be able to bring a computer to the patient's bedside for both reading patient records and recording encounters. These workstations combined with single sign-on can create significant efficiencies.[17] One concern about the use of mobile technology in clinical settings is infection control, particularly in intensive care units. Protocols for cleaning mobile devices must be followed so that these devices do not spread infections.[18] Security of mobile devices is also a concern which is why some hospitals prefer to provide secure devices rather than permit bringing your own device (BYOD).

Once these concerns are addressed, communication between medical team members can be enhanced by mobile devices. For instance, the pager has been the primary mobile device to communicate alerts to medical team members for years. New smartphone technology allows for not only standard alerts but also the ability to respond to a message by placing an order or other action.[19] Data collection is another task which can be enabled efficiently through mobile devices. Whether it is a nurse recording vital signs at the bedside on a tablet or devices in the ICU, such as, a smart IV pump or wireless blood pressure cuff transmitting data directly into the EHR, data collection is becoming more wireless and mobile. With the pervasive use of digital imaging in healthcare, viewing imaging results on a tablet has some advantages including the easy ability to zoom in but also take the images anywhere in the clinical setting. There are limitations based on the size of the screen and some images are better viewed on large desktop screens.[20]

One final example of the use of mobile technology in healthcare is for teleconsultations including teleradiology and telestroke services. While these are discussed in more detail in the chapter on telemedicine, these services over mobile technologies can enable real time discussion of diagnosis and treatment from an ambulance, plane or remote location.[21]

mHEALTH IN THE HOME SETTING

There are a growing number of use cases for mobile technology in remote patient monitoring. With the incentives from valued-based care, such as, bundled payment models, monitoring post-hospitalization or surgery is becoming more common. One study demonstrated

> **Featured App: PTSD Coach**
>
>
>
> This app was developed for the US Department of Veteran Affairs due to the increased risk of PTSD after deployment. The app educates about the symptoms and treatment of PTSD. It has tools for PTSD screening and allows a patient to track their symptoms over time. There are links for additional support. Initial evaluation of the app was positive and reported in the literature.[22] A follow on randomized and controlled study of community trauma survivors showed good acceptance, but PTSD symptom improvement was not statistically significant between the app and the control group.[23]
>
> It is free to download from the iTunes Store and Google Play.[24]

> **Case Study: TEXT ME Trial (2011-2013)**
>
> The Tobacco, Exercise and Diet Messages (TEXT ME) study was a single blind, randomized trial, enrolling 710 Australian patients with documented heart disease.
>
> About half the patients received 4 educational text messages per week about lifestyle changes for six months. A control group did not receive text messages.
>
> Those patients who received text messaging had statistically lower LDL cholesterol levels as the primary end point and lower systolic blood pressures, lower smoking rates and increases in physical activity as the secondary endpoints.
>
> While statistically significant, the only result that was clearly clinically significant was the reduction of smoking from 44% to 26%). It is not known if these changes will persist over time.[25]

the safety of using remote monitoring for heart failure patients with defibrillators. This multicenter trial in Europe showed that office visits could be reduced without compromising patient safety.[26]

Diabetes is probably the most common condition to use mobile technology to monitor and treat. Virtual diabetes prevention programs with live coaching, social discussions and the ability to record progress have been shown to be as effective as in person programs.[27] For Type 1 Diabetes (insulin-dependent), both continuous glucose monitors and insulin pumps have emerged as a new standard of care. Both have the capability of transmitting data to an app to help the patient balance blood sugar levels and insulin requirements. More recently the concept of an artificial pancreas led to enabling the two devices to communicate and therefore keep blood glucose levels in range even at night.[28] This began as an open source project[29] but is also being commercialized.

Another use case uses fitness trackers to monitor movement and fatigue of those with multiple sclerosis.[30] Cardiac rehabilitation is now being delivered to veterans using a smart phone app.[31] Hypertension control has also been demonstrated through the home use of a wireless blood pressure cuff.[32] Pulmonary diseases, such as, asthma are being monitored by the use of smart wireless inhalers.[33]

Interfaces with electronic medical records through applications programming interfaces (APIs) are being developed to better integrate data from devices and apps into the clinical workflow. Apple HealthKit and Apple ResearchKit are two such APIs which help integrate into clinical care and research and discussed in a later section.[34]

mHEALTH FOR WELLNESS AND SPORTS

Fitness trackers are the most common mobile devices and are usually worn on the wrist. The associated apps allow the user to track steps, distance, heart rate, sleep, and diet. But do these devices, by themselves, create significant health behavior change? To date, many studies have tested this but without success. However, some of the research challenges include failure to define the effect size in digital health, the short duration of the studies and the changes in technology over time.[35] Smartwatches as well as sport-specific devices can be integrated with more sophisticated GPS and other tools.

mHEALTH IN LOW AND MIDDLE-INCOME COUNTRIES

Use of mHealth in these countries has grown as wireless networks have become available and, in some cases, more prevalent than wired networks. Many low-income countries have "leap frogged" or skipped the land line communication phase and gone straight to wireless networks. Use cases include texting health reminder messages for pregnant women, sending information to community health workers educational snippets about current treatments and epidemics and community health workers reporting on current outbreaks and requests for assistance.

A World Health Organization 2011 report *mHealth. New horizons for health through mobile technologies* reported that while these countries were quick to adopt mHealth initiatives, they were often not studied well and were challenged by other health priorities and data security. Too often, mHealth studies were generated on a short-term grant such that the study numbers are small with no larger follow-on studies.[36]

MOBILE TECHNOLOGY FOR RESEARCH AND DEVELOPMENT

Apple iOS

Apple HealthKit: Apple created the HealthKit, a software development kit (SDK) for iOS and the WatchOS. These application programming interfaces (APIs) integrate a new app with the phone's health and activity data. Extensive videos, sample code and guides are included on the website.[37]

CareKit: is the open-source SDK to help patients manage their own health.[38] For example, Corrie Health app from Johns Hopkins University monitors a patient's recovery from a heart attack by tracking vital signs, medications and physical activity.[39] Johns Hopkins is using this app as part of the Myocardial Infarction Combined Device Recovery Enhancement (MICORE) Study.[40]

Apple ResearchKit: is an open-source framework that enables an iOS app to be used for medical research.[38] The workflow for research, such as enrollment, consent and patient surveys is part of the Kit. ResearchKit is integrated with HealthKit so that health and activity parameters (steps, calories burned and heart rate) can be included in the study. For example, an app could be developed to measure activity in a population at risk (e.g. Parkinson's Disease) and that data, along with patient input is made available to researchers. Such an app has already been developed (mPower) by the University of Rochester and available in the iTunes App Store.[41] (Figure 13.4)

Figure 13.4: mPower app for Parkinson research

Android OS

Google Fit: is an app but also includes an Android software SDK, similar to Apple HealthKit for building apps using the activity sensor's application programming interface (API). Data is transmitted to a central web repository known as Google Fitness Store.[42]

ResearchStack: is a SDK framework for creating research apps using the Android OS, analogous to the Apple ResearchKit.

Research Droid: is an Android port of the ResearchKit launched in November 2015 to add Android smartphones to the research platform. The first app America Walks was developed that is available on both Android and iOS and can integrate with ResearchKit.[43] A study (sponsored by TrialX) started and ended in 2016 to collect data from volunteers.[44]

REGULATORY REQUIREMENTS

In the United States, the Food and Drug Administration (FDA) for many years did not have a clear statement on the regulation of mobile health. Now some clarification has come to distinguish between devices and apps which are used for treatment or clinical decision support and those which have not.[45] Those devices and apps which are directly involved in treatment or that provide clinical decision support are regulated and must got through the FDA approval process. At the same time, the FDA is encouraging innovation through a new digital health software precertification pilot program with selected companies participating.[46]

mHEALTH CHALLENGES

Distraction: mobile technology, in particular smartphones and specifically text messaging can be distracting when combined with activities such as driving an automobile. Several studies have confirmed that mobile technology is also distracting in a medical environment.[47] Some would argue that distraction due to smartphones increases medical malpractice.[48]

Privacy/security: clearly, mobile technology represents a security risk for patient data and the healthcare system's network.[49]

Source of infection: there is evidence that mobile devices contain bacteria that could be harmful in certain locations such as the operating room.[50]

Interference with medical equipment: there is a potential risk of smartphone electromagnetic radiation on pacemakers and other cardiac equipment. Some argue that phones should be kept at a distance of greater than 1 meter to be safe.[51]

Lack of quality control: Due to the proliferation of mobile apps many argued that there needed to be a means to judge the quality of apps, particularly mHealth apps. In 2015, an Australian expert panel convened and developed the 23-item scale Mobile App Rating Scale (MARS). This scale has excellent internal consistency (alpha = .90) and interrater reliability intraclass correlation coefficient (ICC = .79).[52]

Lack of evidence: despite more than 500 mHealth studies, there is a paucity of high quality evidence to show that mHealth positively impacts patient outcomes. The lack of evidence in this area mirrors what was reported about HIT in general and the section on evidence-based health informatics stated in the chapter on Evidence Based Medicine. For example:

- Direito et al. performed a systematic review with meta-analysis of mHealth technologies and physical activity (PA). They concluded that these technologies had only small effects on PA and sedentary behavior.[53]
- McCarroll et al. conducted a systematic review to determine the effect of mHealth interventions on health eating in adults and determined that the studies were of low quality.[54]

While the tech industry and developers would like to "scale up" the number of apps, they must first be shown to be effective and there should be agreement as to how the apps should be rated with research. The lay public is likely overly optimistic about the effect of healthcare apps as well, due to their wide acceptance worldwide. Many of the studies conducted were pilot studies with no follow up or studies adequately powered to show a significant outcome. Most studies focused on text messaging with mixed results and did not focus on other capabilities of smartphones. Tomlinson, et al. discusses several recommendations regarding what it will take to scale up mHealth.[55] Fiordelli et al. and Kumar et al. elaborate on the state of mHealth research and what we know so far.[56-57]

Healthcare organizations need to develop a mobile technology policy.[58] Policies should include:

- Bring your own device (BYOD) policies to be sure external phones match security policies, encryption, WiFi security, firewall protection, etc.
- For healthcare system owned phones determine who owns the data
- Use separate phones for strictly business use. Be able to locate phones via GPS
- Create "no phone zones", where appropriate

FUTURE TRENDS IN MHEALTH

One can only anticipate greater innovation in this field. First, we will see more reliable devices and sensors. Also, the further miniaturization of sensors will progress. Mobile health is already entering the realm of the Internet of Things (IOT) where Bluetooth devices communicate with each other as well as the mobile phone platform. The uses of sensors on the body will expand to implantables and ingestables which are already in testing. Finally, more comprehensive personalized health dashboards with appear, giving actionable alerts and progress reports. Analytics and artificial intelligence will become pervasive in mHealth promoting new methods of precision medicine.

MHEALTH RESOURCES

- *mHealth Resources to Strengthen Health Programs*. This article includes resources for implementation of mHealth in less developed countries, an e-learning course, online guide, evidence database and high-impact practices.[59]
- *mHealth Evidence*. This web site archives more than 8300 digital health evidence sources with a search engine with standard and advanced search capabilities.[60]

- NIH Fogarty International Center. *Mobile Health information and resources*. This NIH website includes news and general information about mHealth.[61]

RECOMMENDED READING

- *The Asthma Mobile Health Study, a large scale clinical observational study using ResearchKit*. This article is important because this asthma study (recruitment, consent and enrollment) was carried out entirely by smartphone. Data flowed between the 7,593 participants and the researchers. The authors do point out that this approach may be limited by selection bias, low retention rates, reporting bias and data security.[34]
- *Guidelines for reporting of health interventions using mobile phones: mobile health evidence reporting and assessment (mERA) checklist*. The WHO mHealth Technical Evidence Review Group developed a 16-item check list to standardize mHealth reporting and research.[62]
- *What is the economic evidence for mHealth? A systematic review of economic evaluations of mHealth solutions*. This is a systematic review of the economic evidence for mHealth. They used the Consolidated Health Economic Evaluation Reporting Standards (CHEERS) guidelines to evaluate the 39 studies accepted. Most studies were conducted in upper and upper middle-income countries. Using the CHEER guidelines, 74.3% of studies reported the mHealth initiative as cost-effective or economically beneficial.[63]
- *Mobile text messaging for Health: a systematic review of reviews*. This systematic review confirmed the usefulness of text messaging interventions for diabetes, weight loss, smoking, physical activity and medication compliance. However, they recommended longer studies and more research into cost-effectiveness and intervention details.[64]

KEY POINTS

- Mobile technology has become a routine aspect of the personal and professional lives of most healthcare workers
- mHealth is a worldwide movement, not limited to just the United States
- There is a myriad of medical and health & fitness apps available
- While mHealth presents many new advantages, it is associated with many challenges

CONCLUSION

The field of Health Informatics has seen the evolution of mobile technology evolve from primitive personal digital assistants to sophisticated smartphones and tablets. mHealth is a movement to apply these new technologies to health care, education and research. The evidence for mHealth benefit is limited for multiple reasons. Larger and better designed studies are needed.

While the number of medical apps available is astronomical, the percent that are used regularly by either patients or clinicians is limited. Smartphones have multiple advantageous characteristics for education and communication, but they are also associated with numerous challenges.

With the persistent popularity of smartphone technology, they are here to stay but this mandates intelligent smartphone policies by healthcare organizations.

REFERENCES

1. WHO Global Observatory for eHealth. New horizons for health through mobile technologies. Geneva. WHO;2011:112
2. Mobile Fact Sheet. Pew Research. http://www.pewinternet.org/fact-sheet/mobile/ (Accessed November 2, 2017)
3. Thomas Friedman, Thank You For Being Late, Farrar, Straus & Giroux, November 2016.
4. Andrew D Trister, E Ray Dorsey & Stephen H Friend. Smartphones as new tools in the management and understanding of Parkinson's disease, npj Parkinson's Disease, https://www.nature.com/articles/npjparkd20166
5. Personal Connected Health Alliance, Continua Design Guidelines. http://www.pchalliance.org/continua-design-guidelines (Accessed November 2, 2017)
6. Text4Baby, https://www.text4baby.org/ (Accessed November 2, 2017)
7. Text2Quit app, https://www.text2quit.com (Accessed November 5, 2017)

8. Bar-Shain DS, Stager MM, Runkle AP et al. Direct Messaging to Parents/Guardians to Improve Adolescent Immunizations. Journal of Adolescent Health 56:5, May 2015, S21-26.
9. Agoola S, Jethwani K, Lopez L, et al. Text to Move: A Randomized Controlled Trial of a Text-Messaging Program to Improve Physical Activity Behaviors in Patients With Type 2 Diabetes Mellitus. Journal of Medical Internet Research, 2016;18(11):e307 http://www.jmir.org/2016/11/e307/ (Accessed November 3, 2017)
10. Cheng C, Brown C, New T, et al. SickleREMOTE: A two-way text messaging system for pediatric sickle cell disease patients. http://ieeexplore.ieee.org/document/6211602/ (Accessed November 2, 2017)
11. Guidi G, Pollonini L, Dacso CC, Iadanza E. A multi-layer monitoring system for clinical management of Congestive Heart Failure. BMC Med Inform Decis Mak. 2015;15 Suppl 3:S5.
12. Office of the National Coordinator for Health IT, PGHD Policy Framework, https://www.healthit.gov/sites/default/files/Draft_White_Paper_PGHD_Policy_Framework.pdf (Accessed November 2, 2017)
13. Müller A, Schweizer J, Helms TM, et al, Telemedical Support in Patients with Chronic Heart Failure: Experience from Different Projects in Germany. International Journal of Telemedicine and Applications. Volume 2010 (2010), Article ID 181806.
14. Baig MM, Hossenini HG, Moquemm AA, Mizra F, Linden M. A Systematic Review of Wearable Patient Monitoring Systems – Current Challenges and Opportunities for Clinical Adoption. Journal of Medical Systems, July 2017, 41:115.
15. Berthene A. There are many more mobile health apps, but consumer interest flags. October 17, 2016. Digital Commerce. https://www.digitalcommerce360.com/2016/10/17/many-mobile-health-apps-consumer-interest-flags/ (Accessed November 2, 2017)
16. AppBrain, Most popular Google Play categories. https://www.appbrain.com/stats/android-market-app-categories (Accessed February 2, 2018)
17. Gellert GA, Crouch JF, Gibson LA, et al. Clinical impact and value of workstation single sign-on. International Journal of Medical Informatics. Volume 101, May 2017, Pages 131-136
18. Kurtz SL. Identification of low, high, and super gelers and barriers to hand hygiene among intensive care unit nurses. American Journal of Infection Control, Volume 45, Issue 8, 1 August 2017, Pages 839–843.
19. Khanna RR, Wachter RM, Blum M. Reimagining Electronic Clinical Communication in the Post-Pager, Smartphone Era. JAMA. 2016;315(1):21–22.
20. Hirschorn DS, Choudhri AF, Shih G, et al. Use of Mobile Devices for Medical Imaging. J Am Coll Radiol 2014;11:1277-1285.
21. Kim PT, Falcone RA. The use of telemedicine in the care of the pediatric trauma patient. Seminars in Pediatric Surgery, Volume 26, Issue 1, February 2017, Pages 47-53.
22. Kuhn E, Greene C, Hoffman J, et al. Preliminary evaluation of PTSD Coach, a smartphone app for post-traumatic stress symptoms. Mil Med 2014;179(1):12-18.
23. Miner A, Kuhn E, Hoffman JE et al. Feasibility, acceptability and potential efficacy of the PTSD Coach app: A pilot randomized controlled trial with community trauma survivors. Psychological Trauma: Theory, Research, Practice and Policy 2016;8(3):384-392
24. Department of Veterans Affairs, Mobile App: PTSD Coach, https://www.ptsd.va.gov/public/materials/apps/PTSDCoach.asp (Accessed 11/5/2017)
25. Chow CK, Redfern J, Hillis GS, et al. Effect of Lifestyle-Focused Text Messaging on Risk Factor Modification in Patients With Coronary Heart Disease. Journal of the American Medical Association; 2015 Sep 22;314(12):1255.
26. Boriani G, Da Costa A, Quesada A, et al. Effects of remote monitoring on clinical outcomes and use of healthcare resources in heart failure patients with biventricular defibrillators: results of the MORE-CARE multicentre randomized controlled trial. European Journal of Heart Failure, Volume 19, Issue 3, March 2017, Pages 416–425.
27. Castro Sweet CM, Chiguluri V, Gumpina R, et al, Outcomes of a digital health program with human coaching for a diabetes risk reduction in a medical population. Journal of Aging and Health, January 2017, 1-19.
28. Forlenza GP, Deshpande S, Ly TT, et al. Application of Zone Model Predictive Control Artificial Pancreas During Extended Use of Infusion Set and Sensor: A Randomized Crossover-Controlled Home-Use Trial. Diabetes Care 2017 Jun; dc170500.
29. Do It Yourself Artificial Pancreas System https://diyps.org/about/ (Accessed November 2, 2017)
30. Powell DJH et al. Tracking daily fatigue fluctuations in multiple sclerosis: ecological momentary assessment provides unique insights. Journal of Behavioral Medicine. October 2017, Volume 40, Issue 5, pp 772–783.
31. Harzand A, Witbrodt B, Davis-Watts M, et al. Feasibility of a Smartphone-Delivered Cardiac Rehabilitation Program Amongst Veterans, Journal of the American College of Cardiology, Volume 69, Issue 11, Supplement, 21 March 2017, Page 2559.

32. Dandillaya R, Neumann M, Anderson S, et al. Evaluating the 30-Day Effects of a Comprehensive Remote Patient Monitoring, Shortened Provider Feedback Interval, and Patient Engagement and Education Program on Hypertensive Patients. Journal of Healthcare Information Management, Fall 2013, 27(4), 28–35.
33. Merchant RK. Effectiveness of Population Health Management Using the Propeller Health Asthma Platform: A Randomized Clinical Trial. The Journal of Allergy and Clinical Immunology: In Practice. Volume 4, Issue 3, May–June 2016, Pages 455–463.
34. Chan YY, Wang P, Rogers L, et al. The Asthma Mobile Health Study, a large-scale clinical observational study using ResearchKit. Nature Biotechnology 35, 354–362 (2017).
35. Mookherji S, Mehl G, Kaonga N, Mechael P. Unmet Need: Improving mHealth Evaluation Rigor to Build the Evidence Base. J Health Commun. 2015;20(10):1224-9.
36. World Health Organization. mHealth. New horizons for health through mobile technologies. Global Observatory for eHealth series – volume 3. 2011, www.who.int/goe/publications/goe_mhealth_web.pdf (Accessed November 2, 2017)
37. Apple HealthKit: Develop health and fitness apps that work together https://developer.apple.com/healthkit/ (Accessed November 2, 2017)
38. Apple ResearchKit and CareKit. https://www.apple.com/researchkit/ (Accessed November 2, 2017)
39. Johns Hopkins Corrie Health app, https://itunes.apple.com/us/app/corrie-health/id1212463532?mt=8 (Accessed November 2, 2017)
40. Clinical Research Study at Johns Hopkins: MICORE, https://corriehealth.com/#micore (Accessed November 2, 2017)
41. Sage Bionetworks, Parkinson mPower study app, https://itunes.apple.com/us/app/parkinson-mpower-study-app/id972191200 (Accessed November 2, 2017)
42. Google Fit Platform Overview, https://developers.google.com/fit/overview (Accessed November 2, 2017)
43. ResearchDroid – Android port of ResearchKit, http://appliedinformaticsinc.com/researchdroid/ (Accessed November 2, 2017)
44. TrialX, America Walks Study, http://trialx.com/americawalksstudy/ (Accessed November 2, 2017)
45. FDA, Mobile Medical Applications, https://www.fda.gov/MedicalDevices/DigitalHealth/MobileMedicalApplications/ (Accessed November 2, 2017)
46. FDA selects participants for new digital health software precertification pilot program https://www.fda.gov/NewsEvents/Newsroom/PressAnnouncements/ucm577480.htm (Accessed November 2, 2017)
47. Preetinder S Gill, Ashwini Kamath, Tejkaran S Gill, Distraction: an assessment of smartphone usage in health care work settings, Risk Manag Healthc Policy. 2012; 5: 105–114.
48. Little KS. Cell Phones and Medical Malpractice. March 5, 2015. Healthcare Law Blog. https://www.healthcarelaw-blog.com/2015/03/cell-phones-medical-malpractice.html (Accessed November 2, 2017)
49. Warwick A. How to formulate an effective smartphone security policy. ComputerWeekly.com. 2012. http://www.computerweekly.com/feature/How-to-formulate-an-effective-smartphone-security-policy . (Accessed November 2, 2017)
50. Brady RR, Fraser SF, Dunlop MG, Paterson-Brown S, Gibb AP. Bacterial contamination of mobile communication devices in the operative environment. J Hosp Infect. 2007 66(4):397–398.
51. JP Attri, R Khetarpal, V Chatrath, and J Kaur, Concerns about usage of smartphones in operating room and critical care scenario, Saudi J Anaesth. 2016 Jan-Mar; 10(1): 87–94.
52. Stoyanov SR, Hides L, Kavanagh DJ, Zelenko O, Tjondronegoro D, Mani M. Mobile app rating scale: a new tool for assessing the quality of health mobile apps. JMIR mHealth uHealth. JMIR Publications Inc.; 2015 Mar 11;3(1):e27
53. Direito A, Carraça E, Rawstorn J, Whittaker R, Maddison R. mHealth Technologies to Influence Physical Activity and Sedentary Behaviors: Behavior Change Techniques, Systematic Review and Meta-Analysis of Randomized Controlled Trials. Ann Behav Med. 2017;51(2):226–39.
54. McCarroll R, Eyles H, Ni Mhurchu C. Effectiveness of mobile health (mHealth) interventions for promoting healthy eating in adults: A systematic review. Vol. 105, Preventive Medicine. 2017. p. 156–68.
55. Tomlinson M, Rotheram-Borus MJ, Swartz L, Tsai AC. Scaling Up mHealth: Where Is the Evidence? PLoS Med [Internet]. Public Library of Science; 2013 Feb 12.
56. Fiordelli M, Diviani N, Schulz PJ. Mapping mHealth Research: A Decade of Evolution. J Med Internet Res [Internet]. 2013 May 21;15(5):e95.
57. Kumar S, Nilsen WJ, Abernethy A, Atienza A, Patrick K, Pavel M, et al. Mobile health technology evaluation: The mHealth evidence workshop. Vol. 45, American Journal of Preventive Medicine. 2013. p. 228–36.
58. Sweeney P. 10 best IT practices for smartphone security. TechNewsWorld. 2010, https://www.technewsworld.com/story/70826.html . (Accessed November 2, 2017)

59. mHealth resources to strengthen health programs. Global Health: Science and Practice. 2014;2(1):130-131
60. K4Health, mHealth Evidence, https://www.mhealthevidence.org/ (Accessed November 2, 2017)
61. NIH Fogarty International Center. Mobile Health Information and Resources. https://www.fic.nih.gov/RESEARCHTOPICS/Pages/MobileHealth.aspx (Accessed November 2, 2017)
62. Agarwal S, LeFevre AE, Lee J et al. Guidelines for reporting of health interventions using mobile phones: mobile health evidence reporting and assessment (mERA) checklist. BMJ. 2016;352;i1174.
63. Irbarren SJ, Cato K, Falzon L, Stone PW. What is the economic evidence for mHealth? A systematic review of economic evaluations of mHealth solutions. PLOS. 2017.
64. Hall AK, Cole-Lewis H, Bernhardt JM. Mobile text messaging for health: a systematic review of reviews. Ann Rev Pub Health 2015;36:393-415

14

Evidence-Based Medicine and Clinical Practice Guidelines

ROBERT E. HOYT • WILLIAM R. HERSH

"The great tragedy of Science - the slaying of a beautiful hypothesis by an ugly fact"
—Thomas Huxley (1825-1875)

LEARNING OBJECTIVES

After reading this chapter the reader should be able to:

- Explain the definition and origin of evidence-based medicine
- Define the benefits and limitations of evidence-based medicine
- Describe the evidence pyramid and levels of evidence
- Discuss the process of using evidence-based medicine to answer a medical question
- Compare the most important online and smartphone evidence-based medicine resources
- Describe the interrelationship between clinical practice guidelines, evidence-based medicine and electronic health records
- Define the processes required to create and implement a clinical practice guideline

INTRODUCTION

Evidence-Based Medicine (EBM) is included in a textbook on health informatics because information technology has the potential to improve decision making through online medical resources, electronic clinical practice guidelines, electronic health records (EHRs) with decision support, online literature searches, digital statistical analysis and online continuing medical education (CME). This chapter is devoted to finding the best available evidence and discussing one of its end products, clinical practice guidelines. According to the Centre for Evidence-based Medicine, EBM can be defined as:

"a systematic approach to clinical problem solving which allows the integration of the best available research evidence with clinical expertise and patient values"[1]

Furthermore, in *Crossing the Quality Chasm,* the Institute of Medicine (IOM) states:

"Patients should receive care based on the best available scientific knowledge. Care should not vary illogically from clinician to clinician or from place to place"[2]

In other words, the IOM is saying that every effort should be made to find the best answers and that these answers should be standardized and shared among clinicians. Such standardization implies that clinical practice should be consistent with the best available evidence that would apply to most patients. This is easier said than done because clinicians are independent practitioners and interpret patient findings and research results differently. It is true that many questions cannot be answered by current evidence, so clinicians may have to turn to subject matter experts. In addition, clinicians lack the time and the tools to seek the best evidence. Furthermore, greater than 1,800 citations are added to MEDLINE every day, making it impossible for a practicing clinician to stay up-to-date with the medical literature. Likewise, interpreting this evidence requires expertise and knowledge that not every clinician has. One does not have to look very far to see how evidence changes recommendations, e.g. bed rest is no longer recommended for low back pain[3] or following a spinal tap (lumbar puncture); routine activity is recommended instead.[4]

Three pioneers are closely linked to the development of EBM. Gordon Guyatt coined the term EBM

283

in 1991 in the American College of Physician (ACP) Journal Club.[5] The initial focus of EBM was on clinical epidemiology, methodology and detection of bias. This created the first fundamental principle of EBM: not all evidence is equal; there is a hierarchy of evidence that exists. In the mid-1990s, it was realized that patients' values and preferences are essential in the process of decision making and addressing these values has become the second fundamental principle of EBM, after the hierarchy of evidence. Archie Cochrane, a British epidemiologist, was another early proponent of EBM. The Cochrane Collaboration was named after him as a tribute to his early work. The Cochrane Collaboration consists of review groups, centers, fields, methods groups and a consumer network. Review groups, located in 13 countries, perform systematic reviews based on randomized controlled trials. As of mid-2017 the Cochrane Collaboration had completed over 7284 subject reviews and 2548 protocol reviews. The Cochrane Database of Sytematic Reviews has a high impact factor of 6.1.[6-7] The rigorous reviews are performed by volunteers, so efforts are slow. David Sackett is another EBM pioneer who was hugely influential at The Centre for Evidence-based Medicine in Oxford, England and at McMaster University, Ontario, Canada. EBM has also been fostered at McMaster University by Dr. Brian Haynes who is the Chairman of the Department of Clinical Epidemiology and Biostatistics and the editor of the American College of Physician's (ACP) Journal Club.

The first randomized controlled trial was published in 1948.[8] For the first time, subjects who received a drug were compared with similar subjects who would receive another drug or placebo and the outcomes were evaluated. Subsequently, studies became *"double blinded,"* meaning that both the investigators and the subjects did not know whether they received an active medication or a placebo. Until the 1980s evidence was summarized in review articles written by experts. However, in the early 1990s, systematic reviews and meta-analyses became known as a more focused, objective, and rigorous way to summarize evidence and the preferred way to present the best available evidence to clinicians and policy makers. Since the late 1980s more emphasis has been placed on improved study design and true patient outcomes research. It is no longer adequate to show that a drug reduces blood pressure or cholesterol; it should demonstrate an improvement in patient-important outcomes such as reduced strokes or heart attacks.[9]

Despite some reluctance by the US to embrace EBM universally, the US federal government has established multiple Evidence-based Practice Centers to conduct systematic reviews of topics in clinical medicine, social and behavioral science and economics.[10] More recently, nine US medical societies participated in a 2012 initiative known as Choosing Wisely that lists 45 low-value medical tests and therapies that are strongly discouraged, based on best evidence.[11]

IMPORTANCE OF EBM

Learning EBM is like climbing a mountain to gain a better view. One might not make it to the top and find the perfect answer, but individuals will undoubtedly have a better vantage point than those who choose to stay at sea level. Reasons for studying EBM resources and tools include:
- Current methods of keeping medically or educationally up-to-date do not work
- Translation of research into practice is often very slow
- Lack of time and the volume of published material results in information overload
- The pharmaceutical and medical device industries bombard clinicians and patients every day; often with misleading or biased information. They also heavily influence research and publications. Issues resulting from their influence include: treating questionable or early diseases before evidence is in; overpowering studies so there is a statistical but not clinically significant outcome; establishing inclusion criteria so that patients most likely to respond to treatment are included; using surrogate and not clinical endpoints and only selecting studies with positive results.[12]
- Much of what is considered the *"standard of care"* in every day practice has yet to be challenged and could be wrong

Without proper EBM training clinicians will not be able to appraise the best information resulting in poor clinical guidelines and wasted resources.

Traditional Methods for Gaining Medical Knowledge

- Continuing Medical Education (CME). Traditional CME is desired by many clinicians, but the evidence shows it to be highly ineffective and does not lead to changes in practice. In general, busy clinicians are looking for a non-stressful evening away from their practice or hospital with food and drink provided.[13-14] Much CME is provided free by pharmaceutical companies with their inherent biases. Better educational methods must be developed. A recent study demonstrated that online CME was at least comparable, if not superior to traditional CME.[15]

- Clinical Practice Guidelines (CPGs). This will be covered in more detail, later in this chapter. Unfortunately, just publishing CPGs does not in and of itself change how medicine is practiced and the quality of CPGs is often variable and inconsistent.
- Expert Advice. Experts often approach a patient in a significantly different way compared to primary care clinicians because they deal with a highly selective patient population. Patients are often referred to specialists because they are not doing well and have failed treatment. For that reason, expert opinion needs to be evaluated with the knowledge that their recommendations may not be relevant to a primary care population. Expert opinion therefore should complement and not replace EBM.
- Reading. Most clinicians are unable to keep up with medical journals published in their specialty. Clinicians can only devote a few hours each week to reading. All too often information comes from pharmaceutical representatives visiting the office. Moreover, recent studies may contradict similar prior studies, leaving clinicians confused as to the best course.

EBM Steps to Answering Clinical Questions

The following are the typical steps a clinician might take to answer a patient-related question:
- The physician sees a patient and generates a clinical well-constructed question. Here is the **PICO** method, developed by the National Library of Medicine:
  o **P**atient or problem: what is the patient group of interest? Elderly? Gender? Diabetic?
  o **I**ntervention: what is being introduced, a new drug or test?
  o **C**omparison: with another drug or placebo?
  o **O**utcome: what needs to be measured? Mortality? Hospitalizations? A web-based PICO tool has been created by the National Library of Medicine to search Medline. This tool can be placed as a short cut on any computer.[16]
  o It has been recently suggested to add a T and S to PICO (i.e., PICOTS) to indicate the **T**ype of study that would best answer the PICO question and the **S**etting where it would take place.
- Seek the best evidence for that question via an EBM resource or PubMed.
- Critically appraise that evidence using tools mentioned in this chapter. Examine internal and external validity and the potential impact of an intervention
- Apply the evidence to your patient considering patient's values, preferences and circumstances[15]

There are many more detailed treatises of EBM; probably the best and oldest is the textbook Evidence-Based Medicine: How to Practice and Teach It, by Straus, Glasziou, Richardson and Haynes, now in its fourth edition.[17]

Terminology Used in Answering Clinical Questions

- Evidence appraisal: When evaluating evidence, one needs to assess its validity, results and applicability.
- Validity: Validity means is the study believable? If apparent biases or errors in selecting patients, measuring outcomes, conducting the study, or analysis are present, then the study is less valid. If the study is poorly designed, it will have poor *internal validity*. Table 14.1 lists some common sources of research bias.[18]
- Results: Results should be assessed in terms of the magnitude of treatment effect and precision (narrower confidence intervals or statistically significant results indicate higher precision).
- Applicability: Also called *external validity*, applicability indicates that the results reported in the study can be generalized to the patients of interest.[19]

Most Common Types of Clinical Questions

- Therapy question. This is the most common area for medical questions and the primary one discussed in this chapter
- Prognosis question
- Diagnosis question
- Harm question
- Cost question

THE EVIDENCE PYRAMID

The pyramid in Figure 14.1 represents the different types of medical studies and their relative ranking. The starting point for research is often animal studies and the pinnacle of evidence is the meta-analysis of randomized trials. With each step up the pyramid our evidence is of higher quality associated with fewer articles published.[20] Although systematic reviews and meta-analyses are the most rigorous means to evaluate a medical question, they are expensive, labor intensive, and their inferences are limited by the quality of the evidence of the original studies.
- Case reports/case series. Consist of collections of reports on the treatment of individual patients without control groups; therefore, they have much less scientific significance.

Table 14.1: Common Sources of Research Bias

Categories	Bias Types	Explanation
Selection bias	1. Incidence-prevalence 2. Loss to followup 3. Publication	1. Bias that occurs when disease risk is estimated based on data collected at a time point on survivors, rather than on incident cases 2. Patients lost to followup may have a different outcome of interest so can skew results 3. Positive outcome studies are more likely to be published

Information bias	1. Misclassification
	   o Recall
	   o Interviewer
	   o Observer
	   o Regression dilution
	2. Lead time
		1. Imperfect classification of exposure status or outcome
		   a. Imprecise memory
		   b. Interviewer gets answers that support a preconceived hypothesis
		   c. Observer's knowledge of the exposure status influences assessment
		   d. A variable that shows extreme value on first measurement will be less extreme on repeat measures
		2. Outcome difference based on when in the disease course the diagnosis was made

- Case control studies. Study patients with a specific condition (retrospective or after the fact) and compare with people who do not. These types of studies are often less reliable than randomized controlled trials and cohort studies because showing a statistical relationship does not mean that one factor necessarily caused the other.
- Cohort studies. Evaluate (prospectively or followed over time) and follow patients who have a specific exposure or receive a particular treatment over time and compare them with another group that is similar but has not been affected by the exposure being studied. Cohort studies are not as reliable as randomized controlled studies, since the two groups may differ in ways other than the variable under study.
- Randomized controlled trials (RCTs). Subjects are randomly assigned to a treatment or a control group that received placebo or no treatment. The randomization assures that patients in the two groups are balanced in both known and unknown prognostic factors, and that the only difference between the two groups is the intervention being studied. RCTs are often "*double blinded*" meaning that both the investigators and the subjects do not know whether they received an active medication or a placebo. This assures that patients and clinicians are less likely to become biased during the conduct of a trial, and the randomization effect remains protected throughout the trial. RCTs are considered the gold standard design to test therapeutic interventions.
- Systematic reviews. Defined as protocol-driven comprehensive reproducible searches that aim at answering a focused question; thus, multiple RCTs are evaluated to answer a specific question. Extensive literature searches are conducted (usually by several different researchers to reduce selection bias of references) to identify studies with sound methodology; a very time-consuming process. The benefit is that multiple RCTs are analyzed, not just one study. Standardized systematic review instruments, such as the Jadad scale can be used to evaluate the quality of individual RCTs.[21] Another popular rating tool is PRISMA (Preferred Reporting Items for Systematic Reviews and Meta-analyses), a 27-item check list.[22]
- Meta-analyses. Defined as the quantitative summary of systematic reviews that take the systematic review a step further by using statistical techniques to combine the results of several studies as if they were one large single study.[15] Meta-analyses offer two advantages compared to individual studies. First, they include a larger number of events, leading to more precise (i.e., statistically significant) findings. Second, their results apply to a wider range of patients because the inclusion criteria of systematic reviews are inclusive of criteria of all the included studies.[19]

Figure 14.1: The Evidence Pyramid

Table 14.2: Suggested studies for questions asked

Type of Question	Suggested Best Type of Study
Therapy	RCT > cohort > case control > case series
Diagnosis	Prospective, blind comparison to a gold standard
Harm	RCT + cohort > case control > case series
Prognosis	Cohort study > case control > case series
Cost	Economic analysis and modeling

Table 14.2 lists the suggested studies for different questions asked. This chapter will deal primarily with therapy questions so note that RCTs are the suggested study of choice.[23] Studies that don't randomize patients or introduce a therapy along with a control group are referred to as observational studies (case control, case series and cohort) and are usually retrospective in nature. By their nature, RCTs are prospective and not retrospective.

Evidence of harm should be derived from both RCTs and cohort study designs. Cohort studies have certain advantages over RCTs when it comes to assessing harm: larger sample size, longer follow up duration, and more permissive inclusion criteria that allow a wide range of patients representing a real-world utilization of the intervention to be included in the study.

Levels of Evidence (LOE)

Several methods have been suggested to grade the quality of evidence, which on occasion, can be confusing. The most up-to-date and acceptable framework is the GRADE (Grading of Recommendations, Assessment, Development and Evaluation).[24-25] More than 100 organizations, from 19 countries endorse or use GRADE, such as the BMJ, Cochrane Collaboration and UpToDate. A variety of EBM tools are available on their website. Fiftheen GRADE articles on EBM have been published in the Journal of Clinical Epidemiology from 2011-2012. The following is a description of the levels of evidence in this framework:

- Level 1: High quality evidence (usually derived from consistent and methodologically sound RCTs)
- Level 2: Moderate quality evidence (usually derived from inconsistent or less methodologically sound RCTs; or exceptionally strong observational evidence)
- Level 3: Low quality evidence (usually derived from observational studies)
- Level 4: Very low-quality evidence (usually derived from flawed observational studies, indirect evidence or expert opinion)

In this framework, RCTs start with a level 1 and observational studies start with a level 3. Factors that could decrease the level of quality include: design limitations, excessive heterogeneity, imprecision (wide confidence intervals) and high probability of publication bias. The rationale for this rating system reflects the rigor of the RCTs and the strong inference they provide. For example, a recent systematic review and meta-analysis reported that seven observational (non-randomized) studies demonstrated a beneficial association between chocolate consumption and the risk of cardiometabolic disorders.[26] The highest levels of chocolate consumption were associated with significant reduction in cardiovascular disease and stroke compared with the lowest levels. Although these results seem impressive at face value, it is implausible that the effect of chocolate consumption is that profound (37% and 29% reduction in the risk of cardiovascular disease and stroke). This magnitude of effect rivals the best available drugs and interventions used to prevent these diseases. Observational studies like these, have likely exaggerated the magnitude of benefit due to many factors (i.e., bias and confounding). It is possible that chocolate users are healthier, wealthier, more educated or have other characteristics that make them have lower incidence of disease. The opposite is also possible. Therefore, our confidence in estimates of effects generated from observational studies is lower than that of randomized trials. Hence, one derives evidence with different quality rating. Furthermore, it is important to recognize that the quality of evidence can be upgraded or downgraded if additional criteria based on study methodology and applicability is available.

RISK MEASURES AND TERMINOLOGIES

Overall, therapy trials are the most common area of research and ask questions such as, is drug A better than drug B or placebo? To determine what the true effect of a study is, it is important to understand the concept of risk reduction and the number needed to treat. These concepts are used in studies that have dichotomous outcomes (i.e., only two possible answers such as dead or alive, improved or not improved); which are more commonly utilized outcomes. The chapter will define these concepts and then present an example for illustration.
1. Risk describes the probability an adverse event will occur.
2. Odds is the ratio of the probability that an event will occur to the probability that it will not occur.[27]

Risk can be converted to odds and vice versa with the following formulas:
Risk = Odds/1+ odds
Odds= risk / (1-risk)

Example of risk measures

Amazingstatin is a drug that lowers cholesterol. If a physician treats 100 patients with this drug and five of them suffer a heart attack over a period of 12 months, the risk of having a heart attack in the treated group would be 5/100= 0.050 (or 5%). The odds of having a heart attack would be 5/95= 0.052. In the control group, if he or she treats 100 patients with placebo and seven suffer heart attacks, the risk in this group is 7/100=0.070 or 7% and the odds are 7/93=0.075.

Notice that the risk in the experimental group is called experimental event rate (EER) and the risk in the control group is called control event rate (CER). To compare risk in two groups, the following terms are used:

Relative Risk is the ratio of two risks as defined above. Thus, it is the ratio of the event rate of the outcome in the experimental group (EER) to the event rate in the control group (CER).
- RR = EER/CER

Relative Risk Reduction is the difference between the experimental event rate (EER) and the control event rate (CER), expressed as a percentage of the control event rate.
- RRR = (EER-CER)/CER

Absolute Risk Reduction is the difference between the EER and the CER.
- ARR = EER-CER

(Note that "difference" is not the same as subtracting CER from EER. For example, if the EER is 1.5 and the CER is 2.0, the difference is .5, not -.5)

Number Needed to Treat is the number of patients who must receive the intervention to prevent one adverse outcome.[28]
- NNT = 1/ARR (or 100/ARR, if ARR is expressed as a percentage instead of a fraction)

Odds Ratio is the ratio of odds (instead of risk) of the outcome occurring in the intervention group to the odds of the outcome in the control group.

On Amazingstatin, 5% (EER) of patients have a heart attack after 12 months of treatment. On placebo 7% (CER) of patients have a heart attack over 12 months
RR = 5% /7% = 0.71
RRR = (7% - 5%) /7% = 29%
ARR = 7% - 5% = 2%
NNT = 100/2 = 50

In summary, on average, 50 patients must be treated with Amazingstatin over 12 months to prevent one heart attack. As calculated above, the odds for the intervention and control group respectively are 0.052 and 0.075; the odds ratio (*OR*) = 0.52/0.075 = 0.69.

Comments

RR and OR are very similar concepts and if the event rate is low, their results are almost identical. These results show that this drug cuts the risk of heart attacks by 29% (almost by a third), which seems like an impressive effect.

However, the absolute reduction in risk is only 2% and therefore 50 patients need to be treated to prevent one adverse event. Although this NNT may be acceptable, using RRR seems to exaggerate our impression of risk reduction compared with ARR. Most of what is written in the medical literature and the lay press cites the RRR. Unfortunately, very few studies offer NNT data, but it is very easy to calculate if the ARR specific to your patient is known. Nuovo et al. noted that NNT data was infrequently reported by five of the top medical journals despite being recommended.[29] In another interesting article, Lacy and co-authors studied the willingness of US and UK physicians to treat a medical condition based on the way data was presented. Ironically, the data was the same but presented in three different ways.[30]

Examples of Using RRR, ARR and NNT

A full-page article appeared in a December 2005 Washington Post newspaper touting the almost 50% reduction of strokes by a cholesterol lowering drug. This presented an opportunity to look at how drug companies usually advertise the benefits of their drugs. Firstly, in small print, the reader notes that patients must be diabetic with one other risk factor for heart disease to see benefit. Secondly, there are no references. The statistics are derived from the CARDS Study published in the Lancet in Aug 2004.[31] Stroke was reported to occur in 2.8% in patients on a placebo and 1.5% in patients taking the drug Lipitor. The NNT is therefore 100/1.3 or 77. So, on average, a physician would have to treat 77 patients for an average of 3.9 years (the average length of the trial) to prevent one stroke. This doesn't sound as good as *"cuts the risk by nearly half."* Now armed with these EBM tools, look further the next time a miraculous drug effect is advertised.

Number Needed to Harm (NNH) is calculated similarly to the NNT. If, for example, Amazingstatin was associated with intestinal bleeding in 6% of patients compared to 3% on placebo, the NNH is calculated by dividing the ARR (%) into 100. For our example, the calculation is 100/.03 = 33. In other words, the treatment of 33 patients with Amazingstatin for one year resulted, on average, in one case of intestinal bleeding because of the treatment. Unlike NNT, the higher the NNH, the better.

The Case of Continuous Variables and effect size. The results of studies (effect measures) described so far (i.e., RR, OR, ARR) are used when outcomes are dichotomous (such as dead or alive, having a heart attack or not, etc.). However, outcomes can also be continuous or numerical (e.g., blood cholesterol level). These outcomes are usually reported as a difference in the means of two study groups. This difference has a unit, which in the cholesterol example, is mg/dL. In addition to the mean difference, results would also include some measure that describes the spread or dispersion of measurements around the mean (i.e., standard deviation, range, interquartile range or a confidence interval).

If the metrics of continuous variables do not have intuitive intrinsic meaning (e.g., a score on a test or a scale), the *effect size* can be standardized (i.e., difference in means is divided by the standard deviation; which makes the data measured in standard deviation units). This process allows the comparison of students taking different tests, or tests taken in different years, or comparing the results of studies that used different scales as their outcomes. This is possible because all these measurements are standardized (have the same unit, which is standard deviation unit). A commonly used effect size is Cohen's d, which is a standardized difference in means. It is interpreted arbitrarily as a small, moderate or large effect, if d was 0.2, 0.5 or 0.7; respectively. In addition to knowing that a result is statistically significant, calculating the effect size gives one an idea of how big the difference is. Online effect size calculators are available.[32]

Confidence Intervals. Most results published in journals will include confidence intervals that give the reader an idea of the precision of the results. In other words, if the result of interest is a mean of 5.4 kilograms and the 95% confidence intervals are 3.9 - 6.9, this means there is a 95% chance the true result lies somewhere between 3.9 and 6.9 (and 5% would fall outside this range). Be leary of results with wide confidence intervals as this frequently means the sample size was too small. Also, if the confidence intervals include zero (example -3.0 to 30) one can't be sure an intervention had a positive or negative effect.[33] The formula to calculate CIs is complicated, but many online calculators exist.[34]

Cost of Preventing an Event (COPE). Many people reviewing a medical article would want to know what the cost of the intervention is. A simple formula exists that sheds some light on the cost: COPE = NNT x number of years treated x 365 days x the daily cost of the treatment. Using our example of Amazingstatin = 50 x 1 x 365 x $2 or $36,500 to treat 50 patients for one year to prevent one heart attack. COPE scores can be compared with other similar treatments.[35]

LIMITATIONS OF THE MEDICAL LITERATURE AND EBM

Because evidence is based on information published in the medical literature, it is important to point out some

of the limitations researchers and clinicians must deal with on a regular basis:
- There is a low yield of clinically useful articles in general.[36]
- Conclusions from randomized drug trials tend to be more positive if they are from for-profit organizations.[37]
- Up to 16% of well publicized articles are contradicted in subsequent studies.[38] A more recent review of articles published from 2001 to 2010 in just the New England Journal of Medicine concluded that 40% represented reversal of prior recommendations.[39]
- Even systematic reviews have their limitations. An evaluation of over one thousand reviews in the Cochrane Library revealed that 44% of treatments were likely to be beneficial but in only 1% was no further research recommended. Similarly, they found that 49% of interventions were not determined to be either helpful or harmful.[40] Another review of systematic reviews and meta-analyses reported most were in specialty journals, most dealt with therapies, funding resources were not reported in 40% and about half did not state the article was a systematic review or meta-analysis in the title or abstract.[41]
- Peer reviewers are *"unpaid, anonymous and unaccountable"* so it is often not known who reviewed an article and how rigorous the review was.[42]
- Many medical studies are poorly designed:[43]
    o The recruitment process was not described.[44]
    o Inadequate power (size) to make accurate conclusions. In other words, not enough subjects were studied.[45]
    o Studies published in high-impact journals attract a lot of attention but are often small randomized trials with results that may not be duplicated in future studies. This may be positive publication bias.[46]
    o Studies with negative results (i.e., results that are not statistically significant) are not always published or take more time to be published, resulting in *"publication bias."* To prevent this type of bias the American Medical Association advocates mandatory registration of all clinical trials in public registries. Also, the International Committee of Medical Journal Editors requires registration as a condition to publish in one of their journals. However, they do not require publishing the results in the registry at this time. Registries could be a data warehouse for future mining and some of the well-known registries include:
        § ClinicalTrials.gov
        § WHO International Clinical Trials Registry
        § Global Trial Bank of the American Medical Informatics Association
        § Trial Bank Project of the University of California, San Francisco[47]

Although EBM is considered a highly academic process towards gaining medical truth, numerous challenges exist:
- Different evidence rating systems by various medical organizations
- Different conclusions by experts evaluating the same study
- Time intensive exercise to evaluate existing evidence
- Systematic reviews are limited in the topics reviewed (over 7,000 in the Cochrane Library in 2017) and are time intensive to complete (6 to 24 months). Often, the conclusion is that current evidence is weak and further high-quality studies are necessary
- Randomized controlled trials are expensive. Drug companies tend to fund only studies that help a current non-generic drug they would like to promote
- Results may not be applicable to every patient population; i.e. external validity or generalizability
- Some argue that we should aim at *"evidence-based health,"* instead of EBM. This would entail correcting many societal structural deficiencies, in addtion to health care, requiring substantial financial obligations[48]
- Some view EBM as *"cookbook medicine"*[49]
- There is not good evidence that teaching EBM changes behavior[50]

Other Approaches

EBM has had both strong advocates and skeptics since its inception. One of its strongest proponents Dr. David Sackett published his experience with an "Evidence Cart" on inpatient rounds in 1998. The cart contained numerous EBM references but was so bulky that it could not be taken into patient rooms.[51] Since that article, multiple, more convenient EBM solutions exist. While there are those EBM advocates who would suggest the sole use of EBM resources, many others feel that EBM *"may have set standards that are untenable for practicing physicians."*[52-53]

Dr. Frank Davidoff believes that most clinicians are too busy to perform literature searches for the best evidence. He believes that healthcare needs "Informationists" who are experts at retrieving information.[54] To date, only clinical medical librarians (CMLs) have the formal training to take on this role. At large academic centers CMLs join the medical team on inpatient rounds and attach pertinent and filtered articles to the chart. As an example, Vanderbilt's Eskind Library has a Clinical Informatics Consult Service.[55-56] The obvious drawback is that CMLs are only available at large medical centers and are unlikely to research outpatient questions. As covered in the chapter on EHRs, some vendors embed high quality resources, such as UpToDate in their software or have options using infobuttons.

According to Slawson and Shaughnessy clinicians must become an "*information master*" to sort through the "*information jungle*." They define the usefulness of medical information as:

Usefulness = $\frac{Validity \times Relevance}{Work}$

Only the clinician can determine if the article is relevant to his/her patient population and if the work to retrieve the information is worthwhile. Slawson and Shaughnessy also developed the notion of looking for "*patient oriented evidence that matters*" (POEM) and not "*disease oriented evidence that matters*" (DOEM). POEMS look at mortality, morbidity and quality of life, whereas DOEMS tend to look at laboratory or experimental results. They point out that it is more important to know that a drug reduces heart attacks or deaths (POEM), rather than just reducing cholesterol levels (DOEM).[57] This school of thought also recommends that clinicians not read medical articles blindly each week but should instead learn how to search for patient-specific answers using EBM resources.[58] This also implies that physicians are highly motivated to pursue an answer, have adequate time and have the appropriate training. See case study below for example of EBM being applied to a clinical scenario.

EVIDENCE-BASED HEALTH INFORMATICS (EBHI)

EBHI is not a separate field, it represents the application of EBM tools to the field of health informatics. Dr. Elske Ammenwerth, a major proponent of EBHI defined this approach in 2006 as the "*conscientious, explicit and judicious use of current best evidence when making decisions about information technology in healthcare.*"[59] While the quality of health informatics research has improved in the past decade, the overall report card for most studies is mixed, regardless of which technology

---

EBM Case Study

People with blockage of the carotid artery are at risk of stroke and death. They can be treated via surgery (called endarterectomy) or a less invasive procedure (putting a stent in the blocked area by going through the arteries, i.e., without surgery). The choice of procedure is controversial.

The evidence
A systematic review and meta-analysis appraised the quality of the totality of existing evidence in this area. They found 13 randomized controlled trials that enrolled a total of 7,484 patients. The methodological quality of the trials was moderate to high. Compared with carotid endarterectomy, stenting was associated with increased risk of stroke (relative risk [RR], 1.45; 95% confidence interval [CI], 1.06-1.99) and decreased risk of myocardial infarction (MI) caused by surgery (RR, 0.43; 95% CI, 0.26- 0.71). For every 1,000 patients opting for stenting rather than endarterectomy, 19 more patients would have strokes and 10 fewer would have MIs.

Patients values, preferences and context
Patients vary in their values such as aversion (fear) of stroke vs death and their fear of surgery and surgical complications such as scars in the neck and anesthesia. Patients also vary in their surgical risk (e.g., those with history of heart disease may prefer less invasive procedure to avoid prolonged anesthesia).

Guidelines
Due to the different impact of these procedures on the different outcomes, the guidelines were nuanced and stratified and allowed patients values and preferences, age, surgical and anatomical risk factors to be used in decision making. This example highlights the importance of patients' values and preferences as the second principle of EBM

References
Murad MH, Shahrour A, Shah ND, Montori VM, Ricotta JJ. A systematic review and meta-analysis of randomized trials of carotid endarterectomy vs stenting. J Vasc Surg. 2011 Mar;53(3):792-7. Epub 2011 Jan 8

Ricotta JJ, Aburahma A, Ascher E, et al. Updated Society for Vascular Surgery guidelines for management of extracranial carotid disease: executive summary. J Vasc Surg. 2011 Sep;54(3):832-6.

is being studied.⁶⁰ There are at least three reasons why published research studies in health informatics have not been fully evidence-based:

- **Early Hype**. In multiple other chapters the overly optimistic predictions regarding the impact of HIT on healthcare quality, safety, proficiency and cost reduction is pointed out. Many of these predictions were based on expert opinions or modeling and not high-quality research. The hype was not isolated to HIT vendors and techno-enthusiasts; it was shared by academia and the federal government. It was aggravated by *"technology pressure"* or the natural tendency to try to fit new technologies into healthcare, even when the benefits have not been proven. This tends to raise expectations and may cause governments to introduce technology friendly policies, prior to having all the facts. Early success stories were widely broadcast, even though many of the early innovations came from several medical centers with a track record for home grown successful technology.[61-62]

- **Methodological challenges**. Early research studies frequently suffered from internal validity (quality of study design and execution) and external validity (whether results are generalizable to other locations and patients) issues. Most health information technology (HIT) studies reported are observational and retrospective in nature. Many are before/after studies. This distinction is important because cause and effect are difficult to prove with observational studies, compared to prospective RCTs. Randomization and blinding are difficult with health information technology. As an example, randomizing physicians to electronic prescribing (vs. paper prescribing) is difficult to implement and often impractical. In an observational study, physicians who volunteer to try electronic prescribing are likely *"early adopters"* and not representative of average physicians, which could skew the results. Alternate methods of randomization are feasible and desired. For example, *"cluster randomization"* would be a practical methodology in this situation. With this method, several clinics or hospitals can be randomized as a whole practice to electronic prescribing whereas other clinics or hospitals can be randomized to paper prescribing. HIT interventions are complicated in nature and one could argue represent a technosocioeconomic experience. Early studies tended to have small sample sizes, short term outcomes, inadequate endpoints, inadequate cost data and few comments about negative effects.

  Clearly, there are HIT innovations that are popular and save time such as drug look-up apps for mobile technology, patient portals and voice recognition but they have been poorly studied so there is a lack of good qualitative and quantitative data about their overall effect.

  There are several articles that focus on the methodological challenges of HIT research along with recommendations.[63-65]

  Dr. Ammenwerth has been instrumental in developing guidelines for *evaluating* health informatics (GEP-HI) and *reporting* health informatics studies (STARE-HI).[66-67]

- **The failure to anticipate unintended consequences related to HIT adoption**. Weiner coined the term *"e-iatrogenesis"* in 2007 to describe adverse events related to technology.[68] Sittig and Singh divided unintended consequences into: technology unavailable; technology malfunctions and technology functions but there is human error (e.g. e-presribing works properly but clinician entered wrong drug dose).[69] Additional aspects of unintended consequences that include patient safety issues are as follows:

  o The Joint Commission issued a Sentinel Event alert in 2008 to alert healthcare workers that 25% of medication errors were related to a technology issue.[70]

  o Alert fatigue may cause drug and lab test alerts to be ignored.[71]

  o Alarm fatigue is as big an issue as alert fatigue. This is discussed in more detail in the chapter on patient safety.[72]

  o Distraction while using mobile devices and social media and issue while on the job.[72]

  o Upcoding with EHR use could increase healthcare costs and raise thorny ethical/legal issues.[73]

  o HIT may raise, not lower long term healthcare costs.[74]

  o Privacy and security issues are on the increase due to widespread HIT adoption. This is addressed in the chapter on Healthcare Privacy and Security.

In 2015 Dr. Ammenwerth published further EBHI recommendations:

- Establish health IT study registries
- Improve publication quality
- Create incentives for publishing negative studies, to counter publication bias
- Create a health IT systems taxonomy
- Improve indexing of health IT evaluation papers
- Migrate from meta-analysis to meta-summaries; synthesizing both qualitative and quantitative studies
- Include health IT evaluation competencies in curricula

- Create frameworks for evidence-based implementation
- Establish post-marketing surveillance for health IT[75]

The end result of this convergence of factors could be widespread negativism towards HIT, increased medical errors and cost and decreased governmental and payer-based funding. Hopefully, with better research over time one will have fewer questions and more answers.

EBM RESOURCES

Dr. Ammenwerth has been instrumental in promoting EBHI and creating a web-based repository (EVALDB) of over 1500 health informatics interventions archived.[76] In addition, she and Michael Rigby published an ebook on EBHI in 2016 that is a free download.[77] There are many first-rate online medical resources that provide EBM type answers. They are all well referenced, current and written by subject matter experts. Several include the level of evidence (LOE). These resources can be classified as **filtered** (an expert has appraised and selected the best evidence, e.g., up-to-date or **unfiltered** (non-selected evidence, e.g., PubMed). For the EBM purist, the following are considered traditional or classic EBM resources:

- Clinical Evidence[78]
  o British Medical Journal product with two issues per year
  o Sections on EBM tools, links, training and articles
  o Evidence is oriented towards patient outcomes (POEMS)
  o Very evidence-based with single page summaries and links to national guidelines
  o Available in paperback (Concise), CD-ROM, online or PDA format
- Cochrane Library[79]
  o Database of systematic reviews. Each review answers a clinical question
  o Database of review abstracts of effectiveness (DARE)
  o Controlled Trials Register
  o Methodology reviews and register
  o Fee-based
- Cochrane Summaries[80]
  o Part of the Cochrane Collaboration
  o Reviews can be accessed for a fee, but abstracts are free. A search for low back pain in 2011, as an example, returned 393 reviews (abstracts)
- EvidenceAlerts[81]
  o Since 2002 BMJ Updates has been filtering all of the major-medical literature. Articles are not posted until they have been reviewed for newsworthiness and relevance; not strict EBM guidelines
  o Users can go to their site and do a search or choose to have article abstracts e-mailed on a regular basis
  o These same updates are available through www.Medscape.com
- ACP Journal Club[82]
  o Bimonthly journal that can be accessed from OVID or free if a member of the American College of Physicians (ACP)
  o Over 100 journals are screened but very few articles end up being reviewed
  o They have a searchable database and email alerting system
- Essential Evidence Plus[83]
  o Physician oriented content that is fee-based
  o Offers daily patient-oriented evidence that matters (POEMS) (easy to read synopses) emailed to subscribers
  o Essential evidence plus search tool researches EBM topics, EBM guidelines (CPGs), POEMS, Cochrane Systematic Reviews, National Guideline Clearinghouse CPGs, and decision and diagnostic calculators
- TRIP Database has a search engine that using three different strategies to determine a search score[84]
- OVID can search the Cochrane Database of Systematic Reviews, DARE, ACP Journal Club and Cochrane Controlled Trials Register at the same time. Also includes Evidence-based Medicine Reviews.[85]
- SUMSearch. Free site that searches Medline, National Guideline Clearing House and DARE[86]
- Bandolier. Free online EBM journal; used mainly by primary care doctors in England. Provides simple summaries with NNTs. Resource also includes multiple monographs and books on EBM that are easy to read and understand.[87]
- Centre for Evidence-based Medicine is a comprehensive EBM site presented by Oxford University.[88]
- Best Bets (best evidence topics) lists topics of interest to primary care and emergency department clinicians. Hosted by the Emergency Department at the Manchester Royal Infirmary, UK.[89]
- Evidence-based Health Care is a very good EBM resource repository from the Health Sciences Library at the University of Colorado.[90]
- Google. Inserting "evidence-based" with any search question will yield multiple results.[91]
- The NNT web site provides NNT and NNH for multiple medical conditions. In addition to therapy

reviews they provide probabilities for diagnosis related conditions.[92]
- MDCalc is a web-based calculator site based on EBM. Helpful for those looking for examples of common clinical calculations.[93]
- EBM for Mobile Technology:
  - MedCalc 3000 calculators are both web based and available for smartphones. EBM Stats includes approximately 50 EBM calculators to include NNT, NNH, etc. Fee-based app for iPhone and Android operating systems.[94]

CLINICAL PRACTICE GUIDELINES

The Institute of Medicine in 2011 defined clinical practice guidelines (CPGs) as:

> "statments that include recommendations intended to optimize patient care that are informed by a systematic review of evidence and an assessment of the benefits and harms of alternative care options"[95]

CPGs take the very best evidence-based medical information and formulate an approach to treat a specific disease or condition. If one considers evidence as a continuum that starts by data generated from a single study, appraised and synthesized in a systematic review, CPGs would represent the next logical step in which evidence is transformed into a recommendation. Many medical organizations use CPGs with the intent to improve quality of care, patient safety and/or reduce costs. Information technology assists CPGs by expediting the search for the best evidence and linking the results to EHRs and smartphones for easy access. Two areas in which CPGs may be potentially beneficial include disease management and quality improvement strategies, covered in other chapters. As 83% of Medicare beneficiaries have at least one chronic condition and 68% of Medicare's budget is devoted to the 23% who have five or more chronic conditions, CPGs can play an important role in improving care and lowering costs.[96] There is some evidence that guidelines that address *multiple comorbidities* (concurrent chronic diseases) do work. As an example, in one study of diabetics, there was a 50% decrease in cardiovascular and microvascular complications with intensive treatment of multiple risk factors.[97]

Despite evidence to suggest benefit, several studies have shown poor CPG compliance by patients and physicians. The well publicized 2003 RAND study in the New England Journal of Medicine demonstrated that *"overall, patients received 54% of recommended care."*[98-99] In another study of guidelines at a major teaching hospital there was overuse of statin therapy (cholesterol lowering drugs). Overuse occurred in 69% of primary prevention (to prevent a disease) and 47% of secondary prevention (to prevent disease recurrence or progression), compared to national recommendations.[100]

It should be emphasized that creating or importing a guideline is the easy part because hundreds have already been created by a variety of national and international organizations. Implementing CPGs and achieving buy-in by all healthcare workers, particularly physicians, is the hard part.

DEVELOPING CLINICAL PRACTICE GUIDELINES

Ideally, the process starts with a panel of content and methodology experts commissioned by a professional organization. As an example, if the guideline is about preventing venous thrombosis and pulmonary embolism, multi-disciplinary content experts would be pulmonologists, hematologists, pharmacists and hospitalists.

Methodology experts are experts in evidence-based medicine, epidemiology, statistics, cost analysis, etc. The panel refines the questions, usually in PICO (patient, intervention, comparison and outcome) format, that was discussed earlier in this chapter. A systematic literature search and evidence synthesis takes place. Evidence is graded, and recommendations are negotiated. Panel members have their own biases and conflicts of interest that should be declared to CPG users. Voting is often needed to build consensus since disagreement is a natural phenomenon in this context.

The CPG development process has been standardized by several organizations. The National Academy of Medicine published 8 CPG development standards in 2011.[101] The Guideline International Network (GIN) published (2012), eleven CPG development standards shown in table 14. 3 (modified from ref)[102]

The Strength of Recommendations

Guideline panels usually associate their recommendations by a grading that describes how confident they are in their statement. Many organizations use the GRADE method for evaluating strength of evidence, discussed earlier in this chapter. Ideally, panels should separately describe their confidence in the evidence (the quality of evidence, previously described) from the strength of the recommendation. The reason for this separation is that there are factors other than evidence that may affect the strength of recommendation. These factors include: (1) how closely balanced are the benefits and harms of

Table 14.3: Key CPG components of a high-quality CPG

CPG development panel should be diverse, with health professionals, subject matter experts and patients
CPGs should describe the process used to reach a consensus among the CPG team
CPG team should disclose any financial and nonfinancial conflicts of interest
CPGs should specify its objectives and scope
CPGs should clearly and fully describe the methods used for the guideline development
CPG developers should use systematic evidence review methods to evaluate evidence
CPG recommendation should be clearly stated, with benefits, harms and costs
CPGs should use rating systems about strength of evidence and CPG quality
CPG review by external stakeholders should be conducted before publication
CPG should include an expiration date and/or describe the update process
CPG should disclose any financial support received evidence review or recommendations

the recommended intervention, (2) patients' values and preferences, and (3) resource allocation.

For example, even if there is very high-quality evidence from randomized trials showing that warfarin (a blood thinner) decreases the risk of stroke in some patients, the panel may issue a weak recommendation considering that the harms associated with this medicine are substantial. Similarly, if high quality evidence suggests that a treatment is very beneficial, but this treatment is very expensive and only available in very few large academic centers in the US, the panel may issue a weak recommendation because this treatment is not easily available or accessible.

Application to Individuals

A physician should consider a strong recommendation to be applicable to all patients who are able to receive it. Therefore, physicians should spend his/her time and effort on explaining to patients how to use the recommended intervention and integrate it in their daily routine.

On the other hand, a weak recommendation may only apply to certain patients. Physicians should spend more time discussing pros and cons of the intervention with patients, use risk calculators and tools designed to stratify patients' risk to better determine the balance of harms and benefit for the individual. Weak recommendations are the optimal condition to use decision aids, which are available in written, videographic and electronic formats and may help in the decisionmaking process by increasing knowledge acquisition by patients and reduce their anxiety and decisional conflicts.

Appraisal and Validity of Guidelines

There are multiple tools suggested to appraise CPGs and determine their validity. In fact, one systematic review reported 38 appraisal tools in the English medical literature.[103] These tools assess the process of conducting CPGs, the quality and rigor of the recommendations and the clarity of their presentation. AGREE II (Appraisal of Guideline Research and Evaluation) is an instrument with six domains and 23 items, whereas, iCAHE is simpler with 14 quality-related questions.[104]

The following list includes some of the attributes that guidelines users (clinicians, patients, policy makers) should seek to determine if a CPG is valid and has acceptable quality:

- Evidence-based, preferably linked to systematic reviews of the literature
- Considers all relevant patients groups and management options
- Considers patient-important outcomes (as opposed to surrogate outcomes)
- Updated frequently
- Clarity and transparency in describing the process of CPGs development (e.g., voting, etc.)
- Clarity and transparency in describing the conflicts of interests of the guideline panel
- Addresses patients' values and preferences
- Level of evidence and strength of recommendation are given
- Simple summary or algorithm that is easy to understand

- Available in multiple formats (print, online, smartphone, etc.) and in multiple locations
- Compatibility with existing practices
- Simplifies, not complicates decision making[105]

Barriers to Clinical Practice Guidelines

Attempts to standardize medicine by applying evidence-based medicine and clinical practice guidelines have been surprisingly difficult due to multiple barriers:

- **Practice setting**: inadequate incentives, inadequate time and fear of liability. An early study estimated that it would require 7.4 hours/working day just to comply with all the US Preventive Services Task Force recommendations for the average clinician's practice![106]
- **Co-morbidities**: CPGs generally discuss one disease entity, but in reality, many adults have multiple chronic diseases. For example, the average adult type 2 diabetic, also has hypertension, obesity and hypercholesterolemia. That could potentially mean 4 CPGs.
- **Contrary opinions**: local experts do not always agree with CPG or clinicians hear different messages from drug detail representatives
- **Sparse evidence**: there are several medical areas in which the evidence is of lower quality or sparse. Guideline panels in these areas would heavily depend on their expertise and should issue weak recommendations (e.g. suggestions) or no recommendations if they did not reach a consensus. These areas are problematic to patients and physicians and are clearly not ready for quality improvement projects or pay-for-performance incentives. For years, diabetologists advocated tight glycemic control of patients with type 2 diabetes; however, it turned out from results of recent large randomized trials that this strategy does not result in improved outcomes.[107]
- More information is needed why clinicians don't follow CPGs. Persell et al. reported that 94% of the time when clinicians chose an exception to the CPG it was appropriate. Three percent were inappropriate and 3% were unclear.[108]
- **Knowledge and attitudes**: there is a lack of confidence to either not perform a test (malpractice concern) or to order a new treatment (don't know enough yet). Information overload is always a problem.[109-110]
- **CPGs can be too long, impractical or confusing**. One study of Family Physicians stated CPGs should be no longer than two pages.[111-113] Most national CPGs are 50 to 150 pages long and don't always include a summary of recommendations or flow diagram
- **Where and how should CPGs be posted?** What should be the format? Should the format be standardized?
- **Less buy-in if data reported is not local since physicians tend to respond to data reported from their hospital or clinic.**
- **No uniform level of evidence (LOE) rating system**
- **Too many CPGs posted on the National Guideline Clearinghouse**. For instance, a non-filtered search in June 2017 by one author for "type 2 diabetes" yielded 378 CPGs. The detailed search option helps filter the search significantly.[113]
- **Lack of available local champions to promote CPGs**
- **Excessive influence by drug companies:** A survey of 192 authors of 44 CPGs in the 1991 to 1999-time frame showed:
  o 87% had some tie to drug companies
  o 58% received financial support
  o 59% represented drugs mentioned in the CPG
  o 55% of respondents with ties to drug companies said they did not believe they had to disclose involvement[114]
- **Quality of national guidelines**: National guidelines are not necessarily of high quality. A 2009 review of CPGs from the American Heart Association and the American College of Cardiology (1984 to Sept 2008) concluded that many of the recommendations were based on a lower level of evidence or expert opinion, not high-quality studies.[115]
- **No patient input**. At this point patients are not normally involved in any aspect of CPGs, even though they receive recommendations based on CPGs. In an interesting 2008 study, patients who received an electronic message about guidelines experienced a 12.8% increase in compliance. This study utilized claims data as well as a robust rules engine to analyze patient data. Patients received alerts (usually mail) about the need for screening, diagnostic and monitoring tests. The most common alerts were for adding a cholesterol lowering drug, screening women over age 65 for osteoporosis, doing eye exams in diabetics, adding an ACE inhibitor drug for diabetes and testing diabetics for urine microalbumin.[116] It makes good sense that patients should be knowledgeable about national recommendations and should have these guidelines written in plain language and available in multiple formats. Also, because many patients are highly "connected" they could receive text messages via cell phones, social

networking software, etc., to improve monitoring and treatment.

INITIATING CLINICAL PRACTICE GUIDELINES

Examples of Starting Points:

- High cost conditions: heart failure
- High volume conditions: diabetes
- Preventable admissions: asthma
- There is variation in care compared to national recommendations: deep vein thrombophlebitis (DVT) prevention
- High litigation areas: failure to diagnose or treat
- Patient safety areas: intravenous (IV) drug monitoring

The Strategy

- Leadership support is crucial
- Use process improvement tools such as the Plan-Do-Study-Act (PDSA) model
- Identify gaps in knowledge between national recommendations and local practice
- Locate a guideline champion who is a well-respected clinical expert.[117] A champion acts as an advocate for implementation based on his/her support of a new guideline
- Other potential team members:
  o Clinician selection based on the nature of the CPG
  o Administrative or support staff
  o Quality Management staff
- Develop action plans
- Educate all staff involved with CPGs, not just clinicians
- Pilot implementation
- Provide frequent feedback to clinicians and other staff regarding results
- Consider using the checklist for reporting clinical practice guidelines developed by the 2002 Conference on Guideline Standardization (COGS).[118]

CLINICAL PRACTICE GUIDELINE EXAMPLE

There have been thousands of CPGs created and disseminated but far fewer have been studied, in terms of impact and even fewer have been significantly successful. Figure 14.2 represents a 2013 study reported from Kaiser Permanente Northern California (KPNC) for hypertensive control. Note that control of hypertension increased from a baseline of 43% in 2001 to 80% in 2009. The national averages are also presented. It is important to realize that Kaiser has had a sytem wide EHR since 2005 and that they have developed multiple enterprise evidence-based CPGs. Furthermore, because everyone has the same leaderhip and information technology system, it is easier to get everyone on the team on the same page. This study is presented later in this chapter under *"current knowledge."*[119]

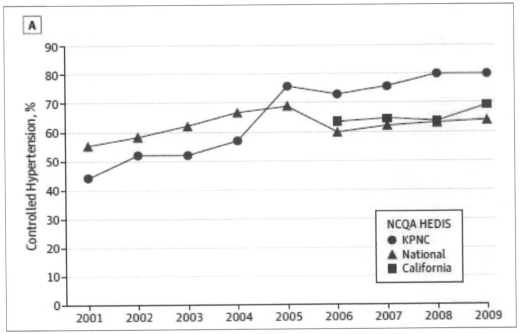

Figure 14.2: Results from 2013 Study using hypertension CPG

ELECTRONIC CLINICAL PRACTICE GUIDELINES

CPGs have been traditionally paper-based and often accompanied by a flow diagram or algorithm. With time, more are being created in an electronic format and posted on the Internet or Intranet for easy access. Zielstorff outlined the issues, obstacles and future prospects of online practice guidelines in an early review.[112] What has changed since then is the ability to integrate CPGs with smartphones and electronic health records.

CPGs on smartphones: These mobile platforms function well in this area as each step in an algorithm is simply a tap or touch of the screen. In Figures 14.3 and 14.4 programs are shown that are based on national guidelines for cardiac risk and cardiac clearance. Figure 14.3 depicts a calculator that determines the 10-year risk of heart disease based on serum cholesterol and other risk factors. A prostate cancer risk calculator was developed in the Netherlands and was validated by clinical research.[120] Many excellent guidelines for the smartphone exist that will be listed later in this chapter.

Figure 14.4: Rotterdam Prostate Cancer Risk Calculator

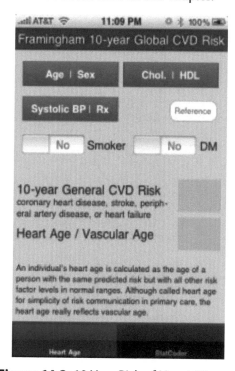

Figure 14.3: 10 Year Risk of Heart Disease

Web-Based Risk Calculators: Many of the CPGs available on a mobile platform and are also available online. While these are not CPGs exactly, they are based on population studies and are felt to be part of EBM and can give direction to the clinician. As an example, some experts feel that aspirin has little benefit in preventing a heart attack unless your 10-year risk exceeds 20%. The following is a short list of some of the more popular online calculators:

- ATP III Cardiac risk calculator: estimates the 10-year risk of a heart attack or death based on your cholesterol, age, gender, etc.[121]
- FRAX fracture risk calculator: estimates the 10-year risk of a hip or other fracture based on all the common risk factors for osteoporosis. The calculator considers a patient's bone mineral density score, gender and ethnicity.[122]
- GAIL breast cancer risk assessment tool: estimates a patient's risk of breast cancer, again, based on known and accepted risk factors.[123]
- Stroke risk calculator: based on the Framingham study it predicts 10-year risk of a stroke based on known risk factors.[124]
- Risk of stroke or death for new onset atrial fibrillation: the calculator is also based on the Framingham study and calculates five-year risk of stroke or death.[125]

Electronic Health Record CPGs

Although not all electronic health records have embedded CPGs, there is definite interest in providing local or national CPGs at the point of care. CPGs embedded in the EHR are clearly a form of decision support. They can be linked to the diagnosis or the order entry process. In addition, they can be standalone resources available by clicking, for example, an "info-button." Clinical decision support provides treatment reminders for disease states that may include the use of more cost-effective drugs. Institutions such as Vanderbilt University have integrated more than 750 CPGs into their EHR by linking the CPGs to ICD-9 codes.[126] The results of embedded CPGs appear mixed. In a study by Durieux using computerized decision support reminders, orthopedic surgeons showed improved compliance to guidelines to prevent deep vein thrombophlebitis.[127] On the other hand, studies by Tierney, failed to demonstrate improved compliance to guidelines using computer reminders for hypertension, heart disease and asthma.[128-130] Clinical decision support, to include order sets is discussed in more detail in the chapters on electronic health records, clinical decision support systems and patient safety.

There are other ways to use electronic tools to promulgate CPGs. In an interesting paper by Javitt, primary care clinicians were sent reminders on outpatient treatment guidelines based only on claims data. Outliers were located by using a rules engine (Care Engine) to compare a patient's care with national guidelines. They could show a decrease in hospitalizations and cost because of alerts that notified physicians by phone, fax or letter. This demonstrates one additional means of changing physician behavior using CPGs and information technology not linked to the electronic health record.[132] Critics might argue that claims data are not as accurate, robust or current as actual clinical results.

Software is now available (Symmetry® EBM Connect®) that can compute compliance with guidelines automatically using administrative data. The program translates guidelines from text to algorithms for 20 disease conditions and therefore would be much more efficient than chart reviews. Keep in mind it tells users if, for example, LDL cholesterol was ordered, not the actual results.[131]

CLINICAL PRACTICE GUIDELINE RESOURCES

Web-based CPGs

National Guideline Clearinghouse. This program is an initiative of the Agency for Healthcare Research and Quality (AHRQ) and is the largest and most comprehensive of all CPG resources. Features offered:
- Includes about 2500+ guidelines
- There is extensive search engine filtering i.e. one can search by year, language, gender, specialty, level of evidence, etc.
- Abstracts are available as well as links to full text guidelines where available
- CPG comparison tool
- Forum for discussion of guidelines
- Annotated bibliography
- They link to 17 international CPG resource sites[113]

National Institute for Health and Clinical Excellence (NICE)

- Service of the British National Health Service
- Approximately 100 CPGs are posted and dated
- A user-friendly short summary is available as well as a lengthy guideline, both in downloadable pdf format
- Podcasts are available[133]

Guidelines International Network

- Not-for-profit organization that began in 2002
- 152 members from 47 countries
- Membership fee includes access to an extensive CPG library[134]

Agency for Healthcare Research and Quality (AHRQ)

- 1 of 12 agencies within the Department of Health and Human Services (HHS)
- AHRQ supports health services research initiatives that seek to improve the quality of health care in America
- AHRQ's funds evidence practice centers that conduct evidence appraisal and reviews to support the development of clinical practice guidelines[135]

Health Team Works (formerly Colorado Clinical Guidelines Collaborative)

- Free downloads available for Colorado physicians and members of CCGC
- As of June 2017, they have 16 CPGs available
- Guidelines are in easy to read tables, written in a pdf format
- References, resources and patient handouts are available[136]

Institute for Clinical Systems Improvement (ICSI)

- Collaboration of three major health plans in Minnesota to improve healthcare quality
- Their web site includes about 30 CPGs
- Each CPG has a main algorithm with hyperlinked steps
- They also have order sets and patient resources. Some are for members only
- Evidence-based and rated CPGs
- Executive summary with date of publication[137]

Smartphone-based CPGs

Most CPGs can be downloaded for the iPhone or iPad through the iTunes Store or the Android Market.

For further information about medical apps, readers are referred to the chapter on mobile technology. The following are a sample of CPGs available for smartphones:

- NCCN Clinical Practice Guidelines in Oncology (NCCN Guidelines™) are available for iPhone and Android.[138]
- Skyscape has multiple free CPGs available for download and 150+ fee-based CPGs. For example, Pediatric Clinical Practice Guidelines & Policies provides access to more than 30 clinical practice guidelines and more than 380 policy statements, clinical reports and technical reports.[139]
- mTBI Pocket Guide provides evidence-based information about traumatic brain injury (TBI) and is available on the Android Market.[140]
- ePSS is an app available for all operating systems, developed by the US Preventive Services Task Force. Preventive medicine guidelines are presented based on age, gender, smoking status, etc.[141]

RECOMMENDED READING

Several recent articles are posted that address EBM and CPGs

- *Improving Adherence To Otitis Media Guidelines With Clinical Decision Support And Physician Feedback* is a 2013 cluster-randomized study of adherence to CPGs for acute otitis media (AOM) and otitis media with effusion (OME), using EHR-CDS and monthly physician feedback. Researchers found that clinical decision support (CDS) and feedback both improved CPG compliance but they were not additive.[142]
- *Childhood Obesity: Can Electronic Medical Records Customized With Clinical Practice Guidelines Improve Screening And Diagnosis?* Researchers wanted to know if CPGs that are part of an EHR improve recording of BMI, growth chart completion, risk score questionnaire completion and coding for obesity. In this before/after study there was an increase in all parameters, but the number of children reported with obesity was still below the known rates of obesity for this community.[143]
- *Use Of Health IT For Higher-Value Critical Care.* Authors advocated using CPGs in EHRs to risk stratify patients, particularly with non-cardiac illnesses, for admission to the critical care unit.[144]
- A *"Smart Heart Failure Sheet: Using Electronic Medical Records To Guide Clinical Decision Making.* The authors report their experience with an embedded CPG Developed at the Beth Israel Deaconess Medical Center. The resource is highly educational for both the physician and patient. The smart sheet automatically uploads lab and imaging pertinent to heart failure diagnosis and treatment. It appears in the EHR after adding heart failure to the problem summary list or demonstrating a low ejection fraction by echocardiography. The program also allows a clinician to see all of his/her patients with heart failure, along with flow charts, etc. No outcome data has been published.[145]
- *Improved Blood Pressure Control Associated with a Large Scale Hypertension Program.* This Kaiser-Permanente Northern California (KPNC) study looked at blood pressure control based on reported HEDIS measures from 2001-2009 in California. After implementing a hypertension CPG and creating a hypertension registry for the entire region they also instituted a polypill (single pill containing several blood pressure medications). Follow-up visits were by medical assistants. The end result was to see control rise from 43% to 80%; a percentage considerably higher than the national average (55% in 2001, 64% in 2009). Also, see Figure 14.2.[119]
- *Why Randomized Controlled Trials are Needed to Accept New Practices: 2 World Views* and *The Necessity for Clinical Reasoning in the Era of Evidence-based Medicine.* Both of these articles appeared in a late 2013 issue of the Mayo Clinic Proceedings. They highlight the healthly controversy between those who believe clinicians must have evidence before they proceed and those who accept that the evidence is lacking or mixed so one must employ a good clinical reasoning.[146-147]

FUTURE TRENDS

The field of EBM continues to evolve. Methodologists continue to identify opportunities to improve our understanding and interpretation of research findings. It is anticipated that more standardization of reporting and more transparency. The Appraisal of Guidelines for Research & Evaluation (AGREE II) is a web-based tool that rates the quality of CPGs with 23 items covering 6 quality domains.

Two methodology studies help refine our knowledge base:

- Trials are often stopped early when extreme benefits are noted in the intervention group. The rationale for stopping enrollments of participants is that it is "unethical" to continue randomizing patients to the placebo arm because researchers are depriving them from the benefits of the intervention. However, it was found that stopping trials early for benefit exaggerates treatment effect by more than 30%; simply because the trial is stopped at a point of extreme benefit that is clearly made extreme by chance. Such exaggeration leads to the wrong conclusions by patients and physicians embarking on comparing the pros and cons of a treatment and leads to the wrong decisions by policymakers. In fact, stopping early may be unethical from a societal and individual point of view.[148]

- The second recent advancement in methodology relates to the finding that authors who have financial affiliation with the industry are three times more likely to make statements that are favorable to the sponsored interventions. It is very plausible that this bias is subconscious and unintentional; nevertheless, as readers of the literature, one should recognize the potential and implications of this bias.[149]

Advances with CPGs will be related to better integration with a variety of HIT and more research into those factors that improve CPG compliance

KEY POINTS

- Evidence-based Medicine (EBM) is the academic pursuit of the best available answer to a clinical question
- The two fundamental principles of EBM are: (1) a hierarchy of evidence exists (i.e., not all evidence is equal) and (2) evidence alone is insufficient for medical decision making. It should rather be complemented by patient's values, preferences and circumstances.
- Health information technology will hopefully improve medical quality, which is primarily based on EBM.
- There are multiple limitations of both EBM and the medical literature.
- The average clinician should have a basic understanding of EBM and know how to find answers using EBM resources.
- Clinical Practice Guidelines (CPGs), based on evidence-based medicine, are the roadmap to standardize medical care.
- CPGs are valuable for chronic disease management or to measure quality of care.

CONCLUSION

Knowledge of EBM is important for those involved with patient care, quality of care issues or research. Rapid access to a variety of online EBM resources has changed how clinicians practice medicine. Despite its shortcomings, an evidence-based approach helps healthcare workers find the best possible answers. Busy clinicians are likely to choose commercial high-quality resources, while academic clinicians are likely to select true EBM resources. Ultimately, EBM tools and resources will be integrated with electronic health records as part of clinical decision support.

The jury is out regarding the impact of CPGs on physician behavior or patient outcomes. Busy clinicians are slow to accept new information, including CPGs. Whether embedding CPGs into EHRs will result in significant changes in behavior that will consistently result in improved quality, patient safety or cost savings remains to be seen. It is also unknown if linking CPGs to better reimbursement (pay-for-performance) will result in a higher level of acceptance. While it is being determined how to optimally improve healthcare with CPGs, most authorities agree that CPGs need to be concise, practical and accessible at the point of care. Every attempt should be made to make them electronic and integrated into the workflow of clinicians.

ACKNOWLEDGEMENT

We thank Dr. M. Hassan Murad and Dr. Brian Haynes for their early contributions to this chapter.

REFERENCES

1. Evidence-based Medicine: What it is, what it isn't. https://www.cebma.org/wp-content/uploads/Sackett-Evidence-Based-Medicine.pdf (Accessed June 1, 2017)
2. Crossing the Quality Chasm: A new health system for the 21th century (2001) The National Academies Press https://www.nap.edu/catalog/10027/crossing-the-quality-chasm-a-new-health-system-for-the?gclid=Cj0KEQjw9r7JBRCj37PlltTskaMBEiQAKTzTfHXAyIfKvrd3PH3ffVO_ZLrx9FCWUqXPKvv-LFgDV3AaAjv88P8HAQ (Accessed June 1, 2017)
3. Choosing Wisely. http://www.choosingwiselycanada.org/materials/treating-lower-back-pain/ (Accessed June 1, 2017)

4. Teece I, Crawford I. Bed rest after spinal puncture. BMJ https://www.ncbi.nlm.nih.gov/pmc/articles/PMC2660496/ (Accessed June 1, 2017)
5. Guyatt GH. Evidence-based medicine. ACP J Club 1991;114:A16
6. Bothwell LE, Podolsky SH. The Emergence of the Randomized, Controlled Trial. N Engl J Med 2016 Aug 11;375(6):501–4. Available from: http://www.nejm.org/doi/10.1056/NEJMp1604635
7. The Cochrane Library http://www.cochranelibrary.com/cochrane-database-of-systematic-reviews/index.html (Accessed May 31, 2017)
8. Medical Research Council. Streptomycin treatment of pulmonary tuberculosis. BMJ 1948;2:769-82
9. Gandhi GY, Murad MH, Fujiyoshi A, et al. Patient-important outcomes in registered diabetes trials. JAMA. Jun 4 2008;299 (21):2543-2549
10. Agency for Healthcare Quality and Research. Effective Healthcare Program. www.effectivehealthcare.ahrq.gov (Accessed June 1, 2017)
11. Choosing Wisely. www.choosing wisely.org (Accessed June 1, 2017)
12. Greenhalgh T, Howick J, Maskrey N, Evidence-based Medicine Renaissance Group. Evidence-based medicine: a movement in crisis? BMJ. 2014;348:g3725. http://www.bmj.com/content/348/bmj.g3725 (Accessed May 31, 2017)
13. Davis DA et al. Changing physician performance. A systematic review of the effect of continuing medical education strategies. JAMA 1995; 274: 700-1.
14. Sibley JC. A randomized trial of continuing medical education. N Engl J Med 1982; 306: 511-5.
15. Fordis M et al. Comparison of the Instructional Efficacy of Internet-Based CME with Live Interactive CME Workshops. JAMA 2005;294:1043-1051
16. National Library of Medicine PICO http://askmedline.nlm.nih.gov/ask/pico.php (Accessed June 1, 2017)
17. Evidence-Based Medicine How to practice and teach it, by Straus, Glasziou, Richardson and Haynes. Fourth Edition. Churchill Livingstone, Elsevier. 2011. Toronto, CA.
18. Tripepi G, Jager KJ, Dekker FW, Wanner C, Zoccali C. Bias in clinical research. Kidney Int 2008;73(2):148–53. http://linkinghub.elsevier.com/retrieve/pii/S008525381552958X (Accessed May 31, 2017)
19. Centre for Evidence-based Medicine http://www.cebm.net (Accessed June 1, 2017)
20. Haynes RB. Of studies, syntheses, synopses and systems: the "4S evolution of services for finding the best evidence." ACP J Club 2001;134: A11-13
21. Jadad AR, Moore RA, Carroll D et al. Assessing the quality of reports of randomized clinical trials: is blinding necessary? Control Clin Trails. 1996;17:1-12
22. Bigby M. Understanding and Evaluating Systematic Reviews and Meta-analyses. Indian J Dermatol. 2014;59(2):134–9. http://www.ncbi.nlm.nih.gov/pubmed/24700930 (Accessed May 31, 2017)
23. The well built clinical question. University of North Carolina Library http://www.hsl.unc.edu/Services/Tutorials/EBM/Supplements/QuestionSupplement.htm (Accessed June 1, 2017)
24. Guyatt GH, Oxman AD, Vist G, Kunz R, Falck-Ytter Y, Alonso-Coello P, Schünemann HJ. GRADE: an emerging consensus on rating quality of evidence and strength of recommendations. BMJ 2008;336:924-926
25. GRADE. http://gradeworkinggroup.org/ (Accessed May 31, 2017)
26. Buitrago-Lopez A, Sanderson J, Johnson L, Warnakula S, Wood A, Di Angelantonio E, Franco OH. Chocolate consumption and cardiometabolic disorders: systematic review and meta-analysis. BMJ. 2011 Aug 26;343:d4488. doi: 10.1136/bmj.d4488.
27. Risk and Odds. http://handbook.cochrane.org/chapter_9/9_2_2_1_risk_and_odds.htm (Accessed May 31, 2017)
28. Henley E. Understanding the Risks of Medical Interventions Fam Pract Man May 2000;59-60
29. Nouvo J, Melnikow J, Chang D. Reporting the Number Needed to Treat and Absolute Risk Reduction in Randomized Controlled Trials JAMA 2002;287:2813-2814
30. Lacy CR et al. Impact of Presentation of Research Results on Likelihood of Prescribing Medications to Patients with Left Ventricular Dysfunction. Am J Card 2001;87:203-207
31. Collaborative Atorvastatin Diabetes Study (CARDS) Lancet 2004;364:685-96
32. Online effect size calculator. http://www.socscistatistics.com/effectsize/Default3.aspx (Accessed May 31, 2017)
33. Confidence Intervals. www.onlinestat book.com (Accessed June 1, 2017)
34. Online confidence interval calculator. http://www.socscistatistics.com/confidenceinterval/Default3.aspx (Accessed May 31, 2017)
35. Maharaj R. Adding cost to number needed to treat: the COPE statistic. Evidence-based Medicine 2007;12:101-102
36. Haynes RB. Where's the Meat in Clinical Journals? ACP Journal Club Nov/Dec 1993: A-22-23
37. Als-Neilsen B, Chen W, Gluud C, Kjaergard LL. Association of Funding and Conclusions in Randomized Drug Trials. JAMA 2003; 290:921-928

38. Ioannidis JPA. Contradicted and Initially Stronger Effects in Highly Cited Clinical Research JAMA 2005;294:218-228
39. Prasad V, Vandross A, Toomey C et al. A Decade of Reversal: An Analysis of 146 Contradicted Medical Practices. Mayo Clin Proc 3013; 88(8):790-798
40. El Dib RP, Attallah AN, Andriolo RB. Mapping the Cochrane evidence for decision making in health care. J Eval Clin Pract 2007;13(4):689-692
41. Moher D, Tetzlaff J, Tricco AC, Sampson M, Altman DG. Epidemiology and reporting characteristics of systematic reviews. PLoS Med 2007;4(3):e78. http://www.ncbi.nlm.nih.gov/pubmed/17388659 (Accessed May 31, 2017)
42. Kranish M. Flaws are found in validating medical studies The Boston Globe August 15, 2005 http://www.boston.com/news/ nation/articles/2005/08/15/flaws_are_found_in_validating_medical_studies/ (Accessed June 12, 2007)
43. Altman DG. Poor Quality Medical Research: What can journals do? JAMA 2002;287:2765-2767
44. Gross CP et al. Reporting the Recruitment Process in Clinical Trials: Who are these Patients and how did they get there? Ann of Int Med 2002;137:10-16
45. Moher D, Dulgerg CS, Wells GA. Statistical Power, sample size and their reporting in randomized controlled trials JAMA 1994;22:1220-1224
46. Siontis KC, Evangelou E, Ionnidis JP. Magnitude of effects in clinical trials published in high impact general medical journals. Int J Epidemiol 2011;40(5):1280-1291
47. Evidence-based Medicine. Clinfowiki. www.clinfowiki.org (Accessed June 1, 2017)
48. Moskowitz D, Bodenheimer T. Moving from evidence-based medicine to evidence-based health. J Gen Intern Med 2011;26(6):658-660
49. Straus SE, McAlister FA Evidence-based Medicine: a commentary on common criticisms Can Med Assoc J 2000;163:837-841
50. Dobbie AE et al. What Evidence Supports Teaching Evidence-based Medicine? Acad Med 2000;75:1184-1185
51. Sackett DL, Staus SE. Finding and Applying Evidence During Clinical Rounds: The "Evidence Cart" JAMA 1998;280:1336-1338
52. Grandage K et al. When less is more: a practical approach to searching for evidence-based answers. J Med Libr Assoc 90(3) July 2002
53. Schilling LM et al. Resident's Patient Specific Clinical Questions: Opportunities for Evidence-based Learning Acad Med 2005;80:51-56
54. Davidoff F, Florance V. The Informationist: A New Health Profession? Ann of Int Med 2000;132:996-999
55. Giuse NB et al. Clinical medical librarianship: the Vanderbilt experience Bull Med Libr Assoc 1998;86:412-416
56. Westberg EE, Randolph AM. The Basis for Using the Internet to Support the Information Needs of Primary Care JAMIA 1999;6:6-25
57. Slawson DC, Shaughnessy AF, Bennett JH. Becoming a Medical Information Master: Feeling Good About Not Knowing Everything J of Fam Pract 1994;38:505-513
58. Shaughnessy AF, Slawson DC and Bennett JH. Becoming an Information Master: A Guidebook to the Medical Information Jungle J of Fam Pract 1994;39:489-499
59. Ammenwerth E. Is there sufficient evidence for evidence-based informatics? 2006. www.gmds2006.de/Abstracts/ 49.pdf (Accessed September 25, 2013)
60. Keizer NF, Ammenwerth E. The quality of evidence in health informatics: how did the quality of health care IT evaluation publications develop from 1982 to 2005? IJMI. 2008;77:41-39
61. Chaudry B. Systematic review: impact of health information technology on quality, efficiency and cost of medical care. Ann Int Med 2006;144(10):742-752
62. Orszag Pl. Evidence on the costs and benefits of health information technology. CBO. July 24, 2008. www.cbo.gov (Accessed July 31, 2008)
63. Shcherbatykh I, Holbrook A, Thabane L, et. al. Methodologic issues in health informatics trials: the complexities of complex interventions. J Am Inform Med Assoc 2008;15:575-580
64. Ammenwerth E, Schnell-Inderst, Siebert U. Vision and challenges of Evidence-Based Health Informatics: A case study of a CPOE meta-analysis. IJMI 2010;79:e83-e88
65. Liu JL, Wyatt JC. The case for randomized controlled trials to assess the impact of clinical information systems. J Am Med Inform Assoc. 2011;18:173-180
66. Nykanen P, Brender J, Ammenwerth E et al. Guideline for good evaluation practice in health informatics (GEP-HI). IJMI 2011;80:815-827
67. Talmon J, Ammenwerth E, Brender J et.al. STARE-HI---Statement on reporting of evaluation studies in Health Informatics. IJMI 2009;78:1-9
68. Weiner J. "e-iatrogenesis": The most critical unintended consequence of CPOE and other HIT. J Am Med Inform Assoc. 2007;14:387-388
69. Sittig D, Singh H. Defining health information errors: new developments since to Err is Human. Arch Int Med. 2011;171(14):1279-1282
70. The Joint Commission. Sentinel Alert Series. No.42. December 11, 2008. http://www.jointcommission.org/assets/1/18/SEA_42.pdf (Accessed September 25, 2013)
71. Singh H, Spitzmueller C, Petersen N, et.al. Information overload and missed test results in electronic health record-based settings. JAMA.

71. ...2013. March 4. Online first. (Accessed April 4, 2013)
72. Top 10 Health technology hazards for 2013. Health Devices. Vol. 41(11). November 2012. www.ecri.org (Accessed December 1, 2012)
73. Schulte F. How doctors and hospitals have collected billions in questionable Medicare fees. www.publicintegrity.org September 15, 2012. (Accessed January ,6 2013)
74. Adler-Milstein J, Gree CE, Bates DW. A survey analysis suggests that electronic health records will yield revenue gains for some practices and losses for many. Health Affairs. 2013;32(3):562-570
75. Ammenwerth E. Evidence-based health informatics: how do we know what we know? Methods Inf Med 2015 http://methods.schattauer.de/en/contents/archivestandard/issue/special/manuscript/24631/download.html (Accessed May 27, 2017)
76. Evaluation Database (EVALDB) http://evaldb.umit.at (Accessed June 1, 2017)
77. Evidence Based Health Informatics: Promoting Safety and Efficiency Through Scientific Mehtods and Ethical Policy. http://ebooks.iospress.nl/volume/evidence-based-health-informatics-promoting-safety-and-efficiency-through-scientific-methods-and-ethical-policy (Accessed May 27, 2017)
78. Clinical Evidence www.clinicalevidence.com (Accessed June 1, 2017)
79. Cochrane Library. www.cochranelibrary.com (Accessed June 1, 2017)
80. Cochrane Review http://www.cochrane.org (Accessed June 1, 2017)
81. EvidenceAlerts http://plus.mcmaster.ca/evidencealerts (Accessed June 1, 2017)
82. ACP Journal Club. www.annals.org/aim/journal-club (Accessed June 1, 2017)
83. Essential Evidence Plus www.essentialevidenceplus.com (Accessed June 1, 2017)
84. Trip Database www.tripdatabase.com (Accessed June 1, 2017)
85. OVID http://gateway.ovid.com (Accessed June 1, 2017)
86. SUMSearch www.sumsearch.org (Accessed June 1, 2017)
87. Bandolier www.bandolier.org.uk (Accessed June 1, 2017)
88. Centre for Evidence-based Medicine www.cebm.net (Accessed June 1, 2017)
89. Best Bets. www.bestbets.org (Accessed June 1, 2017)
90. Evidence-based Healthcare http://hs libraryguides.ucdenver.edu/ebpml (Accessed June 1, 2017)
91. Google Search. www.google.com (Accessed September 25, 2013)
92. The NNT www.thennt.com (Accessed June 1, 2017)
93. MDCalc. www.mdcalc.com (Accessed June 1, 2017)
94. MEDCalc 3000. www.ebmcalc.com/pubapps/nav.htm (Accessed June 1, 2017)
95. Clinical Practice Guidelines We Can Trust. Washington, D.C.: National Academies Press; 2011 http://www.nap.edu/catalog/13058 (Accessed June 5, 2017)
96. O'Connor P. Adding Value to Evidence-based Clinical Guidelines JAMA 2005;294:741-743
97. Gaede P. Multifactorial intervention and cardiovascular disease in patients with type 2 diabetes NEJM 2003;348:383-393
98. McGlynn E . Quality of Health Care Delivered to Adults in the US RAND Health Study NEJM Jun 26, 2003
99. Crossing the Quality Chasm: A new Health System for the 21th century 2001. IOM. https://www.nap.edu/catalog/10027/crossing-the-quality-chasm-a-new-health-system-for-the (Accessed June 1, 2017)
100. Abookire SA, Karson AS, Fiskio J, Bates DW. Use and monitoring of "statin" lipid-lowering drugs compared with guidelines Arch Int Med 2001;161:2626-7
101. Clinical Practice Guidelines We Can Trust. National Academies Press; 2011 http://www.nap.edu (Accessed May 31, 2017)
102. Qaseem A, et al. Guidelines International Network: Toward International Standards for Clinical Practice Guidelines. Ann Intern Med National Academies Pr, Washington, DC; 2012 ;156(7):525. http://annals.org/article.aspx?doi=10.7326/0003-4819-156-7-201204030-00009 (Accessed June 5, 2017)
103. Siering U, Eikermann M, Hausner E, Hoffmann-Eßer W, Neugebauer EA. Appraisal Tools for Clinical Practice Guidelines: A Systematic Review. PLoS One. CMA; 2013;8(12):e82915. http://dx.plos.org/10.1371/journal.pone.0082915 (Accessed June 2, 2017)
104. Kredo T, Bernhardsson S, Machingaidze S, Young T, Louw Q, Ochodo E, et al. Guide to clinical practice guidelines: The current state of play. Int J Qual Heal Care. 2016;28(1):122–8.
105. Oxman A, Flottorp S. An overview of strategies to promote implementation of evidence-based health care. In: Silagy C, Haines A, eds Evidence-based practice in primary care, 2nd ed. London: BMJ books 2001
106. Yarnall KSH, Pollak KL, Østbye T et al. Primary Care: Is There Enough Time for Prevention? Am J Pub Health 2003;93 (4):635-641
107. Montori VM, Fernandez-Balsells M. Glycemic control in type 2 diabetes: time for an evidence-based about face? Ann Intern Med 2009;150 (11):803-808
108. Persell SD, Dolan NC, Friesema EM et al. Frequency of Inappropriate Medical Exceptions

to Quality Measures. Ann Intern Med 2010;152:225-231
109. Grol R, Grimshaw J. From Best evidence to best practice: effective implementation of change in patient's care Lancet 2003;362:1225-30
110. Legare F, O'Connor AM, Graham ID. et. al. Primary health care professionals' views on barriers and facilitators to the implementation of the Ottawa Decision Support Framework in practice. Pat Ed Couns. 2006;63:380-390
111. Wolff M, Bower DJ, Marabella AM, Casanova JE. US Family Physicians experiences with practice guidelines. Fam Med 1998;30:117-121
112. Zielstorff RD. Online Practice Guidelines JAMIA 1998;5:227-236
113. National Guideline Clearinghouse www.guideline.gov (Accessed June 1, 2017)
114. Choudry NK et al. Relationships between authors of clinical practice guidelines and the pharmaceutical industry JAMA 2002;287:612-7
115. Tricoci P, Allen JM, Kramer JM et al. Scientific Evidence Underlying the ACC/AHA Clinical Practice Guidelines JAMA 2009;301(8):831-841
116. Rosenberg SN, Shnaiden TL, Wegh AA et al. Supporting the Patient's Role in Guideline Compliance: A Controlled Study. Am J Manag Care 2008;14 (11):737-744
117. Stross JK. The educationally influential physician – Journal of Continuing Education Health Professionals 1996; 16: 167-172)
118. Shiffman RN, Shekelle P, Overhage JM et al. Standardized Reporting of Clinical Practice Guidelines: A Proposal form the Conference on Guideline Standardization. Ann Intern Med 2003;139:493-498
119. Jaffe MG, Lee GA, Young JD et al. Improved Blood Pressure Control Associatied with A Large Scale Hypertension Program. JAMA 2013;310(7):699-705
120. Pereira-Azevedo N, Osório L, Fraga A, Roobol MJ. Rotterdam Prostate Cancer Risk Calculator: Development and Usability Testing of the Mobile Phone App. JMIR Cancer. 2017;3(1):e1. http://www.ncbi.nlm.nih.gov/pubmed/28410180 (Accessed June 1, 2017)
121. ATP III Risk www.cvriskcalculator.com (Accessed June 1, 2017)
122. Frax Calculator https://www.sheffield.ac.uk/FRAX/tool.jsp (Accessed June 1, 2017)
123. Gail Breast Cancer Risk https://www.cancer.gov/bcrisktool/ (Accessed June 1, 2017)
124. Stroke Risk Calculator. Cleveland Clinic. https://my.clevelandclinic.org/stroke-risk-calculator (Accessed June 1, 2017)
125. Stroke or death due to atrial fibrillation http://www.zunis.org/FHS%20Afib%20Risk%20Calculator.htm (Accessed June 1, 2017)
126. Giuse N et al. Evolution of a Mature Clinical Informationist Model JAIMA 2005;12:249-255
127. Durieux P et al. A Clinical Decision Support System for Prevention of Venous Thromboembolism: Effect on Physician Behavior JAMA 2000;283:2816-2821
128. Tierney WM et al. Effects of Computerized Guidelines for Managing Heart Disease in Primary Care J Gen Int Med 2003;18:967-976
129. Murray et al. Failure of computerized treatment suggestions to improve health outcomes of outpatients with uncomplicated hypertension: results of a randomized controlled trial Pharmacotherapy 2004;3:324-37
130. Tierney et al. Can Computer Generated Evidence-based Care Suggestions Enhance Evidence-based Management of Asthma and Chronic Obstructive Pulmonary Disease? A Randomized Controlled Trial Health Serv Res 2005;40:477-97
131. Javitt JC et al. Using a Claims Data Based Sentinel System to Improve Compliance with Clinical Guidelines: Results of a Randomized Prospective Study Amer J of Man Care 2005;11:93-102
132. Welch, PW et al. Electronic Health Records in Four Community Physician Practices: Impact on Quality and Cost of Care. JAMIA 2007;14:320-328
133. National Institute for Health and Clinical Excellence www.nice.org.uk (Accessed June 1, 2017)
134. Guidelines International Network http://www.g-i-n.net/ (Accessed June 5, 2017)
135. Agency for Health Care Research and Quality (AHRQ) https://www.ahrq.gov/professionals/clinicians-providers/guidelines-recommendations/index.html (Accessed June 1, 2017)
136. Health Team Works www.health teamworks.org (Accessed June 1, 2017)
137. Institute for Clinical Systems Improvement. www.icsi.org (Accessed June 1, 2017)
138. National Comprehensive Network https://www.nccn.org/professionals/physician_gls/f_guidelines.asp (Accessed June 5, 2017)
139. Skyscape www.skyscape.com (Accessed June 1, 2017)
140. mTBI Pocket Guide https://t2health.dcoe.mil/apps/mtbi (Accessed June 5, 2017)
141. Electronic Preventive Sevices Selector http://epss.ahrq.gov/PDA/index.jsp (Accessed (June 1, 2017)
142. Forrest CB, Fiks AG, Bailey LC et.al. Improving adherence to otitis media guidelines with clinical decision support and physician feedback. Pediatrics. 2013;131(4):e1071-1081
143. Savinon, C, Taylor JS, Canty-Mitchell J et.al. Childhood obesity: Can electronic medical records customized with clinical practice guidelines improve screening and diagnosis? J Am Acad Nurse Pract. 2012;24(8):463-471

144. Chen LM, Kennedy EH, Sales A et. al. Use of health IT for higher-value critical care. NEJM 2013;368(7):594-597
145. Battaglia L, Aronson MD, Neeman N et al. A "Smart" heart failure sheet: Using electronic medical records to guide clinical decision making. Am J Med 2011;124(2):118-120
146. Prasad V. Why Randomized Controlled Trials Are Needed to Accept New Practices: 2 Medical Word Views. Mayo Clin Proc 2013;88(10):1046-1050
147. Sniderman AD, LaChapelle KJ, Rachon NA, Furberg CD. The Necessity for Clinical Reasoning in the Era of Evidence-based Medicine. Mayo Clin Proc 2013;88(10):1108-1114
148. Bassler RD, Briel M, Murad MH et al. Stopping Randomized Trials Early for Benefit and Estimation of Treatment Effects: Systematic Review and Meta-Regression Analysis. JAMA 2010;303 (12):1180-7
149. Wang AT, McCoy CP, Murad MH. Association Between Affiliation and Position on Cardiovascular Risk with Rosiglitazone: Cross Sectional Systematic Review. BMJ 2010. March 18:340.c1344. doi:10.1136/bmj.c1344 (Accessed April 10, 2010)

15

Information Retrieval from Medical Knowledge Resources

WILLIAM R. HERSH

LEARNING OBJECTIVES

After reading this chapter, the reader should be able to:

- Enumerate the basic biomedical and health knowledge resources in books, journals, electronic databases, and other sources
- Describe the major approaches used to indexing knowledge-based content
- Apply advanced searching techniques to the major biomedical and health knowledge resources
- Discuss the major results of information retrieval evaluation studies
- Describe future directions for research in information retrieval

INTRODUCTION

One of the most important applications of health informatics is *information retrieval* (IR), sometimes called *search*. IR is the field concerned with the acquisition, organization, and searching of knowledge-based information, which is usually defined as information derived and organized from observational or experimental research.[1-2] Although IR in biomedicine traditionally concentrated on the retrieval of text from the biomedical literature, the purview of content covered has expanded to include newer types of media that include images, video, chemical structures, gene and protein sequences, and a wide range of other digital media of relevance to biomedical education, research, and patient care. With the proliferation of IR systems and online content, even the notion of the library has changed substantially, with the emergence of the *digital library*.[3-4]

Figure 15.1 shows an overview of the components of search systems. The overall goal of the IR process is to find *content* that meets a person's information needs. This begins with the posing of a *query* to the IR system. A *search engine* matches the query to content items through metadata. There are two intellectual processes of IR. *Indexing* is the process of assigning metadata to content items, while *retrieval* is the process of the user entering his or her query and retrieving content items.

The use of IR systems by clinicians, patients, and others has become essentially ubiquitous. It is estimated that among individuals who use the Internet in the United States, over 80 percent have used it to search for personal health information.[5] Virtually all physicians use the Internet.[6-7] Furthermore, access to systems has gone beyond the traditional personal computer and extended to new devices, such as smartphones and tablet devices. Other evidence points to the importance of IR and biomedicine. One researcher defines biology as an "*information science*."[8] Clinicians can no longer keep up with the growth of the literature, as an average of 75 clinical trials and 11 systematic reviews are published each day.[9]

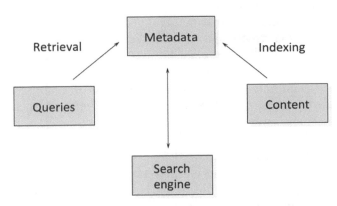

Figure 15.1: Components of information retrieval systems

IR tends to focus on knowledge-based information, which is information based on scientific research and in distinction to patient-specific information that is generated in the care of patient. Knowledge-based information is typically subdivided into two categories. Primary knowledge–based information (also called primary literature) is original research that appears in journals, books, reports, and other sources. This type of information reports the initial discovery of health knowledge, usually with either original data or reanalysis of data (e.g., systematic reviews and meta-analyses). Secondary knowledge–based information consists of the writing that reviews, condenses, and/or synthesizes the primary literature. The most common examples of this type of literature are books, monographs, and review articles in journals and other publications. Secondary literature also includes opinion-based writing such as editorials and position or policy papers. It also encompasses clinical practice guidelines, narrative reviews, and health information on Web pages. In addition, it includes the plethora of pocket-sized manuals that were formerly a staple for practitioners in many professional fields. Secondary literature is the most common type of literature used by physicians. Secondary literature also includes the growing quality of patient/consumer-oriented health information that is increasingly available via the Web.

Profound changes have taken place in the publishing of knowledge-based information in recent years. Virtually all scientific journals are published electronically now. In addition, there is great enthusiasm for electronic availability of journals, as evidenced by the growing number of titles to which libraries provide access. When available in electronic form, journal content is easier and more convenient to access. Furthermore, since most scientists have the desire for widespread dissemination of their work, they have incentive for their papers to be available electronically. Not only is there the increased convenience of redistributing reprints, but research has found that freely available on the Web have a higher likelihood of being cited by other papers than those that are not.[10] As citations are important to authors for academic promotion and grant funding, authors have incentive to maximize the accessibility of their published work.

The technical challenges to electronic scholarly publication have been replaced by economic and political ones.[11-12] Printing and mailing, tasks no longer needed in electronic publishing, comprised a significant part of the "added value" from publishers of journals. There is still however value added by publishers, such as hiring and managing editorial staff to produce the journals and managing the peer review process. Even if publishing companies as they are known were to vanish, there would still be some cost to the production of journals. Thus, while the cost of producing journals electronically is likely to be less, it is not zero, and even if journal content is distributed "free," someone must pay the production costs. The economic issue in electronic publishing, then, is who is going to pay for the production of journals.[12] This introduces some political issues as well. One of them centers around the concern that much research is publicly funded through grants from federal agencies such as the National Institutes of Health (NIH) and the National Science Foundation (NSF). In the current system, especially in the biomedical sciences (and to a lesser extent in other sciences), researchers turn over the copyright of their publications to journal publishers. The political concern is that the public funds the research and the universities carry it out, but individuals and libraries then must buy it back from the publishers to whom they willingly cede the copyright. This problem is exacerbated by the general decline in funding for libraries.

Some proposed models of *"open access"* scholarly publishing to keep the archive of science freely available.[13-15] The basic principle of open access publishing is that authors and/or their institutions pay the cost of production of manuscripts up front after they are accepted through a peer review process. After the paper is published, it becomes freely available on the Web. Since most research is usually funded by grants, the cost of open access publishing should be included in grant budgets. The uptake of publishers adhering to the open access model has been modest, with the most prominent being *Biomed Central* (BMC, www.biomedcentral.com) and the *Public Library of Science* (PLoS, www.plos.org).

Another model that has emerged is *PubMed Central* (PMC, http://pubmedcentral.gov). PMC is a repository of life science research articles that provides free access while allowing publishers to maintain copyright and even optionally keep the papers housed on their own servers. A lag time of up to 6 months is allowed so that journals can reap the revenue that comes with initial publication. The NIH now requires all research funded by its grants to be submitted to PMC, either in the form published by publishers or as a PDF of the last manuscript prior to journal acceptance (http://publicaccess.nih.gov). Publishers have expressed concern that copyrights give journals more control over the integrity of the papers they publish.[16] An alternative approach advocated by non-commercial (using professional society) publishers is the *DC Principles for Free Access to Science* (www.dcprinciples.org), which advocates reinvestment of revenues in support of science, use of open archives such as

PMC as allowed by business constraints, commitment to some free publication, more open access for low-income countries, and no charges for authors to publish.

CONTENT

It is useful to classify the information to gain a better understanding of its structure and function. In this section, we classify content into bibliographic, full-text, annotated, and aggregated categories, although some content does not neatly fit within them.

Bibliographic Content

The first category consists of bibliographic content. It includes what was for decades the mainstay of IR systems: literature reference databases. Also called bibliographic databases, this content consists of citations or pointers to the medical literature (i.e., journal articles). The best-known and most widely used biomedical bibliographic database is MEDLINE, which contains bibliographic references to all the biomedical articles, editorials, and letters to the editors in approximately 5,000 scientific journals. The journals are chosen for inclusion by an advisory committee of subject experts convened by NIH. At present, over 900,000 references are added to MEDLINE yearly. It now contains over 24 million references. A web page devoted to MEDLINE size and searches statistics is at https://www.nlm.nih.gov/bsd/bsd_key.html .

The MEDLINE record may contain up to 49 fields. A clinician may be interested in just a handful of these fields, such as the title, abstract, and indexing terms. But other fields contain specific information that may be of great importance to other audiences. For example, a genome researcher might be highly interested in the Supplementary Information (SI) field to link to genomic databases. Even the clinician may, however, derive benefit from some of the other fields. For example, the Publication Type (PT) field can help in the application of evidence-based medicine (EBM), such as when one is searching for a practice guideline or a randomized controlled trial. MEDLINE is accessible by many means and available without charge via the PubMed system (http://pubmed.gov), produced by the National Center for Biotechnology Information (NCBI, www.ncbi.nlm.nih.gov) of the NLM, which provides access to other databases as well. some other information vendors, such as Ovid Technologies (www.ovid.com), license the content of MEDLINE and other databases and provide value-added services that can be accessed for a fee by individuals and institutions.

MEDLINE is only one of many databases produced by the NLM. There are several non-NLM bibliographic databases that tend to be more focused on subjects or resource types. The major non-NLM database for the nursing field is the *Cumulative Index to Nursing and Allied Health Literature* (CINAHL, CINAHL Information Systems, http://www.ebscohost.com/cinahl/), which covers nursing and allied health literature, including physical therapy, occupational therapy, laboratory technology, health education, physician assistants, and medical records. Another well-known bibliographic database is EMBASE (www.embase.com), which contains over 32 million records and covers many of the same medical journals as MEDLINE but with a more international focus, including more non-English-language journals.

A second, more modern type of bibliographic content is the Web catalog. There are increasing numbers of such catalogs, which consist of Web pages containing mainly links to other Web pages and sites. It should be noted that there is a blurry distinction between Web catalogs and aggregations (the fourth category). In general, the former contains only links to other pages and sites, while the latter include actual content that is highly integrated with other resources. Some well-known Web catalogs include:

- *HealthFinder* (www.healthfinder.gov)—consumer-oriented health information maintained by the Office of Disease Prevention and Health Promotion of the U.S. Department of Health and Human Services.
- *HON Select* (www.hon.ch/HONselect)—a European catalog of quality-filtered, clinician-oriented Web content from the HON foundation.
- *Translating Research into Practice* (TRIP, www.tripdatabase.com)—a database of content deemed to meet high standards of EBM.

An additional modern bibliographic resource is the *National Guidelines Clearinghouse* (NGC, www.guideline.gov). Produced by the Agency for Healthcare Research and Quality (AHRQ), it contains exhaustive information about clinical practice guidelines. Some of the guidelines produced are freely available, published electronically and/or on paper. Others are proprietary, in which case a link is provided to a location at which the guideline can be ordered or purchased. The overall goal of the NGC is to make evidence-based clinical practice guidelines and related abstract, summary, and comparison materials widely available to health care and other professionals.

A final kind of bibliographic-like content consists of RSS feeds, which are short summaries of Web content, typically news, journal articles, blog postings, and other content. Users set up an RSS aggregator, which can

be though a Web browser, email client, or standalone software, configured for the RSS feed desired, with an option to add a filter for specific content. There are two versions of RSS (1.0 and 2.0) but both provide:
- Title—name of item
- Link—URL to content
- Description—a brief description of the content

Full-text Content

The second type of content is full-text content. A large component of this content consists of the online versions of books and periodicals. As already noted, most traditionally paper-based medical literature, from textbooks to journals, is now available electronically. The electronic versions may be enhanced by measures ranging from the provision of supplemental data in a journal article to linkages and multimedia content in a textbook. The final component of this category is the Web site. Admittedly, the diversity of information on Web sites is enormous, and sites may include every other type of content described in this chapter. However, in the context of this category, "Web site" refers to the vast number of static and dynamic Web pages at a discrete Web location.

Electronic publication of journals allows additional features not possible in the print world. Journal Web sites may provide supplementary data of results, images, and even raw data. A journal Web site also allows more dialog about articles than could be published in a "Letters to the Editor" section of a print journal. Electronic publication also allows true bibliographic linkages, both to other full-text articles and to the MEDLINE record.

The Web also allows linkage directly from bibliographic databases to full text. PubMed maintains a field for the Web address of the full-text paper. This linkage is active when the PubMed record is displayed, but users may be met by a "paywall" if the article is not available for free. Many sites allow both access to subscribers or a pay-per-view facility. Many academic organizations now maintain large numbers of subscriptions to journals available to faculty, staff, and students. Other publishers, such as Ovid, provide access within their own password-protected interfaces to articles from journals that they have licensed for use in their systems.

The most common secondary literature source is traditional textbooks, which have essentially made a complete transition to publication in electronic form. A common approach with textbooks is "bundling", sometimes with linkages across the bundled texts. An early bundler of textbooks was Stat!Ref (Teton Data Systems, www.statref.com) that, like many, began as a CD-ROM product and then moved to the Web. Stat!Ref offers over 400 textbooks and other resources. Most other publishers have similar aggregated their libraries of textbooks and other content. Another collection of textbooks is the NCBI Bookshelf, which contains many volumes on biomedical research topics (http://www.ncbi.nlm.nih.gov/books).

Electronic textbooks offer additional features beyond text from the print version. While many print textbooks do feature high-quality images, electronic versions offer the ability to have more pictures and illustrations. They also can provide sound and video. As with full-text journals, electronic textbooks can link to other resources, including journal references and the full articles. Many Web-based textbook sites also provide access to continuing education self-assessment questions and medical news. Finally, electronic textbooks let authors and publishers provide more frequent updates of the information than is allowed by the usual cycle of print editions, where new versions come out only every 2 to 5 years.

As noted above, Web sites are another form of full-text information. Probably the most effective provider of Web-based health information is the U.S. government. Not only do they produce bibliographic databases, but the NLM, AHRQ, the National Cancer Institute (NCI), Centers for Disease Control (CDC), and others have also been innovative in providing comprehensive full-text information for health care providers and consumers. One example is the popular CDC Travel site (http://www.cdc.gov/travel/). Some of these will be described later as aggregations, since they provide many different types of resources.

Many commercial biomedical and health Web sites have emerged in recent years. On the consumer side, they include more than just collections of text; they also include interaction with experts, online stores, and catalogs of links to other sites. There are also Web sites, either from medical societies or companies, that provide information geared toward health care providers, typically overviews of diseases, their diagnosis, and treatment; medical news and other resources for providers are often offered as well.

Other sources of on-line health-related content include encyclopedias, the body of knowledge, and Weblogs or blogs. A well-known online encyclopedia with a great deal of health-related information is Wikipedia, which features a distributed authorship process whose content has been found to reliable[17-18] and frequently shows up near the top in health-related Web searches.[19] A growing number of organizations have a body of knowledge, such as the American Health Information Management

Association (AHIMA, http://library.ahima.org/bok/). Blogs tend to carry a stream of consciousness, but often high-quality information is posted within them.

Annotated Content

The third category consists of annotated content. These resources are usually not stored as freestanding Web pages but instead are often housed in database management systems. This content can be further subcategorized into discrete information types:
- Image databases—collections of images from radiology, pathology, and other areas
- Genomics databases—information from gene sequencing, protein characterization, and other genomic research
- Citation databases—bibliographic linkages of scientific literature
- EBM databases—highly structured collections of clinical evidence
- Other databases—miscellaneous other collections

A great number of biomedical image databases are available on the Web. These include:
- Visible Human http://www.nlm.nih.gov/research/visible/visible_human.html
- Lieberman's eRadiology http://eradiology.bidmc.harvard.edu
- WebPath http://library.med.utah.edu/WebPath/webpath.html
- Pathology Education Instructional Resource (PEIR) www.peir.net
- DermIS www.dermis.net
- VisualDX www.visualdx.com

Many genomics databases are available on the Web. The first issue each year of the journal *Nucleic Acids Research* (NAR) catalogs and describes these databases and is now available by open access means.[20] NAR also maintains an ongoing database of such databases, the Molecular Biology Database Collection (http://www.oxfordjournals.org/nar/database/cap/). Among the most important of these databases are those available from NCBI.[21] All their databases are linked among themselves, along with PubMed and OMIM, and are searchable via the GQuery system (http://www.ncbi.nlm.nih.gov/gquery/).

Citation databases provide linkages to articles that cite others across the scientific literature. The earliest citation databases were the *Science Citation Index* (SCI, Thompson-Reuters) and *Social Science Citation Index* (SSCI, Thompson-Reuters), which are now part of the larger *Web of Science*. Two well-known bibliographic databases for biomedical and health topics that also have citation links include SCOPUS (www.scopus.com) and Google Scholar (http://scholar.google.com). These three were recently compared for their features and coverage.[22] A final citation database of note is CiteSeer (http://citeseerx.ist.psu.edu/), which focuses on computer and information science, including biomedical informatics.

Evidence-based medicine (EBM) databases are devoted to providing annotated evidence-based information. Some examples include:
- *The Cochrane Database of Systematic Reviews*—one of the original collections of systematic reviews (www.cochrane.org)
- *Clinical Evidence*—an "evidence formulary" (www.clinicalevidence.com)
- *UptoDate*—content centered around clinical questions (www.uptodate.com)
- *InfoPOEMS*—"patient-oriented evidence that matters" (www.essentialevidenceplus.com)

There is a growing market for a related type of evidence-based content in the form of clinical decision support order sets, rules, and health/disease management templates. Publishers include EHR vendors whose systems employ this content as well as other vendors such as Zynx (www.zynxhealth.com) and Thomson Reuters Cortellis (http://cortellis.thomsonreuters.com).

There is a variety of other annotated medical content. The ClinicalTrials.gov database (www.clinicaltrials.gov) began as a database of clinical trials sponsored by NIH. In recent years it has expanded its scope to a register of clinical trials[23-24] and to containing actual results of trials.[25-26] Another important database for researchers is NIH RePORTER (http://projectreporter.nih.gov/reporter.cfm), which is a database of all research funded by NIH.

Aggregated Content

The final category consists of aggregations of content from the first three categories. The distinction between this category and some of the highly-linked types of content described above is admittedly blurry, but aggregations typically have a wide variety of different types of information serving the diverse needs of users. Aggregated content has been developed for all types of users from consumers to clinicians to scientists.

Probably the largest aggregated consumer information resource is *MedlinePlus* (http://medlineplus.gov) from the NLM. MedlinePlus includes all the types of content previously described, aggregated for easy access to a given topic. MedlinePlus contains health topics, drug information, medical dictionaries, directories, and other resources. Each topic contains links to health information from the NIH and other sources deemed credible by its selectors. There are also links to current health news (updated daily), a medical encyclopedia, drug references, and directories, along with a preformed PubMed search, related to the topic.

Aggregations of content have also been developed for clinicians. Most of the major publishers now aggregate all their content in packages for clinicians. Another well-known group of aggregations of content for genomics researchers is the model organism databases. These databases bring together bibliographic databases, full text, and databases of sequences, structure, and function for organisms whose genomic data have been highly characterized. One of the oldest and most developed model organism databases is the Mouse Genome Informatics resource (www.informatics.jax.org).

INDEXING

As described at the beginning of the chapter, indexing is the process of assigning metadata to content to facilitate its retrieval. Most modern commercial content is indexed in two ways:

1. Manual indexing—where human indexers, usually using a controlled terminology, assign indexing terms and attributes to documents, often following a specific protocol.
2. Automated indexing—where computers make the indexing assignments, usually limited to breaking out each word in the document (or part of the document) as an indexing term.

Manual indexing is done most commonly for bibliographic databases and annotated content. In this age of proliferating electronic content, such as online textbooks, practice guidelines, and multimedia collections, manual indexing has become either too expensive or outright unfeasible for the quantity and diversity of material now available. Thus, there are increasing numbers of databases that are indexed only by automated means. Before covering these types of indexing in detail, let us first discuss controlled terminologies.

Controlled Terminologies

A controlled terminology contains a set of terms that can be applied to a task, such as indexing. When the terminology defines the terms, it is usually called a vocabulary. When it contains variants or synonyms of terms, it is also called a thesaurus. Before discussing actual terminologies, it is useful to define some terms. A concept is an idea or object that occurs in the world, such as the condition under which human blood pressure is elevated. A term is the actual string of one or more words that represent a concept, such as *Hypertension* or *High Blood Pressure*. One of these string forms is the preferred or canonical form, such as Hypertension in the present example. When one or more terms can represent a concept, the different terms are called synonyms.

A controlled terminology usually contains a list of terms that are the canonical representations of the concepts. If it is a thesaurus, it contains relationships between terms, which typically fall into three categories:

- Hierarchical—terms that are broader or narrower. The hierarchical organization not only provides an overview of the structure of a thesaurus but also can be used to enhance searching (e.g., MeSH tree explosions that add terms from an entire portion of the hierarchy to augment a search)
- Synonym—terms that are synonyms, allowing the indexer or searcher to express a concept in different words
- Related—terms that are not synonymous or hierarchical but are somehow otherwise related. These usually remind the searcher of different but related terms that may enhance a search

The MeSH terminology is used to manually index most of the databases produced by the NLM.[27] The latest version contains over 26,000 subject headings (the word MeSH uses the canonical representation of its concepts). It also contains over 170,000 synonyms to those terms, which in MeSH jargon are called entry terms. In addition, MeSH contains the three types of relationships described in the previous paragraph:

- Hierarchical—MeSH is organized hierarchically into 16 trees, such as Diseases, Organisms, and Chemicals and Drugs
- Synonym—MeSH contains a vast number of entry terms, which are synonyms of the headings
- Related—terms that may be useful for searchers to add to their searches when appropriate are suggested for many headings

The MeSH terminology files, their associated data, and their supporting documentation are available on the NLM's MeSH Web site (http://www.nlm.nih.gov/mesh). There is also a browser that facilitates exploration of the terminology (http://www.nlm.nih.gov/mesh/MBrowser.html). Figure 15.2 shows a slice through the MeSH hierarchy for certain cardiovascular diseases.

There are features of MeSH designed to assist indexers in making documents more retrievable. One of these is subheadings, which are qualifiers of subject headings that narrow the focus of a term. In Hypertension, for example, the focus of an article may be on the diagnosis, epidemiology, or treatment of the condition. Another feature of MeSH that helps retrieval is check tags. These

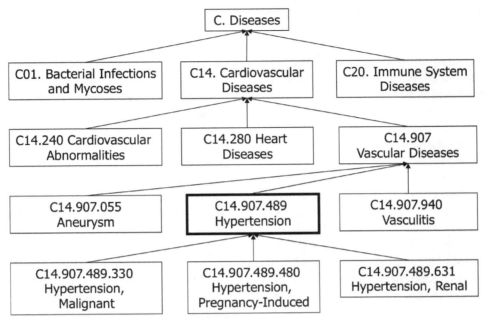

Figure 15.2: "Slice" through MeSH hierarchy

are MeSH terms that represent certain facets of medical studies, such as age, gender, human or nonhuman, and type of grant support. Related to check tags are the geographical locations in the Z tree. Indexers must also include these, like check tags, since the location of a study (e.g., *Oregon*) must be indicated. Another feature gaining increasing importance for EBM and other purposes is the publication type, which describes the type of publication or the type of study. A searcher who wants a review of a topic may choose the publication type *Review* or *Review Literature*. Or, to find studies that provide the best evidence for a therapy, the publication type *Meta-Analysis*, *Randomized Controlled Trial*, or *Controlled Clinical Trial* would be used.

MeSH is not the only thesaurus used for indexing biomedical documents. Several other thesauri are used to index non-NLM databases. CINAHL, for example, uses the CINAHL Subject Headings, which are based on MeSH but have additional domain-specific terms added. EMBASE has a terminology called EMTREE, which has many features similar to those of MeSH (http://www.embase.com/info/helpfiles/emtree-tool/emtree-thesaurus).

Manual Indexing

Manual indexing is most commonly done for bibliographic and annotated content, although it is sometimes for other types of content as well. Manual indexing is usually done by means of a controlled terminology of terms and attributes. Most databases utilizing human indexing usually have a detailed protocol for assignment of indexing terms from the thesaurus. The MEDLINE database is no exception. The principles of MEDLINE indexing were laid out in the two-volume MEDLARS Indexing Manual.[28-29] Subsequent modifications have occurred with changes to MEDLINE, other databases, and MeSH over the years. The major concepts of the article, usually from two to five headings, are designed as main headings, and designated in the MEDLINE record by an asterisk. The indexer is also required to assign appropriate subheadings. Finally, the indexer must also assign check tags, geographical locations, and publication types. Although MEDLINE indexing is still manual, indexers are aided by a variety of electronic tools for selecting and assigning MeSH terms.

Few full-text resources are manually indexed. One type of indexing that commonly takes place with full-text resources, especially in the print world, is that performed for the index at the back of the book. However, this information is rarely used in IR systems; instead, most online textbooks rely on automated indexing (see below). One exception to this is MDConsult (now Clinical Key), which uses back-of-book indexes to point to specific sections in its online books.

Manual indexing of Web content is challenging. With billions of pages of content, manual indexing of more than a fraction of it is not feasible. On the other hand, the lack of a coherent index makes searching much more difficult, especially when specific resource types are being sought. A simple form of manual indexing of the

Web takes place in the development of the Web catalogs and aggregations as described earlier. These catalogs contain not only explicit indexing about subjects and other attributes, but also implicit indexing about the quality of a given resource by the decision of whether to include it in the catalog.

While most Web content is indexed automatically (see below), one approach to manual indexing has been to apply metadata to Web pages and sites, = exemplified by the Dublin Core Metadata Initiative (DCMI, www.dublincore.org).[30] The goal of the DCMI has been to develop a set of standard data elements that creators of Web resources can use to apply metadata to their content. The DCMI was recently approved as a standard by the National Information Standards Organization (NISO) with the designation Z39.85. It is also a standard with the International Organization for Standards (ISO), ISO Standard 15836:2009. The specification has 15 defined elements:

- DC.title - name given to the resource
- DC.creator - person or organization primarily responsible for creating the intellectual content of the resource
- DC.subject - topic of the resource
- DC.description - a textual description of the content of the resource
- DC.publisher - entity responsible for making the resource available in its present form
- DC.date - date associated with the creation or availability of the resource
- DC.contributor - person or organization not specified in a creator element who has made a significant intellectual contribution to the resource but whose contribution is secondary to any person or organization specified in a creator element
- DC.type - category of the resource
- DC.format - data format of the resource, used to identify the software and possibly hardware that might be needed to display or operate the resource
- DC.identifier - string or number used to uniquely identify the resource
- DC.source - information about a second resource from which the present resource is derived
- DC.language - language of the intellectual content of the resource
- DC.relation - identifier of a second resource and its relationship to the present resource
- DC.coverage - spatial or temporal characteristics of the intellectual content of the resource
- DC.rights - rights management statement, an identifier that links to a rights management statement, or an identifier that links to a service providing information about rights management for the resource

There have been some medical adaptations of the DCMI. The most developed of these is the Catalogue et Index des Sites Médicaux Francophones (CISMeF, www.cismef.org).[31] A catalog of French-language health resources on the Web, CISMeF has used DCMI to catalog over 40,000 Web pages, including information resources (e.g., practice guidelines, consensus development conferences), organizations (e.g., hospitals, medical schools, pharmaceutical companies), and databases. The Subject field uses the French translation of MeSH but also includes the English translation. For Type, a list of common Web resources has been enumerated.

While Dublin Core Metadata was originally envisioned to be included in Hypertext Markup Language (HTML) Web pages, it became apparent that many non-HTML resources exist on the Web and that there are reasons to store metadata external to Web pages. For example, authors of Web pages might not be the best people to index pages or other entities might wish to add value by their own indexing of content. A standard for cataloging metadata is the Resource Description Framework (RDF).[32] A framework for describing and interchanging metadata, RDF is usually expressed in Extensible Markup Language (XML), a standard for data interchange on the Web. RDF also forms the basis of what some call the future of the Web as a repository not only of content but also of knowledge, which is also referred to as the Semantic Web.[32] Dublin Core Metadata (or any type of metadata) can be represented in RDF.

Manual indexing has several limitations, the most significant of which is inconsistency. Funk and Reid evaluated indexing inconsistency in MEDLINE by identifying 760 articles that had been indexed twice by the NLM.[33] The most consistent indexing occurred with check tags and central concept headings, which were only indexed with a consistency of 61 to 75 percent. The least consistent indexing occurred with subheadings, especially those assigned to non-central-concept headings, which had a consistency of less than 35 percent. A repeat of this study in more recent times found comparable results.[34] Manual indexing also takes time. While it may be feasible with the large resources the NLM has to index MEDLINE, it is probably impossible with the growing amount of content on Web sites and in other full-text resources. Indeed, the NLM has recognized the challenge of continuing to have to index the growing body of biomedical literature and is investigating automated and semi-automated means of doing so.[35]

Automated Indexing

In automated indexing, the indexing is done by a computer. Although the mechanical running of the automated indexing process lacks cognitive input, considerable intellectual effort may have gone into development of the system for doing it, so this form of indexing still qualifies as an intellectual process. In this section, we will focus on the automated indexing used in operational IR systems, namely the indexing of documents by the words they contain.

Some might not think of extracting all the words in a document as "indexing," but from the standpoint of an IR system, words are descriptors of documents, just like human-assigned indexing terms. Most retrieval systems actually use a hybrid of human and word indexing, in that the human-assigned indexing terms become part of the document, which can then be searched by using the whole controlled term or individual words within it. Most MEDLINE implementations have always allowed the combination of searching on human indexing terms and on words in the title and abstract of the reference. With the development of full-text resources in the 1980s and 1990s, systems that allowed only word indexing began to emerge. This trend increased with the advent of the Web.

Word indexing is typically done by defining all consecutive alphanumeric sequences between white space (which consists of spaces, punctuation, carriage returns, and other non-alphanumeric characters) as words. Systems must take particular care to apply the same process to documents and the user's query, especially with characters such as hyphens and apostrophes. Many systems go beyond simple identification of words and attempt to assign weights to words that represent their importance in the document.[36]

Many systems using word indexing employ processes to remove common words or conflate words to common forms. The former consists of filtering to remove *stop words*, which are common words that always occur with high frequency and are usually of little value in searching. The stop word list, also called a negative dictionary, varies in size from the seven words of the original MEDLARS stop list (and, an, by, from, of, the, with) to the list of 250 to 500 words more typically used. Examples of the latter are the 250-word list of van Rijsbergen,[37] the 471-word list of Fox,[38] and the PubMed stop list.[39] Conflation of words to common forms is done via *stemming*, the purpose of which is to ensure words with plurals and common suffixes (e.g., -ed, -ing, -er, -al) are always indexed by their stem form.[40] For example, the words cough, coughs, and coughing are all indexed via their stem cough. Both stop word removal and stemming reduce the size of indexing files and lead to more efficient query processing.

A commonly used approach for term weighting is TF*IDF weighting, which combines the inverse document frequency (IDF) and term frequency (TF). The IDF is the logarithm of the ratio of the total number of documents to the number of documents in which the term occurs. It is assigned once for each term in the database, and it correlates inversely with the frequency of the term in the entire database. The usual formula used is:

$$IDF(term) = \log \frac{\text{number of documents in database}}{\text{number of documents with term}} + 1 \quad (1)$$

The TF is a measure of the frequency with which a term occurs in a given document and is assigned to each term in each document, with the usual formula:

$$TF(term, document) = \text{frequency of term in document} \quad (2)$$

In TF*IDF weighting, the two terms are combined to form the indexing weight, WEIGHT:

$$WEIGHT(term, document) = TF(term, document) * IDF(term) \quad (3)$$

Another automated approach to precomputing metadata about documents involves the use of link-based methods, which is best known through its use by the Google search engine (www.google.com). This approach gives weight to pages based on how often they are cited by other pages. The *PageRank algorithm* is mathematically complex but can be viewed as giving more weight to a Web page based on the number of other pages that link to it.[41] Thus, the home page of the NLM or a major medical journal is likely to have a very high PR, whereas a more obscure page will have some lower PR. General-purpose search engines such as Google and Microsoft Bing use word-based approaches and variants of the PageRank algorithm for indexing. They amass the content in their search systems by "crawling" the Web, collecting and indexing every object they find on the Web. This includes not only HTML pages, but other files as well, including Microsoft Word, Portable Document Format (PDF), and images.

Word indexing has several limitations, including:
- Synonymy—different words may have the same meaning, such as high and elevated. This problem may extend to the level of phrases with no words in common, such as the synonyms hypertension and high blood pressure.
- Polysemy—the same word may have different meanings or senses. For example, the word lead can refer to an element or to a part of an electrocardiogram machine.

- Content—words in a document may not reflect its focus. For example, an article describing hypertension may make mention in passing to other concepts, such as congestive heart failure (CHF) that are not the focus of the article.
- Context—words take on meaning based on other words around them. For example, the relatively common words high, blood, and pressure, take on added meaning when occurring together in the phrase high blood pressure.
- Morphology—words can have suffixes that do not change the underlying meaning, such as indicators of plurals, various participles, adjectival forms of nouns, and nominalized forms of adjectives.
- Granularity—queries and documents may describe concepts at different levels of a hierarchy. For example, a user might query for antibiotics in the treatment of a specific infection, but the documents might describe specific antibiotics themselves, such as penicillin.

RETRIEVAL

There are two broad approaches to retrieval. Exact-match searching allows the user precise control over the items retrieved. Partial-match searching, on the other hand, recognizes the inexact nature of both indexing and retrieval, and instead attempts to return the user content ranked by how close it comes to the user's query. After general explanations of these approaches, we will describe actual systems that access the different types of biomedical content.

Exact-Match Retrieval

In exact-match searching, the IR system gives the user all documents that exactly match the criteria specified in the search statement(s). Since the Boolean operators AND, OR, and NOT are usually required to create a manageable set of documents, this type of searching is often called Boolean searching. Furthermore, since the user typically builds sets of documents that are manipulated with the Boolean operators, this approach is also called set-based searching. Most of the early operational IR systems in the 1950s through 1970s used the exact-match approach, even though Salton was developing the partial-match approach in research systems during that time.[42] In modern times, exact-match searching tends to be associated with retrieval from bibliographic and annotated databases, while the partial-match approach tends to be used with full-text searching.

Typically, the first step in exact-match retrieval is to select terms to build sets. Other attributes, such as the author name, publication type, or gene identifier (in the secondary source identifier field of MEDLINE), may be selected to build sets as well. Once the search term(s) and attribute(s) have been selected, they are combined with the Boolean operators. The Boolean AND operator is typically used to narrow a retrieval set to contain only documents with two or more concepts. The Boolean OR operator is usually used when there is more than one way to express a concept. The Boolean NOT operator is often employed as a subtraction operator that must be applied to another set. Some systems more accurately call this the AND NOT operator. The Boolean operators are depicted graphically in Figure 15.3.

- AND – only content items that have all terms

- OR – content items that have any term

- NOT – content items with one term but not other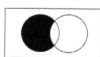

Figure 15.3: Boolean operators

Some retrieval systems allow terms in searches to be expanded by using the wild-card character, which adds all words to the search that begin with the letters up until the wild-card character. This approach is also called truncation. Unfortunately, there is no standard approach to using wild-card characters, so syntax for them varies from system to system. PubMed, for example, allows a single asterisk at the end of a word to signify a wild-card character. Thus, the query word can* will lead to the words cancer and Candida, among others, being added to the search.

Partial-Match Retrieval

Although partial-match searching was conceptualized very early, it did not see widespread use in IR systems until the advent of Web search engines in the 1990s. This is most likely because exact-match searching tends to be preferred by *"power users"* whereas partial-match searching is preferred by novice searchers. Whereas exact-match searching requires an understanding of Boolean operators and (often) the underlying structure of databases (e.g., the many fields in MEDLINE), partial-match searching allows a user to simply enter a few terms and start retrieving documents.

The development of partial-match searching is usually attributed to Salton,[42] who pioneered the approach in the 1960s. Although partial-match searching does not exclude the use of non-term attributes of documents, and for that matter does not even exclude the use of Boolean operators, the most common use of this type of searching is with a query of a small number of words, also known as a natural language query.[43] Because Salton's approach was based on vector mathematics, it is also referred to as the vector-space model of IR. In the partial-match approach, documents are typically ranked by their closeness of fit to the query. That is, documents containing more query terms will likely be ranked higher, since those with more query terms will in general be more likely to be relevant to the user. As a result, this process is called relevance ranking. The entire approach has also been called lexical–statistical retrieval.

The most common approach to document ranking in partial-match searching is to give each a score based on the sum of the weights of terms common to the document and query. Terms in documents typically derive their weight from the TF*IDF calculation described above. Terms in queries are typically given a weight of one if the term is present and zero if it is absent. The following formula can then be used to calculate the document weight across all query terms:

$$\text{Document weight} = \sum_{\text{all query terms}} \text{Weight of term in query} * \text{Weight of term in document} \quad (4)$$

This may be thought of as a giant OR of all query terms, with sorting of the matching documents by weight. The usual approach is for the system to then perform the same stop word removal and stemming of the query that was done in the indexing process. (The equivalent stemming operations must be performed on documents and queries so that complementary word stems will match.) One problem with TF*IDF weighting is that longer documents accumulate more weight in queries simply because they have more words. As such, some approaches "normalize" the weight of a document.

A variety of other variations to the basic partial-matching retrieval approach have been developed. One important addition is *relevance feedback*, a feature allowed by the partial-match approach, permits new documents to be added to the output based on their similarity to those deemed relevant by the user. This approach also allows reweighting of relevant documents already retrieved to higher positions on the output list. The most common approach is the modified Rocchio equation employed by Buckley et al.[44] In this equation, each term in the query is reweighted by adding value for the term occurring in relevant documents and subtracting value for the term occurring in non-relevant documents. There are three parameters, α, β, and γ, which add relative value to the original weight, the added weight from relevant documents, and the subtracted weight from non-relevant documents, respectively. In this approach, the query is usually expanded by adding a specified number of query terms (from none to several thousand) from relevant documents to the query. Each query term takes on a new value based on the following formula:

$$\begin{aligned}\text{New query weight} =\ & a * \text{Original query weight} \\ & + b * \frac{1}{\text{number of relevant documents}} * \sum_{\text{all relevant documents}} \text{weight in document} \\ & - g * \frac{1}{\text{number of nonrelevant documents}} * \sum_{\text{all nonrelevant documents}} \text{weight in document}\end{aligned} \quad (5)$$

When the parameters, α, β, and γ, are set to one, this formula simplifies to:

$$\begin{aligned}\text{New query weight} =\ & \text{Original query weight} \\ & + \text{Average term weight in relevant documents} \\ & - \text{Average term weight in nonrelevant documents}\end{aligned} \quad (6)$$

Several IR systems offer a variant of relevance feedback that finds similar documents to a specified one. PubMed allows the user to obtain "related articles" from any given one in an approach similar to relevance feedback but which uses a different algorithm.[45] A number of Web search engines allow users to similarly obtain related articles from a specified Web page.

RETRIEVAL SYSTEMS

There are many different retrieval interfaces, with some of the features reflecting the content or structure of the underlying database.

As noted above, PubMed is the system at NLM that searches MEDLINE and other bibliographic databases. Although presenting the user with a simple text box, PubMed does a great deal of processing of the user's input to identify MeSH terms, author names, common phrases, and journal names (described in the on-line help system of PubMed). In this automatic term mapping, the system attempts to map user input, in succession, to MeSH terms, journals names, common phrases, and authors. Remaining text that PubMed cannot map is searched as text words (i.e., words that occur in any of the MEDLINE fields). Figure 15.4 shows the PubMed search results screen. The system allows a basic search and then provides access to a wealth of features around the results. The left-hand side of the screen allows setting of limits, such as to study type (e.g., randomized controlled trial), species (e.g., human or others), and age group (e.g., aged ->65 years). The right-hand side provides filters for free full text article and reviews, as well as other features that include the details of the search. As in most bibliographic systems, users can search PubMed by building search sets and then combining them with Boolean operators to tailor the search. This is called the "advanced search" or "search builder" of PubMed, as shown in Figure 15.5. PubMed also has a specialized query interface for clinicians seeking the best clinical evidence (called Clinical Queries) as well as several "apps" that allow access via mobile devices (e.g., iOS or Android).

Another recent addition to PubMed is the ability to sort search results by relevance ranking rather than the long-standing default reverse-chronological ordering. Choosing this option leads to MEDLINE records being sorted based on a formula that includes IDF, TF, a measure for which field in which the word appears (more for title and abstract), and a measure of recency of publication.[46]

Another common entry point for scientific articles and other content is Google Scholar. The search system provides access to a subset of content derived from the larger Google database but limited to scientific resources. The output is sorted by the number of citations to each retrieved item.

A growing number of search engines allow searching over many resources. The general search engines Google, Microsoft Bing, and others allow retrieval of any types of documents they index via their Web crawling activities. Other search engines allow searching over aggregations of various sources, such as NLM's GQuery (https://www.ncbi.nlm.nih.gov/search/), which allows searching over all NLM databases and other resources in one simple interface.

EVALUATION

There has been a great deal of research over the years devoted to evaluation of IR systems. As with many areas of research, there is controversy as to which approaches to evaluation best provide results that can assess searching

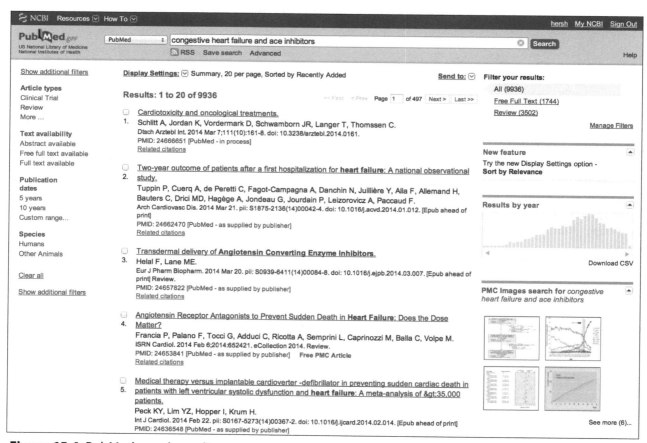

Figure 15.4: PubMed search results

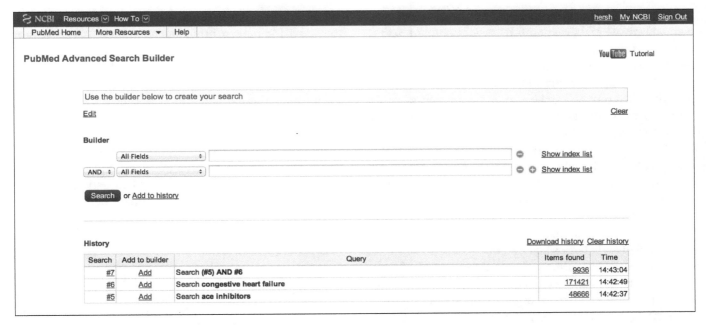

Figure 15.5: PubMed advanced search

in the systems they are using. Many frameworks have been developed to put the results in context. One of those frameworks organized evaluation around six questions that someone advocating the use of IR systems might ask:[47]

1. Was the system used?
2. For what was the system used?
3. Were the users satisfied?
4. How well did they use the system?
5. What factors were associated with successful or unsuccessful use of the system?
6. Did the system have an impact?

A simpler means for organizing the results of evaluation, however, groups approaches and studies into those which are system-oriented, i.e., the focus of the evaluation is on the IR system, and those which are user-oriented, i.e., the focus is on the user.

System-Oriented Evaluation

There are many ways to evaluate the performance of IR systems, the most widely used of which are the relevance-based measures of recall and precision. These measures quantify the number of relevant documents retrieved by the user from the database and in his or her search. Recall is the proportion of relevant documents retrieved from the database:

$$Recall = \frac{number\ of\ retrieved\ and\ relevant\ documents}{number\ of\ relevant\ documents\ in\ database} \quad (7)$$

In other words, recall answers the question, for a given search, what fraction of all the relevant documents have been obtained from the database?

One problem with Equation (7) is that the denominator implies that the total number of relevant documents for a query is known. For all but the smallest of databases, however, it is unlikely, perhaps even impossible, for one to succeed in identifying all relevant documents in a database. Thus, most studies use the measure of relative recall, where the denominator is redefined to represent the number of relevant documents identified by multiple searches on the query topic.

Precision is the proportion of relevant documents retrieved in the search:

$$Precision = \frac{number\ of\ retrieval\ and\ relevant\ documents}{number\ of\ documents\ retrieved} \quad (8)$$

This measure answers the question, for a search, what fraction of the retrieved documents are relevant?

One problem that arises when one is comparing systems that use ranking versus those that do not is that non-ranking systems, typically using Boolean searching, tend to retrieve a fixed set of documents and as a result have fixed points of recall and precision. Systems with relevance ranking, on the other hand, have different values of recall and precision depending on the size of the retrieval set the system (or the user) has chosen to show. Often, we seek to create an aggregate statistic that combines recall and precision. Probably the most common approach in evaluative studies is the mean average precision (MAP), where precision is measured at every point at which a relevant document is obtained, and the MAP measure is found by averaging these points for the whole query.

A good deal of evaluation in IR is done via challenge evaluations, where a common IR task is defined, and a test collection of documents, topics, and relevance judgments are developed. The relevance judgments define which documents are relevant for each topic in the task, allowing different researchers to compare their systems with others on the same task and improve them. The longest running and best-known challenge evaluation in IR is the Text REtrieval Conference (TREC, http://trec.nist.gov/), which is organized by the U.S. National Institute for Standards and Technology (NIST, www.nist.gov). Started in 1992, TREC has provided a testbed for evaluation and a forum for presentation of results. TREC is organized as an annual event at which the tasks are specified and queries and documents are provided to participants. Participating groups submit "runs" of their systems to NIST, which calculates the appropriate performance measure(s). TREC is organized into tracks geared to specific interests. A book summarizing the first decade of TREC grouped the tracks into general IR tasks:[48]

- Static text—ad hoc
- Streamed text—routing, filtering
- Human in the loop—interactive
- Beyond English (cross-lingual)—Spanish, Chinese, and others
- Beyond text—optical character recognition (OCR), speech, video
- Web searching—very large corpus
- Answers, not documents—question-answering
- Domain-specific—genomics, legal

While TREC has mostly focus on general-subject domains, there have been several tracks that have focused on the biomedical domain. The first of these was the Genomics Track, which focused on the retrieval of articles well as question-answering in this domain.[49] A second track to do focused on retrieval from medical records, with a task devoted to identifying patients who might be candidates for clinical studies based on criteria to be discerned from their medical records.[50] Additional tasks have focused on retrieval for clinical decision support[51] and precision medicine.[52]

Some researchers have criticized or noted the limitations of relevance-based measures. While no one denies that users want systems to retrieve relevant articles, it is not clear that the quantity of relevant documents retrieved is the complete measure of how well a system performs.[53-54] Hersh[55] has noted that clinical users are unlikely to be concerned about these measures when they simply seek an answer to a clinical question and are able to do so no matter how many other relevant documents they miss (lowering recall) or how many non-relevant ones they retrieve (lowering precision).

What alternatives to relevance-based measures can be used for determining performance of individual searches? Harter admits that if measures using a more situational view of relevance cannot be developed for assessing user interaction, then recall and precision may be the only alternatives. Some alternatives have focused on users being able to perform various information tasks with IR systems, such as finding answers to questions.[56-60] For several years, TREC featured an Interactive Track that had participants carry out user experiments with the same documents and queries.[61] Evaluations focusing on user-oriented evaluation of biomedical IR will be described in the next section.

User-Oriented Evaluation

A number of user-oriented evaluations have been performed over the years looking at users of biomedical information. Most of these studies have focused on clinicians.

One of the original studies measuring searching performance in clinical settings was performed by Haynes et al.[62] This study also compared the capabilities of librarian and clinician searchers. In this study, 78 searches were randomly chosen for replication by both a clinician experienced in searching and a medical librarian. During this study, each original ("novice") user had been required to enter a brief statement of information need before entering the search program. This statement was given to the experienced clinician and librarian for searching on MEDLINE. All the retrievals for each search were given to a subject domain expert, blinded with respect to which searcher retrieved which reference. Recall and precision were calculated for each query and averaged. The results showed that the experienced clinicians and librarians achieved comparable recall in the range of 50%, although the librarians had better precision. The novice clinician searchers had lower recall and precision than either of the other groups. This study also assessed user satisfaction of the novice searchers, who despite their recall and precision results said that they were satisfied with their search outcomes. The investigators did not assess whether the novices obtained enough relevant articles to answer their questions, or whether they would have found additional value with the ones that were missed.

A follow-up study yielded some additional insights about the searchers.[63] As was noted, different searchers tended to use different strategies on a given topic. The different approaches replicated a finding known from other searching studies in the past, namely, the lack of overlap across searchers of overall retrieved citations as well as relevant ones. Thus, even though the novice searchers had lower recall, they did obtain a great many relevant citations not retrieved by the two expert searchers. Furthermore, fewer than 4 percent of all the relevant citations were retrieved by all three searchers. Despite the widely divergent search strategies and retrieval sets, overall recall and precision were quite similar among the three classes of users.

Recognizing the limitations of recall and precision for evaluating clinical users of IR systems, Hersh and coworkers have carried out several studies assessing the ability of systems to help students and clinicians answer clinical questions. The rationale for these studies is that the usual goal of using an IR system is to find an answer to a question. While the user must obviously find relevant documents to answer that question, the quantity of such documents is less important than whether the question is successfully answered. In fact, recall and precision can be placed among the many factors that may be associated with ability to complete the task successfully.

The first study by this group using the task-oriented approach compared Boolean versus natural language searching in the textbook Scientific American Medicine.[59] Thirteen medical students were asked to answer 10 short-answer questions and rate their confidence in their answers. The students were then randomized to one or the other interface and asked to search on the five questions for which they had rated confidence the lowest. The study showed that both groups had low correct rates before searching (average 1.7 correct out of 10) but were mostly able to answer the questions with searching (average 4.0 out of 5). There was no difference in ability to answer questions with one interface or the other. Most answers were found on the first search to the textbook. For the questions that were incorrectly answered, the document with the correct answer was actually retrieved by the user two-thirds of the time and viewed more than half the time.

Another study compared Boolean and natural language searching of MEDLINE with two commercial products, CD Plus (now Ovid) and KF.[60] These systems represented the ends of the spectrum in terms of using Boolean searching on human-indexed thesaurus terms (Ovid) versus natural language searching on words in the title, abstract, and indexing terms (KF). Sixteen medical students were recruited and randomized to one of the two systems and given three yes/no clinical questions to answer. The students were able to use each system successfully, answering 37.5 percent correctly before searching and 85.4 percent correctly after searching. There were no significant differences between the systems in time taken, relevant articles retrieved, or user satisfaction. This study demonstrated that both types of systems can be used equally well with minimal training.

A more comprehensive study looked at MEDLINE searching by medical and nurse practitioner (NP) students to answer clinical questions. A total of 66 medical and NP students searched five questions each.[64] This study used a multiple-choice format for answering questions that also included a judgment about the evidence for the answer. Subjects were asked to choose from one of three answers:
- Yes, with adequate evidence.
- Insufficient evidence to answer question.
- No, with adequate evidence.

Both groups achieved a pre-searching correctness on questions about equal to chance (32.3 percent for medical students and 31.7 percent for NP students). However, medical students improved their correctness with searching (to 51.6 percent), whereas NP students hardly did at all (to 34.7 percent).

This study also attempted to measure what factors might influence searching. A multitude of factors, such as age, gender, computer experience, and time taken to search, were not associated with successful answering of questions. Successful answering was, however, associated with answering the question correctly before searching, spatial visualization ability (measured by a validated instrument), searching experience, and EBM question type (prognosis questions easiest, harm questions most difficult). An analysis of recall and precision for each question searched demonstrated a complete lack of association with ability to answer these questions.

Two studies have extended this approach in various ways. Westbook et al. assessed use of an online evidence systems and found that physicians answered 37% of questions correctly before use of the system and 50% afterwards, while nurse specialists answered 18% of questions correctly and also 50% afterwards.[65] Those who had correct answers before searching had higher confidence in their answers, but those not knowing answer initially had no difference in confidence whether their answer turned out to be right or wrong. McKibbon and Fridsma[66] performed a comparable study of allowing physicians to seek answers to questions with resources they normally use employing the same questions as Hersh et al.[64] This study found no difference in answer correctness before or after using the search system. Clearly these studies show a variety of effects with different IR systems, tasks, and users.

Pluye et al.[67] performed a qualitative study assessing impact of IR systems on physician practice. The study identified 4 themes mentioned by physicians:
- Recall—of forgotten knowledge.
- Learning—new knowledge.

- Confirmation—of existing knowledge.
- Frustration—that system use was not successful.

The researchers also noted two additional themes:
- Reassurance—that system is available.
- Practice improvement—of patient-physician relationship.

The bulk of more recent physician user studies have focused on ability to users to answer clinical questions. Hoogendam et al. compared UpToDate with PubMed for questions that arose in patient care among residents and attending physicians in internal medicine.[68] For 1305 questions, they found that both resources provided complete answers 53% of the time, but UpToDate was better at providing partial answers (83% full or partial answer for UpToDate compared to 63% full or partial answer for PubMed).

A similar study compared Google, Ovid, PubMed, and UpToDate for answering clinical questions among trainees and attending physicians in anesthesiology and critical care medicine.[69] Users were allowed to select which tool to use for a first set of four questions to answer, while 1-3 weeks later they were randomized to only a single tool to answer another set of eight questions. For the first set of questions, users most commonly selected Google (45%), followed by UpToDate (26%), PubMed (25%), and Ovid (4.4%). The rate of answering questions correctly in the first set was highest for UpToDate (70%), followed by Google (60%), Ovid (50%), and PubMed (38%). The time taken to answer these questions was lowest for UpToDate (3.3 minutes), followed by Google (3.8 minutes), PubMed (4.4 minutes), and Ovid (4.6 minutes). In the second set of questions, the correct answer was most likely to be obtained by UpToDate (69%), followed by PubMed (62%), Google (57%), and Ovid (38%). Subjects randomized a new tool generally fared comparably, with the exception of those randomized from another tool to Ovid.

Another study compared searching UpToDate and PubMed Clinical Queries at the conclusion of a course for 44 medical residents in an information mastery course.[70] Subjects were randomized to one system for two questions and then the other system for another two questions. The correct answer was retrieved 76% of the time with UpToDate versus only 45% of the time with PubMed Clinical Queries. Median time to answer the question was less for UpToDate (17 minutes) than PubMed Clinical Queries (29 minutes). User satisfaction was higher with UpToDate.

Fewer studies have been done assessing non-clinicians searching on health information. Lau et al. found that use of a consumer-oriented medical search engine that included PubMed, MedlinePLUS, and other resources by college undergraduates led to answers being correct at a higher rate after searching (82.0%) than before searching (61.2%).[71-72] Providing a feedback summary from prior searches boosted the success rate of using the system even higher, to 85.3%. Confidence in one's answer was not found to be highly associated with correctness of the answer, although confidence was likely to increase for those provided with feedback from other searchers on the same topic.

Despite the ubiquity of search systems, many users have skill-related problems when searching for information. van Duersen assessed a variety of computer-related and content-related skills from randomly selected subjects in the Netherlands.[73] Older age and lower educational level were associated with reduced skills, including use of search engines. While younger subjects were more likely to have better computer and searching skills than older subjects, they were more likely to use non-relevant search results and unreliable sources in answering health-related questions. This latter phenomenon has also been seen outside the health domain among the "millennial" generation, sometimes referred to as "digital natives."[74]

FUTURE DIRECTIONS

The above evaluation research shows that there is still plenty of room for IR systems to improve their abilities. In addition, there will be new challenges that arise from growing amounts of information, new devices, and other new technologies.

There are also other areas related to IR where research is ongoing in the larger quest to help all involved in biomedicine and health—from patients to clinicians to researchers—better use information systems and technology to improve the application of knowledge to improve health. This has resulted in a research taking place in several areas related to IR, which include:
- Information extraction and text mining—usually through the use of natural language processing (NLP) to extract facts and knowledge from text. These techniques are often employed to extract information from the EHR, with a wide variety of accuracy as shown in a recent systematic review.[75] Among the most successful uses of these techniques have been studies to identify diseases associated with genomic variations[76-77]
- Summarization—Providing automated extracts or abstracts summarizing the content of longer documents[78-79]
- Question-answering—Going beyond retrieval of documents to providing actual answers to questions, as exemplified by the IBM Corp. Watson system,[80] which is being applied to medicine[81]

> **KEY POINTS**
>
> - There are many biomedical and health knowledge resources online available in bibliographic databases, journals and other full-text resources, Web sites, and other sources.
> - Bibliographic content is likely to be indexed using controlled vocabularies assigned by humans.
> - Full-text and other resources are likely to be indexed via extraction of words.
> - The major approaches to searching biomedical and health knowledge resources include exact-match searching using sets and Boolean operators and partial-match searching on words using relevance ranking.
> - System-oriented evaluation studies tend to focus on performance of search systems and usually involvement measurement of the relevance-based measures of recall and precision.
> - User-oriented evaluation studies tend to compare users and their abilities to complete tasks using retrieval systems.

CONCLUSION

There is no doubt that considerable progress has been made in IR. Seeking online information is now done routinely not only by clinicians and researchers, but also by patients and consumers. There are still considerable challenges to make this activity more fruitful to users.

RECOMMENDED READING

Hersh, WR (2009). *Information Retrieval: A Health and Biomedical Perspective* (3rd Edition). New York, NY, Springer.

REFERENCES

1. Hersh W. *Information Retrieval: A Health and Biomedical Perspective (3rd Edition)*. New York, NY: Springer; 2009.
2. Hersh W. Information Retrieval and Digital Libraries. In: Shortliffe E, Cimino J, eds. *Biomedical Informatics: Computer Applications in Health Care and Biomedicine (Fourth Edition)*. New York, NY: Springer; 2014:613-641.
3. Lindberg D, Humphreys B. 2015 - the future of medical libraries. *New England Journal of Medicine*. 2005;352:1067-1070.
4. Witten I, Bainbridge D, Nichols D. *How to Build a Digital Library, Second Edition*. San Francisco: Morgan Kaufmann; 2010.
5. Fox S, Duggan M. *Health Online 2013*. Washington, DC: Pew Internet & American Life Project; January 15, 2013.
6. Purcell K, Brenner J, Rainie L. Search Engine Use 2012. Washington, DC: Pew Internet & American Life Project; March 9, 2012.
7. Anonymous. From Screen to Script: The Doctor's Digital Path to Treatment. New York, NY: Manhattan Research; Google;2012.
8. Insel T, Volkow N, Li T, Battey J, Landis S. Neuroscience networks: data-sharing in an information age. *PLoS Biology*. 2003;1:E17.
9. Bastian H, Glasziou P, Chalmers I. Seventy-five trials and eleven systematic reviews a day: how will we ever keep up? *PLoS Medicine*. 2010;7(9):e1000326.
10. Björk B, Solomon D. Open access versus subscription journals: a comparison of scientific impact. *BMC Medicine*. 2012;10:73.
11. Hersh W, Rindfleisch T. Electronic publishing of scholarly communication in the biomedical sciences. *Journal of the American Medical Informatics Association*. 2000;7:324-325.
12. Sox H. Medical journal editing: who shall pay? *Annals of Internal Medicine*. 2009;151:68-69.
13. Neylon C. Science publishing: Open access must enable open use. *Nature*. 2012;492:348-349.
14. VanNoorden R. Open access: The true cost of science publishing. *Nature*. 2013;495:426-429.
15. Wolpert A. For the sake of inquiry and knowledge--the inevitability of open access. *New England Journal of Medicine*. 2013;368:785-787.
16. Drazen J, Curfman G. Public access to biomedical research. *New England Journal of Medicine*. 2004;351:1343.
17. Giles J. Internet encyclopaedias go head to head. *Nature*. 2005;438:900-901.
18. Nicholson D. *An evaluation of the quality of consumer health information on Wikipedia* [Capstone]. Portland, OR: Medical Informatics & Clinical Epidemiology, Oregon Health & Science University; 2006.
19. Laurent M, Vickers T. Seeking health information online: does Wikipedia matter? *Journal of the*

19. *American Medical Informatics Association.* 2009;16:471-479.
20. Galperin M, Cochrane G. The 2011 Nucleic Acids Research Database Issue and the online Molecular Biology Database Collection. *Nucleic Acids Research.* 2011;39(suppl1):D1-D6.
21. Sayers E, Barrett T, Benson D, et al. Database resources of the National Center for Biotechnology Information. *Nucleic Acids Research.* 2011;39(suppl1):D38-D51.
22. Kulkarni A, Aziz B, Shams I, Busse J. Comparisons of citations in Web of Science, Scopus, and Google Scholar for articles published in general medical journals. *Journal of the American Medical Association.* 2009;302:1092-1096.
23. DeAngelis C, Drazen J, Frizelle F, et al. Is this clinical trial fully registered? A statement from the International Committee of Medical Journal Editors. *Journal of the American Medical Association.* 2005;293:2927-2929.
24. Laine C, Horton R, DeAngelis C, et al. Clinical trial registration: looking back and moving ahead. *Journal of the American Medical Association.* 2007;298:93-94.
25. Zarin D, Tse T, Williams R, Califf R, Ide N. The ClinicalTrials.gov results database--update and key issues. *New England Journal of Medicine.* 2011;364:852-860.
26. Zarin D, Tse T. Trust but verify: trial registration and determining fidelity to the protocol. *Annals of Internal Medicine.* 2013;159:65-67.
27. Coletti M, Bleich H. Medical subject headings used to search the biomedical literature. *Journal of the American Medical Informatics Association.* 2001;8:317-323.
28. Charen T. *MEDLARS Indexing Manual, Part I: Bibliographic Principles and Descriptive Indexing, 1977.* Springfield, VA: National Technical Information Service; 1976.
29. Charen T. *MEDLARS Indexing Manual, Part II.* Springfield, VA: National Technical Information Service; 1983.
30. Weibel S, Koch T. The Dublin Core Metadata Initiative: mission, current activities, and future directions. *D-Lib Magazine.* 2000;6.
31. Darmoni S, Leroy J, Baudic F, Douyere M, Piot J, Thirion B. CISMeF: a structured health resource guide. *Methods of Information in Medicine.* 2000;9:30-35.
32. Akerkar R. *Foundations of the Semantic Web: XML, RDF & Ontology.* Oxford, England: Alpha Science International Ltd; 2009.
33. Funk M, Reid C. Indexing consistency in MEDLINE. *Bulletin of the Medical Library Association.* 1983;71:176-183.
34. Marcetich J, Rappaport M, Kotzin S. Indexing consistency in MEDLINE. Paper presented at: MLA 04 Abstracts2004; Washington, DC.
35. Aronson A, Mork J, Gay C, SM.Humphrey, Rogers W. The NLM Indexing Initiative's Medical Text Indexer. Paper presented at: MEDINFO 2004 - Proceedings of the Eleventh World Congress on Medical Informatics2004; San Francisco, CA.
36. Salton G. Developments in automatic text retrieval. *Science.* 1991;253:974-980.
37. vanRijsbergen C. *Information Retrieval.* London, England: Butterworth; 1979.
38. Fox C. Lexical Analysis and Stop Lists. In: Frakes W, Baeza-Yates R, eds. *Information Retrieval: Data Structures and Algorithms.* Englewood Cliffs, NJ: Prentice-Hall; 1992:102-130.
39. Anonymous. Stopwords. In: Anonymous, ed. *PubMed Help.* Bethesda, MD: National Library of Medicine; 2007.
40. Frakes W. Stemming Algorithms. In: Frankes W, Baeza-Yates R, eds. *Information Retrieval: Data Structures and Algorithms.* Englewood Cliffs, NJ: Prentice-Hall; 1992:131-160.
41. Brin S, Page L. The anatomy of a large-scale hypertextual Web search engine. *Computer Networks and ISDN Systems.* 1998;30:107-117.
42. Salton G, McGill M. *Introduction to Modern Information Retrieval.* New York: McGraw-Hill; 1983.
43. Salton G, Fox E, Wu H. Extended Boolean information retrieval. *Communications of the ACM.* 1983;26:1022-1036.
44. Buckley C, Salton G, Allan J. The effect of adding relevance information in a relevance feedback environment. Paper presented at: Proceedings of the 17th Annual International ACM SIGIR Conference on Research and Development in Information Retrieval1994; Dublin, Ireland.
45. Wilbur W, Yang Y. An analysis of statistical term strength and its use in the indexing and retrieval of molecular biology texts. *Computers in Biology and Medicine.* 1996;26:209-222.
46. Anonymous. PubMed Help. Bethesda, MD: National Library of Medicine; 2014: http://www.ncbi.nlm.nih.gov/books/NBK3827/ (Accessed August 8, 2017)
47. Hersh W, Hickam D. How well do physicians use electronic information retrieval systems? A framework for investigation and review of the literature. *Journal of the American Medical Association.* 1998;280:1347-1352.
48. Voorhees E, Harman D, eds. *TREC: Experiment and Evaluation in Information Retrieval.* Cambridge, MA: MIT Press; 2005.
49. Hersh W, Voorhees E. TREC genomics special issue overview. *Information Retrieval.* 2009;12:1-15.

50. Voorhees E. The TREC Medical Records Track. Paper presented at: Proceedings of the International Conference on Bioinformatics, Computational Biology and Biomedical Informatics 2013; Washington, DC.
51. Roberts K, Simpson M, Demner-Fushman D, Voorhees E, Hersh W. State-of-the-art in biomedical literature retrieval for clinical cases: a survey of the TREC 2014 CDS track. *Information Retrieval Journal*. 2016;19:113-148.
52. Roberts K, Demner-Fushman D, Voorhees E, et al. Overview of the TREC 2017 Precision Medicine Track. Paper presented at: The Twenty-Sixth Text REtrieval Conference (TREC 2017) Proceedings 2017; Gaithersburg, MD.
53. Swanson D. Historical note: information retrieval and the future of an illusion. *Journal of the American Society for Information Science*. 1988;39:92-98.
54. Harter S. Psychological relevance and information science. *Journal of the American Society for Information Science*. 1992;43:602-615.
55. Hersh W. Relevance and retrieval evaluation: perspectives from medicine. *Journal of the American Society for Information Science*. 1994;45:201-206.
56. Egan D, Remde J, Gomez L, Landauer T, Eberhardt J, Lochbaum C. Formative design-evaluation of Superbook. *ACM Transactions on Information Systems*. 1989;7:30-57.
57. Mynatt B, Leventhal L, Instone K, Farhat J, Rohlman D. Hypertext or book: which is better for answering questions? Paper presented at: Proceedings of Computer-Human Interface 92 1992.
58. Wildemuth B, deBliek R, Friedman C, File D. Medical students' personal knowledge, searching proficiency, and database use in problem solving. *Journal of the American Society for Information Science*. 1995;46:590-607.
59. Hersh W, Hickam D. An evaluation of interactive Boolean and natural language searching with an on-line medical textbook. *Journal of the American Society for Information Science*. 1995;46:478-489.
60. Hersh W, Pentecost J, Hickam D. A task-oriented approach to information retrieval evaluation. *Journal of the American Society for Information Science*. 1996;47:50-56.
61. Hersh W. Interactivity at the Text Retrieval Conference (TREC). *Information Processing and Management*. 2001;37:365-366.
62. Haynes R, McKibbon K, Walker C, Ryan N, Fitzgerald D, Ramsden M. Online access to MEDLINE in clinical settings. *Annals of Internal Medicine*. 1990;112:78-84.
63. McKibbon K, Haynes R, Dilks CW, et al. How good are clinical MEDLINE searches? A comparative study of clinical end-user and librarian searches. *Computers and Biomedical Research*. 1990;23(6):583-593.
64. Hersh W, Crabtree M, Hickam D, et al. Factors associated with success for searching MEDLINE and applying evidence to answer clinical questions. *Journal of the American Medical Informatics Association*. 2002;9:283-293.
65. Westbrook J, Coiera E, Gosling A. Do online information retrieval systems help experienced clinicians answer clinical questions? *Journal of the American Medical Informatics Association*. 2005;12:315-321.
66. McKibbon K, Fridsma D. Effectiveness of clinician-selected electronic information resources for answering primary care physicians' information needs. *Journal of the American Medical Informatics Association*. 2006;13:653-659.
67. Pluye P, Grad R. How information retrieval technology may impact on physician practice: an organizational case study in family medicine. *Journal of Evaluation in Clinical Practice*. 2004;10:413-430.
68. Hoogendam A, Stalenhoef A, Robbé P, Overbeke A. Answers to questions posed during daily patient care are more likely to be answered by UpToDate than PubMed. *Journal of Medical Internet Research*. 2008;10(4):e29.
69. Thiele R, Poiro N, Scalzo D, Nemergut E. Speed, accuracy, and confidence in Google, Ovid, PubMed, and UpToDate: results of a randomised trial. *Postgraduate Medical Journal*. 2010;86:459-465.
70. Ensan L, Faghankhani M, Javanbakht A, Ahmadi S, Baradaran H. To compare PubMed Clinical Queries and UpToDate in teaching information mastery to clinical residents: a crossover randomized controlled trial. *PLoS ONE*. 2011;6:e23487.
71. Lau A, Coiera E. Impact of web searching and social feedback on consumer decision making: a prospective online experiment. *Journal of Medical Internet Research*. 2008;10(1):e2.
72. Lau A, Kwok T, Coiera E. How online crowds influence the way individual consumers answer health questions. *Applied Clinical Informatics*. 2011;2:177-189.
73. vanDeursen A. Internet skill-related problems in accessing online health information. *International Journal of Medical Informatics*. 2012;81:61-72.
74. Taylor A. A study of the information search behaviour of the millennial generation. *Information Research*. 2012;17(1).
75. Stanfill M, Williams M, Fenton S, Jenders R, Hersh W. A systematic literature review of automated clinical coding and classification systems. *Journal of the American Medical Informatics Association*. 2010;17:646-651.

76. Denny J. Mining Electronic Health Records in the Genomics Era. *PLOS Computational Biology.* 2012;8(12):e1002823.
77. Denny J, Bastarache L, Ritchie M, et al. Systematic comparison of phenome-wide association study of electronic medical record data and genome-wide association study data. *Nature Biotechnology.* 2013;31:1102-1111.
78. Mani I. *Automatic Summarization.* Amsterdam: John Benjamins; 2001.
79. Fiszman M, Rindflesch T, Kilicoglu H. Summarization of an online medical encyclopedia. Paper presented at: MEDINFO 2004 - Proceedings of the Eleventh World Congress on Medical Informatics2004; San Francisco, CA.
80. Ferrucci D, Brown E, Chu-Carroll J, et al. Building Watson: an overview of the DeepQA Project. *AI Magazine.* 2010;31(3):59-79.
81. Ferrucci D, Levas A, Bagchi S, Gondek D, Mueller E. Watson: beyond Jeopardy! *Artificial Intelligence.* 2012;199-200:93-105.

16

Medical Imaging Informatics

ROBERT E. HOYT • JOHN D. GRIZZARD

LEARNING OBJECTIVES

After reading this chapter the reader should be able to:

- Describe the history behind digital radiology and the creation of picture archiving and communication systems (PACS)
- Enumerate the benefits of digital radiology to clinicians, patients and hospitals
- List the challenges facing the adoption of picture archiving and communication systems
- Describe the difference between computed and digital radiology
- Outline the field of medical imaging informatics
- Understand new imaging technologies such as web PACS and mobile imaging viewer

INTRODUCTION

The field of medical imaging informatics has been slowly evolving over the past three decades and is a subspecialty under biomedical informatics. However, others consider imaging informatics as a subspecialty under Radiology. As an information science, it studies every facet of imaging: acquisition, storage, interpretation and sharing to improve patient care. The field tends to include radiologists and scientists involved with medical physics. Imaging informaticians must understand how imaging data moves throughout the medical enterprise and how it interacts with electronic health records, voice recognition dictation systems, computer-aided diagnosis software, health information organizations, etc. Specialists in this field must also have a good understanding of workflow, networks, security, data quality, hardware and software similar to the skill set needed for electronic health records. The supporting group for the field is the Society for Imaging Informatics in Medicine or SIIM.[1] More information about the history of medical imaging informatics is presented in detail by Branstetter.[2] While Teleradiology could be discussed in this chapter it has been included in the chapter on Telemedicine.

This chapter will discuss the field of medical imaging informatics and the various technologies such as picture archiving and communication systems (PACS) that have revolutionized the field of Radiology.

Definitions

Medical Imaging Informatics (MII): According to the Society for Imaging Informatics in Medicine MII "is *the study and application of processes of information and communications technology for the acquisition, manipulation, analysis and distribution of medical image data.*"[1]

Picture Archiving and Communication Systems (PACS): is a medical imaging technology which provides economical storage of, and convenient access to, images from multiple modalities.[2-3]

Biomedical Imaging Informatics (BII) tends to be broader than MII and includes radiology, pathology, dermatology, and ophthalmology. It is a sub-field of biomedical informatics and can include cellular and molecular imaging. Biomedical imaging informatics is "*a discipline that focuses on improving patient outcomes through the effective use of images and imaging-derived information in research and clinical care.*" Due to the broader nature of BII in also includes interpretation technologies such as machine learning and natural language processing.[4]

In this chapter, we will focus primarily on MII and its close relationship with Radiology.

History of Transitioning to PACS

Digital imaging appeared in the early 1970's by pioneers such as Dr. Sol Nudelman and Dr. Paul Capp. The first reference to PACS occurred in 1979 when Dr. Lemke in Berlin published an article describing the functional concept. In 1983, a team led by Dr. Steven Horii at the University of Pennsylvania began working on the data standard digital imaging and communications in medicine (DICOM) (see chapter on data standards) that would facilitate image sharing. The US Army Medical Research and Materiel Command installed the first large scale PACS in the US in 1992.[5] The University of Maryland hospital system was the first to go "filmless" in 1999.[6]

Medical imaging has progressed along a pathway very similar to conventional photography in that there has been a gradual shift from analog images printed on film to digital images captured on electronic media. This transformation has occurred slowly over time and has been made possible by variety of technical innovations. The initial impetus for this change came about as the result of the development of digital imaging technologies; specifically, computed tomography, ultrasound, and magnetic resonance imaging in the 1970's and 80's. These modalities resulted in digital images that were displayed on monitors at dedicated workstations attached to the source devices. Images could be printed onto film, but the underlying technology was that of digital image acquisitions.

It quickly became apparent that reviewing images at a computer monitor in this "softcopy" format had significant advantages over the prevailing film-based technology in use at the time. Specifically, images could be viewed without delay as soon as they were prepared by the scanner, without the need for film processing. In addition, if the scanner manufacturer supplied an additional workstation, images could be viewed at a remote location in the radiology department as soon as they were available. Developing costs, and the time required for image development would vanish, and storage of films would become greatly simplified. In addition, image retrieval would also be greatly facilitated. (It used to be said that if a radiologist had been a bad person during life and was sent to hell, he would spend eternity looking through film jackets for old studies).

The transition to a completely filmless radiology department was impeded by the extensive initial costs that were involved. Although, in general, films would no longer be printed, printing would remain part of the process as referring physicians would often request a copy of the studies. Therefore, it was difficult to go completely filmless, and so a small fraction of residual printing costs would remain.

Conventional radiographs were initially obtained in the usual fashion using xray film and were then digitized using film scanners in order to make digital viewing of the images possible. Eventually computed radiography and digital radiographs became available but meant that many conventional radiology rooms would have to be significantly upgraded. Computer-based image archiving would also be necessary, requiring significant expense. Lastly, in order to link the various imaging technologies with the image archive, a comprehensive and fast network would need to be built. Although going filmless had the advantages of decreasing printing costs and increasing speed, these initial capital outlays were formidable. However, over time, the advantages of digital filming became sufficiently attractive, and the extensive upfront capital costs of doing away with film and moving to completely digital imaging diminished as computer hardware and network technology rapidly evolved, and PACS systems gradually started to become economically viable.

An additional obstacle to the widespread adoption of PACS technology was that initially the scanner vendors had proprietary imaging formats. That is, a CT scan performed on one manufacturer's equipment could not be viewed using another manufacturer's imaging workstation. Over time, it became apparent that a uniform imaging and communications strategy was required. The DICOM standard was developed to facilitate image sharing and transmission, and this development greatly facilitated the adoption of PACS.

DICOM was developed by the National Electrical Manufacturers Association, along with the American College of Radiology, and provided a mechanism for the accurate handling, storing, printing, and transmitting of digital image information for medical images. The standard enables the integration of imaging equipment, image archiving storage systems, imaging workstations, printers, and network hardware from a variety of different manufacturers to be combined into a picture archiving and communication system. In other words, instead of each manufacturer and piece of equipment speaking a different and unique language, all the manufacturers and equipment makers were now speaking the same language. Initial iterations of the DICOM standard were developed beginning in the mid-and late 1980's, but in 1993 the DICOM 3.0 standard was released, and was found to be very robust, and widespread adoption soon followed.[7]

Hospitals and radiology groups have made the transition from analog to digital radiography. To their credit, radiologists have pushed for this change for years but have had to wait for better technology and financial support from their healthcare organizations. Early pioneers understood that a digital system would mean no more bulky film jackets, frequently lost films and slow retrieval. The technology is now mature and widely accepted but cost is still an issue at smaller healthcare organizations.

PACS are made possible by faster processors, higher capacity disk drives, higher resolution monitors, more robust hospital information systems, better servers and faster network speeds. PACS are also frequently integrated with voice recognition systems to expedite report turnaround. PACS usually have a central server that serves as the image repository and multiple client computers linked with a local or wide area network. Images are stored using the DICOM data standard. Input into PACS can also occur from a DICOM compliant CD or DVD brought from another facility or teleradiology site via satellite. An historical perspective of the development of PACS in the United States is chronicled in this reference.[8]

It is important to point out that a few facilities with digital systems or PACS still print hard copies or have some non-digital services. This could be due to physician resistance, lack of resources or the fact that it has taken longer for certain imaging services such as mammography to go digital. Most radiology departments have all modalities stored on a comprehensive or *Full PACS*--which means that images from ultrasonography (US), magnetic resonance imaging (MRI), positron emission tomography (PET), computed tomography (CT), routine radiography and endoscopy are stored and viewed on the system. *Mini-PACS*, on the other hand, are more limited and process images from only one or two modalities.[9]

As an example, cardiologists will often adopt a mini-PACS, and will use it to display only echocardiography and cardiac catheterization images.

PACS Key Components (see Figure 16.1)

- Digital acquisition devices: the devices that are the sources of the images. Digital angiography, fluoroscopy and mammography are the newcomers to PACS. CT, MRI and ultrasound scanners have always been inherently digital
- The Network: ties the PACS components together—that is, it is the pathway for image transmission from the scanners to the image archive, and from there to the radiologist at a reading station.
- Database server: high speed and robust central computer to process information. This answers the request of the reading radiologist to provide the images at his/her workstation.
- Archival server: responsible for storing images. A server enables short term (fast retrieval) and long term (slower retrieval) storage. HIPAA requires separate back up, usually off-site to prevent data loss in a disaster situation
- Radiology Information system (RIS): system that maintains patient demographics, scheduling, billing information and interpretations
- Workstation or soft copy display: contains the software and hardware to access the PACS. Replaces the standard light box or view box. This is where the radiologist reviews the imaging study and dictates his diagnostic report.
- Teleradiology: the ability to remotely view images at a location distant from the site of origin removed[10]

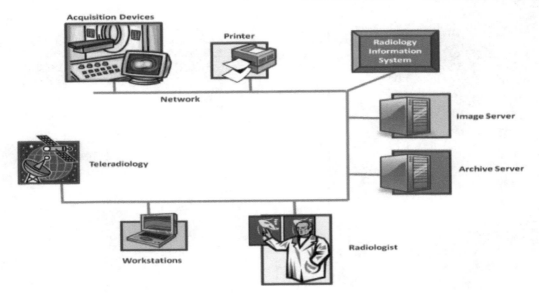

Figure 16.1: PACS Key Components

Types of Digital Detectors

- Computed radiography (CR): after x-ray exposure to a special cassette, a laser reader scans the image and converts it to a digital image. The image is erased on the cassette, so it can be used repeatedly. (Figure 16.2)
- Digital radiography (DR): does not require an intermediate step of laser scanning.[10]

Figure 16.2: Computed Radiography

TYPICAL PACS WORKFLOW

As already noted, a PACS should interface with both the HIS and RIS. Typically, the patient is identified in the HIS and an order created that is sent to the RIS via an HL7 protocol (HL7 and its members provide a framework and related standards for the exchange, integration, sharing, and retrieval of electronic health information). Orders will go to the imaging device via the DICOM protocol and the image is created in DICOM format and sent to the PACS server. Images are stored on the image archive, and the reading physician (radiologist) is notified of a pending study. The study is then read by the radiologist at a computer workstation using high-resolution monitors and viewing software available from a variety of different vendors. (see Figure 16.3 of typical PACS screen)

Viewing software typically allows comparison of the present examination with any prior imaging studies so that interval changes can be detected. In this instance, it is remarkably easy for the radiologist to directly compare the present examination with multiple prior examinations, as these are easily sorted using the computer software provided by the workstation vendor. Therefore, comparison can be made to multiple prior studies without having to endlessly search through a film jacket to find the studies. These studies can be viewed side-by-side, or above and below one another, depending on the radiologist's preference. In addition, the PACs workstations allow linking of the studies such that image locations that are similar on two different studies performed at two different times can be reviewed in unison. These workstations allow manipulation of the brightness and contrast of the images, and also facilitate measurement of the densities of objects seen in the images in order to detect such things as fluid or calcification. The images can be magnified and zoomed up for better evaluation of small and fine structures. Many workstations allow visualization of the image data set with multi-planar reformations. These allow one to view the images simultaneously from the front, from the side, and in the axial or standard imaging plane, and to cross register at the views with one another. Most diagnostic monitors are still grayscale as the majority of the imaging modalities render their images in grayscale, and grayscale monitors are relatively less expensive compared to color. Newer

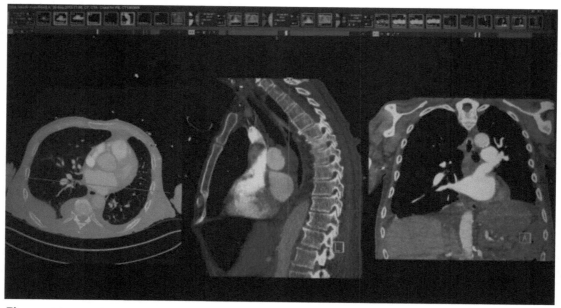

Figure 16.3: PACS Screen (Courtesy Dr. J. Grizzard)

"medical monitors" have 2,048 x 2,560 pixel resolution, and increasingly, color monitors are being adopted to better demonstrate useful color-rendered 3D images.

At the completion of the physician's detailed review of the images, the radiologist, often using voice recognition software, generates a diagnostic report. The dictation is then reviewed, and any corrections made if necessary. The report is then stored on the PACS server linked to the images and is also sent back to the HIS via an HL7 message, so it can be viewed as part of the medical record.

PACS FOR A HOSPITAL DESKTOP COMPUTER

The AGFA IMPAX 6.3 PACS is an example of a client-server-based system used by the US Navy.[11] The PACS receives HL7 messages from the hospital information system (HIS) and provides diagnostic reports and other clinical notes along with the patient's images. Although resolution is slightly better with special monitors, the quality of the images on the standard desktop monitor is very acceptable for non-diagnostic viewing (see Figure 16.4). Any physician on the network can rapidly retrieve and view standard radiographs, CT scans and ultrasounds. The desktop program is intuitive with the following features:
- Zoom-in feature for close-up detail
- Ability to rotate images in any direction
- Text button to see the report
- Mark-up tool that does the following to the image:
- Adds text
- Has a caliper to measure the size of an object
- Has a caliper to measure the ratio of objects: such as the heart width compared to the thorax width
- Measures the angle: angle of a fracture
- Measures the square area of a mass or region
- Adds an arrow
- Right click on the image and short cut tools appear
- Export an image to any of the following destinations:
- Teaching file
- CD-ROM
- Hard drive, USB drive or save on clipboard
- Create an AVI movie

The following are two scenarios that point out how practical PACS can be for the average primary care physician:

Scenario #1: An elderly man is seen in the emergency room at the medical center over the weekend for congestive heart failure and is now in your office on a Monday morning requesting follow up. The practice is part of the Wonderful Medicine Health Organization, so the physician pulls up his chest x-ray on the office PC.

Scenario #2: A physician is seeing a patient visiting the area with a cough and on his chest x-ray a mass in his left lung is noted. The image is downloaded on a CD (or USB drive) for the patient to take to his distant PCM where he will receive a further work up.

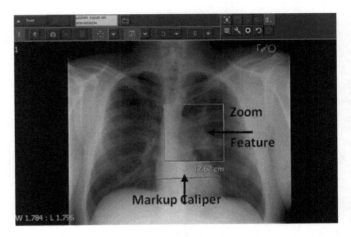

Figure 16.4: Chest X-ray viewed in PACS

PACS EXTENSIONS: WEB BASED IMAGE DISTRIBUTION

Diagnostic imaging plays a significant role in the medical care a patient receives. Reliance on paper and film-based patient records across geographic and institutional borders can decrease the ability for care providers to have immediate access to the patient's entire medical record and imaging history without the implementation of a health information exchange system. Similarly, having patient images present only within a single health care system limits what would otherwise be a potentially widely available resource. Additionally, both patients and referring physicians increasingly request the widespread distribution of images, which can reduce the need for duplicate studies, and allow more rapid diagnosis and treatment. The most readily available means for rapidly and widely disseminating medical imaging is via the Internet, using the World Wide Web. Erin Chesson states that *"the power and reach of the web is empowering the health imaging world – completing the loop from radiology to specialist and back to the referring physician and even the patient."* Furthermore, the benefits of web-based technology provide on-demand, online access to electronic images regardless of the location of patient records, reports and images.[12]

Unfortunately, the DICOM imaging format that has enabled the development of PACS and the interoperability of imaging resources from different vendors has served as something of an impediment to the use of the World Wide Web for image distribution. Specifically, DICOM images are not browser compatible -- that is to say, DICOM images cannot be viewed using a standard Internet browser, as can JPEG, GIF, PNG and other file formats. One solution to this problem is for the browser to serve as a link to a server, which can open and display the images, and then stream them to the viewer. In this instance, client software must be present on the viewing computer to allow this functionality. In many respects, most of the PACS vendors have developed these products to "web-enable" their PACS and provide remote viewing. Usually this entails downloading a small application (thin client) from the PACS vendor that enables the remote viewing station to act like a modified PACS workstation. Changes in browser technology will frequently necessitate updating of this client software mini application.

An alternative type of system enables direct viewing in the browser without client software, enabling its use on any computer with Internet functionality. This type of solution is known as a "zero-footprint" web viewer. As stated previously, DICOM is not intrinsically viewable within an Internet browser. Therefore, for a browser to render the images, they must first be converted to an imaging format that is compatible and can be opened by a conventional browser. Heart Information Technologies WebPAX viewer is one such imaging system and is a true zero-footprint web-based PACS. In this system, DICOM images are pre-converted to GIF files (which are browser compatible), which are then embedded in a webpage. Other vendors have also developed similar strategies for delivering DICOM images in a browser compatible format. One of these is DICOM Web Viewer (DWV), an open source zero footprint viewer that uses only JavaScript and html5 to make DICOM images viewable in a browser.[13] To provide readers with more details regarding web based PACS more information is available on the heartit.com web site.[14]

Although both systems confer a tremendous advantage in terms of more widely distributing medical images, the zero-footprint viewer has the additional advantage of not requiring additional software, and without requiring periodic updates as browser technologies change. In addition, no maintenance is required on the computer involved, as no client software has been downloaded or requires maintenance.

Regardless of the solution used, web-enabled or web-compatible PACS operate through the web environment, much like the ASP model electronic health record (EHR), discussed in the chapter on EHRs. Table 16.1 compares the legacy PACS with web-based PACS. According to a PACS vendor web-based PACS *"is an application that uses different web technologies in a very open manner, regardless if the user is on a PC or Mac, using Linux or Windows for the operating system."*[15] Web based PACS are facilitated by a remote server rendering or processing of the images in 2D, 3D and 4D. This requires robust bandwidth and perhaps the end user to use a thin client or "dumb terminal" or "virtual desktops" which reduces costs, is more secure, more reliable and is available from any location with Internet connectivity.[16]

Table 16.1: Legacy PACS compared to web PACS[16]

Legacy PACS	Web PACS
Only available on computers with proper software installed	Available anywhere with Internet access
Upgrades must be manually installed	Upgrades are done centrally or are not necessary
Multiple user interfaces	One user interface
Difficult to integrate with health information exchanges	Easy to integrate with health information exchanges
Difficult to link to multiple EHRs	Easier to link to EHRs
Labor intensive for PACS administrator for maintenance and training	Much less labor intensive for maintenance and training
Could involve multiple operating systems	One operating system
Less likely to be standards-based	Utilizes JPEG compression, DICOM, HL7 and IHE profiles

Its goal is to offer seamless availability to radiologists, referring physicians, clinicians and nursing staff wherever they need images, i.e. at their office, in the electronic health record, at their homes or wherever there is access to a remote, secure computer. To the patient, it means that their physician has access to all of the medical information required to make informed decisions regarding their medical care: recent and previous images and reports, lab results, medication history, and other pertinent information.[16]

For example, a patient with a fractured lumbar spine can enter the emergency department at a medical facility located 90 miles away from a major city. The emergency department (ED) physician there may be undecided about transporting the patient via helicopter for neurosurgery. The availability of web PACS affords medical personnel the technology to contact the specialist who can log in from a home system using the web viewer to analyze the patient's back images for his/her recommendations.[12] For a real-world story about web-based PACS, see case study in the Infobox.

Other medical facilities may belong to a Health Information Organization (HIO) so regional physicians will have access to images from a variety of facilities.[17] The Consolidated Imaging Initiative (CI-PACS) implemented a regional health information exchange system for radiology for rural hospitals. The system offers a shared, standards-based, interoperable PACS in two hospitals. The system also provides access to remote sites and physicians' offices using the link into the hospital's CI-PACS connection.[19] With widespread adoption of EHRs in the US, it is now common place to find EHRs integrated with PACS systems.

Increasingly, web-based patient portals enable patient access to their medical records, and diagnostic imaging is a valuable part of their records. Embedding links to secure web-based diagnostic image repositories within the electronic medical record is an ongoing development at many leading-edge academic medical centers.

Artificial intelligence (AI) and machine learning (ML) have been integral with digital imaging for many years. Computer assisted diagnosis (CAD) is just one aspect of these newer technologies and can be applied to dermatology and retinal images, in addition to radiological images. CAD can be used on a variety of common imaging such as mammography and chest x-rays. The underlying diagnostic algorithms are based on artificial neural networks and serve as a second opinion for radiologists. The sensitivity of CAD in detecting abnormalities is quite high, in most studies.[20]

Deep learning is the newest type of artificial intelligence to be used with medical imaging. It involves a wide range of neural networks that can segment organs, detect lesions and classify tumors. Unlike traditional machine learning, deep learning learns the important features as well as proper weighting to make predictions on new data. The toolkits and libraries necessary for deep learning are discussed in this reference.[21]

MEDICAL IMAGING AND MOBILE TECHNOLOGY

In 2011, the FDA approved the first primary diagnostic radiology application for mobile devices. Performance evaluation reviewed by the FDA consisted of tests for

---

Case Study: Cardiac PACS

"I was at a meeting in San Francisco, California and was contacted by the MRI technologists at my hospital regarding a complicated cardiac MR case," John Grizzard, Associate Professor of Radiology at Virginia Commonwealth University (VCU), recalls. *"The physician covering the service wanted to consult with me regarding a case where there appeared to be a mass in the heart."* Dr. Grizzard, who is also section chief of non-invasive cardiovascular imaging at the VCU Medical Center, was able to open a browser on his notebook computer and log on to his department's WebPAX server. In seconds the entire cardiac MRI study opened up on his screen in the browser — over 800 cinematic motion images of the heart, moving at the patient's actual heart rate. *"I was able to confirm the suspected diagnosis of a cardiac tumor, and did so from three time zones away, using a pretty vanilla laptop computer and a standard web browser. I didn't need any special client software; I used just a regular off the shelf browser, and it worked. The beauty and the difficulty inherent in cardiac imaging is that you need to see the heart move. And you must see it moving in real time — or in a rhythm that approximates the patient's heart rate. Using WebPax, I was able to do this."* Subsequent surgery confirmed the diagnosis of a cardiac mass, so the story does not have an entirely happy ending, but the ability to remotely view minutes-old motion studies of a patient's heart thousands of miles away demonstrates the power of WebPAX, a true zero-footprint web-based PACS that can display DICOM images using any standard internet browser.[18]

measured luminance, image quality (resolution), and noise referenced by international standards and guidelines. This new mobile radiology application provided medical image viewing on the Apple iPhone, iPad, and iPod. The application is called the Mobile MIM™ and its primary purpose is to give radiologists a means to view images away from their work stations. Figure 16.5 shows an image on a mobile device (iPad).[22]

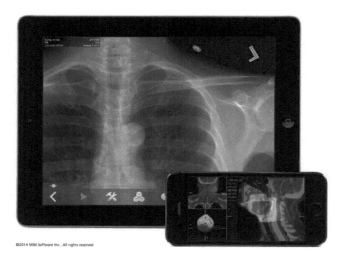

Figure 16.5: Mobile MIM Image (Image courtesy of MIM Software Inc)

ResolutionMD Mobile

In 2011, Calgary Scientific received FDA approval to market ResolutionMD Mobile as a medical imaging diagnostic application. The platform supports several mobile devices and operating systems (iOS and Android). Images are not permanently stored on the mobile devices. ResolutionMD mobile performs on 3/4G wireless.[23]

DICOM Viewers

The following web site lists the currently available free DICOM viewers.[24]

OsiriX is a DICOM viewer for the MAC operating system and one of the earliest viewers available. OsiriXMD is fee based and FDA cleared as a Class II device. 2D, 3D and 4D Images can be uploaded from a CD or USB device.[25]

Mobile Imaging Challenges

In spite of its popularity mobile imaging has several limitations. Images will likely have lower resolution, compared to a dedicated work station. Mobile programs may not permit report generation or editing. Comparing old and new images side by side is generally not possible.[26]

DIGITAL IMAGING ADVANTAGES AND DISADVANTAGES

Advantages

- Replaces a standard x-ray film archive which means a much smaller x-ray storage space; space can be converted into revenue generating services and it reduces the need for file clerks
- Allows for remote viewing and reporting; to also include teleradiology
- Expedites the incorporation of medical images into an electronic health record
- Images can be archived and transported on portable media, e.g. USB drive and Apple's iPhone
- Other specialties that generate images may join PACS such as cardiologists, ophthalmologists, gastroenterologists and dermatologists
- PACS can be web-based and use "service oriented architecture" such that each image has its own URL. This would allow access to images from multiple hospitals in a network.
- Unlike conventional x-rays, digital films have a zoom feature and can be manipulated in innumerable ways
- Improves productivity by allowing multiple clinicians to view the same image from different locations
- Rapid retrieval of digital images for interpretation and comparison with previous studies
- Fewer "lost films"
- Reports are more likely to accompany the digital image
- Radiologists can view an image back and forth like a movie, known as "stack mode"
- Quicker reporting back to the requesting clinician
- Digital imaging allows for computer aided detection (CAD)
- Increased productivity. Several studies have shown increased efficiency after converting to an enterprise PACS. In a study by Reiner, inpatient radiology utilization increased by 82% and outpatient utilization by 21% after transition to a film-less operation.[27] In another study conducted at the University of California Davis Health System, transition to digital radiology resulted in a decrease in the average image search time from 16 to two minutes (equivalent to more than $1 million savings annually in physician's time) and a decrease in film printing by 73% and file clerk full time equivalents (FTEs) dropped by 50% (equivalent to more than $2 million savings annually).[28] The Henry Ford Health Systems film retrieval time dropped from 96 hours to 36 minutes with a net savings of $15 per film.[29]

Disadvantages

- Cost is the greatest barrier, although innovations such as open source and "rental PACS" are alternatives
- New legislation cutting reimbursement rates for certain radiology procedures, thus decreasing capital that could be used to purchase a PACS[30]
- Expense and complexity to integrate with hospital and radiology information systems and EHRs
- Lack of interoperability with other PACSs
- Bandwidth limits may require network upgrades
- Different vendors may use different DICOMS tags to label films
- Viewing digital images is a little slower than routine x-ray films
- Workstations may require upgrades if high resolution monitors are necessary

IMAGING INFORMATICS EDUCATION

Fellowship in Imaging Informatics

A small number (7) of fellowships are available for radiology residents who desire additional informatics training while completing thir clinical residency. Most fellowships are for one year and do not qualify for certification in clinical informatics.[31-32]

Certificate in Imaging Informatics

The American Board of Imaging Informatics (ABII) was founded in 2007 and began administering the board exam in the same year. The Board is the collaborative effort between the Society of Imaging Informatics in Medicine (SIIM) and the American Registry of Radiologic Technologists (ARRT). Certification is particularly pertinent for those who plan to be PACS administrators. Individuals are certified for ten years during which time they must earn and report continuing education credits every two years. Candidates to take the exam must have earned 7 points and minimal criteria as outlined in table 16.2. Further details are available on the ABII web site.[33]

RECOMMENDED READING

The following are articles that discuss issues related to PACS and medical imaging informatics
- The Impact of PACS on Clinician Work Practices in The Intensive Care Unit: A Systematic Review of the Literature. Authors performed a systematic review to determine the impact of PACS on workflow and other issues related to ICU care. Data would suggest that PACS improves efficiency and clinical decision making but may reduce communication between the clinician and the radiologist. They do point out, however, that many articles come from the same institution and no randomized controlled trials have been published so generalizability is limited.[34]
- Imaging Informatics: Essential Tools for The Delivery Of Imaging Services. This review discusses the fact that imaging informatics involves much more than just interpreting digital images; it involves secure storage, delivery, sharing and quality analytics that support research and education. The need for better data standards, standardized reporting and terminologies is also discussed, as well as new standards that will capture and expose image metadata. Radiology clinical decision support is mentioned as a means to reduce inappropriate exam ordering.[35]
- Biomedical Imaging Informatics in The Era Of Precision Medicine: Progress, Challenges, And Opportunities. This review describes the status of the field and challenges such as managing large data sets, developing data standards for interoperability and the need for combined efforts among organizations that deal with medical imaging.[36]
- Biomedical Imaging Informatics. In: Biomedical Informatics: Computer Applications in Health Care and Biomedicine. 2014. Chapter 7 provides much more detail about multiple aspects of BII.[37]

FUTURE TRENDS

Despite its expense PACS has become the de facto standard of care for medical imaging. Making digital images available to all medical staff in a user-friendly manner has been a quantum leap forward. Towards this goal, Stage 2 Meaningful Use required both eligible professionals and hospitals to incorporate (or make accessible) through their electronic health records more than 10% of images ordered. Additionally, there is also a trend towards web based PACS because it is more capable and is a better fit for large healthcare organizations, health information organizations and newer delivery models such as accountable care organizations. This is being supported and facilitated by faster networks, better monitor resolution and more digital imaging. Similarly, there will be better mobile platforms (smartphones and tablets) for viewing images by primary care and specialty physicians, patients and radiologists. Newer image standards are likely such as DICOM GSPS, DICOM SR and

AIM (annotation and image markup) to make image reporting and mining standardized.[35]

Additionally, patients are requesting to view their images in the patient portal that links to web-based image repositories. Also, zero-footprint web viewers are becoming more popular, which means there is no desktop client software to download. The other developments that are ongoing involve integrating a medical record viewer into PACS, so that a single-sign-on gives access to both.

Table 16.2: Criteria for Certificate in Imaging Informatics (Courtesy ABII)

Candidates must have at least 7 points in total AND meet all minimum criteria

Experience	Academic Education from an Accredited* Institution	Certification Credentials	Continuing Education Meeting ABII's Criteria
Minimum 2 points Maximum 4 points	Minimum 0 points Maximum 5 points**	Minimum 0 points Maximum 2 points	Minimum 0 points Maximum 2 points
1 point per 12 months of work experience as a credentialed healthcare imaging professional or healthcare IT professional or healthcare informatics professional Note: Clinical experience completed as part of an imaging informatics training or educational program may be counted in the Experience category or the Academic Education category, but not in both.	1 point: No degree, but at least 30 semester credits or 45 quarter credits 2 points: Associate degree/college diploma (Degree conferred upon completion of at least 60 semester credits or 90 quarter credits) 3 points: Associate degree/college diploma (Degree conferred upon completion of at least 90 semester credits or 135 quarter credits) 4 points: Baccalaureate degree/college diploma (Degree conferred upon completion of at least 120 semester credits or 180 quarter credits) 5 points: Masters or Doctoral degree (Degree conferred upon completion of post-baccalaureate graduate or professional program)	1 point for each Information Technology (IT) or clinical credential See website drop down menu or visit: Qualifying Information-Technology Credentials Qualifying Clinical Credentials	1 point for each 18 hours of continuing education credits in imaging informatics or related disciplines taken within 18 months of the date of application

KEY POINTS

- PACS is the logical result of digitizing x-rays, developing better monitors and medical networks
- PACS is well accepted by radiologists and non-radiology physicians because of the ease of retrieval, quality of the images and flexibility of the platform
- PACS is a type of teleradiology, in that, images can be viewed remotely by multiple clinicians on the same network
- Cost and integration are the most significant barriers to the widespread adoption of PACS
- Web-enabled PACS will promote better interoperability and sharing
- Mobile devices such as smartphones and tablet PCs offer an acceptable alternative for viewing images

CONCLUSION

PACS and digital imaging result from a predictable technological evolution beyond traditional film. For that reason, PACS has become a mainstream technology for moderate to large healthcare organizations. Like electronic health records (EHRs) PACS is an expensive technology to implement, but unlike EHRs, there is greater acceptance by clinicians. EHRs and Health Information Organizations will benefit by being interoperable with web PACS. Healthcare organizations will be looking for ways to interpret and distribute a wide range of images to the entire organization. The technology is moving closer to thin client or zero client web-based PACS for maximum flexibility and interoperability for the enterprise.

REFERENCES

1. Society for Imaging Informatics in Medicine. www.siim.org (WebCite) (Accessed July 2, 2017)
2. Branstetter BF. Basics of Imaging Informatics, part 1 and 2. Radiology 2007; Vol 243: 656-667; Vol 244(1): 78-84
3. Choplin, R., (1992). Picture archiving and communication systems: an overview. Radiographics January 1992 12:127-129 (Accessed August 14 2013)
4. Biomedical Imaging Informatics. AMIA Working Group. https://www.amia.org/programs/working-groups/biomedical-imaging-informatics (Accessed July 1, 2017)
5. Hood MN, Scott H. Introduction to Picture Archive and Communication Systems. J Radiol Nurs. 2006;25:69-74
6. Wiley G. The Prophet Motive: How PACS was Developed and Sold http://www.imagingeconomics.com/library/tools/printengine.asp?printArticleID=200505-01 (Accessed April 14, 2006)
7. DICOM http://dicom.nema.org/ (Accessed July 2, 2017)
8. Huang HK. Short History of PACS: Part 1 USA. Eur J Rad 2011;78:163-176
9. Bucsko JK. Navigating Mini-PACS Options. Set sail with Confidence. Radiology Today. http://www.radiologytoday.net/archive/rt_071904p8.shtml (Accessed January 11, 2007)
10. Samei, E et al. Tutorial on Equipment selection: PACS Equipment overview. Radiographics 2004; 24:313-34
11. AGFA Healthcare. IMPAX 6. www.global.agfahealthcare.com/us/ (Accessed July 2, 2017)
12. Chesson E. Choosing Web-based PACS. Health Imaging.com June 2006 http://www.dominator.com/assets/002/5131.pdf (Accessed October 5, 2010)
13. DICOM web viewer. http://technologyadvice.com/blog/healthcare/5-dicom-viewers/ (Accessed June 28, 2017)
14. Heart IT Imaging. WebPax. http://www.heartit.com/ (Accessed July 2, 2017)
15. Massat M B. Will Web-Based PACS Take Over? Imaging Technology News. January/February 2009 (Accessed October 5, 2010)
16. PACS Vendor Questions the Importance of Web-based vs Non-Web-based Solutions. Dynamic Imaging http://www.dynamic-imaging.com/ pdf/DI-Web-based-WhitePpr.pdf (Accessed October 9, 2010)
17. Task Force to Study Electronic Health Records. Infrastructure Management & Policy Development Workgroup. Health Information Exchange in Maryland. iHealth & Technology March 2005 http://mhcc.maryland.gov/electronichealth/shared/taskforce/february/hinfoexmd021307.pdf (Accessed October 10, 2010)
18. WebPax PACS Solutions. http://heartit.com/solutions/medical-image-management-solutions/webpax-pacs-systems/ (Accessed July 3, 2017)
19. Loux, Stephenie et al. Consolidated Imaging: Implementing a Regional Health Information Exchange System for Radiology in Southern Maine IHealth & Technology. March 2005 (Accessed October 16, 2010)
20. Doi K. Computer-aided diagnosis in medical imaging: historical review, current status and future potential. Comput Med Imaging Graph [Internet]. NIH Public Access; 2007 [cited 2017 Jul 3];31(4–5):198–211. Available from: http://www.ncbi.nlm.nih.gov/pubmed/17349778
21. Erickson BJ, Korfiatis P, Akkus Z, Kline T, Philbrick K. Toolkits and Libraries for Deep Learning. J Digit Imaging [Internet]. Springer International Publishing; 2017 Mar 17 [cited 2017 Jul 3];1–6. Available from: http://link.springer.com/10.1007/s10278-017-9965-6
22. Mobile MIM. https://www.mimsoftware.com/mobile_cloud/mobile_mim (Accessed July 3, 2017)
23. Resolution MD. https://www.calgaryscientific.com/healthcare (Accessed July 5, 2017)
24. Dicom Viewers. https://www.dicom-viewers.com/ (Accessed July 5, 2017)
25. Osirix. OsiriX http://www.osirix-viewer.com (Accessed July 2, 2017)
26. Hirschorn DS, Choudhri AF, Shih G, Kim W. Use of Mobile Devices for Medical Imaging. J ournal Am Coll ege Radiol ege Radiol [Internet]. 2014;11:1277–85. Available from: http://informatics.jacr.org/sites/default/files/12.20-2842.pdf

27. Reiner BI et al. Effect of Film less Imaging on the Utilization of Radiologic Services. Radiology 2000;215:163-167
28. Srinivasan M et al. Saving Time, Improving Satisfaction: The Impact of a Digital Radiology System on Physician Workflow and System Efficiency. J Health Info Man 2006;21:123-131
29. Innovations in Health Information Technology. AHIP. November 2005. www.ahipresearch.org (Accessed January 10, 2007)
30. Phillips J et al. Will the DRA Diminish Radiology's Assets? http://new.reillycomm.com (Accessed June 18, 2011)
31. Society for Imaging Informatics. https://siim.org/page/fellowship_programs (Accessed July 2, 2017)
32. Liao GJ, Nagy PG, Cook TS. The Impact of Imaging Informatics Fellowships. J Digit Imaging. 2016;29(4):438-442
33. American Board of Imaging Informatics. www.abii.org (Accessed July 5, 2017)
34. Hains IM, Georgiou A, Westbrook JL. The Impact of PACS on Clinician Work Practices in The Intensive Care Unit: A Systematic Review of the Literature. J Am Med Inform Assoc. 2012:19(4):506-513
35. Mendelson DS, Rubin DL. Imaging Informatics. Essential tools for the delivery of imaging services. Acad Rad 2013;20(10):1195-1212
36. Hsu W, Markey MK, Wang MD. Biomedical imaging informatics in the era of precision medicine: progress, challenges and opportunities. J Am Med Inform Assoc 2013;20(6):1010-1013
37. Rubin DL, Greenspan H, Brinkley JF. Biomedical Imaging Informatics. In: Biomedical Informatics: Computer Applications in Health Care and Biomedicine. 2014. Editors Shortliffe and Cimino. Springer.

17

Telemedicine

ROBERT E. HOYT • THOMAS R. MARTIN

LEARNING OBJECTIVES

After reading this chapter the reader should be able to:

- State the difference between telehealth and telemedicine
- List the various types of telemedicine consultations, such as teleradiology and teleneurology
- List the potential benefits of telemedicine to patients and clinicians
- Identify the different means of transferring information with telemedicine, such as store and forward
- Enumerate the most significant ongoing telemedicine projects
- List the most common barriers to telemedicine

INTRODUCTION

According to the Office for the Advancement of Telehealth (OAT), Telehealth is defined as:

"the use of electronic information and telecommunications technologies to support long-distance clinical health care, patient and professional health-related education, public health and health administration"[1]

Like e-health, telehealth is an extremely broad term. A review by Oh et al. found 51 definitions for e-health, suggesting that the term is too general to be useful and the same is probably true regarding telehealth.[2] One could argue that Health Information Organizations (HIOs), Picture Archiving and Communication Systems (PACS) and e-prescribing are also examples of telehealth if they exchange healthcare information between distant sites.

Clearly, telehealth is the broader term that incorporates clinical and administrative transfer of information, whereas telemedicine relates to remote transmission or exchange of only clinical information. In this chapter, the term telemedicine is used, instead of telehealth and defined as follows:

"the use of medical information exchanged from one site to another via electronic communications to improve patients' health status"[3]

Telemedicine was theorized in the 1920s when an author from Radio News magazine demonstrated how a doctor might examine a patient remotely using radio and television. Ironically, this was proposed before television was even available (Figure 17.1).[4] The first instance of remote monitoring has been attributed to monitoring the health of astronauts in space in the 1960's.[5] Very rudimentary telemedicine has been conducted using telephone communication for the past fifty years or more. With the advent of the internet and video conferencing many new modes of communication are now available.

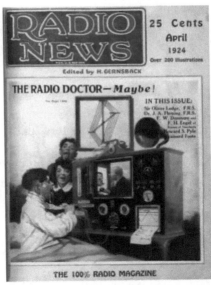

Figure 17.1: Early Telemedicine (Courtesy Radio News)

The goal of telemedicine ultimately is to provide timely and high quality medical care remotely. Telemedicine is becoming increasingly popular for the following reasons: (1) With the rising cost of healthcare worldwide, newer delivery models are appearing that will include telemedicine. In the case of the United States where Medicare will not reimburse for readmission for certain diseases new strategies are needed to prevent readmissions, to include telemedicine. (2) There is a shortage of primary care and intensivist physicians. Moreover, they are maldistributed to urban and not rural areas. Remote delivery of medical care with telemedicine is a partial fix. (3) Additional means are needed to deliver medical care, given the rise in chronic diseases and our graying population. (4) Telemedicine results in improved collaboration among physicians and disparate healthcare organizations. (5) Telemedicine raises patient satisfaction when it results in better access to specialty care, less time lost from work and/or fewer long distant trips to tertiary medical centers.

Like many of the other topics in health informatics covered in this textbook there are multiple interrelationships. Telemedicine can be employed for disease management and as a strategy for improved patient care and communication, thus being part of consumer health informatics. Telemedicine is slowly being integrated with a variety of technologies and platforms such as electronic health records, health information organizations, mobile and picture archiving and communication systems. Due to the pervasive nature of mobile technology, it is also a player in telemedicine.

Telemedicine is part of healthcare reform internationally, in that it aims to improve access to high quality care and education remotely. It can be used for populations at risk, such as rural, indigent and elderly patients. As medical care becomes more patient-centric telemedicine will become part of the patient centered medical home and accountable care organization models.

Recently, telemedicine has been adopted by many major US healthcare delivery systems, such as Saint Joseph's Healthcare, Geisenger and Sentara, to improve access to medical care and hopefully reduce spiraling costs.[6-8] Given a move to bundled and alternative payments, telemedicine remains a potentially useful tool for the informatician. Increasingly payment models are either implicitly or explicitly calling for their use. The Merit Based Incentive Payment System (MIPS) provides opportunities for the use of telehealth services to increase access to care and provide population health outreach. In addition, CMS continues to evaluate waivers for the delployment of telemedicine services as outlined in updates to the comprehensive joint replacement (CJR) bundled payment.[9]

Telemedicine Communication Modes

In this chapter, multiple ways are presented for patients to receive remote care, starting from simple e-mail to complex audio-video teleconferencing. In the past, several years new telemedicine technologies and business models have appeared, with more on the way. Table 17.1 shows several of the communication modes used in telemedicine, along with pros and cons.

Telemedicine Transmission Modes

There are three telemedicine transmission modes:
- **Store-and-forward**. Images or videos are saved and sent later. As an example, a primary care physician takes a picture of a rash with a digital camera and forwards it to a dermatologist to view when time permits. This method is commonly used for specialties such as dermatology and radiology. This could also be referred to as asynchronous communication.
- **Real time**. A specialist at a medical center views video images transmitted from a remote site and discusses the case with a physician. This requires more sophisticated equipment to send images real time and often involves two-way interactive telemonitors. The specialist can see the patient and ask questions. Telemedicine also enables the sharing of images from peripheral devices such as electronic stethoscopes, otoscopes, etc. This would be an example of synchronous communication.
- **Remote monitoring**. A technique to monitor patients at home, in a nursing home or in a hospital for personal health information or disease management.

Telemedicine Categories

In this chapter, Telemedicine is divided into the categories noted below based on current knowledge and initiatives. Virtual patient visits (televisits or e-visits) could be part of telemedicine or consumer health informatics. Virtual visits will be discussed further in the chapter on consumer health informatics.
- Teleconsultations: teleradiology, teledermatology, televisits, etc.
- Telehealth: While some have parsed out unique definitions for each word, ATA treats "telemedicine" and "telehealth" as synonyms and uses the terms interchangeably. In both cases, we are referring to the use of remote health care technology to deliver clinical services. The use of the term telehealth may reference direct to consumer models or care received other care team members operating within a contiumum of care.

Table 17.1: Telemedicine Communication Modes

Communication Mode	Pros	Cons
Patient-Portal secure-messaging	Asynchronous. Able to attach photos. Response can be formatted with template. Could use VoIP. Audit trail is available	Not as personal as live visit. Usually not connected to EHR or other enterprise information but may be in the future
Telephone/Smartphone	Widely available, simple and inexpensive. Real-time	Not asynchronous. Unstructured. No audit trail. Only real-time
Audio-Video	Maximal input to clinician. Can include review of x-rays, etc. Perhaps more personal than just messaging	Currently, most expensive in terms of networks and hardware, but that is changing
EHR "One Click" Integration	Many telehealth solutions operate independently of the medical record. Given the relative simplicity of the underlying technology some EHR vendors may integrate features and functionality directly into platform offerings.	Potentially costly, workflow issues such as scheduling, virtual patient waiting rooms, and licensure issues
"White Labeled" Solutions	Emerging solutions by large organizations or vendors are providing platforms which essentially scale a medical practice or health system to meet demand for virtual patient visits	Requires contractual agreements, issues of vetting of providers, determining appropriate workflow to move a patient from virtual to face to face setting

- Telemonitoring:
  o Telerounding: hospital inpatients
  o Telehomecare: monitoring physiological parameters, activity, diet, etc. at home

TELECONSULTATIONS

Teleconsultation is a worldwide phenomenon because specialists tend to practice in large metropolitan areas, and not in rural areas. Most programs consist of a central medical hub and several rural spokes. Programs attempt to improve access to services in rural and underserved areas, to include prisons. This reduces travel time and lowers the cost for specialists and patients alike. Programs have the potential to raise the quality of care delivered and help educate remote rural patients and physicians. The most commonly delivered services are mental health, dermatology, cardiology and orthopedics. According to the American Telemedicine Association (ATA), as of 2018 there about 200 telemedicine networks and 3,500 practice sites.[10]

Primary Care or Concierge Style Televisits

e-Visits are covered in more detail in the chapter on Consumer Health Informatics. While this section discusses primarily specialty telemedicine it should be noted that virtual audio-visual appointments (televisits) have gained in popularity. For example, *Doctor On Demand* functions largely to offer 24/7 coverage to employees who need after hour access or who do not want to lose time from work. Services include primary care, mental health, preventive health, woman's health and lab services. Many of these vendors contract with employers and insurance companies as a mean of improving access to care.[11]

Teleradiology

The military has taken the lead in this area partly due to the high attrition rate of radiologists and the desire to enhance radiology support for military deployments. By 2007 most Army x-rays became digital, which helped the storage, transmission and interpretation of images. With this newer technology, a computerized tomography (CT) scan performed in Afghanistan can be read at the

Army medical center in Landstuhl, Germany. Another example of military teleradiology can be found on the Navy hospital ships Mercy and Comfort where digital images can be transmitted to shore based medical centers for interpretation or consultation.

In the civilian sector, vRad (formerly NightHawk Radiology Services) helps smaller hospitals by supplying radiology services remotely. All are board certified; most trained in the United States and carry multiple state licenses. They list a staff of 500+ radiologists (70% are subspecialty trained) and interpret seven million studies per year. They offer conventional radiology as well as CT, MRI, Ultrasound and Nuclear Medicine interpretation.[12]

In mid-2013 The American College of Radiologists published Teleradiology Practice Guidelines. The Task Force outlined benefits as well as challenges to the practice.[13]

Another more common but important example of teleradiology is the practice of radiologists reading films after-hours at home. They must have high resolution monitors and high-speed connections to the Internet but with this set up and voice recognition software; they can be highly productive at home. This is becoming the standard practice for radiologists. Instead of driving in or staying at the hospital at night to interpret images, they can deliver interpretations while at home.

Teleneurology

Treatment of stroke with intravenous clot busting drugs has become the standard of care and can result in a small reduction in mortality and increase in the odds of going home and walking better.[14] Many regions lack neurologists to see patients with stroke-like symptoms to determine if they need clot-busting drugs (thrombolytics) or need to be transferred to a higher level of care. This is, in part, due to the increased malpractice risk and decreased reimbursement situation of treating emergency patients. With the advent of telemedicine, the case can be discussed real time and the patient, and their x-rays can be viewed remotely by a stroke specialist. One company, REACH Health developed a web-based solution that includes a complete audio-visual package, so neurologists can view the patient and their head CT (CAT scan). REACH Health was developed by neurologists at the Medical College of Georgia. Because the program is web-based, the physician can access the images from home or from the office. Likewise, the referring hospital only has to have an off-the-shelf web camera, a computer and broadband internet connection. The software can be integrated with EHRs. They conduct an annual Telemedicine Survey and results of the 2017 Survey will be noted later in the chapter.[15] Specialists-on-Call is a Massachusetts based organization that has neurologists on board to handle emergency consults via telemedicine. In 2011, they added telepsychiatry and in 2013 teleintensivists. Their capabilities include the ability to transfer head CT images and bidirectional audio and video conferencing with remote physicians and families. Over 4,000 stroke patients are treated monthly. To accomplish this, they have an infrastructure that consists of a PACS, a call center, an electronic health record and videoconferencing equipment. The cost for this service is not inexpensive and it is unknown if third party payers will eventually reimburse for this service.[16-17] A teleneurology study is reported later in this chapter.

Teleconcussion is a new indication for teleneurology consultation whereby a patient with head trauma is evaluated remotely. An article from the Mayo clinic on this approach is posted in the recommended reading section of this chapter.

Telepharmacy

Like teleradiology, this field arose because of the shortage of pharmacists to review prescriptions. Vendors now sell systems with video cameras to allow pharmacists to approve prescriptions from a remote location. This is very important at small medical facilities or after-hours when there is not a pharmacist on location.[18] A 2011 survey documented that 60% of hospitals have 24-hour review of medication orders by pharmacists and 11% use a telepharmacy company.[19] The North Dakota Telepharmacy Project operates 56 remote sites where pharmacy technicians receive approval for a drug by distant pharmacists via teleconferencing. About 73% of counties are covered with this project, in addition to two counties in Minnesota. This initiative is supported by North Dakota State University. In this manner, a full drug inventory is possible even in small rural communities and the pharmacists still perform utilization reviews and other services remotely.[20]

Telepsychiatry or Telemental Health

The shortage of mental health professionals has helped to drive telemental health (also called telemental health) services. Several studies have indicated that telepsychiatry is equivalent to face-to-face psychiatry for most patients.[21] The American Psychiatric Association promotes telepsychiatry, primarily for remote or underserved areas, using live video teleconferencing. During a telesession, there can be individual or group therapy, second opinions and medication reconciliation. In general, virtual visits help team medicine and patient satisfaction has been good. On the American Psychiatric Association web site, there are valuable telemental health resources.[22] Another telepsychiatry trend that is appearing is the use of free commercial-off-the-shelf (COTS) audiovisual programs, such as Skype. Voyager Telepsychiatry uses this popular program to hold virtual telepsychiatry sessions.[23] One of the most important areas for telepsychiatry is for US military members who return from war with Posttraumatic Stress Disorder (PTSD) and Traumatic Brain Injury (TBI). About 40% of veterans live in rural areas, where transportation may be an issue. The VA has opened three Veterans Rural Health Resource Centers in Iowa, Utah and Vermont to help develop and evaluate telemedicine programs.[24]

A comprehensive resource "The Online Couch" was published in 2012 that categorized technology-enabled treatment of depression into: computer-based cognitive behavioral therapy (CCBT), online counseling, online social networks, mobile platforms, games and virtual reality. The report makes the point that depression is extremely common in the adult population and because there is a shortage of mental health providers, a cost barrier and stigma associated with mental health visits, alternative approaches are needed. CCBT has been endorsed in the United Kingdom for use in the National Health Service, specifically for the treatment of depression (Beating the Blues) and panic/phobic disorders (FearFighter). The University of Pittsburg Medical Center has now also adopted Beating the Blues approach and will use it in the patient centered medical home model. Other technology-related approaches are mentioned, such as interactive voice response, email, chat and video, but are beyond the scope of this chapter. Market challenges, provider perspectives and payments are addressed.[25]

The web site Telemental Health Technologies Compared offers the ability to search for applications focused on private practice, provider networks, the enterprise and the consumer. For example, 90 different technologies are listed under private practice. One platform offers online CCBT for insomnia, depression, anxiety, panic, phobias, obsessive compulsive disorder and addiction, with literature references. This comparison web site makes the point that 38 state counseling boards have policies addressing online counseling and 18 states mandate that telemental services be reimbursed at the same level as face-to-face therapy.[26]

Telemental health networks and organizations have arisen in the past few years. For example, JSA Health Telepsychiatry offers 24/7 coverage by board certified clinicians for emergency departments, rural health clinics, homeless shelters, schools, correctional facilities and cruise ships.[27]

A 2015 review of telepsychiatry reported that telepsychiaty works well for post-traumatic stress disorder (PTSD), depression, and ADHD, and works well for underserved groups such as American Indian, Hispanic, and Asian populations. Despite success they reported barriers that included, personal bias—in leadership, insufficient training, ainconsistent reimbursement, licensing, and prescription policies.[28]

A 2016 systematic review of telepsychiatry demonstrated that overall, patients and clinicians were satisfied with telepsychiatry services. The review also showed that telepsychiatry is not more expensive than face-to-face delivery of mental health services and in fact may be more cost-effective. Lastly, no adverse events were found related to telepsychiatry.[29]

Teledermatology

With the advent of good quality digital cameras and cell phones with medium quality cameras, the concept of teledermatology was born. The Teledermatology Project, created in 2002, has the goal of providing free worldwide dermatology expertise, particularly for third world countries and the underserved. Physicians can easily obtain a teleconsultation and diagnostic and therapeutic advice using the store and forward mode.[30] iDoc24 is a Teledermatology project that began in Sweden for the European Union. It was designed for those patients who were traveling or did not have access to their physician and had a new skin condition. Patients can take a picture of their skin lesion with a digital camera or cell phone (app available) and forward it (can be anonymous) as an attachment to a text message and it would be followed by a response by a dermatologist within 24 hours. The image can also be integrated with the regional personal health portal that is part of the Swedish National Health Service. The goal is to provide better service, answer anonymous requests and decrease overall face-to-face visits to dermatologists.[31]

Direct Derm is a web service for patients to upload pictures of their skin problem and receive recommendations from a team of US board-certified dermatologists within 2 business days. They also offer the ability for physicians to submit patient dermatology issues and photographs. The service costs $85 and is not submitted to third party payers. This platform is supported by the California Healthcare Foundation and Kresge Foundation.[32]

Teledermatology can reduce the need for face-to-face visits with a Dermatologist and can help with education and training of clinicians. The correlation between face-to-face evaluations and store and forward images from cameras or cell phone cameras is high.

For more details on teledermatology, readers are referred to a review article by the California HealthCare Foundation.[33]

Teleophthalmology

The most common indication for teleophthalmology is imaging the retina (teleretinal screening) of patients at risk. Diabetic patients constitute the group most frequently followed due to serious complications, such as blindness related to their diabetes. Some ocular lesions are amenable to photo-coagulation, so early intervention is important. Unfortunately, fewer than 50% of diabetics receive annual screening. Traditionally, patients would schedule an appointment with an ophthalmologist but largely due to the additional expense many patients were not seen. The newer trend is to have patients screened with a digital camera device in a primary care office and the images interpreted as store-and-forward. Newer devices can image the retinal without dilating the pupil, an attractive option for patients. Those patients with any serious pathology are then referred to a retinal expert. A demo video is available by one vendor to describe the process.[34]

Two 2017 studies suggest that screening for retinal issues in a primary care setting is worthwhile. The first study showed an increase in screening for diabetic retinopathy of 16% and wait times for screening reduced about 90%.[35] The second study was part of the North Carolina Diabetic Retinopathy Telemedicine Network. Their study of five clinics showed an increase in screening from 25% to 40%.[36]

TELEMONITORING

Telerounding

This is a new concept developed to help address the shortage of physicians and nurses. Telerounding is being rolled out in facilities with reasonably good reviews, despite obvious criticisms that it further compromises the already strained doctor-patient relationship.

Robot Rounding. A study in 2005 in the Journal of the American Medical Association showed that surgeons could make a second set of rounds using a video camera at the patient's bedside (InTouch Robots). A physician assistant makes the actual rounds, backed up by the attending physician remotely via the robot. Robot units are five and a half feet tall, weigh 220 lbs. and have a computer monitor as a head. The cost is more than $100,000 each or they can be leased for $5,000 monthly plus $5,000 per viewing station. At this time, they are being used in 20 plus hospital systems in the United States. They can move around and can project x-ray results to the patient. Ellison et al. reported on urological patients who either received face-to-face rounds post-operatively or robotic telerounding. They concluded that robotic rounding was safe and well received by patients. Two-thirds of patients stated they would rather see their own physician remotely than a stranger making rounds in person.[37-38] One of the leading companies to offer robots has expanded the use of the robots for telestroke care, hospital and operating room consultations and E-ICU rounding, discussed in the next section. The vendor claims 1600 installations by 2017 and 30 use cases.[39]

E-ICU Rounding. In the United States, it is stated that approximately 35,000 intensivists (physicians who specialize in ICU care) are needed, but only 6,000 exist. Moreover, even though hospital beds are not increasing, ICU beds are. Therefore, remote monitoring makes sense particularly during nighttime hours when physicians might not be present. The Leapfrog Group has advocated care delivered by intensivists for all ICUs as one of its four patient safety recommendations; but this goal remains elusive.[40] Hospitals that use e-ICUs believe there are patient safety and financial benefits, but both need to be proven. An e-ICU service may be less expensive than recruiting full time intensivists. Also, because ICU care can cost $2,500 daily, any cost saving modality that positively affects length of stay or mortality will gain market attention. Avoiding law suits in the ICU also means cost savings. It is estimated that over 100 hospitals now have e-ICU programs, even though there is no reimbursement by insurers.[41]

A few large healthcare systems have created their own eICU systems, but most have used the VISICU platform. It was founded by two intensivists from Johns Hopkins in 1998 and later purchased by Phillips Healthcare division. As of 2017, 11% of hospital systems use this technology. Their approach is to provide two-way video and audio communication, standardization of care, clinical decision support and robust graphical displays of physiological data. They have a research arm with more than 2.5 million patient stays archived. This platform extended support of care outside the ICU in 2007 to include medical surgical floors, emergency departments, step-down units and post anesthesia units.[42]

The cost for e-ICUs is significant considering the uncertain benefits, such as return on investment. Kumar reported costs of eICU programs in 2013 based on the Veterans Health Administration system and estimated that it would cost between $70,000 and $87,000 per ICU-bed for implementation and the first year of support. Based on their review of the literature eICUs could be associated with a several thousand dollar loss or gain per patient admitted to the ICU.[43] In spite of the many potential virtues of the e-ICU, an early article by Berenson et al. expressed the opinion that the actual value of e-ICUs was far from proven and there was a major interoperability issue between the e-ICU software and critical ICU systems like IV fluids and mechanical ventilation.[44] Another early article in by Thomas et al. evaluated the medical care in six ICUs before and after the implementation of an e-ICU system. They concluded that there was not an overall improvement in mortality or length of stay.[45] A more recent meta-analysis by Young et al. showed a decreased mortality and length of stay (LOS) in the ICU but not the overall hospital mortality or length of stay.[46] An article from a single academic medical center reported a lower hospital mortality and length of stay, improved guideline adherence and reduced preventable complications.[47] However, as pointed out in an editorial, it is not known if eICUs can improve care in rural/remote settings and whether hired intensivists on site would be more valuable.

In 2013, a large non-randomized study was reported based on results from 56 ICUs in 19 US healthcare systems. More than 118,000 patients were studied (including a control group). They found that mortality was lower in the eICU group and hospital and ICU LOS was lower. The factors leading to reduced mortality and LOS were intensivists reviewed case within one hour of admission; use of performance data; adherence to ICU EBM/CPGs and quicker alert response times.[48]

The bottom line is that further research is needed to provide the kind of detail necessary to determine the benefit of this type of telemedicine. For example, is the benefit greater for a small hospital with limited ICU expertise compared to a large integrated ICU system with an abundance of intensivists?

Most articles reviewed included the Phillips eICU system, but other healthcare organizations have developed their own solutions. For example, in the Department of Veteran Affairs VISN (Veterans Integrated Service Network) 19, the Denver VAMC serves as the hub for four rural smaller VA hospitals. The telehealth system is ready on call to mobilize a rapid response team when called by a critical care nurse. The goal is to rapidly stabilize a patient, so they can stay local or be transferred safely. This hybrid system does have the advantage of using the same EHR in every hospital and mobile medical carts. Another unique aspect of their system is the ability for outlying surgeons to operate on a patient at the telehealth hub but can participate in virtual rounds remotely as the patient recovers.[49]

Telehomecare and Remote Patient Monitoring

Telehomecare is remote monitoring of the patient at home. One healthcare expert has stated that "*home is the new hub of health*" which implies the home needs to be interoperable with the rest of the healthcare system.[50] It usually involves monitoring vital signs, weights, blood sugars, etc. that can be sent via a wired or wireless mode from homes to physicians' offices, patient portal, EHRs, health information exchanges, etc. While home telemonitoring can also include fitness programs and "*aging in place*" technologies, the chapter will focus on chronic disease management and post-acute care monitoring. The goal is to better educate and monitor patients at home to provide better patient-centric healthcare, while reducing readmissions and unnecessary emergency room visits, thus saving money. There are multiple reasons telemonitoring is burgeoning:

- Chronic diseases are on the rise that will likely increase hospitalizations, readmissions and unnecessary emergency room visits. Measures like home monitoring might decrease this trend. The goal is to intervene immediately, rather than wait till the next appointment.
- Medicare changed reimbursement to home health agencies from the number of visits to a diagnosis-based system, leading to decreased reimbursement for visiting nurses.
- Telemonitoring programs potentially support audio and visual communication with patients at home and therefore can reduce home visits by a nurse or physician. Nurses can make visits only if there is a problem, such as a change in symptoms or vital signs.
- One consulting organization predicts a nursing shortage of 800,000 and a physician shortage of 85,000 to 200,000 by the year 2020.[51]
- Baby boomers are tech savvy and more likely to demand services like telemonitoring.
- Monitoring may be possible using the ubiquitous cell phone and new microsensors.
- Linking home monitoring devices to EHRs with decision support and health information exchanges will increase the functionality of this new technology. The potential to save costs is attractive but elusive and will require high quality confirmatory studies.
- CMS has administered Medicare Medical Home Demonstration projects to test the "*medical home*" and "*hospital at home*" concepts. Medical groups will be paid for coordination of care, health information technology, secure e-mail and telephone consultation and remote monitoring. Details are preliminary and available on the CMS site.[52] Accountable Care

Organizations (ACOs) may also incorporate telemedicine.
- The Affordable Care Act will reduce payments to hospitals deemed to have excessive readmission rates for conditions such as heart failure, acute myocardial infarction (heart attack) and pneumonia. This may help drive more monitoring and technology at home.[53]
- Comprehensive Care for Bone Joint Replacement - the CJR model includes a waiver of the geographic site requirement for any service on the Medicare-approved telehealth list and the originating site requirement only to permit telehealth visits to originate in the beneficiary's home or place of residence.[9]

Many health IT vendors are developing home monitors and new sensors that will transmit information to a physician's office or other healthcare organizations. Programs will be interactive and include patient education for issues such as drug compliance. This data may interface with an electronic health record, health information organization (HIO) or web site for others to evaluate. Some predict that houses (smart homes) will be wired with multiple small sensors known as "motes" that will monitor daily activities such as taking medications and leaving the house. The information would be transmitted to a central organization that would notify the patient and/or family if there was non-compliance or a worrisome trend. Selection of vendors should include some discussion on approaches to shared risk/savings models. While a number of business models exist – for example per member per month approaches for remote monitoring – contractual agreements for larger deployments of telehealth services should include tiered agreements when possible. Also given the rise of external organizations providing "platforms as a service" type offerings, considering should be given to the term of contract and scope of services needed by the contracting organization. In addition, many contemporary used practices employed in EHR selection should be considered when evaluating a homecare solution.

Telemonitoring is actually a process with multiple steps depicted in Figure 17.2.

Home Telemonitoring System Examples

More than 50+ companies offer technology to monitor patients at home and the list continues to grow and include large companies such as Intel and General Electric. Devices can be standalone or be integrated with another system such as an electronic health record or personal health record. Devices connect externally using USB, Bluetooth, ZigBee, telephony (POTS), WiFi and

Figure 17.2: Telemonitoring cyclical process

3G/4G telecommunication networks. The list of available sensors continues to grow. Table 17.2 lists current and future home sensors that assist telemonitoring.

Health Buddy is a FDA approved device that is certified by the National Committee for Quality Assurance and used by the Veterans home telemedicine programs. Health Buddy (Bosch Healthcare) is used by over 12,000 patients and has been shown in one study (of limited design) to increase medication compliance and reduce outpatient visits.[54-55] Additional studies are reported in another section. The Centers for Medicare and Medicaid Services tested the system with about 2,000 patients with chronic diseases and the results are in the section on Telemedicine Studies. Ironically, Bosch Healthcare shuttered Health Buddy in 2015.

HoneyWell Life Care Solutions has over 15,000 monitors currently in use and more than 300,000 patients have been monitored. Features include:
- *Genesis Touch* touch screen device with connectivity to their management platform (LifeStream Manager) via 4G or Wifi. Vital signs can be recorded, and audio-video conferencing can take place. Peripheral devices can be connected via Bluetooth for blood pressure, weigh and oximetry monitoring.
- *Genesis DM* device is designed for disease management of heart failure, diabetes and chronic obstructive lung disease (COPD)
- *LifeStream Connect* is a new option that interfaces with EHRs
- *LifeStream Analytics* analyzes data from the management suite.[56]

Table 17.2: Home telemonitoring sensors

Sensor	Purpose
Weight	Disease management
Blood Pressure	Disease management
Glucose	Disease management
Oximeter	Disease management
Spirometry	Disease management
Temperature	Acute monitoring
Medication tracker	Drug compliance
PT/INR	Anticoagulant monitoring
Home security and other functions	Infrastructure monitoring
Motion detectors/chair and bed sensors	Quality of life monitoring and safety
Fitness	Quality of life monitoring

MyCareTeam: a fee for service diabetic portal developed in cooperation with Georgetown University. Hypertension and weight are also monitored. The application integrates with Allscripts EHR.[57]

WellDoc: a chronic disease management platform intended for sharing data from mobile devices with clinicians and nurse managers. It offers patient coaching for diabetes, cardiovascular disease, respiratory disease, oncology, mental health and wellness. BlueStar is a special type 2 diabetic program reimburseable by insurance companies.[58]

Care Innovations: Intel along with General Electric has entered the telehomecare market with a comprehensive program called Care Innovations. One product line, *ConnectRCM* offers wellness surveys, brain games, medication reminders, biometric data collection and messaging. Another product line, *Connect* Caregiver is a beta program for caregivers and *Connect Guide* is a chronic disease management module.[59]

TELEMEDICINE INITIATIVES AND RESOURCES

The following section provides a sampling of some interesting telemedicine initiatives and resources:

Telehealth Resource Centers: Telehealth Resource Centers (TRCs) have been established to provide assistance, education and information to organizations and individuals who are actively providing or interested in providing medical care at a distance. The charter from the Office for Advancement of Telehealth is to assist in expanding the availability of health care to underserved populations. And because TRCs are federally funded by AHRQ, the assistance provided is generally free of charge.[60]

Center for Connected Health Policy (CCHP): is a nonprofit, nonpartisan organization working to maximize telehealth's ability to improve health outcomes, care delivery, and cost effectiveness.[61]

Informatics for Diabetes Education and Telemedicine (IDEATel): The largest government sponsored telemedicine program in the US. The project evaluated approximately 1,650 computer illiterate patients living in urban and rural New York State. Patients received a home telemedicine unit that consisted of a computer with video conferencing capability, access to a web portal for secure messaging and education and the ability to upload glucose and blood pressure data. These same subjects were assigned a case manager who was under the supervision of a diabetic specialist. They used the Veterans Affairs clinical practice guidelines on diabetes. They were compared to a control group that didn't receive the home monitoring system. The results of this project are reported in the Recommended Reading section.[62]

Global Partnership for Telehealth (GPT): Georgia has 159 counties, many at the poverty level. This network is the first statewide effort to link 36 rural hospitals and clinics with specialists at eleven large urban hospitals.

Project created partnerships among Wellpoint (Blue Cross/Blue Shield) and the state government. Importantly, telemedicine consults were reimbursed as office visits due to a new Georgia law and 20 specialties were felt to be appropriate for telemedicine. In 2012, this initiative had 75,000 visits at 350 locations. The top categories for encounters were wound care, school health and telepsychiatry.[63-64]

Department of Veteran Affairs Telehealth services: The VA is already one of the largest users of telehealth in the US with its VA Video Connect program. This program serviced 67 facilities and 700,000 veterans in 2016. In mid-2017 the Office of American Innovation and the Department of Justice announced an upcoming regulation to provide telehealth services nationally, across state lines as part of a new Anywhere to Anywhere program.[65]

University of Texas Medical Branch at Galveston: Program is the largest telemedicine system in the world with 300 locations and 60,000 annual telemedicine sessions. Sixty per cent of visits deal with a prison population. They also offer specialty services in neurology, addiction medicine and psychiatry.[66]

Massachusetts General Hospital Teleburn Program. MGH specialists offer emergency and long-term consultations, as well as continuing education.[67]

Inova Telepediatrics: Inova connects emergency room physicians with specialty pediatricians to consult on seriously ill children.[68]

Mercy Virtual Care Center. Mercy Health operates in several states and in 2015 opened a 125,000 square foot Virtual Care Center, the largest telemedicine hub in the US. The services offered include: care managment, condition managment, home management, eSitter, Nurse on Call, pharmacotherapy, Teleintensivist, TeleICU, Telesepsis, Telestroke, Utilization management and Wellness.[69]

The Virtual Dental Home Demonstration Project is a California university-based dental initiative in which remote dental hygienists and dental assistants use imaging to connect patients at risk with dentists. Approximately, half can be treated locally, and half are referred for in-person dental visits.[70]

California Central Valley Teleretinal Program: Using a non-proprietary, open source web-based program (EyePACS) images can be forwarded to an ophthalmologist for interpretation. Images are stored on a SQL Server and images are viewed with a web browser. A simple software program on the PC allows for uploading images to the server. There is e-mail notification to the consultant and back to the individual who sent the images. During the grant period they screened 53,000 patients at a cost of $15 per patient.[71-72]

Northwest Telehealth: A consortium of 4 healthcare systems created the Inland Northwest Health Services, located in Spokane, Washington. This initiative has 115 sites using advanced audio-visual technology. They offer the following services: clinical care (15 specialties), teleER, telepharmacy, distance education, administrative and operational planning/coordination.[73]

Federal Communications Commission (FCC): In 2006 they announced a $400 million budget for pilot projects to promote broadband networks in rural areas. The goal was to create networks for public healthcare organizations and non-profit clinicians that would eventually connect to a national backbone. The network could be used for telemedicine or other medical functions in rural areas. In 2007 the FCC created a $417 million fund that supported pilot projects to connect more than 6,000 hospitals, research centers, universities and clinics. Internet2 or the LambdaRail Network was used. Much of the funding came from the Universal Service Fund that derives from a fee added to consumers and telecommunication companies. In 2009, Congress directed the FCC to develop a National Broadband Plan with goal of providing broadband access to every American and funded it as part of the American Recovery and Reinvestment Act (ARRA).

In 2013, The FCC launched Healthcare Connect Fund that dedicated $400 million yearly to support the broadband access nationwide, particularly in rural areas. This will support telemedicine initiatives that require substantial bandwidth. The following applicants are eligible: community health (and mental health) centers, migrant health centers, local health departments, post-secondary educational institutions (including academic centers and medical schools), public or not for profit hospitals and rural health clinics. Their current initiative is known as Connect2Health FCC Task Force which aims to promulgate broadband and healthcare technologies.[74-75] Access to the funds remains an issue of interest to policymakers. Currently, administrative effort may not be paid for out of the funds which has resulted in a number of rejected applications. Policymakers continue to evaluate approaches to better utilize these funds.[76-77]

What is telehealth? Health IT.gov. The ONC site includes the Federal Telehealth Compendium and the Health IT Playbook for Telehealth.[78]

UPMC Telehealth Adoption Model. HIMSS case study discusses the University of Pennsylvania Medical Center telehealth adoption model. Good review of factors behind whether to consider a telehealth initiative.[79]

INTERNATIONAL TELEMEDICINE

In many chapters, international initiatives are included to demonstrate that health information technology is being embraced by both developed and developing nations. This is particularly true regarding mobile technology initiatives. Other countries are facing the same challenges with rising chronic diseases, disparities in healthcare delivery and rising healthcare costs. Cost is a clear barrier but fortunately, the cost for telemedicine interventions is falling.

As more international authors are added telehealth/telemedicine initiatives that are innovative and informative will be highlighted. More information can be gained by visiting the International Society for Telemedicine and eHealth which is linked to the World Health Organization and has 13 working groups.[80]

In a 2016 paper Combi et al. summarizes the telehealth projects in developing countries, as well as challenges and lessons learned.[81]

RECOMMENDED READING

The following are a sample of some of the more interesting and telemedicine articles to appear in the medical literature:

- *Telehealth Interventions to Support Self-Managment of Long-Term Conditions: A Systematic Metareview of Diabetes, Heart Failure, Asthma, Chronic Obstructive Pulmonary Disease, and Cancer.* This 2017 article was a systematic review of all systematic reviews on randomized controlled trials on these conditions from the years 2000-2016. They concluded that blood glucose telemonitoring improved glycemic control only in type 2 diabetes; telemonitoring reduced mortality and hospital admisions for heart failure inconsistently. Results for other conditions was mixed. No harm from any intervention was noted.[85]
- *Care Coordination/Home Telehealth: The Systematic Implementation of Health Informatics, Home Telehealth and Disease Management to Support the Care of Veteran Patients with Chronic Conditions.* The US Department of Veterans Affairs operates perhaps the largest telehomecare networks in the world. This is partly due to the fact that the VA has transitioned from inpatient to outpatient and home care. Also, with so many active duty members returning injured from the war zone they will eventually need telehomecare. Their Care Coordination / Home Telehealth program is also a disease management program. The VA currently runs three programs: telehomecare, teleretinal and a video teleconferencing services that link 110 hospitals and 380 clinics. Data from home devices inputs into the VA's EHR. A study of 17,000 VA home telehealth patients was reported in late 2008. Although the cost per patient averaged $1,600, it was considerably less expensive than in-home care. They utilized individual care coordinators who each managed a panel of 100-150 general medical patients or 90 patients with mental health related issues. They promoted self-management, aided by secure messaging systems and a major goal was early detection of a problem to prevent an unnecessary visit to the clinic or emergency room. 48% were monitored for diabetes, 40% for hypertension, 25% for heart failure, 12% for emphysema and 1% for PTSD. Patient satisfaction was very high. This study showed a 19% reduction in hospitalizations and a 25% reduction in the average number of days hospitalized.[86]
- *IDEA TEL Studies.* The one-year results of the IDEA TEL were published in late 2007 and showed mild improvement in blood sugars, cholesterol and blood pressure compared to the control group. Patient and

---

International Case Study

United Kingdom Department of Health Whole System Demonstrator Program

This is the largest (3000 patients, 177 practices analyzed) cluster randomized telemedicine trial to study the impact of technology (secure messaging, home telemonitoring, etc.) on diabetes, coronary heart disease and chronic lung disease. The study looked at outcomes, use of services, user and professional experiences, etc. Results published in June 2011 BMJ indicate: a reduction in emergency admissions, emergency department visits, elective admissions, reduction in bed days and a reduction in mortality.

Subsequently, another study looked at the cost effectiveness of this telehealth project by analyzing the QALY (quality adjusted life years). They concluded that telehealth added expense to the overall care of patients but did not add to years lived and therefore had a low probability of being cost effective.[82-84]

physician satisfaction were positive but detailed cost data was lacking. Ironically, Medicare claims were higher in the study patients than in the control group, for unclear reasons. The five-year results were published in 2009 and although they showed some statistically significant improvement in blood sugar, cholesterol and blood pressure control, they were of doubtful clinical significance. Importantly, users of this technology had a dropout rate greater than 50%. In 2010, a final report from this group concluded that "telemedicine case management was not associated with a reduction in Medicare claims."[87-89]

- *Telestroke Care.* Teleneurology or telestroke care was evaluated by a study by Meyer in 2008. They compared the outcomes of patients with a possible impending stroke and consultation by telephone, versus full video teleconferencing. Correct treatment decisions were made more frequently (98% versus 82%) for the teleconferencing sessions, but patient outcomes were the same. There was no difference in death rates or hemorrhaging after the clot busting drugs (thrombolytics) were administered.[90] An excellent review article on stroke telemedicine was published by Demaerschalle et al. in the Mayo Clinic Proceedings.[91] The jury is out whether stroke telemedicine is cost effective or a reasonable choice, compared to telephonic consultation.[92]
- Heart Failure Telemedicine. An international meta-analysis of 10 randomized controlled trials looked at remote patient monitoring (RPM) of heart failure patients. They concluded RPM reduced the risk for all-cause mortality and hospitalization for heart failure. The number needed to treat (discussed in evidence-based medicine chapter) was 50 for all-cause mortality and 14 for heart failure hospitalization.[93] Another Cochrane meta-analysis also showed benefit of telemedicine while a 2010 telephone monitoring study reported no benefit; thus again, no consensus.[94-95]
- Effect of Telephone-Administered vs Face-to-Face Cognitive Behavioral Therapy on Adherence to Therapy and Depression Outcomes Among Primary Care Patients: A Randomized Trial. Subjects were randomized to usual CBT or 18 sessions of telephone CBT. Fewer subjects discontinued telephone therapy (21% vs 33%). Overall treatment success was similar but face-to-face CBT was superior at six months.[96]
- Teleconcussion: *An Innovative Approach to Screening, Diagnosis and Management of Mild Traumatic Brain Injury.* The Mayo Clinic presented a case study of a 15-year-old boy in Arizona with a post-concussion syndrome with persistent headaches evaluated by a remote consultant.[97]
- A Randomized Controlled Trial of Telemonitoring in Older Adults with Multiple Health Issues to Prevent Hospitalizations and Emergency Department Visits. Telemonitoring (daily biometrics, symptom reporting and videoconferencing), using Intel Health Guide was compared with usual care in elderly adults in the Mayo Clinic system over a one-year period. There was no difference between the two study groups at one year. The mortality was higher in the telemonitoring group (14.7%) compared to usual care (3.9%) for unclear reasons.[98]
- *Impact of Critical Care Telemedicine Consultations on Children in Rural Emergency Departments.* Study looked at physican-rated quality of care for children seen in emergency department of rural hospitals in California. They found that quality was higher for patients who received telemedicine consults, compared to telephone consultation or no consultation. Telemedicine was associated with more changes in diagnostic and therapeutic approaches and patient satisfaction was high.[99]
- Effectiveness of Telemonitoring Integrated Into Existing Clinical Services On Hospital Admissions for Exacerbation of Chronic Obstructive Pulmonary Disease: Researcher Blind, Multicentre, Randomized Controlled Trial. This well-designed study was reported in 2013 and failed to show a reduction in readmission for COPD one year after randomization or improvement in quality of life. The intervention consisted of a touch screen used for symptom and treatment queries and oxygen saturation measurements.[100]

BARRIERS TO TELEMEDICINE

The barriers to telemedicine are similar to the barriers to all health information technology covered in other chapters. The most significant barriers are as follows:

Limited reimbursement. Most telemedicine networks were created with federal grants. Medicare will reimburse if there is a formal consultation linked by live two-way video teleconferencing and the patient resides in a professional shortage area. Medicare will reimburse physicians, nurse practitioners, physician assistants, nurse midwives, clinical nurse specialists, clinical psychologists and clinical social workers. The originating sites can be offices, hospitals, skilled nursing homes, rural health clinics and community mental health centers. Medicare Part B reimburses for telemedicine services for initial inpatient care, outpatient care, pharmacologic management, end stage renal disease-related visit and psychiatric diagnostic interviews. Clinicians at the remote site submit claims using the correct CPT or HCPCS codes as well as the telemedicine modifier GT. Patients pay 20% of the approved Medicare-approved amount. In 2015, the CPT Editorial Board created a new code for chronic care management (CPT 99490) that does not require face to face visits. This permits asynchronous visits for remote monitoring of chronic conditions and is subject to restrictions (multiple chronic conditions, recent hospital admissions, and use of Certified EHR Technology), for 20-minute visits and is reimbursed by Medicare. Several bills have been introduced to expand Medicare coverage of telemedicine (the most recent is HR 2550 Medicare Telehealth Parity Act of 2017 and S. 2484 The Connect for Health Act) but it is not known if they are likely to pass.

Most states have the ability to cover Medicaid telemedicine care but must comply with state and federal guidelines. State's policies and coverages vary significantly from state to state. Many private insurers don't cover telemedicine, but a few provide the same coverage as face-to-face visits. As of 2017, thirty states mandate that private insurers reimburse for telemedicine services as they do for in-person services.[3] The Center for Connected Health Policy provides maping of coverage and pairity laws across the United States.

The 2017 US Telemedicine Industry Benchmark Survey listed the challenges facing telemedicine. The top four were as follows: Medicare reimbursement, inadequate telemedicine parity laws, Medicaid reimbursement and private payor reimbursement.[101]

Limited research showing reasonable benefit and return on investment. A systematic review of telehealth economics concluded that standard economic evaluation methods were not used therefore the results were not generalizable.[102] A review of 80 systematic reviews of telehealth effectiveness reported 21 were positive, 18 found the evidence promising but limited and 41 reported the evidence is limited and inconsistent.[103] In summary, the studies on telemedicine are mixed and are of low quality. Most studies are based on a small patient population as large randomized controlled trials are expensive. Therefore, results can't be generalized to every population. Similarly, many studies are not conducted over a long period of time, so attrition rates might not be accurately reported. Moreover, there are many flavors of technology (telephone, smartphone, internet, interactive device, etc.) used in telemedicine making comparisons more difficult. It does seem like the addition of a skilled healthcare worker, such as a nurse or pharmacist, is necessary to experience benefits from a telemedicine program. Healthcare organizations that have excellent health IT support as well as disease management teams are the most likely to benefit from telemedicine. In this early stage of telemedicine, the technology by itself does not seem to produce significant benefit.

High cost or the limited availability of high speed telecommunications. Bandwidth issues, particularly in rural areas where telemedicine is most needed. VPN connections slow the process further.

High resolution images or video require significant bandwidth, particularly if x-rays or images or pills must be read by remote clinician. Telepsychiatry may require lower resolution. The following are average file sizes (megabytes): Xray 10 MB, MRI 45 MB, Mammogram 160 MB and 64 slice CT 3,000 MB.[104]

State licensure laws when telemedicine crosses state borders. Some states require participating physicians to have the same state license. In 2011 CMS loosened the requirements for telehealth. The new rule allowed hospitals receiving telehealth services to be privileged and credentialed from the hospital providing telehealth services.[105]

Lack of standards. Currently, there is a paucity of technical standards for telemedicine.

Lack of evaluation by a certifying organization. While there is a lack of certification, organizations such as the American College of Physicians have published guidelines for telemedicine in primary care settings.[106]

Fear of malpractice because of telemedicine. Who is going to evaluate telemonitoring data 24/7?

Ethical and legal challenges. Kluge reviews the challenges faced internationally with telemedicine.[107]

Sustainability is a concern due to an inadequate long-term business plan

Lack of sophistication on the part of the patient, particularly in the elderly and under-educated.

UVA Center for Telehealth

The University of Virginia created this network in 1994 to provide care for rural underserved patients. This network already has more than 40 subspecialists in 85 locations in Virginia. They will now host the new Mid-Atlantic Telehealth Resource Center that will support Delaware, Kentucky, Maryland, North Carolina, the District of Columbia and West Virginia. They are one of 14 Telehealth Resource Centers in the US.[108]

TELEMEDICINE ORGANIZATIONS

Office for the Advancement of Telehealth (OAT): falls under Health Resources and Services Administration (HRSA) that is an agency of the Department of Health and Human Services. Its goal is to promote telemedicine in rural/underserved populations, provide grants, technical assistance and "best practices."[1]

Regional Telehealth Resource Centers: The United States now has 14 telehealth resource centers created to set up a national telehealth network.[60]

American Telemedicine Association (ATA): a non-profit international organization with paid membership that began in 1993. Individual state telemedicine policies are included on their web site. ATA has created a set of telemedicine standards and guidelines covering telemental health, diabetic retinopathy, teleradiology, tele-dermatology, telerehabilitation, telemedicine operations and telepathology. Goals of the ATA are as follows:
- *"Educating government about telemedicine as an essential component in the delivery of modern medical care*
- *Serving as a clearinghouse for telemedicine information and services*
- *Fostering networking and collaboration among interests in medicine and technology*
- *Promoting research and education including the sponsorship of scientific educational meetings and the Telemedicine and e-Health Journal*
- *Spearheading the development of appropriate clinical and industry policies and standards"*[3]

USDA Rural Development Telecommunications Program: The USDA has a program to finance the rural telecommunications infrastructure. In 2007, there were grants and loans totaling $128 million to achieve the goals of broadband access for distant learning and remote medical care. The USDA Rural Development agency has funded several e-ICU programs in the US.[109]

The Agency for Healthcare Research and Quality (AHRQ): AHRQ has funded several telemedicine projects looking at virtual ICUs, telewound projects, cancer management, medication management, heart failure management and others.[110]

FUTURE TRENDS

Telemedicine is a relatively new field created because of the misdistribution of physicians, the need for remote delivery of medical care and the emergence of nascent technologies. Televisits will likely increase if found to be helpful to patients and clinicians for minor illnesses, even if reimbursement lags. Teleconsultation is on the rise worldwide to address access problems for populations at risk: rural, poor, incarcerated, children, elderly and those with multiple chronic diseases. Telemonitoring is complex because it traditionally required sophisticated and expensive technology as well as skilled human intervention to deliver virtual ICU care or home telemedicine. With cell phone cameras, web cams and simple programs such as Skype™ the technology is maturing and more affordable.

KEY POINTS

- Telehealth is a neologism that relates to long distance clinical care, education and administration
- Telemedicine refers to the remote delivery of medical care using technology
- Almost all specialties now have telemedicine initiatives
- Despite the lack of reimbursement, virtual ICUs have gained in popularity because they have perceived benefits
- Telehomecare is a new telehealth initiative that has appeared due to the graying of the US population and the increase in chronic diseases
- Lack of uniform reimbursement, lack of standards and lack of high quality outcome studies have impacted the adoption of telemedicine

---

CONCLUSION

Telemedicine is still in its infancy in the United States and in most areas of the world. The barriers are largely financial due to the high cost to set up the system and the lack of reimbursement in many cases. With the price of telemedicine systems dropping, telemedicine for rural patients is likely more cost-effective than referral to distant urban specialists. If the FCC and ARRA initiatives are successful and/or HIOs flourish, healthcare may have the infrastructure required for telemedicine throughout the United States. Transmission and storage of large images and the ability to compare old and new imaging studies will be greatly aided by LambdaRail and modern web PACS. If future studies prove there is substantial return on investment, then it is a matter of time before more payors support telemedicine. At this time, successful telemedicine programs require an engaged patient and physician, a supportive infrastructure, disease managers and payer reimbursement.

REFERENCES

1. Office for the Advancement of Telehealth http://www.hrsa.gov/telehealth/ (Accessed July 6, 2017)
2. Oh H, Rizo C et al. What is eHealth?: a systematic review of published definitions. J Med Inter Res 2005; 7 (1) e1 http://www.jmir.org/2005/1/e1/ (Accessed September 15, 2013)
3. American Telemedicine Association http://www.atmeda.org (Accessed July 7, 2017)
4. Telemedicine: A guide to assessing telecommunications in health care. Marilyn Field ed. National Academies Press 1996. http://www.nap.edu/catalog.php?record_id=5296 (Accessed September 25, 2013)
5. Puskin DS. HHS Perspective on US Telehealth www.ieeeusa.org/volunteers/committees/mtpc/Saint2001puskin.ppt (Accessed December 6, 2006)
6. St. Joseph Health System introduces telehealth to Orange County, CA. August 8 2012. www.healthcareitnews.com (Accessed August 9, 2012)
7. Miliard M. Sentara to offer telehealth services system wide. August 20 2012. www.healthcareitnews.com (Accessed August 20, 2012)
8. Anderson C. Remote monitoring helps Geisinger cut readmissions. February 29 2012. www.healthcareitnews.com (Accessed February 29, 2012)
9. Comprehensive Care for Joint Replacement. CMS. https://innovation.cms.gov/initiatives/CJR (Accessed July 18, 2017)
10. American Telemedicine Association www.americantelemed.org (Accessed July 7, 2017)
11. Doctor on Demand www.doctorondemand.com (Accessed July 9, 2017)
12. vRad www.virtualrad.com (Accessed July 7, 2017)
13. Silva P, Breslau J, Barr RM et al. ACR White Paper on Teleradiology Practice: A Report from the Task Force on Teleradiology Practice. J ACR 2013;10(8):575-585
14. Saver JL, Fonarow GC, Smith EC et.al. Time to Treatment With Intravenous Tissue Plasminogen Activator and Outcomes From Acute Ischemic Stroke. JAMA 2013;309(23):2480-2488
15. REACH Health www.reachheealth.com (Accessed July 7, 2017)
16. Teleneurology Helps Combat Specialist Shortage, Wait Times. July 17, 2007 www.ihealthbeat.org. (Accessed July 18, 2007)
17. Specialists on call http://specialistsoncall.com/ (Accessed July 8, 2017)
18. ScriptPro Telepharmacy http://www.scriptpro.com/Products/Telepharmacy/ (Accessed September 24, 2013)
19. Pedersen CA, Schneider PJ, Scheckelhoff DJ. ASHP national survey of pharmacy practice in hospital settings: Dispensing and administration—2011. Am J Health Syst Pharm 2012;69(9):768-785
20. North Dakota Telepharmacy Project. www.ndsu.edu/telepharmacy (Accessed July 7, 2017)
21. O'Reilly R, Bishop J, Maddox K et al. Is telepsychiatry equivalent to face to face psychiatry? Results from a randomized equivalence trial. Psychiatr Serv 2007;58(6):836-843
22. American Psychiatry Association http://www.psych.org (Accessed July 7, 2017)
23. Telepsychiatry http://homepsychiatry.com/ (Accessed July 7, 2017)
24. Joch A. Tele-therapy. Government Health IT. www.Healthcareitnews.com November 2008 p.30-31
25. Sarasohn-Kahn J. The Online Couch: Mental Health on the Web. California HealthCare Foundation. June 2012. www.chcf.org (Accessed June 2013)
26. Telemental Health Technologies Compared. www.telementalhealthcomparisons.com (Accessed July 8, 2017)
27. JSA Health Telepsychiatry Services. http://jsahealthmd.com/ (Accessed July 8, 2017)
28. Chan S, Parish M, Yellowlees P. Telepsychiatry Today. Vol. 17, Current Psychiatry Reports. 2015;17(11):89
29. Hubley S, Lynch SB, Schneck C, Thomas M, Shore J. Review of key telepsychiatry outcomes. World J Psychiatry 2016 [cited 2017 Jul 8];6(2):269. http://www.wjgnet.com/2220-3206/full/v6/i2/269.htm (Accessed July 8, 2017)
30. Telederm Project www.telederm.org (Accessed October 25, 2011)
31. iDoc24. www.idoc24.com (Accessed September 24, 2013)
32. Direct Derm. www.directderm.com (Accessed July 8, 2017)
33. Armstrong AW, Lin SW, Liu et al. Store-and-Forward Teledermatology Applications. December 2009. California HealthCare Foundation www.chcf.org (Accessed February 10, 2010)
34. Intelligent Retinal Imaging Systems. http://www.retinalscreenings.com/ (Accessed July 9, 2017)
35. Daskivich LP, Vasquez C, Martinez C, Tseng C-H, Mangione CM, MM K, et al. Implementation and Evaluation of a Large-Scale Teleretinal Diabetic Retinopathy Screening Program in

the Los Angeles County Department of Health Services. JAMA Intern Med [Internet]. American Medical Association; 2017 May 1 [cited 2017 Jul 7];177(5):642. Available from: http://archinte.jamanetwork.com/article.aspx?doi=10.1001/jamainternmed.2017.0204 (Accessed July 5, 2017)
36. Jani PD, Forbes L, Choudhury A, Preisser JS, Viera AJ, Garg S. Evaluation of Diabetic Retinal Screening and Factors for Ophthalmology Referral in a Telemedicine Network. JAMA Ophthalmol [Internet]. SAS Institute; 2017 May 18 [cited 2017 Jul 7];40(3):412–8. Available from: http://archopht.jamanetwork.com/article.aspx?doi=10.1001/jamaophthalmol.2017.1150 (Accessed July 5, 2017)
37. Ellison LM, Nguyen M, Fabrizio MD, Soha A. Postoperative Robotic Telerounding Arch Surg 2007;142(12):1177-1181
38. Robotic Doctor Makes Rounds in Baltimore. www.Ihealthbeat.org Feb 27, 2006. (Accessed March 10, 2006)
39. InTouch Health. www.intouchhealth.com (Accessed July 8, 2017)
40. Leapfrog. www.leapfroggroup.org (Accessed July 8, 2017)
41. Breslow MJ. Effect of a multiple-site intensive care unit telemedicine program on clinical and economic outcomes: An alternative paradigm for intensivist staffing. Crit Care Med 2004;32:31-38
42. Phillips eICU http://www.usa.philips.com/healthcare/product/HCNOCTN503/eicu-program-telehealth-for-the-intensive-care-unit (Accessed July 8, 2017)
43. Kumar G, Falk DM, Bonello RS et al. The Costs of Critical Care Telemedicine Programs: A Systematic Review and Analysis. Chest. 2013;143(1):19-29
44. Berenson RA, Grossman JM, November EA. Does Telemonitoring of Patients—The eICU—Improve Intensive Care? Health Affairs 2009;28(5):w937-947
45. Thomas EJ, Lucke JF, Wueste L et al. Association of Telemedicine for Remote Monitoring of Intensive Care Patients With Mortality, Complications and Length of Stay. JAMA 2009;302(24):2671-2678
46. Young LB, Chan PS, Lu X. Impact of telemedicine intensive care unit coverage on patient outcomes. Arch Intern Med 2011;171(6):498-506
47. Lilly CM, Cody S, Zhao H et al. Hospital Mortality, Length of Stay and Preventable Complications Among Critically Ill Patients Before and After Tele-ICU Reengineering of Critical Care Processes. JAMA 2011;305(1):2175-2228
48. Lilly CM, McLaughlin JM, Zhao H et al. A Multi-Center Study of ICU Telemedicine Reengineering of Adult Critical Care. Chest. 2013. Doi:10.1378/chest.13-1973
49. Hawkins CL. Virtual Rapid Response. The Next Evolution of the Tele-ICU. AACN Advanced Critical Care. 2012;23(3):337-340
50. Sarashon-Kahn, J. The Connected Patient. California Healthcare Foundation. February 2011. www.chcf.org (Accessed September 10, 2011)
51. Healthcare Staffing Growth Assessment. Staffing Industry Strategic Research. June 2005 http://media.monster.com/a/i/ intelligence/pdf (Accessed September 25, 2006)
52. Medicare Medical Home Demonstration http://www.cms.gov/Medicare/Demonstration-Projects/DemoProjectsEvalRpts/ Medicare-Demonstrations-Items/ CMS1199247.html (Accessed September 25, 2013)
53. Affordable Care Act http://www.healthcare. gov/law/index.html (Accessed September 25, 2013)
54. Health Buddy® http://www.bosch-telehealth.com/en/us/products/health_buddy/health_buddy.html (Accessed September 25, 2013)
55. Baker LC, Johnson SJ, Macaulay D, Birnbaum H. Integrated Telehealth and Care Management Program for Medicare Beneficiaries with Chronic Disease Linked to Savings. Health Affairs. 2011; 30 (9), 1689-1697.
56. Honeywell Life Care Solutions www.honeywelllifecare.com (July 8, 2017)
57. MyCareTeam www.mycareteam.com (Accessed July 8, 2017)
58. WellDoc. www.welldoc.com (Accessed July 8, 2017)
59. Care Innovations www.careinnovations.com (Accessed July 8, 2017)
60. Telehealth Resource Centers https://www.telehealthresourcecenter.org/ (Accessed May 5, 2017)
61. Center for Connected Health Policy http://www.cchpca.org/ (Accessed May 5, 2017)
62. IDEATel www.ideatel.org (Accessed September 10, 2007) No longer active.
63. Georgia Partnership for Telehealth. http://www.gatelehealth.org/ (Accessed July 8, 2017)
64. Brewer R, Goble G, Guy P. A Peach of a Telehealth Program: Georgia Connects Rural Communities to Better Healthcare. Winter 2011. Perspectives in Health Information Management. www.perspectives.ahima.org (Accessed October 20, 2011)
65. Slabodkin G. VA to expand telehealth services for veterans nationwide. August 4, 2017. www.healthdatamanagement.com (Accessed August 4, 2017)
66. University of Texas Medical Branch Center for Telehealth Research and Policy. http://telehealth.utmb.edu/index.html (Accessed July 8, 2017)
67. Massachusetts General Hospital Teleburn Program. http://www.massgeneral.org/telehealth/teleburns.

67. aspx http://otn.ca/en/programs/teleburn (Accessed July 8, 2017)
68. Inova Telepediatric Program https://www.inova.org/inova-telemedicine-program/telepediatrics (Accessed July 8, 2017)
69. Mercy Virtual Care Center. http://www.mercyvirtual.net/ (Accessed July 9, 2017)
70. University of the Pacific. http://dental.pacific.edu/departments-and-groups/pacific-center-for-special-care/innovations-center/virtual-dental-home-system-of-care (Accessed July 9, 2017)
71. Diabetic Retinopathy Screening. January 2011. http://www.chcf.org/projects/2009/ diabetic-retinopathy-screening (Accessed September 25, 2013)
72. EyePACS. www.eyepacs.org. (Accessed July 9, 2017)
73. Northwest Telehealth https://www.inhs.info/sub.aspx?id=2244 (Accessed July 9, 2017)
74. Federal Communications Commission www.fcc.gov Accessed July 9, 2017)
75. National Broadband Plan. www.broadband.gov (Accessed July 9, 2017)
76. Landi H. Senators Introduce Bill to Expand Rural Telehealth Services. June 23, 2017. https://www.healthcare-informatics.com/news-item/telemedicine/senators-introduce-bill-expand-rural-telehealth-services (Accessed July 16, 2017)
77. Rheuban KS. Expanding broadband will improve America's health. Richmond Times Dispatch. June 24, 2017 http://www.richmond.com/opinion/their-opinion/guest-columnists/dr-karen-s-rheuban-expanding-broadband-will-improve-america-s/article_1da623a0-731e-5b89-9fb1-326e72eb7bae.html (Accessed July 16, 2017)
78. What is Telehealth? Health IT.gov https://www.healthit.gov/telehealth (Accessed June 20, 2017)
79. UPMC Telehealth Adoption Model -A HIMSS Case Study November 10, 2015. mHealth Summit. HIMSS Connected Health. Natasa Sokolovich and William Fera http://exhibitionfloor.himss.org/mhealth2015/Custom/Handout/Speaker0_Session529_2.pdf (Accessed July 16, 2017)
80. International Society for Telemedicine and eHealth. http://www.isfteh.org/ (Accessed July 9, 2017)
81. Combi C, Pozzani G, Pozzi G. Telemedicine for Developing Countries. Appl Clin Inform [Internet]. 2016;7(4):1025–50. Available from: http://www.schattauer.de/index.php?id=1214&doi=10.4338/ACI-2016-06-R-0089 (Accessed July 15, 2017)
82. Whole System Demonstrator Project http://www.dh.gov.uk/en/Publicationsandstatistics/Publications/PublicationsPolicyAndGuidance/DH_131684 (Accessed July 9, 2017)
83. Steventon A, Bardsley M, Billings J et al. Effect of telehealth on use of secondary care and mortality: findings from the Whole Systems Demonstrator cluster randomized trial. BMJ. 2012;344:e3874
84. Henderson C, Knapp M, Fernandez J-L et al. Cost effectiveness of telehealth for patients with long term conditions (Whole Systems Demonstration Telehealth Questionnaire Study): nested economic evaluation in a pragmatic, cluster randomized controlled trial. BMJ. 2013;346:f1035
85. Hanlon P, Daines L, Campbell C, McKinstry B, Weller D, Pinnock H. Telehealth Interventions to Support Self-Management of Long-Term Conditions: A Systematic Metareview of Diabetes, Heart Failure, Asthma, Chronic Obstructive Pulmonary Disease, and Cancer. J Med Internet Res 2017 May 17;19(5):e172. http://www.jmir.org/2017/5/e172/ (Accessed July 17, 2017)
86. Darkins A, Ryan P, Kobb R et al. Care Coordination/Home Telehealth: the Systematic Implementation of Health Informatics, Home Telehealth and Disease Management to Support the Care of Veteran Patients with Chronic Conditions. Telemedicine and e-Health. 2008;14(10):1118-1126
87. Shea S. The Informatics for Diabetes and Education Telemedicine (IDEATEL) Project. Trans Amer Clin Clim Assoc 2007;118:289-300
88. Shea S, Weinstock RS, Teresi JA et al. A Randomized Trial Comparing Telemedicine Case Management with Usual Care in Older, Ethnically Diverse, Medically Underserved Patients with Diabetes Mellitus: 5 Year Results of the IDEATel Study. JAMIA 2009;16:446-456
89. Palmas W, Shea S, Starren J et al. Medicare payments, healthcare service use and telemedicine implementation costs in a randomized trial comparing telemedicine case management with usual care in medical underserved participants with diabetes. J Am Med Inform Assoc 2010;17:196-202
90. Meyer BC, Raman R, Hemmen T et al. Efficacy of site-independent telemedicine in the STRokEDOC trial: a randomized, blinded, prospective study. August 3 2008 www.thelancet.com/neurology e-publication (Accessed August 10, 2008)
91. Demaerschalk BM, Miley ML, Kiernan TJ et al. Stroke Telemedicine. Mayo Clin Proc 2009;84(1):53-64
92. Berthoid J. Help from afar: telemedicine vs. telephone advice for stroke. ACP Internist April 2009 p. 19
93. Klersy C, De Silvestri A, Gabutti G et al. A meta-analysis of remote monitoring of heart failure patients. J Am Coll Cardiol 2009;54:1683-94
94. Inglis S. Structured Telephone Support or Telemonitoring Programmes for Patients with Chronic Heart Failure. Cochrane Database

of Systematic Reviews 2010; no.8, doi 10.1002/14651858.CD007228.pub 2 (Accessed September 20, 2011)
95. Chaudhry SI, Mattera JA, Curtis JP et al. Telemonitoring in Patients with Heart Failure. NEJM November 16 2010. Doi 10.1056/NEJMoa1010029 (Accessed December 4, 2010)
96. Mohr DC, Ho J, Duffecy J et al. Effect of telephone-administered vs face-to-face cognitive behavioral therapy on adherence to therapy and depression outcomes among primary care patients: a randomized trial. JAMA 2012;307(21): 2278-85
97. Vargas BB, Channer DD, Dodick DW et al. Teleconcussion: An innovative approach to screening, diagnosis and management of mild traumatic brain injury. Telemed eHealth. 2012;18(10):803-806.
98. Takahashi PY, Pecina JL, Upatising B et al. A Randomized Controlled Trial of Telemonitoring in Older Adults with Multiple Health Issues to Prevent Hospitalizations and Emergency Department Visits. Arch Int Med 2012;172(10):773-779
99. Dharmar M, Romano PS, Kuppermann N et al. Impact of Critical Care Telemedicine Consultations on Children in Rural Emergency Departments Crit Care Med 2013. Doi:10.1097/CCM.0b013e31828e9824 Aug 7, 2013
100. Hanley PH, McCloughan L et al. Effectiveness of Telemonitoring Integrated Into Existing Clinical Services On Hospital Admissions for Exacerbation of Chronic Obstructive Pulmonary Disease: Researcher Blind, Multicentre, Randomized Controlled Trial. BMJ 2013; http://www.bmj.com/content/347/bmj.f6070 (Accessed October 31, 2013)
101. REACHHEALTH. April 2017. 2017 U. S. Telemedicine Industry Benchmark Survey. www.reachhealth.com (Accessed July 6, 2017)
102. Bergmo TS. Can economic evaluation in telemedicine be trusted? A systematic review of the literature. BiomedCentral October 2009 (Accessed November 22, 2010)
103. Ekeland AG, Bowes A, Flottorp S. Effectiveness of Telemedicine: A Systematic Review of Reviews. Int J Med Inform 2010;79:736-771
104. Health Care Broadband in America. www.broadband.gov (Accessed January 3, 2011)
105. AAMI. CMS telemedicine restrictions eased. http://www.aami.org/news/2011/ 051211.cms.html (Accessed June 4, 2011)
106. Daniel H, Sulmasy LS, et al. Policy Recommendations to Guide the Use of Telemedicine in Primary Care Settings: An American College of Physicians Position Paper. Ann Intern Med 2015 Nov 17 ;163(10):787. http://annals.org/article.aspx?doi=10.7326/M15-0498 (Accessed July 9, 2017)
107. Kluge EW. Ethical and legal challenges for health telematics in a global world: Telehealth and the technological imperative. IJMI.2011; 80 (2):e1-e5
108. University of Virginia Center for Telehealth https://uvahealth.com/services/telemedicine-telehealth-services (Accessed October 15, 2013)
109. USDA Rural Development. www.rd.usda.gov (Accessed July 8, 2017)
110. Agency for Healthcare Quality and Research. www.ahrq.gov (Accessed July 9, 2017)

18

Bioinformatics

ROBERT E. HOYT • WILLIAM R. HERSH • INDRA NEIL SARKAR

LEARNING OBJECTIVES

After reading this chapter the reader should be able to:

- Define bioinformatics, translational bioinformatics, and other bioinformatics-related terms
- State the importance of bioinformatics in future medical treatments and prevention
- Describe genomics and its important implications for health care
- List major private and governmental bioinformatics databases and projects
- Enumerate several bioinformatics projects that involve electronic health records
- Describe the application of bioinformatics in genetic profiling of individuals and large populations

INTRODUCTION

This chapter is focused on "bioinformatics," the study of data and information as it relates to knowledge within the context of the life sciences. Bioinformatics traces its formal beginning to 1970, when the term was first introduced in scientific literature.[1] In many ways, bioinformatics has evolved in parallel with health informatics. Significant advances in bioinformatics have given rise to contemplation of its applications within the context of biomedicine and health (the sub-discipline of biomedical informatics referred to as "*translational bioinformatics*.")

Definitions

The chapter begins with common definitions and the next section provides a short primer on genomics, which underpins many of the concepts used for bioinformatics within the context of health.

Bioinformatics has been defined as, "*the field of science in which biology, computer science and information technology merge to form a single discipline.*"[2] Bioinformatics makes use of fundamental aspects of computer science (such as databases and artificial intelligence) to develop algorithms for facilitating the development and testing of biological hypotheses, such as: finding the genes of various organisms, predicting the structure or function of newly developed proteins, developing protein models and examining evolutionary relationships.[3-4] A related term is *computational biology*, which refers to the computational aspects of molecular biology. *Translational bioinformatics* focuses on the "*development of storage, analytic and interpretive methods to optimize the transformation of increasingly voluminous biomedical data into proactive, predictive, preventive and participatory health.*"[5] Simply put, translational bioinformatics is the specialization of bioinformatics for human health.

Bioinformatics is sometimes said to work with the various "omes" and "omics." They include:

- Genomics - the study of genetic material in an organism (e.g., the genes that may be associated with a disease).
- Proteomics - the study at the level of proteins (e.g., through the components, structure, and functions).
- Pharmacogenomics - the study of genetic material in relationship to drug targets.
- Metabolomics - the study of genes, proteins or metabolites.
- Interactome - biomolecular pathways and interactions of proteins
- Microbiome - microorganisms inhabiting an individual
- Exposome – environmental factors to which an organism is exposed
- Bibliome – the literature of science

- Metagenomics - the analysis of genetic material derived from complete microbial communities harvested from natural environments.[6]

In addition, bioinformatics studies the relationship between the *genotype*, which is the genetic information that is associated with biological function[7] and the *phenotype*, which is the observable characteristic, structure, function and behavior of a living organism. Examples of phenotypes include hair color, height, and development of diseases. The phenome refers to the total phenotypic traits.

GENOMIC PRIMER

The human body has about 100 trillion cells and each one contains a complete set of genetic information (chromosomes) in the nucleus; exceptions are eggs, sperm, and red blood cells. Humans have a pair of 23 chromosomes in each cell that includes an X and Y chromosome for males and two Xs for females. Offspring inherit one pair from each parent. Chromosomes are listed approximately by size with chromosome 1 being the largest and chromosome 22 the smallest. Organisms have differing numbers of chromosomes (e.g., our closest extant primate relatives, chimpanzees, have 24 pairs). Chromosomes consist of double twisted helices of deoxyribonucleic acid (DNA). DNA is composed of four sugar-based building blocks ("nucleotides": adenine [A], thymine [T], cytosine [C], and guanine [G]) that are generally found in pairs (following "Watson-Crick" pairing templates: A-T, C-G). DNA is often referred to as the "*blueprint for life.*" As such, a given organism's DNA encodes its full complement of proteins essential for cellular function. Some of the encoding of DNA also enables it to control the expression of proteins or affect how other portions of DNA may be decoded based on a biological context (e.g., to accommodate for faulty DNA decoding or DNA damage that may be encountered due to environmental phenomena). Genes are regions on chromosomes that encode instructions, which may result in proteins that then in turn enable biological functions. The process of decoding genes involves transcribing the DNA into ribonucleic acid (RNA) and then translation into amino acids that form the building blocks for proteins (Figure 18.1). Collectively, the complete set of genes is referred to as the "genome" (based on the combination of the terms "gene" and "chromosome").

It is estimated that humans have between 20,000 and 30,000 genes and that genomes are about 99.9% similar between individuals. Variations in genomes between individuals are known as single nucleotide polymorphisms (SNPs) (pronounced "snips"). There are three general types of alterations: single base-pair changes, insertions or deletions of nucleotides, and reshuffled DNA sequences. As an example, one individual might have a chromosome with the sequence TGGC, while another might have the sequence TAGC. Each of these is referred to as an allele. Although SNPs are common, their significance is complex and difficult to decipher.[8-10]

Another type of genetic variation is copy number variations (CNVs). These are repeats of DNA sequences of 50 nucleotides or longer. There may be many CNVs in an individual's genome. Anywhere from 4.8% to 9.5% of the human genome is CNVs. Some of these copies of genomes are deleterious, others are not. When they are deleterious, they are so-called unbalanced rearrangements, involving either loss or gain of segments of the genome.[11]

An additional cause of genetic variation is *epigenetics*, which is the variation in the phenotype or gene expression that is caused by mechanisms other than DNA sequence differences.[12] The molecular mechanisms for epigenetics are beginning to be unraveled, such as DNA methylation.[13] Epigenetics shows that there is influence of the environment on the expression of genes, and therefore leading to genetic variation.

A great deal of progress has been made with genetic testing and our understanding of the human genome and genetic variations. Genome-wide associations studies (GWAS) look at associations between genomic variants and traits of the phenotype.[14] The variations or SNPs discovered are said to be associated with the disease, but true cause and effect cannot be ascertained.[15] Similarly, phenome-wide association studies (PheWAS) are being carried out comparing genes to disease associations, most recently using the electronic health record for phenotypical information.[16]

Genetic material can be obtained from blood, saliva, skin and hair samples. Full genome sequencing has historically been an expensive, time-consuming, and complicated process, although the cost of full genome sequencing has dropped to approximately $1000 (US). SNP-based genomic profiling is available now for less than $200 US (e.g., 23andMe). This cost differential is largely because SNP genotyping analyzes about 0.1 – 0.2% of the genome in contrast to every single nucleotide. Even more cost effective is ultra-low-coverage (ULC) sequencing techniques that analyze the same 10-20% of the genome and cost $60 US. SNP-based genotyping has more specificity since it uses a technique the seeks to identify SNPs from a library of *a priori* selected SNPs of interest (using a technology called "microarrays");

by contrast ULC techniques may be more sensitive but may be difficult to ensure reproducibility. Therefore, ULC techniques may serve the role of SNP discovery, whereas SNP-based genotyping can serve the role of SNP identification.

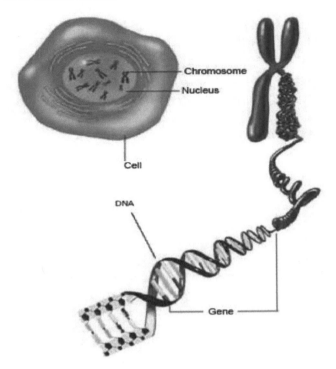

Figure 18.1: Genes (Courtesy of National Institute of General Medical Sciences, National Institutes of Health)

Often, we do not want to know just the genome sequence but the amount that the genes are actually expressed.[17] Some call the genome itself a "parts list," whereas what we want to know is how much those parts are used. One of the early techniques for measuring gene expression was the gene microarray. These are chips that allow measurement of many different biological substances, not only DNA and RNA, but also protein. When nucleotides are measured there are 25-base strands. Since most genes are longer than 25 bases, there are more than one spot per gene. And in fact, a typical microarray chip has up to 1 to 2 million of these spots, so many different genes or transcriptions of RNA of those genes can be measured on microarrays. The technology has now become very easy to mass produce. The initial application of microarrays was to measure gene expression, i.e., the mRNA that had been transcribed. But other applications of microarrays have emerged, including the sequencing of unknown DNA. Microarrays are being supplanted by a newer technology called RNA-seq, which measures mRNA more accurately.

IMPORTANCE OF TRANSLATIONAL BIOINFORMATICS

Besides diagnosing the 3,000 to 4,000 hereditary diseases that are currently known, bioinformatics techniques may be helpful to discover future drugs targets, develop personalized drugs based on genetic profiles and develop gene therapies to treat diseases with a strong genomic component, such as cancer. One approach that has been explored to enable gene therapies involves the use of genetically altered viruses that carry human DNA. This approach, however, has not been definitely shown to work and has not been for general use by the FDA. Recent years has seen significant advances in genome editing using a technology called CRISPR/Cas9 gene editing. However, while manipulation of genomes in other organisms, such as microbes, has shown promise for energy production ("bio-fuels"), environmental cleanup, industrial processing, and waste reduction, clinical applications are still nascent in their design and approval. Nonetheless, the advent of technologies, such as CRISPR/Cas9, suggest that genome editing will become a viable path for treatment of genetically based conditions within the next decade.[18]

This chapter will deal primarily with translational bioinformatics (TBI), the identified area of focus in bioinformatics that is focused primarily on the use of bioinformatics approaches to address challenges in biomedicine and health. A significant goal of TBI is to enable bi-directional crossing of the translational barrier between the research bench and the bed in the medical clinic. With growing genome-wide and population-based research data sets, more genotype-phenotype associations are being uncovered that potentially can detect and treat diseases with a genetic component earlier. Such associations may also help create tailor made drugs for higher efficacy. Figure 18.2 demonstrates the bidirectional nature of data and information flow between bioinformatics and health informatics. The emergence

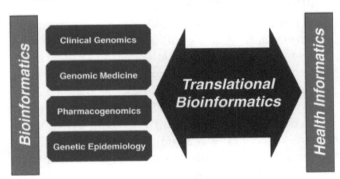

Figure 18.2: Translational bioinformatics (Adapted from Sarkar et al)[19]

of translational bioinformatics is primarily due to the rapid advances in both sub-disciplines, and the realization of the potential to leverage biological data within the context of clinical care. In other words, a variety of advances in bioinformatics, such as faster and cheaper DNA sequencing, and more widespread adoption of electronic health records have made this possible.

Pharmacogenomics is an illustrative example of how translational bioinformatics can be used within the context of pharmaceutical development to make use of genomic information for better drug discovery and utilization. Drug companies are faced with the huge expense of drug development, the long road to producing a new drug and expiring patents. Drug failures are common and can be due to complex combination of a lack of clinical efficacy, side effects and commercial issues. Unfortunately, animal models are often inadequate for the development and evaluation of drugs for treating human conditions. It is thus the goal to use genetic information for:
- New indications for an old drug (repurposing)
- New targets for existing drugs (e.g., treatment of tongue cancer using RET inhibitors)
- Drugs to work better in certain patient groups (gender, age, race, ethnicity, etc.) with possible genetic variants
- Knowing ahead of time what drugs to avoid due to higher incidence of side effects that are genetically modulated
- Develop clinical decision support in electronic health records based on pharmacogenomics [20-21]

Multiple projects are underway to integrate genetic and clinical data that will be discussed later in the chapter. Electronic health records (EHRs) and health information exchange (HIE) efforts, which are rapidly becoming ubiquitous, thanks in large part to federal incentives, are poised to contribute massive amounts of patient information (including demographic, laboratory, and clinical data). It is important to also note that in addition to genomic and clinical data, environmental data may offer valuable insights into the understanding and eventual treatment of disease.

Another important application of translational bioinformatics is in cancer genomics. There have been so-called hallmarks papers that describe the state of the knowledge on how cancer cells proliferate and evade death by a number of mechanisms within living organisms.[22] When whole genome sequencing is applied to tumor cells, it shows that they undergo genomic changes that give them the ability to proliferate and metastasize within living organisms.[23] It has also been discovered that cancer gene mutations, within tumors, are heterogeneous, so cancer cells evolve and attain this ability to proliferate outside the normal defense mechanisms of organisms.[24] Another important discovery is that some genomic changes representing common tumor types occur in different cancer locations, so that the same mechanism may occur in more than one type of cancer.[25-26] There has also been improved understanding of gene function, and of course this may aid in the development of better treatments.[27]

BIOINFORMATICS PROJECTS AND CENTERS

The Human Genome Project

One of the greatest accomplishments in biomedicine was the completion of the Human Genome Project (HGP). This international collaborative project, sponsored by the US Department of Energy and the National Institutes of Health, was started in 1990 and finished in 2003. In the process of acquiring the human genome (as a complete set of DNA sequences, encompassing all 23 chromosomes), genome sequences for several other key organisms ("model" organisms) were also acquired. These included the *Escherichia coli* bacterium, fruit fly (*Drosophila melanogaster*), and house mouse (*Mus musculus)*. By mid-2007 about three million differences (SNPs) had been identified in human genomes. Appreciating the potential significant societal impact, the HGP also addressed the ethical, legal and social issues associated with the project. Since the completion of the HGP, attention is now more focused on the development of approaches to analyze and learn from volumes of data representing increasing numbers of individuals.[11,28-29] These analyses include the annotation of information associated with disease onto chromosomes. Figure 18.3 displays the DNA sequencing of just chromosome number 12. Huge relational databases are necessary to store and retrieve this information. New technologies continue to emerge that reduce the necessity to sequence an entire human genome, such as DNA arrays (gene chips) that help speed the analysis and comparison of DNA fragments.[30] The cost of the HGP was close to $3 billion; but over time, costs have dramatically dropped for genetic analysis.[7]

National Human Genome Research Institute (NHGRI)

NHGRI is an NIH institute that provides many educational resources on their web site. Like other NIH institutes, they conduct and fund research within their intramural division, as well as support extramural research with external partners. Their health section has

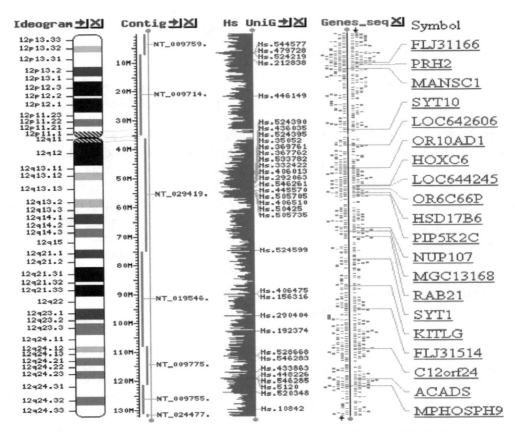

Figure 18.3: Chromosome 12 (Courtesy of the National Library of Medicine)

multiple resources for patients and healthcare professionals with emphasis on the Human Genome Project. The "Issues in Genetics" section covers important controversies in policy, legal and ethical issues in genetic research. They include a large glossary (200+) of genetics-related definitions, also available as a software app for the iPhone and iPad.

In 2003, NHGRI launched the Encyclopedia of DNA Elements (ENCODE) Project. ENCODE is comprised of a consortium of laboratories with the goal to study and characterize the functional elements of the human genome. All ENCODE data are free for research purposes. In 2012, 1640 data sets were published, which continue to produce controversy. For example, ENCODE researchers posited that 80% of the human genome is active and performing a role (and thus not "junk" DNA as has been previously thought).[31]

Human Microbiome Project (HMP)

It is estimated that less than 0.01% of microbes on Earth have been cultured, characterized, and sequenced. As an exception, the complete genome for the common human parasite *Trichomonas vaginalis* was reported in 2007 in the journal *Science*.[32] The HMP is an NIH-sponsored initiative that catalogued the myriad of organisms that co-exist with humans and heretofore have been rarely studied (e.g., flora from oral, nasal, skin, and the gastrointestinal tract). It is important to note that microbial cells on the human body outnumber human cells by a factor of 10 to 1. Initial efforts were aimed at identifying the microbiome in health patients. More recently, extensive work has been done to identify the microbiome in multiple disease states with results too comprehensive to cite in this chapter. Three areas the HMP is currently focusing on include pregnancy and pre-term birth, onset of inflammatory bowel disease and onset of type 2 diabetes.

The HMP used metagenomics, as explained in the definitions section. As detailed on the HMP web site their goals were as follows:
- Determine whether individuals share a core human microbiome
- Understand whether changes in the human microbiome can be correlated with changes in human health
- Develop new technological and bioinformatics tools needed to support these goals
- Address the ethical, legal and social implications raised by human microbiome research[6]

Human Variome Project

This is an international non-governmental organization that began in 2006 with the goal to create systems and standards for storage, transmission and use of genetic variations to improve health. Rather than catalogue "normal" genomes they focus on the abnormalities that cause disease. Another aspect of their vision is to provide free public access to their databases.[33]

The PhenX Project

The goal of this project is to identify 15 high-quality, well established measures and standards for each of 21 research domains. Standardization is important so that phenotypical, risk factors and environmental exposures can be compared. For example, if everyone used a common set of standards data could be more readily compared or combined to gain more statistical power.[34-35]

1000 Genomes Project

This is an international initiative with the goal to catalogue and study the genomes of 2500 individuals from 26 populations, looking for genetic variations that occur at a frequency of about 1%. The data will be free for researchers and hosted on Amazon's Web Services. The project produced 200 terabytes of data during its active phase from 2008 to 2015.[36]

Pediatric Cancer Genome Project

St. Judes Children's Hospital-Washington University created this initiative to combat childhood cancer. Data generated from 800 subjects will be offered free to researchers. As an example of a positive result from this project, two gene variations associated with 50% of low grade gliomas (brain tumors) were identified.[37]

Global Alliance for Genomics and Health

In June 2013, a Global Alliance was formed to share genetic and clinical information. The Alliance was formed by 500 international medical and research organizations. They develop standards for sharing genetic information, information technology platforms with open standards and patient consent policies.[38]

Pharmacogenomics Knowledge Base (PharmGKB)

This Stanford University based resource catalogues the relationships between genes, diseases and drugs. There are sections on drugs, pathways, dosing guidelines and drug labels. Information is downloadable.[39]

Framingham Heart Study SHARe Genome-Wide Association Study

In 2007, the Framingham Heart Study began a new phase by genotyping 17,000+ subjects as part of the FHS SHARe (SNP Health Association Resource) project. The SHARe database is located at NCBI's dbGaP and will contain 550,000 SNPs and a vast array of phenotypical (combined characteristics of the genome and environment) information available in all three generations of FHS subjects. These will include measures of the major risk factors such as systolic blood pressure, total, LDL and HDL cholesterol, fasting glucose, and cigarette use, as well as anthropomorphic measures such as body mass index, biomarkers such as fibrinogen and C-reactive protein (CRP) and electrocardiography (EKG) measures such as the QT interval. Because of this initiative, they have been able to publish multiple articles on genetic associations and heart disease.[40]

The Mayo Clinic Bipolar Disorder Biobank

Researchers at the Mayo clinic and other institutions are analyzing the genetic and clinical information on 2000 patients in their biobank to determine genetic aspects of bipolar disorder. It is hoped that data generated from this project will lead to earlier and better treatment of this mental health disorder.[41]

Informatics for Integrating Biology and the Bedside (i2b2)

i2b2 is a National Institutes of Health National Center for Biomedical Computing initiative located at Harvard Medical School. The Center has developed open source software that will enable investigators to mine existing clinical data for research. Specifically, they will develop a scalable computational framework to speed up translation of genetic findings into healthcare. There are multiple member institutions, including international ones. The project was designed to allow users to query a system-wide de-identified repository for a set of patients meeting certain inclusion or exclusion criteria. On the web site, users can download client-software, client-server software and the source code.[42] The i2b2 infrastructure has been shown to be generalizable to multiple sites for a range of clinical conditions.[43]

The Observational Health Data Sciences and Informatics (OHDSI)

OHDSI is a multi-stakeholder, international collaboration that aims to make the best use of available large health data sets. Health data exist in a variety of file formats and often require standardization for both use and supporting data integration tasks. From the perspective of translational bioinformatics, OHDSI represents a major initiative that provides datasets that can be integrated with biological data. For example, within the context of studying drugs, it is useful to know reported adverse events. While the US FDA does make available data collected from its adverse event reporting system (FAERS), these data are challenging to use for large scale analyses. AEOLUS is an artifact of an OHDSI initiative (LAERTES) that standardizes FAERS data such that it can be used within a range of studies, including those involving translational bioinformatics.[44]

All of Us

Conceptualized at the end of the Obama administration, the All of Us program (which is a key element of the Precision Medicine Initiative) is a White House initiative to develop a US cohort of individuals that will support the development and evaluation of next-generation, personalized treatments. The All of Us program is built around the fundamental tenets of precision medicine, i.e., the right treatment for the right patient at the right time. In this way, the success of this endeavor, which is just formally launching at the writing of this text, will result in a paradigm shift in how biological data can be used to inform health care in a meaningful and efficient manner.[45]

The Cancer Genome Atlas (TCGA)

The suite of conditions that comprise cancer have a strong genomic component. The TCGA project is a joint initiative between the National Human Genome Research Institute and the National Cancer Institute, both of the National Institutes of Health, that aims to provide one of the largest collections of cancer-related genomic data. Currently, the initiative focuses on 33 cancers and the TCGA provides access to genomic data associated with tumor samples gathered from more than 11,000 patients.[46]

Bioinformatics Data and Information Resources

The bioinformatics community has produced a wide variety of knowledge-based information resources that not only provide access to research results but also facilitate scientific discovery information from genomics. Much of this activity is led by the National Center for Biotechnology Information (NCBI), which is part of the National Library of Medicine. A catalog of NCBI and other resources is provided annually in the open-access journal *Nucleic Acids Research*. In fact, there is an annual database issue that covers the myriad of databases available. And there's also a web server issue that covers accessible resources that can be accessed over the Internet. NCBI was created in 1988 and hosts dozens of databases associated with biomedicine, including the popular MEDLINE and GenBank databases. NCBI provides access to sequences from over 500,000 organisms (via GenBank), including the complete genomes of thousands of organisms (via NCBI Genome). Genomes represent both completely sequenced organisms and those for which sequencing is still in progress. Popular NCBI databases, which are linked by a common interface (Entrez), are listed in Figure 18.4. On the Genome project web site one can search for specific genes or proteins from different species. Figure 18.5 shows the result of a search for the tumor protein TP53.

The NCBI site also provides access to BLAST+ (new Basic Local Alignment Search Tool) that enables the identification of significantly related (based on a "expectation" value or "e-value") nucleotide or protein sequences from within the protein and nucleotide databases.[47] Magic BLAST is a recently developed tool that enables rapid searching for related sequences (e.g., as might arise from a full sequence of a human being) to a reference genome.[48] Also available on the NCBI site are databases where one can find data about a gene (Gene), its location on the chromosome (MapViewer), its variants (dbVar), and data about patients who have the gene or its variants (dbGaP).

GenBank

This database was established in 1982 and is the NIH sequence database that is a collection of all publicly available DNA sequences. Along with EMBL (Europe) and DDBJ (Asia), GenBank is a member of the International Nucleotide Sequence Database Consortium (INSDC), which provides free access to sequence data from nearly anywhere with an Internet connection. Interestingly, many biological and medical journals now require submission of sequences to a database prior to publication, which can be done with NCBI tools such as BankIt.[49]

The Online Mendelian Inheritance in Man (OMIM)

Originally an NCBI database but now a standalone resource, Online Mendelian Inheritance in Man (OMIM) is a database of genetic data and human genetic disorders. It was originally developed by Johns Hopkins University and Dr. Victor McKusick, a pioneer in genetic metabolic abnormalities. It includes an extensive reference section linked to PubMed that is continuously updated.[50]

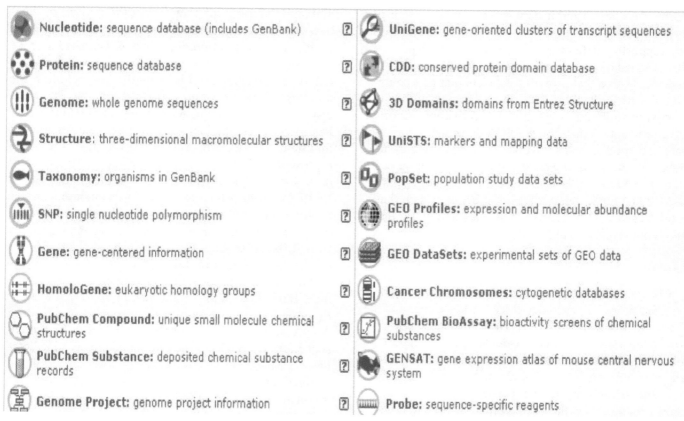

Figure 18.4: NCBI Databases (Courtesy National Library of Medicine)

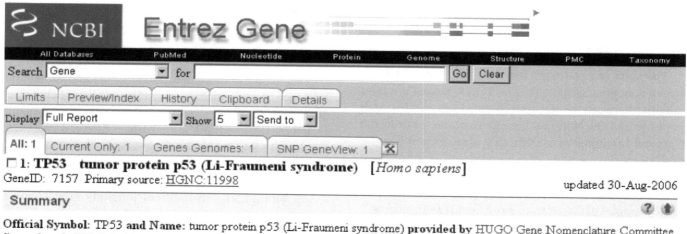

Figure 18.5: Entrez search for tumor protein (Courtesy National Library of Medicine)

World Community Grid

This project was launched by IBM in 2004 and simply asked people to donate idle computer time. By 2007 over 500,000 computers were involved in creating a super-computer used in bioinformatics. Projects include Help defeat Cancer, Fight AIDS@Home, Genome Comparison and Human Proteome Folding projects. This grid promises to greatly expedite biomedical research by analyzing complex databases more rapidly because of this grid.[51]

PERSONAL GENOMICS

The availability of population-based genetic data, the decreasing cost for human genome determination and the availability of commercial personal genetic testing companies provide greater personal uses of genomic data.

Population Studies: There are several ongoing initiatives that will leverage genomic data in the context of population studies. For instance, Oracle Corporation has partnered with the government of Thailand to develop a database to store medical and genetic records. This initiative was undertaken to offer individualized "tailor made" medications and to offer bio-surveillance for future outbreaks of infectious diseases such as avian influenza.[52] Not all such initiatives have been successful. Perhaps the best known is DeCODE Genetics Corporation, which aimed to collect disease, genetic and genealogical data for the entire population of Iceland; however, it filed for chapter 11 bankruptcy in 2009.[53] Nonetheless, DeCODE continues the development of personal genomics based solutions, largely in partnership with organizations such as Pfizer.

Decreasing Cost of Human Genome Determination: Coinciding with the completion of the HGP, the NHGRI has kept track of the cost to perform DNA sequencing of an entire human genome over the past decade. As Figure 18.6 indicates the cost has dropped from an initial cost of $100,000,000 to a current cost of less than $1,000 per genome in 2017. Notably, the decrease in cost of genome sequence is exceeding Moore's Law (attributed to Intel co-founder Gordon Moore, and states that the cost of computing power will be halved every 18 months based on advances in technology).[11]

Personal Genetics Testing. Many patients may want to know their own genetic profile, even if the consequences are uncertain. The following are examples of personal genetics companies ("direct to consumer genomics"):

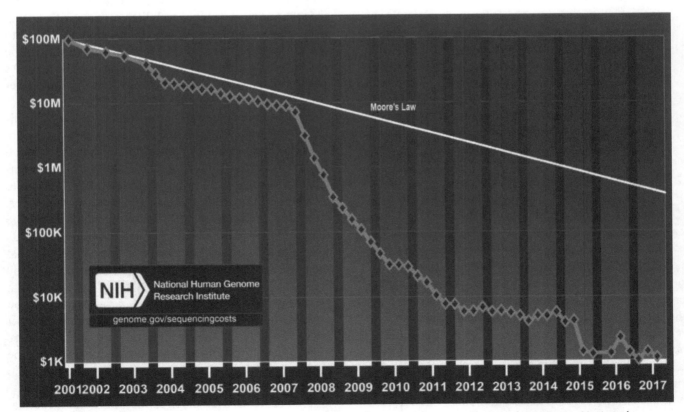

Figure 18.6: Cost per Genome over time (Courtesy National Human Genome Research Institute, National Institutes of Health)[11]

- AncestryDNA is a separate service offered by Ancestry.com. Their analysis will determine ethnicity estimates and will identify remote relatives. Saliva samples are needed, the cost is $79 and the turnaround time is six to eight weeks.[54]
- 23andMe is a direct to consumer online genetic testing company. For $199 they send a testing kit to homes based on analyzing saliva with a turnaround time of four to six weeks. Currently, they look for 240 diseases, multiple carrier states and drug response conditions (a substantial increase in the last two years). They also offer an analysis of ancestry based on the genetic profile.[55] In 2010 a genome wide association study (GWAS) was published that used this technology and showed that patient questionnaire results correlated well with genetic results. Additionally, they were able to describe five new genotype-phenotype associations: freckling, photic sneeze reflex, hair curl and failure to smell asparagus.[56] Google's co-founder Sergey Brin was one of the early funders of 23andMe, focusing on a project through this company to study the genetic inheritance of Parkinson's disease. They hope to recruit 10,000 subjects from various organizations and offer a discount price for complete analysis. In late 2013 the FDA instructed the company to stop performing genetic analyses for medical conditions until they receive 510(k) (pre-market) clearance, which they subsequently received.[57] In 2017, they offered a Health and Ancestry analysis for $199 and an Ancestry analysis for $99. The Health analysis includes genetic risk (e.g. celiac disease, macular degeneration, etc.), wellness reports, trait reports (e.g. eye color, skin pigment) and carrier status reports (e.g. polycystic kidney disease, cystic fibrosis, etc.).
- Myriad™ specializes in genetic testing for cancers with a hereditary component, such as breast, endometrial, melanoma, ovarian, colon, prostate, gastric and pancreatic cancer.[58] A sentinel Supreme Court decision took place in 2013 that determined that Myriad could not patent BRCA gene testing.[59]

As pointed out by Harold Varmus (American Nobel-prize winner, who was a former director of the NIH, and the current director of the NCI), personal genetics *"is not regulated, lacks external standards for accuracy, has not demonstrated economic viability or clinical benefit and has the potential to mislead customers."*[60] For genetics to enter the mainstream, new technologies and specialties will need to be developed and numerous ethical questions will arise. Just finding the abnormal gene or SNP is the starting point. Understanding the sensitivity and specificity of genetic tests within clinical contexts will be essential for them to be accepted. In general, patients may not be willing to undergo major procedures (e.g., a prophylactic mastectomy or prostatectomy to prevent cancer) unless the genetic testing is nearly perfect. It is also important that genetic counseling be available to help patients understand the implication of genetic susceptibility tests (versus genetic guarantee of disease, such as the mutations associated with Huntington's disease).

Additionally, the Genetic Information Nondiscrimination Act of 2008 was passed to protect patients against discrimination by employers and healthcare insurers based on genetic information. Specifically, the Act prohibits health insurers from denying coverage to a healthy individual or charging that person higher premiums based solely on genetic information and bars employers from using individuals' genetic information when making decisions related to hiring, firing, job placement, or promotion.[61]

Many obstacles face the routine ordering of genetic tests by the average patient. Ioannidis et al. pointed out that for genetic testing to be reasonable several facts must be true. The disease of interest must be common. Even with breast cancer, when seven established genetic variants are evaluated, they only explain about 5% of the risk for the cancer. If the disease (e.g., Crohn's disease) is rare, then the test must be highly predictive. For genetic testing to be relevant one should have an effective treatment to offer, otherwise there is little benefit. The test must be cost effective, as many currently are too expensive. As an example, screening for sensitivity to the blood thinner warfarin (Coumadin) makes little sense now due to cost.[62]

A 2010 Lancet journal commentary warned of additional concerns. Whole-genome sequencing will generate a tremendous amount of information that the average physician and patient will not understand without extensive training. At this point, health care lacks adequate numbers of geneticists and genetic counselors that understand the implications of data being made available thanks to continued advances in biotechnology. Patients will need to sign an informed consent to confirm that many of the findings will have unclear meaning. They will have to deal with the fact that they may be found to be carriers of certain diseases that *may* have impact on childbearing, etc. Genetic testing may cause many further tests to be ordered, thus leading to increased healthcare expenditures. As more information about whole-genome sequencing is gained, more patients will desire it but who will pay for it? And, can the costs be justified?[63]

Two other articles drive home additional practical points. When the risk of cardiovascular disease based on the chromosome 9p21.3 abnormality was evaluated in white women, it only slightly improved the ability to predict cardiovascular disease above standard, well-accepted risk factors.[64] Meigs et al. looked at whether multiple genetic abnormalities associated with Type 2 diabetes would be predictive of the disease. They found that the score based on 18 genetic abnormalities only slightly improved the ability to predict diabetes, compared to commonly accepted risk factors.[65]

For more information regarding future bioinformatics trends, readers are referred to the review paper by Altman and Miller.[66]

GENOMIC INFORMATION INTEGRATED WITH ELECTRONIC HEALTH RECORDS

Eventually, the patient's genetic profile may be an additional data field within the electronic health record. Recently, gene variants have been identified for diabetes, Crohn's disease, rheumatoid arthritis, bipolar disorder, coronary artery disease and multiple other diseases.[67] There are several forward-looking initiatives that have started on the path to integrate genomic data with traditional clinical data, for example:

- In 2006, the Veterans Affairs health care system began collecting blood to generate genetic data that it will link to its EHR. The goal is to bank 100,000 specimens as a pilot project and link this information to new drug trials. The new voluntary program was officially launched in 2011 and is known as the Million Veteran Program (MVP). MVP will link genetic, military exposure, health and lifestyle into a single database.[68]
- Kaiser Permanente created the Research Program on Genes, Environment and Health (RPGEH) to determine their medical history, exercise and eating habits. They now have a biobank of 200,000 patients, along with their survey and EHR data. In 2014, Kaiser extended this program to multiple Kaiser pr0grams in the US, with the goal of including 500,000 members. Because the average age of participants is 65 it is anticipated that excellent information about aging will be generated. For example, they are measuring telomere length (the tips of chromosomes) that is thought to correlate with aging. This NIH funded initiative was completed in 15 months and access to data is now available.[69-70]
- The Electronic Medical Records and Genomics (eMERGE) Network is a consortium of 10 healthcare organizations with significant investments in both EHR and genomic analytics across the United States. The National Human Genome Research Institute organizes this network, with additional funding from the National Institute of General Medical Sciences. An important theme is electronic health records are a vital resource for complex genomic analysis of disease susceptibility and patient outcomes in diverse patient populations. The October 2013 issue of *Genetics in Medicine* is devoted to discussing the progress of eMERGE. It is pointed out that for EHRs to support genomic information they must: 1. Store data in structured format 2. Data must be standards based 3. Phenotypic information must also be stored as structured data 4. Data must be available for use by rules engines 5. EHRs must be able to display information needed by the clinician based on phenotypic and genotypic data. All these requirements have challenges that must be addressed. At the top of the list is adequate training of clinicians so they can deal with genomic data and privacy protections of the data. Importantly, will the clinical decision support for interpretation be part of commercial EHRs or reside in a data warehouse?[71-72]
- Vanderbilt University recently published a strong correlation between their genetic biorepository known as BioVU (genotype) with clinical information (phenotype) obtained from their electronic health record. The diseases studied were rheumatoid arthritis, multiple sclerosis, Crohn's disease and type 2 diabetes.[73]
- Mount Sinai BioMe Project has collected genetic profiles from about 26,000 patients in its biobank so they can link it with clinical data in the EHR and their clinical data warehouse. As of early 2016, they had enrolled 32,000 patients, with a goal of 100,000 patients.[74]
- The Mayo Clinic Biobank began in 2009 with the goal of including genetic information on 50,000 of their patients. In 2013, they reported that they were a little over half way there. While their biobank is smaller than other organizations they can collate data from more than 15 years in their EHR. Their research focus is very broad and not targeted to one disease. The volunteer participation rate is high at 29% and their patient population is highly educated, compared to the general population.[75]

SNOMED CT is making changes to its codes to include genetic information. Organizations such as Partners HealthCare, IBM, Cerner and data mining vendors are all gearing up to add genetic information to what is currently known about patients and integrate that with electronic health records.[76]

The Agency for Healthcare Research and Quality (AHRQ) is developing computer-based clinical decision support tools to help clinicians use genetic information to treat conditions with a strong genetic component, such as breast cancer. Such tools that could be integrated into EHRs are: whether women with a family history of breast cancer need BRCA1/BRCA2 testing and which women who already have breast cancer may benefit from additional genetic testing.[77]

It is surprising that family history is often overlooked by clinicians and that it usually does not exist as computable data for analysis. Electronic health records have varying approaches for collecting this information in a common computable format, challenging their use for clinical decision support. Family history data are generally entered as unstructured text that can be of varying quality (based on provider-patient interviews). Input of family history was a menu objective for stage 2 meaningful use. Data standards have been developed so family history can be part of EHRs and PHRs, to be shared.[78] There is a government sponsored free web tool available for the public to record their family history using the newest data standards. In this way, the results can be saved as a XML file and shared by EHRs and PHRs. The site, *My Family Health Portrait*, is available for English or Spanish speaking patients, is easy to use and does not store any patient information on the site. Instead, patients can store the XML file on their personal computers.[79] The program is open source and downloadable from this site.[80] A consortium in North Carolina has developed a self-administered family history tool for the collection of data and creation of clinical decision support for primary care physicians. The tool, known as MeTree will collect information on 48 conditions with a genetic component but provide clinical decision support for only five common conditions. Researchers will study how this information impacts appropriate testing as well as implementation hurdles.[81]

In late 2013, a family history tool for pediatricians was released as a collaborative effort by several medical organizations. Data collected is structured and can be inputted from handheld devices, such as tablets. The program includes 35 genetic-related conditions with associated clinical decision support tools (actionable information) for pediatricians to help children at risk. The pedigree created includes first, second and third-degree relatives. The program utilizes HL7 standards but does not currently integrate with EHRs.[82]

Hoyt, et al. was able to analyze a digital family history with neural networks and thereby create a predictive model of diabetes, hypertension and heart disease in first degree relatives.[83-84]

For further information about the role of EHRs and genomics readers are referred to these citations.[85-87]

RECOMMENDED READING

The following are several recent and interesting articles related to bioinformatics:

- *Invited Commentary: Genetic Prediction for Common Diseases: Will Personal Genomics Ever Work?* The author is responding to an article about atrial fibrillation where including 3 gene variants did not improve on prediction. The point is made that thousands of gene variants have been discovered but few improve predictability of disease for several reasons discussed in the commentary.[88]
- *Risk models for progression to advanced age-related macular degeneration using demographic, environmental, genetic and ocular factors.* Much is known about the genetic and non-genetic risk factors for age-related macular degeneration. The authors build prediction models considering multiple factors which one day could be included in EHRs.[89]
- *Genetic testing behavior and reporting patterns in electronic medical records for physicians trained in a primary care specialty or subspecialty.* This is the first article to report on the ordering of genetic tests trends by multiple physicians at a large academic medical center. Tests were ordered primarily on childbearing women, primarily by internal medicine and ob-gyn physicians. Twenty gene tests accounted for 88% of the volume, with tests for cystic fibrosis and prothrombin variants being the most common. They point out that the genetic tests had no standard reporting format and appeared as free text in EHRs.[90]
- *The CLIPMERGE PGx Program: clinical implications of personalized medicine through electronic health records and genomic-pharmacogenomics.* The article describes the pilot program that will study the pharmacogenomics of 1500 patients at Mount Sinai that are part of the Biome Biobank program. They note that about 100 pharmacogenomic variants have been found and are associated with FDA drug information. The CLIPMERGE PGx program will create an external system that integrates with their EHR and includes a risk assessment engine that generates alerts (clinical decision support) in the EHR.[91]
- *Evaluation of Family History Information Within Clinical Documents and Adequacy of HL7 Clinical Statement and Clinical Genomics Family History Models for Its Representation: A Case Report.* The authors evaluated family history in clinical documents to determine adequacy of existing models. They found that HL7 Clinical Genomics Family History Model and HL7 Clinical Statement Models

would represent most family histories, but refinements are needed.[92]
- *Systematic Comparison of phenome-wide association study of electronic medical record data and genome-wide association study data.* The study reported in 2013 was conducted using data from five institutions in the eMERGE network. They tested 3,144 SNPs for an association with 1358 diseases (phenotypic information) using data from electronic health records on almost 14,000 patients. They replicated prior associations and discovered new ones, particularly single loci SNPs associated with multiple diseases (pleiotrophy).[93]
- *Genetic data and electronic health records: a discussion of ethical, logistical and technological considerations.* This is an excellent early 2014 summary of the issues/challenges extant with potential incorporation of genetic data into electronic health records. They reiterate the need for standardized content, genetic clinical decision support, compression and storage of massive genetic data.[94]

FUTURE TRENDS

Given the rapidly evolving nature of bioinformatics, multiple trends seem likely. First, new gene associations will continue to be reported. For example, in 2013 researchers in Switzerland described the CRTC1 polymorphism that is associated negatively with BMI and fat mass in psychiatric and non-psychiatric patients.[95] Second, the cost to perform complete genome sequencing will continue to drop but it must be accompanied by competent analysis to be meaningful. It is likely newer technologies will continue to improve and make both sequencing and interpretation cost effective in the next decade. Third, the time to complete sequencing will continue to shorten. As a result, infectious diseases and birth defects can be diagnosed in several days.[96-97] Fourth, companies such as 23andMe will continue to offer more tests each year and at a lower price such that patient demand may exceed the ability to know what to do with the data, making this a "disruptive technology". They will also require FDA clearance. Fifth, there will be more decision support associated with both family history and genomic information that correlates with phenotypical information, risk factors and lab results. Lastly, with integration of robust data from multiple sources within the electronic health record there will be a better understanding of what factors turn genes on or off, in epigenetics.[98]

KEY POINTS

- Traditionally, bioinformatics has had limited direct impact on clinical medicine, but translational bioinformatics is bridging this gap
- Advances in biotechnology (such as genome sequencing) are introducing a treasure trove of genetic information that will enable deeper understandings of the manifestation of disease as well as the development of a new cadre of therapeutics over the next decade
- The inclusion of genetic profiles is being contemplated for electronic health records
- At this time, direct to consumer genetic testing is still in its early stages, and cannot be used as a replacement for traditional clinical tests (but may be used to complement)

CONCLUSION

Bioinformatics may seem foreign to many clinicians. The promise of translational bioinformatics is to transform biological knowledge (such as can be inferred from genomic data) into clinically actionable items. The success of translational bioinformatics will not be realized until clinicians can access and clinically interpret data that tells them who should be screened for certain conditions and which drugs are effective in which patients as part of day-to-day practice. In the meantime, biomedical scientists and companies will continue to add to the many genetic databases, develop genetic screening tools and get ready for one of the newest revolutions in medicine. The American Health Information Community (AHIC) recommended that the federal government should prepare for the storage and integration of genetic information into many facets of health care.[99] Their recommendations will initiate the necessary dialogue that must take place to prepare for bioinformatics to align with the practice of medicine. But, as pointed out by Dr. Varmus *"the full potential of a DNA-based transformation of medicine will be realized only gradually, over the course of decades."*[60]

REFERENCES

1. Hogeweg P. The Roots of Bioinformatics in Theoretical Biology. PLoS Comput Biol 2011;7(3): e1002021. doi:10.1371/journal.pcbi.1002021 http://journals.plos.org/ploscompbiol/article?id=10.1371/journal.pcbi.1002021 (Accessed August 20, 2017)
2. NCBI. A Science Primer. www.ncbi.nlm.nih.gov/About/primer/bioinformatics.html (Accessed August 20, 2013) (Archived)
3. NIH Working Definition of Bioinformatics and Computational Biology July 17, 2000 http://www.bisti.nih.gov/docs/CompuBioDef.pdf (Accessed August 24, 2013) (Archived)
4. Bioinformatics Overview. Bioinformatics Web www.geocities.com/bioinformaticsweb/?200630/ (Accessed August 24, 2013) (Archived)
5. Butte AJ, Shah NH. Computationally translating molecular discoveries into tools for medicine: translational bioinformatics articles now featured in JAMIA. J Am Med Inform Assoc. 2011;18(4):352-353
6. Metagenomics. Human Microbiome Project. The NIH Common Fund. http://commonfund.nih.gov/hmp/overview.aspx (Accessed October 22,2017)
7. Genetics Home Reference. http://ghr.nlm.nih.gov/glossary=phenotype (Accessed August 14, 2013)
8. Feero WG, Guttmacher AE, Collins FS. Genomic Medicine—An Updated Primer. NEJM 2010;362:2001-2011
9. National Institute of General Medical Sciences. The New Genetics. http://publications.nigms.nih.gov/thenewgenetics (Accessed October 22, 2017)
10. Genome: The autobiography of a species in 23 chapters. Matt Ridley. Harper Perennial. 2006
11. Zarrei, M, MacDonald, JR, et al. (2015). A copy number variation map of the human genome. *Nature Reviews Genetics*. 16: 172-183.
12. Heard, E and Martienssen, RA (2014). Transgenerational epigenetic inheritance: myths and mechanisms. *Cell*. 157: 95-109.
13. Reik, W (2007). Stability and flexibility of epigenetic gene regulation in mammalian development. *Nature*. 447: 425-432.
14. Price, AL, Spencer, CC, et al. (2015). Progress and promise in understanding the genetic basis of common diseases. *Proceedings of the Royal Society. Biological sciences*. 282: 20151684. http://rspb.royalsocietypublishing.org/content/282/1821/20151684 (Accessed October 15, 2017)
15. National Human Genome Research Institute. www.genome.gov (Accessed October 22, 2017)
16. Bush, WS, Oetjens, MT, et al. (2016). Unravelling the human genome-phenome relationship using phenome-wide association studies. *Nature Reviews Genetics*. 17: 129-145.
17. Goodwin, S, McPherson, JD, et al. (2016). Coming of age: ten years of next generation sequencing technologies. *Nature Reviews Genetics*. 117: 333-351.
18. Komaroff AL. Gene Editing Using CRISPR. Why the Excitement? Viewpoint. August 22/29 2017. JAMA. https://jamanetwork.com/journals/jama/fullarticle/2646800 (Accessed September 20, 2017)
19. Sarkar IN, Butte AJ, Lussier YA et al. Translational bioinformatics: linking knowledge across biological and clinical realms. J Am Med Inform Assoc 2011; 18:354-357
20. Hanahan, D and Weinberg, RA (2011). Hallmarks of cancer: the next generation. *Cell*. 144: 646-674.
21. Chin, L, Hahn, WC, et al. (2011). Making sense of cancer genomic data. *Genes & Development*. 25: 534-555.
22. Gerlinger, M, Rowan, AJ, et al. (2012). Intratumor heterogeneity and branched evolution revealed by multiregion sequencing. *New England Journal of Medicine*. 366: 883-892.
23. Hoadley, KA, Yau, C, et al. (2014). Multiplatform analysis of 12 cancer types reveals molecular classification within and across tissues of origin. *Cell*. 158: 929-944.
24. Lu, C, Xie, M, et al. (2015). Patterns and functional implications of rare germline variants across 12 cancer types. *Nature Communications*. 6: 10086. http://www.nature.com/ncomms/2015/151209/ncomms10086/full/ncomms10086.html
25. Tyner, JW (2014). Functional genomics for personalized cancer therapy. *Science Translational Medicine*. 6: 243fs226. http://stm.sciencemag.org/content/6/243/243fs26
26. Altman RB, Kroemer HK, McCarty CA et al. Phamacogenomics: will the promise be fulfilled? Nat Rev Genet 2011;12:69-73
27. Buchan NS, Rajpal DK, Webster Y et al. The role of translational bioinformatics in drug discovery. Drug Discovery Today. 2011;16(9/10):426-434
28. Human Genome Project http://www.genome.gov/10001772 (Accessed October 22, 2017)
29. NCBI Human Genome Resources www.ncbi.nlm.nih.gov/genome/guide/human/ (Accessed October 22, 2017)
30. DNA Micro-Arrays http://en.wikipedia.org/wiki/Dna_array (Accessed August 4, 2013)
31. National Human Genome Research Institute https://www.genome.gov/ (Accessed October 22, 2017)
32. Carlton JM et al. Draft genome sequence of the sexually transmitted pathogen Trichomonas vaginalis. Science 2007;315:207-212
33. Human Variome Project. www.humanvariomeproject.org (Accessed October 22, 2017))

34. The PhenX Project https://www.phenxtoolkit.org/ (Accessed October 22, 2017)
35. Hamilton CM, Strader LC, Maiese D et al. The PhenX Toolkit: Get the Most From Your Measures. Am J Epid 2011;174(3):253-260
36. 1000 Genomes Project http://www.1000genomes.org/ (Accessed September 22, 2017)
37. St. Jude - Washington University Pediatric Cancer Genome Project https://www.stjude.org/research/pediatric-cancer-genome-project.html (Accessed October 22, 2017)
38. Global Alliance for Genomics and Health. https://www.ga4gh.org/ (Accessed October 22, 2017)
39. Pharmacogenomics Knowledge Base http://www.phsarmgkb.org/ (Accessed September 17, 2011)
40. Framingham SNP Health Association Resource http://www.ncbi.nlm.nih.gov/projects/gap/cgi-bin/study.cgi?id=phs000007 (Accessed October 22, 2017)
41. Mayo Clinic Bipolar Disorder Biobank http://mayoresearch.mayo.edu/mayo/research/bipolar-disorder-biobank/index.cfm (Accessed August 1, 2013)
42. Informatics for Integrating Biology & the Bedside https://i2b2.org (October 22, 2017)
43. Cincinnati's Children's Hospital i2b2 Data Warehouse https://i2b2.cchmc.org/ (Accessed October 22, 2017)
44. Observational Health Data Sciences and Informatics (OHDSI). https://www.ohdsi.org/ (Accessed October 22, 2017)
45. All of Us. https://allofus.nih.gov/about/about-all-us-research-program (Accessed October 22, 2017)
46. The Cancer Genome Atlas https://cancergenome.nih.gov/abouttcga/overview (Accessed October 22, 2017)
47. National Center for Biotechnology Information (NCBI) http://www.ncbi.nlm.nih.gov/ (Accessed October 22, 2, 2013)
48. NCBI Insights. Magic-BLAST 1.3.0 released with new features and improvements. September 26, 2017) https://ncbiinsights.ncbi.nlm.nih.gov/tag/magic-blast/ (Accessed October 10, 2017)
49. GenBank www.ncbi.nlm.nih.gov/Genbank/ (Accessed October 22, 2017)
50. The Online Mendelian Inheritance in Man. www.omim.org (Accessed October 22, 2017)
51. World Community Grid www.worldcommunitygrid.org (Accessed October 22, 2017)
52. Seguin B, Hardy BJ, Singer PA et al. Universal health care, genomic medicine and Thailand: investing in today and tomorrow. Nature Reviews – Genetics. 2008;9: S14-S19
53. DeCODE genetics https://www.decodeme.com (Accessed October 22, 2017)
54. AncestryDNA. http://dna.ancestry.com/ (Accessed October 22, 2017)
55. 23andMe www.23andme.com (Accessed October 22, 2017)
56. Eriksson N, Macpherson JM, Tung JY et al. Web-based, Participant Driven Studies Yield Novel Genetic Associations for Common Traits. PLoS Gen 2010;6(6) e1000993
57. Google Co-Founder To Back DNA Database Study on Parkinsons. March 12, 2009. www.ihealthbeat.org (Accessed March 12, 2009)
58. Myriad www.myriad.com (Accessed October 22, 2017)
59. Supreme Court of the United States. June 1, 2013. http://www.supremecourt.gov/opinions/12pdf/12-398_1b7d.pdf (Accessed August 2, 2013)
60. Varmus H. Ten Years On—The Human Genome and Medicine. NEJM 2010;362:2028-2029
61. Hudson, KL, Holohan JD, Collins FS. Keeping Pace with the Times—the Genetic Information Nondiscrimination Act of 2008. NEJM 2008;358:26612663
62. Ioannidis JPA. Personalized Genetic Prediction: Too Limited, Too Expensive or Too Soon? Editorial. Annals of Internal Medicine 2009;150(2):139-141
63. Samani NJ, Tomaszewski M, Schunkert H. The personal genome—the future of personalized medicine? Lancet 2010;375:1497-1498
64. Paynter NP, Chasman DI, Buring JE et al. Cardiovascular Disease Risk Prediction With and Without Knowledge of Genetic Variation at Chromosome 9p21.3. Annals of Internal Medicine 2009;150(2):65-72
65. Meigs JB, Shrader P, Sullivan LM et al. Genotype Score in Addition to Common Risk Factors for Prediction of Type 2 Diabetes. NEJM 2008;359(21):2208-2219
66. Altman RB, Miller KS. 2010 Translational bioinformatics year in review. J Am Med Inform Assoc 2011;18:358-366
67. Pennisi, E. Breakthrough of the Year: Human Genetic Variation. Science 2007;318 (5858):1842-1843
68. Million Veteran Program http://www.research.va.gov/mvp/ (Accessed October 22, 2017)
69. Kaiser Seeks Member's Genetic Info for Database. www.ihealthbeat.org February 15, 2007 (Accessed February 16, 2007)
70. Kaiser Permanente Research Program on Genes, Environment and Health (RPGEH) https://divisionofresearch.kaiserpermanente.org/genetics/rpgeh aspx (Accessed October 22, 2017)
71. Electronic Medical Records and Genomics (eMERGE) Network https://www.mc.vanderbilt.edu/victr/dcc/projects/acc/index.php/About#About_the_eMERGE_Network (Accessed October 23, 2017)
72. Multiple articles. Genetics in Medicine. October 2013 (Accessed November 22, 2013)

73. Ritchie MD, Denny JC, Crawford DC et al. Robust Replication of Genotype-Phenotype Associations across Multiple Diseases in an Electronic Health Record. Am J Hum Gen 2010;86:560-572
74. BioMe Project. http://icahn.mssm.edu/research/ipm/programs/biome-biobank/pioneering (Accessed October 23, 2017)
75. Olson JE, Ryu Euijung E, Johnson KJ et al. The Mayo Clinic Biobank: A Building Block for Individualized Medicine. May Clin Proc. 2013;88(9):952-962
76. Kmiecik T, Sanders D. Integration of Genetic and Familial Data into Electronic Medical Records and Healthcare Processes. http://www.surgery.northwestern.edu/dos-contact/infosystems/Kmiecik%20Sanders%20Article.pdf (Accessed June 28, 2009)
77. AHRQ Launches Project on Computer-Based Genetic Tools. September 23, 2008 www.ihealthbeat.org (Accessed September 24, 2008)
78. Ferro WG. New tool makes it easy to add crucial family history to EHRs. Perspectives. ACP Internist May 2009 p. 6
79. Family History http://familyhistory.hhs.gov (Accessed August 23, 2013)
80. National Cancer Institute Gforge http://gforge.nci.nih.gov/projects/fhh (Accessed August 25, 2013)
81. Orlando LA, Hauser ER, Christianson C et al. Protocol for implementation of family history collection and decision support into primary care using a computerized family health history system. BMC Health Services Research 2011;11:264 www.biomedcentral.com/1472-6963/11/264 (Accessed August 26, 2013)
82. Family History Tool for Pediatric Providers. Genetics in Primary Care. www.geneticsinprimarycare.org (Accessed November 25, 2013)
83. Hoyt R, Linnville S, Chung Hui-Min, Hutfless B, Rice C. Digital Family History for Data Mining. Perspectives in Health Information Management Fall 2013
84. Hoyt R, Linnville S, Thaler S, Moore J. Digital Family History Data Mining with Neural Networks: A Pilot Study. Perspectives in Health Information Management. Online. Winter 2016
85. Kohane IS. Using EHRs to drive discovery in disease genomics. Nature reviews genetics. 2011;12:417-428
86. Ullman MH, Matthew JP. Emerging landscape of genomics in the EHR for personalized medicine. 2011;32(8):512-516
87. Shoenbill K, Fost N, Tachinardi U et al. Genetic data and electronic health records: a discussion of ethical, logistical and technological considerations. J Am Med Inform Assoc 2013;0:1-10
88. Ioannidis JP. Invited Commentary: Genetic Prediction for Common Diseases: Will Personal Genomics Ever Work? Arch Int Med 2012;172(9):744-745
89. Seddon JM, Reynolds R, Yu Y et al. Risk models for progression to advanced age-related macular degeneration using demographic, environmental, genetic and ocular factors. Ophthalmology;2011;118(11):2203-2211
90. Ronquillo J, Li C, Lester W. Genetic testing behavior and reporting patterns in electronic medical records for physicians trained in a primary care specialty or subspecialty. J Am Med Inform Assoc 2012;19(4):570-574
91. Gottes man O, Scott SA, Ellis SB et al. The CLIPMERGE PGx Program: clinical implications of personalized medicine through electronic health records and genomic-pharmacogenomics Clin Pharm Ther 1 May 2013. Doi:10.1038/clpt.2012.72
92. Melton GB, Raman N, Chen ES, Sarkar IN et al. Evaluation of Family History Information Within Clinical Documents and Adequacy of HL7 Clinical Statement and Clinical Genomics Family History Models for Its Representation: A Case Report J Am Med Inform Assoc 2010;17(3):337-340
93. Denny JC, Bastarache L, Ritchie MD et al. Systematic comparison of phenome-wide association study of electronic medical record data and genome-wide association study data. Nat Biotech 2013; Online 24 November. Doi:10.1038/nbt.2749
94. Shoenbill K, Fost N, Tachinardi U et al. Genetic data and electronic health records: a discussion of ethical, logistical and technological considerations. J Am Med Inform Assoc. 2014;21:171-180
95. Choong E, Quteineh L, Cardinaux JR et al. Influence of CRTC1 Polymorphisms on Body Mass Index and Fat Mass in Psychiatric Patients and th General Adult Population. JAMA Psych Online August 7, 2013. E1-E10. (Accessed August 12, 2013)
96. Saunders CJ, Miller NA, Soden SE et al. Rapid whole-genome sequencing for genetic disease diagnosis in neonatal intensive care units. Sci Transl Med 2012;4(154): 154ra135.doi: 10.1126/scitranslmed.3004041
97. Torok ME, Ellington MJ, Marti-Renom MA et al. Whole-Genome Sequencing for Rapid Susceptibility Testing of M.tuberculosis. NEJM 2013;369(3):290-292
98. Epigenetics. http://learn.genetics.utah.edu/content/epigenetics/ (Accessed September 20, 2013)
99. HHS considers adding genetic information to EHRs. HealthImagingNews June 12, 2008. www.healthimaging.com (Accessed June 12, 2008)

19

Public Health Informatics

BRIAN E. DIXON • SAURABH RAHURKAR

LEARNING OBJECTIVES
After reading this chapter the reader should be able to:

- Define public health informatics
- Explain the importance of informatics to the practice of public health and the role of informatics within a public health agency
- Define and distinguish the various forms of public health surveillance systems used in practice
- List several common data sources used in the field of public health for surveillance
- Differentiate between local and global public health informatics systems and efforts
- Explain how clinics and hospitals participate in surveillance systems and how population level data are used to support clinical care
- Explain the significance of informatics systems to support surveillance for early detection of bioterrorism, emerging diseases and other health events and
- Understand the workforce needs and competencies of public health informaticians

INTRODUCTION

Public health is defined as *"the science and art of preventing disease, prolonging life, and promoting health through the organized efforts and informed choices of society, organizations, public and private communities, and individuals."*[1] Whereas physicians and care delivery organizations focus on the health of individuals, public health focuses on the health of populations and communities. Historical efforts in public health centered on vaccination against diseases like polio and measles as well as management of outbreaks of disease, including influenza and tuberculosis.[2] In the 21st century, the systems and agencies responsible for protecting and promoting the health of populations are faced with numerous challenges beyond infectious diseases, such as obesity, the opioid epidemic, uninsured populations without access to care, toxic environments, health disparities, antimicrobial and antibiotic resistant diseases, hospital-borne infections and bioterrorism.[3] To address these challenges, public health organizations conduct a range of activities across three, broad core functions – assessment, policy development and assurance:[4]

- Assessment – Public health agencies spend most of their time and resources on investigations of potential threats to the public's health. Activities include testing and monitoring of water quality, laboratory examination of diseases carried by mosquitoes, tracking food-borne illnesses, testing for environmental hazards (e.g. soil lead levels), monitoring for potential bioterrorism threats, and tracing the contacts for individuals exposed to diseases as well as hazardous chemicals.
- Policy Development – Public health agencies also create policies and regulations to protect the health of populations. For example, children may be required to have certain immunizations before they can attend school to prevent disease outbreaks that would harm children and disrupt family life. Smoking may be prohibited in certain places (e.g., public buildings, restaurants) to protect people from the harmful effects of secondhand smoke. Agencies use the evidence they gather from their investigations as well as the scientific literature to advocate for policies that state and federal legislative bodies ultimately adopt for the health of populations and communities.

- Assurance – Once laws and regulations are passed to protect health, public health agencies are tasked with assuring compliance with them. Local health departments may perform housing inspections to assure that landlords comply with rules concerning pest control. Restaurant inspectors typically work for local health authorities, and they assure that those who prepare food wash their hands, wear gloves, and take other precautions to prevent the spread of disease. Assurance also involves evaluating the effectiveness of policies as well as programs on health outcomes. For example, evaluations of public policies that prohibit smoking show improve outcomes, including reduced admissions for acute coronary syndrome and reduced mortality from smoking-related illnesses.[5]

Public health surveillance (PHS), or the systematic collection, analysis, interpretation and dissemination of health-related data, is the bedrock of public health practice.[2] This is because the surveillance systems capture and manage the volumes of data and information necessary to support the three core functions of public health. The notion of PHS can be traced as far back as the Renaissance.[6]

The rise of scientific thought gave rise to the use of mortality and morbidity data in public health in Europe, which spread to the Americas with European settlers. Dedicated public health positions were created to monitor and quarantine or block ships as necessary during the black death epidemic in the 14th century.

PHS in the United States historically focused on infectious diseases. Evidence of PHS can be found in mid-18th century Rhode Island where tavern keepers performed the role of reporting communicable diseases among patrons.[7] Just over a century later, physicians in Massachusetts were voluntarily submitting weekly reports in a standard postcard-reporting format. The first national level PHS activities in the United States started in 1850 when the federal government published national mortality statistics.

While the practice of PHS developed, and became more systematic over time, the mode of data collection and monitoring remained largely unchanged with physicians or teams of trained officers collecting data. Although this form of surveillance works well for some community level data (e.g., census data, immunizations), due to the long time taken to collect this information it is not useful in outbreak detection. Just as clinical medicine has benefited from the proliferation of information systems, as well as informatics approaches to improve health care processes and outcomes, public health in the 21st century also stands to benefit from informatics.

In the following sections, we discuss the role of informatics in public health and formally define public health informatics. Next, we describe some of the various information systems commonly found in public health agencies at local, state, and national levels. We also discuss connecting information systems across national borders to support global public health surveillance. Furthermore, we discuss the role of hospitals, clinics, and health systems in exchanging information with governmental public health agencies to support public and population health.

Definitions

- Information system (IS): A collection of technical and human resources that support the storage, computing, distribution, and communication of information required by all or some part of an enterprise such as a public health agency.
- Public health (PH): *"the science and art of preventing disease, prolonging life and promoting health through the organized efforts and informed choices of society, organizations, public and private, communities and individuals."*[1] Typically public health is performed in the context of a governmental agency that seeks to prevent disease and promote healthy behaviors. Furthermore, public health professionals, researchers, and educators work with communities to implement interventions aimed at improving the health of populations.
- Public health informatics (PH Informatics): *"the systematic application of information and computer science and technology to public health practice, research and learning...."*[8] Informatics synthesizes theories and practices from the computer and information sciences as well as the behavioral and management sciences into methods, tools and concepts that lead to information systems that impact health.
- Public health surveillance (PHS): *"the ongoing systematic collection, analysis, and interpretation of health-related data essential to the planning, implementation and evaluation of public health practice, closely integrated with the timely dissemination of these data to those who need to know. The final link in the surveillance chain is the application of these data to prevention and control."*[9]
- Public health reporting: The process in which clinics or health care delivery organizations, or even patients, report data and information about an individual disease episode (or case) to a public health authority. For example, many states in the

U.S. require that information about individual cancer cases be reported to statewide cancer databases.
- Social determinants of health (SDOH): *"the structural determinants and conditions in which people are born, grow, live, work and age."*[10] The SDOH include factors like socioeconomic status, education, physical environment, employment, and social support networks, as well as access to health care.

THE ROLE OF INFORMATICS IN PUBLIC HEALTH

In the words of Johann Wolfgang von Goethe, *"Knowing is not enough; we must apply. Willing is not enough, we must do."* The increasing volumes of data and information generated in the conduct of public health surveillance and practice must be captured, managed, shared, and then applied to impact the health of populations. This is where informatics can play a role in support of public health. Public health informatics (referred to as PH Informatics) is a subset of the broader discipline of health informatics previously defined as *"the systematic application of information and computer science and technology to public health practice, research and learning."*[8]

While accurate, the definition of PH Informatics does not capture the essence of what informatics' contributes to public health practice, research and learning. Informatics facilitates both the technical (e.g., computing systems, interfaces, protocols) as well as the human (e.g., governance, privacy, work process) aspects of collecting, managing, sharing and using data in the context of public health. As such, PH Informatics can and should be pervasive across all core function of public health; it is not the same as the information technology (IT) department within a public health organization.[11] Furthermore, PH Informatics is enhancing how public health professionals draw actionable insights from the data and information gathered by public health agencies.

Several policy initiatives such as the HITECH act, the Affordable Care Act, and MACRA (described in other chapters) have changed the way health data is collected. Incentives from the meaningful use program have brought about widespread use of electronic health records and health information exchange. The proliferation of digital health information and accompanying technological advances shifted the role of health departments from consumers to brokers of information. To carry out the core functions of public health in this environment, health departments need to be informatics-savvy.

What do health departments need to do to be informatics-savvy? A framework from the Public Health Informatics Institute (PHII) focuses on three core elements that an informatics-savvy health department should have: an informatics vision and strategy, a competent and skilled workforce, and a well-designed, effectively used information system.[12-13]

- An informatics-savvy health department should have a strategic vision for how informatics and information technology should support the practice of public health. These health departments can collect, evaluate and securely exchange information electronically to improve public and community health outcomes. Leadership is essential in creating an information savvy health department; ideal leaders would highlight the importance of informatics in achieving public health's goals at the organization level. Furthermore, a shared vision across the health department may motivate a shift towards achieving informatics capabilities at the organizational level.
- Informatics-savvy health departments have competent and skilled public health professionals. Everyone in a public health agency needs to know something about informatics and information systems. The health agency should have an experienced informatician in a leadership position who can provide overall guidance and direction on informatics. Some health departments dub this position the Chief Public Health Informatics Officer. Beyond an informatician, health departments need epidemiologists, restaurant inspectors, environmental health officers, and public health nurses who can interact with and utilize the information systems deployed by the agency to collect, manage, and share data. Consequently, those in health agencies who manage the people who use the systems, including the Health Commissioner or Health Officer at the top of the organization, should also know something about informatics and technology.
- LaVenture et al. (2015)[13] suggest that informatics-savvy health departments require *"a disciplined approach to design and use of information systems that effectively support agency program objectives."* These health departments will utilize informatics best practices, including standard systems development lifecycle approaches as well as technical and semantic standards, to procure, optimize and use information systems across the public health spectrum in support of the organization's mission: to protect and serve the health of populations. Information systems should, for example, ensure confidentiality and interoperability while enabling analytics and visualization in addition to standard reporting. Above all, information systems should

deliver value to the public health agency by supporting the work done by its professional staff.

INFORMATION SYSTEMS TO SUPPORT PUBLIC HEALTH FUNCTIONS

Public health is organized into a set of core functions.[4] The functions include: surveillance, assessment, etc. Information systems can be utilized to support each of the core functions. In the following sections, we describe several core functions of public health and provide examples of how informatics supports them.

Public Health Surveillance

Public health surveillance (PHS) is essential to understanding the health of a population. Surveillance is the continuous, systematic collection, analysis and interpretation of health-related data needed for the planning implementation and evaluation of public health practice. Additionally, PHS also involves timely dissemination of the information gleaned from the collected data to agencies and organizations responsible for preventing and controlling disease and injury. Until recent years, PHS was primarily paper-based. However, with the increasing shift towards eHealth, PHS has embraced the field of public health informatics.[14] To monitor disease events in a large population one needs interoperable technologies such as standards-based networks, databases and reporting software. Current electronic surveillance systems employ complex information technology and embedded statistical methods to gather and process large amounts of data and to display the information for networks of individuals and organizations at all levels of public health. Public health surveillance serves to:
- Estimate the significance of the problem
- Determine the distribution of illness
- Outline the natural history of a disease
- Detect epidemics
- Identify epidemiological and laboratory research needs
- Evaluate programs and control measures
- Detect changes in infectious diseases
- Monitor changes in health practices and behaviors
- Assess the quality and safety of health care, drugs, devices, diagnostics and procedures and
- Support planning.[15]

PHS is primarily of two types: indicator-based and event-based.[16] Indicator-based surveillance refers to the monitoring of a specific disease/health condition, or a class of diseases/health conditions that are of interest to public health. Examples include mortality, morbidity, environmental indicators (water quality, air quality), prescriptions, syndromes etc. Non-communicable (chronic) illnesses, such as diabetes, depression and hypertension, are tracked using indicator-based methods; something of high interest in health system in addition to public health agencies. Data sources for indicator-based surveillance include population health surveys as well as individual cases reported by health care providers, laboratories, community clinics, etc. For example, public health laws often require providers to report cases of influenza where the patient dies, usually from complications of the disease. Hypertension and other chronic illnesses are usually identified through population-based surveys that ask people about their health status.

Event-based surveillance monitors data from specific events where a large number of people gather in one place. Examples include music concerts, the Olympic games, natural disasters, or the Muslim pilgrimage referred to as the Hajj. These events are of public health importance as large populations experience similar exposures in a relatively short time frame. The objective of event-based surveillance is to identify, in real-time, emerging threats to the public's health, such as an increased number of cases of influenza among Hajj pilgrims or a cluster of poison monoxide cases stemming from poor ventilation in a nightclub during a concert. Data sources for event-based surveillance includes reports, stories, rumors and any other information about health events that could be a serious risk to public health.

Both indicator- and event-based surveillance focus on identifying a trend (e.g., increased prevalence, decreased incidence) or an emerging threat to public health (e.g., unknown disease, outbreak of known disease) using data from a variety of sources. The potential trend or emerging threat is initially referred to as a signal. Epidemiologists at a health agency then work to verify the signal using additional, detailed data they gather from the same sources and/or additional sources. For example, over-the-counter sales of anti-diarrhea medication and increased call volume at a local poison control center might signal an outbreak of a food-borne illness, such as *shigellosis*. However, to confirm this epidemiologists would want to corroborate the signal with information from electronic health records and potentially map the addresses of the people who called with those who purchased the medications. Epidemiologists might also want to ask people who reported illness questions about where they recently ate or purchased groceries. If the data are confirmed to indicate an outbreak, then the health department would notify the public and work to identify additional people affected. They would further work to identify the source and prevent further exposure to what caused the disease.

Case Study

Legionnaire's disease (LD) is a severe pneumonia caused by a waterborne bacterium that most frequently affects susceptible populations including those 50 and older, former smokers, those with chronic disease, or immunosuppressed individuals. LD is a reportable condition, and 85% of all LD outbreaks are associated with water system exposures.[17] In 2015, New York City (NYC) experienced an outbreak of 438 potential LD cases reported by the clinical laboratory system as well as health care professionals.[18] Reported cases are followed up with interviews of patients or close relatives to identify potential exposures. Moreover, automated detection systems enable the health department to detect clusters of disease by time and location. Based on information gathered from surveillance through reported cases and automated cluster detection, an outbreak was identified in the Bronx with a higher than normal incidence of LD. Physicians do not usually treat pneumonia patients for LD and thus do not obtain sputum samples for culture. Therefore, the health department encouraged physicians in relevant zip codes to consider LD as a possible diagnosis and to collect respiratory samples. The health department also reached out to the NYC Chief Medical Examiner's office to request autopsies on individuals that died due to unexplained respiratory illness. Using knowledge of the known factors associated with LD along with data from multiple sources, including reported cases and autopsy reports, a cooling tower was hypothesized to be the cause of the outbreak. Clinical staff in internal medicine, pediatrics, geriatrics, primary care, infectious diseases, emergency medicine, family medicine, laboratory medicine, and infection control, were requested to consider LD as a potential diagnosis in all individuals presenting with community acquired pneumonia and to perform the relevant tests. Samples were taken from cooling towers in the outbreak zone followed by decontamination of towers that tested positive for *legionella*. Using epidemiological data on cases and satellite imagery, the source of the outbreak was verified. To provide public assurance, the state of NY expanded free testing of all cooling towers in the Bronx area.

Additionally, the department of health conducted numerous communication and community engagement activities to allay any panic among residents in the outbreak zone.

The 2015 outbreak of LD which lasted from July till August was the largest in NYC history with 138 cases and 16 fatalities.

In the aftermath of the outbreak, legislation was passed known as Local Law 77 that required all cooling towers to be registered with the department of buildings and to be inspected once every 90 days.

The 2015 NYC LD outbreak demonstrates how public health relies on data gathered from multiple sources, requires rapid integration of data to identify the source of a health threat, and works in partnership with health care professionals and policymakers to address public health challenges.

To identify signals, a bidirectional flow of information between health care delivery and public health is essential. As such, to identify a sustainable, consistent, nationwide approach that would allow the utilization of EHR data to advance public health surveillance, Digital Bridge was created in 2016. The Digital Bridge initiative[19] was funded by the Robert Wood Johnson Foundation (RWJF) and included the PHII as well as the Deloitte consulting group as partners. Additional stakeholders[20] included EHR vendors, laboratories, health care payers, key entities in the public health community, the CDC, public health associations and health care delivery systems. Digital Bridge's initial focus as identified by the stakeholders was on automating the generation and transmission of case reports from clinical care to public health agencies using EHR data, also called Electronic Case Reporting (eCR). The goal of eCR was to provide complete and accurate signals to public health agencies to allow for early detection of outbreaks. The automated nature of eCR reduces burden of reporting on healthcare providers and responds directly to the needs local and state partners. In doing so, eCR directly links health care to population and public health.

Types of Surveillance Systems

There exist a range of surveillance systems designed to support the two types of PHS. These systems can be classified based on data collection purpose and design. Table 19.1 demonstrates the more common varieties of surveillance systems found in practice.

There are many similarities between surveillance systems and information systems, which represent a collection of technical and human resources that provide

Table 19.1: Types of Surveillance Systems

Surveillance System	Description	Examples
Case-based surveillance systems	Collect data on individual cases of a disease with previously determined case definitions or criteria outlined for person, time, place, clinical & laboratory diagnosis; Analyze case counts and rates, trends over time and geographic clustering patterns; Historically, case-based surveillance systems have been the focus of most public health surveillance.	National Notifiable Disease Surveillance System (NNDSS)
Syndromic surveillance systems	Collect data on clusters of symptoms and clinical features of an undiagnosed disease or health event in near real-time allowing for early detection, rapid response mobilization and reduced morbidity and mortality; Data can be obtained from specific information systems such as those in emergency departments using HL7 Admission, Discharge and Transfer (ADT) messages as well as existing sources such as insurance claims, school and work absenteeism reports, over the counter (OTC) medication sales, consumer driven health inquiries on the Internet, mortality reports and animal illnesses or deaths; Geographic and temporal aberration and geographic clustering analyses are performed with real-time syndromic surveillance data; Syndromic surveillance systems can also be used to track longitudinal data and monitor disease trends.	Biosurveillance Common Operating Network (BCON) National Syndromic Surveillance Platform
Sentinel surveillance systems	Collect and analyze data from designated health care facilities selected for their geographic location, medical specialty, and ability to accurately diagnose and report high quality data. They include health facilities or laboratories in selected locations that report all cases of a certain health event or disease to analyze trends in the entire population. While these are useful to monitor and identify suspected health events or diseases, they are less reliable in assessing the magnitude of health events on a national level as well as rare events since data collection is limited to specific geographic locations.	PulseNet FoodNet ILINet
Behavioral surveillance systems	Collect data on health-risk behaviors, preventative health behaviors, and health care access in relation to chronic disease and injury. Also collect data on social determinants of health (SDOH), including race/ethnicity, education level, household income, employment status, disability, emotional support, home ownership SDOH data can be used to derive concepts such as poverty, socioeconomic status, stress, social support, social exclusion, early life experiences. Analyze the prevalence of behaviors as well as the trends in the prevalence of behaviors over time. Information is most commonly collected by personal interview or examination Most acute when conducted regularly, every 3 to 5 years	Behavioral Risk Factor Surveillance System (BRFSS) National Health Interview Survey (NHIS) Pregnancy Risk Assessment Monitoring System (PRAMS)

Table 19.2: Comparison of Information Systems with Surveillance Systems in the Context of Public Health

	Public Health Information System	Public Health Surveillance System
General Definitions	A collection of technical and human resources that support the storage, computing, distribution, and communication of information required by all or some part of a public health agency	A collection of interacting components, including subsystems, that facilitate the systematic collection, analysis, and interpretation of health-related data essential to the planning, implementation and evaluation of public health practice
Similarities	Like a surveillance system, an IS involves a collection of discrete, complex components; in both cases, humans are components of the system.	Like an IS, surveillance systems focus on the capture, management, sharing and use of data and information to support core business processes in public health.
Differences	An IS must include some aspect of computing in which data or information are analyzed using machines.	A surveillance system does not require computing; surveillance can occur on paper using manual data entry and capture techniques. This occurs frequently in low resource settings.

the storage, computing, distribution, and communication of information required by all or some part of an enterprise like a public health agency. In fact, all surveillance systems could be classified as information systems (even though the reverse is not true). In Table 19.2, we compare the concept of an information system to that of a surveillance system in the context of public health.

Syndromic Surveillance

Syndromic surveillance systems detect early symptoms of a disease before a diagnosis based on further testing is established. For example, if multiple individuals complain of stomach symptoms over a short period of time, one might suspect there is an outbreak of gastroenteritis. The main goal of syndromic surveillance is early detection of clusters of illness allowing rapid mobilization of public health response. However syndromic surveillance systems do not identify the cause of the outbreak, rather they provide comparison of trends which allows public health officials to initiate outbreak investigation techniques. Signals from syndromic surveillance systems must be investigated further because, as described in a recent review of syndromic surveillance methods, current techniques are not very accurate.[21]

More than 80% of state health departments across the U.S. report some level of syndromic surveillance activity.[22] While the precise methods and systems used for syndromic surveillance vary across states, many are adopting syndromic surveillance systems to augment their case- and event-based surveillance programs.

Most syndromic surveillance systems rely upon chief complaint (the reason for visiting an emergency department provided by the patient) data received electronically from emergency department (ED) information systems. In addition to data from electronic health records, public health officials can also incorporate the following types of data into their syndromic surveillance systems: unexplained deaths, insurance claims, school absenteeism, work absenteeism, over the counter medication sales, Internet based health inquiries by the public and animal illnesses or deaths.[23]

Data captured by syndromic surveillance systems are grouped into "bins" or categories of similar data. For example, complaints about an "upset stomach" or "tummy ache" can be grouped together to represent two cases of gastrointestinal symptoms. These symptom groups are commonly referred to as syndromes by epidemiologists. The syndrome categories most commonly monitored are:

- Botulism-like illnesses
- Febrile (fever) illnesses (influenza-like illnesses)
- Gastrointestinal (stomach) symptoms
- Hemorrhagic (bleeding) illnesses
- Neurological syndromes
- Rash associated illnesses
- Respiratory syndromes
- Shock or coma

Originally developed to detect bioterrorist attacks as well as seasonal influenza, syndromic surveillance systems are used in practice to detect a growing array of community-level events and disease patterns. In recent times, syndromic surveillance has been used to detect morbidity due to inclement weather. Specifically, Leonardi et al. used data from calls made to the National Health Service's nurse-led helpline in England to monitor the effects of a heat wave in 2003.[24] Calls made to the helpline increased, showing a marked increase for heat/sun stroke. Jossern et al. analyzed emergency department (ED) data to examine effects related to a heat wave in 2004. A higher frequency of conditions such as dehydration, hyperthermia, malaise, hyponatremia, renal colic and renal failure were seen during the hot period.[25] Additionally, the EDs observed an increase in visits by elderly patients. Similar results were found by Schaffer et al. using data from ED visits, ambulance calls, and mortality in Australia.[26] In both cases, syndromic surveillance was useful in identifying the elderly as a vulnerable population to rising temperature.

Ambulatory electronic health records (EHR) systems are a potentially rich source of data that can be used to track disease trends and biosurveillance. EHR systems contain both structured (e.g. ICD-10 coded) data as well as narrative free text. Hripcsak et al. assessed the value of outpatient EHR systems to identify influenza-like illnesses and gastrointestinal infectious illnesses from Epic® EHR data from 13 community health centers.[27] The first system analyzed structured EHR data and the second used natural language processing (MedLEE processor) of narrative data. The two systems were compared to influenza lab isolates and to a verified ED surveillance system based on *"chief complaint."* The results showed that for influenza-like illnesses the structured and narrative data correlated well with proven cases of influenza and ED data. For gastrointestinal infectious diseases, the structured data correlated very well but the narrative data correlated less well. They concluded that EHR structured data was a reasonable source of biosurveillance data.[27]

Figure 19.1 depicts typical syndromic surveillance data at a national level. Facilities that report data are indicated geographically by red pins. Trends for influenza-like illness and asthma symptoms are plotted over a 10-month period for two states: Texas and Arizona. Whereas asthma symptoms fluctuated normally over time with a slight decrease towards the end of the period, influenza-like illness symptoms steadily rose from December through February then fell almost to zero indicating a seasonal pattern. Graphs like this can serve as *"dashboards"* for flu season monitoring as well as periodic review of other trends in symptoms. Sudden increases or decreases would trigger further investigation by epidemiologists at a health agency.

National Syndromic Surveillance Program

The National Syndromic Surveillance Program (NSSP) in the United States supports the development of surveillance systems that focus on collection and timely exchange of syndromic data.[28] In addition to

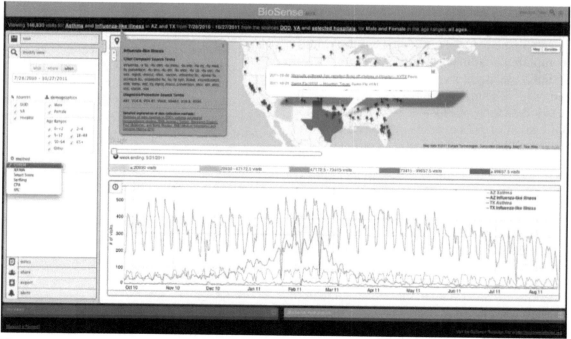

Figure 19.1: Data view in BioSense

providing a nationwide situation awareness, the NSSP promotes public health preparedness to disease outbreaks and bioterrorism-related events. The NSSP features two major components: 1) the BioSense technical platform enabling analysis of syndromic surveillance data, and 2) a collaborative "Community of Practice" that functions across local, state and federal levels of public health.

BioSense is a cloud-based information system created by the U.S. Centers for Disease Control and Prevention (CDC) that provides public health officials with standardized tools and procedures to collect, store, analyze and exchange syndromic data across regional and jurisdictional boundaries. It was created in 2003 through the Public Health Security and Bioterrorism Preparedness Act, primarily as a system for the early detection and monitoring of bioterrorism-related events. In 2010, BioSense was redesigned to integrate existing syndromic surveillance systems and allow for better regional sharing of information. The BioSense 2.0 platform allowed state and local health departments to share and access syndromic data in support of the nation's meaningful use program. While state and local public health agencies could view the detailed data provided by hospitals and clinics in their jurisdictions, only de-identified data could be viewed by the CDC and other jurisdictions.

In 2014, the CDC embarked upon building a third generation of the Biosense platform depicted in Figure 19.2. The goal was to provide a web-based clearinghouse where data could be stored, searched and analyzed from and by multiple parties; decreasing the need for local health departments to purchase additional expensive information technologies.[29] To facilitate rapid evaluation and analysis of syndromic data, the BioSense platform offers integrated tools such as ESSENCE, SQL tools, and R studio within a cloud-based computing environment hosted by the CDC.

The NSSP Community of Practice (NSSP CoP) was developed in partnership with the International Society for Disease Surveillance (ISDS) in 2016.[30-31] It is an active collaboration between the CDC, state and local health departments and voluntary members that include academia, government agencies, non-profit organizations and other stakeholders to advance the science and practice of syndromic surveillance. The NSSP CoP does this by engaging stakeholders in the continuous evaluation and ongoing development of the BioSense platform. The stakeholder input and collaborative approach ensures that the BioSense platform remains cognizant and adaptive of the differing user needs and changes in the technology over time.

CASE MANAGEMENT SYSTEMS

Once public health agencies have captured data on a reported case of disease, they need to persist those data as well as augment them with additional information they gather from disease investigations. To manage data

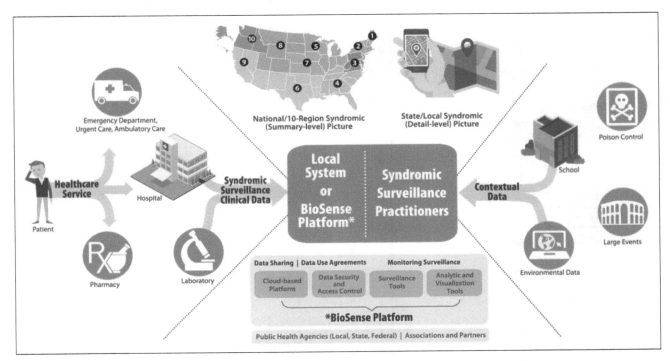

Figure 19.2: The Biosense platform

on individuals with disease, public health agencies use case management systems. The NEDSS Base System (NBS) is an example of a case management system.[32] This system, developed by CDC and available as open source software, is used in over 20 states. A screenshot showing the dashboard a user sees upon login to the system is presented in Figure 19.3.

Case management systems function like that of EHR systems used in clinics and hospitals. The system focuses on documenting information about individual cases of disease. Workers at the health agency enter data such as the name, date of birth, gender, race, etc. about individuals. They also enter information about health care facilities, laboratory test results, treatments, vaccinations, and other details about a particular case (or suspected case). Disease investigators at the health department follow up on suspected cases reported from health care providers and document whether the case is confirmed or not. Clusters of confirmed cases can indicate an outbreak of a disease. Health authorities seek to follow up on suspected cases to both confirm them and ensure individuals with disease receive appropriate treatment at a health care facility. Case management systems therefore document not only the presence of disease but also what actions were taken by health care providers and the community to treat illness and stop an outbreak.

Case management systems utilize informatics standards to both receive and send information. Many health departments utilize an interface to automatically import electronic lab reports (ELRs) (HL7 ORU messages) received from laboratory information systems. The interfaces automatically populate fields in the case management system like patient name, lab test name, lab result, etc. Interfaces also exist to export data from the state to the CDC to enable national-level monitoring of disease trends. Case management systems can also export data into files that can be analyzed by statistical software packages such as SAS and R.

THE PUBLIC HEALTH INFORMATION NETWORK (PHIN)

The Prevention and Public Health Fund, as part of the Affordable Healthcare Act of 2010, in conjunction with the Health Information Technology for Economic and Clinical Health (HITECH) Act allowed the public health infrastructure to move into the eHealth era. Driven by the mission to prevent, reduce and treat disease, these initiatives focus on developing interoperable public health information systems that are beneficial to the healthcare of all Americans.[33-34]

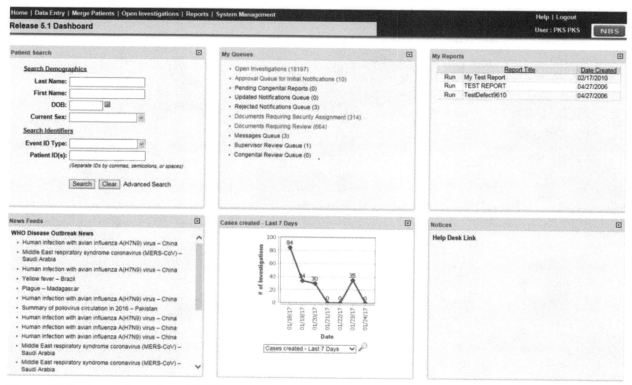

Figure 19.3: Screenshot of the NEDSS Base System

The Public Health Information Network (PHIN) is a CDC initiative established to facilitate fully capable and interoperable public health information systems.[35] Through the PHIN, the CDC provides a set of tools and resources aimed at improving public health preparedness across local, state and federal jurisdictions as well as to improve health information exchange between clinical healthcare and public health.

The PHIN focuses on sharing critical health data between the different levels of public health and clinical healthcare by supporting the development of standards, specifications and an architecture for health information exchange. Further, it monitors the health information exchange capabilities for state and local health departments. Moreover, the PHIN assists local and health departments meet its requirements by providing technical assistance as well by advancing supportive policy. Tools provided by the PHIN include, the PHIN messaging system (MS), the message quality framework (MQF), the vocabulary access and distribution system (VADS), the public health directory (DIR), guides for public health reporting and standards for data exchange.[36] For more information about these data standards, readers are referred to the chapter on data standards.

MEANINGFUL USE AND PUBLIC HEALTH

Integral to a vision of a robust public health information infrastructure is the need for EHR systems as well as other eHealth information systems to interoperate with surveillance systems and other types of public health information systems.[37] Each stage of the Meaningful Use (MU) program managed by the U.S. Centers for Medicare and Medicaid Services (CMS) has several core and menu requirements for eligible professionals (EPs) and eligible hospitals (EHs) that intersect with public health:

Stage 1 & Stage 2

- The capability to electronically transmit immunization data to immunization registries or immunization information systems. EPs or EHs must test the ability to transmit a HL7 message to a local public health agency.
- The capability to electronically transmit reportable lab results (as determined by state or local law). EHs only must use HL7 2.5.1 standard, LOINC and SNOMED-CT to test the ability to transmit electronic messages from the lab to public health agencies.
- The capability to electronically transmit syndromic surveillance data from an EHR. EPs and EHs must test the ability to transmit HL7 messages of syndromic surveillance data to public health agencies, which may include input into the BioSense platform.

Stage 2 only

- The capability of EPs to report cancer cases to a state registry from a certified EHR.
- The capability of EPs to report specific cases to a non-cancer state registry from a certified EHR.[37]

Stage 3 only

EPs must meet two of the following measures while EHs must meet four:[38]
- Report to Immunization Registries
- Report to syndromic surveillance
- Report reportable cases
- Report to Public Health Registry
- Report to Clinical Data Registry
- Electronic reporting of laboratory results

In addition to Meaningful Use, the Medicare Access and Chip Reauthorization Act (MACRA) also incentivizes public health reporting. The MACRA primarily repeals the Medicare Sustainable Growth Rate (SGR) methodology used to update physician fee schedules; in other words, it changes how care providers are paid through the Medicare Part B reimbursements. MACRA gets rid of the fee-for-service (FFS) model which rewards volume rather than quality and introduces the quality payment program which rewards high quality and effective care. Eligible clinicians and groups can participate in this program through two tracks: Merit-Based Incentive Payment System (MIPS) and the Advanced Alternate Payment Models (APMs).

Under MIPS provider care quality is determined by 4 performance measures: 1) quality; 2) clinical practice improvement activities; 3) resource use or cost; and 4) advancing care information or meaningful use of CEHRT. Starting in January of 2017, advancing care information replaces the Medicare EHR incentive program. As part of the new advancing care information measure, MACRA reduces the number of public health registries to which providers must report. While providers are only required to report to immunization registries, increased performance (e.g., higher reimbursements) could be earned by providers for reporting to additional public health registries.

GEOGRAPHIC INFORMATION SYSTEMS (GIS)

Epidemiologists often characterize data by place, time and person. As early as 1855, Dr. John Snow created a simple map to show where patients with cholera lived in

London in relation to the drinking water source in the Soho District of London. Using his hand drawn map and basic epidemiological investigation techniques, much of which are still used today, he determined the source of the epidemic to be a common water pump. Epidemiology, public health surveillance and indeed the field of public health have improved significantly since the pioneering work of Snow and others after him. Much of this transformation has been the result of the emergence and proliferation of advanced computing technologies, the Internet and other automated information systems that have facilitated the amalgamation of large datasets to map out disease patterns.

A Geographic Information System (GIS) is a system of hardware, software and data used for the mapping and analysis of geographic data. GIS provides access to large volumes of data; the ability to select, query, merge and spatially analyze data; and visually display data through maps. GIS can also provide geographic locations, trends, conditions and spatial patterns. Spatial data has a specific location such as longitude-latitude, whereas attribute data is the database that describes a feature on the map.

GIS maps are created by adding layers. Each layer on a GIS map has an attribute table that describes the layer. The data can be of two types: *Vector* or *Raster*. *Vector* data appears as points, lines or polycons (enclosed areas that have a perimeter like parcels of land). *Raster* data utilizes aerial photography and satellite imagery as a layer. Using GPS and mobile technology, field workers can enter epidemiologic data to populate a GIS. This geospatial visualization has been useful in tracking infectious diseases, public health disasters and bioterrorism.[39-40]

With the recent shift in public health focus towards chronic diseases, GIS has also been used to monitor chronic diseases and social and environmental determinants of health for public health policy. In early 2011, the Centers for Disease Control and Prevention launched a new project, Chronic Disease GIS Exchange.[41] Designed for public health professionals and community leaders, GIS experts will use as an information exchange forum to network and collaborate with the goal of preventing heart disease, stroke and other chronic diseases. Data and information shared in this forum will be used in documenting the disease burden, informing policy decisions, enhancing partnerships and facilitating interventions from the use of GIS data.[42-44] Figure 19.4 shows a GIS display of diabetes prevalence by State. The image comes from the Robert Wood Johnson Foundation's *"The State of Obesity"* website that provides data, policy analysis and information on the prevalence and burden of obesity across the United States.[45]

Virtually all surveillance systems used in modern public health practice have a GIS component that allows for the mapping of disease trends or events giving public health practitioners the ability to deploy resources to

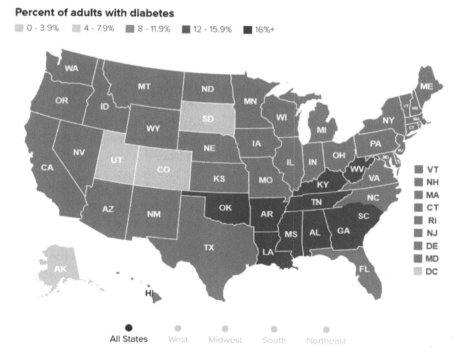

Figure 19.4: GIS display of diabetes prevalence by State (Courtesy, the Robert Wood Johnson Foundation)

monitor health status over time or control outbreaks. Key variables can be input by zip code, latitude, longitude, that help public health disease investigators narrow down the source of the problem.

COMMON TYPES AND SOURCES OF PUBLIC HEALTH DATA

High quality population data need to be collected, managed and shared to achieve the aims of public health. Data originate as fragmented, discrete observations about individuals (e.g., a systolic blood pressure reading, a weight, a height). When systematically collected, normalized, and managed in a database, data from multiple individuals transforms into data about a population. Data can also be transformed into information (e.g., combine the height and weight to become a body mass index).

Public health agencies systematically collect, store, manage, exchange, and use several kinds of population health data, including the following:
- Vital statistics, including information about births and deaths
- Infectious or communicable diseases, including tuberculosis, chlamydia and HIV
- Chronic or non-communicable diseases, including diabetes, hypertension, depression
- Immunizations administered, especially those for children and vulnerable populations
- Cancer cases, including tumor diagnosis and treatment information
- Injuries, including motor vehicle accidents, sports-related injuries and traumatic brain injuries
- Poisonings, including drug overdoses and opioid abuse
- Environmental data, including air quality and water quality
- Crime statistics, such as the types of crimes in various geographic areas
- Behavioral data, including nutrition, smoking, and alcohol use data
- Social determinant data, including poverty rates, education levels, and vacant housing
- Nursing home and assisted living facility data

Public health agencies employ a range of methods to gather data and information about populations. One of the most common methods used is a population-based survey, such as the Behavioral Risk Factor Surveillance System (BRFSS).[46] Population-based surveys are given to a representative sample of a population (e.g., 500 people in a county with 100,000 residents) where the participants are randomly drawn from every type of potential resident (e.g., people over 65 years of age, people of Asian heritage). The responses from the survey represent the views, behaviors and health of the entire population.

Agencies also receive data from electronic health records and other types of information systems used in health care delivery organizations. For example, infectious disease data often come from laboratory information systems via electronic lab reports (ELRs).[47] Disease investigators at a public health agency might also be given direct access to an EHR system to gather detailed information on a confirmed disease case following the receipt of an ELR. Injury data on individuals who were hurt or died from a motor vehicle accident are extracted from an emergency department information system and transmitted electronically to the public health agency, usually via an Internet-based form. Immunization data are transmitted electronically from EHR systems using a specific type of HL7 message.[48] In the case of cancer, specially trained registars collect data on newly diagnosed cancer cases and treatment as well as outcomes for existing cancer cases. These individuals work in hospitals or clinics to collect information and report them to public health agencies.[49] Data collected by public health can also come from community organizations as well as other branches of government such as the housing authority or criminal justice system.

Data collected or managed by public health agencies are used for a variety of purposes. Epidemiologists monitor trends in the prevalence (how many people have diabetes in a state?) as well as incidence (how many new cases of breast cancer were there since last year?). Agencies also use the information to assess health behaviors, such as the proportion of individuals who smoke regularly or go to the gym at least once per week. When combined or integrated, the information managed by public health agencies can be used to examine the impact of a potential public policy or provide insights into the health of a neighborhood.[50] Such data can be useful to providers as well as payers, community organizations, and people looking to move into a community from another county or state.

GLOBAL PUBLIC HEALTH INFORMATICS

Public health threats can originate from any nation. Since the start of the twenty-first century, the world has experienced outbreaks of SARS, MERS-COV, Ebola and Zika viruses in Asia, the Middle East, Africa and the Americas, respectively. Transmission of infectious disease is aided by the fact that travel across nations and regions is both quick and efficient. Therefore, real-time collaboration and data sharing across national borders by governmental public health agencies, health care delivery organizations, and non-governmental organizations is critical to protect the global public's health. Furthermore, the regular collection and timely reporting of health indicators, including mortality and morbidity, from each nation is critical to understanding the global prevalence and burden of disease. In this section, we highlight the key policies, organizations, and information systems that support collaboration of global public health informatics activities around the world.

International Public Health Regulations

The third edition of the World Health Organization's International Health Regulations (IHR) was published in 2005.[51] This legally-binding agreement provides a framework for the management of international public health emergencies, while also addressing the capacity of participating nations to detect, evaluate, alert, and respond to public health events. The IHR specifies operational procedures for disease surveillance, notification and reporting of public health events and risks as well as for the coordination of international response to those events. For example, in response to the Zika virus outbreak, the World Health Organization (WHO) published a handbook to assist nations maintain *"sanitary standards at international borders at ports, airports, and ground crossings (points of entry)."*[52]

While each nation is responsible for its internal surveillance system, the IHR facilitates sharing of data and information across surveillance systems and networks. When connected, these national information and communications technologies form a global platform for decision support.[53] The IHR further enables WHO to mobilize human experts who can bring technical expertise and skills as well as logistical support to nations during a public health emergency. Since the start of the Ebola outbreak in West Africa, the WHO's Global Outbreak Alert and Response Network (GOARN) deployed a multidisciplinary workforce of 895 experts, including doctors, nurses, infection control specialists, logisticians, laboratory specialists, as well as communication, anthropology and social mobilization experts, emergency management and public health professionals.[54]

A visualization of the entities and scope of global collaboration is presented in Figure 19.5. Significant effort is required not only of local public health agencies but also a wide range of governmental (e.g., homeland security, municipal) and non-governmental organizations (e.g., health care providers, charity organizations, faith-based organizations) to coordinate a response to a natural disaster or emerging infectious disease.

Global Public Health Organizations

The leading international public health entity is the World Health Organization (WHO). Organized in 1948 as an agency of the United Nations (UN), WHO directs and coordinates public health efforts worldwide. WHO and its Member States collaborate with other UN agencies, non-governmental organizations, and the private sector to:

- **Foster health security:** Through its surveillance and disaster/epidemic response systems, WHO works to identify and curb outbreaks of emerging or epidemic-prone diseases.
- **Promote health development:** Through this objective WHO works to increase access to life-saving

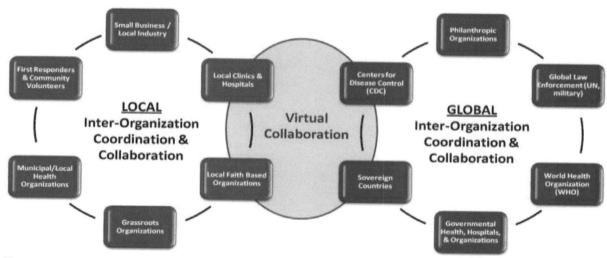

Figure 19.5: Coordination between local and global public health organizations

and health-promoting interventions, particularly in poor, disadvantaged, or vulnerable groups. WHO's health development efforts focus on the treatment of communicable and non-communicable disease (e.g. diabetes), prevention and treatment of tropical diseases (e.g. malaria), women's health issues, and healthcare within African nations.
- **Strengthen health systems:** In poor and medically underserved areas, WHO endeavors to strengthen and supplement existing health systems. Activities include providing trained healthcare workers, access to essential drugs, and assistance in collecting vital health information.[55]

In 2005, the 58th World Health Assembly, recognizing the need to incorporate emerging health information technologies into WHO and Member States, adopted an eHealth strategy resolution. The resolution urged Member States to plan for appropriate eHealth services in their countries. That same year WHO established a Global Observatory for eHealth (GOe) to study eHealth, including its evolution and impact on health within countries. The objectives of the GOe are to:[56]

- Provide relevant, timely, and high-quality evidence and information to support national governments and international bodies in improving policy, practice, and management of eHealth
- Increase awareness and commitment of governments and the private sector to invest in, promote, and advance eHealth
- Generate knowledge that will significantly contribute to the improvement of health using ICT
- Disseminate research findings through publications on key eHealth research topics as a reference for governments and policy-makers

The GOe principally focuses on surveying Member States about their individual efforts to adopt and use information systems for public health. Analysis of this information provides a critical benchmark for global adoption and use of health IT systems as well as their impact on health outcomes. For example, in the third survey of Member States the GOe analyzed how eHealth systems impacted universal health coverage – a high priority for the WHO. The survey examined how eHealth policies and information systems were enabling delivery of health services to remote populations and underserved communities through telehealth or mHealth as well as facilitative the training of the health workforce using eLearning.[57]

Another useful resource from the GOe is the Directory of eHealth Policies.[58] This online directory contains a collection of national eHealth policies and strategies published by Member States. This resource is designed to support the development of eHealth strategies by governments through easy access to existing policy and strategy documents worldwide. It further provides insight into which countries have existing national strategies. For example, the federal health IT strategic plan from the U.S. Office of the National Coordinator for Health Information Technology is in the directory along with links to other American eHealth strategy documents.

In 2012, the WHO named the National Health and Family Planning Commission of the People's Republic of China as the WHO Collaborating Centre for Health Information and Informatics. In this role, this arm of the Chinese ministry of health supports *"comprehensive monitoring of the global, regional and country health situation, trends and determinants, using global standards, and leadership in the new data generation and analyses of health priorities."*[59]

Within the United States, there are several national public health organizations that contribute significantly to the development, implementation and use of global public health informatics systems:

- In 2008, the U.S. Centers for Disease Control and Prevention (CDC) established the **Global Public Health Informatics Program (GPHIP)** within its Division of Global Health Protection.[60] The GPHIP collaborates with an array of partners to promote data-driven decision making and apply informatics best practices around the world. Its partners include the President's Emergency Plan for AIDS Relief (PEPFAR), U.S. Agency for International Development (USAID), among others. Working with ministries in China, Saudi Arabia, and Swaziland, the GPHIP has supported the introduction of telehealth, mHealth and electronic laboratory reporting systems. The GPHIP further serves as the WHO Collaborating Centre for Public Health Informatics.
- **PATH** is a non-governmental organization that seeks to accelerate innovation across five areas: vaccines, drugs, diagnostics, devices, as well as system and service innovations.[61] In addition to supporting eHealth use broadly, PATH also supports the use of social media to raise awareness of disease threats and promote healthy behaviors in populations.
- The **Task Force for Global Health** is a non-governmental organization that works both in the U.S. and internationally to strengthen the public health infrastructure by training healthcare workers in how to detect and respond to disease outbreaks and by improving the use of information to protect and promote health.[62]
- The **Institute for Health Metrics and Evaluation (IHME)** is an independent global health research

center at the University of Washington. The institute seeks to systematically gather, normalize and share health indicators from 195 countries to make available a longitudinal data resource for policymaking and public health decision support.[63]

In addition to the organizations detailed here, several philanthropic organizations, including the Bill and Melinda Gates Foundation as well as the Rockefeller Foundation, President's Emergency Plan for AIDS Relief (PEPFAR) and USAID, have provided financial support and resources to support the implementation and use of information systems in support of public health in numerous countries.

Examples of Global Public Health Systems

While there are many integrated surveillance systems, we highlight three examples that represent the range of systems used in practice. One system gathers data from around the world to inform WHO about emerging health threats. Another gathers in-country data from remote sensors to inform multiple stakeholders about pending natural disasters. A third system collects health indicators from multiple nations to examine the prevalence and burden of both infectious and chronic diseases.

The **Global Public Health Intelligence Network (GPHIN)** was developed by the Public Health Agency of Canada to electronically monitor infectious disease outbreaks. Approximately 40 percent of the outbreaks investigated by WHO each year come from the GPHIN. This network "*is a secure, Internet-based 'early warning' system that gathers preliminary reports of public health significance in seven languages on a real-time, 24/7 basis.*"[64] GPHIN "*continuously and systematically crawls web sites, news wires, local online newspapers, public health email services and electronic discussion groups for key words.*"[65] Although originally developed to detect infectious disease outbreaks, GPHIN now scans for food and water contamination, exposure to chemical and radioactive agents, bioterrorism, and natural disasters. It uses automated analysis to process the gathered data to alert human analysts to conduct additional review of any serious issues or trends. These data are then made available to WHO/GOARN and other subscribers through a web-based application and to the public through the WHO website.

The **Global Burden of Disease (GBD)**, an online resource published by the Institute for Health Metrics and Evaluation at the University of Washington, gathers and quantifies data on hundreds of diseases, injuries, and risk factors from nations around the globe, so that health systems can be improved and disparities can be eliminated.[67] Collected and analyzed by a consortium of more than 2,300 researchers in more than 130 countries, the data capture premature death and disability from more than 300 diseases and injuries in 195 countries, by age and sex, from 1990 to the present, allowing comparisons over time, across age groups, and among populations. In Figure 19.6, two treemaps produced from the GBD compares causes of death in high income nations with those in low income nations. Chronic conditions, such as ischemic heart disease (IHD) and stroke, are the primary causes of death in high income countries, accounting for over 30% of all deaths. On the other hand, low income countries are burdened principally by communicable diseases such as lower respiratory infection (LRI), diarrhea, and malaria.

CHALLENGES IN GLOBAL PUBLIC HEALTH INFORMATICS

While we highlight some of the many efforts to strengthen the public health infrastructure using information and communications technologies in the recent past, there were several systemic failures during the Ebola outbreak from 2014-2016 that illustrate more work is required by government and health care organizations in the U.S. and globally to achieve a fully integrated,

---

Malaysia: Early Warning and Risk Navigation Systems

eWARNS is Malaysia's Early Warning and Risk Navigation Systems for natural disasters including rainfall, flash flood, soil erosion, landslide, tidal wave, and forest fire. Remote Sensing and Transmission Units (RSTU) placed throughout the country are used to predict floods and other natural disasters. Each RSTU collects rainfall data, senses the impact of the rain, and transmits the data via the internet to a receiving unit. The RSTU also acts as a web-server allowing the 'remote panel' to be viewed via the internet. The system alerts the public to real time risk levels and forecasts via SMS text messaging on their mobile phones. Information on daily rainfall, erosivity index, and erosion hazards are also available on the website.[66]

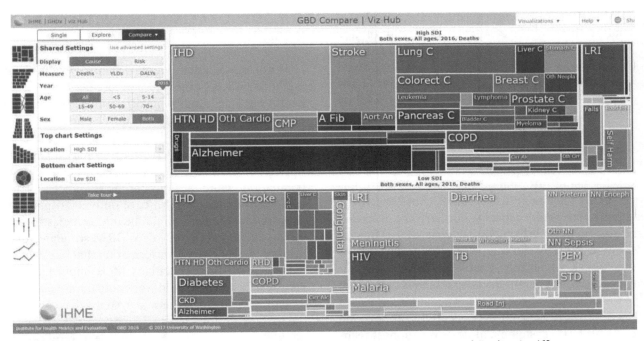

Figure 19.6: Treemaps produced from the GBD (Institute for Health Metrics and Evaluation)[68]

coordinated response to emerging health threats. Recent publications[69-70] by public health scholars and professionals outline several areas that need strengthening for the world to be ready for the next epidemic:

- **Surveillance systems need to integrate zoonotic with human data.** More than 60% of emerging infectious disease events that affected humans in recent decades began as zoonoses, or diseases that affect animals.[71] Health surveillance systems capture data on humans, animals and the environment.[72] Such systems have just begun to be adopted globally, particularly in Europe. Yet few nations, including the U.S., have policies that promote their development, adoption and use. Information on human and animal health remains fragmented; few systems integrate human and environment data. Understanding complex relationships, such as why Ebola has thinned populations of chimpanzees and gorillas in Central Africa long before the human outbreak in West Africa,[73] is essential to understanding the epidemiology of infectious diseases and their threat to the world.
- **Global capacity for public health responses must be strengthened.** As of 2016, just one-in-five WHO Member States had fully implemented the IHR. Therefore 80% of the world lacks full capacity to handle an outbreak like that of Ebola or Zika. Beyond tracking the spread of disease, nations need to prepare healthcare workers for dealing with emerging infections. For example, healthcare workers in Africa were at far higher risk from contracting Ebola than the general public Moreover, even in the U.S., there was a significant lack of knowledge about how to care for someone infected with Ebola. Attention to the health care delivery aspects of preparedness must be addressed worldwide.
- **Communication is rapidly changing in a globally connected society.** Communication during the Ebola crisis was itself a disaster. People in the U.S. showed up in hospitals complaining of Ebola symptoms before the first patient from West Africa arrived on U.S. soil. In Africa, families did not understand how their cultural practices were putting them at risk for infection. Communication from public health authorities with appropriate, timely updates on the situation lagged behind newer communications channels, including cable news networks and social media. As a result, fear and miscommunication spread quickly in Africa as well as the U.S. Some of the miscommunication came from healthcare workers on the ground in Ebola affected countries. As Carney and Weber put it, *"When CNN becomes a more trusted news source on a public health crisis than the [CDC], we have a major problem."* [72]
- **Better diagnostic tools are needed to improve detection of disease.** Early diagnosis is key to identifying cases and stopping the spread of disease. At the time of the outbreak, Ebola diagnostic tools could not accurately detect the virus during its incubation period or at the beginning of the symptomatic phase.

Furthermore, there did not exist any biomarker-based analysis that could be conducted at the point-of-care or even a central public health laboratory to assist in screening geographic populations. Improved methods for detection of diseases, available at low cost, are needed to assist in scaling efforts to rapidly detect the spread of infections. Laboratory testing devices that can output data using vocabulary standards like LOINC would also be useful.[74]

The issues raised in the wake of the Ebola crisis are addressable using informatics approaches. New information systems and analytics methods could be used to strengthen public health capacity. Furthermore, informaticians also focus on information flow as well as human workflow (especially the interaction between humans and machines), which may enable policies, organizational processes, and communication methods to improve during times of public health crisis. The same approaches used to improve patient safety in operating rooms and clinics using EHR systems and decision support technologies could be adapted to improve how ministries of health conduct knowledge management around emerging health threats. This is a key contribution public health informatics can make to population health globally.

PUBLIC HEALTH INFORMATICS WORKFORCE

The information technologies in use within public health agencies require a diversity of human expertise for development, implementation, management, analysis, and exchange of data. While all public health workers require some knowledge of IT depending on the demands of their position, public health agencies need informaticians. Informaticians are public health professionals who apply informatics methods to enhance surveillance, prevention, preparedness, health promotion, and the other aspects of public health practice. They provide leadership within public health agencies to develop and execute informatics strategies that enhance data quality, interface with clinical information systems, integrate data from external sources, and analyze population health data.

Currently informaticians account for just over one percent of the public health workforce.[75] To adequately support public health organization adoption and use of information technologies in the future, the number of public health informaticians needs to increase. Public health professional organizations support this expansion of the workforce, and academic programs are expanding the number of informatics-related courses offered to public health students. There are also a range of training programs in public health informatics offered through the CDC[76] as well as the U.S. National Library of Medicine.[77]

THE ROLE OF CLINICS, HOSPITALS AND HEALTH SYSTEMS

While this chapter focuses principally on informatics methods and systems that support the work of public health agencies, public health informatics is relevant to health systems, hospital, clinics and other health care delivery settings. As emphasized in the chapter, frontline health care workers often play a central role in detecting disease and documenting data that drive population health systems for surveillance. Therefore, it is important for health care delivery organizations to work closely with public health agencies to aim for *"write once, read many"* or WORM strategies that can minimize dual entry of information that has value to all those who care for populations. More importantly, health care system actors should work collaboratively to define and implement processes and systems that can facilitate the daily workflow in care delivery organizations using population health data. For example, awareness of past diagnosis of methicillin resistant staph aureus (MRSA) delivered from a public health registry to an infection control practitioner in a health system can facilitate proper isolation upon admission to a hospital. Awareness of a local salmonella outbreak could help facilitate better screening and diagnosis in emergency departments and urgent care clinics. This is sometimes referred to as public health decision support.[78] Given the existence of standards to facilitate data exchange from public health authorities to clinics and hospitals, working together should enable us to improve care for many types of populations in the future.

RECOMMENDED READING

The following are samples of recent interesting articles related to public health informatics:
- *Public Health, Population Health, and Epidemiology Informatics: Recent Research and Trends in the United States.*[79] *This article surveys recent advances in the field of public health informatics within the United States. PH Informatics studies continue to describe new technologies that are impacting public health practice, especially PHS. Furthermore, the workforce in PH Informatics is expanding. However, there exist several gaps preventing further integration between clinical and public health information systems.*
- *Getting from Here to There: Health IT Needs for Population Health.*[80] *Existing health IT systems were designed for organizations that are structurally,*

operationally, and culturally focused on individual care delivery, rather than improving health for a population. Opportunities exist to align health IT resources and population health management strategies to fill the gaps among technological capabilities, use and the emerging demands of population health. To realize this alignment, healthcare leaders must think differently about the types of data their organizations need, the types of partners with whom they share information, and how they can leverage new information and partnerships for evidence-based action.

- *Electronic Health Records and Meaningful Use in Local Health Departments: Updates From the 2015 NACCHO Informatics Assessment Survey.* [81] This article provides a snapshot of information system adoption and usage in local health departments within the United States. While nearly half of local health departments use an EHR system to capture primary health data, only a quarter conduct informatics training for their workforce. Larger health departments or those with support from their state health department were more likely to have EHR systems. Smaller and rural health departments lag behind their urban peers in adopting and using information systems. The article points out gaps by geography as well as system functionality available in most health departments. These data suggest where states and the nation should invest going forward to strengthen the public health information infrastructure.

FUTURE TRENDS

As highlighted in this chapter, public health agencies increasingly use a variety of information systems to facilitate the core functions of PH. These systems are capturing and managing ever growing volumes of "big data" from a variety of sources, including clinical, government and commercial sectors. In the future, public health agencies will continue to develop, adopt and use new information systems, especially since the distribution of these systems is not equal across local, state and national levels. In addition, public health informatics research and practice will likely focus on the following macro-level trends:

1. Precision Health – One of the fastest growing areas of research and practice for health informatics is precision medicine. Precision medicine leverages clinical, genomic, and other types of data to inform the care provided to individuals. For example, cancer patients can receive tailored treatment regimens based on their genomic profile and medical history. Public health agencies desire to utilize precision techniques to inform population health within the health system as well as communities.

2. Social Determinants of Health – The National Academy of Medicine (NAM) recommends that EHR systems capture *"psychosocial vital signs"* also referred to as measures of the Social Determinants of Health or SDOH.[82] Public health agencies have access to a range of data on SDOH, especially from population health surveys as well as programs that focus on serving vulnerable populations. In partnership with health care delivery organizations, public health agencies could leverage SDOH data to inform policies, research, and care delivery. In the future, greater interoperability between clinical and public health information systems could facilitate achievement of the NAM goals without necessarily requiring duplicate data entry across the health system.

3. Social Media Surveillance – Public health agencies are interested in leveraging the exploding volumes of data and information shared rapidly via social media services such as Twitter. Recently the San Francisco Department of Public Health experimented with analyzing restaurant reviews posted on Yelp.com to predict health code violations.[83] While the models performed adequately, low specificity indicates they implicated several restaurants that did not violate health code. More experimentation and analysis is likely to occur in the future as health agencies try to figure out how best to leverage social media information. Yet skepticism will be necessary because utilizing social media information is not a panacea. For example, early reports on Google's Flu Trends product looked promising, yet the service turned out to be flawed and the service was terminated[84]

> **KEY POINTS**
>
> - Public health informatics is an important sub-category of health informatics
> - Public health reporting is a part of meaningful use stages 1-3 and will be a component of advancing care informatics, which replaces meaningful use starting in 2017
> - Public health surveillance is a core function of the science and practice of public health that informatics supports through enhanced data collection, management, sharing and application in the context of both infectious and chronic diseases
> - Informatics supports more than surveillance, including immunization administration, food safety and coordination of care during outbreaks.
> - Public health informatics is an important part of health systems not only in the U.S. but many other nations that have established health information infrastructures

CONCLUSION

Public health is concerned with the health of populations, instead of individuals. To effectively monitor the health of large populations and trends in public health activities, public health agencies require robust health information systems that can interoperate with equivalent systems deployed in care delivery organizations. A robust public health information infrastructure therefore requires data standards, a variety of electronic information systems and health information exchange networks that allow for both clinical and public health organizations. Policies such as the HITECH Act and Affordable Care Act in the United States are creating incentives that lay the foundation for a robust health information infrastructure. Similar policies and strategies in other nations are also laying a foundation. Yet achieving the broad vision of a robust, interconnected learning health system that integrated genomic, clinical, social determinants, and environmental data to inform care delivery, policy and the population's health will require more work.

While we have highlighted some of the excellent work happening in local and state health departments as well as national ministries of health, many health agencies lack even basic electronic systems. The most recent report from the National Association of City and County Health Officials in the U.S. suggests that less than half of local health departments use an EHR and only one-quarter of departments conducted some kind of informatics training in the past 12 months.[85] Moreover, the adoption and use of electronic systems is not equitable across health agencies. Smaller and rural health departments tend to lack access to both information systems and informatics expertise. We have more work to do to train public health informaticians who can work at agencies at all levels to increase adoption and use of the information systems described in this chapter. We further must advocate for better funding of public health agencies to enable the implementation of these systems across the health system. Finally, we need to encourage more partnership between clinical and public health organizations to interconnect and interoperate their systems to enable analytics, decision support and precision health at scale – across the entire spectrum of health services offered within a community or nation.

REFERENCES

1. Winslow, Charles-Edward Amory (1920 Jan 9). "The Untilled Fields of Public Health". Science 51 (1306): 23–33. doi:10.1126/science.51.1306.23. PMID 17838891. (Accessed August 10, 2017)
2. Lee LM, Thacker SB. The cornerstone of public health practice: public health surveillance, 1961--2011. MMWR Surveill Summ. 2011 Oct 7;60 Suppl 4:15-21.
3. Institute of Medicine. The Future of the Public's Health in the 21st Century. Washington, DC: The National Academies Press; 2003. 536 p.
4. Core Functions of Public Health and How They Relate to the 10 Essential Services. Centers for Disease Control and Prevention. https://www.cdc.gov/nceh/ehs/ephli/core_ess.htm (Accessed September 27, 2017)
5. Frazer K, Callinan JE, McHugh J, van Baarsel S, Clarke A, Doherty K, Kelleher C. Legislative smoking bans for reducing harms from secondhand smoke exposure, smoking prevalence and tobacco consumption. Cochrane Database of Systematic Reviews 2016, Issue 2. Art. No.: CD005992. DOI: 10.1002/14651858.CD005992.pub3
6. Declich S, Carter AO. Public health surveillance: historical origins, methods and evaluation. Bulletin of the World Health Organization. 1994;72(2):285.

7. CDC MMWR Reports https://www.cdc.gov/mmwr/preview/mmwrhtml/su6103a2.htm (Accessed August 10, 2017)
8. Magnuson JA, O'Carroll P. Introduction to Public Health Informatics. In: Magnuson JA, Fu JPC, editors. Public Health Informatics and Information Systems. Health Informatics: Springer London; 2014. p. 3-18.
9. Centers for Disease Control. Comprehensive Plan for Epidemiologic Surveillance. Atlanta: US Department of Health and Human Services, Public Health Service; 1986.
10. Michael Marmot et al., Closing the Gap in a Generation: Health Equity through Action on the Social Determinants of Health. The Lancet 372, no. 9650 (Nov. 8, 2008):1661–1669.
11. Dixon BE, McFarlane TD, Dearth S, Grannis SJ, Gibson PJ. Characterizing Informatics Roles and Needs of Public Health Workers: Results From the Public Health Workforce Interests and Needs Survey. J Public Health Manag Pract. 2015 Nov-Dec;21 Suppl 6, Public Health Workforce Interests and Needs Survey (PH WINS):S130-S40.
12. LaVenture M, Brand B, Ross DA, Baker EL. Building an informatics-savvy health department: part I, vision and core strategies. J Public Health Manag Pract. 2014 Nov-Dec;20(6):667-9.
13. LaVenture M, Brand B, Ross DA, Baker EL. Building an informatics-savvy health department II: operations and tactics. J Public Health Manag Pract. 2015 Jan-Feb;21(1):96-9.
14. Krishnamurthy, R. & St. Louis, M. (2010). Informatics and the management of surveillance data. In Principles and Practice of Public Health Surveillance. Third edition, Oxford Press. Eds. Lee LM, Teutsch SM, Thacker SB, St. Louis ME.
15. Teutsch, S. (2010). Considerations in planning a surveillance system. In Principles and Practice of Public Health Surveillance. Third edition, Oxford Press. Eds. Lee LM, Teutsch SM, Thacker SB, St. Louis ME.
16. CDC Global Health https://www.cdc.gov/globalhealth/healthprotection/gddopscenter/how.html (Accessed September 26, 2017)
17. CDC MMWR https://www.cdc.gov/mmwr/volumes/66/wr/mm6622e1.htm (Accessed September 26, 2017)
18. Chamberlain, A. T., Lehnert, J. D., & Berkelman, R. L. (2017). The 2015 New York City Legionnaires' Disease Outbreak: A Case Study on a History-Making Outbreak. Journal of Public Health Management and Practice, 23(4), 410-416. doi:10.1097/PHH.0000000000000558
19. Centers for Disease Control and Prevention https://www.cdc.gov/ehrmeaningfuluse/docs/ehr-vendors-collaboration-initiative/2017-01-17-digitalbridge_cdc_ehr_-vendor_final.pdf (Accessed September 26, 2017)
20. Digital Bridge http://www.digitalbridge.us/about/partners/ (Accessed September 26, 2017)
21. Mathes RW, Lall R, Levin-Rector A, Sell J, Paladini M, Konty KJ, et al. (2017) Evaluating and implementing temporal, spatial, and spatio-temporal methods for outbreak detection in a local syndromic surveillance system. PLoS ONE 12(9): e0184419. https://doi.org/10.1371/journal.pone.0184419
22. Reynolds T, Gordon S, Soper P, Buehler J, Hopkins R, Streichert L. Syndromic Surveillance Practice in the United States 2014: Results from a Nationwide Survey. Online Journal of Public Health Informatics. 2015;7(1):e90. doi:10.5210/ojphi.v7i1.5756.
23. Henning, K. (2004). Overview of syndromic surveillance: What is syndromic surveillance? MMWR, 53(Suppl), 5-11.
24. Leonardi GS, Hajat S, Kovats RS, Smith GE, Cooper D, Gerard E. Syndromic surveillance use to detect the early effects of heat-waves: an analysis of NHS direct data in England. Sozial-und Präventivmedizin. 2006 Jul 1;51(4):194-201
25. Josseran L, Caillère N, Brun-Ney D, Rottner J, Filleul L, Brucker G, Astagneau P. Syndromic surveillance and heat wave morbidity: a pilot study based on emergency departments in France. BMC medical informatics and decision making. 2009 Feb 20;9(1):14.
26. Schaffer A, Muscatello D, Broome R, Corbett S, Smith W. Emergency department visits, ambulance calls, and mortality associated with an exceptional heat wave in Sydney, Australia, 2011: a time-series analysis. Environ Health 2012:11(1):3
27. Hripcsak G, Soulakis, ND, Li L et al. Syndromic Surveillance Using Ambulatory Electronic Health Records. JAMIA 2009;16:354-361
28. CDC National Syndromic Surveillance Program https://www.cdc.gov/nssp/overview.html (Accessed September 26, 2017)
29. BioSense https://www.cdc.gov/nssp/biosense/index.html (Accessed August 10, 2017)
30. Syndromic Surveillance https://www.syndromicsurveillance.org/about/index.html (Accessed September 26, 2017)
31. CDC National Syndromic Surveillance Program https://www.cdc.gov/nssp/community.html (Accessed September 26, 2017)
32. https://www.cdc.gov/nbs/overview/index.html (Accessed September 26, 2017)
33. Public Health Informatics Institute. (2010, June 30). Finding common ground: Collaborative requirements development for public health information systems. http://www.phii.org/ (Accessed August 10, 2017)

34. Foldy, S. (2011, February 11). Public health and the Health IT for Economic & Clinical Improvement (HITECH) Act: CDC's roles. www.cdc.gov/phin/library/about/IRGC_on_HITECH_20110211.pdf (Accessed August 10, 2017)
35. CDC State of CDC Information Technology. March 27, 2009. https://www.cdc.gov/od/ocio/state/state-of-cdc-it-v508.pdf (Accessed September 26, 2017)
36. Public Health Information Network https://www.cdc.gov/phin (Accessed September 26, 2017)
37. Shapiro JS, Mostashari F, Hripcsak G et al. Using Health Information Exchange to Improve Public Health. Am J Pub Health 2011;101:616-623
38. Public Health & Electronic Health Records Meaningful Use: Overview https://www.cdc.gov/nssp/events/documents/2.-sanjeev-tandone_nssp_meeting_feb_2017-v002.pdf (Accessed September 26, 2017)
39. Geographic Information Systems www.esri.com/what-is-gis/ (August 10, 2017))
40. Wiki GIS: The GIS Encyclopedia. Retrieved from http://wiki.gis.com/wiki/index.php (Accessed August 10, 2017)
41. CDC. Chronic Disease GIS Exchange https://www.cdc.gov/dhdsp/maps/gisx/index.html (Accessed September 20, 2017)
42. Ghirardelli A, Quinn V, Foerster S. Using geographic information systems and local food store data in California's low-income neighborhoods to inform community initiatives and resources. Am J Pub Health 2010;100(11):2156-2162
43. Centers for Disease Control and Prevention, National Center for Chronic Disease Prevention and Health Promotion, Division for Heart Disease and Stroke Prevention. (2011). Chronic Disease GIS Exchange. http://www.cdc.gov/DHDSP/maps/GISX/index.html (Accessed August 10, 2017)
44. Diabetes Interactive Atlas https://www.cdc.gov/diabetes/data/county.html (Accessed August 10, 2017)
45. State of Obesity https://stateofobesity.org/diabetes/ (Accessed September 20, 2017)
46. Behavioral Risk Factor Surveillance System https://www.cdc.gov/brfss/index.html (Accessed September 14, 2017)
47. Dixon BE, McGowan JJ, Grannis SJ. Electronic laboratory data quality and the value of a health information exchange to support public health reporting processes. AMIA Annu Symp Proc. 2011;2011:322-30.
48. Rajamani S, Roche E, Soderberg K, Bieringer A. Technological and Organizational Context around Immunization Reporting and Interoperability in Minnesota. Online Journal of Public Health Informatics. 2014 12/15;6(3):e192.
49. Jabour AM, Dixon BE, Jones JF, Haggstrom DA. Data Quality at the Indiana State Cancer Registry: An Evaluation of Timeliness by Cancer Type and Year. Journal of Registry Management. 2016 Winter;43(4):168-173.
50. Dixon BE, Zou JF, Comer KF, Rosenman M, Craig JL, Gibson PJ. Using Electronic Health Record Data to Improve Community Health Assessment. Frontiers in Public Health Services and Systems Research. 2016 Oct;5(5):50-6.
51. Fidler DP, "From International Sanitary Conventions to Global Health Security: The New International Health Regulations" Chinese Journal of International Law vol 4, issue 2, p 325-392. http://chinesejil.oxfordjournals.org/content/4/2/325.full (Accessed August 10, 2017)
52. WHO Vector surveillance and control of ports, airports and ground crossing http://www.who.int/ihr/publications/9789241549592/en/ (Accessed September 12, 2017)
53. WHO A global system for alert and response http://www.who.int/ihr/alert_and_response/en/ (Accessed September 12, 2017)
54. WHO Partners: Global Outbreak Alert and Response Network http://www.who.int/csr/disease/ebola/partners/en/ (Accessed September 26, 2017)
55. World Health Organization. http://www.who.int (Accessed August 10, 2017)
56. WHO Global Observatory for eHealth http://www.who.int/goe/en/ (Accessed September 12, 2017)
57. WHO Atlas of eHealth country profiles 2015 http://www.who.int/goe/publications/atlas_2015/en/ (Accessed September 12, 2017)
58. WHO Directory of eHealth policies http://www.who.int/goe/policies/en/ (Accessed September 12, 2017)
59. WHO Collaborating Centre for Health Information and Informatics http://apps.who.int/whocc/Detail.aspx?cc_ref=CHN-112&cc_title=informatics& (Accessed September 12, 2017)
60. CDC Global Public Health Informatics Program https://www.cdc.gov/globalhealth/healthprotection/gphi/index.html (Accessed September 12, 2017)
61. PATH. Leading innovation in global health https://www.path.org/about/index.php (Accessed September 12, 2017)
62. The Task Force for global health https://www.taskforce.org/our-work (Accessed September 12, 2017)
63. HealthData https://www.healthdata.org/ (Accessed September 12, 2017)
64. Global Public Health Intelligence Network (GPHIN). Public Health Agency of Canada. http://www.phac-aspc.gc.ca/index-eng.php (Accessed December 2, 2013) or https://gphin.canada.ca/cepr/

aboutgphin-rmispenbref.jsp?language=en_CA (Accessed August 16, 2017)
65. Heymann, DL. Dealing with Global Infectious Disease Emergencies in Gunn SWA, et.al. Understanding the Global Dimensions of Health. 2005, 169-180
66. Early Warning and Risk Navigation Systems(EWARNS). www.ewarns.com.my/index.php?im=about (Accessed August 10, 2017)
67. HealthData. Global Burden of Disease http://www.healthdata.org/gbd/about (Accessed September 12, 2017)
68. Institute for Health Metrics and Evaluation (IHME). GBD Compare. Seattle, WA: IHME, University of Washington, 2017. Available from http://vizhub.healthdata.org/gbd-compare. (Accessed September 26, 2017.)
69. Jacobsen KH, Aguirre AA, Bailey CL, Baranova AV, Crooks AT, Croitoru A, et al. Lessons from the Ebola Outbreak: Action Items for Emerging Infectious Disease Preparedness and Response. EcoHealth. 2016 March 01;13(1):200-12.
70. Carney TJ, Weber DJ. Public Health Intelligence: Learning From the Ebola Crisis. Am J Public Health. 2015 Sep;105(9):1740-4.
71. Jones KE, Patel NG, Levy MA, Storeygard A, Balk D, Gittleman JL, Daszak P (2008) Global trends in emerging infectious diseases. Nature 451:990–993.
72. Stark, K. D., Arroyo Kuribrena, M., Dauphin, G., Vokaty, S., Ward, M. P., Wieland, B., & Lindberg, A. (2015). One Health surveillance - More than a buzz word? Prev Vet Med, 120(1), 124-130. doi:10.1016/j.prevetmed.2015.01.019
73. Ryan SJ, Walsh PD (2011) Consequences of non-intervention for infectious disease in African great apes. PLoS One 6:29030.
74. Vreeman DJ, Hook J, Dixon BE. Learning from the crowd while mapping to LOINC. J Am Med Inform Assoc. 2015 Nov;22(6):1205-11.
75. Dixon BE, McFarlane TD, Dearth S, Grannis SJ, Gibson PJ. Characterizing Informatics Roles and Needs of Public Health Workers: Results From the Public Health Workforce Interests and Needs Survey. J Public Health Manag Pract. 2015 Nov-Dec;21 Suppl 6, Public Health Workforce Interests and Needs Survey (PH WINS):S130-S40.
76. Public Health Informatics Fellowship Program https://www.cdc.gov/phifp (Accessed September 15, 2017)
77. National Library of Medicine Biomedical Informatics and Data Science Research Training Programs https://www.nlm.nih.gov/ep/GrantTrainInstitute.html (Accessed September 15, 2017)
78. Dixon BE, Gamache RE, Grannis SJ. Towards public health decision support: a systematic review of bidirectional communication approaches. J Am Med Inform Assoc. 2013 May 1;20(3):577-83.
79. Massoudi BL, Chester KG. Public Health, Population Health, and Epidemiology Informatics: Recent Research and Trends in the United States. Yearb Med Inform. 2017;26(01):241-7
80. Vest JR, Harle CA, Schleyer T, Dixon BE, Grannis SJ, Halverson PK, et al. Getting from here to there: health IT needs for population health. The American journal of managed care. 2016 Dec;22(12):827-9.
81. Williams KS, Shah GH. Electronic Health Records and Meaningful Use in Local Health Departments: Updates From the 2015 NACCHO Informatics Assessment Survey. Journal of Public Health Management and Practice. 2016 Nov-Dec;22(Suppl 6):S27-S33.
82. Hripcsak G, Forrest CB, Brennan PF, Stead WW. Informatics to support the IOM social and behavioral domains and measures. J Am Med Inform Assoc. 2015 Jul;22(4):921-4.
83. Schomberg JP, Haimson OL, Hayes GR, Anton-Culver H. Supplementing Public Health Inspection via Social Media. PloS one. 2016;11(3):e0152117.
84. Lazer D, Kennedy R. What we can learn from the epic failure of google flu trends. Wired. October 1, 2015. https://www.wired.com/2015/10/can-learn-epic-failure-google-flu-trends/ (Accessed September 26, 2017)
85. FrontlineSMS:Medic. http://medic.frontlinesms.com/ (Accessed August 10, 2017)

20

eResearch

JOHN SHARP

LEARNING OBJECTIVES

After reading this chapter the reader should be able to:

- Understand the scope of eResearch and Clinical Research Informatics within the clinical research workflow
- Describe the use of EHR data in various phases of research including research originating from EHR data
- Conceptualize how informatics tools can be utilized in recruiting subjects for clinical research
- Detail how informatics supports the ongoing management of clinical trials
- Review the new trends in big data, real-time analytics and data mining

INTRODUCTION

Simply stated, eResearch refers to the use of information technology to support research. Within the past ten years, there has been a dramatic shift from paper-based records in research to almost completely electronic. Paper case report forms being transposed into spreadsheets or early database programs are rapidly disappearing. Now every aspect of clinical research is supported by informatics tools. Several factors enabled this rapid change: availability of open source programming, major support from the National Center for Research Resources of the National Institutes of Health for informatics, consolidation of field of clinical research informatics with the American Medical Informatics Association (AMIA), and academic medical centers' move toward securing patient data as a result of HIPAA and HITECH. These forces accelerated the move toward informatics permeating clinical research. But the most significant change is the adoption of electronic health records. In a perspective from the New England Journal of Medicine titled *"Evidence Based Medicine in the EHR Era"* the authors give examples of how an electronic cohort of patient data in the electronic health record (EHR) can be used in clinical decision support. They conclude: *"the growing presence of EHRs along with the development of sophisticated tools for real-time analysis of de-identified data sets will no doubt advance the use of this data driven approach to health care delivery."*[1] There is no doubt that health informatics and specifically eResearch will have a major impact on evidence-based medicine in the future. In fact, there are now informatics solutions for every phase of the research process. This chapter will explore the current state of these tools and their usefulness in promoting clinical research.

PREPARATORY TO RESEARCH

The first step for a researcher with a question is to research the literature. It has been well documented that the medical literature is growing at a rate which overwhelms the practicing physician and the clinical researcher. Informatics tools are increasingly needed to assist with sorting through the literature and creating a reasonable background for any study. Fortunately, PubMed offers an array of tools which can be utilized on the site or integrated into a website or application using web services and RSS feeds. Entrez Programming Utilities provide a catalog of XML scripts as well as Perl scripts and other tools for custom extraction of medical journal data. A mobile version is also available.[2] The National Library of Medicine sponsors App contests to improve searches and create visualizations to improve data analysis. Google Scholar provides a broader database search which includes PubMed but also other scientific and academic publications (scholar.google.com). Google Books provides access to excerpts of books and allows searches through published works as well (books.google.com). For more details on online medical resources and medical information retrieval, please see additional chapters.

ClinicalTrials.gov provides the researcher with a search of all registered clinical trials within the U.S. As with PubMed, ClinicalTrials.gov provides an open API (Application Programming Interface) for linking and XML for connecting through web services.[3] For a wider search, the World Health Organization (WHO) provides a search tool which incorporates international trials.[4] Both PubMed and WHO now have mobile versions of the clinical trial search tools.

Research collaboration networks have seen significant growth in recent years. Research networks are typically web-based applications which include features similar to other social networks, such as a personal profile, opportunities to connect with others with similar interests and the ability to post status updates. Often research networks have personal profiles of researchers pre-populated with publications (thanks to integration with PubMed) and clinical trials (integration with ClinicalTrials.gov) and grants through the NIH Exporter.[5] With these rich data sources, some research networks have created semantic connections between researchers (vivoweb.org). However, most research networks provide search tools to enable finding connections between those with common interests. Three tools stand out, although many have been developed:

- **Vivo**. Vivo is an open source tool developed at Cornell University and a semantic web application (common framework that allows data to be shared and reused across application)[6]
- **Harvard Profiles Catalyst**. This is an open source community of over 130-member institutions with built-in network analysis and data visualization tool[7]
- **SciVal Experts**. This commercial solution also has modules to find research funding and measure benchmarks.[8]

Other available tools generate National Institutes of Health (NIH) biosketches (https://www.ncbi.nlm.nih.gov/sciencv/) and add publications and grants dynamically (see Figure 20.1). Since research networks are relatively new, there is not substantial evidence of their effectiveness beyond anecdotal examples.

In addition to finding collaborators, clinical researchers would like to know the feasibility of their studies before they initiate them. One approach enabled by electronic health record data is doing queries to evaluate adequate pools of patients to be recruited into the study. This requires a clinical data repository from EHR data using a query tool to search de-identified clinical information. By modifying inclusion and exclusion criteria, a researcher can find the appropriate cohort for recruitment based on a reasonable recruitment rate. There are already successful examples of this that have saved years of unsuccessful or under recruited studies.[9]

Electronic grant submission is now common for government agencies. Through the Office of Extramural Research at the NIH, grant submission and award management are all web-based. Forms are completed online and uploaded to the site, email alerts are available about the posting of new grants, and grant awards are posted online at the site. In addition, some Clinical Trial Management Systems (CTMS) discussed later are integrated with the NIH electronic Research Administration Commons (eRA), which allows institutions to centrally manage grant submissions to the NIH (https://era.nih.gov/).

STUDY INITIATION

Informatics has a role in the initiation of studies as well. Volunteer recruitment can be enabled over the Internet. Two approaches to volunteer recruitment are ResearchMatch and TrialX.

ResearchMatch provides a way to connect patients seeking clinical trials and researchers seeking volunteers[10] (see Figure 20.2). Volunteers can create an account and indicate what their health issues are that may match with clinical trials. Researchers from institutions affiliated with the network can enter the clinical trials and contact information by completing an online form. Then the researcher can search volunteers and email them an invitation to participate. The volunteer can accept or decline to receive more information.[11]

TrialX is a commercial venture which allows the volunteer to search clinical trials from ClinicalTrials.gov. Based on search terms, the user can see how closely their search matches available trials and then select a trial and email the investigator by registering on the site. Researchers can also register to list their trials and organizations can partner with TrialX to create custom listings of their studies.[12] Yet another model is a social network built around volunteering for clinical trials. ArmyOfWomen provides that platform and has provided thousands of volunteers for dozens of trials, initially for breast cancer but now for a variety of conditions.[13]

Online recruitment of subjects using social media is an emerging trend. Information on clinical trials using major social media outlets like Facebook and Twitter are new ventures. A transition from traditional advertising to

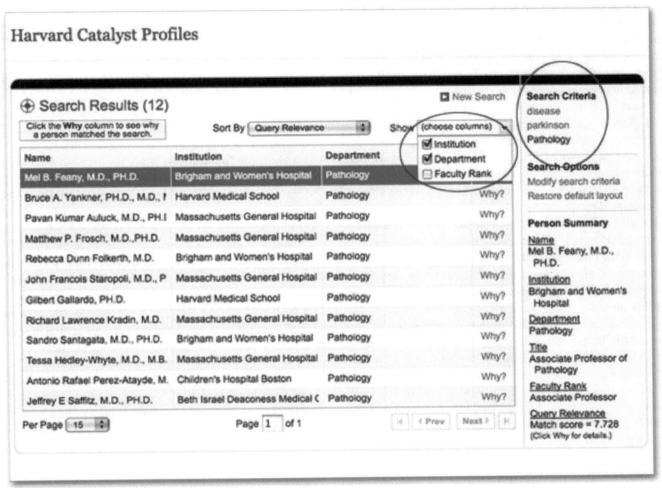

Figure 20.1: Example of a search from Harvard Profiles Catalyst

Figure 20.2: ResearchMatch program (Courtesy ResearchMatch)

online promotion of clinical trials is growing but many Institutional Review Boards (IRBs) are unfamiliar with this approach and need education to promote acceptance and establish standards for appropriate use. Another promising use of social media is provider groups and researchers developing relationships with online patient networks. These groups of ePatients are receptive to clinical trials and partnerships with researchers. A successful partnership was documented between women who have a rare cardiac condition and the Mayo Clinic. The women, whom already had an online community, were eager to participate in trials.[14] Patient social networks are already collecting data on their treatments and so the word about new clinical trials travels quickly.[15] Many healthcare organizations still caution patients and employees from using social media; in this context, patients should be cautioned that information on clinical trials communicated through social media must be evaluated like other online health content, with a critical mind.

Recruitment of subjects through the capabilities of the electronic health record has two possible modes. First, the EHR can be used to find cohorts of eligible patients and create patient contact lists for recruitment.[16] Second, clinical trial alerts can be embedded within the EHR based on diagnoses, lab tests or other patient characteristics. The alert would typically remind the provider that their patient may be eligible for a clinical trial and who to contact.[17-18]

STUDY MANAGEMENT AND DATA MANAGEMENT

There are several informatics tools which support study management and particularly managing research data. Clinical trial management systems (CTMS) are now common in academic medical centers. The purpose of these tools is to manage the planning, preparation, performance, and reporting of clinical trials. A CTMS has multiple functions in study management including: budget management, study calendar of patient visits, and creating electronic case report forms (eCRFs).[18] These tools can be open source or commercially available products.[19]

Some applications provide eForms or eCRFs with a focus on study data management. These tools enable the building of web-based forms for research without the support of programmers. Probably the most widely distributed tool is Research Electronic Data Capture (REDCap), which was developed at Vanderbilt University. REDCap provides a secure, web-based application based on PHP and MySQL, which can be installed locally and provides an online designer for creating data collection instruments. REDCap also provides a method for controlling user rights and user access groups as well as maintaining an audit trail[20-21] (see Figure 20.3).

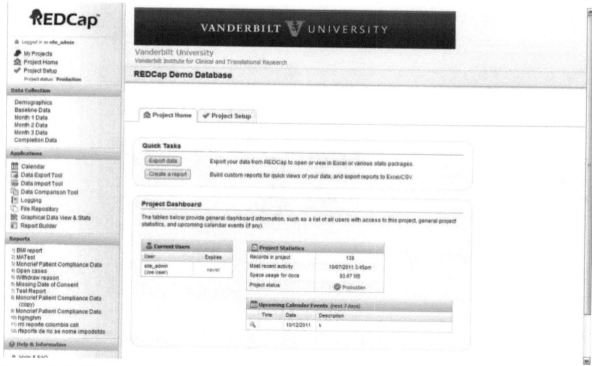

Figure 20.3: REDCap program (Courtesy Vanderbilt University)

OpenClinica is another example of a data management system. It is an open source tool which provides the ability to submit and extract data, manage protocols and other study administration tools. It enables compliance with Good Clinical Practice (GCP) and regulatory guidelines such as FDA regulations for electronic databases. OpenClinica provides a free community edition and a licensed enterprise addition.[22] CAISIS Cancer Data Management System, developed at Memorial Sloan Kettering Cancer Center, is an open source .NET application which provides eForms for study data collection. CAISIS has an active open source community supporting and enhancing the application. There are also some tools within CAISIS to import data from clinical systems.[23]

Integration of EHR data into clinical trials provides an efficient method to add routine data into the study database. While this feature is rarely available within commercial EHRs, the data from EHRs or clinical data warehouses can be exported on study patients and then imported into study data management systems. The challenge is selecting the appropriate data, such as lab results from study visits, and exporting only that data. Some commercial data management systems have tools to automate this process. An important part of secondary use of EHR data for research should include a validation process to ensure that data which was collected in clinical care is appropriate for a research study or registry.[25]

EHR data can be used exclusively to produce a variety of study types. For instance, epidemiologic research, studying population characteristics or trends, can be extracted from EHRs containing large groups, such as, from regional or national health systems. Biosurveillance studies are also enabled by EHR data. With daily or near real-time data on large populations, outbreaks of new infections or other disease trends can be tracked. Biosurveillance using EHR data has also been shown as a method of diagnosing strep pharyngitis in real time.[26]

Identification of risk factors has been demonstrated through the use of EHR data. For instance, a study from Harvard demonstrated the ability to rapidly identify risk of stroke associated with diabetes medication using signal detection analysis.[27] Another study, from Cleveland Clinic, used EHR data to predict six-year mortality risk in type 2 diabetes.[28] The Archimedes Model developed by David Eddy, provides predictive modeling for diabetes.[29] In addition to predictive studies, EHR data has been used in identifying post-operative complications, medication adherence and triggered adverse event reporting.[30-31] From these and other uses, it is clear that decision support is increasingly being supported by EHR data.

Comparative effectiveness research (CER) is of increasing interest related to healthcare reform and research sponsored by the Agency for Healthcare Research and Quality (AHRQ).[32] EHR data can answer some questions that clinical trials cannot and can often do so more quickly. Hoffman and Podgurski propose using EHR data to develop personalized comparisons of treatment effectiveness, applying the rich clinical data to decision support in a personalized medicine approach.[35] Observational studies, which infer (but not prove) causation from EHR data, can examine large cohorts who receive different treatments and then evaluate the outcomes and costs associated with each. A study of diabetes management of 27,207 patients demonstrated the comparative effectiveness of using EHRs as opposed to paper records; showing greater improvement in disease outcomes for those managed with EHRs.[33] The National Academy of Medicine (previously known as the Institute of Medicine) has developed a substantial workshop summary on the "*Infrastructure Required for Comparative Effectiveness Research*" which includes not only better research design, but a move from "siloed" evidence based medicine to "*semantically integrated, information-based medicine*" which requires "*a substantial informatics platform to interpret, query and explore clinical data.*"[34]

What to do about data that is not routinely collected in EHRs? For instance, what about disease specific information which may be helpful in populating a disease registry? The solution is the use of smart forms within the EHR which are specific to a specialty clinic or treatment protocol. These forms must be designed with care to gather discrete clinical observations and judgments while being easy to complete in a busy clinical environment. Back-end integration with EHR data structure is essential.[35]

Collection of research data using medical devices is another informatics challenge. With more medical devices being integrated with the EHR or generating their own databases, a significant amount of new clinical monitoring data is available for research. Whether these are EKG monitors, automated anesthesia records, implanted devices[36] or activity sensors[37] data collection from medical devices provides a method to quickly acquire research data for analysis.

Patient Reported Outcomes (PROs) is another area of growing emphasis in clinical research with the National Institutes of Health developing a program called PROMIS to focus attention on it. PROs "*is the term used to denote health data that is provided by the patient through a system of reporting.*"[38] In the context of PROs, the use of tablet devices is gaining popularity as a method for collecting patient reported data at the point of care, such as, in the study of pain[39] or cognitive impairment.[40] Tablets also have broader uses, including social networking and cataloging relevant articles for research.[41] Studies in patient reported outcomes are now being funding by the Patient Centered-Outcome Research Institute (PCORI), a private organization funded by the U.S. government to promote outcomes research which includes PRO.[42]

On a similar vein, the use of Patient Generated Health Data (PGHD) from apps, devices and questionnaires provides another method for data collection for research.[43]

DATA MANAGEMENT SYSTEMS FOR FDA REGULATED STUDIES

The unique requirements of the Food and Drug Administration (FDA) for data management for studies of new drugs and devices present challenges for informatics. The regulation 21 CFR Part 11: Electronic Records, Electronic Signatures[44] sets a high bar for implementing data management systems and their validation. In addition to selecting a system which is compatible with the regulatory requirements, significant validation test cases must be developed and executed. While this area is typically the purview of drug/device companies or contract research organizations, academic medical centers often require this capability to support early stage, investigator-initiated studies. Commercial systems such as PhaseForward [45] and Oracle Clinical [46] dominate this market, but open source tools like OpenClinica can also be validated in compliance with these regulations. Remote Data Capture (RDC) is a term often used for these systems which enable secure data collection over multiple study sites for large clinical trials.

INTERFACES AND QUERY TOOLS

In recent years, Clinical Data Repositories and Registries using EHR data have been developed at many academic medical centers. A review by Weiner et al. discusses four such systems with a variety of features.[47] One more broadly adopted tool, supported through the National Center for Research Resources (NCRR), is i2b2 (Informatics for Integrating Biology and the Bedside) which enables the secure storage and query of EHR and other data.[48] A query tool developed in the United Kingdom called TrialViz allows for searching by phenotype and data quality.[49]

Stanford University is creating their own clinical data repository called the Stanford Translational Research Integrated Database Environment (STRIDE).[50] This repository has five functions: *"Anonymized Patient Research Cohort Discovery, Electronic Chart Review for Research, IRB-Approved Clinical Data Extraction, Biospecimen Data Management, Data Management and Research Registries."* Registries will become an even more important tool to track patients with chronic and rare diseases.

To support these large clinical data repositories, tools which support data mapping, semantic ontologies, and natural language processing have been developed. The National Center for Biomedical Ontology provides a repository of tools through its Bioportal for medical ontology standards and mapping (see Figure 20.4).[51]

Figure 20.4: Bioportal Ontology Search (Courtesy NCBO)

Wynden et al. note that the two main challenges in maintaining an integrated data repository for research are, *"the ability to gain regular access to source clinical systems and the preservation of semantics across systems during the aggregation process."*[52] Natural Language Process (NLP) is required when one seeks to mine clinical text notes, such as encounter notes, operative notes, radiology reports and discharge summaries. Many centers are developing such systems, such as cTAKES developed by the Mayo Clinic.[53] It can examine notes and extract data elements based on structured vocabularies, such as LOINC® (Logical Observation Identifiers Names and Codes) for laboratory values.[54]

Health information exchange (HIE) is another technology which has potential for clinical research. Although developed primarily to enable care across health systems and states with various EHR implementations, it can be used in a de-identified mode to mine data for state or national trends including public health research.

Web services continue to expand in their support of many of the technologies noted above. Particularly the use of APIs, such as FHIR (Fast Healthcare Interoperability Resources) that uses REST and the registry i2b2.[55]

The category of big data is now being defined in healthcare, not just business. Big data is typically defined in the multiple terabyte or petabyte range and creates unique management problems in traditional relational databases. Often, this scale of data requires cloud computing solutions for storage and analysis. A new focus on NOSQL databases and a group of tools developed by the Apache Foundation is called Hadoop. *"The Apache Hadoop software library is a framework that allows for the distributed processing of large data sets across clusters of computers using a simple programming model."*[56] While some of the initial applications of these NOSQL databases are in genomics, other research applications, such as, exploring PACS (radiology images)[57] and multisite clinical trials may be future applications.[58] New analytic tools for large sets of EHR data are enabling data exploration. Hadoop is becoming essential to healthcare because 80% of healthcare information is unstructured data.[59] The practice of combining phenotypic from the EHR and genomic data is relatively new but shows promise is researching specific diseases with genetic markers.[60]

DATA ANALYSIS

While Clinical Research Informatics has traditionally left the statistical analysis tools to their biostatistical partners, with the wealth and volume of clinical data now available, some role in data analysis is appropriate. With tools like The R Project for Statistical Computing, an open source statistical package,[61] there is the potential for integration of the statistical package with the data repository.[62] Tools like REDCap provide access to their API (Application Programming Interface) to connect directly to statistical programs. SAS also provides for integration of patient data from a variety of sources with tools for data cleaning, standardization and exploration.[63]

Data visualization has progressed beyond simple charts and graphs to a part of informatics which enables the researcher to see data patterns as part of data exploration and planning for analysis. Data visualization in research is in its early stages, so new approaches for how to visualize data need to be created and standardized. But when done well, visualization can help detect errors in the data and explore relationships.[64] The selection of visualization tools is key, and informaticists can aid in the selection of these tools as they do with other software. Tools like Tableau,[65] Acesis[66] and functions embedded in statistical packages like SAS should be considered.

Real time analytics are also helpful tools for dealing with large datasets and clinical decision support. Real time analytics is the provision of analyzed data relatively instantly to support decision making. While this approach is relatively new in medicine, IBM's Watson project is proposing to provide this kind of service. This is closely tied to predictive analytics based on clinical data including discrete data, text and unstructured data.[67]

RECOMMENDED READING

The following articles are recommended for supplemental reading on e-research:

- *Evidence Generating Medicine:Redefining The Research-Practice Relationship To Complete The Evidence Cycle.* The authors review the technical, regulatory, fiscal and socioeconomic challenges facing clinical research. They maintain the relationship between clinical medicine and research be bi-directional. That is, not only should research results drive the practice of medicine but the practice of medicine should drive research.[68]
- *Clinical Research Informatics: A Conceptual Perspective.* A conceptual model of clinical research informatics (CRI) is presented. The authors used the model to discuss 18 articles that were devoted to CRI in one issue of JAMIA.[69]
- *Time To Integrate Clinical And Research Informatics.* The authors plead the case to combine clinical and research informatics in order to improve patient care and create a *"learning healthcare system."* They also outline known *"bottlenecks"* associated with the potential integration.[70]
- *A Survey Of Informatics Platforms That Enhance Distributed Comparative Effectiveness Research Using Multi-Institutional Heterogeneous Clinical Data.* Authors discuss what is needed in order for there to be effective comparative effectiveness research (CER) among disparate research organizations. They note that there are six large informatics platforms for CER being studied and they identified six steps towards successful CER among multi-institutions.[71]
- *Electronic Health Record–Enabled Research in Children Using the Electronic Health Record for Clinical Discovery.* The authors present a systematic approach to the use of EHR data in research including data validation and the development of research cohorts.[72]

FUTURE TRENDS

The future of eResearch is leading toward the nationwide learning healthcare system as described by the National Academy of Medicine. With the number of tools in active use as described in the chapter, further use and enhancement of these informatics resources combined with the broad adoption of EHRs, make huge amounts of clinical data available for analysis and further discovery. Research networks will enable collaboration that was not possible a decade ago. Research volunteer recruitment, which has been chronically low, can see new opportunities through web-based tools and social media. Study and data management, tied to paper records for so long, are now freed in a digital form for secondary use. Biosurveillance can detect new outbreaks in hours instead of weeks. Data poor registries now have the opposite challenge – large data and how to store and manage it. E-Research will enable researchers to reduce the time from *"the creation and validations of new biomedical knowledge and translation of that knowledge into practice."*[73]

KEY POINTS

- eResearch and Clinical Research Informatics have a role within every aspect of the clinical research workflow
- EHR data can be effectively utilized in clinical trials, registries, public health studies and can include research originating from EHR data
- Informatics tools are effective in recruiting subjects for clinical research
- Informatics supports the ongoing management of clinical trials including study calendars, data management, grant management and subject recruitment and consent.
- New trends include: big data, real-time analytics and data mining

CONCLUSION

The emergence of clinical research informatics as a field within bioinformatics has been made possible by major advances in technology and institutional support. Every aspect of clinical research now has a set of tools to support its processes. A mix of commercial-off-the-shelf tools, software-as-a-service applications (SaaS) and open source tools developed at academic medical centers have enabled this transformation. The growing availability of EHRs nationally is just beginning to make a contribution to clinical research and is poised to become a standard method for comparative effectiveness and population-based research. New devices, such as, tablets and smart phones, and the ability to obtain data from medical devices, increase the amount of data available for research. Data analysis and visualization tools enable researchers to quickly turn the data into usable information. eResearch is now maturing as a field of informatics.

REFERENCES

1. Frankovich J., Longhurst C.A., Sutherland S.M. Evidence Based Medicine in the EHR Era. N Engl J Med 2011; 365:1758 – 1759
2. National Center for Biotechnology Information: Entrez Programming Utilities Help: http://www.ncbi.nlm.nih.gov/books/NBK25500/ (Accessed October 1, 2017)
3. ClinicalTrials.gov http://clinicaltrials.gov/ct2/info/linking (Accessed October 1, 2017)
4. World Health Organization: International Clinical Trials Registry Platform Search Portal http://apps.who.int/trialsearch/ (Accessed October 1, 2017)
5. National Institutes of Health: ExPORTER https://exporter.nih.gov/ (Accessed October 1, 2017)
6. VIVO http://www.vivoweb.org (Accessed October 1, 2017)
7. Profiles Research Networking Software http://profiles.catalyst.harvard.edu/ (Accessed October 1, 2017)
8. SciVal https://www.elsevier.com/solutions/scival (Accessed October 1, 2017)
9. Reddel HK et al, Assessment of US Electronic Medical Records to Guide Feasibility and Design of the NOVELTY Study. American Journal of Respiratory and Critical Care Medicine 2017;195:A2032
10. Research Match https://www.researchmatch.org (Accessed October 1, 2017)
11. Harris PA, Scott KW, Lebo L, Hassan N, Lighter C, Pulley J. ResearchMatch: A National Registry to Recruit Volunteers for Clinical Research. Academic Medicine, 87:1, 1-8, 2012.
12. TrialX http://trialx.com (Accessed October 1, 2017)
13. Army of Women http://www.armyofwomen.org/ (Accessed October 1, 2017)

14. Tweet MS, Gulati R, Aase LE, Haynes SN. Spontaneous Coronary Artery Dissection: A Disease-Specific, Social Networking Community-Initiated Study. Mayo Clinic Proceedings September 2011 vol. 86 no. 9 845-850.
15. Wicks P, Massagli M, Frost J, Brownstein C, Okun S, Vaughan T, Bradley R, Heywood J. Sharing Health Data for Better Outcomes on PatientsLikeMe. Med Internet Res. 2010 Apr-Jun; 12(2): e19.
16. Pickett M, Sharp JW. Research Recruitment in Anesthesia Using EMR Data. American Medical Informatics Association Clinical Research Informatics Summit, 2011.
17. Embi PJ, Jain A, Clark J, Bizjack S, Hornung R, HarrisCM. Effect of a clinical trial alert system on physician participation in trial recruitment. Arch Intern Med 2005 Oct 24; 165(19): 2272-7.
18. Baum S. Penn Medicine to expand use of EMR software that finds clinical trial candidates, MedCity News, Oct. 31, 2011, http://www.medcitynews.com/2011/10/penn-medicine-to-expand-pilot-using-emr-app-for-clinical-trial-candidates/ (Accessed October 1, 2017)
19. Leroux H, McBride S, Gibson S. On selecting a clinical trial management system for large scale, multi-centre, multi-modal clinical research study. Stud Health Technol Inform. 2011; 168: 89-95.
20. Geyer J, Myers, K, Vander Stoep A, McCarty C, Palmera N, DeSalvo A, Implementing a low-cost web-based clinical trial management system for community studies: a case study. Clin Trials October 2011 vol. 8 no. 5 634-644.
21. Research Electronic Data Capture (REDCap) https://projectredcap.org/ (Accessed October 1, 2017)
22. Harris P, Taylor R, Thielke R, Payne J, Gonzalez N, Conde J, Research electronic data capture (REDCap) - A metadata-driven methodology and workflow process for providing translational research informatics support, J Biomed Inform. 2009 Apr;42(2):377-81.
23. OpenClinica http://www.openclinica.com (Accessed October 1, 2017)
24. Caisis http://www.caisis.org/ (Accessed October 1, 2017)
25. Navaneethan SD, Jolly SE, Schold JD, Arrigain S, Saupe W, Sharp J, Lyons J, Simon JF, Schreiber MJ Jr, Jain A, Nally JV Jr. Development and validation of an electronic health record-based chronic kidney disease registry. Clin J Am Soc Nephrol. 2011 Jan;6(1):40-9
26. Fine AM, Nizet V, Mandl KD. Improved diagnostic accuracy of Group A Streptococcal Pharyngitis with use of real-time Biosurveillance. Ann Intern Med 2011; 155:345-352.
27. Brownstein JS, Murphy SN, Goldfine AB, Grant RW, Sordo M, Gainer V, Colecchi JA, Dubey A, Nathan DM, Glaser JP, Kohane IS. Diabetes Care 33:526-531, 2010.
28. Wells BJ, Jain A, Arrigain S, Yu C, Rosenkrans WA, Kattan MW. Predicting 6-year mortality risk in patients with Type 2 Diabetes. Diabetes Care 31:2301-2306, 2008.
29. Stern M, Williams K, Eddy D, Kahn R. Validation of prediction of diabetes by the Archimedes Model and comparison with other predicting models. Diabetes Care 31: 1670-1671, 2008.
30. Murff HJ, FitzHenry F, Matheny ME, Gentry N, Kotter, KL, Crimin K,, Dittus RS, Rosen AK, Elkin PL, Brown SH, Speroff T. Automated identification of postoperative complications within an electronic medical record using natural language processing. JAMA 2011;306(8):848-855.
31. Linder JA, Haas JS, Iyer A, Labuzetta MA, Ibara M, Celeste M, Getty G, Bates DW. Secondary use of electronic health record data: spontaneous triggered adverse drug event reporting. Pharmacoepidemiol Drug Saf. 2010 Dec;19(12):1211-5.
32. Agency for Healthcare Research and Quality (AHRQ): Comparative Effectiveness Research Grant and ARRA Awards https://grants.nih.gov/recovery/ (Accessed October 1, 2017)
33. Hoffman S, Podgurski A. Improving health care outcomes through personalized comparisons of treatment effectiveness base on electronic health records. Journal of Law, Medicine & Ethics, Fall 2011, 425-436.
34. Cebul RD, Love TE, Jain AK, Herbert CJ. Electronic health records and the quality of diabetes care. N Engl J Med 365; 9, 2011:825-833.
35. Olsen LA, Grossmann C, McGinnis JM. Learning What Works: Infrastructure Required for Comparative Effectiveness Research. Institute of Medicine, The National Academies Press, 2011, p.35.
36. Olsha-Yehiav M, Palchuk MB, Chang FY, Taylor DP, Schnipper JL, Linder JA, Li Q, Middleton B. Smart forms: Building condition-specific documentation and decision support tools for ambulatory EHR. AMIA Annu Symp Proc. 2005: 1066.
37. Greenlee R, Magid D, Go A, Smith D, Reynolds K, Gurwitz J, Cassidy-Bushrow A, Jackson N, Glenn K, Hammill S, Kadish A, Varosy P, Suits M, Garcia-Montilla R, Vidaillet H, Masoudi F. PS2-15: Linking Disparate Data Sources to Evaluate Implantable Cardioverter Defibrillator Outcomes in the Cardiovascular Research Network: Initial Lessons. Clin Med Res. 2011 Nov;9(3-4):151.
38. Stanley KG, Osgood ND, The Potential of Sensor-Based Monitoring as a Tool for Health Care, Health

39. PROMIS: What Patient Related Outcomes (PROs) Are http://www.nihpromis.org/Patients/PROs (Accessed October 1, 2017)
40. Minton O, Strasser F, Radbruch L, Stone P. Identification of Factors Associated with Fatigue in Advanced Cancer: A Subset Analysis of the European Palliative Care Research Collaborative Computerized Symptom Assessment Data Set. Journal of Pain and Symptom Management, 2011.03.025.
41. Kim H. Exploring technological opportunities for cognitive impairment screening. ACM CHI Conference on Human Factors in Computing Systems, 2011. http://dl.acm.org/citation.cfm?id=1979512 (Accessed October 1, 2017)
42. FDA: Code of Federal Regulations Title 21 http://www.accessdata.fda.gov/scripts/cdrh/cfdocs/cfcfr/cfrsearch.cfm?cfrpart=11 (Accessed October 1, 2017)
43. Accenture Federal Services, Conceptualizing a Data Infrastructure for the Capture, Use, and Sharing of Patient-Generated Health Data in Care Delivery and Research through 2024. https://www.healthit.gov/sites/default/files/Draft_White_Paper_PGHD_Policy_Framework.pdf (Accessed October 1, 2017)
44. Patient-Centered Outcomes Research Institute. https://www.pcori.org/ (Accessed October 1, 2017)
45. Oracle: Phase Forward http://www.phaseforward.com/ (Accessed October 1, 2017)
46. Oracle Clinical http://www.oracle.com/us/products/applications/health-sciences/e-clinical/clinical/ (Accessed October 1, 2017)
47. Weiner MG, Lyman JA, Murphy S, Weiner M. Electronic health records: high-quality electronic data for higher-quality clinical research. Informatics in Primary Care, Volume 15, Number 2, June 2007, pp. 121-127(7).
48. Murphy SN, Gainer V, Mendis M, Churchill S, Kohane I. Strategies for maintaining patient privacy in i2b2. J Am Med Inform Assoc. 2011 Oct 7.
49. Tate AR, Beloff N, Al-Radwan B, Wickson J, Puri S, Williams T, Van Staa T, Bleach A. Exploiting the potential of large databases of electronic health records for research using rapid search algorithms and an intuitive query interface. J Am Med Inform Assoc. 2013 Nov 22. doi: 10.1136/amiajnl-2013-001847.
50. Stanford Medicine Cohort Discovery Tool http://med.stanford.edu/researchit/tools/cohort-tool.html (Accessed October 1, 2017)
51. BioPortal http://bioportal.bioontology.org/ (Accessed October 1, 2017)
52. Wynden R, Weiner MG, Sim I, Gabriel D, Casale M, Carini S, Hastings S, Ervin D, Tu S, Gennari JH, Anderson N, Mobed K, Lakshminarayanan P, Massary M, Cucina RJ. Ontology Mapping and Data Discovery for the Translational Investigator. AMIA Summits Transl Sci Proc 2010, 2010:66-70.
53. http://ctakes.apache.org (Accessed October 1, 2017)
54. Logical Observation Identifiers Names and Codes (LOINC) http://loinc.org/ (Accessed October 1, 2017)
55. Wagholikar KB et al. SMART-on-FHIR implemented over i2b2. Journal of the American Medical Informatics Association, Volume 24, Issue 2, 1 March 2017, Pages 398–402.
56. Apache Hadoop http://hadoop.apache.org/ (Accessed October 1, 2017)
57. Apache Hadoop http://hadoop.apache.org/ (Accessed October 1, 2017)
58. ACM Digital Library: An application architecture to facilitate multi-site clinical trial collaboration in the cloud http://dl.acm.org/citation.cfm?id=1985511 (Accessed October 1, 2017)
59. Pallavi Poojary, Big Data In Healthcare: How Hadoop Is Revolutionizing Healthcare Analytics. https://www.edureka.co/blog/hadoop-big-data-in-healthcare (Accessed October 1, 2017
60. Denny JC, Bastarache L, Ritchie MD, et al. Systematic comparison of phenome-wide association study of electronic medical record data and genome-wide association study data. Nat Biotechnol. 2013 Nov 24;31(12):1102-1111. doi: 10.1038/nbt.2749.
61. The R Project for Statistical Computing http://www.r-project.org/ (Accessed October 1, 2017)
62. Hothorn T, James D A, Ripley BD. R/S Interfaces to Databases. DSC 2001 Proceedings of the 2nd International Workshop on Distributed Statistical Computing. http://www.ci.tuwien.ac.at/Conferences/DSC-2001/Proceedings/HothornJamesRipley.pdf (Accessed October 1, 2017)
63. SAS: Healthcare Analytics https://www.sas.com/en_us/industry/health-care-providers.html (Accessed October 1, 2017)
64. Fox P, Hendler J. Changing the Equation on Scientific Data Visualization. Science 331:705-708, 11 February 2011.
65. Tableau http://www.tableau.com/ (Accessed October 1, 2017)
66. Acesis http://www.acesis.com (Accessed October 1, 2017)
67. IBM: Predictive Analytics for Healthcare https://www.ibm.com/blogs/watson-health/6-ways-leverage-predictive-modeling-phm/ (Accessed October 1, 2017)
68. Embi PT, Payne PR. Evidence generating medicine: redefining the research-practice relationship

69. Kahn MG, Weng C. Clinical Research Informatics: a conceptual perspective. J Am Med Inform Assoc. 2012;19:236-242. Open Access.
70. Katzan IL, Rudick RA. Time to integrate clinical and research informatics. Sci Transl Med 2012;4(162):162fs41
71. Sittig D, Hazlehurst Bl, Brown J et al. A survey of informatics platforms that enhance distributed comparative effectiveness research using multi-institutional heterogeneous clinical data. Med Care. 2012;50:S49-S59
72. Sutherland SM et al. Electronic Health Record–Enabled Research in Children Using the Electronic Health Record for Clinical Discovery. Pediatric Clinics, April 2016, Volume 63, Issue 2, Pages 251–268
73. Friedman CP, Wong AK, Blumenthal D. Achieving a Nationwide Learning Health System. Science Translational Medicine 2:57, 1-3, 10 November 2010.

to complete the evidence cycle. Med Care. 2013;51(8suppl3):581-591

21

International Health Informatics

ALISON FIELDS • CHRIS PATON • GLEBER NELSON MARQUES
NAOMI MUINGA • STEVE MAGARE • ROBERT HOYT

"Every 10 seconds we lose a child to hunger. This is more than HIV/AIDS, malaria and tuberculosis combined."
—Josette Sheeran President and CEO Asia Society

LEARNING OBJECTIVES

After reading this chapter the reader should be able to:

- Describe innovative international eHealth projects
- Differentiate between different national strategic approaches to health informatics
- Detail the way economic and infrastructure issues impact health informatics projects in low and middle income countries (LMIC)
- Describe how mobile health (mHealth) technology is enabling developing countries to access healthcare information in the absence of formal infrastructure

INTRODUCTION

It is worth reflecting on the considerable progress the international community has achieved in a relatively short period of time. In regions where data are available, the uptake of electronic health records (EHRs) and other health information systems (HISs), has grown considerably in the past 10 years.

In this chapter, the progress and challenges in informatics from multiple developed and developing countries will be discussed. While that is the stated purpose of the chapter, we should mention that the basic healthcare systems in these countries vary considerably. For example, Australia, Canada, France, Germany, Switzerland and the US typically use the fee-for-service model, whereas the Netherlands, New Zealand, Norway and the UK use a combination of capitation and fee-for-service. Additionally, primary care practices tend to be smaller in size throughout the Netherlands, France, Germany and Switzerland, but much larger in countries such as Australia, Canada, and the United States.[1]

The higher levels of adoption have been realized by smaller, more developed countries that have implemented nationwide eHealth strategies. European countries including Norway, Sweden and Denmark have advanced national health IT (NHIT) systems that have been largely successful. In Australasia, New Zealand is in the process of implementing its second 5-year Health IT Plan.[2] However, some countries have shown that neither considerable financial support nor national strategy guarantees successful implementation of a health informatics program.[3]

Many developed countries such as Norway, the Netherlands, New Zealand, the United Kingdom and Australia are now approaching 100% adoption of EHRs. Table 21.1 displays the EHR adoption rate by primary care physicians in 10 developed or high-income countries (HIC), based on a 2012 survey. The table also displays how often online tools are used or are available to patients.[1]

Even with such evidence of great advancement and adoption it is important to recognize that with great strides come growing pains and sometimes great setbacks. For example, some countries such as Denmark have found that rapid adoption comes with its own set of challenges, such as interoperability issues and fragmented patient care.[4] England and Australia suffered strains to eHealth and dismantled their original plans and systems. Both however, are moving forward with new initiatives to serve their populations.

HEALTH INFORMATICS IN EUROPE

Health Informatics in Europe has developed in an irregular fashion with countries usually adopting their own plans and national strategies without a homogenous Pan-European approach. Despite differing frameworks, there has been some uniformity in the development of interoperability standards.[5-6]

European EHR Standards

The ISO EN 13606 standard is defined by the European Committee for Standardization (CEN) that covers all major standards in Europe (not just healthcare standards) and the International Standards Organization (ISO).[7] This standard is designed to achieve "semantic interoperability," meaning different computer systems should not only be able to both read and write data to each other but also understand the meaning of the messages by means of a look-up reference that defines all the different types of data included in the standard.[8]

By using semantic interoperability standards computer systems can more intelligently use data from other systems. For example, a hospital EHR system could use data like blood pressures from a clinic system in its decision support system to issue an alert when inpatient blood pressures values are higher than expected.[8]

In this chapter, the acronym ICT (information and communications technology) as a broad IT umbrella, under which HIT (health information technology) will be used.

Denmark

Denmark is widely held to be a leader in Health Informatics across Europe.[9] It was an early adopter of systems providing patients access to medical records and the centralized government offered healthcare providers a range of services that assisted the adoption of health IT systems.[10] Part of the country's success is undoubtedly due to the universal healthcare system that is publicly financed.[11]

Early Adoption and Acceptance

Denmark began their foray into eHealth practices with modest steps beginning in the 1980's with subsidizing the electronic transmission of medical claims in primary care. These early steps encouraged infrastructure development and a foundation for subsequent advancements in HIT.[10] Successive programs were also aided by the 1968 establishment of Denmark's central citizen registry. This

Table 21.1: International EHR and Online Tool Adoption

Country	2012 EHR Adoption	Email Access	Online Appointments	Online Refills
Australia	92	20	8	7
Canada	56	11	7	6
France	67	39	17	15
Germany	82	45	22	26
Netherlands	98	46	13	63
New Zealand	97	38	13	25
Norway	98	26	51	53
Switzerland	41	68	30	48
United Kingdom	97	35	40	56
United States	69	34	30	36

had allowed citizens to become incrementally comfortable with their personal information being held privately and securely in a high quality and readily accessible electronic manner.[12]

Medcom

MedCom is a not-for-profit publicly financed company owned by the Danish government that provides the infrastructure to allow the secure transfer of digital messages between all patients, hospitals, labs and pharmacies, such as, "discharge letters, referrals, lab test orders, e-prescriptions and insurance reimbursement." The system has been very successful with most healthcare documents now being transmitted in digital format.[13-14]

Interoperability

While national standards provide the framework for all EHR systems, there are 5 regions that each subscribe to their own EHR systems but each one is interoperable on a national level. By 2017 there will be 2 regions that will be using Epic EHR, further reducing the number of EHR systems down to 4 for the whole country.[15]

Sundhed.dk

Sundhed.dk is the official health portal for Denmark that allows patients to access their health data, make appointments and access informational resources, such as, disease guides. They can also see waiting list times, ratings of public hospitals, and access online patient support groups. Doctors and other healthcare professionals can also access patient data through the "e-Journalen" and "Shared Medication Record." These tools essentially allow access to health data that was previously more difficult to access.[16] Figure 21.1 displays how Sundhed.dk is organized.

Healthy mHealth Sector

Denmark has conditions that has made mobile health (mHealth) thrive, including at least 70% smartphone use with more than 90% considering themselves regular users.[17] Denmark has been listed as the first choice for mHealth business based on readiness, reputation, practitioner acceptance, and level of digitization.[18] In September 2015, the Danish Digitization Authority reported that it would invest $3.3 million in 5 new telehealth projects to address home physiotherapy to reduce readmissions, home heart monitoring, a virtual endocrinology outpatient clinic, acute care monitoring for elders, and communication between type 1 diabetes patients and healthcare staff.[19] An active and engaging health tech environment has allowed such innovations as Monsenso's app for the treatment and analysis of mental health disorders and "Virtual Rehabilitation" for physiotherapy.[20-21]

Strategy for Success

Factors that have contributed to globally recognized success in Denmark include universal healthcare, regional record systems, adhering to national standards of interoperability, unique patient identifiers, small population size, higher incomes, and high population access to the Internet.[2,22]

United Kingdom (UK)

HIT Strategy

Like Denmark, HIT has been guided by a National Health Service (NHS). Patients are assigned unique identifiers, and electronic prescriptions are widely used. However, this appears to be where the similarities end. In

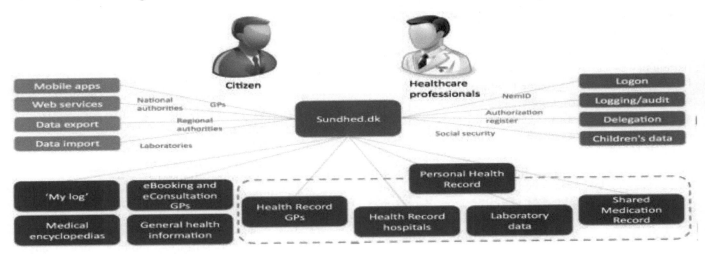

Figure 21.1: Sundhed.dk organization

1998, the NHS initially planned to provide EHRs, online access to health services, and a national system to develop and coordinate HIT services.[23] However, subsequent implementation has been a costly strain.

The National Programme for IT

Perhaps the best-known health IT project in Europe is UK's National Programme for IT (NPfIT). This was an ambitious program initiated by former Prime Minister Tony Blair to create a centrally managed system of IT to cover the whole of the NHS in England.[24] At the core of the project was the development of the "Spine", a central database that would hold the records of some 50 million patients.[25] The project commissioned several large IT companies to provide EHR and information systems to hospitals across the country.

The program has been largely viewed as a failure, with over $12.7 billion pounds spent over the course of 9 years with little software or infrastructure delivered, while functioning only at "base functionality."[24,26] Multiple impediments have been cited, including haste, architecture issues, and lack of motivation, leadership and technical expertise.[24,27-28]

NHS Identifier

Unique patient identifiers are important in enhancing patient safety and preventing adverse events.[29] In the UK, the NHS number is a unique patient identifier used to access health services and to match individuals to their clinical records. With accuracy and completeness of this identifier ranging from 90%-99% across all levels of care it is routinely used to correctly identify patients. From a research perspective, the NHS number is of relevance as it is often used to link clinical data to other datasets, such as socioeconomic or mortality data.[30] Unlike many other countries, the NHS number is consistent and highly unlikely to change over the lifetime of an individual.[31]

However, there is a growing need in the UK to assure patient privacy and confidentiality, thus identifiable data is pseudonymised, or replaced with codes that are only recognizable to treating physicians and not researchers. The UK Data Protection Act of 1998 limits the sharing of confidential data to only those instances where explicit consent has been provided and is rigorously applied across the UK to prevent unethical use of identifiable data through legislation. These restrictions while aimed at maintaining confidentiality and privacy of personal information, have also been criticized for causing many research endeavors to be carried out less efficiently or not at all.[32]

TeleHealth in the UK

The UK aspires to increase the use of telecommunication technologies to monitor patients at a distance.[33] This has been seen as a cost-effective alternative to long-term institutional care. Providing care in the home with an intention to support self-management of patient care requirements is one-way telehealth. Driven by the Department of Health, the UK undertook a clinical trial (Whole Systems Demonstrator Programme) which was possibly the largest telehealth and telecare trial in the world aimed at testing the efficacy of telehealth use.[34] In an early analysis of results on chronic conditions, such as diabetes and heart failure, the study showed an overall improvement in health measures and a significant reduction in costs, emergency visits, hospital admissions and mortality rates.

However, other recent studies have showed no significant effect on patient outcomes with the use of home monitoring, as compared to usual care.[35] Most studies of teleHealth interventions did not adequately define nor address the meaning or differences in "usual care" and trials failed to consider the new innovations in technology such as smartphones, tablets and the Internet.

Healthspace

Healthspace was envisioned in 2006 to be a patient health record web portal that would allow patients to access and edit their own records, create their own health data, provide a platform for secure communication to providers, and make appointments.[36] The program suffered from low strategic importance, limited functionality and as described by Dr. Charles Gutteridge, the National Director for Informatics at the Ministry of Health, " just too difficult."[37] By December 2012 Healthspace was officially shut down as a result of low interest, as less than 0.01% signed up for comprehensive accounts.[38]

Post NPfIT

In 2011 the UK government announced that NPfIT would be dismantled in favor of a system of devolved procurement to hospitals, provided that local programs adhere to "nationally specified technical and professional standards."[39] Currently, practices and hospitals are encouraged to utilize electronic records and forms, but there is no requirement to do so, and medical records (from hospitals, general practitioners and specialists) are not integrated.[40]

Despite a lack of a requirement for a specific EHR or definitive interoperability standards, 96% of physicians use some sort of electronic patient record in their practice; 38% can share patient summaries and test results with physicians outside of their practice and 85% receive electronic notification of improper drug dosage or interaction.[40]

In 2014, the NHS released the Five Year Forward View Plan outlining the future of their health care system. Included in the brief document was the creation of a National Information Board to increase transparency, expand health apps, advance the number of interoperable EHRs, and bolster the use of technologies, such as smartphones by staff and patients.[41]

The NHS published "Personalized Health and Care 2020" in 2014 and it built upon the Five Year Forward Plan. They referred to it as a "Framework for Action," rather than a formalized plan or strategy.[39] Follow up "roadmaps" were then published in June 2015 outlining achievements and regional conferences were scheduled throughout July 2015 for patients and practitioners to provide feedback on the direction of the roadmap, barriers to delivery, and incentives provided to achieve success.[42]

The Interoperability Handbook was published by the NHS in September 2015 as a resource to aid interoperability, standards and implementation.[43]

To add to the challenges experienced by the UK, the Cambridge University System implemented Epic EHR, but experienced a myriad of difficulties.[44]

Positives in the mHealth market

Despite difficulties in the overall eHealth design and implementation, the UK has received recognition as an excellent country for mHealth.[18] Examples of UK's mHealth programs include: iPlato's Patient Care Messaging service; mHealth Assist that provides a platform for communication, information, and support for those with chronic conditions; and at one-time PersonalTechMD's exploration of a new wearable to aid dementia patients, their families, and practitioners.[45-47]

Germany

Early Adopter of Medical Informatics

Germany was once a leader in advancing Health Informatics. In 1949 Gustav Wagner formed the world's first professional organization for Medical Informatics in Germany.[48] Since then, Germany has gone on to build a mature eHealth system with 92% electronic records and a strong medical technology sector. Despite their strengths, Germany still struggles to improve in the areas of health information exchange, percentage of General Practitioners (GPs) with websites and e-prescribing.[49]

German Smart Card eGK

A critical component of Germany's eHealth infrastructure is the national Electronic Health Card (Elektronische Gesundheitskarte OR eGK) (Figure 21.2).[50] The original card was designed for 72 million customers, doctors, hospitals, and pharmacies. It was intended to hold insurance and prescription information and optionally house data on drug intake, chronic diseases, blood type, operations, lab results, and a disease diary.[51]

Figure 21.2: German Smart Card

However, the massive and expensive (1.7 billion EUR) roll out of the universal e-card was virtually suspended early in 2010 due to difficulties surrounding the complex nature of the e-card plan, data privacy concerns, and opposition from providers.[52]

The eGK was relaunched in October 2011 to six German states and the general population in 2013.[53] The initially ambitious aim of including medical data and e-prescribing information has been scaled back, but the card is designed to support future national eHealth projects.[54] The card facilitates identification and access to services, such as, an electronic patient file, diagnoses, labs and x-rays, and e-prescribing. There is also an added layer of security that requires a patient to enter their PIN number for the e-card to operate. Since the introduction of the updated eGK 95% of the population has been equipped with the new cards.[55]

Unleashing Interoperability

In 2014 Germany released the "Digital Agenda 2014-2017" underscoring the need for increased interoperability, as well as eHealth innovations. To that end, they announced the implementation of a standards framework at the federal level to increase interoperability.[56] As part of the restructuring the new "eHealth Initiative" was announced as an independent working party to address the needs defined in the Digital Agenda. The two cornerstones are a study to determine concrete ways to remove interoperability barriers between their 200 different HIT systems and the creation of a national telemedicine portal.[57] The original eHealth infrastructure leaned heavily on the eGK, however the Digital Agenda, the eHealth Initiative, and the eHealth Act all aim to establish a singular telematics infrastructure to function in conjunction with the electronic health card.[58]

In addition to announcing these changes, the draft bill also introduces incentives for physicians and hospitals to cooperate with new innovations, such as, creating emergency data or electronic discharge letters.[59]

mHealth App uncertainty

The mHealth market for apps in Germany is mixed and considered the most controversial of the EU countries.[18] There have been pockets of positive news for apps, such as, Caterna Vision Therapy for amblyopia treatment, the first reimbursable app.[60]

France

Early Groundwork

As early as 1978, the French National Social Security System planned an evolutionary leap to use smart cards and electronic care sheets. It took 20 years for an initial roll out of the Sesame Vitale card but during that time a strong backbone of optical fiber was installed.[61]

DMP

The Dossier Médical Personnel (DMP) is a nationwide online patient health information project that was initially launched in 2004, halted in 2006, and re-launched in 2011.[62] The record includes: patient history, allergies, medication history and lab results. Records are accessible on the Internet and patients can choose which providers can view their records and allow full access or read- only status.[63] It was designed to aid in the coordination of care between providers, reduce duplication of actions and documentation and prevent drug and treatment interactions. Providers are also required to use office EHR software that is interoperable with DMP.[64]

Like HealthSpace, the DMP project is voluntary (opt-in) and has not garnered much use so far. As of December 2013, less than 1% of the population had a DMP file and the cost thus far is reportedly more than $210 million EUR. However, unlike HealthSpace, the project continues to move forward.[38]

Carte Vitale Card

Carte (Sesame) Vitale is also a staple in the French eHealth system. The original card was introduced in 1998, strictly for insurance billing purposes, but was envisioned for varied purposes. However, after several generations, the Vitale 2 cards continue to function primarily as a paperless billing and reimbursement vehicle. It is important to note that its use has been linked to reduced administrative costs which in turn have reduced overall treatment costs.[65]

The CPS Smartcard

The CPS card is France's healthcare professional card that contains data on the provider's identity, profession, specialty, and hospital or facility affiliation. They afford a provider the ability to sign and send electronic forms, add and edit EHRs, and access a secure messaging platform.[66]

Potential mHealth growth

Like Germany, France is considered a complicated market for mHealth. Low adoption has plagued mHealth in France, however the market is poised for large potential growth in the future.[18]

Health Informatics Systems Interoperability Framework (HIS-IF)

HIS-IF adapts the international Integrating the Healthcare Enterprise (IHE) profiles to the French context. The main purpose of implementing the HIS-IF across the country is to provide hospitals and clinics an avenue to communicate with the DMP system and to enable the secure creation and storage of personal health records (PHRs). It is a central framework that addresses both technical and semantic interoperability; thus, allowing vendors to concentrate on details and specialty functions while removing the guesswork for providers wanting to purchase compatible products.[67]

HEALTH INFORMATICS IN AUSTRALASIA

Australia and New Zealand (NZ) have both undergone significant developments in Health IT over the past decade. The focus in NZ has been on the development and implementation of a NHIT Plan. In Australia, the National eHealth Transition Authority (NeHTA) has implemented several projects, the highest profile of which was the Personally Controlled Electronic Health Record (PCEHR) that later evolved into My Health Record.

New Zealand

New Zealand (NZ) has been a well-recognized provider of eHealth services since the 1980s when regional hospitals began integrating electronic administrative systems.

Their initial steps towards eHealth began in 1992 with the creation of unique patient identifiers and an interoperability framework.[68]

NZ has a top-down approach to standardization and interoperability and the government is directly involved with the development of HIT, runs the Health Information Standards Organization (HISO), and updates guidelines regularly. The NHIT Plan aims to implement common health IT platforms across each region of NZ for managing patient administration and clinical information. Each of their 4 regions can choose their own EHR, repository, and support systems but are required to integrate with the National Health Index systems.[69]

Currently, their use of electronic means for provider follow-up and preventative care reminders is 92% and physician use of EHR is nearly 100%, while also ranking high in areas of patient centered care and coordinated care. Patient portals are also accessible to view medical records, set appointments, and renew prescriptions.[11]

There are multiple different add-on platforms that the country has introduced in recent years with great success. One such program is the national child IT platform to track important health milestones for children.[70]

Orion Health

Orion Health is an NZ based eHealth technology company founded in 1993 that has become a global force in providing EHRs and other healthcare solutions. They provide services to most health districts in NZ and their products have been implemented in over 30 countries.[71]

Interoperability

While their use of eHealth in individual settings is optimal, data sharing between facilities and specialties is limited.[11] Concerned by the lack of functionality on a national scale, the Minister of Health in 2015 announced a plan to move to a single countrywide EHR system with portal and mobile capabilities. In April 2016, the Ministry released the New Zealand Health Strategy 2016 with one of its five goals being a "smart system." They specifically plan to support a universal EHR, patient portals, analytics and a health app formulary.[72]

mHealth Innovation

NZ is reportedly leading the way with mHealth. The University of Auckland conducted the first study of text messages for: smoking cessation, advice for pregnant women and families, diabetes patient self-management, and a youth-line for the text friendly teen population to help with issues such as bullying, relationships and sex, drugs, and abuse and violence.[73] Another interesting app is "Beating the Blues," a mental health treatment program for depression and anxiety that is available through a patient's GP.[74]

Australia

Australia's eHealth plans have been slowly developing since 1993 when they created the National Health Information Agreement (NHIA) "to develop, collect, and exchange uniform health data, information, and analysis tools." From 1993 until 2007 they laid the foundation for nationwide eHealth in 1999 with their Health Information Action Plan, in 2005 by forming the National E-Health Transition Authority (NEHTA), and in 2007 with legislative changes to update codes and guidelines.[75]

In 2010 AU passed the Healthcare Identifiers Act to assign identifiers to individuals, providers, and organizations.[76] The PCEHR was established in 2012 by the NEHTA; designed to used interoperability standards such as HL7 and the Cross Enterprise Document Sharing (XDS) profile from Integrating the Health Enterprise (IHE) to allow medical professionals to share a variety of clinical information with patients and each other.[77]

However, their progress has not been without growing pains. PCEHR was plagued with low participation, high cost, and eventually replacement. Some practitioners felt that the program had been hijacked and bogged down by lawyers and legislators, did not serve the needs of clinicians or patients, lacked government support, and suffered multiple access and functionality issues.[78] PCEHR was like the DMP in France in that it had a significantly lower number of participants along with a reportedly hefty price tag.

My Health Record system

In 2015, it was announced that the PCEHR would undergo an overhaul. It was modified to become My Health Record, an online PHR. It can store medication summaries, claims history, hospital discharge summaries, imaging reports, organ donation statuses and pathology reports. As of May 2016, 2.7 million patients and 8500 providers had joined the service. The physician incentive program will be tied to participation in the new record system. It is too early to know the impact of this system on patients and clinicians.[79]

mHealth in the AU

mHealth is in its infancy but has potential, as there are an estimated 15 million Australians (~63%) with smartphones.[80] mHealth technology is growing and thriving within AU and is seeing successful exportation as well. Telstra's ReadyCare is a teleHealth service launched in 2015, offering connection to a GP through a 1-800 number or an app. Patients will be able to discuss their issues, upload photos, and receive treatment including prescriptions.[81] Another mHealth system is KinetiGraph for Parkinson's disease patients that is worn like a wristwatch and logs activity and reminds patients when to take their medications.[82]

HEALTH INFORMATICS IN AFRICA

There are a wide range of health informatics initiatives underway in Africa that are quite different from initiatives in Europe and the Americas. Due to a lack of IT infrastructure and financial resources, many projects make use of open source software and mobile technology (mHealth).

Prioritizing eHealth in Africa

Historically, Africa's health agenda was dominated by donor supported programs that mainly focused on a narrow and high-profile disease specific areas, such as HIV, TB, or malaria.[83] The interventions frequently used standalone HIT to support their programs and to track individuals, monitor the useful indicators, and to store the health information they collected. The interventions were often ad hoc, in response to epidemics, had a short to medium term outlook, and subsisted on inadequate ownership from beneficiary governments. While the investments had impact and uplifted the health status of populations across many African countries, there was poor coordination, which in-turn resulted in a replication of efforts and disjointed approaches to eHealth.[84-86]

A significant advantage realized was that resources that would have otherwise been spent on tackling crucial healthcare issues, were now funded by international bodies, thus making these resources available to African governments for alternative investment. Investments in HISs fell lower on the list of priorities as policy makers leaned towards other national issues that would maximize the general populations' state of wellbeing, rather than investing in eHealth which was not seen as a direct and immediate need.

In those rare cases where these funds were invested in eHealth, most initiatives rarely went past the pilot phase and were not well implemented or documented. This made it difficult for countries to learn from the experiences of other similar countries.[87] Furthermore, most programs lacked a comprehensive evaluation mechanism.

These limitations led to some failures, and a lag in eHealth adoption, compared to developed countries.

Current state of eHealth

Over the past 5-10 years, many African countries have made significant strides and rolled out more coordinated health information systems. National strategies to standardize eHealth have now been developed across Africa. These efforts can be seen in multiple African countries. With governments and relevant institutions such as Ministries of Health playing more pivotal roles, the progress achieved is beginning to be felt.[88]

This has not been without its challenges and failures in eHealth implementations are more heavily felt in resource constrained environments. In the past couple of years, significant progress has been made towards standardizing developments in eHealth. Ad-hoc project specific approaches are being gradually refined and initiatives are taking a more systematic approach by establishing government backed eHealth departments, dedicated IT infrastructure and funding.

Less expensive and more sustainable open source software (OSS) systems, such as OpenMRS and other health applications have been adopted widely. These technologies are being used extensively in public facilities and remote communities to record health events and connect rural populations to skilled health workers. Local customization of such widely available open source solutions and the proliferation of mobile phones minimizes the financial outlay and are proving to be a preferred and more sustainable strategy over the long-term.

Open MRS

The project is backed by I-TECH at the University of Washington and was co-founded by the Regenstrief Institute at the University of Indiana and several partner organizations across Africa. OpenMRS is now being developed into a full EHR solution that can be used by both hospitals and clinics in Africa, although it is primarily still used for smaller clinics. The most recent features include patient summaries, vital sign capture, outpatient or inpatient capture, and the ability to add diagnoses using a coded or non-coded methodology (Figure 21.3).[92]

DHIS: District Health Information System2

One of the most significant health informatics successes in Africa in recent years has been the adoption of the open source District Health Information System (DHIS).[89] This cloud-based service allows countries such as Kenya, Tanzania, Uganda, Rwanda, Ghana,

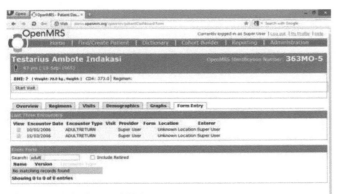

Figure 21.3: Open MRS

and Liberia to manage complex data collected from healthcare facilities across their respective countries (Figure 21.4).

DHIS2 is the current version of the open source system which allows anyone to download, install and change the software for free. The project is coordinated with the University of Oslo in Norway and the Health Information Systems Programme (HISP).[90] DHIS2 pulls in data from local hospitals, either by manually entering statistics about the functions of the hospitals (how many patients, what kind of treatments, etc.) or by integrating DHIS2 with an EHR system, such as OpenMRS. Once in the system, users can use DHIS2 to explore the data, run reports, and generate visualizations.

mHealth in Africa

African countries have experienced a rapid adoption of mobile technologies in recent years. The introduction of mobile phones with SMS capabilities has enabled many highly successful initiatives, from reminding patients when to attend HIV clinics to sending out regular advice for expectant mothers.

As 3G network connectivity continues to spread and tablet PCs and smartphones increase in adoption, new opportunities are emerging in developing countries for a wireless Internet infrastructure that has the potential to "leapfrog" the huge investment in fiber-optic and copper networks that have been installed in the developed world over the last few decades. With 4G, and 5G networks just around the corner, there is a significant opportunity to develop bandwidth intensive healthcare solutions (EHR, imaging, and video-conferencing) for a relatively low cost compared to the vast sums invested in hard-wired networks.

In addition, the introduction of low cost satellite Internet connections such as those offered by Facebook's Internet.org, SpaceX, and Project Loon offer intriguing new ways to accelerate connectivity in developing nations and may be able to bridge the gap in connectivity in areas where 3G networks are not currently available.[93-95]

Mobile Phone Subscriptions

The International Telecommunications Union (ITU) has collated statistics on comparative mobile phone adoption in developed and developing countries. (Figure 21.5)

The proportion of users in the developing world is accelerating quickly. The number of people using their phones and tablet PCs to access the Internet is also increasing more rapidly in developing countries than the developed world.[96]

Current mHealth Options

Medic Mobile

Medic Mobile is a non-profit organization based in San Francisco that operates in 15 sub-Saharan African countries. It developed a platform for delivering SMS

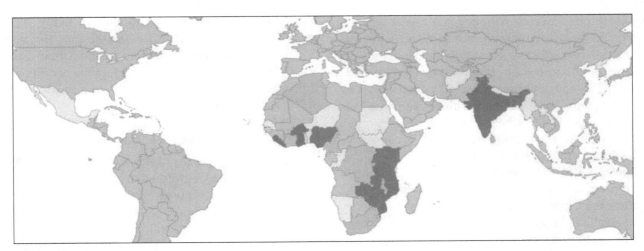

Figure 21.4: DHIS2 Adoption

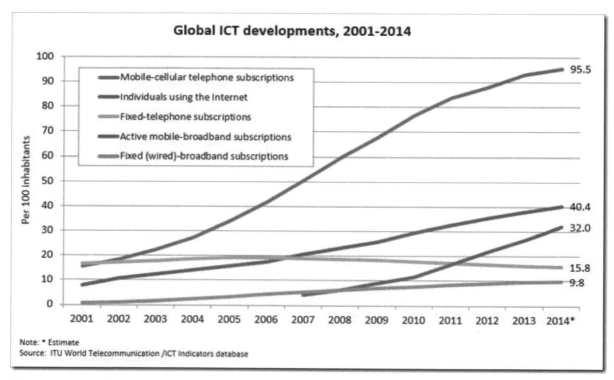

Figure 21.5: Global ICT use

(text) messages based on the open source FrontlineSMS software that was originally developed by conservation charities in Africa. Medic Mobile uses the software, combined with a suite of EHR modules, to give healthcare workers the ability to use SMS to help with vaccination efforts, keep track of patients, send out appointment reminders, and conduct research.[97-98]

Vodafone and mHealth Alliance. Vodafone and United States Agency for International Development (USAID) have been working with partner organizations on several mHealth projects in low resource settings through the mHealth Alliance initiative. In Africa, the Alliance has been involved in several initiatives:

- **GAVI The Vaccine Alliance** is a 3-year partnership between Vodafone and health ministries across sub-Saharan Africa to use mobile phones to improve rates of immunization. Proposed methods include "alerting mothers to the availability of vaccinations by text message, enabling health workers to access health records and schedule appointments through their phones, and helping health facilities in remote locations monitor (vaccine) stocks."[99-101]
- **Vaccines in Mozambique.** Vodafone has been working with the global healthcare provider GSK, Save the Children, and the Mozambique Ministry of Health to implement a project using mobile technology to raise awareness of vaccination among expectant mothers.[102]

- **SMS for Life.** Vodafone, Novartis and other partners have developed an mHealth initiative in Tanzania with Roll Back Malaria.[103] The SMS for Life project aims to use mobile SMS messages to help healthcare workers keep track of malaria drug stock levels across three districts in Tanzania. This initiative has achieved a high reporting compliance rate in Tanzania and Kenya.[104]
- **Comprehensive Community Based Rehabilitation in Tanzania (CCBRT)** is a partnership between the Vodafone Foundation and m-Pesa, the mobile money transfer and microfinancing service, to address the problem of obstetric fistula that causes maternal incontinence post-childbirth in many African women. The CCBRT hospital uses the m-Pesa system to enable mothers to fund their travel to hospitals for surgery to correct the condition. The project aims to enable 3,000 women a year to get access to treatment.[105-106]

Mobile Alliance for Maternal Action

Mobile Alliance for Maternal Action (MAMA) is a successful mHealth project that has been working with a wide range of partner organizations, including the United National Foundation, USAID, Johnson and Johnson, the mHealth Alliance, and BabyCenter. MAMA aims to improve healthcare provisions for new and expectant mothers in low-resource settings using SMS messages containing health advice for them and their newborn children.[107]

As can be seen by the range of projects previously described, many different types of organizations are using relatively low-tech solutions to improve healthcare provision in Africa. As the technology improves, it will be interesting to see how new features of smartphones such as high-resolution cameras, fingerprint identification systems, and faster and more powerful applications will be used in future projects.

According to Grices's conversational maxims, the nature of SMS messaging other than for the very simplest of instruction may be open to multiple interpretations.[104,108] Because text messaging is not privy to essential nonverbal cues, they may lack indications of urgency or importance. The way a message is framed can also affect whether a person is receptive to making a behavior change or not.[109] It is crucial that text messages are written in the most appropriate way for the population, including ethical and cultural considerations. Despite challenges in establishing text message content, a recent study showed simple text reminders increased women's attendance of breast cancer screenings, as well as in adherence to taking HIV/ART medication.[84, 110]

However, overall reviews performed for interventions applied in more developed countries, such as in the use of mobile text messaging in self-management of long term illnesses, reminders to attend clinical appointments, and messages to communicate results of medical investigations, have showed only a moderate effect on outcomes or behavior.[111-113] Existing literature has proven the efficacy of text messaging for appointments or drug reminders in adult populations.[111, 114-115]

Kenya

Kenya's eHealth strategy was developed through a consultative process and is anchored by Vision 2030.[116-117] The strategy promotes the delivery of efficient health services enabled by ICT and the implementation of the official standards and guidelines which were developed to harmonize the various donor-funded implementations. The country continues to embrace open source solutions with use of OpenMRS and DHIS2, described earlier. A 2011 study showed that proprietary and open source systems are used in almost equal measure. Proprietary systems are mainly used in private hospitals and faith-based organizations which mainly purchase off the shelf proprietary solutions to manage their facilities.[118]

Kenya, like many LMIC, has "leapfrogged" most traditional eHealth activities. New mobile or cloud systems have been adopted to replace paper systems without investment in PC or mainframe systems, routinely adopted by HIC. Wireless technologies have been shown to reduce the administrative burden on health care workers and resulted in improved patient care.[119] Web and telephone-based health consultations, such as an e-consultation service Sema Doc launched by a leading mobile phone company, is also gaining traction in urban and semi-urban areas.[120-121]

Systematic EHR use outside of donor funded health programs remains lacking and where they are used, the systems tend to be immature. Kenya continues to face limitations in eHealth and ICT financing, medical personnel shortages, regulation, and legislation to protect privacy and security of patient data.[122]

These factors have considerable impact on healthcare and by extension, eHealth. The Kenyan government's national strategy "Towards a digital Kenya" aims by the year 2017, to make broadband available to all health centers country wide.[123] This will assist greatly in positioning local/rural facilities to realize the potential of eHealth and benefit from teleconsultations through telemedicine.

South Africa

Like Kenya, the South African health information system was fragmented and faced challenges of interoperability where automated systems existed. This then led to the development of an eHealth Strategy to be used as a guide in integrating HIT for health delivery. South Africa through the National Department of Health (NDoH) commissioned the development of a NEMRS.[124]

Rwanda

Rwanda is one of the poorest, smallest, and most densely populated countries in Africa.[125-126] Nevertheless, it has made significant progress in deploying a NHIS driven by a strategic plan for eHealth, championed by the Ministry of Health and Sanitation. The strategy stresses the need for governance in the development of eHealth infrastructure to aid in the efforts against proliferation of fragmented and piecemeal HISs, placing system interoperability high on the list of priorities.[127-128] It is based on an eHealth Enterprise Architecture Framework which functions as a broad roadmap to ensure interoperability between all databases within the health sector. Rwanda is yet to meet any of its Millennium Development Goals (MDG) targets, but it has, over the last decade, halved the maternal mortality rate and made significant progress in all the other MDG indicators. It is believed to be on track to achieve most of the MDGs by the end of the decade.[127,129-130] The Government's vision to further utilize technology in healthcare, has also made Rwanda a pioneer in integrating technology into the healthcare system, both regionally and also in Africa.[131] Regional health experts and officials from the East African Community (EAC) have recognized this progress and made it the lead for eHealth and technology use in healthcare.[132] Rwanda has also adopted DHIS2 and OpenMRS which are seen as more established and stable open source web based platforms. With the OpenMRS platform being piloted successfully in 24 health facilities in 2013, it is being extended nationwide.

Various mHealth initiatives are also important components of this strategy which feed into Rwanda's Health Management Information System. Maternal and newborn health, HIV, and malaria epidemics have been a central target for Rwanda's HIS efforts. The focus is now gradually shifting from infectious/communicable diseases to non-communicable and chronic conditions. Community health workers (CHWs), numbering over 45,000 operating at the village level, provide first line health care delivery.[130] CHWs increase efficiency in reporting disease incidences and delivering care and advice to the populace. To do this they have embraced low cost but effective mHealth solutions such as:

RapidSMS

RapidSMS is an interactive two-way short messaging service (SMS) based information system developed to support the documentation of pregnancies and related events including emergencies within the community. RapidSMS has proved effective in preventing maternal deaths with CHWs being central in collecting maternal indicators and recording births for purposes of aggregation and central storage.[133-134]

TRACnet

Treatment and Research AIDS Centre and other extensions, such as TRACnet Plus, accessed largely through the mobile phone and also through the Internet, have been used to capture health events.[135] These systems enable practitioners in remote HIV/AIDS clinics to communicate with administrative units to manage patient information and submit reports. The systems have enabled real time access to critical tracking indicators, such as HIV transmission patterns, drug stock levels, etc. and have reduced the time for result reporting.[136] TRACnet is currently being extended beyond HIV to capture over 23 other indicators for communicable diseases. This includes monthly monitoring of infectious diseases including TB and malaria. TRACnet SMS messages also feed centrally into the electronic system for integrated disease surveillance in Rwanda.[137] Criticisms of the technology include inconsistencies in recording of clinical data and lack of recent studies needed to gauge the accuracy and quality of the data collected.[138]

SIScom

This program, accessed primarily by mobile phone, allows CHWs involved with maternal health to send patient data via text messages and report on any incidents during the pregnancy to the district hospitals. This system has enabled better patient/maternal follow-up, improved mother and child health, and reduced mortality rates significantly.[139]

HEALTH INFORMATICS IN ASIA

China

History

China holds the distinction of having not only the world's largest population but also the largest elderly population.[140] These factors have put pressure on the government, healthcare organizations, and providers to adopt some form of digitized health care. Computerization of administrative and computer data in hospitals began in the late 1980's with single computer usage.[141] During the 1990s China began increasing IT development in multiple industries through several "Golden Projects." In 1995 China's Ministry of Public Health announced the "Golden Health Project" aimed specifically at healthcare to increase connectivity of hospitals, medical research, and medical education.[142-143]

The Chinese government announced another national health reform plan in 2009 to improve medical coverage to 90% of the population, increase access to essential drugs at lower prices through providers outside of hospitals, and increase use of local healthcare centers as primary contact points rather than larger central hospitals. Increased adoption and interoperability of HIT is an integral part of addressing all the other aspects of their plan.[144] In 2012, the MOH released the 12th five-year plan that included the establishment of a NHIS.[145]

EHR Adoption

As part of the 2009 reform, a nationwide EHR was announced with coverage goals of 30% rural and 50% urban by 2011.[146] As of 2014, China was still short of their original goals with 20% rural and 30% urban adoption rates. However, they are planning continued expansion of EHR to 50% rural and 50% urban by 2020.[145]

Interoperability

While most healthcare providers have some form of EHR, a significant number of them are not interoperable and there is still no unified national system in place to integrate patient records. While hospitals are linked to insurance for payment administration, they are frequently not linked to one another, even in the same region and when owned by the same entity. Currently, patients do not use EHRs or online services for patient access, appointment booking, prescription refills, and messaging.[147]

mHealth

Mobile phone ownership in China tops 1.17 billion devices and subscription levels are high at 92% of the population, incorporating even rural portions of China. The role of mHealth is expanding into primary care as a means to bridge the gap in rural and underserved areas. Providers are currently allowed to offer electronic consultation, without treatment or prescribing medication.[148]

China is home to nearly 100 million diabetic patients leading to the development and ample funding of mHealth apps and plugins to monitor and manage glucose levels, dietary intake, and medications including Weitang and Dnurse.[149-150] Other mHealth app sectors looking to take advantage of the market potential are electronic consultations, intelligent mobile medical devices, and rural child immunization.[151] To address the growing need for mHealth that specifically addresses the nuances necessary for the Chinese market, the Stanford Center at Peking University launched "Digital Health Bootcamps" in 2015. These events bring together hospitals, programmers, designers, marketers, and business experts to develop mHealth apps.[152]

Patients are responding to the convenience and ease of mHealth technologies with a rise in use, from 66 million people in 2014 to 138 million in 2015.[153] High mobile use, rising need, and increased funding will ensure that growth will continue for apps, wearables, and monitoring devices.[148]

Telemedicine

China is now exploring telemedicine solutions to address a quality and distribution gap between urban and rural areas.[154] With the continued increase in their already high population of elderly, telemedicine is also seen as a way to care for their aging patients while reducing pressure at overburdened urban hospitals, already hindered by provider shortages.[155] Several companies have stepped up to offer online platforms including Haodiafu, Chunyu Doctor, and DXY while doctors are ramping up their use of online platforms to communicate with peers and patients.[156]

Cultural complexity and Social makeup

The Chinese culture is infused with a reverence for tradition, conservatism, and caution. While these values may hold their community together, they may also be impeding their progress towards innovation and acceptance of new health technologies.[157] In addition, the large rural makeup of the population and providers has been a barrier to adoption and use of EHR, even when available. Lack of understanding, comfort with electronics, and security concerns has prevented wider adoption of HIT.[146]

Traditional Chinese Medicine (TCM)

Healthcare in China is a chimera with a blend of TCM practices coexisting with Western medicine. While there has been some embrace of Western medicine, use of TCM hospitals and clinics is on the rise, accounting for 17.9% of all visits (530 million) and one third of the total medical industry financially.[158] Hsu et al. noted barriers to HIT implementation by TCM practitioners, including lack of adoption and limited computer experience, concerns surrounding patient confidentiality, trepidation over potential interference with practice workflow, and reluctance to share propriety techniques and formulas.[159] Additionally, current HIT designed for Western medicine does not appear to be "plug and play" with TCM. The use and interface of the two may be similar; however, "TCM EMRs required a different logic and very different terminology."[160]

Limited research

One of the obstacles potentially reducing progress is a lack of research into health technology assessment and usability. Research in these areas ensures accurate policy, funding, creation, and application of new health technologies.[161] There is a shortage of staff to conduct research and an even more limited number of those are clinically well trained.[162] In addition, there has not been a streamlined or coordinated effort to assess and monitor the past or future application of HIT.[161]

Market Complexity

Translational issues, both language and cultural, have been cited by large companies such as Epic, Accenture, and KLAS Enterprises as additional impediments to the expansion of HIT in China.[160] The U.K. Digital Health Demonstration Centre is also trying to take some of the guesswork out of importing HIT services into China,

Japan

History

Japan began building the base infrastructure of digital health in the 1970's to address billing administration and management.[163] The 1980s saw the formation of the Japan Association for Medical Informatics (JAMI).[164] In the 1990s Health Information Exchanges (HIEs) were formed to focus on data sharing between hospitals, clinics, and laboratories, but not regions.[163] The 2000s ushered in Electronic Medical Records (EMRs) as a part of their "e-Japan Strategy."[165]

Current Systems

One exemplary healthcare system in the Nagasaki Prefecture is AjisaiNet. It is Japan's largest health care network linking 27 hospitals and 434 clinics, pharmacies, and labs for over 38,000 patients.[166] Each provider determines level of sharing such as providing access to lab images while privatizing physician notes. It is considered a vanguard system that may pave the way to greater regional and eventually national coverage.[166-168]

Japan's "My Hospital Everywhere" was launched in 2013 to allow patients and providers the ability to store and access EHRs across the country.[168] Despite EHR initiatives, as of 2011/2012 large 400+ bed hospitals had only a 50% EMR adoption rate, while small 20-199 bed clinics only reported 13% EHR adoption.[169] The lack of uniformly and nationally available health records was clear after the 2011 earthquake off the coast of Japan and resulting tsunamis. Hospital facilities were destroyed, paper records were unavailable, and providers and patients alike were in an upheaval.[170]

Unique Identifiers

Beginning in 2016 a new citizen identification numbering system will be rolled out in Japan. While not the initial and primary purpose of the system, medical record creation and access is planned for launch in 2018.[171]

mHealth

Japan is seemingly slow in adapting and adopting mHealth measures.[172] The current bulk of mHealth in Japan is providing connectivity between seniors in remote rural areas sending health data such as weight, activity levels, and blood pressure to their providers.[173] One such movement is the joint venture between the Japan Post and Apple to provide the elderly with iPads enhanced with a variety of health apps to address senior specific topics.[174]

Telemedicine

In Japan telemedicine services has mostly been limited to chronically ill patients in very rural areas. A major obstacle is health insurance guidelines that restrict practitioner's earnings to a fraction of their normal fee when providing telemedicine services.[175]

Issues

One major hindrance to the growth of eHealth in Japan has been the stringent regulations placed on providers that prevent discussing specific matters of health or disease via phone or email.[176] Additionally, lack of continuous funding appears to have been a limiting factor in the spread of breadth and depth of HIT.[163] Despite the fact that growth and adoption of digital health has been limited in Japan, there has been a movement for health care funding and management to shift from public to private to pursue profit and efficiency.

India

Healthcare in India is a paradox. They have well trained physicians and can offer medical tourism for foreign clients. However, despite having 398 medical schools they do not have enough primary care physicians for rural areas. The private sector health organizations provide about 80% of outpatient and 60% of inpatient care, but overall about 70% of healthcare consists of out of pocket spending. India is moving towards universal coverage and as such they passed a draft National Health Policy in 2015. The reality is that universal healthcare would likely cost $23 billion US dollars over the next four years, to serve one sixth of the world's population. This is very unlikely given the fact that India only spends about $61 US dollars per capita and healthcare amounts to only 4% of the gross domestic product (GDP). Universal health coverage is not the only priority as India also faces widespread poverty and poor sanitation. India is delivered by a mix of private and public health services. The country has a wide range of healthcare demands, from very poor rural areas that lack adequately trained healthcare workers, to dense urban areas with booming populations placing a strain on existing healthcare services in the public and private sectors. Although the country has seen rapid economic development in recent years, healthcare provision for most India's inhabitants is still very poor by international standards with high levels of infant mortality, poor sanitation and infectious disease control.[177-179]

National eHealth Authority (NeHA)

Although EHRs have been adopted by many private hospitals and clinics in India, they are largely focused around billing and patient administration and lack interoperability tools that would enable continuity of care. To address this problem, the government of India produced the "EMR/EHR Standards for India" report and is in the process of setting up a new National eHealth Authority to guide institutional adoption of international standards such as SNOMED-CT.[180]

SMARTHealth

SMARTHealth India is a mHealth initiative from the George Institute for Global Health, based in Sydney Australia, the UK, India, and China. The project uses smartphones and tablet PCs to give community health workers (CHWs) the tools to help diagnose and manage chronic diseases, such as diabetes and heart disease. This means that tasks currently performed by doctors can be "task-shifted" to CHWs while still maintaining high standards of care using clinical decision support and communication tools on the mobile devices.[181]

Sana Mobile

Sana Mobile is a volunteer organization based at the Massachusetts Institute of Technology in the US. They have partnered with a range of organizations including universities, NGOs, health organizations and social enterprises to create several projects aimed at using mobile phones to provide healthcare services in LMICs.

The Sana Mobile open source platform allows CHWs to communicate with medical specialists using text, audio, video, and photos for real-time decision support. One of the largest implementations of the Sana Mobile systems has been to help detect oral cancer in rural south India. CHWs can use their phones to take photos of the inside of the patient's mouth and send the images or video to a doctor who can review it and communicate back to the CHW.

The open source Sana platform uses an Android mobile phone application linked to a web interface created using the Django Python Web Framework that is, in turn, linked to the Java OpenMRS open source EMRs.[182]

Singapore

History

Singapore has been forerunner in digitizing national health since their launch of the Patient Master Index (PMI) in 1985 that connected 5 large hospital sites to store and access minimal patient data.[183] By 1995, the next installment was the National Patient Master Index that included limited demographics, allergies, and a small number of specific medical alerts.[184] Singapore announced their intent to create one of the first NEHRs in 2009, and launched the first phase in 2011.[185-186] The current iteration includes analytics for research and online patient access.[187]

In 2015, they launched the multifaceted HealthHub portal, available on PC and mobile.[188] It provides patient access to some health records, appointment booking, health articles, provider/facility directory, and offers deals or rewards at wellness centers, recreation facilities, and food and beverage establishments. Residents access the portal via their Singapore Personal Access or "SingPass"; a government issued account system launched in 2003 for accessing over 600 online eservices.[188-189]

EHR Adoption

Overall use of NEHRs has been ramping up with over 11,400 users and 600,000 record searches in March 2016, a 63% increase in users and a 3-fold increase in searches from the previous year.[190] Adoption of EHRs is higher by public practitioners, pharmacies, and labs. GPs have expressed their hesitance lies in potential interference with daily operations, increased costs, and unfamiliarity with technology.[191]

Interoperability

Singapore's Ministry of Health Holdings has guided interoperable development by establishing the framework for health IT, coordinating continuity with vendors, involving providers early in the process, and engaging in monitoring and management at each phase.[190-191] Connectivity of patient information has also reportedly been less difficult in Singapore as current programs are built on previously established systems.[193]

mHealth

Residents engage in high Internet (70%) and social media (60%) usage driving a blooming mHealth market.[194] Many programs are government and health facility sponsored to track patient health between appointments or for patients to diary and report daily aspects of diet, water intake, and physical activity.[195] There are apps currently under development that will relay health stats for those with conditions such as hypertension or kidney failure, monitor them for abnormalities, and alert providers if intervention is required.[196]

Telemedicine

With a blend of a rapidly aging population and high level of chronic disease, Singapore has been open to exploring options that reduce demand on their finite healthcare resources.[197-198] To this end, they have been gradually adjusting guidelines since the 90's to increase implementation of telehealth in their country.[199-201] While not as rapidly adopted as EHRs, telehealth implementation is ramping up in a variety of specialties: ophthalmology, physical rehabilitation, elder care, and cardiology.[202-205]

Potential Issues

At present, typical patient access is limited to read-only summaries of care but increased patient literacy, use of apps and wearables, and virtual care is driving demand for greater access. There is a growing division between patients and providers with 80% of patients wanting full access while only 17% of doctors reporting the same desire.[206] Some challenges noted in Singapore are also some of the driving forces behind their HIT agenda. The "Silver Tsunami" of increasing elderly, prevalence of chronic disease, and health staff shortage are all aspects that need to be addressed both by and for the advancement of HIT.[207-208]

Bright Future

Singapore has many advantages that have contributed to the successful implementation of HIT including small size, incremental and programmatic sharing of health data, high physician technology literacy, and patient and provider demand for access.[209-210] The country also has the benefit of positive global perception for being a recognized leader in health efficiency, fostering an environment that encourages innovation and expansion of e-health, proper legal structure to protect research and intellectual property, and strong cooperative ties between government and international private industry.[210-212]

HEALTH INFORMATICS IN SOUTH AMERICA

Although a relatively late adopter of HIT internationally, there are now several interesting health informatics initiatives occurring across Central and South America.

Brazil

The Brazilian National Health System

In the 1960s and 1970s, Brazilian health information systems (HISs) were mostly distributed across both public and private health organizations. It was after the 1980's that the Brazilian health community established the need for an integrated NHIS, as noted in some of the Brazilian National Health Conferences.[213] The main change was the Brazilian Federal Constitution, which established the rights for public health in 1989 to all Brazilian people, as a fundamental and constitutional right, to be provided by the Government in the form of a Unified Health System (bras. SUS) or national health system.[213-214]

The continental length, the huge regional asymmetries concerning social and technological development, and the coexistence of private and public health entities results in a complex mosaic scenario that has created the main challenges in the Brazilian system.[214-215] Many health information systems have been developed and, in the 1990's, the Informatics Department of SUS, called DATASUS, was created.[216] The continuing social and scientific development as well as the economic growth and stability achieved by Brazil in this period have brought some of the necessary conditions to support the engagement of more ambitious national plans in the context of the decentralized Brazilian health system. In this period, some institutions have recommended changes to national health information policies, such as the Interagency Web of Information for Health (bras. RIPSA), the National Web of Health Information (bras. RNIS), the Health Informatics Brazilian Society (bras. SBIS), Brazilian Association of Telemedicine and Telehealth (bras. ABTms) and the National Card for Health project (SUS Card). (see Figure 21.6) SBIS is the Brazilian delegate at IMIA (international Medical Informatics Association).[217]

Figure 21.6: The Brazilian Health card

The SUS Card project has been an essential element for the integration and management of Brazilian NHITS throughout the decentralized health entities, like the German Card. In 1999, it was implemented as a pilot project with around 13 million people.[218]

The first steps were to register all Brazilians, living in metropolitan and remote regions. Many challenges were found during this prolonged beginning and the methods of implementing the SUS Card program have been continuously reviewed and updated. There are Apple and Android smartphone apps that can access the SUS card. Access to the card enables users to view weight, blood pressure, glucose, allergies, vaccines, doctor visits and medications. This mHealth strategy has been improved by the development of the e-SUS program, from which e-booking will be available and personal health information will be integrated and available among the different health entities.[219]

The Brazilian National Plan for Health Information and Informatics

The Brazilian National Policy for Information and Informatics in Health (PNIIS) was endorsed by the United Nations Program for Development.[220-222]

In 2011, the Health Ministry established the basic standards for interoperability and for health information. The chosen reference model for EHRs was the OpenEHR platform.[221] The architecture of the clinical document is specified by HL7 CDA and SNOMED-CT standards; image results use the DICOM standard and the ISBT 128 standard is used for encoding information related to blood, tissues, and human organs.[220] In 2015, the PNIIS was institutionalized by the Health Ministry.[222] The principles of the PNIIS includes an individual unified NHIS, with guarantees of health information access at no cost, as well as security and privacy. For this purpose, a national public EHR system was developed, known as the patient electronic chart (bras. PEP), which has been freely distributed for public health entities. Because of technological infrastructure asymmetries, two versions are available. The PEP is integrated with the SUS Card and e-SUS.[223]

The Brazilian Telehealth Network aims to connect health professionals to physicians of reference at health institutions for specialized referral. Teleconsultation and referral in real time or offline messages, second formative opinions, and e-learning courses are in place, although it depends on the technological infrastructure available. There are 46 centers of telehealth in 23 Brazilian states. In 2016, there were about 580,868 tele-diagnoses, 566,371 e-learning activities and 110,944 tele-consulting; a modest result considering the user population around 170 million. The limited IT infrastructure and services available in the countryside regions seem to be the main limiting factor. Additionally, a federal health evidence-based Portal was created to motivate and disseminate evidence-based health practices in cooperation with OPAS and OMS.[223]

Collective Health Data in Brazil

The main HISs of DATASUS are organized into the following categories: 1) National Registers; 2) Outpatient; 3) Epidemiologic; 4) Hospital; 5) Social; 6) Financial; 7) Life Events; and 8) Regulation.[213] These information systems involve a basic healthcare system, outpatient information system, pregnancy tracking system, immunization programs, renal replacement management system, system for registration and monitoring of hypertension and diabetes patients (HIPERDIA), system for management of transplants of organs and tissues, systems for cancer surveillance, epidemiologic data surveillance, and several others. DATASUS also provides a national health data warehouse called TABNET, which is open for public access. The interface provides graphical visualizations, using charts and colored maps and statistical analysis. This system has been extended to be used also for support the data collect of other health initiatives such as the Health Academy Program for physical activities, enabling the easier creation of forms to collect health data from different public health programs. All the health information systems have been constantly updated to gradually increment the interoperability.[223-228]

Preparing people for the use of HIT and its standards in Brazil

Other important guidelines in the PNIIS are those relating to HIT training. The strategy has included efforts from the Ministry of Telecommunications for IT infrastructure and data communication, and from the Ministry of Science and Technology, to stimulate universities and institutes to meet the demand for research and development in HIT.

Finally, the Ministry of Education took part in the national plan and supports the national distance e-learning project called Brazil Open University (bras. UAB). The UAB courses are mostly supported by Brazilian universities and the Federal Government and Municipalities (UNASUS project). E-learning plays an important role for training of SUS staff and make available courses on information and health informatics relevant for the national strategies. It is freely accessed by health professionals and staff, health careers students and public health managers.[229] All these different health strategies, programs and systems can be used by State government and Municipalities accordingly to its local health strategy and the available technological infrastructure.[223]

Argentina

Argentina has a rich history of academics, including computer science. However, unlike many countries it has had decades of political instability that dramatically interfered with the evolution of the sciences. Under more democratic circumstances the Latin American School of Higher Education in Informatics was created in 1985 but later closed in 1990 due to insufficient financial support.

Interest in computer science and informatics continued, despite setbacks. The program Conectar Igualdad was established to distribute netbook computers to all students ages 10-12 to make them more competitive in technology. To date several million computers have been distributed.[230]

Argentina was the main participant in a new Medical Informatics online course delivered in the 2009-time frame. Oregon Health and Science University partnered with the Hospital Italiano of Buenos Aires to translate and adapt an AMIA 10x10 course. Most of the students participating were healthcare professionals from Argentina.[231]

As of 2014 there were 139 private and public institutions of higher education that offer either undergraduate or associate degrees in informatics and 9 doctoral programs.[232]

Uruguay

Uruguay has a national EHR system that was implemented in 2006 under the auspices of the "National Integrated Health System." This system was mandated for all healthcare providers from 2007 onwards and integrated prior private and public partners into one healthcare system.

Uruguay uses HL7v3 CDA documents and IHE profiles to enable interoperability between hospital systems. In addition to their national EHR project, they also have a Vital Statistics project, a Maternal and Child Health Program, a Perinatal Information System, the Aduana Program (for child health up to 2 years), and Electronic Death Certificate (CD-e).[233]

In 2008, President Vazquez initiated the Ceibal Project to distribute simple wireless laptops, known as XO computers to every child.[235] This is part of the "One Laptop Per Child" program that has distributed more than 2 million computers in 42 countries.[235]

Bolivia

Text To Change (TTC) is a social enterprise with offices in Uganda, Amsterdam and Boliva that offers text-messaging programs for a variety of healthcare projects. Examples of their programs include anti-smoking, financial, farming and handwashing educational programs. It now runs campaigns in 17 countries across the world. TTC is using relatively simple, low-cost technology to support public health campaigns that reach many thousands of people.[236]

HEALTH INFORMATICS IN NORTH AMERICA

Canada

Paving the way for Canadian HIT

In 1994, the National Forum on Health was launched, and the recommendation was made to move toward a NHITS.[237] In addition, the Canadian Advisory Council on Health Infostucture began recommending a nationally unified health information system or "information highway" in the late 1990's.[238] To this end, $500 million was budgeted in 2001 to Health Infoway to develop and implement EHRs. During 2003 and 2004 an additional $700 million was granted to increase interoperability, address rural adoption, and develop a Pan-Canadian health surveillance program.[239]

Canada Health Infoway

Canada Health Infoway is an independent, not-for-profit, organization whose membership is derived from deputy ministers of the federal government, territories, and provinces. The board of directors is comprised of a variety of community advisors including those from public health, technology, legal, and finance sectors.[240]

With a substantial investment of public funds, both progress and transparency are paramount. Infoway regularly monitors progress of projects against a "Benefits Evaluation Framework" and publishes both individual and annual reports on overall progress.[241]

The goals of Infoway were to accelerate the implementation of a national EHR and to develop an EHR blueprint for the creation of individual EHRs that suit the needs of differing regions and specialties.[111] Infoway encourages participation and adherence to the Blueprint by reimbursement through a "gated funding" approach, or the reimbursement of up to 75% of eligible costs for conforming products based on milestones achieved.[242]

Infoway released the first version of the Blueprint in 2003 to deal with data standards and a second version in 2006. Version 2 was released to optimize eHealth and EHR projects by better defining both form and function, expanding into addressing telehealth and disease surveillance, and aligning projects with the Privacy and Security Architecture.[243]

Interoperability and Adoption

Despite having a unified board to oversee development and an architectural blueprint, there has been a lack of interoperability amongst the systems. While Infoway is funding the production of eHealth products, it may not be monitoring them sufficiently. This has led to 11 separate jurisdictions with different standards and different software. The result has been poor interoperability between the provinces.[244]

By 2014, EHR adoption by primary care physicians had reached an average rate 77%, 99% for image digitization, 81% for electronic lab test results, and 98% for hospital access to telehealth applications.[240, 245]

Patient Access to Personal Health Records

While 8 in 10 Canadians have expressed a desire to access their health care records and test results online, reportedly only about 6% currently have the ability.[246] To bridge the gap, some regions and specific hospitals are beginning to provide their patients with various online health options, such as RelayHealth, a web-based patient portal.[246]

mHealth

Canadians are gradually beginning to embrace the addition of mHealth into the traditional medical setting. Recent reports indicate that while 74% of Canadian's have accessed health information online only 15% currently use a mHealth app.[247]

An interesting mHealth initiative being promoted is a joint effort between Saint Elizabeth Healthcare and Samsung to provide 5,000 home health workers and 500 administrators with tablets and mHealth apps to manage and monitor patients in variety of home programs from physiotherapy to wound, cancer, and palliative care. Home workers will be able to optimize patient care by tracking the most efficient routes, securely inputting patient data, and scheduling follow up appointments.[248]

Mexico

Mexico's healthcare system has public institutions and private providers, all under the Ministry of Health. The public-sector accounts for about 44% of the population while the rest are served by the System for Social Protection in Health, a health insurance program for the poor. Health care is delivered through the 32 state Health Services.[249]

Mexico was the first Latin American country to introduce an EHR on a larger scale at the Mexican Institute of Social Security (IMSS), responsible for administering social security and healthcare benefits. The EHR was introduced in 2003 and consisted of multiple linked databases. A study by Perez-Cuevas et al. demonstrated that this basic EHR could evaluate the care of type 2 diabetics using EHR data. However, the study was based on only four clinics and two of the clinics could not provide A1c data.[250]

The state of Colima introduced another EHR system known as SAECCOL in 2005, based on Microsoft SQL server databases and Visual Basic. This was a joint project between the US Agency for International Development (USAID) and the Mexican National Public Health Institute. The modular approach had limitations, such as no inpatient modules. Hernandez-Avila et al. pointed out that many of the EHRs in Mexico were developed prior to an official Mexican EHR standard of 2010 so there is great disparity in terms of functionality and interoperability. His published evaluation of the program in 2012 was not dissimilar from early studies from developed countries. In other words, the same challenges faced by the US (funding, standards, lack of training, privacy, resistance to change, etc.) were experienced in this program.[249]

RESOURCES

eHealth Strategy Toolkit

The World Health Organization (WHO) and the International Telecommunications Union (ITU) have been working to help LMIC achieve higher levels of ICT adoption through the development of eHealth strategies and plans. In 2011, they jointly published the eHealth Strategy Toolkit that gives advice and guidance to developing countries as they begin the process of adopting EHR and other HIS.

Although the toolkit provides good advice on developing a national eHealth vision, it also points out the need to develop a concrete eHealth plan and then to continue to monitor progress through regular assessments and re-evaluations of the plan.[251]

IMIA: International Medical Informatics Association

The International Medical Informatics Association (IMIA) is an organization that brings together health and biomedical informatics organizations from around the world. It describes itself as an "association of associations." Its stated goals are as follows:

- "promote informatics in health care and research in health, bio, and medical informatics.
- advance and nurture international cooperation.
- to stimulate research, development, and routine application.
- move informatics from theory into practice in a full range of health delivery settings, from physician's office to acute and long-term care.
- further the dissemination and exchange of knowledge, information, and technology.

- promote education and responsible behaviour.
- represent the medical and health informatics field with the World Health Organization and other international professional and governmental organizations."

IMIA hosts the MedInfo conference series, sponsors the publication of several leading health informatics journals, and hosts several Working Groups on a variety of health informatics topics with researchers contributing from around the world.[252]

WHO Atlas of eHealth Country Profiles

In 2015, the WHO published the results of the third global eHealth survey of 125 countries. This is an excellent up-to-date resource for statistics regarding populations, physician and nurse density, hospitals, life expectancy, etc. Sections for each country include: eHealth foundations, legal frameworks for eHealth, telehealth, EHRs, use of eLearning in health sciences, mHealth, social media and big data.[253]

CHALLENGES AND BARRIERS

Lack of Infrastructure

LMICs generally lack the funds to develop high speed Internet access using fiber optic cabling, so they migrate towards mHealth solutions with smartphones. Despite high penetration of this technology, it is not scalable for many of the HIT/ICT solutions currently available.

Clinical Adoption

One of the most significant barriers has been clinician buy-in of new systems. Even in environments where systems are provided at no cost to providers, such as the UK's National Project for Information Technology or Canada's Health Infoway, switching from a paper-based workflow to electronic workflow has been difficult.

Interoperability

Interoperability between different EHR systems remains a significant challenge internationally. Countries that have achieved success in exchanging clinical documents, such as Denmark (MedCom) are exceptions, rather than the rule. Issues such as over-ambitious targets for "semantic interoperability" using the HL7 version 3 standard have hampered government-directed standards adoption programs. On the other hand, issues with inconsistent vendor and provider adoption of HL7 version 2 continue to present problems, although several countries have successfully adopted HL7 version 2 for e-Prescribing and e-Discharge Summaries. The new HL7 FHIR standard appears to offer some help with these issues offering a solution that is somewhere between the v2 flexibility and v3 semantic interoperability. Several countries are currently investigating FHIR for national adoption.[254]

Hardware limitations

LMICs struggle to find the resources to fund basic computers and Internet access for the masses. Two exceptions are Ceibal and Conectar Igualidad projects, discussed previously.[230,234]

Software limitations

Adoption of open source solutions has been attractive to LMICs but the reality is that none of these systems are a complete or perfect solution.[255] The OpenEHR framework has been recently adopted by several countries as an attempt to encourage interoperability and accurate data sharing. Although endorsed by the EU (through the CEN 13606 standard), OpenEHR has largely been adopted outside of the EU, with the framework adopted for national registries in Australia, Brazil and Sweden.[221]

Lack of Leadership

Conversion to eHealth/EMR represents a significant paradigm shift, resulting in strained relationships between clinicians and healthcare administrators. Effective leadership is clearly needed to realize a smooth transition. HIS implementation and use is highly knowledge driven and to be successfully implemented, will need support and buy-in from stakeholders both vertically and horizontally. Leaders should be able to motivate and inspire personnel to enhance cooperation and also develop synergies with decisional bodies at local, national, regional, and even global levels. A national eHealth Strategy with consistent funding is mandatory to move forward.

Cost effectiveness

Cost effectiveness of HIT in general has remained a significant issue hampering the adoption of EHRs internationally.

Workforce development

In order to deliver and maintain complex information systems there needs to be a well-trained workforce. AMIA has maintained a Global Partnership Program, funded by the Bill and Melinda Gates Foundation.[256]

Lack of evidence

In 2001 Mitchell said of EMR implementation in resource-limited situations, they are a "descriptive feast but an evaluation famine."[255] The reality is that most articles about HIT/ICT, in all clinical settings, are of low quality, in terms of inadequate number of patients studied, poor endpoints, poor documentation of unintended consequences, etc.[257] This is covered in more detail in the section on the chapter on Evidence Based Medicine.

Sustainability

Too many HIT/ICT initiatives in LMICs rely on outside funding for "pilot projects", so that when the project concludes there are no funds for sustainment. In the 2014 review by Luna, sustainability was discussed in the light of five necessary factors: effectiveness, efficiency, financial viability, reproducibility, and portability. These are difficult factors in LMICs when faced with civil strife and limited resources.[258]

Conflicting Priorities

Despite the potential for HIT/ICT initiatives to improve the delivery and documentation of healthcare, it must compete with more basic needs, such as electricity, sanitation, and clean water.

FUTURE TRENDS

Health Informatics adoption will continue to mature in European countries and other developed economies as they transition to a fully digital provision of health data storage and a range of services for patients and healthcare professionals aimed at making the business of healthcare more efficient and effective. Political and professional issues are likely to continue with areas such as privacy of healthcare information and professional autonomy continuing to be debated as new systems are rolled out.

In the developing world, it is likely that national HIT systems will start being adopted that may favor open source systems over the existing range of North American and European commercial systems. The developing world has been leading the way in the range and scope of mobile device usage and it is likely that these mHealth systems will also mature and become integrated with hospitals and government systems rather than to continue as stand-alone projects largely funded by overseas aid.

International agreements on interoperability standards continue to be a significant issue although new standards, such as HL7 FHIR, may prove to be a breakthrough that enables countries that have adopted a "best of breed" approach to funding EHR systems to share data across the healthcare sector.

KEY POINTS

- High income countries (HIC) and low and middle-income countries (LMIC) have adopted substantially different approaches to creating their health information infrastructure
- Mobile technology is playing an important role in health informatics in developing countries
- Cultural and political differences are reflected in the different approaches to national and international health informatics initiatives
- Open source software is an important alternative for LMIC

CONCLUSION

We are living in an age of rapid adoption of information technology in many industries and healthcare is no exception. By looking internationally at the various approaches and projects taken by different countries it seems that William Gibson was correct in stating that the future is here, but not yet evenly distributed.[259] Some of that future is visible in developing countries in their rapid adoption of mobile technology and some is present in well-funded and coordinate national projects such as the MedCom system in Denmark.

REFERENCES:

1. Schoen C, Osborn R, Squires D et al. A Survey of Primary Care Doctors in Ten Countries Shows Progress in Use of Health Information Technology, Less in Other Areas. Health Affairs. 2012; 31(12):2805-2816
2. McDonald K. Health IT Board working on next five-year plan for NZ. PulseIT. November 16, 2014. http://www.pulseitmagazine.com.au/index.php?option=com_content&view=article&id=2167:health-it-board-working-on-next-five-year-plan-

for-nz&catid=49:new-zealand-ehealth&Itemid=274 (Accessed October 18, 2015)
3. European Commission. Commission publishes four reports of eHealth Stakeholder Group. April 11, 2014. https://ec.europa.eu/digital-single-market/en/news/commission-publishes-four-reports-ehealth-stakeholder-group (Accessed October 19, 2015)
4. Kierkegaard, P. (2013) eHealth in Denmark: A Case Study. Journal of Medical Systems, 37 (6).
5. European Federation for Medical Informatics. www.efmi.org (Accessed October 20, 2015)
6. I2-Health. Borderless Communication for a Healthy Europe. http://www.i2-health.eu/ (Accessed October 19, 2015)
7. Austin T, Sun S. Evaluation of ISO EN 13606 As a Result of Its Implementation in XML. Health Informatics Journal. 2013;19(2):264-280
8. Martínez-Costa C, Menárguez-Tortosa M, Fernández-Breis JT. An approach for the semantic interoperability of ISO EN 13606 and OpenEHR archetypes. J Biomed Inform. 2010;43(5):736-746.
9. Studzinski J. HIMSS Europe. Three Current EHealth Trends in the German Speaking Region. http://www.himss.eu/node/6916 - .VfKu3m1mvSw.linkedin (Accessed October 22, 2015)
10. Protti D, Johansen I. Widespread adoption of information technology in primary care offices in Denmark: a case study. Commonwealth Fund. March 2010. http://www.commonweathfund.org/ (Accessed October 25, 2015)
11. 2014 International Profiles of Healthcare Systems. January 2015. Commonwealth Fund. http://www.commonwealthfund.org/ (Accessed June 7, 2016)
12. Civil Registration and Vital Statistics. World Health Organization. 2013. http://www.who.int/healthinfo/civil_registration/crvs_report_2013.pdf (Accessed October 20, 2015)
13. Sundek. dk https://www.sundhed.dk/ (Accessed October 30, 2015)
14. Country Brief: Denmark. October 2010. http://www.academia.edu/1400740/Country_Brief_Denmark (Accessed November 3, 2015)
15. MedCom. http://medcom.dk/om-medcom (Accessed November 1, 2015)
16. Kierkegaard, P. Interoperability after deployment: persistent challenges and regional strategies in Denmark. Int. J Quality in Healthcare. 2015. http://intqhc.oxfordjournals.org/content/early/2015/02/25/intqhc.mzv009.full (Accessed November 2, 2015)
17. Wickland, E. mHealth in Europe: a mixed bag. MobiHealthNews. June 2, 2015. http://www.mobihealthnews.com/ (Accessed November 10, 2015)
18. EU Countries' mHealth App Market. 2015. Research2Guidance. http://www.digitalezorg.nl/digitale/uploads/2015/07/research2guidance-EU-Country-mHealth-App-Market-Ranking-2015.pdf (Accessed November 1, 2015)
19. Wickland, E. Denmark Kicks Off 5 Telehealth Projects. MobiHealthNews. September 22, 2015. http://www.mobilhealthnews.com/(Accessed October 28, 2015)
20. Monsenso Aps. http://www.monsenso.com/ (Accessed November 1, 2015)
21. Bennett, J. In Denmark Home Healthcare Rehab working well using Kinect. InternetMedicine. November 24, 2013. http://www.internetmedicine.com/(Accessed November 20, 2015)
22. Nielson, C, Branebjerg J, Marcussen C, et al. Strategic Intelligence monitoring of personal health systems. Phase 2. Country Study Denmark. European Commission. 2012. http://is.jrc.ec.europa.eu/pages/TFS/documents/SIMPHSCountrystudyDenmarkfinalrev2.pdf (Accessed October 22, 2015)
23. Executive Summary. Information for Health. An Information Strategy for the Modern NHS 1998-2005. http://webarchive.nationalarchives.gov.uk/+/www.dh.gov.uk/en/Publicationsandstatistics/Publications/PublicationsPolicyAndGuidance/DH_4002944 (Accessed November 2, 2015)
24. Campion-Awwad, Hayton A, Smith L, et al. The National Programme for IT in the NHS. A Case Study. http://www.cl.cam.ac.uk/~rja14/Papers/npfit-mpp-2014-case-history.pdf (Accessed October 20, 2015)
25. NHS. A Guide to the National Programme for Information Technology. 2005 http://www.providersedge.com/ehdocs/ehr_articles/A_Guide_to_the_National_Programme_for_Information_Technology.pdf (Accessed October 30, 2015)
26. Dismantling the National Programme for IT. Gov. Uk. September 2011. https://www.gov.uk/government/news/dismantling-the-nhs-national-programme-for-it (Accessed November 5, 2015)
27. Currie, W. Translating Health IT Policy Into Practice in the UK NHS. Scan J of IS. 2014;26(2):3-26
28. Maughan, A. Six reasons why the NHS National Programme for IT failed. Computer Weekly. http://www.computerweekly.com/opinion/Six-reasons-why-the-NHS-National-Programme-for-IT-failed (Accessed October `5, 2015)
29. Everybody's Business—strengthening health systems to improve health outcomes. WHO 2007. http://apps.who.int/iris/handle/10665/43918 (Accessed October 24, 2015)
30. Wallace P, Delaney B, Sullivan F. Unlocking the research potential of the GP electronic care record. Br J Gen Pract. 2013;63(611):284-285.
31. Gill L, Goldacre M. English national record linkage of hospital episode statistics and death registration

31. records. Unit of Health-Care Epidemiology, Oxford University, Oxford. 2003. http://nchod.uhce.ox.ac.uk/NCHOD Oxford E5 Report 1st Feb_VerAM2.pdf (Accessed February 20, 2016)
32. Smyth RL. Regulation and governance of clinical research in the UK. BMJ. 2011;342:d238
33. Department of Health. Raising the Profile of Long Term Conditions Care A Compendium of Information. Department of Health; 2008. http://webarchive.nationalarchives.gov.uk/20130107105354/http://www.dh.gov.uk/prod_consum_dh/groups/dh_digitalassets/documents/digitalasset/dh_082067.pdfr (Accessed February 15, 2016)
34. Department of Health. Whole System Demonstrator Programme Headline Findings – December 2011. Department of Health; .http://www.gov.uk/ (Accessed February 15, 2016)
35. Cartwright M, Hirani SP, Rixon L, et al. Effect of telehealth on quality of life and psychological outcomes over 12 months (Whole Systems Demonstrator telehealth questionnaire study): nested study of patient reported outcomes in a pragmatic, cluster randomized controlled trial. BMJ. 2013;346:f653.
36. Department of Health: The National Programme for IT in the NHS: 2006-2007. http://www.publications.parliament.uk/pa/cm200607/cmselect/cmpubacc/390/390.pdf (Accessed November 11, 2015)
37. Greenhalgh T, Stramer B, Bratant T et al. The Devil's in the Details. 7 May 2010 https://www.ucl.ac.uk/news/scriefullreport.pdf (Accessed October 28, 2015)
38. De Lusignan S, Seroussi B. A Comparison of English and French approaches to providing access to summary care records: scope, consent, cost. Stud Health Tech Inform 2013;186:61-65
39. NHS. Personalised Health and Care 2020. November 2014. https://www.gov.uk/government/uploads/system/uploads/attachment_data/file/384650/NIB_Report.pdf (Accessed November 1, 2015)
40. Thompson, S. International Profiles of Healthcare Systems. 2013. Commonwealth Fund. http://www.commonwealthfund.org/ (Accessed October 15, 2015)
41. NHS. Five Year Forward View. October 2014. http://www.england.nhs.uk/wp-content/uploads/2014/10/5yfv-web.pdf (Accessed October 25, 2015)
42. Gov. UK. Plans to improve digital services for the health and care sector. National Information Board. 19 June 2015. http://www.gov.uk/ (Accessed October 20, 2015)
43. NHS. Interoperability Handbook. September 2015. http://www.england.nhs.uk/ (Accessed November 15, 2015)
44. Monegain, B. Epic EHR adds to UK hospital's financial mess. HealthcareITNews. September 28, 2015. http://www.healthcareitnews.com/ (Accessed November 12, 2015)
45. iPlato. http://www.iplato.net/ (Accessed December 1, 2015)
46. Tunstall Healthcare. www. Tunstall.co.uk (Accessed December 1, 2015)
47. Personaltechmd. http://www.personaltechmd.com/ (Accessed December 1, 2015)
48. Raghavulu, V. Prasad, A. Role of computer science in healthcare. Int J Sci Engin Tech Res 2014;3(46):9386-9387
49. Currie W, Seddon J. A cross-national analysis of eHealth in the EU: some policy and research directions. Inf Man 2014;51(6):783-797
50. Stroetmann KA, Artmann J, Giest S. Country Brief: Germany. October 2010. eHealth Strategies http://www.ehealth.strategies.eu/ (Accessed November 15, 2015)
51. Tuffs A. Germany plans to introduce electronic health card. BMJ. 2004;329:7458
52. Hoeksma J. Germany suspends ehealth card project. Digital Health. 19 January 2010. http://www.digitalhealth.net/ (Accessed October 28, 2015)
53. Germany and the challenge of rolling out eHealth on a large scale. 15 May 2013. http://www.esante.gour.fr/ (Accessed November 15, 2015)
54. The Health Systems and Policy Monitor. Germany. 2014. http://www.hspm.org/countries/Germany (Accessed October 28, 2015)
55. Viactiv. https://www.viactiv.de/english/electronic-health-card-egk/ (Accessed October 28, 2015)
56. The Federal Government. Digital Agenda 2014-2017. http://www.digitale-agenda.de/ (Accessed November 20, 2015)
57. European Commission. Hillenius G. Germany's digital agenda reshuffles country's eHealth policy. January 9, 2015. https://joinup.ec.europa.eu/ (Accessed November 20, 2015)
58. Federal Ministry. Act on secure digital communication and applications in the healthcare system. September 29, 2015. http://www.bmg.bund.de/ (Accessed October 20, 2015)
59. Bertlemann H, Dinger F, Schreiber L. Germany: New healthcare reforms 2015. Health Law Pulse. August 19, 2015. http://www.thehealthlawpulse.com/ (Accessed November 1, 2015)
60. Germany Trade and Invest. German mHealth market outlook. http://www.gtai.de/ (Accessed October 25, 2015)

61. Sesam Vitale. Smart Card Alliance. http://d3nrwezfchbhhm.cloudfront.net/pdf/Sesam_Vitale.pdf (Accessed December 5, 2015)
62. Overview of the national laws on EHR member states. National report for France. Milieu Law and Policy Consulting. January 2014. http://ec.europa.eu/health/ehealth/docs/laws_france_en.pdf (Accessed December 1, 2015)
63. De Lusignan S, Ross P, Shifrinm P, Seroussi B. Comparison of approaches to providing patients access to summary care records across old and new Europe: an exploration of facilitators and barriers to implementation. Stud Health Tech Inform 2013;192(1):397-401
64. Dossier Medical Personnel. http://www.dmp.gouv.fr/web/dmp/ (Accessed November 15, 2015)
65. Sesam-Vitale Evaluation. A contribution to French sustainability policy. Gemalto. http://www.gemalto.com/brochures-site/download-site/Documents/gov_sesam_vitale.pdf (Accessed November 5, 2015)
66. What is a CPS card? 2014 http://esante.gouv.fr/en/services/espace-cps/what-a-cps-card (Accessed November 4, 2015)
67. Health Informatics Systems Interoperability Framework (HIS-IF). 6 May 2015. http://esante.gouv.fr/en/node/2053 (Accessed November 20, 2015)
68. Gray B, Bowden T, Johansen I et al. Electronic Health Records: An International Perspective on "Meaningful Use". The Commonwealth Fund November 2011. http://www.commonwealthfund.org/ (Accessed November 20, 2015)
69. Park Y, Atalag K. Current National Approach to Healthcare ICT Standardization: Focus on the Progress in New Zealand. Health Inform Res 2015;21(3):144-151
70. McDonald K. HiNZ 2015: National child health IT platform scores early wins. Pulse IT. 22 October 2015. http://www.pulseitmagazine.com.au/ (Accessed November 30, 2015)
71. OrionHealth. https://orionhealth.com/ (Accessed November 24, 2015)
72. Ministry of Health. New Zealand Health Strategy 2016. http://www.health.govt.nz/ (Accessed June 13, 2016)
73. Smartphones changing the face of mobile health. National Health IT Board. 9 April 2015. http://www.healthitboard.health.govt.nz/ (Accessed December 15, 2015)
74. Beating the Blues. http://www.beatingtheblues.co.nz/ (Accessed December 15, 2015)
75. Bartlett C, Boehncke K. E-health: enabler for Australia's health reform. November 2008. Booz & Co. http://www.health.gov.au/ (Accessed December 20, 2015)
76. Healthcare Identifiers Act of 2010. http://www.legislation.gov.au/ (Accessed December 20, 2015)
77. Australian Government Department of Health. Personally Controlled EHR for all Australians. 14 September 2012. http://www.health.gov.au/ (Accessed December 11, 2015)
78. McDonald K. Problems plaguing PCEHR provider portal. Pulse IT. 26 February 2014. www.pulseitmagazine.com.au (Accessed December 20, 2015)
79. My Health Record. https://myhealthrecord.gov.au/ (Accessed June 13, 2016)
80. Is Australia ready for mobile health? Healthcare Innovation. August 18, 2015. http://www.enterpriseinnovation.net/ (Accessed December 20, 2015)
81. Telstra. http://www.telstra.com.au/ (Accessed January 2, 2016)
82. Personal KineticGraph. Global Kinetics Corporation. http://www.globalkineticscorporation.com/ (Accessed January 2, 2016)
83. Van de Maele N, Evans D, Tan-Torres T. Development assistance for health in Africa: are we telling the right story? Bulletin for the WHO. 2013;91:483-490
84. Sharma P, Agarwal P. Mobile phone text messaging for promoting adherence to antiretroviral therapy in patients with HIV infection. The WHO Reproductive Health Library; Geneva: World Health Organization 2012(3)
85. Department of State, 2012. PEPFAR Blueprint :Creating an AIDS-free Generation,
86. Frasier, H., May, M.A. & Wanchoo, R., 2008. e-Health Rwanda Case Study, Available at: http://ehealth-connection.org/files/resources/Rwanda + Appendices.pdf. (Accessed December 20, 2015)
87. Afarikumah E. Electronic health in ghana: Current status and future prospects. Online J. Public Health Inform 2014, May;5(3):230
88. Were MC, Siika A, Ayuo PO, Atwoli L, Esamai F. Building Comprehensive and Sustainable Health Informatics Institutions in Developing Countries: Moi University Experience. Studies in health technology and informatics. 2015;216:520.
89. Manya A, Braa J, Øverland LH, et al. National roll out of district health information software (DHIS 2) in Kenya, 2011--central server and cloud based infrastructure. http://www.ist-africa.org/Conference2012 (Accessed December 20, 2015)
90. Braa J, Humberto M. Building collaborative networks in Africa on health information systems and open source software development--experiences from the HISP/BEANISH http://www.ist-africa.org/Conference2007 (Accessed December 20, 2015)
91. Wolfe BA. The openmrs system: Collaborating toward an open source EMR for developing countries. AMIA Annu. Symp. Proc 2006:1146

92. OpenMRS. http://www.openmrs.org/ (Accessed March 23, 2016)
93. Levy S. Zuckerberg explains internet. org, Facebook's plan to get the world online. August 26, 2013. http://www.wired.com/ (Accessed May 14, 2016)
94. SpaceX founder files with government to provide Internet service from space. The Washington Post. 2015 http://www.washingtonpost.com/business/economy/spacex-founder-files-with-government-to-provide-internet-service-from-space/2015/06/09/db8d8d02-0eb7-11e5-a0dc-2b6f404ff5cf_story.html. (Accessed May 14, 2106)
95. Handwerk B. Google's Loon Project Puts Balloon Technology in Spotlight. Natl Geogr Mag. June 18, 2013.
96. The World in 2014. http://www.itu.int/en/ITU-D/Statistics/Documents/facts/ICTFactsFigures2014-e.pdf (Accessed May 14, 2016)
97. Medic Mobile. http://medicmobile.org/ (Accessed June 10, 2016).
98. FrontlineSMS | FrontlineCloud. http://www.frontlinesms.com/ (Accessed June 10, 2016).
99. GAVI. http://www.gavi.org/ (Accessed June 10, 2016)
100. Gerber T, Olazabal V, Brown K, Pablos-Mendez A. An agenda for action on global e-health. Health Aff. 2010;29(2):233-236.
101. Saxenian H, Cornejo S, Thorien K, et al. An analysis of how the GAVI alliance and low-and middle-income countries can share costs of new vaccines. Health Aff. 2011;30(6):1122-1133.
102. Vodafone. http://www.vodafone.com/ (Accessed June 4, 2016)
103. Roll Back Malaria http://www.rollbackmalaria.org/ (Accessed May 10, 2016)
104. Githinji S, Kigen S, Memusi D, Nyandigisi A, Mbithi AM, Wamari A, Muturi AN, Jagoe G, Barrington J, Snow RW, Zurovac D. Reducing stock-outs of life saving malaria commodities using mobile phone text-messaging: SMS for life study in Kenya. PLoS One. 2013 Jan 17;8(1):e54066
105. Siddle K, Vieren L, Fiander A. Characterising women with obstetric fistula and urogenital tract injuries in Tanzania. Int Urogynecol J Pelvic Floor Dysfunct. 2014;25(2):249–55.
106. M-pesa. http://www.safaricom.co.ke/personal/m-pesa (Accessed May 10, 2016)
107. Coleman J. Monitoring MAMA: Gauging the Impact of MAMA South Africa. J Mob Technol Med. 2013;2(4s):9.
108. Grice HP, Cole P, Morgan JL. Syntax and semantics. Logic and conversation. 1975;3:41-58
109. Rothman AJ, Salovey P, Antone C, Keough K, Martin CD. The Influence of Message Framing on Intentions to Perform Health Behaviors. J Exp Soc Psychol. 1993;29(5):408-433
110. Kerrison RS, Shukla H, Cunningham D, et al. Text-message reminders increase uptake of routine breast screening appointments: a randomised controlled trial in a hard-to-reach population. Br J Cancer. 2015;112(6):1005-1010.
111. de Jongh T, Gurol-Urganci I, Vodopivec-Jamsek V, et al. Mobile phone messaging for facilitating self-management of long-term illnesses. Cochrane Database Syst Rev. 2012;(12):CD007459.pub2.
112. Gurol-Urganci I, de Jongh T, Vodopivec-Jamsek V, Atun R, Car J. Mobile phone messaging reminders for attendance at healthcare appointments. Cochrane database Syst Rev 2013. http://www.ncbi.nlm.nih.gov/pubmed/24310741 (Accessed June 1, 2016)
113. Car J, Gurol-Urganci I, De Jongh T, Vodopivec-Jamsek V, Atun R. Mobile phone messaging reminders for attendance at scheduled healthcare appointments. Cochrane Database of Systematic Reviews. 2008;(4). doi:10.1002/14651858.CD007458.
114. Shet A, De Costa A, Kumarasamy N, et al. Effect of mobile telephone reminders on treatment outcome in HIV: evidence from a randomised controlled trial in India. BMJ. 2014;349:g5978.
115. Pop-Eleches C, Thirumurthy H, Habyarimana JP, et al. Mobile phone technologies improve adherence to antiretroviral treatment in a resource-limited setting: a randomized controlled trial of text message reminders. AIDS. 2011;25(6):825-834.
116. Ogara E, Magana O. eHealth Strategy 2011 - 2017. 2011. http://www.isfteh.org/files/media/kenya_national_ehealth_strategy_2011-2017.pdf (Accessed June 8, 2016)
117. Vision 2030. http://www.vision2030.go.ke/ (Accessed June 10, 2016)
118. MOH. Kenya EMR Review Towards Standardization Report. September 7, 2011. https://www.ghdonline.org/ (Accessed June 11, 2016)
119. West D, Branstetter DG, Nelson SD, Manivel JC, Blay J-Y, Chawla S, et al. How Mobile Devices are Transforming Healthcare. BrookingsEdu 2012;18(16):1–38
120. Sema Doc. http://hellodoctor.co.ke/ (Accessed June 10, 2016)
121. Piette J, Lun K, Moura L, et al. Impacts of e-health on the outcomes of care in low- and middle-income countries: where do we go from here? Bull World Health Organ. 2012;90(5):365-372.
122. Ministry of ICT. The Kenya National ICT MasterPlan : Towards a Digital Kenya. 2014. https://www.kenet.or.ke/sites/default/files/Final ICT Masterplan Apr 2014.pdf (Accessed June 10, 2016)
123. Nisingizwe MP, Iyer HS, Gashayija M, et al. Toward utilization of data for program management and evaluation: quality assessment of five years of health

management information system data in Rwanda. Global health action. 2014;7
124. Department of Health South Africa, 2012. National eHealth Strategy, South Africa 2012/13-2016/17. Department of Health, Republic of South Africa, pp.1–36
125. Springer, 2012. Foundations of Health Informatics Engineering and Systems Z. Liu & A. Wassyng, eds., Berlin, Heidelberg: Springer Berlin Heidelberg
126. NISR, 2015. NATIONAL INSTITUTE OF STATISTICS OF RWANDA March 2015 Gross Domestic Product – 2014 Gross Domestic Product and its structure In 2014 , (March), pp.1–8
127. Farmer, P.E. et al., 2013. Reduced premature mortality in Rwanda: lessons from success. Bmj, 346(jan18 1), pp.f65–f65.
128. Crichton, R. et al., 2013. An architecture and reference implementation of an open health information mediator: Enabling interoperability in the Rwandan health information exchange. Lecture Notes in Computer Science (including subseries Lecture Notes in Artificial Intelligence and Lecture Notes in Bioinformatics), 7789 LNCS, pp.87–104.
129. UNECA. MDG Report 2014. Assessing Progress in Africa toward the Millennium Development Goals. www.undp.org (Accessed June 4, 2016)
130. Perry, H. Case Studies of Large-Scale Community Health Worker Programs. Appendix. www.mchip.net (Accessed June 4, 2016)
131. Frasier H, May M, Wanchoo R. e-Health Rwanda case study. Am Med http://ehealth-connection.org/files/resources/Rwanda + Appendices.pdf (Accessed May 15, 2016)
132. Ventures Africa & Iruobe, E., 2015. Rwanda tasked with pioneering eHealth for East Africa. http://venturesafrica.com/rwanda-tasked-with-pioneering-ehealth-for-east-africa/ (Accessed June 8, 2016)
133. Perry HB, Zulliger R, Rogers MM. Community health workers in low-, middle-, and high-income countries: an overview of their history, recent evolution, and current effectiveness. Annual review of public health. 2014 Mar 18;35:399-421.
134. Aranda-Jan CB, Mohutsiwa-Dibe N, Loukanova S. Systematic review on what works, what does not work and why of implementation of mobile health (mHealth) projects in Africa. BMC public health. 2014 Feb 21;14(1):188
135. Källander K, Tibenderana JK, Akpogheneta OJ, et al. Mobile health (mHealth) approaches and lessons for increased performance and retention of community health workers in low-and middle-income countries: a review. Journal of medical Internet research. 2013;15(1):e17.
136. Kizito K, Adeline K, Baptiste K, Anita A. TRACnet: A National Phone-based and Web-based Tool for the Timely Integrated Disease Surveillance and Response in Rwanda.2013 http://www.ncbi.nlm.nih.gov/pmc/articles/PMC3692857/ (Accessed May 21, 2016)
137. United Nations, 2007. TRACnet , Rwanda : Fighting Pandemics through Information Technology. www.un.org (Accessed June 5, 2016)
138. Svoronos T, Jillson IA, Nsabimana MM. TRACnet's absorption into the Rwandan HIV/AIDS response. International Journal of Healthcare Technology and Management. 2008 Jan 1;9(5-6):430-45.
139. Leuchowius K. Report on the health care sector and business opportunities in Rwanda. 2014 http://www.swecare.se/Portals/swecare/Documents/Report-on-the-Health-Care-Sector-and-Business-Opportunities-in-Rwanda-Sep2014-vers2.pdf (Accessed May 22, 2016)
140. Sun R, Cao H, Zhu Z et al. Current aging research in China. Protein Cell 2015;6(8):314-321
141. CHIMA. The white paper on China's hospital information systems. May 2008. http://cdn.medicexchange.com/images/whitepaper/chinas hospital information systems.pdf?1294036467 (Accessed January 5, 2016)
142. Walton G. China's Golden Shield: corporations and the development of surveillance technology in China. Rights and Democracy. 2001
143. Grace Yu. "China HIT Case Study." in Health Information and Technology and Policy Lab HIT Briefing Book, ed. Claire Topal and Kaleb Brownlow (Seattle, WA: National Bureau of Asian Research, 2007)
144. Lei J, Sockolow P, Guan P et al. A comparison of EHRs at two major Peking university hospitals in China to US meaningful use objectives. BMC MI and Decision Making. 28 August 2013. http://www.bmcmedicinformdecismak.biomedcentral.com/ (Accessed September 20, 2016)
145. Parikh H. Overview of EHR systems in BRIC nations. Clinical leader. April 15, 2015. http://www.clinicalleader.com/ (Accessed September 22, 2016)
146. He P, Yuan Z, Liu G et al. An evaluation of a tailored intervention on village doctors use of electronic health records. BMC Health Serv Res. 2014. 14:217
147. International Health Care System Profiles. http://international.commonwealthfund.org/features/ehrs/ (Accessed January 15, 2016)
148. Xiaohui Y, Han H, Jiadong D et al. mHealth in China and the United States. Center for Tech Innovations at Brookings. http://www.brookings.edu/~/media/research/files/reports/2014/03/12-mHealth-china-united-states-health-care/mHealth_finalx.pdf (Accessed January 20, 2016)
149. Yoo E. Diabetes management platform Weitang raises series B from Yidu cloud. Tech Node. January

4, 2016. http://www.technode.com/ (Accessed January 20, 2016)

150. Custer C. Chinese smartphone glucometer DNurse raises millions. Tech In Asia. February 10, 2015. http://www.techinasia.com/ (Accessed January 20, 2016)

151. Chen L, Wang N, Du X et al. Effectiveness of a smartphone app on improving immunization of children in rural Sichuan Province, China: study protocol for a paired cluster randomized controlled trial. BMC Public Health. 2014. DOI:10.1186/1471-2458-14-262

152. China Digital Health Boot Camp. http://scpku.fsi.standford.edu/ (Accessed January 25, 2016)

153. Zhihua L. Healthcare at your fingertips. China Daily. February 22, 2016. http://www.chinadaily.com.cn/ (Accessed February 25, 2016)

154. Cusano D. Can digital health solve China's healthcare quality and distribution problems? TeleCare. July 23,2015. http://www.telecareaware.com/ (Accessed January 20, 2016)

155. Sun J, Guo Y, Wang X et al. mHealth for aging China: opportunities and challenges. Aging and Disease. 2016;7(1):53-67

156. Jourdan A. Digital doctors: China sees tech cure for healthcare woes. Reuters. October 14, 2014. http://www.reuters.com/ (Accessed January 20, 2016)

157. E-Commerce Security: Advice From Experts. M. Khosrow-Pour Editor. Cybertech Publishing. 2004

158. Traditional chine medicine hospitals growing. The State Council. The Peoples Republic of China. January 14, 2016. http://English.gov.cn (Accessed February 3, 2016)

159. Hsu W, Chan E, Zhang Z et al. A survey to investigate attitudes and perceptions of Chinese medicine professionals in HIT in Hong Kong. Eur J Int Med 2015;7(1):36-46

160. Richie J. China's EMR Market. EMR & HIPAA. October 13, 2013 http://www.emrandhipaa.com/ (Accessed February 5, 2016)

161. Lei J, Xu L, Meng Q et al. The current status of usability studies of IT in China: a systematic review.2014. http://www.hindawi.com/journals/bmri/2014/568303/ (Accessed February 5, 2016)

162. Zhang Y, Tang Z. Health technology assessment in China. Health Affairs. 2013;32(2):438

163. Abraham C, Nishihara E, Akiyama. Transforming healthcare with information technology in Japan: A review of policy, people, and progress. Int J Med Inform. 2011; 80(3): 157-170.

164. Okada M, Yamamoto K, Kawamura T. Health and medical Informatics education in Japan. IMIA Yearb Med Inform. 2004; 193-198.

165. Sonoda T. Evolution of electronic medical record solutions. Fujitsu Sci Tech J. 2011; 47(1): 19-27.

166. Juhr M, Haux R, Suzuki T, Takabayaski K. Overview of recent trans-institutional health network projects in Japan and Germany. J Med Syst. 2015; 39(5): 50.

167. Ezaki H. President's message. National Hospital Organization Nagasaki Medical Center. http://www.nagasaki-mc.jp/content/en/ (Accessed September 20, 2016)

168. Organization for Economic Co-operation and Development. OECD health policy studies: Strengthening health informatics infrastructure for health care quality governance. Paris, France: OECD Publishing; 2013

169. Yoshida Y, Imai T, Ohe K. The trends in ENR and CPOE adoption in Japan under the national strategy. Int J Med Inform. 2013; 82(10): 1004-1011

170. Yokobori Y, Sakai T, Takeda T, Oi T. Post Earthquake Health Information Management in Japan—the Challenges. Perspectives in Health Information Management. 2015; 1: 1-13.

171. Matsuda R. The Japanese health care system. In: Mossialos E and Wenzl M, ed. 2015 International profiles of health care systems. New York, NY: The Commonwealth Fund; 2016: 107-114.

172. Essany M. mHealth: Ample room for growth in Japan. mHealthwatch. January 3, 2013 http://mhealthwatch.com/mhealth-ample-room-for-growth-in-japan-19265/. (Accessed September 5, 2016)

173. Montgomery S. 8 Examples of how mHealth is increasing access to healthcare around the world. Nuviun Digital Health. February 20, 2014. http://nuviun.com/content/mHealth. (Accessed September 4, 2016)

174. Apple Press. Japan Post group, IBM and Apple deliver iPads and custom apps to connect elderly in Japan to services, family and community. Apple. April 30, 2015 https://www.apple.com/pr/library/2015/04/30Japan-Post-Group-IBM-and-Apple-Deliver-iPads-and-Custom-Apps-to-Connect-Elderly-in-Japan-to-Services-Family-and-Community.html. (Accessed September 1, 2016)

175. Japan Times Staff. Medical industry signals interest in telemedicine. Japan Times. March 16, 2016. http://www.japantimes.co.jp/news/2016/03/16/national/social-issues/medical-industry-signals-interest-telemedicine/#.V2wZUPkrLAW. (Accessed August 4, 2016)

176. Abd Ghani MK, Bali RK, Marshall IM, Wickramasinghe NS. Electronic health records approaches and challenges: A comparison between Malaysia and four East Asian countries. Int J Electron Healthc. 2008; 4(1): 78-104.

177. Reddy KS. India's Aspirations for Universal Health Coverage. NEJM 2015;373:1-5

178. Healthcare Industry in India. India Brand Equity Foundation. www.ibef.org (Accessed September 23, 2016)
179. Deloitte. 2016 Global health care outlook: Battling costs while improving care. London: Deloitte Touche Tohmatsu; 2016: 1-28.
180. Ministry of Health and Family Welfare. EHR Standards for India. August 2013. http://www.mohfw.nic.in/showfile.php?lid=1672 (Accessed January 20, 2016)
181. Smart Health India. The George Institute. http://www.georgeinstitute.org/philanthropic-opportunities/smart-health-india (Accessed June 25, 2016)
182. Sana. http://sana.mit.edu (Accessed May 1, 2016)
183. Ooi LC, Tan P. Out IT journey: One Patient-One Record. In: Earn LC, Satku K, eds. Singapore's Health Care System: What 50 Years Have Achieved. Hackensack, NJ: World Scientific Publishing Co; 2016: 337-350.
184. Tan, LT. National Patient Master Index in Singapore. Int J Biomed Comput. 1995; 40(2): 89-93.
185. Accenture Staff. Accenture implements nationwide electronic health record system in Singapore. Accenture Newsroom. June 20, 2011 https://newsroom.accenture.com/industries/health-public-service/accenture-implements-nationwide-electronic-health-record-system-in-singapore.htm. (Accessed July 10, 2016)
186. U.S. International Trade Administration. Singapore Type of market: Moderate/Growing. In: 2015 ITA Health IT Top Markets Report. 2015: 31-34.
187. Liu C, Haseltine W. The Singapore health care system. In: Mossialos E and Wenzl M, ed. 2015 International profiles of health care systems. New York, NY: The Commonwealth Fund; 2016: 143-151.
188. Singapore Ministry of Health. Fact sheet on HealthHub. Ministry of Health. October 18, 2015. https://www.moh.gov.sg/content/dam/moh_web/PressRoom/20151018_HealthHub%20Factsheet.pdf. (Accessed July 10, 2016)
189. Singapore Government Staff. eCitizen: eServices. eCitizen.gov. https://www.ecitizen.gov.sg/eServices/Pages/Default.aspx#tabs-3. (Accessed July 10, 2016)
190. Singapore Ministry of Finance. MOF Committee of Supply speech 2016 by Senior Minister of State for Finance MS Sim Ann. Ministry of Finance. April 16, 2011 http://www.mof.gov.sg/news-reader/articleid/1628/parentId/59/year/2016. (Accessed July 10, 2016)
191. Song, GGY. GPs utilization of NEHR. Ministry of Health. February 12, 2015 https://www.moh.gov.sg/content/moh_web/home/pressRoom/Parliamentary_QA/2015/gps-utilisation-of-nehr.html. (Accessed July 10, 2016)
192. Shehadi R, Tohme W, Bitar J, Kutty S. Anatomy of an eHealth ecosystem. Beirut: Booz & Co; 2011.
193. Accenture. Connected health: The drive to integrated healthcare delivery. http://www.himss.eu/sites/default/files/Accenture-Connected-Health-Global-Report-Final-Web.pdf. (Accessed July 10, 2016)
194. Deloitte. Healthcare and life sciences predictions 2020: A bold future. London: Deloitte Touche Tohmatsu; 2015: 1-32.
195. Goh G, Tan NC, Malhotra R, et al. Short-term trajectories of use of a caloric-monitoring mobile phone app among patients with Type 2 diabetes mellitus in a primary care setting. J Med Internet Res. 2015; 17(2): e33.
196. Wee L. Dr. apps on call 24/7. Singapore General Hospital. March 22, 2012. http://www.sgh.com.sg/about-us/newsroom/News-Articles-Reports/Pages/DrAppsoncall247.aspx. (Accessed July 10, 2016)
197. Chin CW, Phua KH. Long-term care policy: Singapore's experience. J Aging Soc Policy. 2016; 28(2): 113-129.
198. HIMSS Asia Pacific. Transforming health through IT. HIMSS Asia Pac. http://www.himssasiapac.org/sites/default/files/HIMSSAP_ExclusiveArticles_CaseStudy_ContinuityofCareforElderly.pdf. (Accessed July 10, 2016)
199. Singapore Ministry of Health. National Telemedicine Guidelines. Ministry of Health. January 2015 https://www.moh.gov.sg/content/dam/moh_web/Publications/Guidelines/MOH%20Cir%2006_2015_30Jan15_Telemedicine%20Guidelines%20rev.pdf. (Accessed July 10, 2016)
200. Singapore Medical Council. Ethical code and ethical guidelines. Health Professionals. July 12, 2011 http://www.healthprofessionals.gov.sg/content/dam/hprof/smc/docs/publication/SMC%20Ethical%20Code%20and%20Ethical%20Guidelines.pdf. (Accessed July 10, 2016)
201. Kwang TW. Smart hospitals, telehealth and EHR in Singapore. Enterprise Innovation. September 28, 2015 http://www.enterpriseinnovation.net/article/smart-hospitals-telehealth-and-ehr-singapore-1526708088. (Accessed July 10, 2016)
202. Hassan NJ. More needs to be done to ensure telemedicine is used in right setting: Experts. Channel News Asia. January 20, 2016 http://www.channelnewsasia.com/news/singapore/more-needs-to-be-done-to/2441808.html. (Accessed July 10, 2016)
203. Senthilingam M, Stevens A. The doctor will not see you now: How Singapore is pioneering telemedicine. CNN. November 3, 2015. (Accessed July 10, 2016)
204. Wee L. See your doctor over a webcam. Singapore General Hospital. February 23, 2012 http://www.sgh.com.sg/about-us/newsroom/News-Articles-Reports/

Pages/Seeyourdoctoroverawebcam.aspx. (Accessed July 10, 2016)
205. Riistama J., Pauws S, Tesanovic A., et al. First of a kind telehealth implementation study in Singapore: Methodology and challenge to tailor to Asian healthcare system. Circ Cardiovasc Qual Outcomes. 2015; 8: A370.
206. Accenture. Consumers and doctors in Singapore increasingly divided on who should have access to a patient's electronic health record, Accenture survey finds. Accenture. May 18, 2016.
207. Association of Chartered Certified Accounts. Global perspectives on health challenges. 2012. ACCA Global. http://www.accaglobal.com/content/dam/acca/global/PDF-technical/health-sector/tech-tp-gpohc.pdf. (Accessed July 10, 2016)
208. Rowlands D. Guest editorial: National eHealth record systems – the Singapore experience. September 10, 2012Pulse IT Magazine. http://www.pulseitmagazine.com.au/asia-pacific-health-it/1139-guest-editorial-national-ehealth-record-systems-the-singapore-experience. (Accessed July 10, 2016)
209. National eHealth Transition Authority Ltd. Evolution of eHealth in Australia: Achievements, lessons, and opportunities. Sydney: NEHTA; 2016.
210. Greene W. Health tech innovation blooms in Singapore. Techonomy. April 18, 2016 http://techonomy.com/2016/04/health-tech-innovation-blooms-in-singapore/. (Accessed July 10, 2016)
211. Lawrence S. Philips, Singapore to jointly invest in population health management companies targeting Asia. January 13, 2016 Fierce Biotech. http://www.fiercebiotech.com/medical-devices/philips-singapore-to-jointly-invest-population-health-management-companies. (Accessed July 10, 2016)
212. Bloomberg Briefs Staff. 2015 most efficient health care (Table). Bloomberg Briefs. 2015 http://www.bloombergbriefs.com/content/uploads/sites/2/2015/11/health-care.pdf. (Accessed July 10, 2016)
213. Morais RM, Costa AL, Gomes EJ. Information Systems Sus: a Historical Perspective and Policies of Computing and Information. Nucleus. 2014;11(1):239–56
214. Paim J, Travassos C, Almeida C, et al. The Brazilian health system: History, advances, and challenges. Lancet. 2011;377(9779):1778–97
215. Victora CG, Barreto ML, do Carmo Leal M, et al. Health conditions and health-policy innovations in Brazil: the way forward. Lancet. 2011;377(9782):2042–53.
216. Presidency of the Republic. http://www.planalto.gov.br/ccivil_03/decreto/1990-1994/D0100.htm (Accessed December 20, 2015)
217. Duarte LS, Pessoto UC, Guimarães RB, et al. Regionalization of health in Brazil: an analytical perspective. Saúde e Soc. SciELO Brasil; 2015;24(2):472–85.
218. Cunha RE Da. Cartão Nacional de Saúde: os desafios da concepção e implantação de um sistema nacional de captura de informações de atendimento em saúde. Cien Saude Colet. 2002;7(4):869–78.
219. Health Ministry launches digital version of SUS card. Portal Saude. www.portalsaude.saude.gov.br (Accessed June 17, 2016)
220. Brazilian Health Ministry. Interoperability and standards in Health Information and Informatics in Brazil. {PORTARIA} {N} 2073 de Agosto 2011 do Ministério da Saúde. 2011. p. 4.
221. OpenEHR. www.openehr.org (Accessed June 1, 2016)
222. Brazilian Health Ministry. Brazilian National Plan for Health Information and Informatics - {PNIIS}. DOU -Brazilian Fed Press - Minist Heal. 2015;Portaria 589(Seção 1):72.
223. Brazilian Health Ministry. Management Comittee for (Brazilian) E-Health Strategy . {Guidelines} Estratégia E-Saúde para o Brasil (en. E-Health Strategy for Brazil), July 12, 2017. p. 1-80. http://portalarquivos.saude.gov.br/images/pdf/2017/julho/12/Estrategia-e-saude-para-o-Brasil.pdf (Accessed August 29, 2017)
224. Prestes IV, de Moura L, Duncan BB, Schmidt MI. A national cohort of patients receiving publicly financed renal replacement therapy within the Brazilian Unified Health System. Lancet. 2013;381:S119.
225. Bittencourt SA, Camacho LAB, do Carmo Leal M. O Sistema de Informação Hospitalar e sua aplicação na saúde coletiva Hospital Information Systems and their application in public health. Cad Saúde Pública. 2006;22(1):19-30.
226. Junior CN, Padoveze MC, Lacerda RA. Governmental surveillance system of healthcare-associated infection in Brazil. Rev Esc Enferm USP. 2014;48(4):657-662.
227. Pires MRGM, Gottems LBD, Vasconcelos Filho JE, Silva KL, Gamarski R. The Care Management Information System for the Home Care Network (SI GESCAD): support for care coordination and continuity of care in the Brazilian Unified Health System (SUS). Cien Saude Colet. 2015;20(6):1805-1814.
228. TABNET http://datasus.saude.gov.br/informacoes-de-saude/tabnet (Accessed June 2, 2106)
229. UNIVERSUS. http://universus.datasus.gov.br/ (Accessed June 1, 2016)
230. Watkins SC. Conectar Igualdad. Argentina's bold move to build an equitable digital future. 9/16/2011. DML Central. http://

dmlcentral.net/conectar-igualdad-argentina-s-bold-move-to-build-an-equitable-digital-future/?variation=b&utm_expid=427257085. oNX7Gn2HRIOm8WhS1GUvRg.1&utm_referrer=https%3A%2F%2Fwww.google.com%2F (Accessed May 20, 2016)
231. Otero P, Hersh W, Luna D. A Medical Informatics Distance Learning Course for Latin America. Methods Inf Med. 2010;49:310–5
232. Wachenchauzer R. The Evolution of Computer Education in Latin America: The Case of Argentina. ACM Inroads. 2014;5(1):70–6.
233. Oreggioni I, Arbulo V, Castelao G et al. Building Up the national integrated health system. Uruguay case study. WHO. September 2015. www.who.int (Accessed June 20, 2016)
234. Ceibel Project http://www.ceibal.edu.uy/ (Accessed May 15, 2016)
235. One Laptop per Child Program http://one.laptop.org/ (Accessed May 10, 2016)
236. Text to Change Mobile www.ttcmobile.com Accessed May 5, 2016
237. Health Canada. Canada's Health Infrastructure. http://www.hc-sc.gc.ca/hcs-sss/eHealth-esante/infostructure/hist-eng.php (Accessed January 24, 2016)
238. Zelmer J, Hagens S. Advancing primary care use of EMRs in Canada. Health Reform Observer.10 October 2014. Volume 2. https://escarpmentpress.org/ (Accessed January 6, 2016)
239. Infoway. The path of progress. Annual Report 2014-2015. https://www.infoway-inforoute.ca/en/component/edocman/2771-annual-report-2014-2015/view-document (Accessed February 2, 2016)
240. Canada Health Infoway. Health Policy Monitor. Torgenson R. October 10, 2006. http://www.hpm.org/ca/b8/2.pdf (Accessed February 20, 2016)
241. Office of the Auditor General of British Columbia. EHR in British Columbia. February 2010. https://www.bcauditor.com/sites/default/files/publications/2010/report_9/report/bcoag-electronic-health-records-ehr.pdf (Accessed February 20, 2016)
242. Canada Health Infoway. EHRS Blueprint: An interoperable EHR Framework. Version 2. March 2006 http://www.cas.mcmaster.ca/~yarmanmh/Recommended/EHRS-Blueprint.pdf (Accessed February 20, 2016)
243. Webster P. E-health progress still poor $2 billion and 14 years later. CMAJ. 2015. Doi:10.1503/cmaj.109-5088
244. Canada Health Infoway. First digital health week launches in Canada. November 10, 2014. http://www.infoway-inforoute.ca/en (Accessed February 20, 2016)
245. Canada Health Infoway. Leaver C, Hagens S. Impact of patients' online access to lab results in British Columbia. May 14, 2014. https://www.cahspr.ca/en/presentation/538486ba37dee8572cd5018f (Accessed February 25, 2016)
246. Mosher D. A framework for patient-centered care coordination. Healthcare Management Forum. 2014. http://hmf.sagepub.com/content/27/1_suppl/S37.full.pdf (Accessed February 28, 2016)
247. Mobile Health Computing Between Clinicians and Patients. Emerging Technology Series. White Paper. April 2014. http://www.infoway-inforoute.ca/en (Accessed March 2, 2016)
248. Samsung Canada and Saint Elizabeth team up to expand mobile innovation. January 14, 2014. http://www.saintelizabeth.com/ (Accessed March 3, 2016)
249. Hernández-Ávila JE, Palacio-Mejía LS, Lara-Esqueda A, et al. Assessing the process of designing and implementing electronic health records in a statewide public health system: the case of Colima, Mexico. J Am Med Inform Assoc 2013;20(2):238–44
250. Pérez-Cuevas R, Doubova S V, Suarez-Ortega M, et al. Evaluating quality of care for patients with type 2 diabetes using electronic health record information in Mexico. BMC Med Inform Decis Mak BioMed Central; 2012 [cited 2016 Jun 24];12(1):50.
251. eHealth Strategy Tool Kit. https://www.itu.int/dms_pub/itu-d/opb/str/D-STR-E_HEALTH.05-2012-PDF-E.pdf (Accessed May 2, 2106)
252. International Medical Informatics Association. www.imia-medinfo.org (Accessed May 3, 2016)
253. WHO Atlas of eHealth Country Profiles. 2015. http://www.who.int/goe/publications/atlas_2015/en/ (Accessed September 15, 2016)
254. HL7 FHIR. https://www.hl7.org/fhir/ (Accessed June 25, 2016)
255. Millard PS, Bru J, Berger CA. Open-source point-of-care electronic medical records for use in resource-limited settings: systematic review and questionnaire surveys. BMJ Open. 2012;2(4)
256. Hersh W, Margolis A, Quiros F, Otero P. Building A Health Informatics Workforce In Developing Countries. Health Aff 2010;29(2):274–7
257. Luna D, Otero C, Marcelo A. Health Informatics in Developing Countries: Systematic Review of Reviews. Yearb Med Inf 2013;8:28-33
258. Luna D, Almerares A, Mayan JC, et al. Health Informatics in Developing Countries: Going beyond Pilot Practices to Sustainable Implementations: A Review of the Current Challenges. Healthc Inform Res World Health Organization; 2014 ;20(1):3
259. William Gibson. Wikiquotes. https://en.wikiquote.org/wiki/William_Gibson (Accessed May 25, 2016)

22

Introduction to Data Science

ROBERT HOYT • DALLAS SNIDER • S. MANTRAVADI

"The ability to take data, to be able to understand it, to process it, to extract value from it, to visualize it, to communicate it's going to be a hugely important skill in the next decades"

—Hal Varian, chief economist Google 2009

LEARNING OBJECTIVES

After reading this chapter the reader should be able to:

- Define the field of data science
- Enumerate the general requirements for data science expertise
- Differentiate between modeling using statistics and machine learning
- Describe the general steps from data wrangling to data presentation
- Discuss the characteristics of big data
- Discuss how data warehousing and relational database systems are important in data science
- Provide examples of how data analytics is assisting healthcare
- Discuss the challenges facing healthcare data analytics

INTRODUCTION

We are surrounded by data every day in our work and in our play. The data we encounter can be very small, such as a phone number (several bytes), or as large as a corporate database (several petabytes). A decade ago data scientists focused primarily on business intelligence (BI), but now this new science is an integral part of all industries. In the business world algorithms were developed to predict customers who would quit membership (churn) and determine what additional products a consumer might purchase (market basket analysis). Even in less scientific fields, such as sports, they are highly dependent on statistics and data analytics.

In this chapter, the focus will primarily be on the impact of data science on the healthcare sector. Healthcare data analytics was also discussed by Bill Hersh in a separate chapter. Given the nascent nature of this field, there is a need for continuous updates and resources available about the importance of data science and health care data analytics. Being able to work effectively and efficiently with data is especially relevant in this age of healthcare technology. The ease of analysis and modeling of complex and free text data is now key and data science needs individuals with both data science and domain (healthcare) specific expertise. Multiple organizations are supporting data science initiatives and education. Details of data science resources are available in Appendix 22.1 of this chapter.

The following are two important definitions:
- Data science *"means the scientific study of the creation, validation and transformation of data to create meaning."*[1] Because data science is relatively new, definitions are still evolving.
- Data analytics is *"the discovery and communication of meaningful patterns in data."* While some would argue for separating data analytics from data mining and knowledge discovery from data (KDD), we will use the terms interchangeably.[2]

Background

The concept of data analytics was outlined by Tukey in 1962 and represented a departure from traditional statistics, which is based on mathematics.[3] The first published use of the term "data science" was in a paper by William Cleveland in 2001, in which he called for expansion of the scope of statistics.[4] Early data scientists worked for some of the most innovative Internet companies, such as Google, Facebook, LinkedIn and Twitter, to assist them with gleaning information from their exponentially growing volumes of data. Data scientists generated new data use cases of data for the consumer and for new business models and products. The advent of the Internet was one of the greatest sources of the data explosion witnessed over the past two decades. For example, in 2014 for every minute there were 4,000,000 Google searches and 2,460,000 shares on Facebook.[5]

In 2013 Cukier and Mayer-Schoenberger aptly named the quantification of data *"datafication."*[6] Almost every Internet activity today can be measured (*datafied or quantified*) and mined. In addition to the meteoric increase in data volume there is also tremendous variety in the data, such as location data (geographic information system (GIS)), survey data, image data, email data, tweet data and sensor data. The data science field has been greatly facilitated by faster computer processor speed, open-source software designed to process large volumes of data and commodity hardware with more expansive storage. This has led to the Big Data era, which will be covered later in the chapter. For more information regarding the history of data science, Gil Press provided a history of data science from 1962 to 2012.[7]

Clearly, data science is a relatively new field with significant recent popularity, as evidenced by a Google Trends worldwide analysis from 2013 to 2017 (Figure 22.1).[8] While there is undoubtedly some hype associated with data science there has been great attention to new courses in data science and the creation of data science centers across the nation. The field is new enough that many details are being worked out among disparate academic departments, such as statistics and computer science.

Data scientists are in great demand today and sought after in all industries. The educational and career aspects of data science will be discussed later in the chapter. Data science has evolved rapidly as a field to include many necessary skill sets required by data scientists. While statistics is the cornerstone of data science, there are many more requirements. The fundamental skill sets and expertise required for data scientists are:
- Mathematics and statistics
- Domain expertise e.g. business and healthcare
- Programming in multiple languages: R, Python, SQL, etc.
- Database management and data warehousing
- Predictive modeling and descriptive statistics
- Machine learning and algorithms
- Big data
- Communication and presentation[9]

Given the relative newness of data science it is not surprising that there is some controversy regarding its exact definition and where it fits into the information sciences. The field of statistics has been keenly aware of this new field for several decades, with some statisticians believing they are data scientists, while others believing the field needs to be expanded to include more computer science. Since the 1962 publication by Tukey, there has been a push to expand the education of statisticians to make it more in line with data science requirements.[10]

Biomedical Informatics and Health Informatics have also struggled with the new field of data science. Some authorities feel that well trained informaticists are data scientists, while others maintain that data science training is different and more multi-disciplinary.[11]

The computer science field has evolved with specialized degrees in data science and information science. Multiple universities now offer degrees in data science and this will be addressed in more detail later in the chapter. Figure 22.2 displays a simple Venn diagram of the

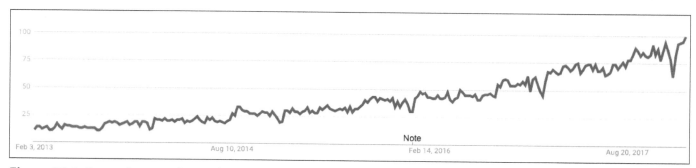

Figure 22.1: Data science web search popularity 2013-2017 (Google Trends)

field. Data science is located at the center, in the overlap section and is the aggregation of the multiple skill sets.

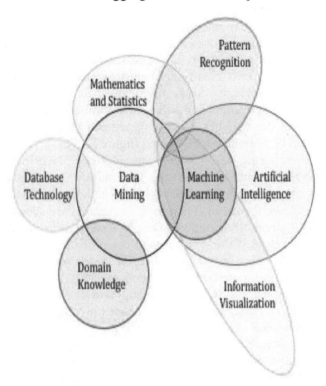

Figure 22.2: Venn diagram of data science

DATA BASICS

Bits and Bytes: Datum is singular and data is plural. In an earlier chapter the definition and concepts of data, information and knowledge were discussed. Data is simply a number without any particular meaning e.g. 10. Information is data with meaning, e.g. a hemoglobin of 10. Knowledge is information that is felt to be true, e.g. a hemoglobin of 14 is normal.

The smallest unit of data is the bit (**b**inary dig**it**) which can be represented as binary choices (0 or 1). A byte consists of eight bits and can provide a potential 256 combinations of data. For example, 0100 0001 represents the capital letter A. Four bytes together would provide more than 4 million possibilities. Because these strings of numbers could be extremely long, the codes can also be displayed as octal (base 8), decimal (base 10) or hexadecimal (base 16). This type of binary coding is important, as computers can rapidly interpret data in this binary format.[12]

With the explosion of data has come the increasing size that challenges us all and is displayed in Table 22.1.[13] What seems like massive datasets today, may seem like medium or small datasets in the future.

Table 22.1: Data sizes

Data Size	Byte Equivalent
Byte (B)	8 bits
Kilobyte (KB)	1,024 bytes
Megabyte (MB)	1,024 Kilobytes
Gigabyte (GB)	1,024 Megabytes
Terabyte (TB)	1,024 Gigabytes
Petabyte (PB)	1,024 Terabytes
Exabyte (EB)	1,048,576 Terabytes
Zettabyte (ZB)	1,073,741,824 Terabytes
Yottabyte (YB)	1,099,511,627,776 Terabytes

Given the many recent "open data" initiatives in existence it is very easy to find and analyze a variety of datasets. In Appendix I of this chapter multiple public-access free datasets are posted.

For data to be machine readable it must be in a format the computer recognizes, such as a text file e.g. a comma separated value (.csv) file. CSV and XLS (Excel) files are excellent formats to populate spreadsheets, the most common starting point for data analytics in any industry. While we will not cover spreadsheets in this chapter in detail, it is strongly recommended that students become proficient in applying formulas, math functions, filters, conditional formatting and pivot tables to them.

STATISTICS BASICS

It is well beyond the scope of this chapter to include an in-depth discussion of statistics, so only the salient concepts that relate to data science will be presented.

Data structure: Data can be classified as *structured* (discrete or fits into a defined field in a database or spreadsheet, e.g. text in name field is a name), *unstructured* (free text) and *semi-structured* (doesn't fit into a database but has a known schema or is tagged, e.g. extensible markup language (XML) file).

Types of data: A classification schema to evaluate types of data would be to categorize it as categorical [nominal or ordinal] (qualitative) or continuous data [interval or ratio] (quantitative). Nominal (also known as categorical) data are *non-numerical* data with no order, such as gender (male or female). Ordinal data are similar, but have order, such as small, medium and large, but there is no numerical measurement difference between the variables. In other words, medium is not twice as large as small. Interval data are *numerical* data with meaningful intervals but no meaningful zero value, such as Celsius temperature. Multiplying interval data does not make sense. Ratio data are numerical data and have order, but also have a meaningful zero value, such as height and weight. Multiplying ratio data does make sense.[14] Data can be discrete, such as the number of discharges (no such thing as ½ discharge) or continuous, such as weight where the number can be fractional (e.g. 140.5 lbs.). Additionally, note that "real numbers" can have decimal points (e.g. 180.2) whereas, integers are whole numbers.

Categorical/nominal data can also be considered qualitative or describing a quality, whereas continuous data tends to be quantitative or numerical data.

Parametric and non-parametric data analysis: Data analysis can also be classified as parametric or non-parametric. Parametric analysis has some form of assumption; is usually associated with a large data sample and the data follows a normal distribution. Non-parametric tests are easier to understand and are usually used for smaller sample sizes. Non-parametric data analytics has minimal assumptions and is used on non-normally distributed data.[15] In general, biological data tends to be non-parametric.

Statisticians generally use non-parametric tests (such as a Chi-square test) for non-normally distributed data and parametric tests (such as t-tests) for normally distributed data.

Data scientists will frequently look at measures of central tendency, such as mean, median and standard deviation to see how the data is distributed. The mean is the sum of values divided by the number of values; whereas the median is the middle of the distribution and the mode is the most common value. The range is the difference between the lowest and highest value. The standard deviation is a measure of the dispersion of data from the mean. Note that many statistical methods focus on comparing the means of two different attributes, but the mean of non-parametric data is meaningless (such as the mean of small, medium and large), so other statistical methods must be used.[14] Table 22.2 shows important characteristics of non-parametric and parametric data.

Central Limit Theorem: There are many reasons why data scientists want to know more about their datasets. The central limit theorem posits that if the number of independent random variables is very large, then the distribution should be normal or bell-shaped (figure 22.3).[16]

Viewing figure 23.3, you can see that 95% of data that are normally distributed falls within 2 standard deviations and 99.7% falls within 3 standards deviations from the mean. Therefore, values that are greater than 3 standard deviations must be looked at critically to determine if the result is real or an error. They are "outliers." An example might be the distribution of the height of all male college students (n=500). The mean might be e.g. 68 inches with a standard deviation of 3 inches. Those students over 77 inches (68 + 9), or under 59 inches (68 - 9) would fall outside the 3 standard deviations. Standard deviation (SD) is calculated by taking the square root of the variance. To calculate variance, you subtract each data value from the mean, square each result, then add them up, then divide by the number of values -1 (you subtract 1 from the total (n) if you are dealing with a sample, but don't subtract by 1 if you are measuring an entire population). If the variance is e.g. 7, then the standard deviation is the square root of 7 or 2.64. The larger the population, the smaller the variance and standard deviation are likely to be. Also, the curve will be narrower, so variance and standard deviation measure how much the values vary, or the spread or dispersion of the data. If the standard deviation is very small, there is very little variance and/or the sample (n) is very large.[17]

Table 22.2: Data types and central tendency measures

Data	Central tendency measures	Comments
1. Non-parametric		
Nominal (categorical)	Mode	Non-numerical data. Mean not used.
Ordinal	Mode or median	Non-numerical data. Mean not used. Median helpful with outliers
2. Parametric		
Interval	Mean, mode and median	Numerical data
Ratio	Mean, mode and median	Numerical data

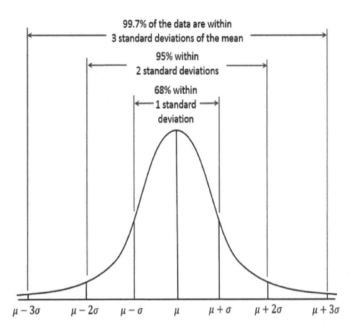

Figure 22.3: Normal distribution (Wikipedia)

Figure 22.4 shows a distribution skewed to the right (positive) with possible outliers.[18]

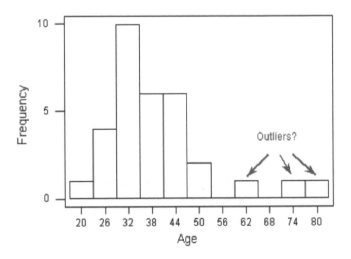

Figure 22.4: Abnormal distribution, skewed to the right with outliers

Dependent and independent variables:

Another important concept regarding data analysis is dependent versus independent variables. The outcome (also known as response, label, target or class) variable you are trying to measure is commonly known as the *dependent* variable, such as "developed diabetes" or "didn't develop diabetes." (binary outcome). While a binary outcome is common, it can include more than two variables and be "multi-class." The predictor (explanatory or feature) variables that predict the dependent variable are known as *independent* variables. For example, in your spreadsheet you may have a column labeled outcome where patients either developed diabetes or didn't (dependent variable). Your analysis examines the other columns such as fasting glucose, BMI, HDL cholesterol, family history of diabetes, blood pressure, etc. (independent variables) to see if they predict outcome. In fact, Wilson conducted such a study in 2007 to predict type 2 diabetes in middle-aged adults using logistic regression to examine common risk factors as predictors of outcome.[19]

Descriptive and Inferential Statistics

These are two common approaches by researchers who use data to answer a question.

Descriptive statistics: this approach *describes* the data with several techniques, such as measures of central tendency that includes mean, median and mode. Mean is the most common measure but is distorted by outliers. The median is the midpoint of a distribution of numbers and is less subject to outliers. The mode is the most common value in a dataset. Table 22.2 describes which measures are used with parametric and non-parametric data. Measures of variation are also used, such as range, variance and standard deviation, discussed in a prior section.

Inferential Statistics: analyzes a random sample to infer conclusions about a population. For example, surveys commonly gather information about a sample to predict characteristics about an entire population. Confidence intervals and hypothesis testing are two important approaches used with inferential statistics.

- **Confidence Intervals (CIs):** measure the uncertainty of the sample. With any survey or sample, one can expect a sampling error or frequently mentioned "margin of error." The most commonly cited confidence level is the 95% confidence level, but other levels such as 90% or 99% could be used. For instance, a 95% confidence interval is a range of values that you can be 95% certain contains the true mean of the population. Using the prior example of 500 male college students with a mean height of 68 inches, a SD of 3 and using 95% confidence intervals, we have 95% confidence that the mean of 68 inches falls between 67.73 - 68.26. The smaller the sample (n) size the wider the confidence intervals. CIs can be determined for numerical and categorical data.

CIs serve multiple purposes:

o an easy to understand visual display of the range of values
o the precision of the estimate
o comparison of CIs between different studies (e.g. in a meta-analysis)
o hypothesis testing, with any value outside the 95% CI can help reject the null hypothesis. Figure 22.5 shows how means and CIs are used to compare three hospitals, using data from the government web site Hospital Compare. On the website the CIs are colored gold and indicate they are compatible with national averages. The CIs are narrow, which is optimal, and the actual results can be seen with a "mouse-over" the mean.[20]
o help "power" a study to determine how many patients a researcher might need to achieve acceptable CIs.[21-22]

- **Hypothesis testing**: measures the strength of evidence to reject or not reject a hypothesis. For example, a researcher has a hypothesis that drug A is better than placebo. The null hypothesis is there is no difference between the mean (m) of the treatment effect of a drug (stated as $H0$: $m1 = m2$) and the alternative hypothesis would be a difference exists (stated as Ha: $m1 \neq m2$). As an illustration, the average or mean systolic blood pressure measured on drug A is m1 and the average or mean systolic blood pressure on placebo is m2. Commonly, a p-value of .05 is used to determine statistical significance. If the p-value is less than .05 the findings are unlikely to have been caused by chance and we reject the null hypothesis that there is no difference between the two treatments.[14] While p-values are one of the most common statistics to be reported, there are serious limitations and many authorities request confidence intervals and effect size be reported as well.

Effect size (ES): ES measures the *magnitude* of the treatment effect. Unlike significance tests such as the p-value, *effect size is independent of sample size*. There are a variety of effect size tests. For example, a common test for comparing two independent means is Cohen's d measure which is simply the difference between two means, divided by the pooled standard deviation. A small effect would be d =.2, medium effect d =.5 and large effect d =.8.[23-24] In spite of the known limitations of only publishing significance values, Chavalarias et al. reported that of the 1000 abstracts they reviewed (1990-2015) p- values were reported in only 15.7%, confidence intervals in 2.3% and effect size in 13.9%.[25] While it is beyond the scope of this chapter, confidence intervals can also be calculated on the effect size.

Type 1 and Type 2 errors: a type 1 error is defined as observing a statistical significance (p <.05) when no real difference actually exists (false positive). Remember, p =.05 means there is a 1 in 20 chance for a spurious result. A type II error is defined when you conclude there is no effect (null hypothesis is not rejected) when one really exists (false negative). The most common cause of a type II error is a sample size that is too small.[14]

DATABASE SYSTEMS

Spreadsheets (flat files), such as Microsoft Excel are ubiquitous in many industries, including healthcare and are a logical starting point for exploring and visualizing data. Flat files are two dimensional (rows and columns) and as such are limited, in terms of data analysis compared to a relational database system (RDBS). Healthcare organizations might use spreadsheets for smaller datasets that require simple analysis (departmental data) and visualization and use RDBSs for larger datasets (hospital wide data).

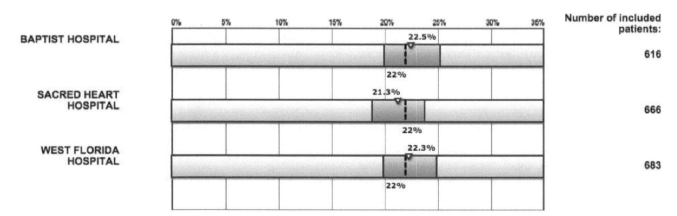

Figure 22.5: Hospital Compare: heart failure readmission rates

Relational Database structure: Data tables are created that link to other tables to store information in a three-dimensional space (data cube). Table 22.3A shows a simple table with patients as the rows (tuples or instances) and the columns (attributes or features) related to admission and discharge dates. This could link to table 22.3B, a demographic table. Patient_ID in table 22.3A will be the "primary key" which will link to the Patient-ID "foreign key" in table 22.3B.

Cardinality (uniqueness) is an important concept for table construction. A one-to-one relationship means that each row of a table relates to only one row in another table. One-to-many means each row in one table may relate to more than one row in another table. Lastly, many-to-many means that each row in a table may relate to many rows in another table and vice versa. In the example above, Table 23.3A can have the same Patient ID multiple times, while Table 22.3B will have patients uniquely defined.

Table 22.3A: Simple database table

Patient ID	Last Name	First Name	DOB
69785747	Jones	Roger	01/02/1956
58585758	Smith	Sally	11/15/1940
36637484	Edwards	Edward	03/22/1938

Traditionally, database systems were designed to adhere to the atomicity, consistency, isolation and durability (ACID) model. Specifically, there needs to be **a**tomicity which is all or none transactions; **c**onsistency or ensuring that if any element of the transaction fails, it all fails; **i**solation means that each transaction is completed separately; **d**urability ensures that each transaction is permanently preserved. However, with the advent of the unstructured NoSQL database models, the ACID model would not pertain. This newer model is known as BASE and will be discussed in the section on Big Data.

Another important concept is "*normalization*" of database tables, primarily to prevent duplication of data and increase database efficiency. There are 5 defined levels of normalization (also known as normal forms); however, in practice just 3 normal forms are commonly used. The first normal form prevents each table row from containing duplicate data. The second normal form prevents the repetition of data within a table's column. The third normal form requires that every column have a dependency on the table's primary key and independent from the remaining non-key columns in the table.[26]

If databases are being created for a specific purpose they have to be "modeled" or organized so that they are designed to answer important business or clinical questions. In other words, how they are designed will help determine what queries or searches are possible. Databases can be separate (federated) and connected by a computer network or single (central) as in central data repository (CDR).[26-27]

Table 22.3B: Connecting demographics table

Patient ID	Hospital ID	Admission Date	Discharge Date
69785747	H445598	2/1/2016	2/12/2016
69785747	H445598	10/3/2014	10/5/2014
58585758	H193240	2/5/2016	2/13/2016
36637484	H148679	2/7/2016	2/14/2016

Database management: Relational data base systems (RDBS) can be simple and PC (client) based, such as Microsoft Access and OpenOffice Base. These programs are easy to use but limited in terms of scalability (limit of 2GB data for MS Access and limited user access). They may be appropriate, for example, for a hospital department. Most larger organizations need a commercial database management system (DBMS), such as Oracle or Microsoft SQL Server. Most significant RDBSs are web accessible and use structured query language (SQL) for extracting information, discussed in another section of this chapter.[14] These database systems are classified as online transactional processing (OLTP) systems, where data manipulations are contained in transactions which can be committed on successful processing or rolled back in case of an error.

Clinical data warehouse (CDW): It is likely that a large hospital system may have multiple database systems that need to be integrated so data can be analyzed and reported. Therefore, organizations frequently establish an enterprise clinical data warehouse. The clinical data warehouse will integrate data from clinical, demographic, financial, insurance, coding and quality-related data systems. They will have, for example, registration, outpatient, inpatient, pharmacy, emergency department and surgical data in one physical location. The data is likely to be both structured and unstructured, to include progress notes, documents and images. While integration of electronic health record data into the warehouse is new in the past 1-2 decades, most the data is unstructured.

The goal is to provide a single platform for analytics so that strategic decisions can be made. Data warehouses are used to generate internal and external reports and to perform predictive and prescriptive analytics (see prior chapter on Healthcare Data Analytics). In addition, the warehouse may interface with clinical decision support systems. Importantly, data warehouses are a great source of data for research.

Many large healthcare systems will also have data marts which are like small warehouses that focus on a single subject area, such as revenue, and are used by a single department. Figure 22.6 displays a prototypical hospital clinical data warehouse.[28] Enterprise clinical data warehouses can be built from multiple data marts which are connected by conformed dimensions, such as patients, providers, facilities and services provided. A data mart is a collection of data that is of interest to a line of business. For a hospital system, the lines of business can include accounting, clinical services, human resources and facility management. The data in a data warehouse is divided into facts and dimensions. Facts are database tables that contain measurements, such as counts and amounts. Dimensions are database tables that contain data that describe the facts, such as name, date of birth and procedure code.

Data warehouses are often associated with online analytical processing (OLAP) that analyzes multidimensional data. OLAP consists of three operations: consolidation (roll-up) or aggregation; drill-down or narrowing a search and slicing and dicing or taking apart the OLAP cube to view from different perspectives. While OLAP helps organize the data, it is not analytical in the usual sense. For that reason, a data mining application must be used. There are programs that combine OLAP with data mining and are known as online analytical mining (OLAM). Most multi-dimensional data models have a star or snowflake schema (architecture). Extract, transform and load (ETL) tools are used to populate and maintain data warehouses. ETL tools allow for the extraction of data from a variety of databases and file types, the transformation of data for cleaning and integration, and then loading the transformed data to the data warehouse.[28] Some of the more popular ETL tools are IBM's DataStage, Microsoft's SQL Server Integration Services and Pentaho's Kettle. These tools provide a graphical user interface, so analysts can quickly assemble data processing flows. Furthermore, ETL tools manage the technical details of connecting to databases and opening files.

Application Programming Interface (API): Internet-based companies such as Google and Amazon have needed a method to transfer and share data with consumers, developers and researchers. The most common method today is via an application programming interface or API which creates a portal for data retrieval. The most common communication standard today for this is RESTful API.[29] Internal APIs can be created for internal customers and external APIs for everyone else. This technology is also used by web and mobile applications. Healthcare as an industry has been slow in adopting this approach. Open, standardized APIs would make data available to more people on more platforms and overcome the challenge of interoperability. Importantly, developers could innovate and create new operational and analytical tools, not in existence today.[30] There is already an evolving strategy to use the Fast

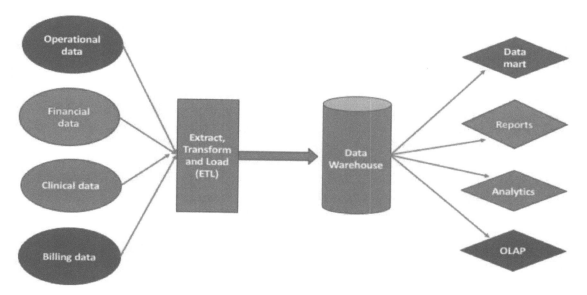

Figure 22.6: Hospital Clinical Data Warehouse

Healthcare Interoperability Resources (FHIR) standard and RESTful APIs to retrieve data from those EHR vendors who have open APIs. The document standards most commonly used are Javascript Object Notation (JSON) and XML. A multidisciplinary group known as the Argonaut Project, that includes some EHR vendors is moving in this direction. They will use new open-source standards (OAuth 2.0 and Open ID) for secure data exchanges.[31] To further support this new direction, the Federal Government is including access to many data sets via open APIs. For example, there is an API catalog for datasets found on Data.Gov.[32]

DATA ANALYTICAL PROCESSES

The general steps undertaken with data analysis are displayed in Figure 22.7, but they must be flexible and iterative. The steps in the process are clearly non-linear; they are multiple interdependent cycles. For example, a researcher might explore data in the exploratory data analysis (EDA) phase with a visualization, then perform a simple analysis with statistical modeling which doesn't reveal anything interesting and return to the EDA step to look for more patterns of interest.

Like evidence-based medicine, data scientists begin with an interesting or important question. The question should be specific, plausible with a potential answer and hopefully solved with available data. If the question is in the form of a hypothesis, the researcher will also have to calculate the sample size of the data to be collected to achieve a meaningful result.

Raw Data

Obviously, the first step is to locate an appropriate data set that hopefully will answer the proposed question. In an ideal world, the data would exist in an acceptable and easy to compute format, such as a text file, spreadsheet or relational database. In reality, the data may be unstructured and messy and therefore may need to be transformed into an acceptable format before the pre-processing stage can begin. Raw data (also known as primary data) is therefore unformatted and in its natural state, prior to cleaning or transforming. Raw data from clinical instruments can be stored in the manufacturer's proprietary data format, which typically is not human readable and requires a program to transform the data.

Data Pre-processing

Prior to the actual analysis, the data will likely need *"wrangling"* or *"munging"*:
- Cleaning: correcting and/or removing inaccurate records
- Integration: combining several sources of data into one spreadsheet or database
- Reduction: consolidating categories to reduce the number of attributes[33]

Exploratory Data Analysis (EDA)

EDA builds on the pre-processing phase. The data is examined in more depth, using multiple steps and tools. The main tool used is descriptive statistics to examine the distribution, mean, mode, range, variance and standard deviation (SD), so the appropriate statistical method or machine learning algorithm for eventual analysis can be chosen. Many data scientists claim that roughly 80% of their time is spent cleaning, processing and learning the data, prior to the actual analysis.[9] Visualization of the data is a major component of EDA. Data scientists may examine non-parametric data using pie and bar charts and parametric data using box plots and histograms. Figure 22.8 shows a boxplot of blood pressure. Note the mean, median and standard deviation. The whiskers define the range or minimum to maximum values. The goal is to look for possible skewed data (non-normal distribution) and outliers. Scatter plots are useful to compare two

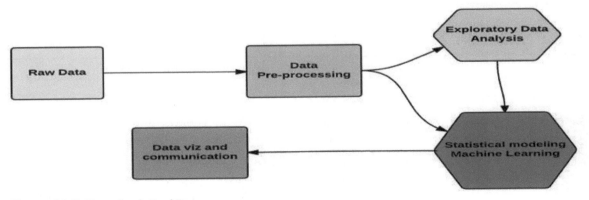

Figure 22.7: Data Analytical Process

variables for a potential linear relationship. For instance, does a rising BMI seem to correlate with serum glucose? Also, look for missing, incorrect or duplicate data. Will the data set you selected answer the proposed question? The following are additional steps involved with EDA: [34]

- Missing data: most datasets are imperfect, which includes missing data. A strategy must be developed to deal with this, to include imputation and other schemes.[35]
- Standardizing and normalizing data: Data column values that are on differing orders of magnitude or are non-normally distributed may need to be "normalized" to prevent one column from biasing or skewing the analytical interpretation. For example, if Column A contains values from 1 to 100 and Column B contains values from 0.0 to 1.0, then these columns must be normalized to put the values in the same range. Software can normalize or convert the numerical variables to a scale from 0 to 1. Another scheme is standardization where the numerical mean is zero with a standard deviation of one. This is valuable when your data tends to be normally distributed and you want to use logistic regression or Bayes. Another approach is to convert the data into z-scores by subtracting the observed value from the mean of the data and dividing by the standard deviation. Additionally, z-scores can tell you if an outlier is greater than 3 standard deviations and should be left out of the model.
- Categorization (binning or bagging): If you use a linear (line-like) model, such as linear regression, then the data should be linear. Converting continuous data into bins (e.g. patient age by decade) is also known as "discretization" and this may help the analysis. Other non-linear techniques, such as decision trees and neural networks are affected less by non-linear data.
- There are times when nominal data must be converted to numerical values, such as 0 and 1; so called "dummy coding." For example, male = 0, female = 1
- Variable selection: a data analyst can use a variety of filters in a machine learning software program to see if excluding attributes impacts the results. In fact, part of building a model is to exclude certain variables and see if they impact outcome.
- Segmentation or grouping. Analysts may analyze only a meaningful subgroup and see if that impacts the results.[36]
- Transformation: using a mathematical method (e.g. log transformation) to convert a skewed distribution to a normal distribution or using a Fourier transform to change from the time domain to the frequency domain.[33]
- Table 22.4 displays which visualization and test should be used, based on the type of dependent and independent variables (categorical or numerical).

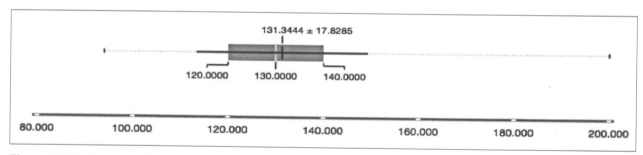

Figure 22.8: Boxplot of blood pressure

Table 22.4: Recommended test and visualization for dependent and independent variables

		Dependent Variable	
		Categorical	Numerical
Independent Variable	Categorical	Visualize with two-way table and percentages. Test with chi-square	Visualize with side-by-side box plots. Test significance with t-test and ANOVA
	Numerical	Test significance with logistic regression, decision tree or Naïve Bayes	Visualize with scatterplot. Test strength and direction with correlation. Test mathematical relation with linear regression

Analyzing the Data Approaches

Prior to discussing analytical methods, we will discuss more about analysis in general. There are *three* general approaches to conduct an analysis: statistical modeling, machine learning and programming languages (see Figure 22.9):

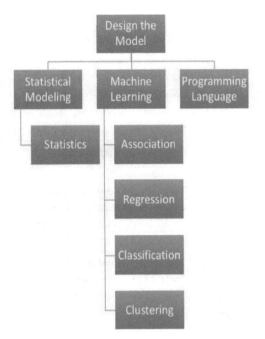

Figure 22.9: Modeling approach

- Statistical modeling is one of the most common approaches and was developed by statisticians. Statistical tools are discussed later in the chapter
- Machine learning approach was developed by computer scientists but is based on mathematics and statistics. Keep in mind that statistical modeling will also address associations, regression and classification, but not clustering. Modeling with machine learning depends on algorithms which are really mathematical formulas to make predictions from data. For example, there is a decision tree algorithm named classification and regression tree (CART) which uses a mathematical formula to calculate a data column and value that best splits the data so as to separate observations based on the dependent variable. All algorithms learn a mapping from input to output. Models used for prediction are called predictive modeling. Different algorithms are compared using the same outcome. This tends to be different from statistical modeling where one statistical method is usually selected.

Supervised machine learning means you already know the classes of data you will be dealing with. For example, you know that the columns (attributes) will be health and demographic factors and the rows are patients.

Unsupervised machine learning implies you don't know the classification of the data. Techniques such as clustering and associations will discover and report groups or patterns of data. Clustering techniques will be discussed in another section.

Semi-supervised machine learning means you know some of the classes of data only, the remainder are unknown. Machine learning is based on algorithms to make predictions. Table 22.5 enumerates various common algorithms used today

- Programming languages. **R language**: The R and Python programming languages are both used in the statistical approach to data analytics, particularly by data scientists. R (statistical computing and graphics) language was developed in 1995 and has gained in popularity ever since. The software is free to download and is supported by a large network and a repository of R functions. This language does include machine learning algorithms. The program tends to be a little slow and there is a substantial learning curve. **Python:** This language was created in 1991 and is one of the easier languages to master. In general, the R language is used more by data scientists and Python is used more by computer scientists; but considerable overlap exists. Python is a better choice if you need to integrate analytics with a web app or database. There is also a machine learning package (scikit-learn), in addition to the statistical approach.[37]

Structured Query Language (SQL): is a universal language to manipulate databases. Its basic functions to modify tables are **CRUD: C**REATE (new table or database), **R**EAD (select or query), **U**PDATE (edit), and **D**ELETE (remove) information in a table in a database. Additional important functions are INSERT and TRUNCATE and many more commands are available. Data must be characterized as data types: **char** (size) or fixed length characters with parameter size in parenthesis; **varchar** (size) or variable length characters with parameter size in parenthesis; **number** (size) or number with a max number of digits in parenthesis; **date** or date value; **number** (size, d) or the number value with a maximum number of digits of size and the precision which is the number of digits to the right of the decimal.[38]

MAJOR TYPES OF ANALYTICS

Descriptive Analytics (not the same as descriptive statistics): analyzes data to look for patterns where no

Table 22.5: Machine Learning Algorithms

Algorithm	Examples	Indication	Benefits	Limitations
Linear			Simple, fast and requires little data	Linear model assumptions
Linear regression		Regression	Widely used	For numerical data only
Logistic regression		Classification	Widely used	For binary categorical data
Linear discriminant analysis		Classification		For more than 2 classes of categorical data
Non-linear				Requires more training data. Slow
Decision Trees	CART, CHAID, C 5.0	Classification and regression	Fast, easy to understand	Risk of overfitting
K-nearest neighbor		Classification and regression	There is no learning so entire training set is used. Simple	May have to experiment with K values to optimize. Works best with small # inputs and no missing data. May need to normalize data
Support vector machine	libSVM	Classification and regression	Powerful, handles high-dimensional data. Used for linear and non-linear models	Uses only numeric or dummy coded categorical data; decision boundary can be difficult to interpret
Bayes	Naive Bayes, Gaussian Naïve Bayes	Classification and regression	Fast, often considered a baseline since it uses probabilities	Assumes input is independent. Input is categorical or binary, but numerical for Gaussian
Neural networks	Multilayer Perceptron	Classification and regression	Powerful, robust and generally more accurate	More complex and more difficult to evaluate results
Unsupervised learning				
Distance based clustering	k-means	Pattern discovery where there are no data labels	Simple and easy to implement	# groups must be selected ahead of time, does not handle concave shapes well
Density based clustering	DBScan	Builds clusters based on the density of data points	Finds clusters with concave shapes	A spatial index is required to run efficiently
Association	aPriori, frequent pattern tree	Set of rules defining a pattern where condition A implies condition B	Simplicity	Can be memory and processor intensive
Ensemble	Bootstrapping, Random forest AdaBoost	Classification and regression	*Uses several algorithms for more accuracy*	Interpretation of the results from multiple models can be challenging

target value or class exists or is used. Therefore, this is unsupervised learning and uses *three* categories of machine learning algorithms:

- *Association rules* are frequently used to associate a purchase of item A with B. They are used to make recommendations and do market basket analysis (MBA). They measure correlation and not causation. Association rules are rated in terms of *support* and *confidence*. Support is defined as the percent of total transactions from a transaction database that a rule satisfies. Confidence measures the degree of certainty of the association. Mining occurs by using the minimal support and confidence levels set by the analyst. *A priori* is one of the most common association rules (algorithms). Multiple rules are generated so you must narrow them down, based on confidence versus support levels established.[28] For example, an association rule might say if the patient was Asian, under age 50 and BMI under 25 then there is no association with diabetes with 95% confidence.
- *Sequence rules* are concerned with the sequence or order of events in a transaction
- *Clustering* identifies patterns or groupings. Hierarchical clustering builds clusters within clusters to assist in comprehending data similarities and reducing the number of representative groupings. Hierarchical clusters can be built from the bottom up (agglomerative) where each observed data point begins in its own cluster and then clusters are merged as the algorithm climbs up the hierarchy. Another way to construct a hierarchical cluster is from the top down (divisive), with all observed data points in one hierarchy and then having the algorithm recursively split the data into smaller clusters. Non-hierarchical clustering uses techniques such as k-means to calculate distances. The number of clusters (k) has to be established ahead of time. Clustering is generally used with numerical data but algorithms, such as FarthestFirst can handle categorical data.[39] Clustering has been successfully used in the medical field to analyze microarray gene data to look for previously unrecognized relationships. Figure 22.10 shows clustering using k-means algorithm where k = 3 in the machine learning software WEKA. The clusters are based on iris flower petal length on the y axis and number of samples on the x axis.[40]

Predictive Analytics (modeling): Creates a model based on a target (dependent) variable, using multiple (independent) variables. Complex machine learning algorithms are not as interpretable or easy to understand. Comments on benefits and limitations of algorithms are

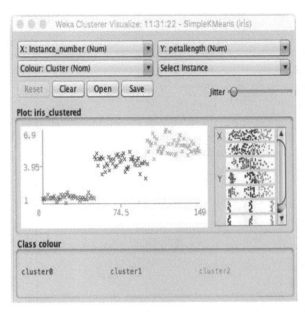

Figure 22.10: Clustering with three identifiable groups

included in table 22.5. Modeling approaches include the following:

- **Regression model** is a linear model that uses supervised learning. *Simple linear regression*: this model displays the mathematical relationship between *two continuous (numerical)* variables. *Multiple linear regression* is for multiple variables. It plots the target or dependent variable (y axis) against an independent variable (x axis). The data is numerical data, such as income or age. The mathematical formula is y = ax + b; where **a** represents the slope of the regression line (increase in y divided by increase in x) and **b** is the y intercept or what the y value is when **x** is zero. The goal is to fit the values as closely to the line as possible. For example, if we use the formula y = 0.425x + 0.785 and we set x = 2, then y = 1.64. Figure 22.11 shows a simple linear regression using this formula. Note that the line does not fit perfectly. The line is based on "the sum of least squared errors" or squaring the difference between the actual value and the value represented on the line, then summing the errors. For instance, when x = 3 and y = 1.3 the residual or distance from the line is 0.7. The optimal line will minimize the sum of the squared errors.[41]

Logistic regression is used with dichotomous or binary outcomes that can be categorical data or dummy coded as 1 or 0. The independent (predictor or explanatory) variables are interval or ratio data. Outcome can be described with odds ratios. As an illustration, the outcome could be lived (1) or died (0) and the independent variables could be age, smoking, diabetes, heart disease, etc. Logistic regression is very useful to answer questions such as what is the probability of developing diabetes (outcome) for x increase in BMI or age (independent variables)? Note, that although we have included logistic regression under regression model, it is used in classification discussed in the next section. Also, note the logistic regression (LR) is commonly used in both the statistical approach and machine learning.

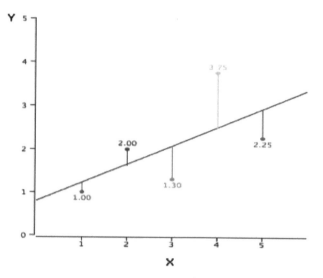

Figure 22.11: Simple linear regression (OnlineStatBook)

- **Classification model** is a non-linear model that uses supervised learning: unlike the regression model, the target variable is categorical (nominal). Common classification algorithms are as follows:
  o **Decision tree** algorithms are now commonly called classification and regression trees (CART). The tree displays root and leaf nodes (outcomes). Examples of decision tree algorithms are C4.5 and J48. Trees can evaluate both nominal data (classification) and continuous data (regression). This is a good way to evaluate attributes, as those with the most information gained will appear first in branching at the root of the tree. Decision trees and regression are the easiest algorithms to explain and present to non-statisticians. These models can be considered a set of IF-THEN rules. Figure 22.12 displays a decision tree looking at contact lens selection. Note, the first branch (root) is based on tear production. IF tear rate is low, THEN no contact lenses. Final decision is no lenses, soft or hard lenses.[40,42] Decision trees can *"overfit"* the data if they learn the model training data too well; leading to poor predictions on new data presented to the model. Trees often need to be "pruned" with techniques, such as, k-fold validation.
  o **Random forests** involve multiple decision tree classifiers (hence the name forest). This is also the reason they are usually organized under ensemble algorithms because they combine several approaches. They work well with multiple predictors and can perform classification and regression. Importantly, a "measure of variable importance" can be studied to determine the

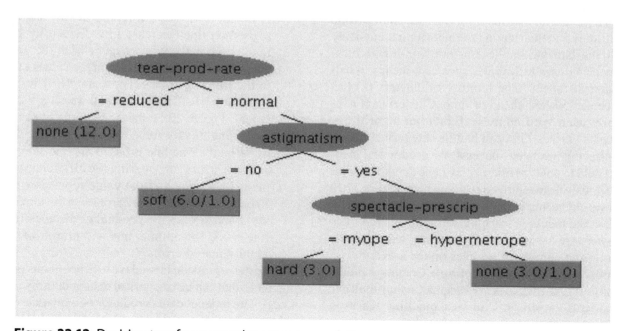

Figure 22.12: Decision tree for contact lens recommendations

relative importance of each variable. Random forests are generally more accurate than standard decision trees but are more complex.[42]

o **Neural networks** (also known as multilayer perceptrons). While the networks work like the human brain, they are statistical models, like logistic regression. Results are powerful but more difficult to interpret, compared to decision trees. There is also more trial and error associated with this approach due to attempts to optimize the algorithm parameters for the best results.[36]

o **Naïve Bayes** uses Bayes theorem to perform classification and regression, based on probability. Specifically, future or posterior probability is based on prior knowledge or prevalence. It is fast and can evaluate binary, categorical and numerical data. The algorithm is called Naïve because it assumes that each input variable is independent. While this is unrealistic, this approach works quite well. Numerical data can be analyzed using a Gaussian or normal distribution Bayes. Probabilities are calculated based on the normal distribution.[36, 42]

o **K-nearest neighbor (KNN)** is an algorithm used for classification and regression. It evaluates the entire training data set, so there is no learning. Predictions are made for a new data point by looking for the "nearest neighbor." There are several ways to evaluate the distance to a data point, with the most common being the Euclidean distance. KNN does not perform as well on large data sets (multiple input variables); due to the "curse of dimensionality."[36,42]

o **Support Vector Machines (SVMs)** is a technique that separates the attribute space with a hyperplane, therefore maximizing the margin between the instances of different classes. The algorithm determines the best coefficients for separation of the numerical data by the hyperplane. SVMs can be used to classify linear and non-linear models. They can also be used for regression. While a little slow, they are highly accurate and less prone to overfitting.[36,42]

o **Ensemble methods** use multiple models to compensate for inherent deficiencies of any one model. The above mentioned random forest is an example of an ensemble method. More robust ensemble methods will use classification algorithms from the different categories of models listed above.

The above techniques can handle "multi-class" classification where they can evaluate more than two target variables. Classification of two target variables is known as binary classification and is common in health informatics. Examples of binary classification target variables would be (benign/malignant), (hypertensive true/hypertensive false), and (Zika virus yes/Zika virus no).[36,43]

Evaluation of the Performance of Predictive Analytics

It is important to have methods to evaluate the predictive ability of these different analytical methods. Traditionally, with the classification model, the data is split into training (67-80%), used to create and train the model, and test data (20-33%), used to test the model. However, with k-nearest neighbor, neural networks and decision trees all the data must be used to create the model. In smaller data sets (e.g. less than 1000 observations) cross validation using K-folds (K most commonly is 10) is the best technique. The advantage of this approach is that all the data is used for both training and testing. Multiple averages can be calculated as well as confidence intervals and standard deviations. Small samples are also helped by bootstrapping, which means when random sampling a value it can be selected (replaced) more than once. A computer can generate many probabilities therefore from a smaller dataset. Most software programs will automatically evaluate performance.

Classification model evaluation: the goal is to establish the overall accuracy of the model or algorithm selected. For example, using a validated dataset for predicting heart disease you know who eventually develops heart disease and you can compare that result with what was predicted using one or more models. This allows you to create a confusion matrix (truth table), that describes the performance of the classification model, by displaying the true positives (TP), false positives (FP), true negatives (TN) and false negatives (FN). Table 22.6 displays a confusion matrix based on heart disease data and generated by the machine learning software program WEKA.[40] From that table you can determine the classification accuracy, classification error, sensitivity, specificity and precision. Table 22.7 displays the most important measures and formulas related to calculating classification performance. Figure 22.13 shows the results of a 10-fold validation using three machine learning algorithms. The classification accuracy (CA) was highest with Naïve Bayes. The precision (positive predictive value) was also highest with Naïve Bayes, as was recall (sensitivity). The F-1 measure is the weighted average of precision and recall. Another way to compare these methods is to plot true positives on the y-axis and false positives (1-sensitivity) on the x-axis. A "perfect" result would be 1 where there are only true positives and no false positives. The curve that is created should be in the upper left area and the area under the curve (AUC) should be in the 70-90% range to be a truly significant finding. Figure 22.14 shows an actual ROC curve with Naïve Bayes showing the greatest AUC (.838).

Another choice is a lift curve which gives similar results. Figures 22.13 and 22.14 were created using the machine learning program Orangeâ, discussed in another section.[44]

Regression model evaluation: The performance is usually measured using R (correlation coefficient) and R^2 (coefficient of determination). R measures the strength and direction of a linear relationship, such that it can be in the range from -1 to +1. R^2 determines how close the data fits the regression line and can vary from 0-100%. 100% indicates that the model explains all the variability of the data around its mean. Additionally, the root mean square error (RMSE) is used as a measure of the fit; the lower the RMSE, the better.[36]

Do the results make sense, it terms of magnitude, direction and uncertainty? The results of the above tests

```
Test & Score
  Settings
    Sampling type: 10-fold Cross validation
    Target class: Average over classes
  Scores
    Method              AUC     CA      F1      Precision   Recall
    Naive Bayes         0.838   0.841   0.819   0.829       0.808
    kNN                 0.652   0.659   0.603   0.625       0.583
    Classification Tree 0.749   0.752   0.722   0.719       0.725
```

Figure 22.13: Ten-fold cross validation results for three machine learning algorithms

Table 22.6: Confusion matrix for Bayes classifier results

	Predicted No	**Predicted Yes**	
Actual No	TN = 130	FP = 20	TN + FP = 150
Actual Yes	FN = 23	TP = 97	FN + TP = 120
	TN + FN = 153	FP + TP = 117	Total cases = 270

Table 22.7: Classification performance measures and formulas

Measure	Formula	Result
Accuracy or correct classification rate	$\dfrac{TP + TN}{Total}$	$\dfrac{97 + 130}{270} = .84$
Misclassification or error rate	$\dfrac{FP + FN}{Total}$	$\dfrac{20 + 23}{270} = .15$
Sensitivity, recall, true positive	$\dfrac{TP}{FN + TP}$	$\dfrac{97}{120} = .81$
Specificity, true negative	$\dfrac{TN}{TN + FP}$	$\dfrac{130}{150} = .87$
Precision	$\dfrac{TP}{TP + FP}$	$\dfrac{97}{117} = .83$

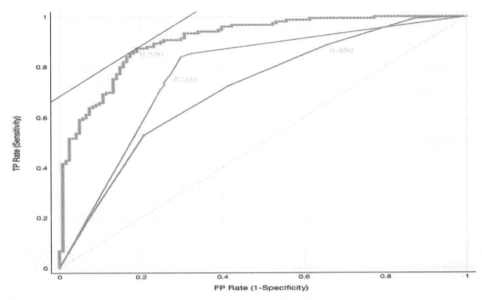

Figure 22.14: Receiver operator characteristic (ROC) curve

may cause the data scientist to tweak the algorithm to see if the optimal predictors were used. Similarly, running the model with fewer parameters to see if that affects the outcome is reasonable. Additionally, it is reasonable to re-run the model without outliers to see if that impacts outcomes.

If a confusion matrix is calculated, a decision must be made regarding the ideal sensitivity and/or specificity desired. As an illustration, if the goal of the algorithm is to not miss a diagnosis of a serious condition, then a high sensitivity is important. If the goal is to be sure a negative test means no disease, then high specificity is important.[45]

PUTTING IT ALL TOGETHER

There are multiple scenarios where predictive modeling in the medical field would be useful. Predicting patients at risk of morbidity (e.g. early sepsis detection), mortality, readmission, high cost and poor compliance are just few of the reasons prediction is important. While retrospective (after the fact) prediction is common, ultimately the goal is to predict serious situations real-time. In the latter scenario, predictive analytics = clinical decision support; which will be discussed in the chapter on clinical decision support.

It is important to understand the concept of modeling. An analogy might be baking a cake is the model and the recipes might be the machine learning algorithms or a statistical approach; with multiple ways to bake a cake. After selecting an appropriate data set, the researcher develops a question or hypothesis. He or she first explores the data set to clean it up and then explore it to look at whether there is missing data, whether the attributes are normally distributed and whether the scales of the attributes are similar or dissimilar. If the outcome variable is nominal, then a classification model is indicated. One option would be a statistical approach. In the case of machine learning (ML) there are a multitude of classification algorithms available to build a model. In addition, more than just raw data can be analyzed. The same algorithms can be used for normalized, standardized, and imputed data, and with data sets where the attributes have been reduced to see if that alters results. The overall results can be compared with t-tests to see if there is any statistical difference between the results. ML results can be analyzed for sensitivity and specificity as well. Once a model has been selected, the algorithms can be further tweaked for the best performance, as they all have customizable options. Algorithm customization is beyond the scope of this chapter. Finally, the model is saved and used for analysis of similar new data.[46] The real test of any model is when it is tested with new and independent data. Case study A illustrates many of these points.

NATURAL LANGUAGE PROCESSING AND TEXT MINING

The last area of data analytics we would like to mention is text mining using natural language processing (NLP). Natural language processing technology, refers to the ability of artificial intelligence to extract information from natural human language. NLP applications are based on the linguistic-artificial intelligence intersection of computer processing that was developed to identify semantics and spoken prose.[49] Health care data, especially electronic health record (EHR) data, is largely unstructured, so ripe for text mining, ready to reap health care information from "troves of unstructured data."[50] Since clinical notes often contain bullet points, lack complete sentences, contain ambiguous words and acronyms, health care data does pose a challenge for natural language processing. Natural language processing and text mining have the potential to address clinical outcomes and billing, as well as improve interoperability during transitions between electronic health records.[51-52] A systematic review reported in 2016 concluded that NLP mining of EHR text resulted in improved case detection rates when combined with ICD-9/-10 coding.[53] The demand for text mining is increasing, and global market for NLP is predicted to increase from $1.1 to $2.7 billion, driven by the quick evolution in NLP technology, reduced cost and increases in healthcare unstructured data.[54]

Text mining has become more achievable with open-source software now readably available. An example would be the R language (discussed in a prior section). For example, there are packages ready-to-install into R that aid in text mining, such as the *tm* package.[55] Other

---

Case Study A: Predictive Modeling With a Diabetes Data Set

A data set hosted on Vanderbilt's Department of Biostatistics web site was used that contains lab and demographic factors on 403 African-American adults from 2 rural counties in Virginia. (See data sets in the data science resource appendix). Thirteen patients were excluded due to missing labs. Sixty patients were diabetic out of the total of 390 or 15%. Diabetes was diagnosed based on a lab test (A1c) of 7 or greater. The outcome (dependent variable) was diabetes or no diabetes. The other 14 attributes (independent variables) were glucose, total cholesterol, HDL cholesterol, Cholesterol/HDL ratio, age, gender, height, weight, BMI (calculated), systolic blood pressure, diastolic blood pressure, waist size, hip size and waist/hip ratio.[47]

Research question: what features (attributes) in the data set best predict the presence or absence of type 2 diabetes in this cohort?

Statistical predictive modeling: IBM Watson Analytics (IBMWA) (discussed in later section) was used. Data was first evaluated with IBMWA's *Explore* option and it was noted that not all variables were normally distributed. Data was then evaluated with IBMWA's *Predict* option. The *single* strongest predictor (90%) was glucose, followed by age and systolic blood pressure (85%). Logistic regression was used and all calculations had significant p values, but glucose also had a large effect size. When a *combination* of factors was analyzed, glucose and weight, glucose and waist/hip and glucose and age all had good predictive values but they were not statistically significant.[48]

Machine Learning predictive modeling: WEKA machine learning software was used. *Exploration*: There are 390 instances (rows or patients) and there was no missing data. The independent variables were first visualized to determine if they were normally distributed and about half were. The scales for the attributes were different, thus normalization of the data should be considered. The outcome variable was not balanced (many more non-diabetics than diabetics) so another adjustment (weighting) may be in order. All attributes were visualized with a scatter plot and the outcome was not clearly separable, another reason the data should be evaluated with different views. *Prediction:* given that the outcome variable was nominal, several common classification algorithms were selected: Random Forest, Logistic Regression, Naïve Bayes and K-Nearest Neighbor. The 4 algorithms were used to analyze 1. Raw data 2. Normalized data 3. Standardized data 4. Top 4 ranked attributes data. The results showed that normalizing and standardizing data did not result in better performance. Reducing the attributes from 14 to 4 had little effect on performance of the model. Logistic Regression was the best performer but not statistically different from Bayes or Random Forest. It successfully classified non-diabetes 97% of the time, and diabetes 91% of the time with a 4% standard deviation. Using just glucose as the primary predictor had high sensitivity and specificity.[40]

Conclusion: The statistical and machine learning approaches resulted in similar results. We used logistic regression in both, but machine learning provided more choices. Glucose was the strongest predictor of diabetes. IBMWA provided preliminary predictions faster than WEKA, but the latter provided more information about accuracy (such as precision, recall, ROC, etc.). The machine learning model selected was logistic regression and we know that in the future we could use glucose only in the model for prediction.

open source software that can be used for text mining are Apache, Stanford's core, and GATE.[56] There is also an open source consortium to advance efforts and collaboration in natural language processing in healthcare.[57] More resources for natural language processing and text mining are available in Appendix I of this textbook. Case study B demonstrates how using NLP on EHR data can help identify heart failure patients.

VISUALIZATION AND COMMUNICATION

Data scientists will likely need to visualize data more than once during the complete data analytical process. Data visualization assists in the understanding and analysis of complex data by placing the data in a context that is easier to perceive. The visualization must present the data accurately and concisely and be free from distractions.[58] They may first visualize data during the EDA phase to look for patterns of interest. After completion of the formal analysis it is quite likely the findings will need to be communicated to others, particularly those working in the C-suite (CEO, COO, etc.). *"A picture is worth a thousand words"* also applies to presenting complicated analytics. A dashboard may need to be created that displays a variety of charts, depending on the nature of the data. Graphs and tables can be created with any spreadsheet software but may not offer as much functionality as data visualization software discussed in another section. Microsoft Excel offers the following graph tools:

- Column (vertical) charts of several varieties to include 3-D
- Bar (horizontal) charts of several varieties to include 3-D
- Pie charts
- Line charts of several varieties to include 3-D
- Scatter charts and Bubble charts
- Area charts
- Sparkline charts which are tiny charts within one cell that summarize data in a line or column[60]

As stated previously, bar graphs are excellent choices for categorical (nominal) data, as are pie charts, if the user is attempting to show the breakdown of components contributing to the whole. Line graphs (run charts) are useful to display numerical data (y axis) over time (x axis). Histograms display continuous numerical data, frequently in bins, such as age groups or decades. Unlike bar graphs the bars are touching. Scatter plots or diagrams are helpful to show the relationship between two numerical variables. Data displayed in this manner

---

Case Study B: Automated ID and Predictive Risk for Heart Failure Patients

Heart failure (HF) admission and readmission are quite common and costly for patients and healthcare systems. Many institutions are using analytics to try to identify patients at increased risk of morbidity, mortality and readmission. Ideally, patients should be identified while still in the hospital and aggressive treatment and education administered.

Research question: can predictive analytics using natural language processing identify patients better than relying on only manual chart reviews?

Predictive analytical approach: Intermountain Healthcare mined free text dictated reports residing in their EHR daily (uploaded to the data warehouse), using natural language processing (NLP). The goal was to identify and risk stratify current patients hospitalized for heart failure (HF) and to determine if adding NLP would improve on manual chart reviews. The statistical model calculated their 30 day all cause readmission risk and 30 mortality risk. The prediction model for identification of HF patients consisted of these predictors: diuretic use, b-type natriuretic peptide level > 200pg/ml, ejection fraction less or equal to 40 in the previous year and ever eligible for CMS or Joint Commission HF core measures.

Results: The addition of NLP to coding increased the identification score from 82.6% to 95.3% and specificity from 82.7% to 97.5%. This resulted in a HF Risk Report reviewed daily in targeted discharge planning sessions and a reduced amount of time for clinicians to review HF patient's risk, compared to the manual method. It should be noted that Intermountain Healthcare tested this strategy at only one hospital and while they reported a significant reduction in 30-day mortality rates and an increase in the percent of patients discharged to home, they did not see a reduction in readmission rates.

Conclusions: NLP improves HF identification rates using EHR free text. Leveraging technology and targeted discharge planning can result in improvement in HF morbidity and mortality.[59]

may help to show a positive or negative relationship between variables. Summary tables are good choices for both categorical and numerical values. For more specific guidance to presenting tables and graphs, readers are referred to the textbook by Horton.[61]

A newer presentation strategy is Infographics or a mashup of graphs, charts and text to make a strong point. Displays can be web-hosted so the data is interactive and available to anyone with Internet access. A variety of visualization software can create infographics as discussed in another section. Piktochart is a company that offers free and paid accounts to create infographics with about 400 templates.[62]

BIG DATA

Clearly, in the data science and analytics fields no term has received as much attention and hype as "Big Data." Even the definitions tend to be controversial. The following are two very different definitions of big data:
- Data so large it can't be analyzed or stored on one computational unit[28]
- Five Vs: the reality is that the definition started with three Vs but has increased to five:
  o Volume: massive amounts of data are being generated each minute
  o Velocity: data is being generated so rapidly that it needs to be analyzed without placing it in a database
  o Variety: roughly 80% of data in existence is unstructured so it won't fit into a database or spreadsheet. There is tremendous variety, in terms of the data that could potentially be analyzed. However, to do this requires new training and tools.
  o Veracity: current data can be "messy" with missing data and other challenges. Because of the very significant volume of data, missing data may be less important than in the past.
  o Value: data scientists now have the capability to turn large volumes of unstructured data into something meaningful. Without value, data scientists will drown in data, not information or knowledge.[63]

Big data is different from "little data" for several reasons. With big data, domain specific knowledge is of major importance due to the unstructured nature of big data. Collaboration between domain specific researchers (health care), computer scientists, statisticians, and data scientists ensures that knowledge is accurately extracted from the copious amounts of unstructured data. Data science is focused on interpretation of the data and dissemination of the results.

Google developed MapReduce in 2004 which is a system to deal with huge datasets. Later the Hadoop Distributed File System (HDFS) was developed which is the open-source version of MapReduce plus the Google File System (GFS). MapReduce has two functions, mapping and reducing which works by distributing the work among many computers (nodes or clusters). Google routinely analyzes data with 1000 computers, as an example. Although Hadoop is "open source" and free to use, Cloudera provides management tools and offers support contracts. By default, the HDFS creates three copies of the data as it is ingested into the file system, which ensures that the data is always backed up in case of hardware failure. The most common approach to analyzing big data is to actually reduce the volume of data down to something more manageable and to distribute the workload and data storage across multiple computers by taking advantage of parallel processing. MapReduce runs on the HDFS and, as its name implies, MapReduce first creates a data map of key-value pairs (tuples) and then reduces the data by combining the key-value pairs into a smaller set of tuples.[64] Apache Hive is the data warehousing system built on Hadoop.[65] Apache Hive allows structured query language statements to be executed to retrieve data from the HDFS, allowing a HDFS novice

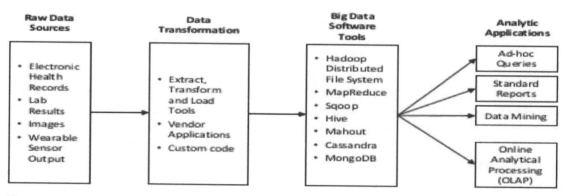

Figure 22.15: Typical big data platform

the ability to query and analyze data without the need to fully understand the complexities of the file system. In this regard, Apache Hive is like relational database query engines. Apache Mahout is a machine learning library that can also be used with Hadoop. (DDS).[66] Figure 22.15 shows a typical big data platform (adapted from [67]).

In a clinical environment, medical professionals collect patient data such as heart rate, blood pressure and temperature, and enter codes for a diagnosis. Also, most test results are reported in numerical quantities. These values are easily stored in columns within a database table row which makes for straightforward data analysis and pattern recognition. Where big data processing enters the picture is when there is a need to analyze any notes that are manually entered. These notes are considered text documents that contain unstructured data. While medical terminology attempts to be standardized, there are still synonyms, shorthand, dialect and language differences that complicate pattern recognition in text documents. It is important that pattern recognition and text mining software can handle the semantics within the text to attain the meaning the author is attempting to convey.

As an alternative to the Hadoop Distributed File System and traditional relational databases, NoSQL databases can als0 store and retrieve large volumes of unstructured data. NoSQL is a general term for a variety of database technologies that were created due to the exponential rise in the volume of generated data, the increase in data access frequency, and the performance and processing needs this surge in data volume requires. Relational databases, with their roots in 1970's technology, were not designed to accommodate the demands of scaling up nor the agility that modern applications require. Furthermore, traditional relational databases were not designed to leverage the inexpensive storage and processing hardware available today. NoSQL database systems are different because their non-relational and distributed architecture allows for the quick, impromptu arrangement and investigation of high-volume, dissimilar data types.

The term NoSQL is a misnomer since many of the NoSQL databases have a structured query language to manipulate the data. For example, Apache Cassandra has its own query language named Cassandra Query Language (CQL) with commands that are like ANSI SQL. Sometimes NoSQL databases are called cloud databases because of their prevalence on cloud service providers, non-relational databases, or Big Data databases. NoSQL databases are specialized in terms of the types of data they were designed to accommodate.[68] Apache Cassandra[69] is designed to store key-value pairs while MongoDB[70] is well suited to store large volumes of unstructured text documents.

While traditional relational databases are governed by the ACID principles (atomicity, consistency, isolation and durability), NoSQL databases use the principles of BASE. In this play on words from chemistry and pH levels, BASE is an acronym for Basically Available, Soft state, Eventual consistency. Basically Available takes advantage of the distributed system architecture so that data is always available even though parts of the system might be offline. Soft state means that the data does not need to be immediately consistent and inconsistencies can be tolerated for specified time periods. Eventually consistent means that after an allotted amount of time, the database will return to a consistent state.[71]

Another application of big data processing in a health care environment is with pattern recognition in digitized images such as x-rays, ultrasounds, CT scans and MRIs. As the resolution and image sizes continue to increase, it is imperative that hardware and processing techniques maintain their ability to efficiently analyze this image data. Image data volume can be reduced using wavelet transforms and other compression algorithms to decrease pattern recognition algorithm processing times.

Healthcare big data analytics faces many challenges. Only large healthcare organizations are likely to be able to afford a big data center and have the expertise to run it. Even when using leased hardware from a cloud service provider to reduce hardware costs, the personnel costs can for big data analytics expertise can be prohibitive. Manipulating large volumes of unstructured data will be difficult, except for those with significant training and experience. Integrating genomic information will eventually become routine, but currently, there are multiple questions about how to use the data. There is inadequate physician training in genetics and a shortage of geneticists to counsel patients. (see chapter on Bioinformatics). Social media data is interesting but "messy" with many missing values. Sensor data, particularly from activity monitors is increasingly common, but how can the data be aggregated and which payer will reimburse for this type of monitoring? Unique challenges are also associated with image, geospatial and temporal data.

According to Gina Neff, the biggest challenge "is social, not technical". She points out that the data needs of physicians, patients, administrators, payers and researchers are very different. Most clinicians are likely to tell you that they have lots of data but very little information and very little time and expertise to analyze these data. She also questions how the analytical results will be incorporated into the daily workflow of busy clinicians. Patients have similar issues and concerns, as evidenced by the number who download mobile apps, but do not use them long term. Patient privacy and security are also an extremely important issue with large datasets.[72]

ANALYTICAL SOFTWARE FOR HEALTHCARE WORKERS

Statistical software programs

- Microsoft Excel with Data Analysis ToolPak. The Toolpak is a free add-in for Excel users. Most of the commonly used statistical tests are offered. There is a steep learning curve, as users must select the optimal test with no guidance or wizards.[73] A target attribute must be compared with one independent variable at a time, which is labor intensive. There is also a data mining add-in for Excel that connects to an existing Microsoft SQL Server Analysis Services, discussed in the data mining section.
- SQL Server Analysis Services has multiple tools for statistical modeling and machine learning that include both supervised and unsupervised learning: clustering, factor analysis, Bayesian and neural networks, time series analysis, association analysis, recommendations, and shopping basket analysis scoring binary outcomes and linear regression. Wizards are available to expedite the process and assist in configuration.[74]
- Statistical Package for the Social Sciences (SPSS). There are many stat packages available, but we will only concentrate on one. SPSS is a family of analytical packages: *SPSS Statistics*: covers all of the basic statistical approaches to include temporal causal modeling, geospatial analytics and R programming integration. *SPSS Modeler*: is a predictive analytics platform that includes machine learning algorithms. *SPSS Analytic Server*: accepts data from Hadoop and the Modeler to analyze big data. *SPSS Social Media Analytics:* analyzes Twitter feeds[75]
- IBM Watson Analytics (IBMWA) was released in 2015 for data analysis by any industry. It is unique, in that it automates the exploration, prediction and visualization of data. The analytical approach is based on advanced statistics (SPSS) and natural language processing (NLP), and not machine learning. NLP tools automatically generate 10 observations about the uploaded dataset with the ability to use NLP to ask additional questions. Data quality is scored and associations are noted. In the background under "statistical details" a user can see the test used. Visualizations are automatically created and can be later used to create dashboards and infographics. An academic program is available for teaching so that an instructor and his/her students can use the program without charge. [45,76] Figure 22.16 shows a prediction result looking for predictors of obesity, based on County Health Ranking Data, for the state of Florida.[77]

Machine Learning/Data Mining

- SQL Server Analysis Services (see prior section)
- WEKA: Developed by the University of Waikato in New Zealand, this software is associated with its own textbook and online course. It differs from

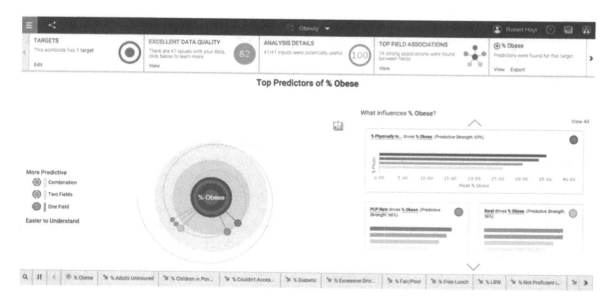

Figure 22.16: Predict option in IBM Watson Analytics

the other programs by not having a graphical user interface that requires users to move operators (widgets) and connectors around to run an analysis. It performs classification, regression, association and clustering. Data visualization is available but is primitive, compared to newer programs.[40]

- KNIME: is a suite of open-source analytical tools. The program supports a large variety of data types and connectors (algorithms and steps). KNIME integrates with programming languages, such as R and Python. More than 1000 modules are available which is both a strength and weakness. Software is available for all operating systems. The user must select the correct operators and sequence for this to work properly. A Quickstart Guide is available.[78]
- RapidMiner. They offer a suite of paid and free machine learning software, available for all operating systems. The free version is RapidMiner Studio, but they also offer a server-based solution intended to be integrated into an enterprise platform. Excellent tutorials are available. This is a very comprehensive system, but you must know which operators to drag and drop, and how they are connected, to get results.[79]
- Orange: Developed by the University of Ljubljana in Slovenia, this program has good tutorials and instructions. Major categories are Data, Classification, Regression, Clustering, and Visualization. Software is available for Windows, Mac and Linux OSs. Probably the most intuitive choice of the free open-source machine learning software. Right clicking operators tells the user what they need to know and gives examples. Figure 22.17 shows a screen shot of a machine learning exercise using three common algorithms.[44]

Data Visualization. Only two data visualization software programs will be presented. Both suggest that they perform analytics but in reality, they don't perform descriptive or predictive analytics, just visualization.

- Tableau: The program offers a suite of web-based and client-based software. There is a free version known as Tableau Public, as well as paid versions. The Public version is non-web based and available for Windows and Mac computers. Input can be Microsoft Access files, Excel or .csv files. Software is very intuitive, with drag and drop functionality. The visualization choices are like Excel, but more elegant and with more options. There is also a free reader for Tableau program output. Tableau offers white papers to improve visualization skills.[80-81]
- Qlik Sense Desktop: This is the free version of the software, which is only available for Windows operating systems. Like Tableau it has an easy drag and drop functionality and multiple chart formats to choose from. They offer developers an open API option.[82]

DATA SCIENCE EDUCATION

DataScience Community is a web site that displays colleges and universities offering a data science -related degree, such as data mining, data analytics and data science in the US and overseas. An early 2018 query revealed 581 institutions worldwide offering programs that focused on data science. The breakdown is quite varied and is as follows:
- Certificate (96); Bachelor (51); Masters (411); Doctoral (22)
- 29% of courses are offered online[83]

KDNuggets is an excellent resource for US and international online courses in analytics, big data, data mining and data science. They also have sections

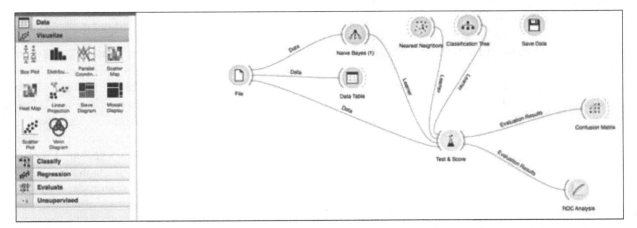

Figure 22.17: Machine learning using Naïve Bayes, nearest neighbor and classification tree

on certification and certificates in analytics and data science.[37] Certifications:
- Certified analytics professional (CAP): CAP has offered certification since 2013. The CAP® credential targets professionals with 5 years of professional analytics experience and a bachelor's degree, or with 3 years of experience and a master's degree.[84]
- Certified Health Data Analyst (CHDA): hosted by AHIMA, the eligibility requirements include: Registered Healthcare Information Technician (RHIT) credential and a minimum of three years of healthcare data experience; Baccalaureate degree and a minimum of three years of healthcare data experience; Registered Healthcare Information Administrator (RHIA) credential; Master's in Health Information Management (HIM) or Health Informatics from an accredited school or Master's or higher degree and one year of healthcare data experience.[85]

In addition, Coursera lists more than 100 low-cost courses related to data science. There are a variety of short courses that would allow someone to see if they have interest or potential in this new area.[86]

The previous paragraphs suggest that there are numerous conventional and non-conventional data science programs available and reflects the overall attention this field has received recently. At the current time, most students will likely opt to get a Master's degree, but have to find a way to gain domain experience, such as internships and employment.

In addition to a plethora of new data science courses, there are also an abundance of new Data Science Centers appearing around the US. An informal search found thirteen major centers with many being multi-disciplinary. This is an encouraging trend because in years past it would have likely that they would have belonged to either computer science or math and statistics departments. Universities are realizing that data science/data analytics affects all industries and therefore affects multiple schools and departments. Perhaps the most striking data science center plan was announced by the University of Michigan in 2015. They anticipate spending $100 million over the next five years to create their Data Science Center for students and faculty and hire 35 new faculty members.[87]

DATA SCIENCE CAREERS

Multiple jobs are available in the data science and analytics field. A late 2017 query on Indeed.com for "data scientist" revealed 3,689 jobs listed. Salary ranged $95K to over $140K.[88] A 2015 O'Reilly Media Data Science Salary Survey surveyed 600 individuals in the data science field. Two thirds of respondents were from the US and only 25% were data scientists, the remainder were employed in a variety of data-related positions. The survey also revealed that 80% of employees in the field were male; having a PhD added $7500 on average and management roles also added to salary estimates.[89] Most data science-related employees had a Master's level degree, but experience, curiosity and problem-solving skills were equally important.[90]

Large healthcare organizations such as Kaiser-Permanente have a large data analytics workforce. According to a Kaiser-Permanente director of information the most important skills are programming, communication and critical thinking. The general data employment categories are:
- Data analysts: technical group who deal with the architecture
- Report analysts: design reports and dashboards related to key performance indicators (KPIs)
- Business systems analysts: work with the operational aspects of business
- Informatics analysts: have strong statistics background and who help others with data analysis [91]

KD Nuggets posts a non-academic and academic data science jobs board. Companies such as Verizon, Aetna, Geisinger Health System, Schwab, Microsoft and Sears are just a sampling of companies seeking data science and analytics expertise.[37]

DATA SCIENCE RESOURCES

See Appendix 22.1 at end of chapter

DATA SCIENCE CHALLENGES

The primary challenge facing the new field of data science is training enough data scientists. According to a 2011 McKinsey Global report there will be a shortage of 140,000 - 190,00 data scientists with significant analytical skills for all industries, by 2018.[92] It is unclear whether a single Master's level training program is adequate. Clearly, some individuals will need additional training and many more will need experience in a domain, such as healthcare or business. Actual exposure to complicated data situations and big data will be critical for success. Due to the shortage of data scientists, it is likely that many organizations will have to rely on a team with multiple individual skill sets, rather than a single individual with all of the required skills. It is possible with better automation and simplification of data analytics; shorter training may be acceptable.

It is also challenging to educate people involved with data to use both statistical and machine learning tools. Both are important, and both should be used to expand the analytical tool set. Classic statistical tools such as linear regression may be too rigid for evaluating biological data that is more complex with more unknowns. This is often better analyzed with machine learning algorithms. Newer data challenges such as image or speech recognition have demanded newer methods, such as neural networks.[93] Many have argued that medical data analytics is unique and not like data from the physical sciences. They cite that the data is more heterogeneous and complex (free text physician, lab, pathology and xray reports); associated with more social, ethical (privacy and security) and legal issues; subject to physician interpretation of results and more difficult to model than other non-healthcare data. Despite the information being electronic, it can still be inaccurate and incomplete.[94]

Another challenge facing the data science field is hype. While there is no doubt that we need to train more people in data analytics, it is unclear how often big data analytics is required. Challenges with big data were covered in that section. Finding correlations, using statistics or machine learning is not difficult, but correlation doesn't prove causation. For example, fire engines don't cause fires, just because they seem to be omnipresent at fires.

FUTURE TRENDS

It is likely that automation of analytics, such as IBM Watson Analytics will become more common place and will allow for more healthcare workers to be involved in data science and analytics. A 2013 survey by KDNuggets sought to find what users thought of the likelihood of automated predictive analytics: 5.1% thought it already existed, while 46% thought it would be 1-10 years off, 30% that it would take another 10-50 years and 19% thought it would never happen.[37]

More and more data analytical platforms are becoming available. In addition to IBM, Amazon and Microsoft, Google began offering the Google Cloud Platform in early 2016. This comprehensive platform offers a myriad of tools (26) that fall into the following categories: compute, storage, networking, big data, machine learning, operations and tools.[95]

In David Donoho's sentinel paper *"50 years of DS (data science)"* he posited the following future trends:
- Open science: more sharing of data and code among researchers to promote reproducibility
- Science as data: verifiable results, available to all researchers
- Data analysis tested empirically: rather than using mathematical models under ideal conditions, analyses will be judged with empirical methods.[9]

We will continue to see refinement in many areas of data science. It is likely that future data scientists will successfully blend statistics with machine learning and not treat them as separate approaches. Not only will the definition of the field likely see changes, the training programs will likely evolve as well. Microsoft began offering a free six-week course in data science in July 2016 as just another example of industry promoting this new field.[96]

Data science will continue to be promoted by the federal government because the field is critical to the future direction of healthcare. Precision medicine is *"identifying which approaches will be effective for which patients based on genetic, environmental and lifestyle factors"* which will require the integration and analysis of extremely large and complex data sets.[97] This will be one approach to achieve a *learning health system* or *"an ecosystem where all stakeholders can securely, effectively and efficiently contribute, share and analyze data."*[98]

Given the focus on big data, one can expect improvement in processing, analyzing and storing huge datasets. More data centers are likely to appear, with specialization by industry.

Moore's Law will continue to drive miniaturization and improvement in speed, memory and storage.[99]

KEY POINTS

- Data science is a new information science field with significant overlap with other fields
- A data scientist must have multiple skills, in addition to mathematics and statistics
- There is a shortage of data scientists worldwide
- Healthcare needs data science and analytics, like all other industries
- Big data brings new promises and new challenges
- We need new tools to automate the process of descriptive, visual and predictive analytics

CONCLUSION

Data science is an exciting new field that is broad and incorporates multiple disciplines. The field mandates multiple skill sets that might not be achieved with only one Master's degree. Given the current shortage of data scientists, many healthcare organizations will have to rely on a team for the required expertise in multiple areas. Despite numerous training programs in data science, it will take a considerable amount of time for there to be adequate data scientists in the pipeline. In addition, most graduates will need experience in the field (domain) to develop expertise. Healthcare organizations need to look at potential internships in the CIO's arena to gain the necessary experience to intelligently analyze healthcare data.

Colleges and Universities are already "ramping up" in the data science field to educate more workers in multiple industries about data analytics, statistics and machine learning. Healthcare workers who are unable to enroll in local data science courses have a plethora of low-cost online national courses available.

It seems likely that we will see more automation of analytics in the future. Data visualization is perhaps the easiest area to automate, so that is why we are seeing vendors expand their influence in this area. It is likely we will see vendors produce new products for predictive analytics, in addition to IBM Watson Analytics. Importantly, new products will need the means to evaluate and report their results. In

healthcare, performance is most commonly reported out as sensitivity, specificity, predictive values, likelihood ratios (LRs) and area under the curve (AUC). It is just a matter of time before we see this type of robust analytical program embedded into healthcare data sets.

APPENDIX 22.1 DATA SCIENCE RESOURCES

Appendix table 1 General datasets

General datasets	Details
• University of California, Irvine Repository: https://archive.ics.uci.edu/ml/datasets.html	Site includes 325 validated datasets covering many domains, different sizes and data types and different analytical methods. These data sets are commonly used for machine learning exercises.
• KDNuggets: www.kdnuggets.com	Site includes 71 data sets available for free download, from various industries.
• The Datahub: https://datahub.io/dataset	Managed by the Open Knowledge Foundation, this site hosts more than 10,000 datasets from most industries.
• Kaggle: www.kaggle.com	Provides free, interesting datasets for various user interests and analysis.
• DATA USA www.datausa.io	Organized by maps, cities/places, jobs and downloads
• Data World https://data.world	New site for creating collaborative data projects with ability to host data and analyze with embedded SQL. Tools available to link to Tableau, R and Python languages and machine learning. Multiple datasets from healthcare and other industries are available. Free for academic use.

Appendix table 2 Healthcare datasets

Healthcare datasets	Details
• Healthcare Cost and Utilization Project: http://www.hcup-us.ahrq.gov/	Includes U.S. longitudinal hospital care data with databases, software and online tools
• Health Data.Gov: http://www.healthdata.gov/content/about	Users can search by data category and format (.csv, .xls, zip, PDF, rdf, JSON, html, txt and API)
• Centers for Disease Control and Prevention: http://www.cdc.gov/nchs/data_access/ftp_data.htm	Includes public use files (PUFs) from surveys from multiple government agencies
• Expert Health Data Programming: http://www.ehdp.com/vitalnet/datasets.htm	Host links to about 45 large data sets
• Health Services Research Information Central:https://www.nlm.nih.gov/hsrinfo/datasites.html#488International	Has extensive health datasets, statistics, international data and data tools
• Vanderbilt Biostatistics Datasets: http://biostat.mc.vanderbilt.edu/wiki/Main/DataSets	Multiple health related data sets are available to download as Excel, ASCII, R and S-Plus files. Also includes links to international data sets.
• MIMIC III Critical Care database: https://mimic.physionet.org/	Site is a repository of more than 40,000 de-identified critical care patient-level data
• CMS Data Navigator: https://dnav.cms.gov/Default.aspx	Expedites the search for Medicare and Medicaid data

Appendix table 3 Free online data science resources

Free online data science resources	Details
• School of Data: http://schoolofdata.org/courses/	Online course covers data fundamentals, data cleaning, exploring data, extracting data and mapping data
• Class Central: https://www.class-central.com/subject/data-science	Offers multiple free data science and big data-related courses
• Data Science Academy: http://datascienceacademy.com/free-data-science-courses/	Aggregates courses from multiple universities
• Udacity: https://www.udacity.com/courses/data-science	Includes data science courses at beginner through advanced levels
• Oregon Health and Science University (OHSU) Dr. Bill Hersch www.informaticsprofessor.blogspot.com	Free Healthcare Data Analytics Course (9/7/2016) OHSU Big Data to Knowledge (BD2K) Open Edu-cational Resources (OERs) Project (10/23/2016)
• IBM Big Data University: https://bigdatauniversity.com	Offers multiple free courses related to data science and big data analytics for beginners and intermediate level learners
• KD Nuggets: www.kdnuggets.com	Check the main "Courses" tab

Appendix table 4 Free online statistics resources

Free online statistics resources	Details
Online textbook: http://www-bcf.usc.edu/~gareth/ISL/	An Introduction to Statistical Learning with Applications in R
Online textbook: http://www-stat.stanford.edu/~tibs/ElemStatLearn/	The Elements of Statistical Learning: Data Mining, Inference, and Prediction, 2nd edition.
Stat Trek: http://stattrek.com/tutorials/free-online-courses.aspx	Online tutorials guide students through the introductory steps of statistics. There are also brief quizzes and calculators to add interest
Biostatistics textbook: http://bolt.mph.ufl.edu	Biostatistics. Open Learning Textbook. University of Florida.
OnlineStatBook: http://onlinestatbook.com/2/index.html and iTunes.apple.com	Excellent Introductory free online reference from David Lane at Rice University. There is also a free e-book for Mac or iOS devices "Introduction to Statistics: An Interactive E-Book.
OpenIntro: www.OpenIntro.org	Three free PDF download books. One book is associated with about 100 free datasets
Statistics How To: www.statisticshowto.com	There is an online textbook as well as a companion e-book for sale. The web site includes calculators and stats tables.
Kaggle: www.kaggle.com	There are forums for those just getting started in data science, as well as information about public data sets. Kaggle also provides job forums for those interested in careers in the data science field. In addition, Kaggle hosts data science competitions for both health and non-health care data.
StatPages: http://statpages.info/	A mega-site for essentially any free online statistical calculator imaginable.

Appendix table 5 Spreadsheet tutorials

Spreadsheet tutorials	Details
University of California at Berkeley: http://multimedia.journalism.berkeley.edu/tutorials/spreadsheets/	Provides the basics on spreadsheets. Based on Google spreadsheets
Google spreadsheets: https://sites.google.com/a/g.risd.org/training/RISD-Video-Tutorials/google-spreadsheet-tutorials	Good introductory video tutorials on Google spreadsheets

Appendix table 6 Programming language tutorials

Programming language tutorials	Details
Tutorials Point: https://www.tutorialspoint.com/tutorialslibrary.htm	Web site offers tutorials in multiple areas of data science to include the R, Python and SQL Tutorial
R Tutorial by Kelly Black: http://www.cyclismo.org/tutorial/R/	Interactive introduction to R from the University of Georgia Department of Mathematics
The Python Tutorial: https://docs.python.org/2/tutorial/	Introductory level instruction on Python
SQLZoo: http://sqlzoo.net/wiki/SQL_Tutorial	Provides SQL tutorials, references, sample databases

Appendix table 7 Web data extraction tools

Web data extraction tools	Details
• Import.io: https://www.import.io/	Commercial site (free trial available) to scrape data from the web into .csv files
• Google Chrome extension scraper: https://chrome.google.com/webstore	Free Google extension that will convert web based data into a format compatible with Google spreadsheets

Appendix table 8 Geo-coding tools

Geo-coding tools	Details
• Geonames: http://www.geonames.org/	Free international geographical database with over 10 million geographical names, maps, etc.
• QGIS (desktop GIS): http://qgis.org/en/site/about/index.html	Open source desktop GIS tool

Appendix table 9 Data science blogs

Data science blogs	Details
• KD Nuggets: http://www.kdnuggets.com/	Lists 90 blogs that cover most aspects of data science
• Informatics Professor: www.informaticsprofessor.blogspot.com	Dr. Bill Hersh's blog on Biomedical and Health Informatics includes topics related to data science and data analytics and free courses offered by OHSU

Appendix table 10 Data science journals

Data science journals	Details
• Data Mining and Knowledge Discovery: http://www.springer.com	Six issues published by Springer each year. Available as open access and non-open access
• Data Science Journal: http://datascience.codata.org/	Peer-reviewed open-access journal
• Journal of Data Science: http://www.jds-online.com/	Publishes international research articles on data science. Online access is free

REFERENCES

1. Data Science Association. http://www.datascienceassn.org/about-data-science (Accessed September 16, 2016)
2. Data mining. Wikipedia. www.wikipedia.org (Accessed January 16, 2016)
3. Tukey JW. The future of data analysis. Annals of Math Stats. 1962;33(1):1-67
4. Cleveland WS. Data science: an action plan for expanding the technical areas of the field of statistics. Int Stat Rev 2001. 69(1):21-26
5. Gunelius S. The Data Explosion in 2014 Minute by Minute-Infographic. July 12, 2014. http://aci.info/2014/07/12/the-data-explosion-minute-by-minute-infographic (Accessed February 23, 2014)
6. Cukier KN, Mayer-Schoenberger V. The Rise of Big Data. Foreign Affairs. May/June 2013. www.foreignaffairs.com (Accessed February 15, 2016)
7. Press G. A very short history of data science. May 28, 2013. www.Forbes.com (Accessed April 15, 2016)
8. Google Trends https://www.google.com/trends/ (Accessed October 15, 2016)

9. Donoho D. 50 years of data science. Presentation at the Tukey Centennial Workshop September 15, 2015. http://courses.csail.mit.edu/18.337/2015/docs/50YearsDataScience.pdf (Accessed October 25, 2015)
10. Baumer B. A Data Science Course for Undergraduates: Thinking with Data. American Statistician. 2015;69(4):334-342
11. Hersh W. 60 years of Informatics: in the context of data science. Informatics Professor Blog. February 1, 2016. http://informaticsprofessor.blogspot.com (Accessed March 1, 2016)
12. Stanton J. Chapter 1 About Data. Introduction to Data Science. 2012 https://ischool.syr.edu/media/documents/2012/3/DataScienceBook1_1.pdf (Accessed January 3, 2016)
13. Geraci R. Data analytics: go big or go home. 2015. Business Week 4436, S1-S6
14. White S. Chapter 1 Introduction to Data Analysis in A Practical Approach to Analyzing Healthcare Data. Third edition. AHIMA Press. 2016. Chicago, IL
15. Nonparametric Statistical Methods—3rd Edition. M. Hollander. 2014. John Wiley & Sons. Hoboken, NJ
16. Wikipedia. Normal Distribution. https://en.wikipedia.org/wiki/Normal_distribution (Accessed February 25, 2016)
17. Standard deviation calculation. Easy calculation. com https://www.easycalculation.com/statistics/standard-deviation.php (Accessed February 26, 2016)
18. Biostatistics. University of Florida. http://bolt.mph.ufl.edu (Accessed March 17, 2016)
19. Wilson WF, Meigs JB, Sullivan L, et al. Prediction of incident diabetes mellitus in middle-aged adults. Arch Intern Med 2007;167:1068-1074
20. Hospital Compare. www.medicare.gov/hospitalcompare (Accessed February 29, 2016)
21. What is a confidence interval? Stat Trek. http://stattrek.com/estimation/confidence-interval.aspx (Accessed February 29, 2016)
22. Confidence interval for mean calculator. Easy calculation.com https://www.easycalculation.com/statistics/confidence-limits-mean.php (Accessed February 29, 2016)
23. Kallnowski P. Understanding confidence intervals (CIs) and effect size estimation. Observer. 2010;23(4). www.psychologicalscience.org (Accessed February 29, 2016)
24. Becker L. Effect size (ES)-effect size calculators. www.uccs.edu/becker/effect-size.html (Accessed February 29, 2016)
25. Chavalarias D, Wallach JD, Ting Li AH et al. Evolution of reporting p values in the biomedical literature, 1990-2015. JAMA 2016;315(11):1141-1148.
26. Database Primer. www.databaseprimer.com (Accessed February 22, 2016)
27. Sheta OE, Eldeen AN. The technology of using a data warehouse to support decision-making in health care. Int J Data Man Sys 2013;5(3):75-86
28. Data Mining: Concepts and Techniques. Han J. Elsevier 2000
29. Learn REST: A RESTful Tutorial. http://www.restapitutorial.com (Accessed February 25, 2016)
30. Huckman R, Uppaluru M. The Untapped Potential of Health Care APIs. Harvard Bus Rev Dec 23, 2015 https://hbr.org/2015/12/the-untapped-potential-of-health-care-apis (Accessed February 26, 2016)
31. The Argonaut Project. http://hl7.org/fhir/2015Jan/argonauts.html (Accessed February 25, 2016)
32. Data.Gov API Catalog. http://catalog.data.gov/dataset?q=-aapi+api+OR++res_format%3Aapi#topic=developers_navigation (Accessed February 25, 2016)
33. Kandel S, Heer J, Plaisant C et al. Research directions in data wrangling: visualizations and transformations for usable and credible data. http://vis.stanford.edu/files/2011-DataWrangling-IVJ.pdf (Accessed January 30, 2016)
34. Doing Data Science, straight talk from the frontline. O'Neil C, Schutt R. O'Reilly Publisher. 2014
35. Sauro J. 7 Ways to Handle Missing Data. June 2, 2015. http://www.measuringu.com/blog/handle-missing-data.php (Accessed January 6, 2016)
36. Analytics in a Big Data World: The Essential Guide to Data Science and Its Application. Baesens B. Wiley 2014
37. KD Nuggets. www.kdnuggets.com (Accessed January 30, 2016)
38. W3 Schools. SQL. http://www.w3schools.com/sql (Accessed January 30, 2016)
39. Chauhan R. Clustering Techniques: A Comprehensive Study of Various Clustering Techniques. Int J Adv Res Comput Sci 2014;5(5): 97-101
40. WEKA. http://www.cs.waikato.ac.nz/ml/weka/downloading.html (Accessed February 25, 2016)
41. OnlineStatBook. http://onlinestatbook.com/2/index.html (Accessed March 1, 2016)
42. Jason Brownlee. Master Machine Algorithms. Discover How They Work and Implement Them from Scratch. 2016. http://machinelearningmastery.com/ (Accessed June 4, 2016)
43. Yoo I, Alafaireet P, Marinov M et al. Data Mining in Healthcare and Biomedicine: A Survey of the Literature. J Med Syst 2012;36:2431-2448
44. Orange. http://orange.biolab.si (Accessed January 3, 2016)
45. Peng, R and Matsui E. The Art of Data Science. LeanPress. 2016 https://leanpress.com/ (Accessed June 2, 2016)

46. Brownlee, J. Machine Learning with WEKA. E-book. 2016. http://machinelearningmastery.com/
47. Vanderbilt Department of Biostatistics. http://biostat.mc.vanderbilt.edu/wiki/Main/DataSets (Accessed March 20, 2016)
48. IBM Watson Analytics. http://www.ibm.com/analytics/watson-analytics/ (Accessed January 4, 2016)
49. Nadkarni PM, Ohno-Machado L, Chapman WW. Natural language processing: an introduction. *Journal of the American Medical Informatics Association: JAMIA*. 2011;18(5):544-551. doi:10.1136/amiajnl-2011-000464
50. Monegain B. Natural Language Processing in High Demand. August 14, 2015. Healthcare IT News. http://www.healthcareitnews.com/news/natural-language-processing-demand (Accessed August 30, 2016)
51. Townsend, H. Natural Language Processing and Clinical Outcomes: The Promise and Progress of NLP for Improved Care. http://bok.ahima.org/doc?oid=106198 (Accessed August 30, 2016)
52. Harris B. 5 benefits of natural language understanding for healthcare. Healthcare IT News. October 9, 2012 http://www.healthcareitnews.com/news/5-benefits-natural-language-understanding-healthcare (Accessed August 30, 2016)
53. Ford E, Carroll JA, Smith HE et al. Extracting information from the text of electronic medical records to improve case detection: a systematic review. JAMIA 2016;23:1007-1015
54. Pennic J. Healthcare Natural Language Processing Market to Reach $2.67B by 2020. HIT Consultant. August 13, 2015. http://hitconsultant.net/2015/08/13/healthcare-natural-language-processing-market-reach-2-67b/ (Accessed August 31, 2016)
55. Feinerer I. Introduction to the tm Package Text Mining in R. July 3, 2015. https://cran.r-project.org/web/packages/tm/vignettes/tm.pdf (Accessed August 31, 2016)
56. Ingersoll G. 5 open source tools for taming text. Opensource.com July 8, 2015. https://opensource.com/business/15/7/five-open-source-nlp-tools (Accessed September 1, 2016)
57. OHNLP Consortium. http://www.ohnlp.org/index.php/Main_Page (Accessed September 1, 2016)
58. E. R. Tufte. *The Visual Display of Quantitative Information*, 2nd ed., Cheshire, CT, USA: Graphics Press, 2001
59. Evans RS, Benuzillo J, Home BD et al. Automated identification and predictive tools to help identify high-risk heart failure patients: pilot evaluation. JAMIA 2016;23:872-878
60. Microsoft Excel 2016. https://products.office.com/en-us/excel (Accessed March 7, 2016)
61. Calculating and Reporting Healthcare Statistics. Third Edition. 2010. Horton LA. AHIMA Press.
62. Piktochart. www.piktochart.com (Accessed March 7, 2016)
63. Marr B. Why only one of the five Vs of big data really matters. March 19, 2015 http://www.ibmbigdatahub.com/blog/why-only-one-5-vs-big-data-really-matters (Accessed April 3, 2016)
64. What is MapReduce? IBM. https://www-01.ibm.com/software/data/infosphere/hadoop/mapreduce/ (Accessed June 10, 2016)
65. Apache Hive. https://hive.apache.org/ (Accessed June 10, 2016)
66. Apache Mahout. http://mahout.apache.org/ (Accessed June 15, 2016)
67. Raghupathi W, Raguhpathi V. Big data analytics in healthcare: promise and potential. Health Info Sci 2014;2(3). www.hissjournal.com/content/2/1/3 (Accessed February 24, 2016)
68. NoSQL. http://nosql-database.org/ (Accessed June 12, 2016)
69. Apache Cassandra. http://cassandra.apache.org/ (Accessed June 10, 2016)
70. MongoDB. http://www.mongodb.com/ (Accessed June 11, 2016)
71. Pritchett D. "Base: An Acid Alternative", ACM Queue, vol. 6, no. 3, July 28, 2008
72. Neff G. Why Big Data Won't Cure Us. Big Data. 2013;1(3):117-123 For further reading we refer readers to a 2013 book Frontiers in Massive Data http://www.nap.edu/download.php?record_id=18374# (Accessed June 12, 2016)
73. Microsoft Excel Analytics ToolPak. https://support.office.com/en-us/article/Use-the-Analysis-ToolPak-to-perform-complex-data-analysis-f77cbd44-fdce-4c4e-872b-898f4c90c007 (Accessed March 7, 2016)
74. SQL Server Analysis Services. https://msdn.microsoft.com/en-us/library/hh231701.aspx (Accessed March 7, 2016)
75. Statistical Package for the Social Sciences (SPSS). http://www-01.ibm.com/software/analytics/spss/ (Accessed March 7, 2016)
76. IBM Watson Analytics Academic Program. https://www.ibm.com/web/portal/analytics/analyticszone/wanew (Accessed January 4, 2016)
77. County Health Ranking http://www.countyhealthrankings.org/ (Accessed February 20, 2016)
78. KNIME. https://www.knime.org (Accessed February 26, 2016)
79. RapidMiner. https://rapidminer.com (Accessed February 26, 2016)
80. Tableau Public https://public.tableau.com/s/ (Accessed January 10, 2016)
81. Tableau Whitepaper. Visual Analysis Best Practices http://www.tableau.com/sites/default/files/media/

82. Qlik Sense Desktop. http://www.qlik.com/products/qlik-sense/desktop (Accessed March 8, 2016)
83. DataScience Community. http://datascience.community/colleges (Accessed April 20, 2018)
84. Certified analytics professional. https://www.certifiedanalytics.org (Accessed March 2, 2016)
85. AHIMA. Certification Health Data Analyst. http://www.ahima.org/certification/chda (Accessed March 2, 2016)
86. Coursera. www.coursera.org (Accessed March 3, 2016)
87. Madhani T. University announces $100 million for data science initiative. September 8, 2015 The Michigan Daily. https://www.michigandaily.com/section/news/university-announces-100-million-data-science-initiative (Accessed March 1, 2016)
88. Indeed. www.indeed.com (Accessed March 1, 2016)
89. O'Reilly Media Data Science Salary Survey (2015). https://www.oreilly.com/ideas/2015-data-science-salary-survey (Accessed February 25, 2016)
90. Harpham B. Career Boost: Break into data science. February 25, 2016. www.infoworld.com (Accessed February 25, 2016)
91. Hersh W. Gimme Some Analytics (We already have it). 2013. Informatics Professor. http://informaticsprofessor.blogspot.com (Accessed February 20, 2016)
92. McKinsey Global Institute. May 2011. Big data: the next frontier for innovation. http://www.mckinsey.com/business-functions/business-technology/our-insights/big-data-the-next-frontier-for-innovation (Accessed February 28, 2016)
93. Breiman, L. Statistical Modeling: The Two Cultures. Stat Science. 2001;16(3):199-231
94. Krzysztof JC, Moore GW. Uniqueness of medical data mining. Art Int Med 2002;26:1-24
95. Google Cloud Platform. https://cloud.google.com (Accessed June 7, 2016)
96. eDX Courses. Microsoft Data Science Professional Project. https://www.edx.org/course/data-science-professional-project-microsoft-dat102x (Accessed July 14, 2016)
97. National Research Council. Towards Precision Medicine: Building a Network for Biomedical Research and a new Taxonomy of Disease. National Academies Press. 2011 www.nap.edu (Accessed October 1, 2016)
98. Connecting Health and Care for the Nation. A Shared Nationwide Interoperability Roadmap. October 2015. www.healthit.gov (Accessed October 1, 2016) Moore's Law. http://www.mooreslaw.org/ (Accessed June 12, 2016)

Index

A

Accountable Care Organization (ACOs)   12, 21, 69, 131, 138, 140-1, 144, 151, 169, 263, 335, 340, 346
ACOs *see* Accountable Care Organization
ADEs *see* adverse drug event
Adverse drug event (ADEs)   78, 116, 176
Agency for Healthcare Research and Quality (AHRQ)   9, 16, 18, 88, 165, 173-4, 176, 178, 183, 196-7, 299, 309-10, 352, 367, 401
AHIMA *see* American Health Information Management Association
AHRQ *see* Agency for Healthcare Research and Quality
AI *see* Artificial intelligence
Algorithms   31, 38, 51, 54, 63, 71, 105, 149, 151-2, 166, 168, 197-8, 298-9, 449-53, 455-6
American Health Information Management Association (AHIMA)   9, 15, 18-21, 254, 256, 311, 462
American Medical Informatics Association (AMIA)   3, 8, 15, 19-20, 85, 164, 237, 253, 397, 426, 428
American Telemedicine Association (ATA)   340-1, 352
AMIA *see* American Medical Informatics Association
Analytics   2, 35, 139-40, 149, 151-2, 154-6, 273, 277, 375, 415, 423, 446, 449, 461-3
APIs *see* application programming interface
Apple HealthKit   274, 276
Application programming interface (APIs)   52, 62-3, 81, 92, 113, 119, 136-7, 170, 274-6, 398, 403, 446, 465
ARRA (American Recovery and Reinvestment Act)   9, 11, 16, 41, 67, 89, 184, 188, 215, 227, 348
Artificial intelligence (AI)   7, 22, 32, 73, 273, 333, 456
ATA *see* American Telemedicine Association

B

Big data   6, 149, 152, 154-5, 174, 202, 391, 403-4, 428, 440, 445, 458-63
Bioinformatics   360
Biomedical Imaging Informatics (BII)   327
Biomedical informatics   3-4, 9, 20, 40, 327, 335, 357
Blockchain   142, 144
Blue Button   135, 261

C

CCDs *see* Continuity of Care Document
CDC *see* Centers for Disease Control
CDS *see* clinical decision support
CDSSs *see* Clinical decision support systems
CDWs *see* clinical data warehouses
Centers for Disease Control (CDC)   6, 9-10, 310, 377, 381-3, 387, 389-90
CER *see* Comparative effectiveness research
CHI *see* consumer health informatics
Chief information officer (CIO)   18, 220
Chief medical information officers (CMIOs)   18-19, 175
CIO *see* chief information officer
Classification   38, 114, 121, 124, 151, 168, 449-50, 452-3, 461
Clinical data   9, 30, 35-7, 114, 125, 137, 139, 150, 152, 154-6, 173, 198, 360, 401, 403-4
Clinical data warehouses (CDWs)   34-7, 401, 445-6
Clinical decision support (CDS)   22, 41, 43, 49, 69-70, 78-9, 81, 83, 86-7, 125-6, 161-79, 276, 299-301, 367-9, 455
Clinical decision support systems   73, 79, 81, 93, 161, 178, 187, 197, 199, 299, 446
Clinical decision support systems (CDSS)   73, 78-81, 83, 88, 93, 161-2, 166, 170, 174, 176, 178, 185, 187-8, 193, 197-201
Clinical informatics   3-4, 18-19, 155, 335
Clinical Practice Guidelines (CPGs)   5, 70, 73-4, 80, 82, 162-3, 165-6, 170, 172, 196, 201, 283-301, 308-9, 347
Clinical research informatics (CRI)   34, 184, 397, 403-4
CMIOs *see* chief medical information officers
Comparative effectiveness research (CER)   11, 36, 93, 155, 288, 401, 403
Computer-Based Patient Record   12, 73, 183-4
Computerized physician order entry (CPOE)   16, 35, 73, 77-9, 83, 85, 88, 161, 165, 174, 176-7, 199
Consumer health informatics (CHI)   3, 12, 253-64, 340-1
Continuity of Care Document (CCDs)   74, 77, 110, 113, 132, 135, 261
CPGs *see* Clinical Practice Guidelines
CPOE *see* computerized physician order entry

D

Data analytics 52, 72, 83, 144, 149-53, 155-6, 202, 439-41, 449, 455, 461, 463-4, 467
Data warehouses, clinical 34-5, 367, 401, 445-6
Databases 2, 6, 53-4, 214, 239, 244, 309, 311-17, 319, 362-3, 384-5, 441, 445-7, 449, 458-9
Datasets 6, 21, 363, 412, 441-4, 447, 453, 458, 463-4
Decision support 20, 43, 73, 76, 78, 80-1, 114, 177, 283, 298, 345, 369, 386, 392, 401
DICOM *see* Digital Imaging and Communications
Digital Imaging and Communications (DICOM) 111, 328, 332, 425
Direct Project 73, 134-6, 140-1
Drug-drug interaction (DDI) 70, 78, 80, 164, 166-7, 170, 199-200

E

EBM *see* evidence-based medicine
EHealth 12, 84, 174, 349, 376, 387, 410, 414-16, 419-20, 422, 426, 428
EHR implementation 70-1, 84, 87-8, 90-1, 184
EHRs *see* electronic health records
Electronic health records (EHRs) 4-9, 11-18, 35, 67-94, 135-42, 161-3, 169-71, 183-5, 188-93, 197-201, 297-301, 332-5, 367-9, 400-4, 423-8
Evidence-based medicine (EBM) 5, 17, 155, 162, 171-2, 283-5, 289-91, 293-4, 298, 300-1, 309, 311, 313, 397, 447

F

Fast healthcare interoperability resources (FHIR) 81, 86, 92, 108, 113, 126, 142-4, 165, 170-1, 178, 403, 428, 447
FDA (Food and Drug Administration) 9, 81, 176, 201, 226, 276-7, 333-4, 346, 359, 366, 402
FHIR *see* Fast healthcare interoperability resources

G

Genetics 162, 358, 360, 362, 365-9, 459, 463
Genomics 4, 113, 150, 153-4, 174, 253, 320, 357, 360, 362-3, 367, 391, 403
Geographic information systems (GIS) 10, 383-4, 440
GIS *see* Geographic information systems

H

Health and Human Services (HHS) 6-7, 9-11, 87, 140-1, 143, 221, 227, 299
Health data 43-4, 104, 214, 227, 256, 271, 363, 375, 401, 411-12, 424-5
Health information exchange (HIE) 2, 6-9, 12, 73, 82-3, 87, 131-4, 136-44, 151, 332, 345, 360, 383, 402, 422
Health information organizations (HIOs) 18, 57, 63, 70-2, 82, 132-44, 327, 333, 335, 337, 339-40, 346
Health Information Privacy and Security 214-27
Health information technology (HIT) 4-5, 7-12, 14-18, 20-2, 29, 41-3, 87-9, 173-4, 176-7, 184, 214-15, 292-3, 410-11, 419-22, 424-5
Health Information Technology for Economic and Clinical Health (HITECH) 9, 11-12, 67, 88-9, 131, 136, 141, 149, 184, 215-16, 375, 382, 397
Health Insurance Portability and Accountability Act (HIPAA) 11-12, 16, 86, 105, 184, 215-20, 223, 227, 329, 397
Health Resources and Services Administration (HRSA) 9-10, 352
Healthcare data 2, 4, 6-7, 15, 30-44, 51, 86, 126, 135, 151-2, 156, 163, 174, 220, 254-5
Healthcare Data Analytics 150-6, 439, 446
Healthcare Information and Management Systems Society (HIMSS) 8, 18, 20-1, 165, 178
HGP *see* Human Genome Project
HHS *see* Health and Human Services 6-7, 9-11, 87, 140-1, 143, 221, 227, 299
HIE *see* health information exchange
HIMSS *see* Healthcare Information and Management Systems Society
HIOs *see* health information organization
HIPAA *see* Health Insurance Portability and Accountability Act
HIT *see* health information technology
HITECH *see* Health Information Technology for Economic and Clinical Health
HITECH Act 7, 11, 14-15, 71, 84, 86, 89, 92, 131, 136, 139, 174, 217-18, 220, 227
HL7 3, 80, 86, 102-3, 106, 108, 113, 126, 142, 165, 169-71, 330, 383, 415
HRSA *see* Health Resources and Services Administration
Human Genome Project (HGP) 8, 360-1, 365

I

IBM Watson Analytics (IBMWA) 456, 460, 463-4
IBMWA *see* IBM Watson Analytics
ICD *see* International Classification of Diseases
ICD-10 18, 74, 80, 115-16, 118, 126, 142, 178, 380
IEEE *see* Institute of Electrical and Electronics Engineers
Information retrieval (IR) 32, 114, 307-8, 323
Institute of Electrical and Electronics Engineers (IEEE) 60-2, 101, 112
International Classification of Diseases (ICD) 82, 115
International Standards Organization (ISO) 101-4, 112, 219, 314, 410
Internet Protocol (IP) 54, 58

IP *see* Internet Protocol
IR *see* information retrieval
ISO *see* International Standards Organization

L

LANs *see* local area networks
Local area networks (LANs) 59

M

Machine learning (ML) 32, 149-51, 153-4, 156, 167, 191-2, 202, 273, 327, 333, 440, 449, 455-6, 460-1, 463-4
Meaningful Use (MU) 21, 67, 72-3, 86, 88-9, 107, 131, 138-9, 141-2, 144, 164, 184, 335, 383, 391
Medical Subject Headings (MeSH) 124, 312-13
MeSH *see* Medical Subject Headings
MHealth 48, 271, 274-6

N

National Center for Biotechnology Information (NCBI) 309, 311, 363
National Coordinator for Health Information Technology 9, 21, 133, 184
National eHealth Transition Authority (NeHTA) 414-15
National Human Genome Research Institute (NHGRI) 360-1, 363, 365, 367
National Institute of Standards and Technology (NIST) 11, 103, 219, 320
National Institutes of Health (NIH) 9, 155, 183, 308-9, 311, 359-60, 363, 365-7, 397-8, 401
National Library of Medicine (NLM) 7, 9, 17, 21, 103, 119, 121, 123, 183-4, 285, 309-12, 314-15, 317, 361, 363
National Provider Identifier (NPI) 89, 106
Nationwide Health Information Network (NwHIN) 11, 13, 133-4, 144
Natural language processing (NLP) 7, 33, 35, 37, 71, 80, 93, 115, 124, 150, 191, 322, 402, 455-7, 460
NCBI *see* National Center for Biotechnology Information
NHGRI *see* National Human Genome Research Institute
NIH *see* National Institutes of Health
NIST *see* National Institute of Standards and Technology
NLM *see* National Library of Medicine
NoSQL databases 403, 459
NPI *see* National Provider Identifier
NwHIN *see* Nationwide Health Information Network

O

Office of the National Coordinator (ONC) 9, 21, 84, 86-8, 103, 105, 110, 113, 133, 135, 140-2, 164, 173-4, 253-4, 256
Office of the National Coordinator for Health Information Technology (ONC) 9
OHSU *see* Oregon Health and Science University
ONC *see* Office of the National Coordinator
Open access 20, 308-9, 311, 467
OpenMRS 416-17, 419-20
Order sets 79-81, 91, 162-5, 171-2, 299
Oregon Health and Science University (OHSU) 220-1, 260, 426, 465, 467

P

PACS *see* Picture Archiving and Communication Systems
PANs *see* Personal Area Network
Patient data 71, 80-1, 87, 114, 134, 138, 141, 164, 169, 214, 216, 220, 239, 277, 419-20
Patient education 5, 73, 76, 137, 172, 346
Patient portals 5, 70, 73, 76, 92, 135, 137, 254, 261, 271, 292, 336, 345, 415
Patient privacy 141, 152, 215, 225, 227, 412, 460
Patient safety 9, 11, 13, 41, 70, 85, 88, 91, 105, 163, 170, 172, 184, 195
Personal Area Network (PANs) 59
Personal health information (PHI) 11, 16, 86-7, 89, 132, 216-18, 220, 307, 340, 425
Personal health record (PHRs) 2, 30-1, 68, 113, 132, 135, 213, 253-4, 260-1, 264, 346, 368, 414, 427
PHI *see* personal health information
PHRs *see* personal health record
PHS *see* Public health surveillance
Picture Archiving and Communication Systems (PACS) 5, 16, 59, 73, 111, 140, 327-32, 334-7, 339, 342, 403
Population health 5, 72-3, 90, 139, 163, 373-5, 390-2
Public health informatics 3, 374-5, 390, 392
Public Health Information System 379, 383, 390-1
Public health surveillance (PHS) 374-9, 384, 390, 392
PubMed 38, 285, 293, 310-11, 316-18, 322, 363, 397-8

R

REDCap *see* Research Electronic Data Capture
Reference Information Model (RIM) 39-40, 107
Regional Health Information Organization (RHIO) 132-3
Research Electronic Data Capture (REDCap) 400, 403
RIM *see* Reference Information Model

S

SDOs *see* standards development organizations
Semantic interoperability 82, 107, 113, 126, 134, 136, 410, 414, 428
Sequoia Project 133

Simple Object Access Protocol (SOAP) 56, 67, 136
Smartphones 8, 13, 21, 48, 51-2, 59, 73, 102, 259-60, 271-3, 277-8, 294, 298, 335-6, 412-13
SNOMED *see* Systematized Nomenclature of Medicine
SNOMED CT 74, 121, 123, 126, 367
SOAP *see* Simple Object Access Protocol
Society for Imaging Informatics and Medicine (SIIM) 9, 327, 335
SQL (structured query language) 53, 155, 440, 445, 449, 458-9
Standards development organizations (SDOs) 102, 164-5
Systematized Nomenclature of Medicine (SNOMED) 3, 118, 121-4, 126

T

Telehealth 62, 93, 259, 262, 339-40, 347-9, 351-2, 387, 424-5, 428
Telemedicine 5, 10, 63, 259, 262, 271, 274, 327, 339-52, 419, 421-2, 424
Text mining 149, 322, 455-7
Translational bioinformatics 357, 359-60

U

Usability 3-4, 10-11, 48, 85, 93, 156, 162, 190, 421

V

Virtual private networks 62, 243
Virtual Private Networks (VPNs) 62, 243
VPNs *see* Virtual Private Networks

W

WANs *see* Wide Area Networks
Web services 47, 56-7, 81, 343, 398, 403
Wide Area Networks (WANs) 61
Wireless Local Area Network (WLANs) 60-1
Wireless networks 48, 60-1, 63, 271, 273, 276
WLANs (Wireless Local Area Network) 60-1
World Health Organization 12, 115-16, 271, 276, 349, 386, 398, 427-8